IB DIPLOMA PROGRAMME

Physics

COURSE COMPANION

Tim Kirk
Neil Hodgson

SECOND EDITION

OXFORD

Great Clarendon Street, Oxford OX2 6DP

Oxford University Press is a department of the University of Oxford. It furthers the University's objective of excellence in research, scholarship, and education by publishing worldwide in

Oxford New York

Auckland Cape Town Dar es Salaam Hong Kong Karachi Kuala Lumpur Madrid Melbourne Mexico City Nairobi New Delhi Shanghai Taipei Toronto

With offices in

Argentina Austria Brazil Chile Czech Republic France Greece Guatemala Hungary Italy Japan South Korea Poland Portugal Singapore Switzerland Thailand Turkey Ukraine Vietnam

Oxford is a registered trade mark of Oxford University Press in the UK and in certain other countries

First published 2010

British Library Cataloguing in Publication Data

Data available

ISBN 9780199139545

10 9 8 7 6 5 4 3 2 1

Printed in Singapore by KHL Printing Company

Paper used in the production of this book is a natural, recyclable product made from wood grown in sustainable forests. The manufacturing process conforms to the environmental regulations of the country of origin.

Acknowledgements

We are grateful to the following to reproduce copyright material

P7tl: EcoPrint/Shutterstock; **P7tb**: Devi/Shutterstock; **P7tr**: Amy Johansson/ Shutterstock; **P7br**: Patrik stollarz/AFP; **P10tl**: James Steidl/Shutterstock; **P10tm**: Stephen Girimont/Shutterstock; **P10tr**: Radu Razvan/Shutterstock; **P22**: Niel Hodgson/Sha Tin College technicians; **P30tl**: Olga Besnard/ Shutterstock; **P30tm**: Afaizal/Shutterstock; **P30tr**: Alexander Kalina/ Shutterstock; **P30bm**: Stefan Matzke/NewSport/Corbis; **P40tl**: Dwphotos/ Shutterstock; **P40tm**: Jason Gower/Shutterstock; **P40tr**: Vladimir Wrangel/ Shutterstock; **P45**: Niel Hodgson; **P46tl**: Niel Hodgson; **P46tm**: Niel Hodgson; **P46tr**: Niel Hodgson; **P57tl**: David Woods/Shutterstock; **P57tm**: Stefanie van der Vinden/Shutterstock; **P57tr**: Ben Smith/Shutterstock; **P64**: Niel Hodgson; **P64**: Niel Hodgson; **P64**: Niel Hodgson; **P64**: Niel Hodgson; **P64**: Niel Hodgson; **P64**: Niel Hodgson; **P68tl**: Stephen McSweeny/Shutterstock; **P68tm**: Mandy Godbehear/Shutterstock; **P68tr**: Galyna Andrushko/ Shutterstock; **P73tl**: Ella/Shutterstock; **P73tm**: Joshua David Treisner/ Shutterstock; **P73tr**: Andras_Csontos/Shutterstock; **P74**: Siepmann Siepmann/Photolibrary; **P74**: Marko_Marcello/Bigstockphoto; **P74**: Tony Gentile/Reuters; **P74**: Stephen Mcsweeny/Shutterstock; **P74**: Jennifer Boggs/ Gettyimages; **P74**: Argironeta/Dreamstime; **P76t**: Niel Hodgson; **P76b**: Niel Hodgson; **P81**: Steve Pyke/Getty Images; **P91**: Jennifer Boggs/Gettyimages; **P98tl**: Dwphotos/Shutterstock; **P98tm**: Tsian/Shutterstock; **P98tr**: Andrejs Pidjass/Shutterstock; **P101**: Spencer Grant/Photo Researchers; **P104**: Nikolay Titov/Shutterstock; **P108m**: DR Larpent/ G.R.E.H.G.E.P./Science Photo Library; **P111t**: Neil Hodgson; **P111m**: Eastwest Imaging/Dreamstime; **P112b**: Vince Cavataio/Photolibrary; **P113**: Bill Brennan/Pacific Stock/ Photolibrary; **P119tl**: Roman Pavlik/Shutterstock; **P119tm**: Jonathan Larsen/ Shutterstock; **P119tr**: Robdigphot/Shutterstock; **P126**: Firenight/Fotolia; **P129l**: Neil Hodgson; **P129r**: Jason Ware/NASA Astronomy Picture of the Day Collection; **P130t**: Nicola vernizzi/Shutterstock; **P130b**: Neil Hodgson; **P138tl**: Simon Krzic/Shutterstock; **P138tm**: Eric Limon/Shutterstock; **P138tr**: Hardtmuth/Shutterstock; **P150tl**: SF photo/Shutterstock; **P150tm**: Kodda/Shutterstock; **P150tr**: Rene Baumgartner/Shutterstock; **P151t**: Niel Hodgson; **P151bl**: Niel Hodgson; **P151br**: Niel Hodgson; **P160tl**: Thomas Skjaeveland/Shutterstock; **P160tm**: Xtrekx/Shutterstock; **P160tr**: Fotohunter/Shutterstock; **P163tl**: PhotostoGo; **P163tr**: Africa Studio/ Shutterstock; **P163bl**: Phase4Photography/Shutterstock; **P163br**: PhotostoGo; **P172t**: PhotostoGo; **P172b**: Ted Pink/Alamy; **P174**: Dan Barba/ Photolibrary; **P175l**: Andrew Lambert Photography/Science Photo Library; **P175r**: Art Directors & TRIP/Alamy; **P178**: Andrew Lambert Photography/ Science Photo Library; **P183tl**: Holger Mette/Shutterstock; **P183tm**: Intraclique LLC/Shutterstock; **P183tr**: Greg Epperson/Shutterstock; **P184t** : John Barry de Nicola/Shutterstock; **P184m**: Eric Gevaert/Shutterstock; **P184b**: The Everett Collection/Rex Features; **P185t**: Elena Elisseeva/ Shutterstock; **P185l**: NASA Goddard Space Flight Center; **P185m**: NASA; **P202l**: Marilyn Barbone/Shutterstock; **P202r**: Marilyn Barbone/Shutterstock; **P206tl**: Joggie Botma/Shutterstock; **P206tm**: Cristi Matei/Shutterstock; **P206tr**: Alex0001/Shutterstock; **P212**: NASA/courtesy of nasaimages.org; **P213**: Johnson Space Center/NASA; **P217tl**: Teresa Levite/Shutterstock; **P217tm**: ID1974/Shutterstock; **P217tr**: Rob Byron/Shutterstock; **P226**: Brent Wong/Dreamstime; **P234tl**: David W Hughes/Shutterstock; **P234tm**: Nobor/ Shutterstock; **P234tr**: Gualtiero Boffi/Shutterstock; **P235**: Ryckaert, David III (1612-61)/Musee des Beaux-Arts Andre Malraux, Le Havre, France/Giraudon/ The Bridgeman Art Library; **P236**: Neil Hodgson; **P236**: Neil Hodgson; **P236**: Neil Hodgson; **P236**: Neil Hodgson; **P237t**: Architecture UK/Alamy; **P237m**: Science Museum/Science & Society; **P237b**: Andrew Lambert Photography/ Science Photo Library; **P238tl**: Science Photo Library; **P238tr**: Popperfoto/ Alamy; **P238bl**: Prof. Peter Fowler/Science Photo Library; **P238bm**: American Institute of Physics/Science Photo Library; **P239t**: A. Barrington Brown/Science Photo Library; **P239b**: Science Photo Library; **P249**: Physics Today Collection/American Institute of Physics/Science Photo Library; **P252t**: NOAO/Science Photo Library; **P252b**: Royal Observatory, Edinburgh/Science Photo Library; **P253t**: David Parker/ Science Photo Library; **P253b**: Hank Morgan/Science Photo Library; **P262tl**: Shcherbakov Sergii/Shutterstock; **P262tm**: Sebastian Kaulitzki/Shutterstock; **P262tr**: Inga Nielsen/ Shutterstock; **P267t**: 1983 University Science Books; "Quantum Chemistry" by Donald A. McQuarrie; **P267b**: Physics World; **P268**; **P269**: Andrew Lambert Photography/ Science Photo Library; **P269**: Andrew Lambert Photography/ Science Photo Library; **P269**: Andrew Lambert Photography/ Science Photo Library; **P269**: Andrew Lambert Photography/ Science Photo Library; **P269**: Andrew Lambert Photography/ Science Photo Library; **P275**: Carl Anderson/Science Photo Library ; **P281tl**: Yvan/Shutterstock; **P281tm**: Algis Balezentis/Shutterstock; **P281tr**: TebNad/Shutterstock; **P290**: Neil Hodgson; **P294t**: Joerg Boethling/Still Pictures; **P294b**: Robin Smith/ Photolibrary; **P298tl**: Elena Elisseeva/Shutterstock; **P298tm**: Martin D. Vonka/Shutterstock; **P298tr**: Bierchen/Shutterstock; **P299**: Beat Bieler/ Shutterstock; **P301**: Neil Hodgson; **P310tl**: Roman Berk/Shutterstock; **P310tm**: Bestseller/Shutterstock; **P310tr**: Erwin Wodicka/Shutterstock; **P314**: Greenpeace/Jeremy Sutton-Hibbert; **P318tl**: Vladfoto/Shutterstock; **P318tm**: Morgan Lane Photography/Shutterstock; **P318tr**: Foto-Ruhrgebiet/ Shutterstock; **P321t**: Robert Galbraith/Reuters; **P321m**: **P321b**: **P324tl**: Mana Photo/Shutterstock; **P324tm**: Tyler Olson/Shutterstock; **P324tr**: Vladislav Gajic/Shutterstock; **P324br**: Jim Parkin/Shutterstock; **P326t**: Wen Zhenxiao/XinHua/Xinhua Press/Corbis; **P326b**: Best View Stock/Photolibrary; **P330tl**: Sergey Kamshylin/Shutterstock; **P330tm**: John Wollwerth/ Shutterstock; **P330tr**: Tatonka/Shutterstock; **343tl**: Georgy Shafeev/ Shutterstock; **343tm**: Dom/Shutterstock; **343tr**: Rasch/Shutterstock.com; **P345**: Andrew Lambert Photography/ Science Photo Library; **P348**: Dr. Tony Brain/Science Photo Library; **P349**: Dr. Jeremy Burgess/Science Photo Library; **P351t**: Giphotostock/Science Photo Library; **P351b**: Sheila Terry/Science Photo Library; **P352**: Neil Hodgson; **P354**: NASA; **P356**: Mazzzur/ Shutterstock; **P356**: Darkgreenwolf/Shutterstock; **P358tl**: Monkey Business Images/Shutterstock; **P358tm**: Ferenc Szelepcsenyi/Shutterstock; **P358tr**: Graeme Dawes/Shutterstock; **P364tl**: Lawrence Gough/Shutterstock; **P364tm**: Aleksi Potov/Shutterstock; **P364tr**: Eric Von Seggern/Shutterstock; **P371**: William Attard McCarthy/Shutterstock; **P371**: Sebastian Kaulitzki/ Shutterstock; **P371**: Wolfgang Kloehr/Shutterstock; **P371**: Peter Michaud/ NASA; **P371**: NASA/courtesy of nasaimages.org; **P371**: NASA/courtesy of nasaimages.org; **P374**: N.A.Sharp, NOAO/NSO/Kitt Peak FTS/AURA/NSF; **P389tl**: Eduardo Rivero/Shutterstock; **P389tm**: Mikhail Nekrasov/ Shutterstock; **P389tr**: Tsian/Shutterstock; **P390**: Celia Peterson/Photolibrary; **P390**: Matilde Gattoni/Photolibrary; **P390**: Jason Larkin/Photolibrary; **P390**: Celia Peterson/Photolibrary; **P403**: MalDix/Shutterstock; **P413tl**: Jocicalek/ Shutterstock; **P413tm**: Andre Blais/Shutterstock; **P413tr**: Oleg Nekhaev/ Shutterstock; **P415t**: Maria Skaldina/Shutterstock; **P415b**: Courtney Keating/ Shutterstock; **P427t**: Andy Z/Shutterstock; **P427m**: Kevin Page/Fotolia; **P427b**: Tobias Machhaus/Shutterstock; **P433tl**: Pavel B/Shutterstock; **P433tm**: Olaru Radian-Alexandru/Shutterstock; **P433tr**: Alexander Gordeyev/Shutterstock; **P449**: K. Sharon/NASA; **P453tl**: Monkey Business Images/Shutterstock; **P453tm**: Andresr/Shutterstock; **P453tr**: Mats/ Shutterstock; **P456**: Science Photo Library; **P457t**: Mehau Kulyk/Science Photo Library; **P457b**: Ntoine Rosset/Science Photo Library; **P460**: Russell D. Curtis/Science Photo Library; **P461**: CNRI/Science Photo Library; **P463**: Antonia Reeve/Science Photo Library; **P466tl**: www.teachingmedicalphysics. org.uk; **P466tm**: www.teachingmedicalphysics.org.uk; **P466tr**: www. teachingmedicalphysics.org.uk; **P466m**: www.teachingmedicalphysics.org. uk; **466bl**: www.teachingmedicalphysics.org.uk; **466bm**: www. teachingmedicalphysics.org.uk; **466br**: www.teachingmedicalphysics.org. uk; **P470**: Michael Taylor/Shutterstock; **P470**: Alice Day/Shutterstock ; **P470**: Morten Kjerulff/Shutterstock ; **P470**: Photodisc/OUP; **P471**: Carl Anderson/ Science Photo Library; **P502tl**: Marikond/Shutterstock; **P502tm**: Gmwnz/ Shutterstock; **P502tl**: Nohab/Shutterstock

We have tried to trace and contact all copyright holders before publication. If notified the publishers will be pleased to rectify any errors or omissions at the earliest opportunity.

Course Companion definition

The IB Diploma Programme Course Companions are resource materials designed to provide students with extra support through their two-year course of study. These books will help students gain an understanding of what is expected from the study of an IB Diploma Programme subject.

The Course Companions reflect the philosophy and approach of the IB Diploma Programme and present content in a way that illustrates the purpose and aims of the IB. They encourage a deep understanding of each subject by making connections to wider issues and providing opportunities for critical thinking.

These Course Companions, therefore, may or may not contain all of the curriculum content required in each IB Diploma Programme subject, and so are not designed to be complete and prescriptive textbooks. Each book will try to ensure that areas of curriculum that are unique to the IB or to a new course revision are thoroughly covered. These books mirror the IB philosophy of viewing the curriculum in terms of a whole-course approach; the use of a wide range of resources; international-mindedness; the IB learner profile and the IB Diploma Programme core requirements; theory of knowledge; the extended essay; and creativity, action, service (CAS).

In addition, the Course Companions provide advice and guidance on the specific course assessment requirements and also on academic honesty protocol.

The Course Companions are not designed to be:

- study/revision guides or a one-stop solution for students to pass the subjects
- prescriptive or essential subject textbooks.

A note on academic honesty

It is of vital importance to acknowledge and appropriately credit the owners of information when that information is used in your work. After all, owners of ideas (intellectual property) have property rights. To have an authentic piece of work, it must be based on your individual and original ideas with the work of others fully acknowledged. Therefore, all assignments, written or oral, completed for assessment must use your own language and expression. Where sources are used or referred to, whether in the form of direct quotation or paraphrase, such sources must be appropriately acknowledged.

How do I acknowledge the work of others?

The way that you acknowledge that you have used the ideas of other people is through the use of footnotes and bibliographies.

Footnotes (placed at the bottom of a page) or endnotes (placed at the end of a document) are to be provided when you quote or paraphrase from another document, or closely summarize the information provided in another document. You do not need to provide a footnote for information that is part of a 'body of knowledge'. That is, definitions do not need to be footnoted as they are part of the assumed knowledge.

Bibliographies should include a formal list of the resources that you used in your work. 'Formal' means that you should use one of the several accepted forms of presentation. This usually involves separating the resources that you use into different categories (e.g. books, magazines, newspaper articles, Internet-based resources, CDs and works of art) and providing full information as to how a reader or viewer of your work can find the same information. A bibliography is compulsory in the extended essay.

What constitutes malpractice?

Malpractice is behaviour that results in, or may result in, you or any student gaining an unfair advantage in one or more assessment component. Malpractice includes plagiarism and collusion.

Plagiarism is defined as the representation of the ideas or work of another person as your own. The following are some of the ways to avoid plagiarism:

- Words and ideas of another person used to support one's arguments must be acknowledged.
- Passages that are quoted verbatim must be enclosed within quotation marks and acknowledged.
- CD-ROMs, email messages, web sites on the Internet, and any other electronic media must be treated in the same way as books and journals.
- The sources of all photographs, maps, illustrations, computer programs, data, graphs, audio-visual, and similar material must be acknowledged if they are not your own work.
- Works of art, whether music, film, dance, theatre arts, or visual arts, and where the creative use of a part of a work takes place, must be acknowledged.

Collusion is defined as supporting malpractice by another student. This includes:

- allowing your work to be copied or submitted for assessment by another student
- duplicating work for different assessment components and/or diploma requirements.

Other forms of malpractice include any action that gives you an unfair advantage or affects the results of another student. Examples include, taking unauthorized material into an examination room, misconduct during an examination, and falsifying a CAS record.

IB mission statement

The International Baccalaureate aims to develop inquiring, knowledgeable, and caring young people who help to create a better and more peaceful world through intercultural understanding and respect.

To this end the IB works with schools, governments, and international organizations to develop challenging programmes of international education and rigorous assessment.

These programmes encourage students across the world to become active, compassionate, and lifelong learners who understand that other people, with their differences, can also be right.

The IB learner profile

The International Baccalaureate aims to develop internationally minded people who, recognizing their common humanity and shared guardianship of the planet, help to create a better and more peaceful world. IB learners strive to be:

Inquirers They develop their natural curiosity. They acquire the skills necessary to conduct inquiry and research and show independence in learning. They actively enjoy learning and this love of learning will be sustained throughout their lives.

Knowledgeable They explore concepts, ideas, and issues that have local and global significance. In so doing, they acquire in-depth knowledge and develop understanding across a broad and balanced range of disciplines.

Thinkers They exercise initiative in applying thinking skills critically and creatively to recognize and approach complex problems, and make reasoned, ethical decisions.

Communicators They understand and express ideas and information confidently and creatively in more than one language and in a variety of modes of communication. They work effectively and willingly in collaboration with others.

Principled They act with integrity and honesty, with a strong sense of fairness, justice, and respect for the dignity of the individual, groups, and communities. They take responsibility for their own actions and the consequences that accompany them.

Open-minded They understand and appreciate their own cultures and personal histories, and are open to the perspectives, values, and traditions of other individuals and communities. They are accustomed to seeking and evaluating a range of points of view, and are willing to grow from the experience.

Caring They show empathy, compassion, and respect towards the needs and feelings of others. They have a personal commitment to service, and act to make a positive difference to the lives of others and to the environment.

Risk-takers They approach unfamiliar situations and uncertainty with courage and forethought, and have the independence of spirit to explore new roles, ideas, and strategies. They are brave and articulate in defending their beliefs.

Balanced They understand the importance of intellectual, physical, and emotional balance to achieve personal well-being for themselves and others.

Reflective They give thoughtful consideration to their own learning and experience. They are able to assess and understand their strengths and limitations in order to support their learning and personal development.

Contents

Introduction

The aim of this book is to encourage and support IB students who are studying physics as part of the IB Diploma Programme. Success in the final examinations does, of course, depend on gaining knowledge and understanding of physics but by the end of the course you should have gained much more than just an ability to answer examination questions. The IB learner profile provides a useful checklist of some of the wider skills you should be aiming to develop.

The goal of physics is no less than an attempt to understand the natural world in its entirety. Although it should be obvious that an IB physics course alone will not achieve this, the insights that physics can provide into the mechanisms that operate in the world around us can be both breathtakingly simple and hugely complex at the same time. If nothing else, we hope that after studying IB physics you gain a love, fascination and respect for the universe in which we live.

There are many excellent physics books available. In writing this course companion we have deliberately focused on some of the wider aspects of subject and attempted to emphasize how the subject links with other subjects within the IB Diploma Programme model.

In this second edition we have included material on all option units. You will find chapters on units E-J (astrophysics, communications, electromagnetic waves, relativity, medical physics and particle physics) and there is a guide to finding material for the other options in Chapter 33.

We have also used this second edition to replace a lot of photographs and diagrams to make them more interesting and relevant. Some chapters have been entirely re-written but all have been updated. End of chapter summaries have been added to help you focus on your learning, and see where it might be useful to consult other resources.

End of chapter questions have also been added to all chapters. Answers to all questions are available on the CD that now comes with the book. The CD also contains a glossary of terms for all units, interactive tests of past examination questions, arranged by topic, and spreadsheets for you to practice data modelling and analysis.

Over the years we have both been fortunate enough to teach fabulous students and we would like to thank those keen and enthusiastic students (you know who you are) for making teaching and learning so rewarding and fun.

The publishers and authors would like to thank Graeme Littler, Jean Godin and Jonathan Allday for their contributions to the first edition, and particularly Mark Sylvester for his work on this second edition.

Individual thanks (from the first edition)

Neil Hodgson: Writing this book has been an extremely challenging process and it has only been possible with the support and best wishes of my colleagues at Sha Tin College, in particular the physics team. Paul Drew read some of the material, giving valuable advice. The team of technicians provided valuable support by photographing experiments and recording primary data. Of course the support and encouragement of my fantastic family; Meher and children, Raoul and Tanya has been invaluable and I would like to thank them for overlooking all my failings as a father and husband during this time. Then, there are the wonderful students at Sha Tin College who provide a purpose and inspiration for this book and I would like to thank them also for their interest, enthusiasm and for making teaching such a worthwhile profession.

Tim Kirk: This book was written while I had the privilege of teaching talented and motivated students at the Godolphin and Latymer School in Hammersmith, and I would like to start by thanking the students there. I would also like to give my heartfelt thanks to my family, friends and colleagues for allowing me the space and time to undertake this project. Without their individual support and understanding, the task would have been impossible. Above all, however, I thank Betsan for her support, patience and encouragement, which was above and beyond the call of duty.

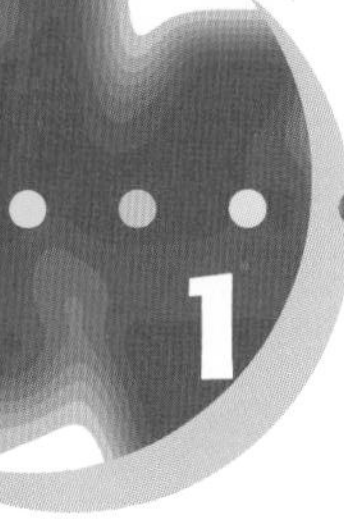

1 Learning in physics

The nature of physics

Physics is an attempt to understand the universe in which we live. Richard Feynman, winner of the Nobel Prize in Physics in 1965, ranks as one of the world's greatest theoretical physicists of recent times. This is how he described physics:

> *Physics is the most fundamental and all-inclusive of the sciences, and has had a profound effect on all scientific development. In fact, physics is the present-day equivalent of what used to be called* natural philosophy, *from which most of our modern sciences arose. Students of many fields find themselves studying physics because of the basic role it plays in all phenomena.* (Feynman et al. 1965: 1)

Ernest Rutherford, another Nobel Prize winner (for chemistry, in 1908), stressed the importance of physics to the progress of science in a more direct way:

> *All Science is either physics or stamp collecting.* (quoted in Birks 1962: 108)

The same sentiment is echoed by Robert Bunsen (of Bunsen burner fame):

> *Ein chemiker der kein Physiker ist, ist gar nichts—A chemist who is not a physicist is nothing at all.* (quoted in Partington 1961: 282)

Perhaps the most succinct description of the physicist's role comes from another Nobel Prize winner, Steven Weinberg. In his Nobel lecture in 1979 he said:

> *Our job in physics is to see things simply, to understand a great many complicated phenomena in a unified way, in terms of a few simple principles.*

But building an understanding of the principles is only the beginning. The astronomer Martin Rees gave the following analogy:

> *The physicist is like someone who's watching people playing chess and, after watching a few games, he may have worked out what the moves in the game are. But understanding the rules is just a trivial preliminary on the long route from being a novice to being a grand master. So even if*

Figure 1 German Chancellor Angela Merkel studied physics at the University of Leipzig

we understand all the laws of physics, then exploring their consequences in the everyday world where complex structures can exist is a far more daunting task, and that's an inexhaustible one I'm sure. (quoted in Wolpert and Richards 1988: 37)

The search for knowledge and understanding is a fundamental part of the human condition, but neither physics nor philosophy has found enough definitive answers for us to be able to stop asking questions. There are so many fundamental puzzles still to solve, and this is part of the thrill and excitement of physics.

The following list (taken from postings on the *Cosmic Variance* blog) summarizes some of the reasons given by students for studying physics:

- To gain a deeper understanding of the fundamental principles that govern our universe and everything in it, while at the same time picking up a broad and eminently useful skillset: the ability to analyse and deconstruct problems, to effectively communicate solutions.
- You learn to see what is important.
- Because you want to succeed in (choose one): business, law, medicine, education, engineering, politics or research in physics.
- There are lots of people in the world who can read and write well. There are far fewer who can think clearly. The world needs more of the latter.
- The thrill of being on the brink of discovery is second only to being madly in love.
- You get to play with cooler, more expensive toys than your friends.
- It is fascinating and fun at the same time.

Approaches used in this book

In addition to introducing much of the material needed for IB Diploma Programme physics (and thus supporting the acquisition of knowledge and improving the ability to be a thinker), the following approaches are used in this book to try to support your learning in physics:

- **Higher level:** These sections contain material studied by higher level (HL) candidates, and are not examined for standard level (SL). However, they will provide useful background information and an extension of ideas to help SL students gain a deeper understanding.
- **The bigger picture:** These sections contain challenging material that goes beyond the curriculum (not just in breadth but in more depth). The aim is to improve real understanding of the issues, and to encourage inquiry.
- **Science tools:** These sections introduce and reinforce the practical techniques and manipulative internal assessment skills that form part of the skill of inquiry.
- **Working with data:** These questions foster and encourage the development of practical skills – particularly DCP and CE (see pages 19 and 23).

- **Rediscovering physics:** The practical situations introduced in these sections provide ideas for experimental work where the outcomes can be predicted from the topics being studied.
- **Investigating physics:** The practical situations introduced in these sections provide ideas for experimental work where outcomes could be unknown.
- **Physics issues:** These sections focus on wider issues in science, science and technology, and internationalism. They address some of the wider aspects of the learner profile, involving physics in the real world, risk, implications, CAS, etc.
- **Thinking about science:** These sections focus on aspects of the scientific method, and highlight opportunities for links to be made with aspects of your studies in theory of knowledge.
- **Questions:** The best way to test, reinforce, and stretch your understanding of physics is to attempt to answer questions. These have been added throughout the text and at the end of chapters.
- **Inquiry:** These sections present short topics for you to research yourself using information readily available in textbooks or the internet. Clear instructions or a series of questions are given to guide the inquiry. The aim is to promote independent learning.
- **Data-based questions:** These encourage learning from the analysis of information, which may involve new information not on the syllabus (see "The bigger picture" above.)
- **Mathematical physics:** Mathematics is a very important tool in physics. These sections highlight useful approaches or necessary skills. Graphical display calculators can be used to support physics learning.

IB Diploma Programme physics can be studied without having to use the mathematical tool called calculus. However, many students learn about calculus in their mathematics courses. If you find the use of calculus notation hard, whenever a new quantity is introduced you should focus instead on the word equations being used.

References

Birks, J.B. 1962. *Rutherford at Manchester*. London. Heywood & Co.

Feynman, R., Leighton, R.B., and Sands, M. 1965. *The Feynman Lectures on Physics*, Vol. 3. Reading, MA. Addison-Wesley.

Partington, J.R. 1961. *A History of Chemistry*, Vol. 4. London. Macmillan.

Wolpert, L. and Richards, A. (eds) 1988. *A Passion for Science*. Oxford. Oxford University Press.

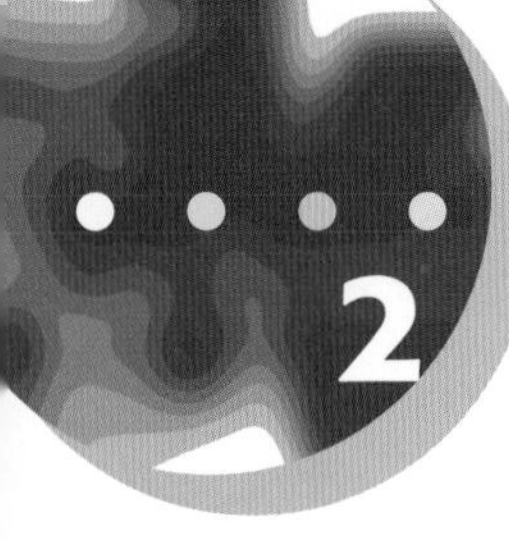

2 Experimental measurements and internal assessment

Estimating the very large and the very small

Physics concerns itself with the study of all natural phenomena, and so the range of possible measurements associated with the study of physics is huge. Everyday objects are, of course, studied in physics, but we also consider much smaller and much larger objects. The range that we consider goes from the very small (the tiny constituents of an atom, for example) up to the very large: individual galaxies, clusters of galaxies, or even the universe itself.

Some things cannot be measured directly. For example, we can estimate the size of a galaxy and the size of a nucleus, but we cannot measure either directly.

Scientific notation (and approximations in terms of orders of magnitude) allows the same basic unit to be used to measure any given quantity (such as mass, length, or time), no matter how large or small. It is important to get a "feel" for the range of the numbers used. You can use the following exercises to estimate the range of possible measurements in the three fundamental quantities of mass, length, and time.

For each of the following lists:

a) Rearrange the list in terms of increasing order of magnitude.

b) Write down a guess for the approximate value for each quantity in terms of the unit used for that measurement.

c) Do some research to find more accurate values. Check the reliability of your data by comparing your values with other students' estimations.

d) Add at least five other measurements of your own choosing that cover the full range of the quantity.

e) Design a poster to illustrate the range of possible measurements for one of the fundamental quantities.

1 **Mass in kilograms:** a person, the Sun, an electron, the observable universe, a car, a grain of rice, a blood cell, a proton, an atom, the Earth, the Milky Way galaxy.

2 **Length in metres:** a person's height, the distance to the Sun, the radius of a proton, the thickness of a piece of paper, the radius of the Earth, the height of the tallest building, the furthest distance travelled by mankind (distance from the Earth to the Moon), a light year, the radius of the observable universe, a blood cell, size of sub-nuclear particles, the diameter of a hydrogen atom.

3 **Time in seconds:** an average human lifespan, the time needed for light to cross a nucleus, an average human reaction time, the time taken for a camera's flash, the age of the universe, the expected life of the Sun, the time needed for light to cross a room, the age of the Earth, the time taken for light to come from the Sun to Earth, your age, the time period for visible light.

1 In each case state the ratio of the two quantities as differences of
orders of magnitude.

a) the diameter of the hydrogen atom compared with its nucleus [1]

b) the distance from the Earth to the Moon compared with the
distance to the Sun [1]

c) the expected life of the Sun compared with an average
human lifespan [1]

d) the age of the universe compared with the expected life of the Sun [1]

e) the mass of a car compared with the mass of an atom [1]

f) the mass of the Earth compared with the mass of a grain of rice. [1]

2 The diameter of the nucleus of a hydrogen atom is of the order of

A 10^{-8} m **B** 10^{-15} m

C 10^{-23} m **D** 10^{-30} m

3 The number of heartbeats of a person at rest in one hour, to the nearest order of magnitude, is

A 10^{1} **B** 10^{2}

C 10^{3} **D** 10^{5}

4 The diameter of a proton is of the order of magnitude of

A 10^{-12} m **B** 10^{-15} m

C 10^{-18} m **D** 10^{-21} m

5 Which of the following gives the approximate ratio of the separation of the molecules in water and in steam at atmospheric pressure?

	Water : Steam
A	1 : 1
B	1 : 10
C	1 : 100
D	1 : 1000

6 The order of magnitude of the weight of an apple is

A 10^{-4} N **B** 10^{-2} N

C 1 N **D** 10^{2} N

Working with data: SI units

Physics is an experimental science. The process of experiment relies upon accurate measurement. To make a measurement, we need to define the units to be used. Many of the units used in science will already be familiar to you (the second, the metre, or the kilogram, for example), but others (such as the henry, or the weber) are not as well known. Many non-standard units are widely used in everyday measurements, but they are best avoided in physics. Some examples of common country-specific units that are widely used are calories, pounds, and gallons.

In order to rationalize the units being used worldwide, the international system of units (*Le Système international d'unités* or the SI system) was developed. It is based around measurement of seven fundamental quantities, with a base unit accurately defined for each one: see Table 1.

Table 1 The SI base units

Quantity	Unit name	Unit symbol
mass	kilogram	kg
length	metre	m
time	second	s
electric current	ampere	A
amount of substance	mole	mol
thermodynamic temperature	kelvin	K
luminous intensity	candela	cd

It would be clumsy to only use the base units in all the combinations that are necessary, so **derived units** can be defined from the base units. It is sometimes possible to use the units of given quantities to predict how different factors may be related.

Table 2 gives the units of many of the quantities that you will meet in your studies. In all cases, the SI derived unit can be worked out from the word equation. Be careful how you use this table, because the equations given refer to particular situations. Whenever you come across a quantity that is new to you, check this list to see if it is mentioned. If it is not here, then add it!

Table 2 Derived units

Quantity	Symbol	Word equation	SI derived unit	SI base unit	Alternative unit
area	A	length × breadth	–	m^2	–
volume	V	length × breadth × height	–	m^3	–
density	ρ (rho)	mass/volume	–	$kg\ m^{-3}$	–
velocity (speed)	v, u	displacement/time (distance/time)	–	$m\ s^{-1}$	–
acceleration	a	change of velocity/time	–	$m\ s^{-2}$	–
concentration		amount of matter/volume	–	$mol\ m^{-3}$	–
momentum	p	mass × velocity	–	$kg\ m\ s^{-1}$	N s
force	F	mass × acceleration (change of momentum/time)	newton (N)	$kg\ m\ s^{-2}$	
pressure (stress)	P	force/area	pascal (Pa)	$kg\ m^{-1}\ s^{-2}$	$N\ m^{-2}$
frequency	f	number of cycles/time	hertz (Hz)	s^{-1}	–
energy (work) (quantity of heat)	Q, W, E	force × distance ($\frac{1}{2}mv^2$ or mgh)	joule (J)	$kg\ m^2\ s^{-2}$	N m
power	P	rate of using energy (rate of doing work)	watt (W)	$kg\ m^2\ s^{-3}$	$J\ s^{-1}$
charge	Q	current × time	coulomb (C)	A s	–
potential difference (p.d.) (voltage) (emf)	V	energy/charge	volt (V)	$kg\ m^2\ s^{-3}\ A^{-1}$	$W\ A^{-1}$ $J\ C^{-1}$
resistance	R	voltage/current	ohm ()	$kg\ m^2\ s^{-3}\ A^{-2}$	$V\ A^{-1}$
magnetic flux	ϕ (phi)	emf × time (magnetic field × area)	weber (Wb)	$kg\ m^2\ s^{-2}\ A^{-1}$	V s
magnetic flux density (magnetic field strength)	B	force/(current × length) (flux/area)	tesla (T)	$kg\ s^{-2}\ A^{-1}$	$Wb\ m^{-2}$
moment of a force	T	force × perpendicular distance	N m	$kg\ m^2\ s^{-2}$	–
heat capacity (entropy)	C (S)	energy/temperature rise	$J\ K^{-1}$	$kg\ m^2\ s^{-2}\ K^{-1}$	–
specific heat capacity	c	heat capacity/mass	$J\ kg^{-1}\ K^{-1}$	$m^2\ s^{-2}\ K^{-1}$	–
electric field strength	E	force/unit charge	$V\ m^{-1}$	$kg\ m\ s^{-3}\ A^{-1}$	NC^{-1}
radioactivity	A	number of disintegrations time	becquerel (Bq)	s^{-1}	–
exposure	X	charge/mass	$C\ kg^{-1}$	$kg^{-1}\ s\ A$	–
absorbed dose of radiation	D	energy/mass	gray (Gy)	$m^2\ s^{-2}$	$J\ kg^{-1}$

Table 2 (continued)

Quantity	Symbol	Word equation	SI derived unit	SI base unit	Alternative unit
dose equivalent (radiation)	H	absorbed dose × factor	sievert (Sv)	$m^2\ s^{-2}$	$J\ kg^{-1}$
intensity	I	power received/area	$W\ m^{-2}$	$kg\ s^{-3}$	–
angle	θ (theta)	arc length/radius	(simple ratio)	–	radians (rad) *
angular velocity	ω (omega)	angle/time	–	s^{-1}	$rad\ s^{-1}$
albedo (reflectivity)	α (alpha)	power absorbed/incident power	(simple ratio)	–	–

*Technically the unit for angle, the radian, is known as a supplementary unit.

SI prefixes and multipliers

To obtain multiples and submultiples of units, standard prefixes are used, as shown in Table 3.

Table 3 SI prefixes and multipliers

Multiplication factor			Prefix	Symbol
1 000 000	=	10^6	mega	M
1 000	=	10^3	kilo	k
100	=	10^2	hecto	h
10	=	10^1	deca	da
0.1	=	10^{-1}	deci	d
0.01	=	10^{-2}	centi	c
0.001	=	10^{-3}	milli	m
0.000 000 1	=	10^{-6}	micro	μ
0.000 000 000 1	=	10^{-9}	nano	n

Thinking about science: Uncertainties

All measurements in science suffer from uncertainty. No matter how hard we try to control all the variables in any situation, some level of experimental error is inevitable. This does not mean all experimental work is flawed, but it *does* mean that we need to consider the uncertainties whenever data are properly analysed.

A complete analysis of experimental data is very time-consuming, and involves a range of statistical techniques that go beyond the IB syllabus. There are simple statistical tests that allow researchers to calculate the likelihood that a given set of experimental readings are significantly different from what might be expected as the results of chance alone. These mathematical tests (such as the *t*-test and the χ^2-test) are often straightforward to apply to data, but should not be used unless one understands their limitations.

Central to many statistical theories is the concept of a **normal distribution**. Consider, for example, a machine designed to manufacture identical metal rods (perhaps for use as clamp stands in physics laboratories). In order to check the operation of the machine, the diameters of all the rods are measured.

A perfect machine would make each rod exactly the same, but a real machine will produce rods that are all slightly different from one another. The variation will be random, with the distribution shown in Figure 1, which plots the number of rods against the measured diameter.

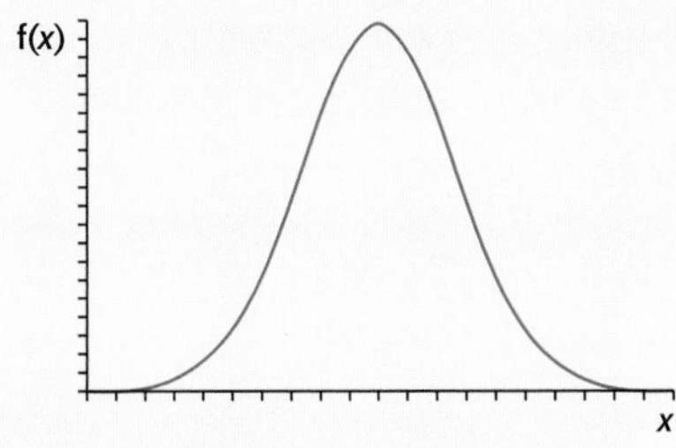

Figure 1 Normal distribution

The area under the curve between any two diameters represents the number of rods in that range of diameters.

This distribution is very common in practical situations. For example, if we measure the height of a large number of adults, the results will fit a normal distribution curve. Any normal distribution is characterized by two numbers: the **mean** ($\bar{x}$) and the **standard deviation, SD (σ)**.

In a normal distribution:

- 68% of samples fall between ±1 SD.
- 95% of samples fall between ±2 SD
- 99.7% of samples fall between ±3 SD.

See Figure 2.

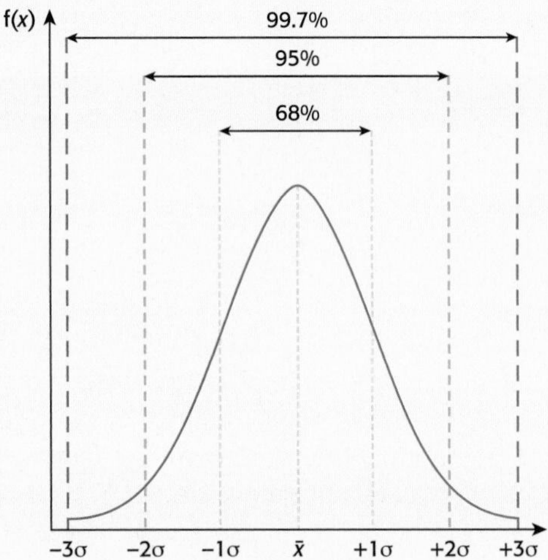

Figure 2 The normal distribution in more detail

We can expect a normal distribution to be produced whenever a given reading is repeated many times and there is the possibility of a **random error** on each reading. Repeating readings will tend to reduce the effect of random errors, because their effect will be "averaged out" over a large number of repeats.

Example of random errors include:

- the readability of a scale (parallax errors in particular: not reading the scale exactly "head on" – sometimes too large, sometimes too small)
- the experimenter's reaction time (if an experiment involves timing the beginning and end of a time interval using a stopwatch, then the recorded time will not be the same as the actual time)
- a meter (with a moving pointer) randomly reading too high or too low as a result of the meter needle having a tendency to "stick".

A full statistical analysis of repeated values would involve calculating the mean and the standard deviation of the recorded data. For IB physics it is usually sufficient to work with uncertainties assessed in terms of a limit of confidence that can be assigned to any given measurement. For example, most students would feel that the confidence limit when using a ruler to measure the length of this book correctly would be about ±1 mm.

It should be obvious from the graph in Figure 2 that a confidence limit is approximately equal to three standard deviations. Ideally, all experimental readings should be repeated several times, and the range of values obtained can be directly used to assign a confidence limit. However, the best practice is to assess practical uncertainties from first principles. See page 77, for an example of how to do this.

We can use these confidence limits to evaluate the results of an experiment. You should already know that the gravitational field strength, g, at the surface of the Earth is approximately 10 N kg^{-1}. Suppose we develop a theory that predicts that g is exactly 10 N kg^{-1}. We devise an experiment to test this theory, and assess the resulting uncertainty. The result of our experiment is $g = 9.2 \pm 1.0$ N kg^{-1}. This value (9.2) is very different from our prediction (10), but the uncertainty range means that the theory could still be correct, because the possible range for g goes from 8.2 up to 10.2 N kg^{-1}.

We now devise a more **precise** experiment – one with a smaller associated uncertainty in the end result. This time the experimental result is $g = 9.85 \pm 0.06$ N kg^{-1}. This second result is closer to our predicted value, but the uncertainty range means that our theory and the experimental result cannot both be correct. So, provided we have not made any errors in the experiment, we have discredited the theory.

Good precision in readings does not necessarily mean the measurements are without error. For example, a meter might have a **zero error**: the zero mark on the scale does not correspond to the position of the needle when the quantity is zero – all readings are in error by the same amount. Simply repeating readings and then taking the average cannot resolve such **systematic errors**. An experiment with small systematic errors is an **accurate** experiment.

Mathematical physics: Dealing with uncertainties

We have highlighted the need to assess the uncertainties associated with any experimental reading. Now we look at the techniques we can use to take these uncertainties into consideration.

Propagation of uncertainties through a calculation

Some simple experiments allow a quantity to be calculated directly from a set of practical measurements. Each measurement will have an associated uncertainty. The most complete way to assess the overall uncertainty of the final result is to do the calculation three times: first using the measured values, and then using the maximum or minimum values to give the maximum and minimum possible answers.

For example, measurements of the length and diameter of a piece of copper wire allow us to calculate the volume of copper in the wire. Say we record the following measurements:

length l = 234 mm ± 2 mm
diameter d = 1.86 mm ± 0.05 mm

The **absolute uncertainty** in the length measurement is ±2 mm, and in the diameter is ±0.05 mm.

So the possible maximum and minimum values of length and diameter are:

l_{max} = 236 mm, l_{min} = 232 mm
d_{max} = 1.91 mm, d_{min} = 1.81 mm

The equation for the volume V of copper is

$$V = \frac{\pi d^2}{4} l$$

So the measured volume, and the possible maximum and minimum values, are

$$V = \frac{\pi(1.86)^2}{4} \times 234\,\text{mm}^3 = 635.816\,\text{mm}^3$$

$$V_{max} = \frac{\pi(1.91)^2}{4} \times 236\,\text{mm}^3 = 676.190\,\text{mm}^3$$

$$V_{min} = \frac{\pi(1.81)^2}{4} \times 232\,\text{mm}^3 = 596.946\,\text{mm}^3$$

Therefore $V_{max} = V + 40.374$ mm³, and $V_{min} = V - 38.870$ mm³, which means that the uncertainty in V is approximately ±40 mm³: that is,

V = 636 ± 40 mm³

This procedure will always work, but can prove very time-consuming. We can use a short cut for quantities that have to be multiplied, divided, or raised to a power. In these situations it helps if we calculate the fractional uncertainty or the percentage uncertainty in each measurement. So, for example, the **fractional uncertainty** in our length measurement is $\pm\frac{2}{234} = \pm 0.0085$, and the **percentage uncertainty** is therefore ±0.85%.

A good way to estimate the overall percentage uncertainty in a quantity is to add all the individual percentage uncertainties in the values we use to calculate it, as follows:

Percentage uncertainty in length $l = \frac{2}{234} \times 100 = \pm 0.9\%$

Percentage uncertainty in diameter $d = \frac{0.05}{1.86} \times 100$
$= \pm 2.7\%$

Therefore percentage uncertainty in $d^2 = 2 \times (\pm 2.7\%) = \pm 5.4\%$

Therefore total uncertainty in volume = (±5.4%) + (±0.9%) = ±6.3%

6.3% of 636 mm³ = 40.068 mm³
That is, V = 636 ± 40 mm³

Note that, when using percentage uncertainties:

- This technique works only with quantities that have to be multiplied or divided. If they have to be added or subtracted, then the maximum and minimum values need to be calculated directly.
- Whether the numbers are multiplied or divided, we still *add* the percentage errors.
- If a reading has to be raised to a power n in calculating the result, the percentage uncertainty in the result is n times the percentage uncertainty in the reading. So, above, the uncertainty in d^2 was twice the uncertainty in d. Similarly, if a reading has to be cubed, the overall percentage uncertainty is three times the uncertainty in the original reading, and if we have to take the square root of a reading (that is, raise it to the power ½), the overall percentage uncertainty is half the value of the original reading.

Graphical representations of uncertainties

A graph visually represents the relationship between two variables. The line of best fit represents the trend shown by the results. In general there will be an uncertainty associated with each point on the graph. We can use **error bars** to show these uncertainties on the graph. For example, the graph in Figure 3 shows the variation of acceleration a with resultant force F.

These results seem to represent a linear relationship. For this to be the case, the error bars for every point must agree with the best-fit line, as shown in Figure 3.

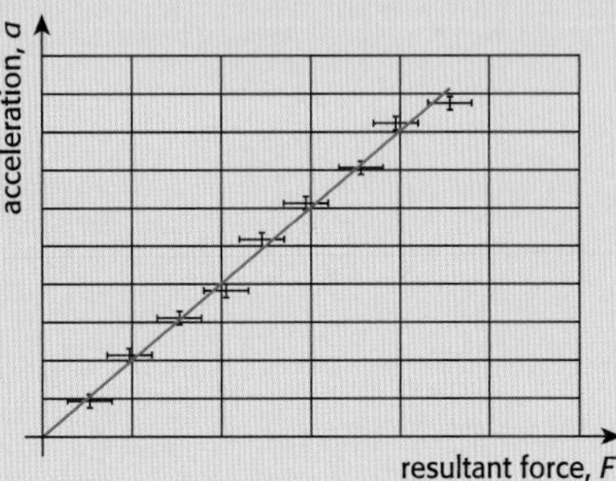

Figure 3 Error bars show the uncertainties in each reading

In many situations, the gradient of a straight-line graph can be used to calculate a constant related to the situation. In the example above the gradient (= acceleration/resultant force) turns out to be the inverse of the mass of the object involved. The range of possible gradients supported by the data is calculated by considering the maximum and minimum gradient lines that can be drawn.

When error bars exist in both the x and y directions, the confidence limits are represented by the rectangle of values that is defined by the two error bars (Figure 4).

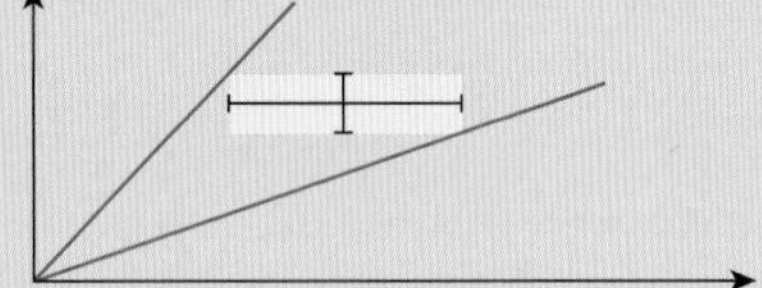

Figure 4 The shaded rectangle shows the range of possible values represented by the two error bars

So, for example, in our acceleration/force case, the maximum and minimum gradient lines would be as shown in Figure 5. We can determine these gradients, and so we can calculate the confidence limits for the gradient.

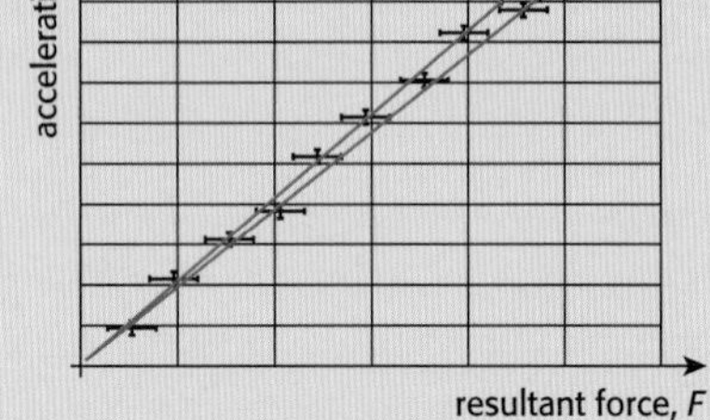

Figure 5 The maximum and minimum gradients of a line passing through the origin

It can be very time-consuming to plot individual error bars for every point. In practice, we often need only add error bars to the first and last data points. And if we know that the graph goes through the origin (because the two quantities are thought to be directly proportional) then it is often sufficient to consider just the point representing the largest values, unless we suspect a systematic error in our readings.

Science tools: general practical skills

Whether you are studying physics at standard or at higher level, 24% of your final marks in the Diploma Programme are awarded for your practical skills. Three quarters of these marks, or 18% of the Diploma, are assessed directly from the practical reports that you submit for marking. So it is important to keep up to date with your practical work, and to learn from earlier mistakes.

The marks for the practical (otherwise known as the internal assessment or IA) are awarded directly by your teacher throughout the course, and strict criteria are published to ensure that the same standards are applied worldwide. A sample of work from every IB school is sent off to check this. Remember: the final marks are awarded as a result of the skills that you can *demonstrate*. For example, you may know how to record raw data, but if you forget to do this properly in a piece of work that is being assessed then it cannot gain full marks.

The following sections focus on each of the five skills that will be assessed, but some general points apply to all of these skills.

- Each skill is given a mark out of 6.
- The first three skills (Design, D; Data collection and processing, DC; and Conclusion and evaluation, CE) are assessed from the written work that you submit.
- Your best two marks for D, DCP, and CE are the ones that count: each of these skills is thus worth a maximum of 12 marks. $3 \times 12 = 36$
- The Manipulative skills (MS) are **summative**. This means they are assessed once only, at the end of the course. The maximum MS mark is 6.
- The Personal skills (PS) are assessed once during the group 4 project. The maximum PS mark is 6.
- The maximum total is out of 48 ($36 + 6 + 6 = 48$).
- Each skill is assessed in terms of three separate **aspects**. The overall mark in each aspect is decided on the extent to which your work matches the criteria for that aspect. The match can be "complete" (2 marks), "partial" (1 mark), or "not at all" (0 marks).
- The assessment of these skills applies to an individual student and not to a team. So although non-assessed practicals can be done with groups of students working together and helping one another, you need to be individually responsible for any tasks that are being assessed.

The following sections look at these practical skills in more detail.

Design (D)

Designing an appropriate experimental technique takes practice, and many students find this skill difficult to master.

All experiments must involve observations. Usually the observations will be quantitative measurements of a given property, such as mass, length, volume, or time. In any experiment, the value of the measured property depends on several variables. For example, the pressure of a sample of gas depends on at least four different variables:

- the volume of the gas
- the temperature of the gas
- the mass of the gas
- the particular gas used.

It is no use changing each of these variables at random and hoping that some pattern will be obvious from the results. We need to discover how each one affects the pressure. For the experiment to be a **fair test**, it needs to be designed so that we can vary only one at a time.

The variable that we choose to control (the one that is *manipulated*) is called the **independent variable**: for example, the volume of the gas.

The variable that we *measure* is called the **dependent variable**: for example, the pressure of the gas.

The variables that are held *constant* throughout the experiment (so that only one thing is affecting the dependent variable) are called the **control variables**: for example, the temperature of the gas, its mass, and the particular gas used.

Your experimental report needs to identify explicitly the dependent variable (measured), the independent variable (manipulated), and the control variables (constants).

Usually, the raw data from an experiment must include four things:

- corresponding values for the independent and dependent variables
- a wide range of values for the independent variable
- repeated readings whenever possible
- readings of all the control variables, when possible, to check that they have been kept constant during the experiment.

The three assessable aspects for the design skill are:

1 defining the problem and selecting variables;
2 controlling variables; and
3 developing a method for collection of data.

Aspect 1: defining the problem and selecting variables

This depends on the context in which the task is set, but for you to demonstrate your ability to define the problem, your teacher can at best give only a general outline of a possible situation to investigate. It is up to you to identify the variables and choose which ones you are going to measure, manipulate, or keep constant. Possible general outlines might be:

- Investigate some physical property of a bouncing ball.
- Investigate the deflection of a cantilever (it is allowable for the teacher to define the dependent variable).

It is up to you to develop this general outline into a specific investigation. If you do not manage to focus the question any further than this, then you have not made progress.

Aspect 2: controlling variables

Your laboratory report must explain your experimental design or method. You need to ensure that the independent variable will be manipulated appropriately, and you must mention how all the other variables are going to be controlled or monitored.

Aspect 3: developing a method for collection of data

The data you collect must be relevant to the problem, and sufficient for you to establish reliable conclusions. This depends on the context, but whenever repeats are easily possible, you must ensure that these take place. If every reading was repeated three times, it would be relatively easy to assess the uncertainty levels, and this technique would also allow large random errors to be immediately identified and checked.

You should aim to have at least ten different values of the independent variable. You may be able to reach valid conclusions on the basis of fewer readings, but it is always best to take as many readings as possible in the time available. If you are plotting average results on a graph, the absolute minimum numbers of readings needed to identify a linear trend is five, but ten would be better.

You also need to consider the data range. In some situations it makes sense to start the experiment by arranging for the dependent variable to take its largest reasonable value and its smallest reasonable value. You can then choose subsequent readings to cover the available range evenly between these two limits.

Table 4 Design

Levels/marks	Aspect 1	Aspect 2	Aspect 3
	Defining the problem and selecting variables	**Controlling variables**	**Developing a method for collection of data**
Complete/2	Formulates a focused problem/research question and identifies the relevant variables.	Designs a method for the effective control of the variables.	Develops a method that allows for the collection of sufficient relevant data.
Partial/1	Formulates a problem/research question that is incomplete **or** identifies only some relevant variables.	Designs a method that makes some attempt to control the variables.	Develops a method that allows for the collection of insufficient relevant data.
Not at all/0	Does not identify a problem/research question **and** does not identify any relevant variables.	Designs a method that does not control the variables.	Develops a method that does not allow for any relevant data to be collected.

Data collection and processing (DCP)

This skill is at the core of any practical work. The data need to be recorded accurately, precisely, and in a way that is not ambiguous. Once the data have been recorded, the resulting information usually has to be processed in order for any conclusions to be drawn. Throughout this process it is important to ensure that uncertainties are considered. They need to be assessed and recorded during the data collection, and they need to be taken into account whenever data are processed.

A very useful approach to data analysis is a graph plot, but sometimes the choice of variables to plot is not straightforward.
By convention, it is usual to plot the independent variable on the x-axis and the dependent variable on the y-axis. In this context time is often considered to be the independent variable, and is therefore plotted on the x-axis.

A line can be added to show the general trend represented by the data points. This **best-fit line** should either be a single smooth curve, or a straight line drawn with a ruler. Never "join the dots": the trend line is an attempt to highlight the general relationship suggested by the data. The graphs in Figure 6 show possible trend lines for different plots of the variation of the dependent variable D with the independent variable I.

Mathematical physics: Graphs of mathematical relationships

Each of the graphs A to H in Figure 6 represents a common mathematical relationship that occurs in physics between a dependent and an independent variable. The graphs are discussed below. Before you read these descriptions, in each case try to describe the mathematical relationship that is suggested by the shape of the graph.

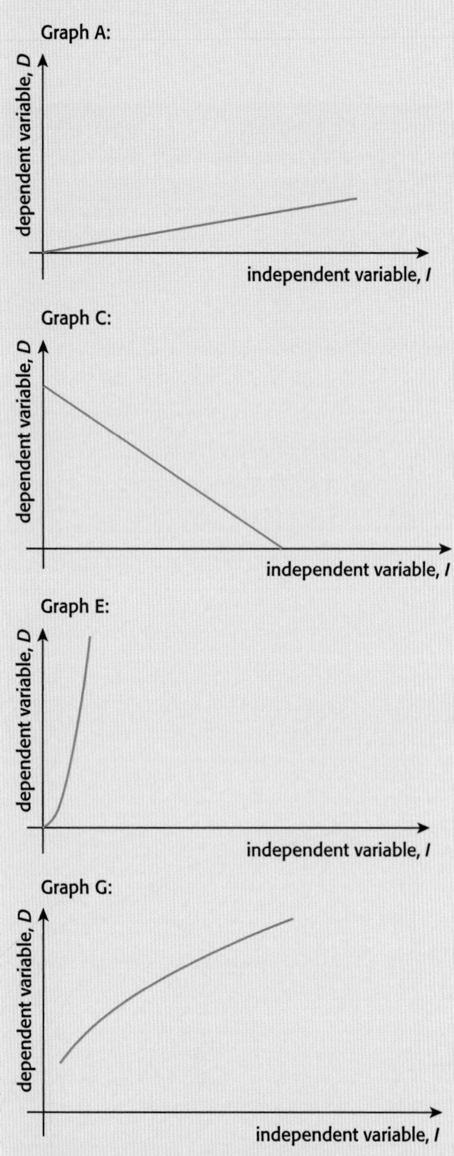

Graph B:
dependent variable, D
independent variable, I
Graph D:
dependent variable, D
independent variable, I
Graph F:
dependent variable, D
independent variable, I
Graph H:
dependent variable, D
independent variable, I

Figure 6 Common shapes for graphs

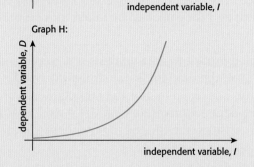

- *Graph A* This graph shows two variables that are **proportional**, because the line is straight and it goes through the origin. A graph with only one of these properties does not show proportionality. The phrase "directly proportional" means the same as "proportional". If the independent variable is doubled, then the dependent variable will also double. There is no such thing as "indirectly proportional". The general mathematical formula for a straight-line graph is $y = mx + c$. In this situation the *y*-intercept (c) is zero, because the line goes through the origin. Thus the equation of this line is $D = \text{constant} \times I$.

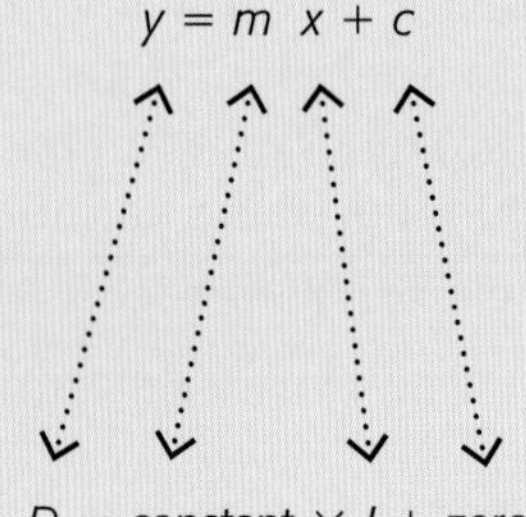

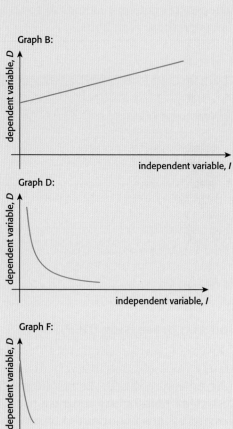

- *Graph B* This graph shows two variables that are *not proportional* because the line is straight *but it* does not go through the origin. There is a linear relationship between the variables. The equation of this line involves two different constants, and is $D = m \times I + c$.
- *Graph C* This graph shows two variables that are **inversely related**: as one increases, the other one decreases. They are not "inversely proportional", because one does not halve when the other one doubles. These variables could also be described as linearly related. This straight line graph has a negative gradient: $D = c - (m \times I)$.
- *Graph D* This graph shows two variables that are **inversely proportional**: as one doubles, the other one halves. Note that the graph does not touch either axis. D is proportional to the inverse of I and vice versa: $D = k \times 1/I$, where k is a constant.
- *Graph E* In this graph the *y*-value is proportional to the square of the *x*-value. Mathematically the relationship is $D = k \times I^2$, where k is a constant.
- *Graph F* The graph hits the *y*-axis but does not cross the *x*-axis. This is an exponential decrease. The mathematical relationship is given by $D = D_0e^{-kI}$: see page 276 for further information.
- *Graph G* In this graph the *y*-value is proportional to the square root of the *x*-value. Mathematically the relationship is $D = k \times \sqrt{I}$ or $I = k' \times D^2$, where k and k' are constants.
- *Graph H* The graph hits the *y*-axis and the slope continues to increase as *x* increases. This is an exponential increase. The mathematical relationship is given by $D = D_0e^{kI}$: see page 276 for an example.

It is useful to plot an appropriate straight-line graph if possible. For example, the time *t* taken for an object to fall, from rest, a distance *s* as a result of gravity is given by the equation

$$s = \frac{1}{2}gt^2$$

where *g* is the acceleration due to gravity. Readings can be taken of the time taken to fall and the distance fallen. A straight plot with *s* on the *y*-axis and *t* on the *x*-axis would look like graph E. It would be more useful to plot *s* on the *y*-axis and t^2 on the *x*-axis. A comparison with the general formula for a straight-line graph shows that this plot would be a straight line. The gradient of the line *m* would be equal to ½*g*, and the intercept on the *y*-axis, *c*, would be zero: that is, the line would go through the origin

$$s = \frac{1}{2}gt^2 + zero$$

$$\updownarrow \quad \updownarrow \quad \updownarrow$$

$$y = mx + c$$

1 The focal length f of a concave mirror, the object distance u, and the corresponding image distance v are related by the equation

$$\frac{1}{u}+\frac{1}{v}=\frac{1}{f}$$

Using the data below, draw a suitable graph in order to find the focal length of the mirror.

u / mm	204	250	345	455	714
v / mm	1000	526	333	270	222

2 An experiment is set up to record the time taken by a ball bearing to fall a known distance (Figure 7). The ball bearing is initially held a measured distance above a surface by an electromagnet. An electronic timer records the time taken between the electromagnet being switched off and the ball bearing hitting the surface. The results are recorded in Table 5 on page 22.

The experimenter assesses the uncertainties as follows:

- Uncertainty in time is ±0.005 s owing to residual magnetism in the ball bearing.
- Uncertainty in distance is ±0.5 cm.

Analyse these data in order to estimate the acceleration of free fall, g. Include an estimate of the uncertainty in your answer.

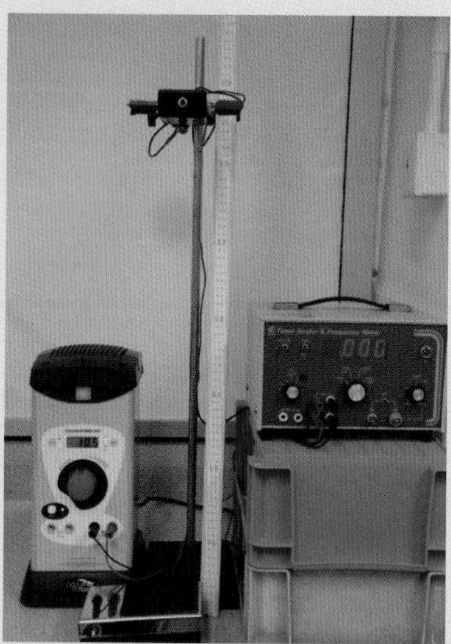

Figure 7 Apparatus to measure the acceleration due to gravity by freefall

Table 5 Experimental data for question 2

Distance, D/cm	Time for free fall, T / seconds			
	T_1	T_2	T_3	T (average)
30.0	0.270	0.269	0.269	0.269
40.0	0.305	0.305	0.304	0.305
50.0	0.333	0.334	0.333	0.333
60.0	0.366	0.365	0.366	0.366
70.0	0.392	0.392	0.392	0.392
80.0	0.413	0.413	0.412	0.413
90.0	0.438	0.438	0.439	0.438
100.0	0.458	0.459	0.457	0.458

Aspect 1: recording raw data

Most of the time you will be working with quantitative data. Whenever such data are recorded it is important to make clear the precise quantity that is being measured, the units it is being measured in, and the uncertainty range for the measurement. It is good practice to ensure that all this information is clearly in the headings for a data table. The example to the right shows a possible table designed for an investigation into the I–V electrical characteristics of a bulb.

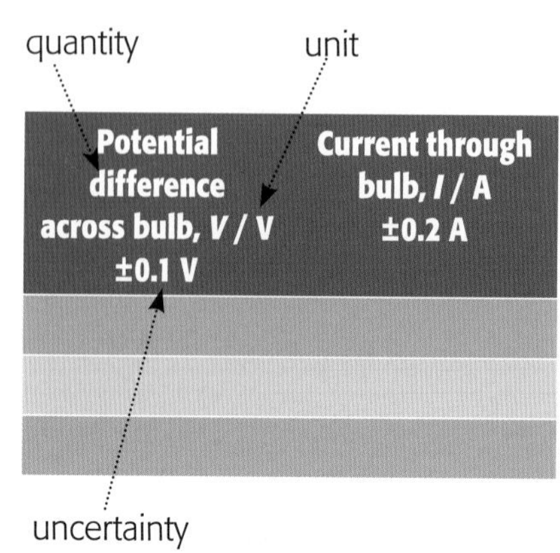

You should discuss in your report your reasons for choosing to assign any particular level of uncertainty, and the number of significant digits you use to record the data should be consistent with the stated uncertainty. In the table it makes sense for there to be a potential difference reading of, say, 4.0 V. It would not make sense to record the reading either as 4.000 (too many significant digits) or as 4 (too few). If the uncertainty is the same throughout an experiment, then the number of decimal places used should always be the same. Average values should also be stated to an appropriate number of significant digits.

Aspect 2: processing raw data

To gain full marks for this aspect you need to be able to process the raw data that you recorded in some mathematical way. Often experiments involve plotting a graph, but a simple direct plot of the variables does not involve any processing. Graphs are ways of presenting data. If the data do not need processing before producing a straight-line graph, at the very least you need to choose to calculate the average of repeated readings, plot these average points, identify a suitable best-fit line, and calculate the gradient and/or the intercept.

Aspect 3: presenting processed data

You need to present the results of any processed data clearly and unambiguously, and you must take uncertainties into consideration by, for example, using the techniques shown on pages 13–16. Graphs need to have: appropriate scales; labelled axes with units; and accurately plotted data points with a suitable best-fit line or curve. Specifically, you need to:

- include uncertainty bars where significant
- explain where uncertainties are not significant
- draw lines of minimum and maximum gradients
- determine the uncertainty in the best straight-line gradient.

Table 6 Data collection and processing

Levels/marks	Aspect 1	Aspect 2	Aspect 3
	Recording raw data	**Processing raw data**	**Presenting processed data**
Complete/2	Records appropriate quantitative and associated qualitative raw data, including units and uncertainties where relevant.	Processes the quantitative raw data correctly.	Presents processed data appropriately and, where relevant, includes errors and uncertainties.
Partial/1	Records appropriate quantitative and associated qualitative raw data, but with some mistakes or omissions.	Processes quantitative raw data, but with some mistakes and/or omissions.	Presents processed data appropriately, but with some mistakes and/or omissions.
Not at all/0	Does not record any appropriate quantitative raw data **or** raw data is incomprehensible.	No processing of quantitative raw data is carried out **or** major mistakes are made in processing.	Presents processed data inappropriately **or** incomprehensibly.

Conclusion and evaluation (CE)

The purpose of most physics experiments is either to measure a physical quantity or to investigate the relationship between two (or more) variables. A good overall conclusion and evaluation of an experiment does three things:

1 It summarises the findings of the experiment, and discusses these findings in the context of the assessed uncertainty limits.
2 In the light of what has been discovered, it critically analyses the experimental procedure to see whether the procedure contains any weaknesses or limitations.
3 It proposes sensible improvements to be made if the experiment is to be repeated in the future. These improvements should aim to correct any weaknesses or limitations already identified.

Aspect 1: concluding

The conclusion of an experiment should be statements of trends or patterns revealed by the data obtained rather than just a restatement of any predicted outcome. Once again, you need to take uncertainties into consideration when justifying any conclusions, and it is good practice always to discuss whether the main sources of error were systematic or random in nature.

If the aim of your experiment is to measure a physical quantity that is already known, you should compare your experimental value with literature values. If you have correctly assessed the uncertainty associated with your experimental procedure, these two values should agree within experimental error. You should fully reference any value that you quote from the literature.

Aspect 2: evaluating procedure(s)

You should comment on the design and method of the investigation as well as on the quality of the data. This should not just be a list of possible weaknesses; it should include some discussion of how significant particular weaknesses turn out to be. It is good practice always to discuss the precision and accuracy of the measurements taken, and to comment on the processes, use of equipment, and management of time.

Aspect 3: improving the investigation

The suggestions for improvement should arise naturally from the weaknesses and limitations that you have identified when evaluating the procedures used, and they should be realistic and appropriate. If, for example, the analysis of uncertainties in a particular experiment assigned ±1% as a result of human reaction time and ±15% to readability issues for the other measurements, then it would not be appropriate to simply suggest using more precise timing equipment.

Table 7 Conclusion and evaluation grid

Levels/ marks	Aspect 1: concluding	Aspect 2: evaluating procedure(s)	Aspect 3: improving the investigation
Complete/2	States a conclusion, with justification, based on a reasonable interpretation of the data.	Evaluates weaknesses and limitations.	Suggests realistic improvements in respect of identified weaknesses and limitations.
Partial/1	States a conclusion based on a reasonable interpretation of the data.	Identifies some weaknesses and limitations, but the evaluation is weak or missing.	Suggests only superficial improvements.
Not at all/0	States no conclusion or the conclusion is based on an unreasonable interpretation of the data.	Identifies irrelevant weaknesses and limitations.	Suggests unrealistic improvements.

Manipulative skill (MS)

By the end of your course you should have gained wide experience of different experimental procedures and approaches. Your teacher will submit a mark to the IB that reflects how far you are able to follow instructions accurately and precisely, carry out a wide range of

1 The diagram below shows the position of the meniscus of the mercury in a mercury-in-glass thermometer.

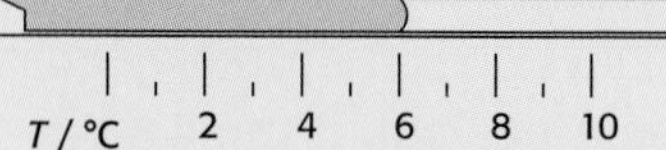

Which of the following best expresses the indicated temperature with its uncertainty?

A (6.0 ± 0.5)°C

B (6.1 ± 0.1)°C

C (6.2 ± 0.2)°C

D (6.2 ± 0.5)°C [1]

N.B. *In this question you need to make a judgement about how accurately you can read the thermometer. The standard "rule" of half a division is too pessimistic.*

2 The time period T of oscillation of a mass m suspended from a vertical spring is given by the expression

$$T = 2\pi\sqrt{\frac{m}{k}}$$

where k is a constant.

Which **one** of the following plots will give rise to a straight-line graph?

A T^2 against m

B $\sqrt{T}$ against $\sqrt{m}$

C T against m

D $\sqrt{T}$ against m [1]

3 An ammeter has a zero offset error. This fault will affect

A neither the precision nor the accuracy of the readings.

B only the precision of the readings.

C only the accuracy of the readings.

D both the precision and the accuracy of the readings. [1]

techniques, and appreciate the importance of approaching all practical work in a safe way.

Table 8 Manipulative skills grid

Levels/marks	Aspect 1: following instructions	Aspect 2: carrying out techniques	Aspect 3: working safely
Complete/2	Follows instructions accurately, adapting to new circumstances (seeking assistance when required).	Competent and methodical in the use of a range of techniques and equipment.	Pays attention to safety issues.
Partial/1	Follows instructions but requires assistance.	Usually competent and methodical in the use of a range of techniques and equipment.	Usually pays attention to safety issues.
Not at all/0	Rarely follows instructions or requires constant supervision.	Rarely competent and methodical in the use of a range of techniques and equipment.	Rarely pays attention to safety issues.

Personal skills (PS)

It is a requirement of the course for all IB Diploma Programme physics students to work together and be involved in a collaborative project across the science disciplines, known as the **group 4 project**. The theme and mechanics of the project will vary between schools and colleges worldwide. The aims are:

- to encourage an understanding of the relationships between scientific disciplines and the overarching nature of the scientific method
- to raise awareness of the moral, ethical, social, economic, and environmental implications of using science and technology
- to develop and apply information and communication technology skills in the study of science
- to develop an appreciation of the possibilities and limitations associated with science and scientists
- to develop an ability to analyse, evaluate, and synthesize scientific information.

The aim of every team involved in the group 4 project is to make scientific progress, but in this context the process is more important than the actual experimental outcomes.

The personal skills mark assesses your self-motivation and perseverance, your ability to work with other scientists as part of a team, and the extent to which you are aware of your own strengths and weaknesses and are able to take responsibility for your own learning.

Table 9 Personal skills grid

Levels/marks	Aspect 1: self-motivation and perseverance	Aspect 2: working within a team	Aspect 3: working within a team
Complete/2	Approaches the project with self-motivation and follows it through to completion.	Collaborates and communicates in a group situation and integrates the views of others.	Shows a thorough awareness of their own strengths and weaknesses and gives thoughtful consideration to their learning experience.
Partial/1	Completes the project but sometimes lacks self-motivation.	Exchanges some views but requires guidance to collaborate with others.	Shows limited awareness of their own strengths and weaknesses and gives some consideration to their learning experience.
Not at all/0	Lacks perseverance and motivation.	Makes little or no attempt to collaborate in a group situation.	Shows no awareness of their own strengths and weaknesses and gives no consideration to their learning experience.

End of chapter summary

Chapter 2 has four main themes: the realm of physics, measurements, uncertainties and practical skills. This last theme is explored separately for measurements, calculated results and graphs. The list below summarises the knowledge and skills that you should be able to undertake after having studied this chapter. Further research into more detailed explanations using other resources and/or further practice at solving problems in all these topics is recommended – particularly the items in bold.

The realm of physics

- State and compare quantities to the nearest order of magnitude including the ranges of magnitude of distances, masses and times that occur in the universe, from smallest to greatest.
- State ratios of quantities as differences of orders of magnitude and estimate approximate values of everyday quantities to one or two significant figures and/or to the nearest order of magnitude.

Measurements

- State the fundamental units in the SI system and distinguish between fundamental and derived units (with examples).
- **Convert between different units of quantities and state units in the accepted SI format (and in scientific notation or in multiples of units with appropriate prefixes).**

Uncertainties

- Describe **and give examples** of random and systematic errors and explain how the effects of random errors may be reduced.
- Distinguish between precision and accuracy in an experiment.
- **Calculate quantities and results of calculations to the appropriate number of significant figures.**
- State uncertainties in measurements as absolute, fractional and percentage uncertainties and use these to determine the uncertainties in results.
- Use error bars in graphs to identify uncertainties and go on to determine the uncertainties in the slope and intercepts of a straight-line graph.

Practical skills

- Understand the detailed techniques that need to be mastered for each of the following practical skills: Design, Data collection and processing, Conclusion and evaluation as well as the manipulative and personal skills necessary to undertake scientific investigations both individually and as part of a team.

Chapter 2 questions

1 This question is about measuring the permittivity of free space ε_0.

The diagram below shows two parallel conducting plates connected to a variable voltage supply. The plates are of equal areas and are a distance d apart.

The charge Q on one of the plates is measured for different values of the potential difference V applied between the plates. The values obtained are shown in the table below. Uncertainties in the data are not included.

V / V	Q / nC
10.0	30
20.0	80
30.0	100
40.0	160
50.0	180

a) Plot a graph of V (x-axis) against Q (y-axis). [4]

b) Draw the line of best fit for the data points. [1]

c) Determine the gradient of your best-fit line. [2]

d) The gradient of the graph is a property of the two plates and is known as *capacitance*. Deduce the units of capacitance. [1]

The relationship between Q and V for this arrangement is given by the expression

$$Q = \frac{\varepsilon_0 A}{d} V$$

where A is the area of one of the plates.

In this particular experiment $A = 0.20\ m^2$ and $d = 0.50$ mm.

e) Use your answer to (c) to determine a value for ε_0. [3]

(Total 11 marks)

2 Data based question. This question is about change of electrical resistance with temperature.

The table below gives values of the resistance R of an electrical component for different values of its

temperature *T*. (*Uncertainties in measurement are not shown.*)

T/°C	1.2	2.0	3.5	5.2	6.8	8.1	9.6
R/Ω	3590	3480	3250	3060	2880	2770	2650

a) Plot a graph to show the variation with temperature *T* of the resistance *R*. Show values on the temperature axis from *T* = 0°C to *T* = 10°C. [3]

b) i) Draw a curve that best fits the points you have plotted. Extend your curve to cover the temperature range from 0°C to 10°C. [1]

ii) Use your graph to determine the resistance at 0°C and at 10°C. [2]

c) On your graph, draw a straight-line between the resistance values at 0°C and at 10°C. This line shows the variation with temperature (between 0°C and 10°C) of the resistance, assuming a linear change. [1]

d) i) Assuming a linear change of resistance with temperature, use your graph to determine the temperature at which the resistance is 3060. [1]

ii) Use your answer in (d)(i) to calculate the percentage difference in the temperature for a resistance of 3060 that results from assuming a linear change rather than the non-linear change. [3]

(Total 11 marks)

3 The Geiger-Nuttall theory of α-particle emission relates the half-life of the α-particle emitter to the energy *E* of the α-particle. One form of this relationship is

$$L = \frac{166}{E^{\frac{1}{2}}} - 53.5.$$

L is a number calculated from the half-life of the α-particle emitting nuclide and *E* is measured in MeV. Values of *E* and *L* for different nuclides are given below. (*Uncertainties in the values are not shown.*)

Nuclide	E / MeV	L	$\frac{1}{E^{\frac{1}{2}}}$/MeV$^{-\frac{1}{2}}$
^{238}U	4.20	17.15	0.488
^{236}U	4.49	14.87	0.472
^{234}U	4.82	12.89	0.455
^{228}Th	5.42	7.78	______
^{208}Rn	6.14	3.16	0.404
^{212}Po	7.39	−2.75	0.368

a) Complete the table above by calculating, using the value of *E* provided, the value of $\frac{1}{E^{\frac{1}{2}}}$ for the nuclide ^{228}Th. Give your answer to three significant digits. [1]

The graph below shows the variation with $\frac{1}{E^{\frac{1}{2}}}$ of the quantity *L*. Error bars have not been added.

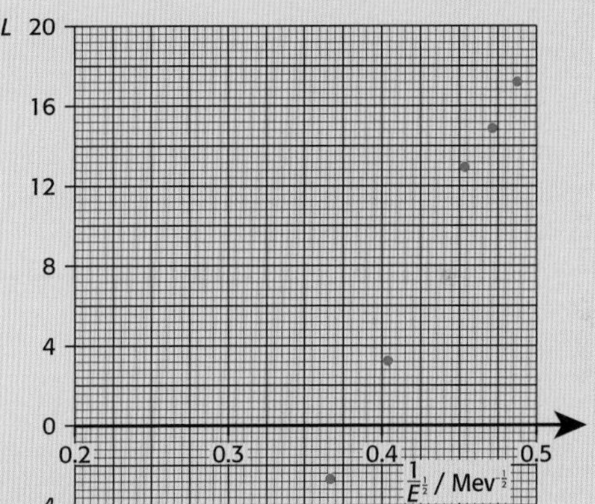

b) i) Identify the data point for the nuclide ^{208}Rn. Label this point R. [1]

ii) On the graph, mark the point for the nuclide ^{228}Th. Label this point T. [1]

iii) Draw the best-fit straight-line for all the data points. [1]

c) i) Determine the gradient of the line you have drawn in (b)(iii). [2]

ii) Without taking into consideration any uncertainty in the values for the gradient and for the intercept on the x-axis, suggest why the graph does **not** agree with the stated relationship for the Geiger-Nuttall theory. [2]

d) On the graph above, draw the line that would be expected if the relationship for the Geiger-Nuttall theory were correct. No further calculation is required. [2]

(Total 10 marks)

4 This question is about thermal energy transfer through a rod.

A student designed an experiment to investigate the variation of temperature along a copper rod when each end is kept at a different temperature.

In the experiment, one end of the rod is placed in a container of boiling water at 100°C and the other end is placed in contact with a block of ice at 0.0°C as shown in the diagram.

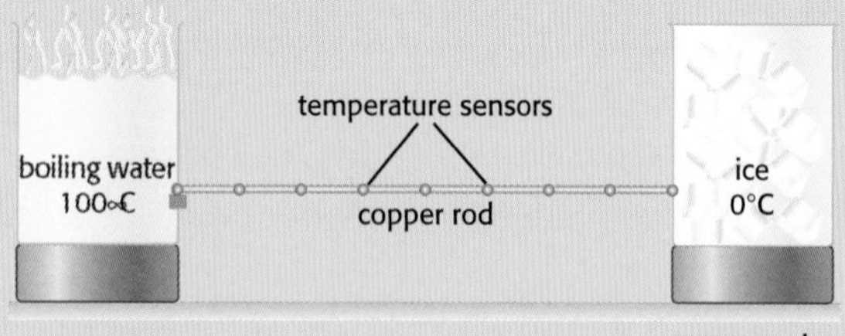

Temperature sensors are placed at 10 cm intervals along the rod. The final steady state temperature θ of each sensor is recorded, together with the corresponding distance x of each sensor from the hot end of the rod.

The data points are shown plotted on the axes below.

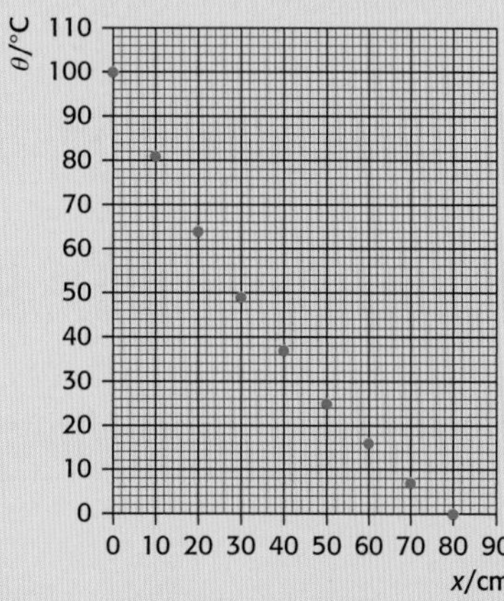

The uncertainty in the measurement of θ is ±2°C. The uncertainty in the measurement of x is negligible.

a) On the graph above, draw the uncertainty in the data points for x = 10 cm, x = 40 cm and x = 70 cm. [2]

b) On the graph above, draw the line of best-fit for the data. [1]

c) Explain, by reference to the uncertainties you have indicated, the shape of the line you have drawn. [2]

d) i) Use your graph to estimate the temperature of the rod at x = 55 cm. [1]

ii) Determine the magnitude of the gradient of the line (the temperature gradient) at x = 50 cm. [3]

e) The rate of transfer of thermal energy R through the cross-sectional area of the rod is proportional to the temperture gradient $\frac{\Delta\theta}{\Delta x}$ along the rod. At x = 10 cm, R = 43W and the magnitude of the temperature gradient is $\frac{\Delta\theta}{\Delta x} = 1.81°\text{C cm}^{-1}$.

At x = 50 cm the value of R is 25 W.

Use these data and your answer to d(ii) to suggest whether the rate R of thermal energy transfer is in fact proportional to the temperature gradient. [3]

(Total 12 marks)

5 As part of a road-safety campaign, the braking distances of a car were measured.

A driver in a particular car was instructed to travel along a straight road at a constant speed v. A signal was given to the driver to stop and he applied the brakes to bring the car to rest in as short a distance as possible. The total distance D travelled by the car after the signal was given was measured for corresponding values of v. A sketch-graph of the results is shown below.

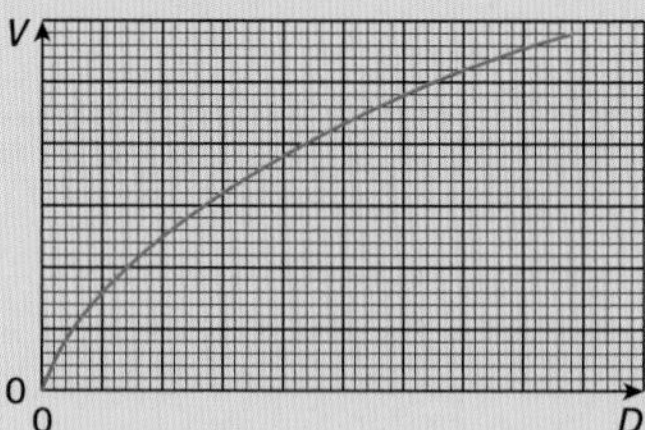

a) State why the sketch graph suggests that D and v are **not** related by an expression of the form

$$D = mv + c,$$

where m and c are constants. [1]

b) It is suggested that D and v may be related by an expression of the form

$$D = av + bv^2,$$

where a and b are constants.

In order to test this suggestion, the data shown below are used. The uncertainties in the measurements of D and v are not shown.

v / m s^{-1}	D / m	$\frac{D}{v}$ /........
10.0	14.0	1.40
13.5	22.7	1.68
18.0	36.9	2.05
22.5	52.9	
27.0	74.0	2.74
31.5	97.7	3.10

i) In the table above, state the unit of $\frac{D}{v}$. [1]

ii) Calculate the magnitude of $\frac{D}{v}$, to an appropriate number of significant digits, for v = 22.5 m s^{-1}. [1]

c) Data from the table are used to plot a graph of $\frac{D}{v}$(y-axis) against v(x-axis). Some of the data points are shown plotted below.

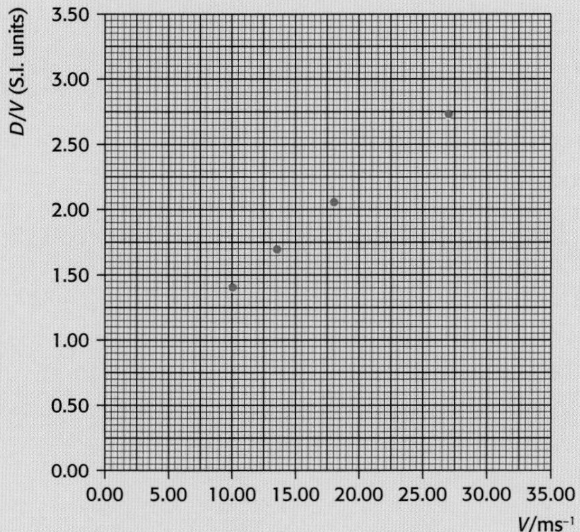

On the graph above,

i) plot the data points for speeds corresponding to 22.5 m s^{-1} and to 31.5 m s^{-1}. [2]

ii) draw the best-fit line for all the data points. [1]

d) Use your graph in (c) to determine

i) the total stopping distance D for a speed of 35 m s^{-1}. [2]

ii) the intercept on the $\frac{D}{v}$ axis. [1]

iii) the gradient of the best-fit line. [2]

e) Using your answers to (d)(ii) and (d)(iii), deduce the equation for D in terms of v.

D = [1]

f) **i)** Use your answer to (e) to calculate the distance D for a speed v of 35.0 m s^{-1}. [1]

(ii) Briefly discuss your answers to (d)(i) and (f)(i). [1]

(Total 14 marks)

6 The resistive force F that acts on an object moving at speed v in a stationary fluid of constant density is given by the expression

$F = kv^2$

where k is a constant.

a) State the derived units of

i) force F. [1]

ii) speed v. [1]

b) Use your answers in (a) to determine the derived units of k. [1]

(Total 3 marks)

3 Kinematics

Kinematics is the study of objects in motion. In the photographs above you can see examples of sportspeople controlling the motion of various objects. This chapter focuses on the accurate and precise measurement of the quantities that can be used to describe the way in which an object moves. Subsequent chapters will analyse these quantities in more detail. The principles involved can be applied in a wide range of situations, from the motion of individual subatomic particles up to complex extended objects such as stars or even galaxies.

Most everyday objects in motion undergo a complex mixture of different types of motion, all occurring together. For example, consider an athlete attempting the high jump. The athlete's overall motion through the air can be considered as a mixture of:

- The translational motion of the **centre of mass** of the athlete in each of three independent directions represented by the x, y, and z-axes. For these axes to be independent of one another they need to be mutually perpendicular, i.e. at 90° to one another, but we are free to choose their direction and/or orientation to fit the specific situation.
- The rotational motion of the athlete around each of three independent axes of rotation that pass through the centre of mass. Again, for these axes to be independent of one another, they need to be mutually perpendicular. It is often convenient to choose the axes of rotation to be the same as the axes used to describe the translational motion, but this is not always necessary.

Figure 1 Three independent axes for translational and rotational notion

1 A bicycle is being ridden in a straight line. Identify the separate translational and rotational motions that fully describe the overall motion of one of the bicycle wheels.

2 A bicycle is being ridden around a corner. The bicycle leans over as it turns to the left. Identify the separate translational and rotational motions that fully describe the overall motion of one of the bicycle wheels.

We shall start by considering the simplest possible situations, with no rotational motion, and translational motion in only one direction,

e.g. an object sliding along a surface in a straight line. We can then extend the ideas to motion in two dimensions, e.g. an object sliding along a surface, and going around a corner (circular motion) or the motion of an object thrown through the air. Three-dimensional motion and rotation about more than one axis are extensions of the same ideas.

To express an overall translational motion in terms of separate motions in independent directions, we need to consider not only the size (i.e. the **magnitude**) of the various quantities, but also their **direction**. Quantities that have both magnitude and direction are called **vector** quantities, whereas quantities that have only magnitude are called **scalar** quantities. Vector and scalar quantities need to be treated in different ways – see page 41.

IB Diploma Programme physics limits the quantitative study of motion to either translational motion or constant rotational motion (circular motion) but not both at the same time. The mathematics of accelerating rotational motion is not included.

Key terms

We need three quantities to describe an object's motion fully in any given situation: its **displacement**, its **velocity,** and its **acceleration**. These are all vector quantities. Two other related scalar quantities are often also used: the object's **distance** and its **speed**. For each of these five quantities it is important to understand the difference between an **average** value over a time interval and the **instantaneous** value at a particular time. Formulae exist that relate these quantities together, but they only apply in certain circumstances.

Many of the equations given below use calculus notation to define the quantities precisely. If your mathematics is not yet up to using this notation, it is sufficient to focus on the word equations and the associated units. If this is the first time you have encountered the definitions, it is best to start with the simplest situation of all: motion in only one direction. You should revisit these definitions once you have studied vector mathematics (see page 41).

Displacement, s

This quantity is the vector equivalent of distance. The displacement $\boldsymbol{s}$ of a particle is the length and direction of the line drawn to the particle from some fixed point (called the **origin**). In SI units the magnitude of the displacement is measured in metres, centimetres, kilometres, etc. For example, the displacement of my place of work from my home is 5.4 km, northeast (Figure 2).

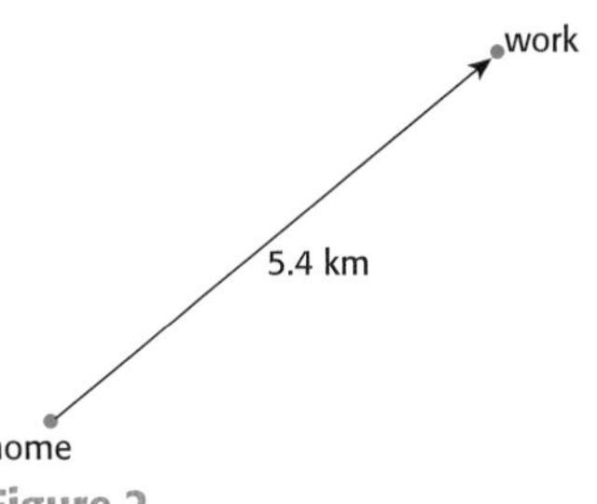

Figure 2

Velocity, v

This quantity is the vector equivalent of the scalar quantity, speed. A moving particle's displacement changes with time. In a time interval Δt, the change in displacement (found by subtracting the initial displacement from the final displacement using vector mathematics) is $\Delta \boldsymbol{s}$. The **average velocity** $\boldsymbol{v}_{av}$ during an interval of time Δt is defined as follows:

$$\boldsymbol{v}_{av} = \frac{\text{change in displacement}}{\text{time taken}} \qquad \boldsymbol{v}_{av} = \frac{\Delta \boldsymbol{s}}{\Delta t}$$

The units of average velocity are m s^{-1}. This quantity is different from **instantaneous velocity**, v, which is the value calculated using the above equation when the change of displacement and the time interval become vanishingly small. This is called the rate of change of displacement. To signify that Δs and Δt are very small they are written as δs and δt.

$$v = \frac{\text{small change in displacement}}{\text{small time taken}} = \text{rate of change of displacement}$$

$$v = \lim_{\delta t \to 0}\left(\frac{\delta s}{\delta t}\right) = \frac{ds}{dt}$$

Speed

Speed is a scalar quantity. It is also measured in m s^{-1}. The average speed is not necessarily equal to the magnitude of the average velocity. The definition is

$$\text{average speed} = \frac{\text{distance travelled along actual path}}{\text{time taken}} = \text{rate of change of distance}$$

The instantaneous speed is the value calculated by the above equation when the time interval becomes vanishingly small, and is always equal to the magnitude of the instantaneous velocity.

Acceleration, a

This is a vector quantity related to the change in a moving particle's velocity with time. In a time interval Δt, the change in velocity is $\Delta \boldsymbol{v}$. The **average acceleration $\boldsymbol{a}_{av}$** is defined as follows:

$$\boldsymbol{a}_{av} = \frac{\text{change in velocity}}{\text{time taken}} \qquad \boldsymbol{a}_{av} = \frac{\Delta \boldsymbol{v}}{\Delta t}$$

The units of the magnitude of average acceleration are m s^{-2}. This quantity is different from **instantaneous acceleration**, $\boldsymbol{a}$, which is the value calculated using the above equation when the change of velocity and the time interval become vanishingly small. This is called the rate of change of velocity.

$$\boldsymbol{a} = \frac{\text{small change in velocity}}{\text{small time taken}} = \text{rate of change of velocity}$$

$$\boldsymbol{a} = \lim_{\delta t \to 0}\left(\frac{\delta v}{\delta t}\right) = \frac{dv}{dt}$$

An important experimental observation is that the acceleration of an object falling in a vacuum near the Earth's surface is independent of its mass. This is called the acceleration of free fall, g. The average value for g at sea level is 9.81 m s^{-2}, or approximately 10 m s^{-2}.

Uniform acceleration equations

Provided an object moves with uniform acceleration, we can manipulate the definitions of average velocity and average acceleration to create a set of five equations. Each of these equations relates four of the five relevant quantities that we would need to fully describe the object's motion:

initial displacement = zero (by definition)
final displacement = $\boldsymbol{s}$

This question is about linear motion.

A police car P is stationary by the side of a road. A car S, exceeding the speed limit, passes the police car P at a constant speed of 18 m s^{-1}. The police car P sets off to catch car S just as car S passes the police car P. Car P accelerates at 4.5 m s^{-2} for a time of 6.0 s and then continues at constant speed. Car P takes a time t seconds to draw level with car S.

a) i) State an expression, in terms of t, for the distance car S travels in t seconds. [1]

ii) Calculate the distance travelled by the police car P during the first 6.0 seconds of its motion. [1]

iii) Calculate the speed of the police car P after it has completed its acceleration. [1]

iv) State an expression, in terms of t, for the distance travelled by the police car P during the time that it is travelling at constant speed. [1]

b) Using your answers to (a), determine the total time t taken for the police car P to draw level with car S. [2]

initial velocity = $\boldsymbol{u}$
final velocity = $\boldsymbol{v}$
acceleration = $\boldsymbol{a}$
time taken = t

In any given situation, if we know any three of these quantities, then we can calculate the other two.

$$v = u + at \qquad s = \frac{u+v}{2}t$$

$$v^2 = u^2 + 2as \qquad s = ut + \frac{1}{2}at^2$$

$$s = vt - \frac{1}{2}at^2$$

Relative velocity

Consider two points A and B, moving with velocities $\boldsymbol{v}_A$ and $\boldsymbol{v}_B$, respectively. The velocity of A relative to B, $\boldsymbol{v}_{AB}$ (otherwise known as the relative velocity of A with respect to B), is the velocity that A appears to have according to an observer who is moving with B. This is calculated by subtracting the velocity of B from the velocity of A:

$$\boldsymbol{v}_{AB} = \boldsymbol{v}_A - \boldsymbol{v}_B \qquad \boldsymbol{v}_{BA} = \boldsymbol{v}_B - \boldsymbol{v}_A$$

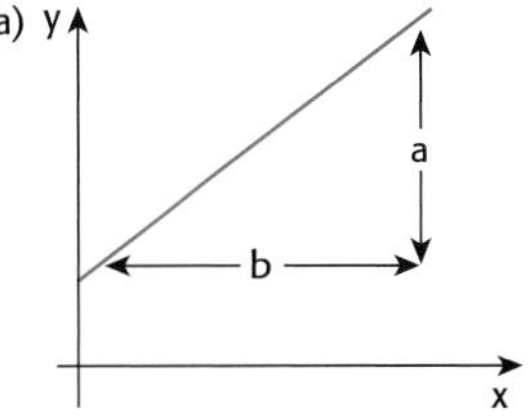

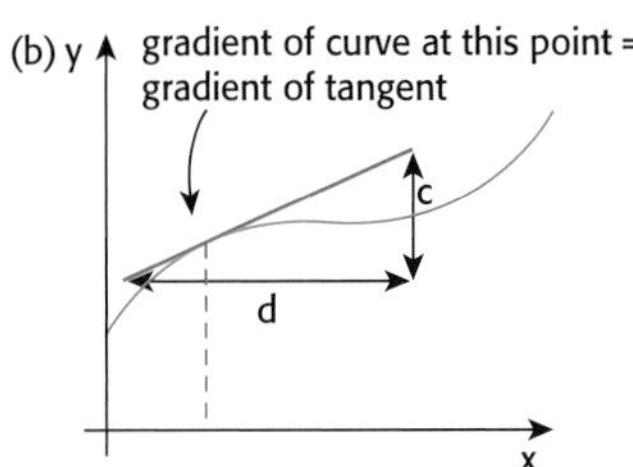

Figure 3 (a) gradient of straight line = $\frac{a}{b}$ (b) gradient of tangent = $\frac{c}{d}$

Graphical work

The definitions of displacement, instantaneous velocity, and instantaneous acceleration correspond to either gradients or areas found from displacement–time graphs or velocity–time graphs.

The gradient is the ratio of the change in the y-axis measurement to the change in the x-axis measurement ("rise over run"). A straight-line graph can be analysed directly (Figure 3a) but a curve may be best analysed by calculating the gradient of the tangent to the line (Figure 3b).

The area under the graph takes into account the units used on the axes. We can calculate simple shapes mathematically, but to determine the area under complicated curves we need to count the number of squares on graph paper (Figure 4).

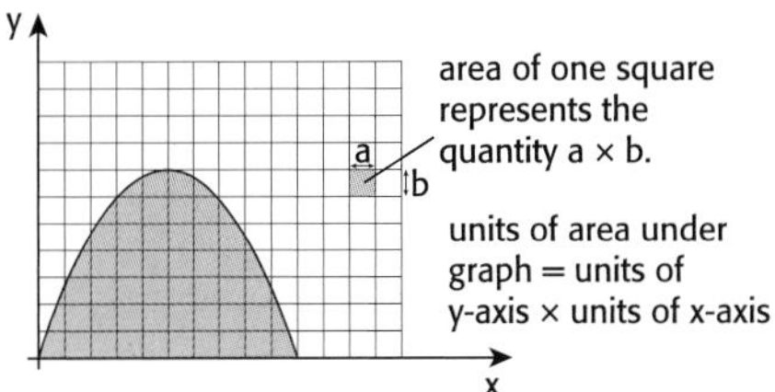

Figure 4 The area under a curve

Displacement–time graphs

For these graphs the slope of the graph at any point represents the instantaneous velocity, and the slope of the line joining any two points on the graph represents the average velocity over the chosen time interval (Figure 5). The area under the graph does not represent any physical quantity.

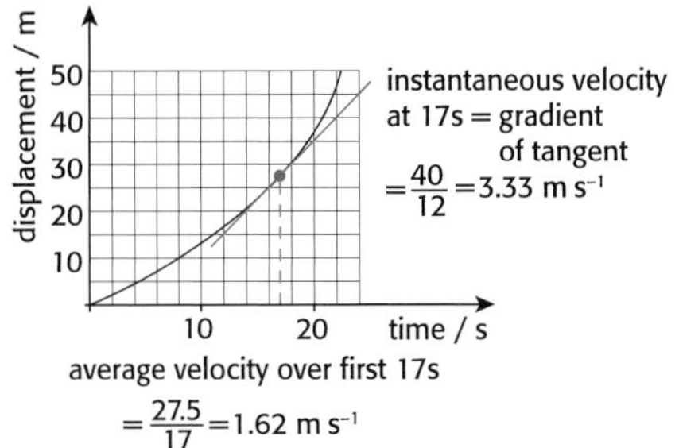

Figure 5 Calculations from a displacement–time graph

Velocity–time graphs

For these graphs the slope of the graph at any point represents the instantaneous acceleration, and the slope of the line joining any two points on the graph represents the average acceleration over the chosen time interval. The area under the graph – between the line and the x-axis – represents the displacement that takes place over the chosen time interval (Figure 6). It is possible for the area to represent a negative number if either the displacement or the velocity is negative.

Acceleration–time graphs

For these graphs the slope of the graph at any point represents the rate of change of acceleration, and the slope of the line joining any two points on the graph represents the average rate of change of acceleration over the chosen time interval. The area under the graph represents the change in velocity that takes place over the chosen time interval (Figure 7).

Graphical display calculators often have powerful built-in tools for determining the gradient of the tangent to a curve or the area under the curve. See pages 87–8 for an example of this for simple harmonic motion.

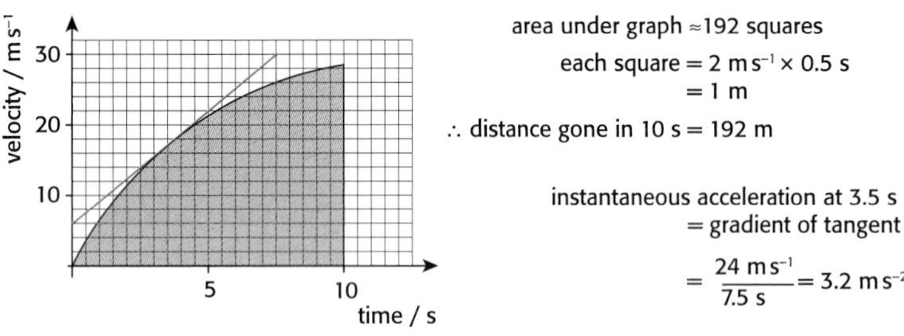

Figure 6 Calculations from a velocity–time graph

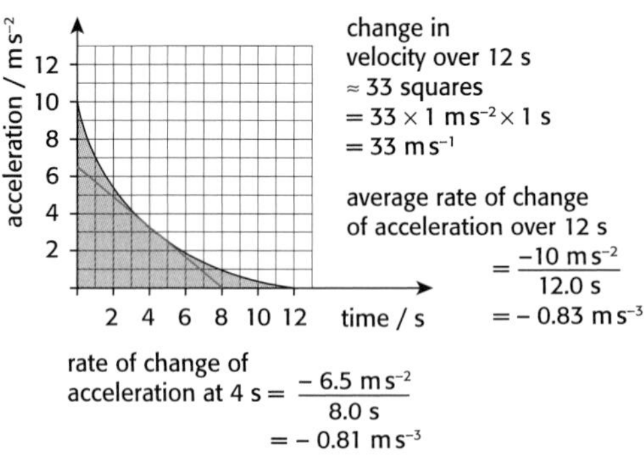

Figure 7 Calculations from an acceleration–time graph

***Mathematical physics:** Modelling falling under gravity*

Predicting the time taken for an object to free-fall a known distance under gravity involves a straightforward application of the uniform acceleration equations. This prediction assumes that, throughout the fall, air resistance can be taken as negligible.

For example, the time taken for an object to free-fall 10 m would be calculated as follows:

initial velocity $\boldsymbol{u}$ = zero

final velocity $\boldsymbol{v}$ = ?

uniform acceleration $\boldsymbol{a}$ = 9.81 m s^{-2}

time taken $\boldsymbol{t}$ = ?

final displacement $\boldsymbol{s}$ = 10 m

$$s = ut + \frac{1}{2}at^2$$

$$\therefore t = \sqrt{\frac{2s}{a}} = \sqrt{\frac{2 \times 10}{9.81}} = 1.43\,\text{s}$$

If, however, the air resistance has a significant effect, the problem is much harder to solve mathematically, even though the principles are still straightforward. Air resistance acts to oppose the motion through the air. At any given instant of time this results in a reduction of the object's

acceleration. The amount of reduction can be taken to be proportional to the speed of the object.

air resistance ∝ speed of object

∴ reduction in acceleration = constant × speed of object

The net result is that the object's acceleration is constantly changing, and this affects the speed of the object, which in turn affects the new value of air resistance.

A powerful mathematical technique for problem-solving in situations like this is to create a model of the changes that are taking place with time. By making the assumption that accelerations will not vary significantly over small time intervals, repeated approximate calculations can be used to predict accurately the variation over significant periods of time. Spreadsheets on computers or graphics calculators allow for a large number of calculations to be performed in a very short time. This is an example of an **iterative technique**, where the results of one calculation are fed into the next one and so on by following a simple rule.

The spreadsheet in Figure 8 shows a possible layout that could be used to check that the process works if friction is negligible. The object starts with a positive displacement, which represents a height measured upwards from the surface of the Earth, and a zero initial velocity. This means the acceleration will be negative. The minus sign represents the fact that the acceleration is downwards.

	A	B	C	D	E	F	G	H	I	J
1	**calculation number**	**Time**	**Displacement**	**Initial velocity**	**Change in velocity**	**Final velocity**	**Average velocity**	**acceleration due to gravity**	−9.80	m s^{-2}
2	0	0.00	10.00000	0.00	−0.10	−0.10	−0.05	**time increment**	0.01	s
3	1	0.01	9.99951	−0.10	−0.10	−0.20	−0.15	**initial displacement**	10.00	m
4	2	0.02	9.99804	−0.20	−0.10	−0.29	−0.25	**initial velocity**	0.00	m s^{-1}
5	3	0.03	9.99559	−0.29	−0.10	−0.39	−0.34			
6	4	0.04	9.99216	−0.39	−0.10	−0.49	−0.44			
7	5	0.05	9.98775	−0.49	−0.10	−0.59	−0.54			
8	6	0.06	9.98236	−0.59	−0.10	−0.69	−0.64			
9	7	0.07	9.97599	−0.69	−0.10	−0.78	−0.74			
10	8	0.08	9.96864	−0.78	−0.10	−0.88	−0.83			
11	9	0.09	9.96031	−0.88	−0.10	−0.98	−0.93			
12	10	0.10	9.95100	−0.98	−0.10	−1.08	−1.03			
13	11	0.11	9.94071	−1.08	−0.10	−1.18	−1.13			
14	12	0.12	9.92944	−1.18	−0.10	−1.27	−1.23			
15	13	0.13	9.91719	−1.27	−0.10	−1.37	−1.32			

Figure 8 A spreadsheet to model acceleration due to gravity using an iterative method

The following points need to be understood:

- All the constants used in the calculation are in column I. Once we have created the spreadsheet, this allows us to change the variables easily and see the effect this has on the final answer. These cells need to be referenced in a direct way (e.g. I2).
- Column A is a record of the number of repeated calculations taking place.
- Column B uses the chosen interval size (the number in I2) to work out the time at the end of this interval, i.e. B3 = B2 + I2
- Column D is the velocity at the beginning of the time interval: D3 = F2
- Column E uses the acceleration and the time interval to work out the change in velocity during the time interval: E2 = I1*I2

- Column F works out the velocity at the end of the time interval:
  F2 = D2 + E2
- Column G works out the average velocity during the time interval:
  G2 = (D2 + F2)/2
- Column C works out the new overall displacement using the displacement at the beginning of the time interval, the average velocity during the time interval: C3 = C2 + G2*I2

This series of calculations is then repeated for the next time interval by "filling down" the equations. The software automatically works out the new formula for each line from that in the line above.

With the values chosen to fit the above situation, the calculations show that the object would be expected to hit the ground (the displacement goes negative if it is below the surface of the Earth) between 1.42 and 1.43 s after release (Figure 9). The value predicted from the uniform acceleration equations is shown in cell I144 for comparison.

Displacement goes negative between 1.42 and 1.43

	A	B	C	D	E	F	G	H	I	J
1	**calculation number**	**Time**	**Displacement**	**Initial velocity**	**Change in velocity**	**Final velocity**	**Average velocity**	**acceleration due to gravity**	−9.80	m s^{-2}
142	140	1.40	0.39600	−13.72	−0.10	−13.82	−13.77			
143	141	1.41	0.25831	−13.82	−0.10	−13.92	−13.87			
144	142	1.42	0.11964	−13.92	−0.10	−14.01	−13.97	**time to hit ground from rest**	1.42857143	
145	143	1.43	−0.02001	−14.01	−0.10	−14.11	−14.06			
146	144	1.44	−0.16064	−14.11	−0.10	−14.21	−14.16			
147	145	1.45	−0.30225	−14.21	−0.10	−14.31	−14.26			
148	146	1.46	−0.44484	−14.31	−0.10	−14.41	−14.36			
149	147	1.47	−0.58841	−14.41	−0.10	−14.50	−14.46			
150	148	1.48	−0.73296	−14.50	−0.10	−14.60	−14.55			
151	149	1.49	−0.87849	−14.60	−0.10	−14.70	−14.65			
152	150	1.50	−1.02500	−14.70	−0.10	−14.80	−14.75			
153	151	1.51	−1.17249	−14.80	−0.10	−14.90	−14.85			
154	152	1.52	−1.32096	−14.90	−0.10	−14.99	−14.95			
155	153	1.53	−1.47041	−14.99	−0.10	−15.09	−15.04			

Figure 9 Calculation of displacement is negative when the object hits the ground

	A	B	C	D	E	F	G	H	I	J
1	**calculation number**	**Time**	**Displacement**	**Initial velocity**	**Change in velocity**	**Final velocity**	**Average velocity**	**acceleration due to gravity**	−9.80	m s^{-2}
2	0	0.00	10.000000000	0.00	−0.10	−0.10	−0.05	**time increment**	0.01	s
3	1	0.01	9.999510000	−0.10	−0.10	−0.20	−0.15	**Initial displacement**	10.00	m
4	2	0.02	9.998042450	−0.20	−0.10	−0.29	−0.24	**Initial velocity**	0.00	m s^{-1}
5	3	0.03	9.995602238	−0.29	−0.10	−0.39	−0.34	**Friction constant**	0.50	
6	4	0.04	9.992194227	−0.39	−0.10	−0.49	−0.44			
7	5	0.05	9.987823255	−0.49	−0.10	−0.58	−0.53			
8	6	0.06	9.982494139	−0.58	−0.10	−0.68	−0.63			
9	7	0.07	9.976211668	−0.68	−0.09	−0.77	−0.72			
10	8	0.08	9.968980610	−0.77	−0.09	−0.86	−0.82			
11	9	0.09	9.960805707	−0.86	−0.09	−0.96	−0.91			
12	10	0.10	9.951691679	−0.96	−0.09	−1.05	−1.00			
13	11	0.11	9.941643220	−1.05	−0.09	−1.14	−1.10			
14	12	0.12	9.930665004	−1.14	−0.09	−1.24	−1.19			

Figure 10 Spreadsheet to include the effects of air friction

A few minor modifications allow the effect of friction to be modelled: see Figure 10.

- Another fixed constant, the friction constant, has been added to cell I5.
- The change in velocity that takes place over the time interval has been modified. The acceleration is not just the acceleration due to gravity. It

needs to be reduced by an amount equal to the product of the friction constant and the initial velocity at the start of the time interval.
E2 = (I1-I5*(D2))*I2.

In this example, the effect is that the time taken to hit the ground has been increased to approximately 1.62 s (Figure 11).

This process gives a good estimate of the time taken, because the chosen time interval is sufficiently small. The equations in column E effectively assume that the frictional force is constant over the chosen time interval. Once the data have been calculated, it is a simple matter to get the computer or a GDC to plot graphs showing the variation with time of either displacement or velocity.

	A	B	C	D	E	F	G	H	I	J
1	**calculation number**	**Time**	**Displacement**	**Initial velocity**	**Change in velocity**	**Final velocity**	**Average velocity**	**acceleration due to gravity**	−9.80	m s^{-2}
158	156	1.56	0.636443763	−10.63	−0.04	−10.68	−10.66			
159	157	1.57	0.529891545	−10.68	−0.04	−10.72	−10.70			
160	158	1.58	0.422892087	−10.72	−0.04	−10.77	−10.74			
161	159	1.59	0.315447626	−10.77	−0.04	−10.81	−10.79			
162	160	1.60	0.207560388	−10.81	−0.04	−10.85	−10.83			
163	161	1.61	0.099232586	−10.85	−0.04	−10.90	−10.88			
164	162	1.62	0.009533577	−10.90	−0.04	−10.94	−10.92			
165	163	1.63	−0.118735909	−10.94	−0.04	−10.99	−10.96			
166	164	1.64	−0.228372229	−10.99	−0.04	−11.03	−11.01			
167	165	1.65	−0.338440368	−11.03	−0.04	−11.07	−11.05			
168	166	1.66	1.448938166	−11.07	−0.04	−11.11	−11.09			
169	167	1.67	1.559863475	−11.11	−0.04	−11.16	−11.14			
170	168	1.68	1.671214158	−11.16	−0.04	−11.20	−11.18			

Figure 11 Air friction increases the time taken to hit the ground

1 Create a spreadsheet that models an object falling under gravity that is affected by air resistance and plots the variations of displacement and velocity.

a) Explain the shapes of the curves when the value of friction is chosen

i) to be significant, and

ii) when it is not.

b) Use your answers to (a) to explain the term "terminal velocity".

c) Explore the effect on the predictions of the model of

i) increasing the chosen time interval, and

ii) decreasing the chosen time interval.

d) Explore the effect of changing the other fixed constants; explain the physical relevance of any change of number.

End of chapter summary

Chapter 3 has one main theme: classical kinematics – the study of objects in motion. The list below summarises the knowledge and skills that you should be able to undertake after having studied this chapter. Further research into more detailed explanations using other resources and/or further practice at solving problems in all these topics is recommended – particularly the items in bold.

- Define displacement, velocity, speed and acceleration and explain the difference between instantaneous and average values of speed, velocity and acceleration.

- Outline the conditions under which the equations for constantly accelerated motion may be applied and solve problems involving the equations of uniformly accelerated motion.
- **Identify the acceleration of a body falling in a vacuum near the Earth's surface with the acceleration *g* of free fall and describe the effects of air resistance on falling objects.**
- Draw and analyse distance – time graphs, displacement – time graphs, velocity – time graphs and acceleration – time graphs. In particular, calculate and interpret the slopes of displacement – time graphs and velocity – time graphs, and the areas under velocity – time graphs and acceleration – time graphs.
- **Determine relative velocity in one and in two dimensions.**
- Use a spreadsheet to model the acceleration due to gravity using an iterative method.

Chapter 3 questions

1 This question is about throwing a stone from a cliff.

Antonia stands at the edge of a vertical cliff and throws a stone vertically upwards.

The stone leaves Antonia's hand with a speed $v = 8.0\text{m s}^{-1}$.

The acceleration of free fall g is 10 m s^{-2} and all distance measurements are taken from the point where the stone leaves Antonia's hand.

a) Ignoring air resistance calculate

i) the maximum height reached by the stone. [2]

ii) time taken by the stone to reach its maximum height. [1]

The time between the stone leaving Antonia's hand and hitting the sea is 3.0 s.

b) Determine the height of the cliff. [3]

(Total 6 marks)

2 This question is about the motion of a bird.

A bird starts from rest on the ground and flies to a tree branch. A simplified graph of the variation of the bird velocity with time is shown below.

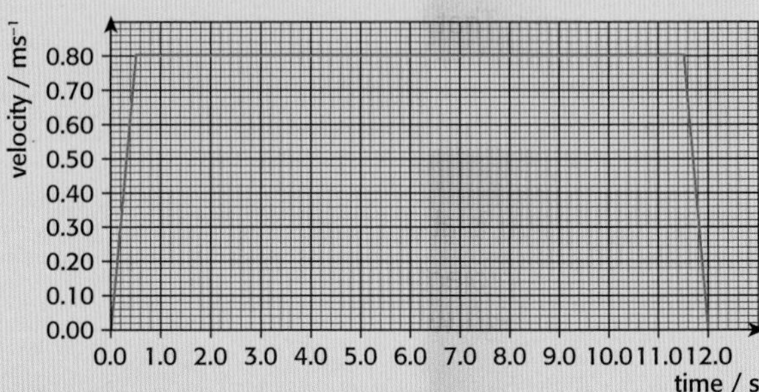

Use the graph to calculate the total vertical distance travelled by the bird. [2]

3 Linear motion

a) Define the term *acceleration*. [2]

b) An object has an initial speed u and an acceleration a. After time t, its speed is v and it has moved through a distance s.

The motion of the object may be summarized by the equations

$$v = u + at,$$
$$s = \frac{1}{2}(v + u)t.$$

i) State the assumption made in these equations about the acceleration a. [1]

ii) Derive, using these equations, an expression for v in terms of u, s and a. [2]

c) The shutter speed of a camera is the time that the film is exposed to light. In order to determine the shutter speed of a camera, a metal ball is held at rest at the zero mark of a vertical scale, as shown below. The ball is released. The shutter of a camera is opened as the ball falls.

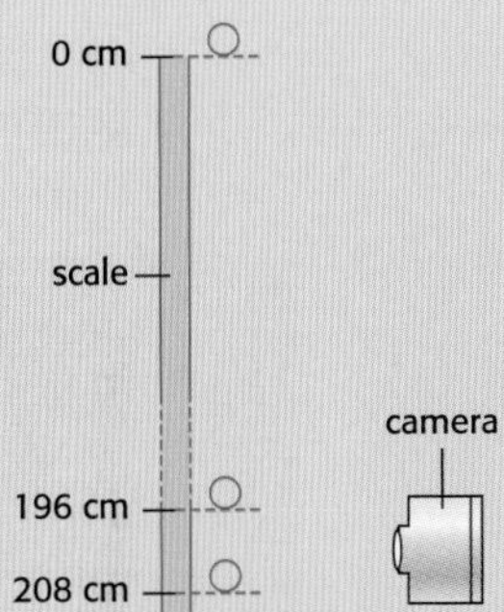

The photograph of the ball shows that the shutter opened as the ball reached the 196 cm mark on the scale and closed as it reached the 208 cm mark. Air resistance is negligible and the acceleration of free fall is 9.81 m s^{-2}.

i) Calculate the time for the ball to fall from rest to the 196 cm mark. [2]

ii) Determine the time for which the shutter was open. That is, the time for the ball to fall from the 196 cm mark to the 208 cm mark. [2]

(Total 9 marks)

4 This question is about projectile motion.

A marble is projected horizontally from the edge of a wall 1.8 m high with an initial speed *V*.

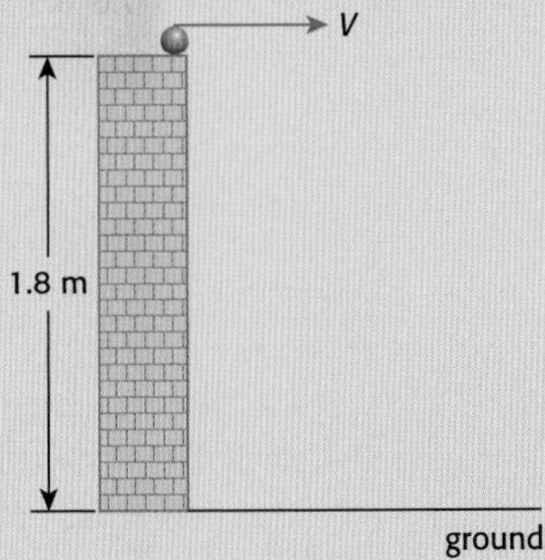

A series of flash photographs are taken of the marble. The photographs are combined into a single photograph as shown below. The images of the marble are superimposed on a grid that shows the horizontal distance *x* and vertical distance *y* travelled by the marble.

The time interval between each image of the marble is 0.10 s.

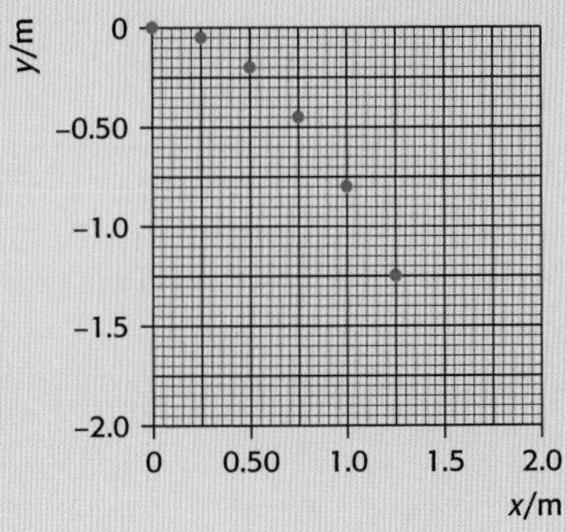

a) On the images of the marble at $x = 0.50$ m and $x = 1.0$ m, draw arrows to represent the horizontal velocity V_H and vertical velocity V_V. [2]

b) Draw a suitable line to determine the horizontal distance *d* from the base of the wall to the point where the marble hits the ground. Explain your reasoning. [3]

c) Use data from the photograph to calculate a value of the acceleration of free fall. [3]

(Total 8 marks)

5 This question is about linear motion.

A car moves along a straight road. At time $t = 0$ the car starts to move from rest and oil begins to drip from the engine of the car. One drop of oil is produced every 0.80 s. Oil drops are left on the road. The position of the oil drops are drawn to scale on the grid below such that 1.0 cm represents 4.0 m. The grid starts at time $t = 0$.

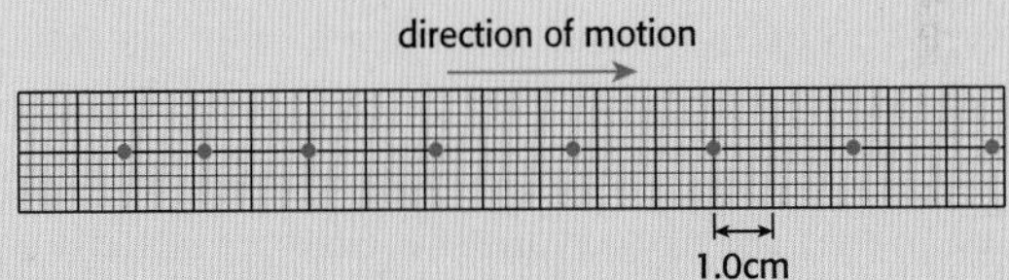

a) i) State the feature of the diagram above which indicates that, initially, the car is accelerating. [1]

ii) On the grid above, draw further dots to show where oil would have dripped if the drops had been produced from the time when the car had started to move. [2]

iii) Determine the distance moved by the car during the first 5.6 s of its motion. [1]

b) Using information from the grid above, determine for the car,

i) the final constant speed. [2]

ii) the initial acceleration. [2]

(Total 8 marks)

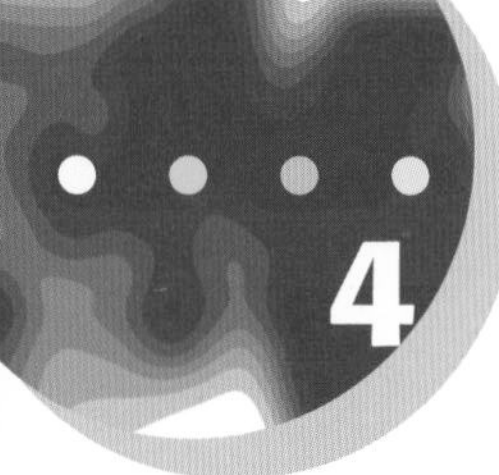

4 Forces and motion

This chapter investigates how forces affect motion. Before you study it, make sure that you have read the earlier material on displacement, velocity, acceleration, and the links between them.

The block in Figure 1 is resting on a table. Experience suggests a force is needed to get the block moving along the surface, and that if the force were removed, the block would stop moving. From these observations, the hypothesis that a constant force causes a constant velocity might seem reasonable. However, this is not the case.

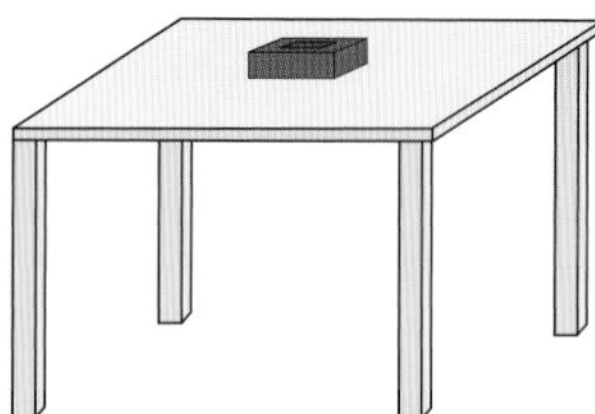

Figure 1 A block resting on a table

Careful observation shows that the block does not come to rest as soon as a force is removed, but that it takes time (and distance). Rather than reject the above hypothesis, some have attempted to modify it. Before the work of Isaac Newton in the 17th century, the accepted model was that the effects of a push continued after the actual event and took time to "run out" (Figure 2). However, this view arose because not all of the forces were being considered. In reality, there are *four* forces acting on the block as it moves along a flat surface: the forward push, friction from the surface, the pull of the Earth's gravity, and the vertical reaction force up from the surface (see Figure 3).

Figure 2 The diagram is from an old manuscript and shows the predicted path of a cannon ball after being fired. It is close to being a straight line until the force "runs out" when the cannon ball falls down nearly vertically. Isaac Newton's work caused this prediction to be altered …

When studying motion experimentally, there are two forces that cannot be removed: gravity and friction. However, it is possible to compensate for each of these forces, or set up situations where their effects can be minimized. For example, if an object is dropped so that it falls freely, then, briefly, there is only the force of gravity acting on it. Unless it is moving very fast or is very light, the effects of air resistance can be ignored. Under these circumstances, experimental measurements show that the resulting motion is uniform acceleration. Resultant force causes acceleration.

Later in the chapter, Newton's laws of motion will be introduced. However, before looking at these, one needs to understand how forces can combine, and why it is important to allow for their directions.

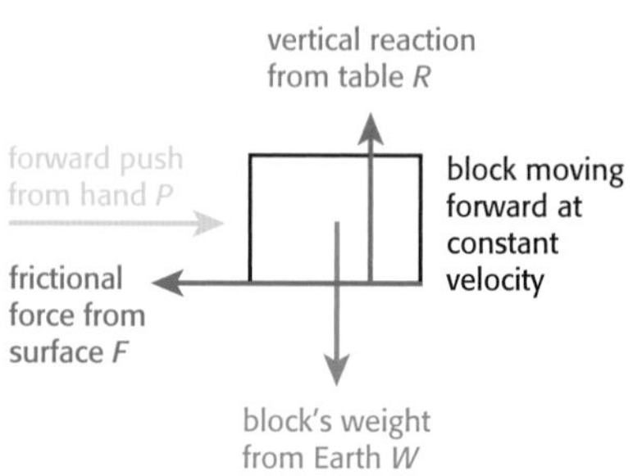

Figure 3 Forces in a block moving at constant velocity

Mathematical physics: Vectors and scalars

All quantities are either **vectors** or **scalars**. Vectors (such as force, velocity, and acceleration) have direction as well as magnitude (size). A change in either magnitude or direction, or both, means that the vector has changed. Scalars (such as mass, density, and energy) just have magnitude. There is no direction to take account of (see Table 1).

Velocity and speed are related, but different. Velocity is a vector, while speed is a scalar. In other words, if you say that a velocity is 10 m s^{-1}, you also have to give a direction (to the right, for example). However, if the speed is 10 m s^{-1}, no direction is implied. If a motorcycle is travelling around a circular track at 10 m s^{-1}, its speed is constant, but its velocity is not, because the direction is changing. And if the velocity is not constant, then the motorcycle has acceleration. Having acceleration does not necessarily mean getting faster or slower.

Table 1

Scalars	Vectors
Volume	Velocity
Density	Acceleration
Speed	Force
Concentration	Displacement
Frequency	Momentum
Energy / work	
Power	
Distance	
Mass	

Mathematically, scalars and vectors need to be handled in different ways.

Scalar addition Scalar addition is the everyday mathematics of numbers: a mass of 3 kg and a mass of 4 kg together give a total mass of 7 kg.

Vector addition A 3 N force and a 4 N force do not necessarily create a total force of 7 N. The maximum that they can create is 7 N, but the minimum is 1 N, and any value between 1 N and 7 N could be possible depending on the directions of the forces. The force produced by the combination is called the **resultant force**.

To add two vectors, one can use the **parallelogram rule** shown in Figure 4. First, the vectors are represented by arrows, with the length of each arrow being in proportion to the magnitude of the vector.

Next, a parallelogram is completed, as in the example below. The diagonal gives the direction and magnitude of the resultant.

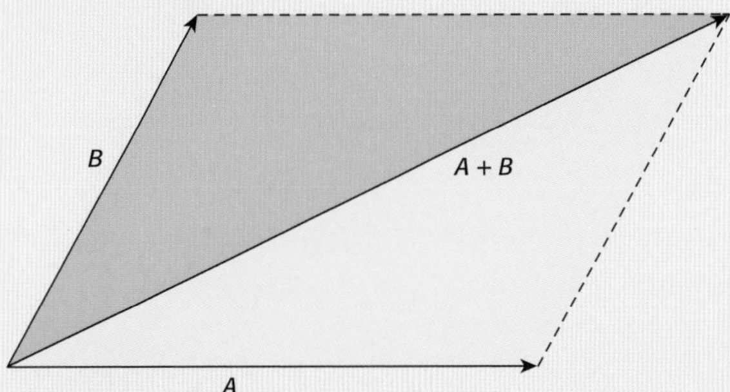

Figure 4 Parallelogram of vectors

If you have more than two vectors to add, the parallelogram rule can be applied with different pairs in turn. Alternatively, the arrows representing the different vectors can be drawn one after another with the tail of one arrow starting from the tip of the proceeding arrow. The arrow joining the tail of the first arrow to the tip of the last arrow represents the resultant vector.

Vector subtraction A negative sign in front of a vector does not alter its magnitude, but means that the direction is reversed. The subtraction of two vectors just involves the addition of a negative vector, as shown in Figure 5. The technique could be used, for example, to calculate the relative velocity between two moving objects.

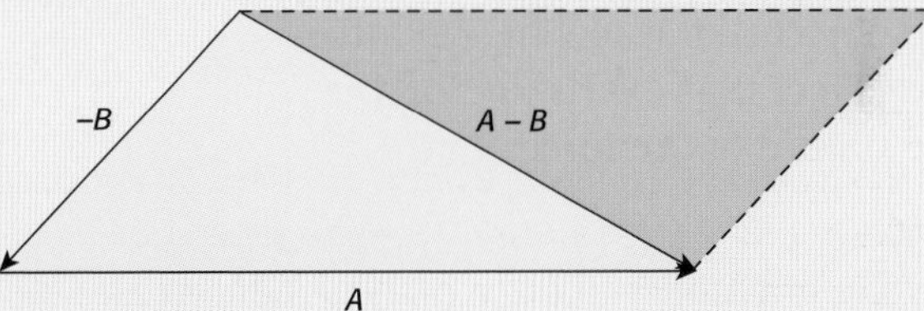

Figure 5 Subtraction of vectors

Vector trigonometry The above techniques involve scale diagrams. However, it is often possible, and certainly more accurate, to use trigonometry to calculate the resultant. This is of course much simpler if right-angled triangles are involved.

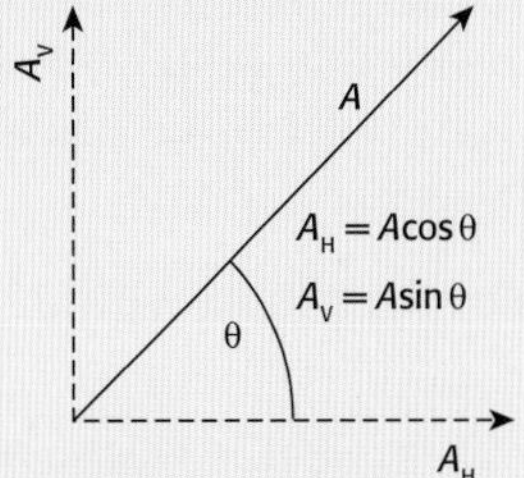

Figure 6 Components of vectors

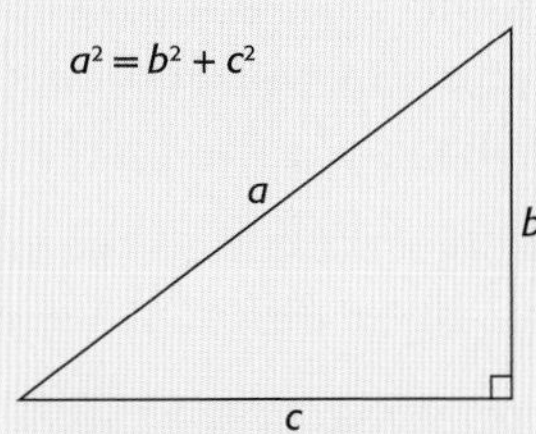

Figure 7 Pythagoras's theorem

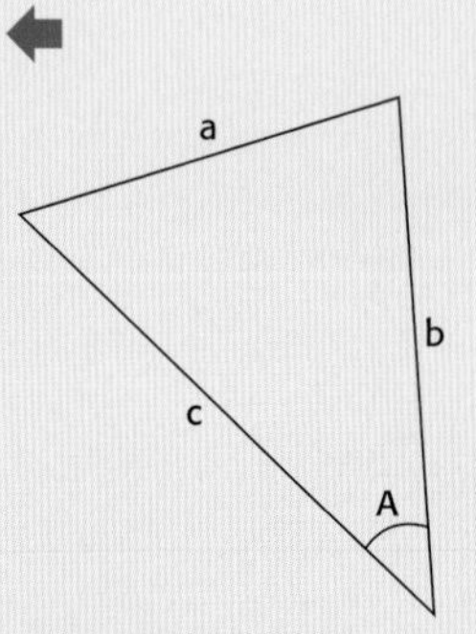

Figure 8 The cosine rule for any triangle is $a^2 = b^2 + c^2 - 2bc \cos A$

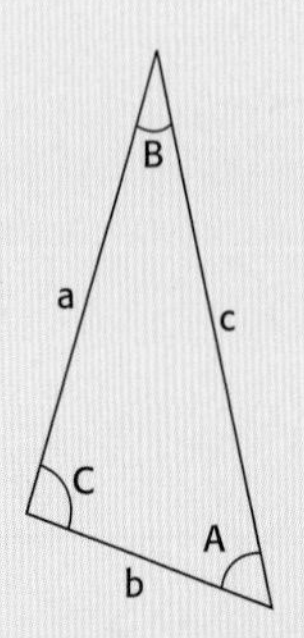

Figure 9 The sine rule for any triangle is $\frac{a}{\sin A} = \frac{b}{\sin B} = \frac{c}{\sin C}$

Resolving into components The techniques used for vector addition can be employed in reverse to split a vector into two (or more) separate vectors, called **components**. If the components are at right angles, they can be treated independently to find out their effects on motion. Analysis of two- or three-dimensional vector problems often involves a separate analysis for each of the different directions. This is followed by a vector addition to find the overall result.

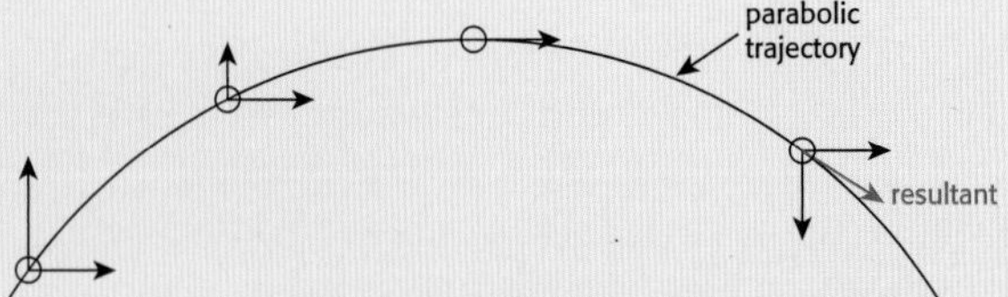

Figure 10 A ball thrown horizontally can be analysed in terms of its independent horizontal and vertical motions. In this example the horizontal velocity is constant whereas the vertical velocity varies.

Further operations In some situations, it is not possible to treat vectors at right angles as separate, because they interact with one another. The motion of a charge in a magnetic field is one example. The full analysis of such situations involves a process called **vector multiplication**. There are two techniques. One produces a **vector product** (or **cross product**), the other a **scalar product** (also known as a **dot product** or **inner product**). Which is chosen depends on the circumstances, but note that, in the second case, vector multiplication produces a *scalar*.

Two scalars can also be multiplied, but that process is much simpler.

Key forces

There are many types of force. This section looks at some of them. A full analysis of any force not only includes its magnitude and direction, but the following as well:

- the object that is feeling the force
- the object that exerts the force
- the nature of the force.

Weight

This is the gravitational force of attraction on an object. It is sometimes confused with mass, but mass is not a force. Mass is sometimes described as the 'amount of matter' in an object, but 'resistance to acceleration' might be more meaningful, as discussed later.

An object's weight and mass are related by this equation:

$$W = mg$$

W is the weight of the object, measured in newtons (N)
m is the mass of the object, measured in kilograms (kg)
g is the gravitational field strength (9.81 N kg^{-1} near the Earth's surface)

An object taken to the surface of the Moon would have the same mass but a different weight.

Normal reaction

When two surfaces are pushed together, there is a repulsion between them that results from the electromagnetic interaction between electrons. This force acts whenever the two surfaces are in contact.

?

A student moves between two points P and Q as shown below.

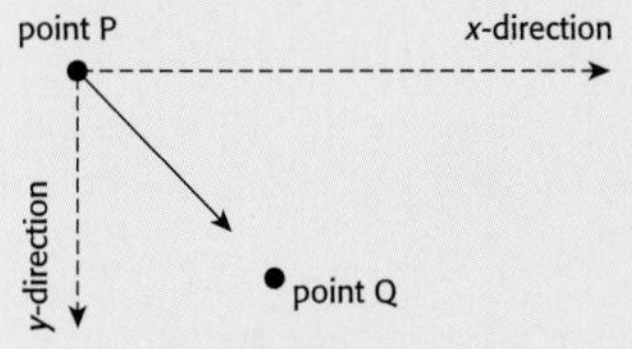

The displacement from P in the x-direction is d_X. The displacement from P in the y-direction is d_Y. The resultant displacement from P is d_R.

Which of the following diagrams shows the **three** displacements from point P?

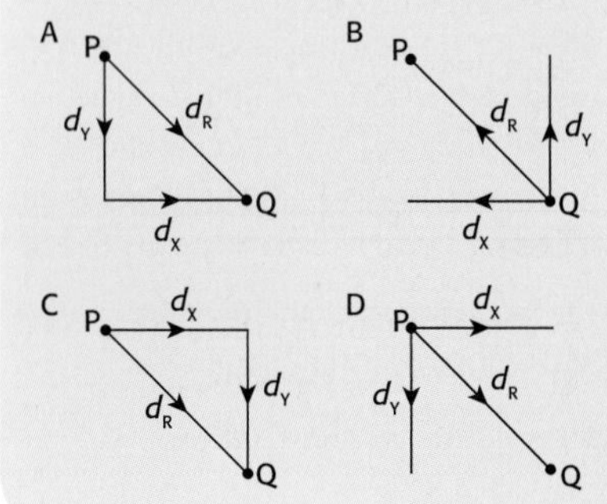

It is called the normal reaction and is at right angles to the surface. For example, a block resting on a table is subject to a normal reaction acting up (see Figure 11).

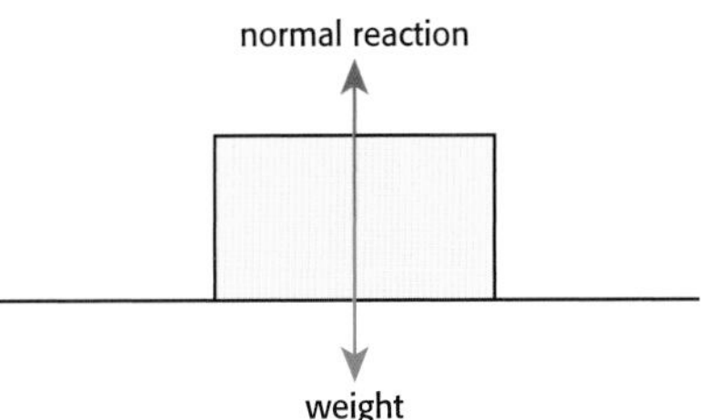

Figure 11 A block at rest on a table

Friction

When two surfaces are in contact, with one sliding over the other, the force that opposes this motion is called friction. Friction is also present before the surfaces start to slide. The resultant force between two surfaces in contact is the combination of the normal reaction and friction. A smooth surface is one with very little friction. Air resistance is a form of friction.

Tension and compression

These are produced by opposing forces. In the case of tension, the force directions are such that the length of the body is increased. With compression, it is decreased.

Figure 12 The above rod is in tension.

Figure 13 The above rod is in compression.

Upthrust

A fluid (liquid or gas) has weight. Because of this, it exerts pressure on any object placed in it. Pressure differences in the fluid cause an upward force on the object. This is called an **upthrust**, or buoyancy force. If it is strong enough, the object will float, like a boat in water, for example.

Lift

Air flowing over an aircraft's wing produces an upward force. This force is known as lift.

Newton's laws of motion

In the late 1600s, Isaac Newton (1643–1727) proposed three laws to describe the nature of forces and the relationship with motion. This section looks at these laws, expressed in their modern form, and some of their implications.

Newton's first law

This expresses the idea that the motion of an object does not change unless a resultant force acts. With no such force, a stationary object will stay still, and a moving object will continue to move at constant velocity (steady speed in a straight line):

$$\text{If } \mathbf{F} = 0, \Delta \mathbf{v} = 0$$

$\mathbf{F}$ is the resultant force acting on the object, measured in newtons (N)
$\Delta \mathbf{v}$ is the change in velocity, measured in m s^{-1}

Note that **bold** text has been used to show vector quantities: for example, $\mathbf{F}$ for force. In diagrams, this isn't necessary because F can indicate the magnitude, while an arrow can be used to show the direction.

Newton's second law

This provides a way of calculating an object's acceleration if the force acting and the mass are known. Expressed as an equation:

$$\boldsymbol{F} = m\,\boldsymbol{a}$$

$\boldsymbol{F}$ is the resultant force acting on the object, measured in newtons (N)
m is the mass of the object, measured in kilograms (kg)
$\boldsymbol{a}$ is the acceleration of the object, measured in m s^{-2}

Note that the resultant force and the acceleration are both vectors and have the same direction.

Although the above equation is very useful, it is not the most fundamental way of stating the law. That is done using the concept of momentum, which is covered later (see pages 47–8).

Relative motion

The first and second laws raise another key idea: being stationary is not essentially any different from having a constant velocity. In both cases, the resultant force on the object must be zero. Put another way, being stationary means having a constant velocity of zero.

A consequence of this idea is that there is no such thing as an absolute frame of reference. If you observe an object moving past you at constant velocity, it is impossible to tell if you are stationary and the object moving, or if it is the other way around – the object is stationary and you are moving. The situations are identical.

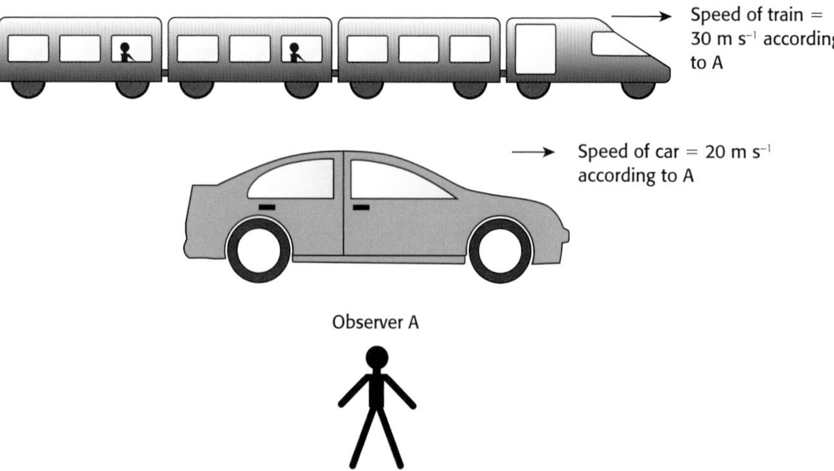

Figure 14 It is impossible to know an object's absolute velocity

According to passengers on the train, A is moving at 30 m s^{-1} and the car is moving at 10 m s^{-1} (both in the same direction). According to the driver of the car the train is moving at 10 m s^{-1} and A is moving at 20 m s^{-1}, in opposite directions.

Another point to note is that forces control only the *changes* in velocity, and not its absolute value. It is impossible to calculate an object's velocity just by knowing the resultant force and the mass. You also need to know the initial conditions – the velocity that the object had before the forces started to act.

Newton's third law

This expresses the idea that forces always involve *two* objects, and always act in pairs. The forces are sometimes called the action and the

?

1 An elevator (lift) is used to either raise or lower sacks of potatoes. In the diagram, a sack of potatoes of mass 10 kg is resting on a scale that is resting on the floor of an accelerating elevator. The scale reads 12 kg.

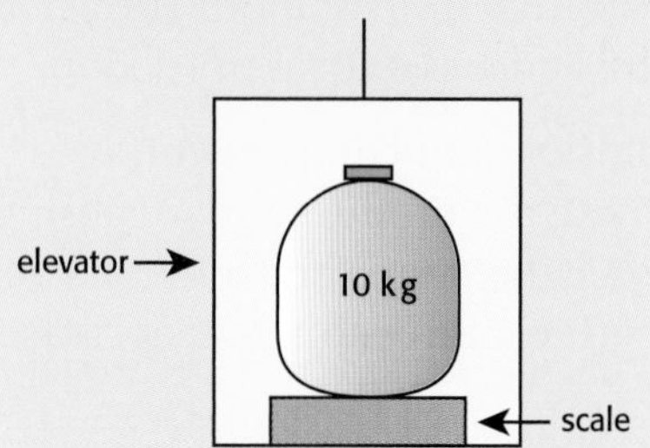

The best estimate for the acceleration of the elevator is

A 2.0 m s^{-2} downwards
B 2.0 m s^{-2} upwards
C 1.2 m s^{-2} downwards
D 1.2 m s^{-2} upwards

2 A light inextensible string has a mass attached to each end and passes over a frictionless pulley as shown.

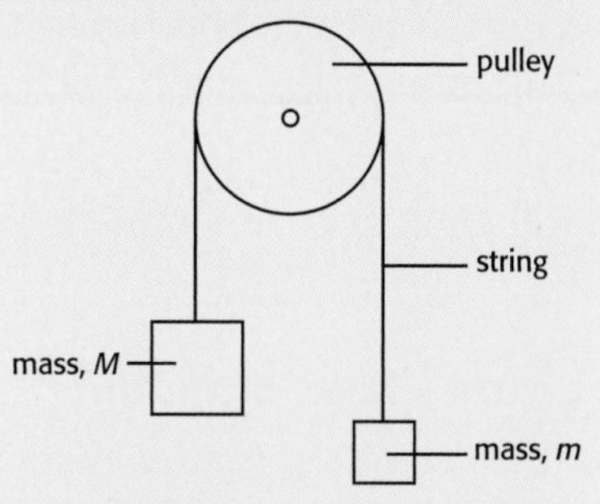

The masses are of magnitudes M and m, where $m < M$. The acceleration of free fall is g. The downward acceleration of the mass M is

A $\dfrac{(M-m)g}{(M+m)}$ **C** $\dfrac{(M+m)g}{(M-m)}$

B $\dfrac{(M-m)g}{M}$ **D** $\dfrac{Mg}{(M+m)}$

reaction, but it does not matter which you call which, because one cannot exist without the other. The forces have these features:

- they act on different objects
- they are equal in magnitude
- they act in opposite directions, along the same line
- they are the same type of force.

The law is sometimes stated like this:

If object A exerts a force $\boldsymbol{F}$ on object B, then object B will exert an equal but opposite force on object A.

$$\boldsymbol{F}_{AB} = -\boldsymbol{F}_{BA}$$

$\boldsymbol{F}_{AB}$ is the force acting on object B from object A.
$\boldsymbol{F}_{BA}$ is the force acting on object A from object B.

For example, your weight is the gravitational pull of the *Earth* on *you*, downwards, So, according to Newton's third law, there must also be another force: the gravitational pull of *you* on the *Earth*, upwards. If you push against a wall, to the left, then the wall must push on you with an equal force, to the right.

Free-body diagrams

Forces act in pairs and always involve *two* objects. However, to solve problems, one needs to concentrate on just *one* object at a time. A free-body diagram is a useful way of doing this. It shows one object in isolation, with all the forces acting on it.

- Start by drawing a diagram representing the general situation, showing all the bodies involved (a body is just another name for an object).
- Choose the one body you are interested in. Draw it removed from its environment. In other words, don't include any of the supports or strings holding it up.
- Draw in *all* the forces acting *on* the body. If, for example, there is a support underneath, this will produce an upward force. So draw in the force (but not the support).
- *Don't* include any forces that are exerted *by* the body on anything else. Newton's third law can be used to help identify all the relevant forces.
- If you want to find the body's acceleration, find the resultant force acting and use $\boldsymbol{F} = m\,\boldsymbol{a}$. In doing this, remember that you can always consider one direction at a time, and then add the components.

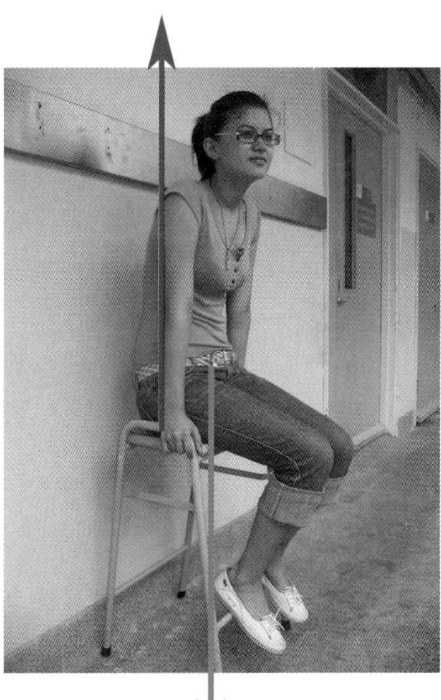

Figure 15 Forces on a physics student sitting on a stool

EXAMPLE 1:

A block moving at constant velocity:
As the block is moving at constant velocity,

$T = F$
$R = W$

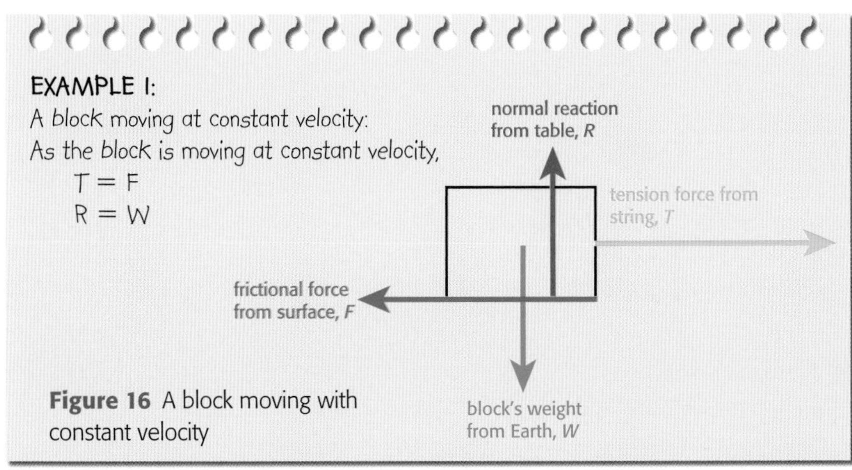

Figure 16 A block moving with constant velocity

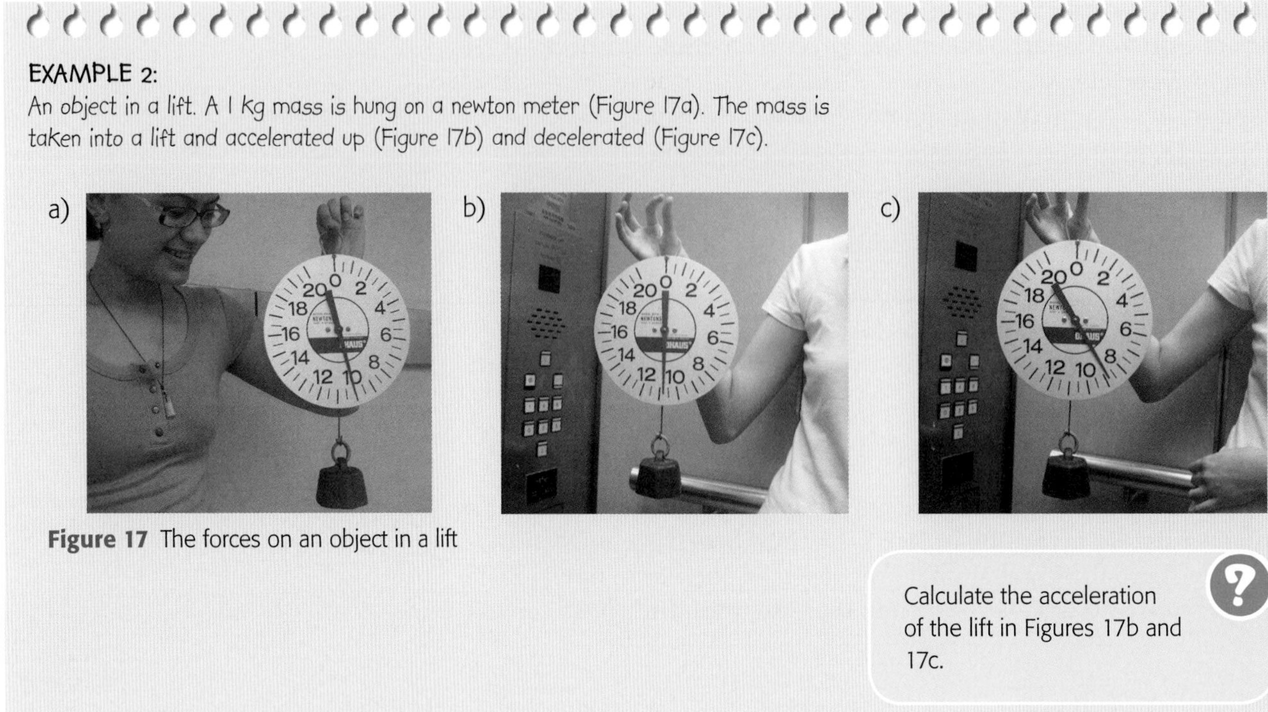

EXAMPLE 2:

An object in a lift. A 1 kg mass is hung on a newton meter (Figure 17a). The mass is taken into a lift and accelerated up (Figure 17b) and decelerated (Figure 17c).

a) b) c)

Figure 17 The forces on an object in a lift

Calculate the acceleration of the lift in Figures 17b and 17c.

Verifying Newton's second law

This section follows the design of an experiment that aims to verify Newton's second law. Although the experiment can provide evidence to support Newton's second law, a detailed analysis of the results shows that the procedure has been based on some false assumptions. Can you spot the mistakes as they are introduced?

The experiment involves the following variables: resultant force, mass, and acceleration. This suggests two possible approaches:

First, the force could be chosen as the independent variable, the acceleration as the dependent variable, and the mass as the controlled variable, which is held constant. The prediction is that the acceleration should be proportional to the force.

Second, the mass could be the independent variable, the acceleration the dependent variable (as before), and the force as the controlled variable, which is held constant. The prediction is that the acceleration should be inversely proportional to the mass.

A possible setup for the experiment is shown in Figure 18.

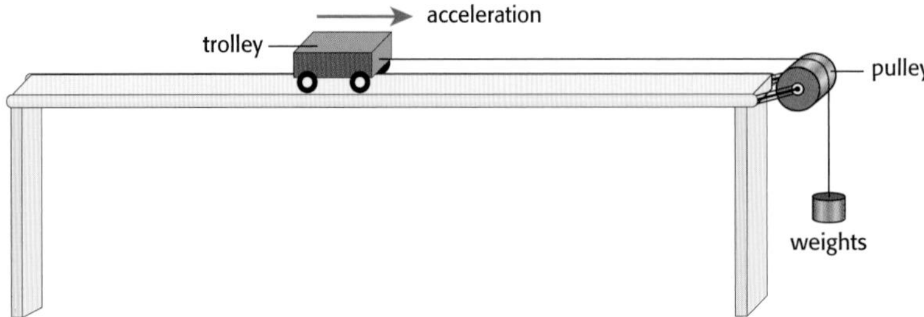

Figure 18 Possible experimental setup to verify Newton's second law

In the first approach, the force is changed by altering the weights hanging from the string. The acceleration of the trolley is recorded electronically using motion sensors. The values of the force and the acceleration are recorded, then the experiment repeated with different forces.

1 What shape of graph would you expect if the measured acceleration and the force were proportional? Explain your answer. [2]

2 The actual results are shown in Figure 19 with uncertainty limits shown for the force values.

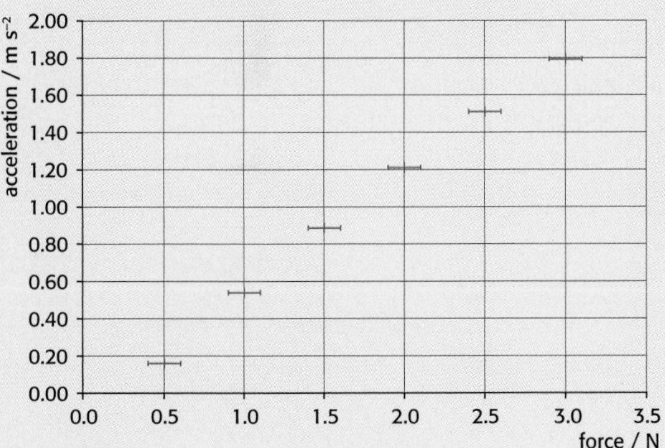

Figure 19

a) Add the best-fit line for this data. [2]

b) Use the graph to explain what is meant by a *systematic* error. [2]

c) Estimate the value of the frictional force acting on the trolley. [1]

d) Estimate the mass of the trolley. [2]

[Total 7 marks]

3 Careful analysis of the graph shows that the last two points are not quite in the same straight line as the first four points. The data is checked by taking further readings. This shows that the weight is *not* proportional to the acceleration as shown in Figure 20.

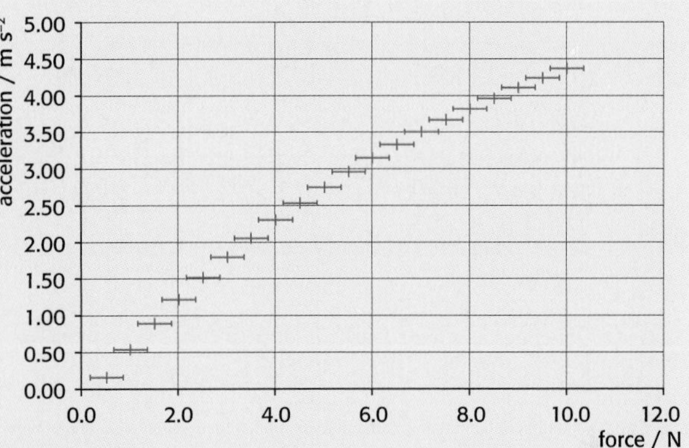

Figure 20

a) Add the best-fit line for this data. [2]

b) i) Draw free-body force diagrams for the trolley **and** for the weight. [4]

ii) Use the diagrams to explain why both the trolley and the weights accelerate together. [4]

c) The value of the weights is continually increased. The acceleration of the trolley would eventually reach a maximum value. Explain why this is so and estimate its value. [3]

d) How could the procedure be modified to improve the experiment? [4]

[Total 17 marks]

Mathematical physics: Force and momentum

Linear momentum, $\boldsymbol{p}$, is a vector quantity defined by the following equation:

$$\boldsymbol{p} = m\,\boldsymbol{v}$$

m is the mass of the object, measured in kilograms
v is the velocity of the objects, measured in m s^{-1}
$\boldsymbol{p}$ is the momentum of the body, measured in kg m s^{-1}

Since 1 N = 1 kg m s^{-2}, an alternative unit for momentum is N s (these two units are equivalent).

This definition provides a more fundamental way of stating Newton's second law:

The rate of change of momentum of a body is proportional to the resultant force acting.

$$F \propto \frac{d}{dt}(m\boldsymbol{v})$$

In SI units, the definition of Newton's second law means that the resultant force and the rate of change of momentum are equal.

$$F = \frac{d}{dt}(m\boldsymbol{v})$$

F is the resultant force, measured in newtons
$m\boldsymbol{v}$ is the momentum of the object, measured in kg m s^{-1}

1 A 1600 kg car is travelling at 30 m s^{-1}. Calculate the average force on the car in each of the following ways that it can be brought to rest:

a) Gently applying the brakes, taking 20 s to stop.

b) An emergency stop taking 2.5 s.

c) Hitting a tree, taking 0.3 s to stop.

2 A 60 g tennis ball moving at 20 m s^{-1} hits a wall horizontally and bounces back along the same path at 18 m s^{-1}. The ball is in contact with the wall for 0.5 s. Calculate:

a) The change in momentum for the ball. *(Be careful – the answer is not 0.12 kg m s^{-1})*

b) The average force that acts on the ball.

$\frac{d}{dt}(m\mathbf{v})$ is the rate of change of momentum in kg m s^{-2} or N

The above equation relates the instantaneous value of force to the instantaneous rate of change of momentum. Over a time interval, Δt, the average force $F_{average}$ is calculated using:

$$F_{average} = \frac{\Delta p}{\Delta t}$$

Impulse

The previous equation can be rearranged to give this:

$$\Delta p = F_{average} \times \Delta t$$

The product of a force and the time for which it acts is called the impulse. The above equation shows that it is equal to the change of momentum. Impulse is a vector quantity.

Analysing a graph

In many practical situations, forces vary significantly over time. If one can record the actual variation of the force, the above relationship allows the overall change in momentum to be calculated.

The graph in Figure 21 represents a force that varies with time. It is reasonable to assume that over a very short time interval, δt, the instantaneous value of the force, F, does not significantly change. The impulse given to the object during this short time interval is the product of F and δt, represented in Figure 21 by the shaded area of the rectangle.

The total impulse is the sum of all the small impulses, added over the time for which the force acts. Mathematically, the way of adding up these impulses is by calculating the area under the graph.

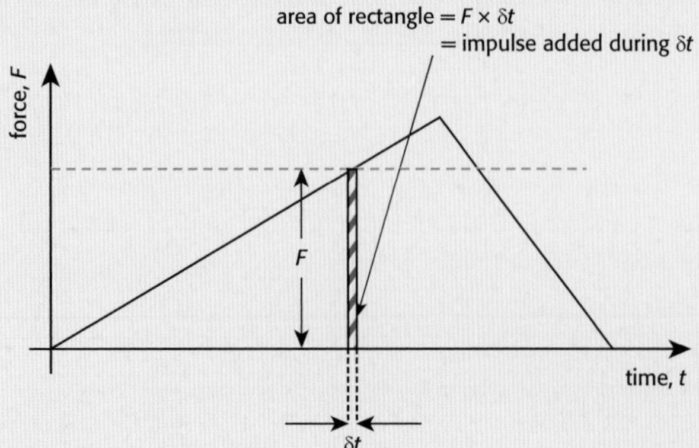

Figure 21 Incremental increase in impulse

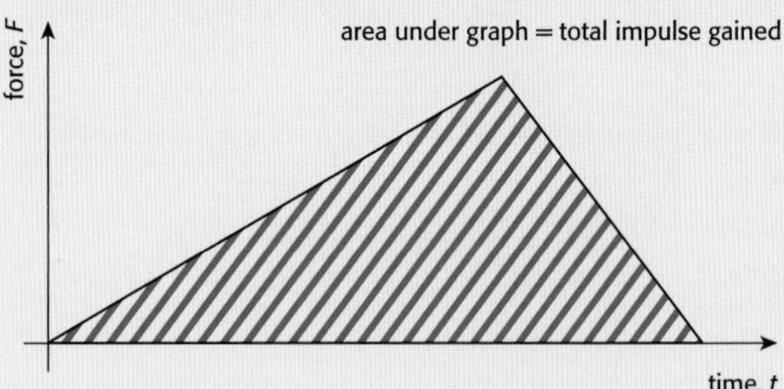

Figure 22 The total impulse is the area under the graph

3 During a tennis serve, a 60g tennis ball is hit with a tennis racket. The initial horizontal velocity of the ball is zero and Figure 23 represents the variation with time of the force on the ball.

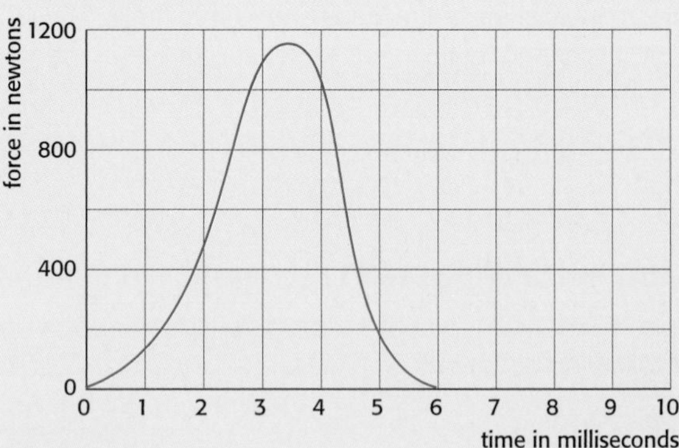

Figure 23

Calculate:

a) The area under the graph

b) The impulse given to the ball

c) The final velocity of the ball.

In many situations, calculating the area under a graph can be achieved either with some simple mathematics or (less accurately) by counting the number of squares and calculating what one square represents. If the variation can be summarized in a mathematical equation, then the area under the graph can be calculated using a process called integration. The mathematics of integration is not required for IB physics, but it is studied in the SL and HL mathematics courses.

$$\text{change in momentum} = \text{impulse} = \int F dt$$

Conservation of momentum

When two objects collide, the momentum of each one changes. From Newton's third law, the forces on the two objects are equal but opposite. They also act for the same time, so the two impulses are equal but opposite. Therefore the change in momentum of one object must be equal but opposite to that of the other. Put another way, the total momentum of the objects is the same after the collision as before. This is an example of the **law of conservation of momentum**:

The overall momentum of an isolated system of interacting objects must be constant.

Here, an "isolated system" means one without any external forces acting on it. It is rare for any system to be completely isolated. However, being a vector, momentum, like force, can be resolved into components. So, if there is no component of a force in one direction, the component of the momentum *in that direction* is constant.

The law of conservation of momentum can sometimes provide a mathematical shortcut to help analyse complex situations. If you know that the total momentum of a system does not change, then you can ignore the internal interactions within it. The law of conservation of energy can be useful for similar reasons.

4 Research a formal mathematical derivation of the law of conservation of momentum. Explain the relevance of the following to the derivation:
   a) Newton's second law
   b) Newton's third law.

5 A 8 kg truck moving at 5 m s^{-1} collides with a stationary 12 kg truck. The collision takes 0.2 s. Afterwards, the two trucks are joined together and move off with a common velocity *v*. Calculate:
   a) the common velocity *v* after the collision
   b) the impulse given to each truck during the collision
   c) the average force acting on each truck during the collision.

Circular motion

Can an object accelerate without getting any faster? The answer is yes, if it is moving in a circle. But before considering this, you need to know how angles and angular rotation are measured.

Radian measure

Angles can of course be measured in degrees but, in studying rotation, the radian is a more useful measure.

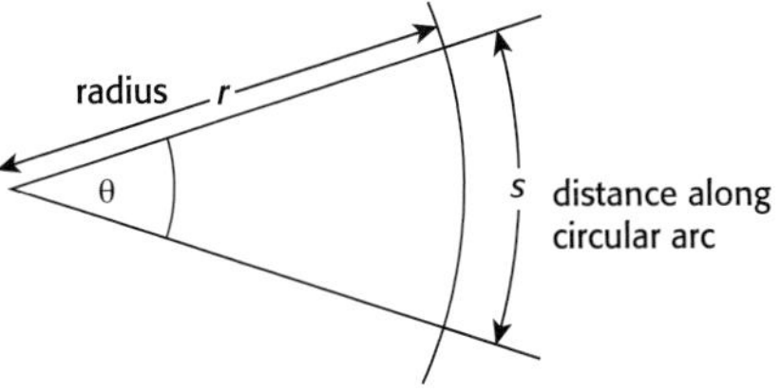

Figure 24 Calculating an angle in radians

The diagram in Figure 24 shows an angle θ.

The value of the angle in radians is found like this:

$$\theta = \frac{s}{r}$$

One radian (symbol rad) is approximately 60 degrees.

For a full circle (360°), $s = 2\pi r$, so θ in this case is 2π radians.

As $360° = 2\pi$ rad, it follows that $180° = \pi$ rad, and $90° = \pi/2$ rad

In each of the above examples, it isn't strictly necessary to include the unit (rad), because the value s/r is a number, with no dimensions.

For small angles (less than about 0.1 rad, or 5°), the arc and two radii in Figure 24 form a shape that approximates to a triangle. Therefore the following relationship applies:

$$\sin\theta \approx \tan\theta \approx \theta$$

Angular velocity

In Figure 25 the object is moving round a circle at constant speed. Put another way, it has uniform circular motion.

The velocity is continuously changing direction. However, at any instant, it is always at right angles to the radius at that position. As the object rotates, this radius makes a different angle with the axes, as in Figure 25.

In a given time interval, the increase in angle is always the same. In other words, there is a constant rate of rotation. This rate called the **angular velocity** is given the symbol ω (omega), and it is defined as:

$$\omega_{\text{av}} = \frac{\text{change in angle turned}}{\text{time taken}}$$

$$\omega_{\text{av}} = \frac{\Delta\theta}{\Delta t}$$

The definition above gives the *average* angular velocity. The following equations express the angular velocity at any instant:

$$\omega = \frac{\text{small change in angle}}{\text{small time taken}} = \text{rate of change of angle}$$

$$\omega = \lim_{\delta t \to 0}\left(\frac{\delta\theta}{\delta t}\right) = \frac{d\theta}{dt}$$

Angular velocity is measured in radians per second: for example, 2 rad s^{-1}. However, as the radian is dimensionless, this could be expressed as 2 s^{-1}.

Linking speed and angular velocity

Look again at the diagram showing the arc of a circle with the angle θ (Figure 26).

Since $\theta = \frac{s}{r}$, it follows that $s = r\theta$

So: $v = \frac{s}{t} = \frac{\theta r}{t} = \omega r$

If the circular motion is uniform, there is no difference between the instantaneous and average values for angular velocity.

Period

This is the time taken for the object to complete one full circle. In a time period T, the object will have rotated through an angle of 2π radians. As the angular velocity is found by dividing the angle by the time:

$$\omega = \frac{2\pi}{T} \quad \text{or} \quad T = \frac{2\pi}{\omega}$$

Centripetal acceleration

Velocity is a vector quantity. So, if its direction is changing, it is not constant. And if it is not constant, the object must have acceleration

Table 2 Conversion between degrees and radians

Angle / degrees	Angle / radians
0	0.00
5	0.09
45	$0.79 = \frac{\pi}{4}$
60	1.05
90	$1.57 = \frac{\pi}{2}$
180	$3.14 = \pi$
270	$4.71 = \frac{3\pi}{2}$
360	$6.28 = 2\pi$

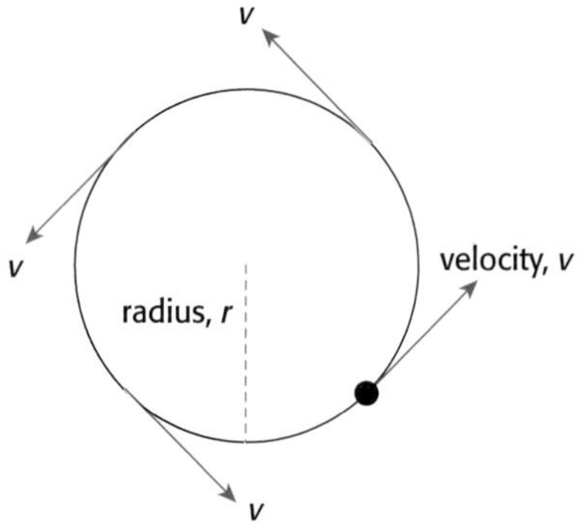

Figure 25 Uniform circular motion

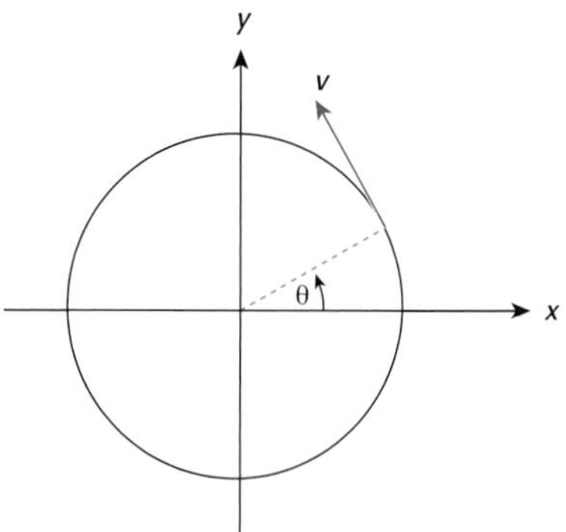

Figure 26 Angles in uniform circular motion

(even though its speed remains steady). The direction of the acceleration can be deduced from the change in velocity that takes place as an object moves around the circle between two points, say A and B (Figure 27).

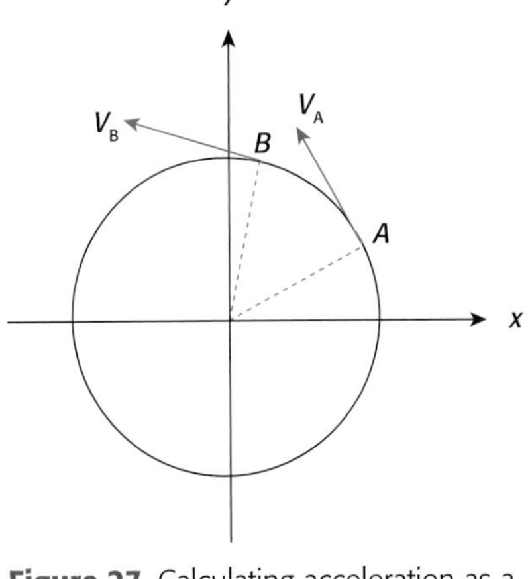

Figure 27 Calculating acceleration as a change in direction of velocity

The vector diagram shows that the average acceleration between A and B is directed towards the centre of the circle (Figure 28). This is also true of the instantaneous acceleration.

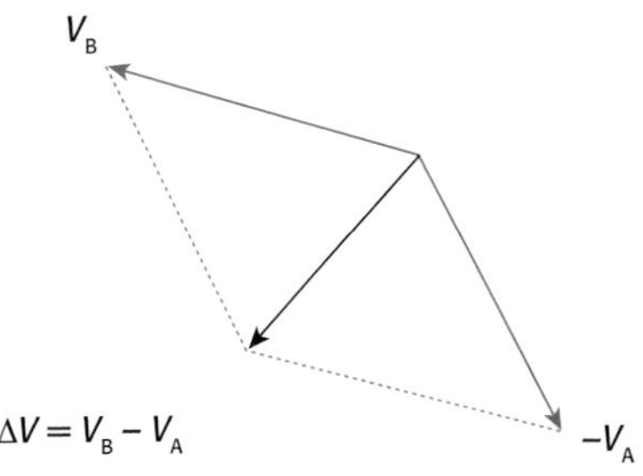

Figure 28 Using vectors to calculate acceleration

The acceleration of an object that is in uniform circular motion is called the **centripetal acceleration**. A more detailed analysis allows one to calculate the instantaneous magnitude of this centripetal acceleration. The result is as follows:

$$a_{\text{centripetal}} = \frac{v^2}{r}$$

This equation can be stated in several different equivalent forms:

$$a_{\text{centripetal}} = r\omega^2 = \frac{4\pi^2 r}{T^2}$$

Centripetal force

Newton's second law tells us that, as the object is accelerating, there must be a resultant force on it. This force is called the **centripetal force** (CPF). As $F = ma$:

$$\text{CPF} = \frac{mv^2}{r} = mr\omega^2 = m\frac{4\pi^2 r}{T^2}$$

The centripetal acceleration is towards the centre of the circle, so the centripetal force must also act in this direction. However, it is important to realize that CPF is not a new or additional force. Something must provide it. Without it, the object would travel in a straight line and not in a circle. Table 3 shows some examples of CPF providers:

Table 3

Situation	Origin of the centripetal force
Mass on the end of a string being whirled around in a horizontal circle	The string exerts a tension force on the mass which provides the necessary CPF
The Earth in orbit around the Sun	The gravitational force of attraction exerted by the Sun on the Earth provides the necessary CPF
A car turning around a corner	The sideways frictional force exerted by the road on the tyres provides the necessary CPF
An electron in a hydrogen atom	The electrostatic attraction exerted by the nucleus on the electron provides the necessary CPF

End of chapter summary

Chapter 4 has three main themes: a comparison between vectors and scalars, forces and dynamics and uniform circular motion. The list below summarises the knowledge and skills that you should be able to undertake after having studied this chapter. Further research into more detailed explanations using other resources and/or further practice at solving problems in all these topics is recommended – particularly the items in bold.

Vectors and scalars

- Distinguish between vector and scalar quantities, and be able to give examples of each.
- Determine the sum or difference of two vectors by a graphical method.
- Resolve vectors into perpendicular components along chosen axes.

Forces and dynamics

- Calculate the weight of a body using the expression $W = mg$.
- Identify the forces acting on an object and draw free-body diagrams representing the forces acting and thus determine the resultant force in different situations.
- State Newton's three laws of motion and be able to give examples and solve problems for each law.
- **State the condition for translational equilibrium and solve related problems.**
- Define *linear momentum* and *impulse*, state the law of conservation of linear momentum and **solve problems with linear momentum and impulse** including being able to determine the impulse due to a time-varying force by interpreting a force–time graph.

Uniform circular motion

- Draw a vector diagram to illustrate that the acceleration of a particle moving with constant speed in a circle is directed towards the centre of the circle.
- Use radian measure to define *angle* and *angular velocity*.
- Apply the expression for centripetal acceleration, identify the force producing circular motion in various situations and thus **solve problems involving circular motion.**

Chapter 4 questions

1 This question is about momentum and the kinematics of a proposed journey to Jupiter.

a) State the law of conservation of momentum. [2]

A solar propulsion engine uses solar power to ionize atoms of xenon and to accelerate them. As a result of the acceleration process, the ions are ejected from the spaceship with a speed of 3.0×10^4 m s^{-1}.

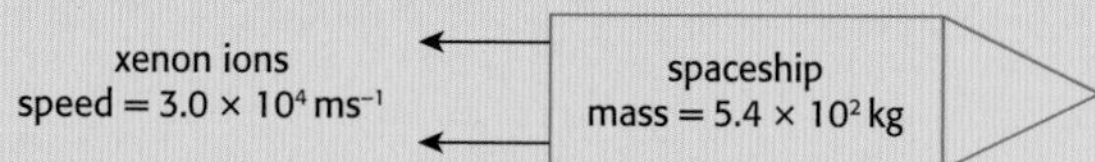

Figure 29

b) The mass (nucleon) number of the xenon used is 131. Deduce that the mass of one ion of xenon is 2.2×10^{-25} kg. [2]

c) The original mass of the fuel is 81 kg. Deduce that, if the engine ejects 77×10^{18} xenon ions every second, the fuel will last for 1.5 years. (1 year = 3.2×10^7 s) [2]

d) The mass of the spaceship is 5.4×10^2 kg. Deduce that the initial acceleration of the spaceship is 8.2×10^{-5} m s^{-2}. [5]

The graph below shows the variation with time t of the acceleration a of the spaceship. The solar propulsion engine is switched on at time $t = 0$ when the speed of the spaceship is 1.2×103 m s^{-1}.

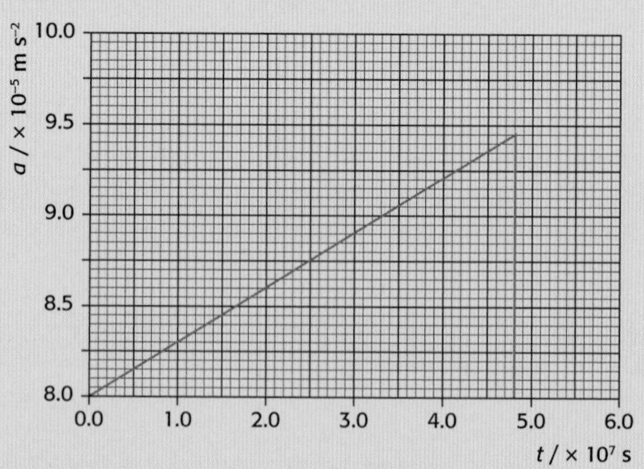

Figure 30

e) Explain why the acceleration of the spaceship is increasing with time. [2]

f) Using data from the graph, calculate the speed of the spaceship at the time when the xenon fuel has all been used. [4]

g) The distance of the spaceship from Earth when the solar propulsion engine is switched on is very small compared to the distance from Earth to

Jupiter. The fuel runs out when the spaceship is a distance of 4.7×10^{-11} m from Jupiter. Estimate the total time that it would take the spaceship to travel from Earth to Jupiter. [2]

(Total 19 marks)

2 This question is about circular motion.

A linear spring of negligible mass requires a force of 18.0 N to cause its length to increase by 1.0 cm.

A sphere of mass 75.0 g is attached to one end of the spring. The distance between the centre of the sphere M and the other end P of the unstretched spring is 25.0 cm, as shown below.

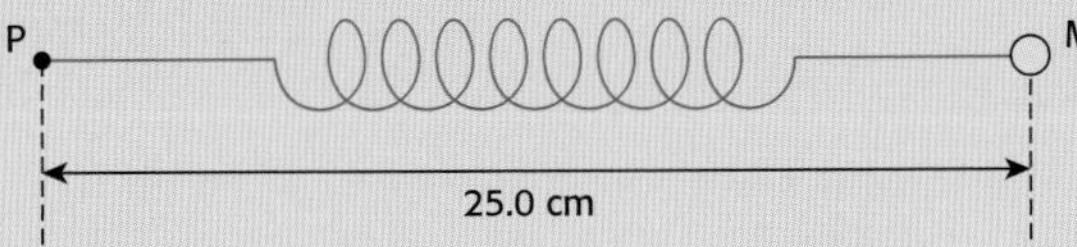

Figure 31

The sphere is rotated at constant speed in a horizontal circle with centre P. The distance PM increases to 26.5 cm.

a) Explain why the spring increases in length when the sphere is moving in a circle. [2]

b) Determine the speed of the sphere. [4]

(Total 6 marks)

3 Linear momentum

a) Define

i) *linear momentum;* [1]

ii) *impulse.* [1]

b) Explain whether momentum and impulse are scalar or vector quantities. [1]

c) By reference to Newton's laws of motion, deduce that when two particles collide, momentum is conserved. [5]

A rubber ball of mass 50 g is thrown towards a vertical wall. It strikes the wall at a horizontal speed of 20 m s^{-1} and bounces back with a horizontal speed of 18 m s^{-1} as shown below.

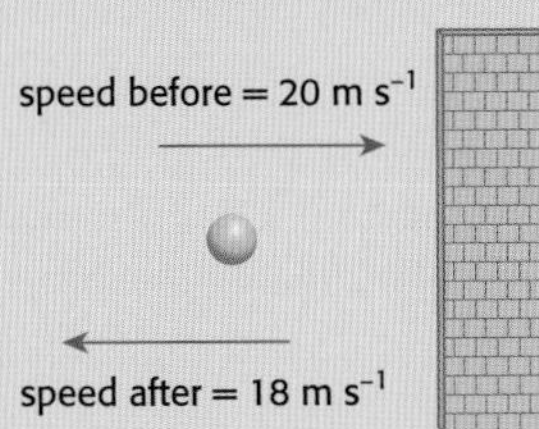

The ball is in contact with the wall for 0.080 s.

d) i) Calculate the change in momentum of the ball. [2]

ii) Calculate the average force exerted by the ball on the wall. [2]

iii) Suggest, in terms of Newton's laws of motion, why a steel ball of the same mass and the same initial horizontal speed exerts a greater force on the wall. [3]

(Total 15 marks)

4 Momentum

a) State the law of conservation of momentum. [2]

b) An ice hockey puck collides with the wall of an ice rink. The puck is sliding along a line that makes an angle of 45° to the wall.

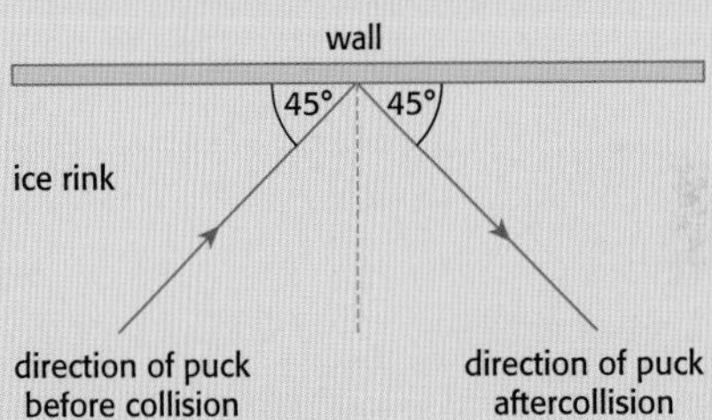

The collision between the wall and the puck is perfectly elastic.

i) State what is meant by an *elastic collision*. [1]

ii) Discuss how the law of conservation of momentum applies to this situation. [2]

c) The diagram below is a scale diagram that shows the vector representing the momentum of the puck before collision.

Scale: 1.0 cm = 0.10 N s

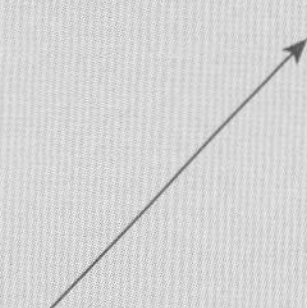

By adding appropriate vectors to the diagram, deduce that the magnitude of the change in momentum of the puck as a result of the collision is 0.71 N s. [4]

d) The sketch-graph below shows the variation with time t of the force F exerted by the wall on the puck.

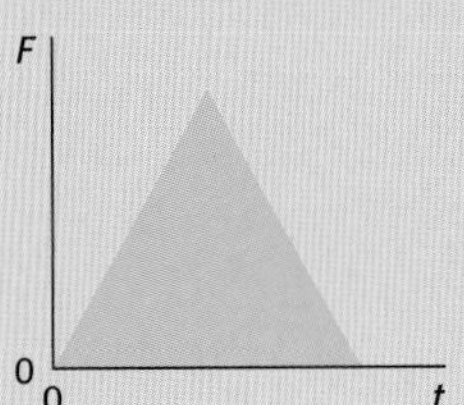

The total contact time is 12 ms. Estimate, explaining your reasoning, the maximum force exerted by the wall on the puck. [3]

(Total 12 marks)

5 Collisions

A large metal ball is hung from a crane by means of a cable of length 5.8 m as shown below.

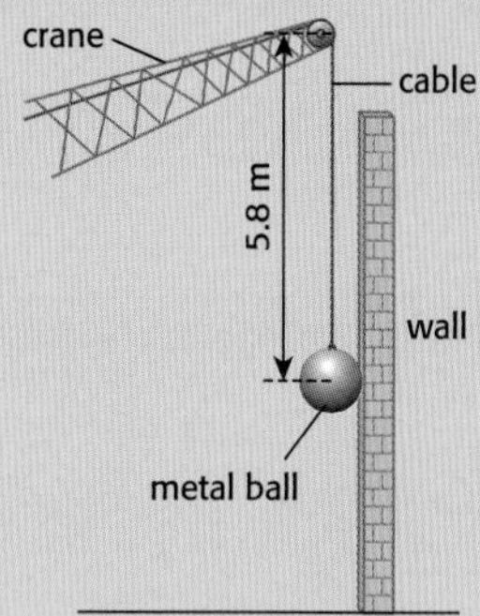

In order to knock down a wall, the metal ball of mass 350 kg is pulled away from the wall and then released. The crane does not move. The graph below shows the variation with time *t* of the speed *v* of the ball after release.

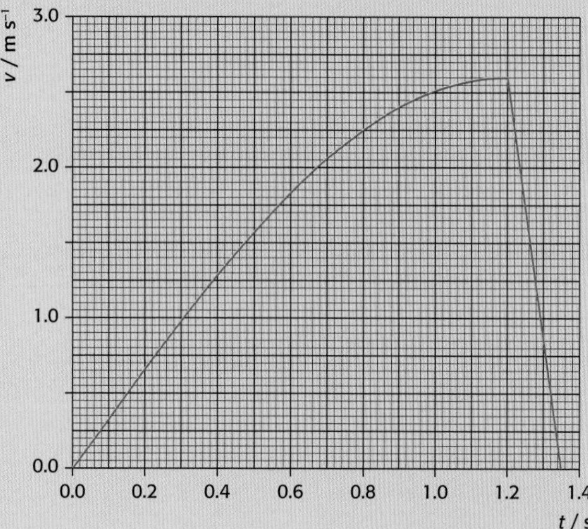

The ball makes contact with the wall when the cable from the crane is vertical.

a) For the ball just before it hits the wall,

i) state why the tension in the cable is not equal to the weight of the ball; [1]

ii) by reference to the graph, estimate the tension in the cable. The acceleration of free fall is 9.8 m s^{-2}. [3]

b) Use the graph to determine the distance moved by the ball after coming into contact with the wall. [2]

c) Calculate the total change in momentum of the ball during the collision of the ball with the wall. [2]

d) i) State the law of conservation of momentum. [2]

ii) The metal ball has lost momentum. Discuss whether the law applies to this situation. [2]

e) During the impact of the ball with the wall, 12% of the total kinetic energy of the ball is converted into thermal energy in the ball. The metal of the ball has specific heat capacity 450 J kg^{-1} K^{-1}. Determine the average rise in temperature of the ball as a result of colliding with the wall. [4]

(Total 16 marks)

6 This question is about a balloon used to carry scientific equipment.

The diagram below represents a balloon just before take-off. The balloon's basket is attached to the ground by two fixing ropes.

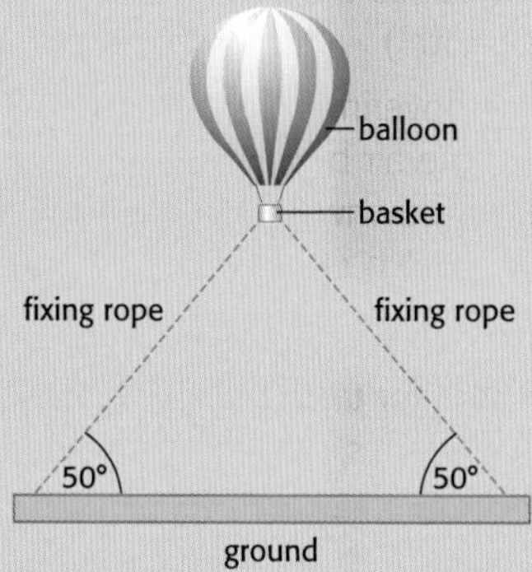

There is a force *F* vertically upwards of 2.15×10^3 N on the balloon. The total mass of the balloon and its basket is 1.95×10^2 kg.

a) State the magnitude of the resultant force on the balloon when it is attached to the ground. [1]

b) Calculate the tension in **either** of the fixing ropes. [3]

c) The fixing ropes are released and the balloon accelerates upwards. Calculate the magnitude of this initial acceleration. [2]

d) The balloon reaches a terminal speed 10 seconds after take-off. The upward force *F* remains constant. Describe how the magnitude of air friction on the balloon varies during the first 10 seconds of its flight. [2]

(Total 8 marks)

7 This question is about Newton's laws of motion, the dynamics of a model helicopter and the engine that powers it.

a) Explain how Newton's third law leads to the concept of conservation of momentum in the collision between two objects in an isolated system. [4]

b) The diagram illustrates a model helicopter that is hovering in a stationary position.

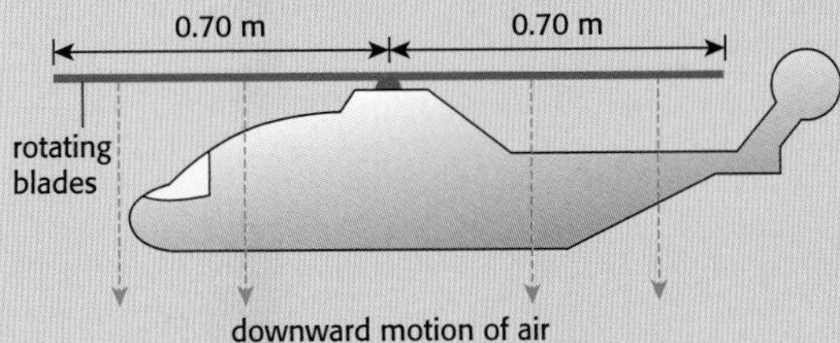

The rotating blades of the helicopter force a column of air to move downwards. Explain how this may enable the helicopter to remain stationary. [3]

c) The length of each blade of the helicopter in (b) is 0.70 m. Deduce that the area that the blades sweep out as they rotate is 1.5 m^2. (Area of a circle = πr^2) [1]

d) For the hovering helicopter in (b), it is assumed that all the air beneath the blades is pushed vertically downwards with the same speed of 4.0 m s^{-1}. No other air is disturbed.

The density of the air is 1.2 kg m^{-3}.

Calculate, for the air moved downwards by the rotating blades,

i) the mass per second; [2]

ii) the rate of change of momentum. [1]

e) State the magnitude of the force that the air beneath the blades exerts on the blades. [1]

f) Calculate the mass of the helicopter and its load. [2]

g) In order to move forward, the helicopter blades are made to incline at an angle θ to the horizontal as shown schematically below.

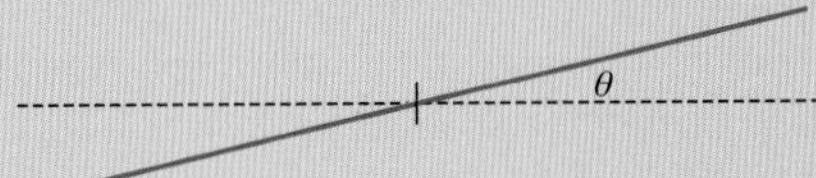

While moving forward, the helicopter does not move vertically up or down. In the space provided below draw a free body force diagram that shows the forces acting on the helicopter blades at the moment that the helicopter starts to move forward. On your diagram, label the angle θ. [4]

h) Use your diagram in (g) opposite to explain why a forward force F now acts on the helicopter and deduce that the initial acceleration a of the helicopter is given by

$$a = g \tan \theta$$

where g is the acceleration of free fall. [5]

i) Suggest why, even though the forward force F does not change, the acceleration of the helicopter will decrease to zero as it moves forward. [2]

(Total 25 marks)

8 This question is about circular motion.

A geo-stationary satellite is one that orbits the Earth in an equatorial plane in the same direction of rotation as that of the Earth and with an orbital period of 24 hours. Since the period of rotation of the Earth is 24 hours, this means that the satellite is stationary relative to a point on the Equator.

a) The diagram below shows a geostationary satellite in orbit about the Earth.

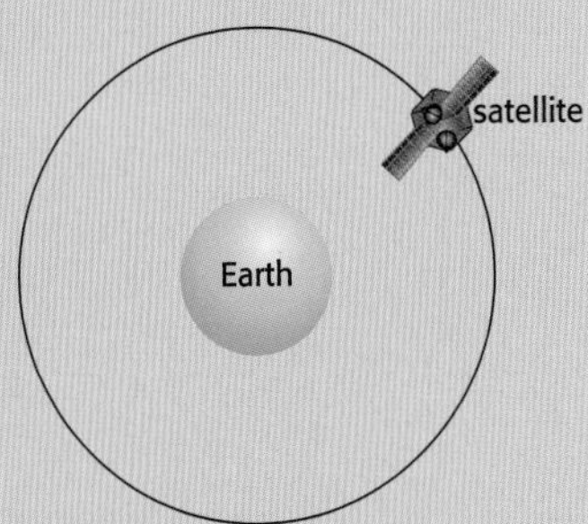

On the diagram above, draw an arrow to show the direction of acceleration of the satellite. [1]

b) State the name of the force causing the satellite's acceleration. [1]

c) The distance of the satellite from the centre of the Earth is 4.2×10^7 m. Calculate the acceleration of the satellite. [3]

(Total 5 marks)

9 This question is about circular motion.

A stone is attached to an inextensible string. The stone is made to rotate at constant speed v in a horizontal circle. Diagram 1 below shows the stone in two positions A and B.

Diagram 1

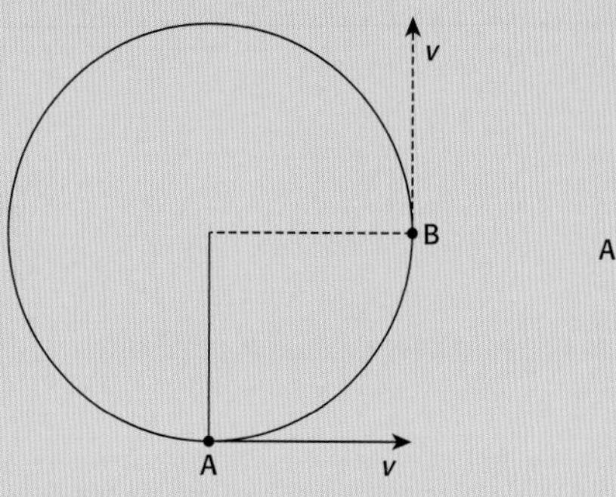

Diagram 2

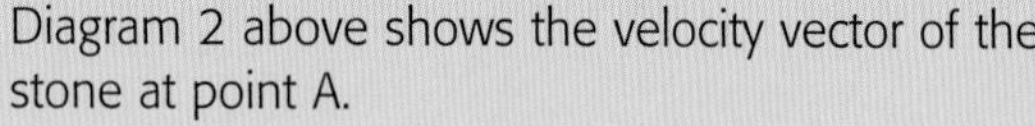

Diagram 2 above shows the velocity vector of the stone at point A.

a) On **diagram 2**, draw vectors to show the change in velocity Δv of the stone from point A to point B. [3]

b) Use your completed diagram 2 to explain why a force, directed towards the centre of the circle, is necessary to cause circular motion. [2]

(Total 5 marks)

5 Energy, work, and power

In physics the three physical quantities of **work, energy,** and **power** have precise mathematical definitions.

The concepts of work and energy provide a useful set of tools that allow us to analyse complicated situations. For example, you should already be able to use Newton's laws to estimate the final velocity v of a 0.8 kg block as it slides from rest down a 30° slope from a height of 1.8 m to ground level (Figure 1).

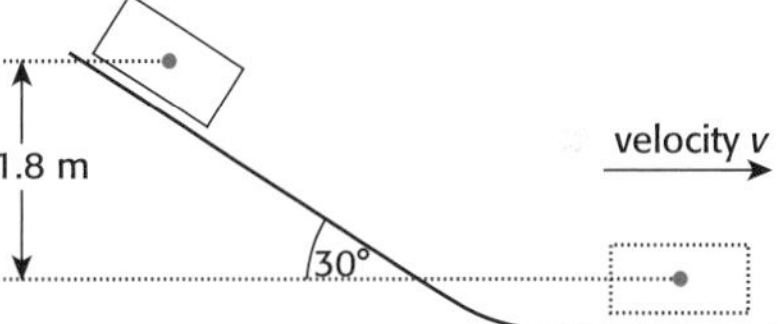

Figure 1 A block slides down a slope

The necessary stages in your calculation are as follows:

1 assume that friction is negligible
2 draw a free-body diagram for the block
3 resolve the forces in the directions (a) down the slope and (b) perpendicular to the slope
4 calculate the resultant force down the slope
5 use $F = ma$ to calculate the acceleration of the block down the slope
6 use trigonometry to calculate the distance down the slope
7 use the constant-acceleration equations to calculate the final velocity after going down the slope.

You should now have calculated that the block is moving down the slope at 6.0 m s^{-1}.

Note: This approach considers only the motion down the slope. We have not yet considered the curved section at the end. The forces during this section must be accelerating the block (it certainly changes direction), and so we might expect it to also change its speed. But if we assume that the effect on its speed will be negligible, then we have a prediction for its final velocity.

We can calculate the same final answer much more quickly using the energy concepts introduced below.

Work and the conservation of energy

Work and energy are related concepts. In any given situation, the work done measures the amount of energy that has been transferred.

Work

In physics, doing work involves moving a force through a distance. Although work is a scalar quantity, force is a vector quantity, and we need to consider the directions when calculating the work done. The simplest situation is when the force and the displacement are in the same direction.

work done, $W = Fs$

In general, however, the force and the displacement are not always in the same direction (Figure 2).

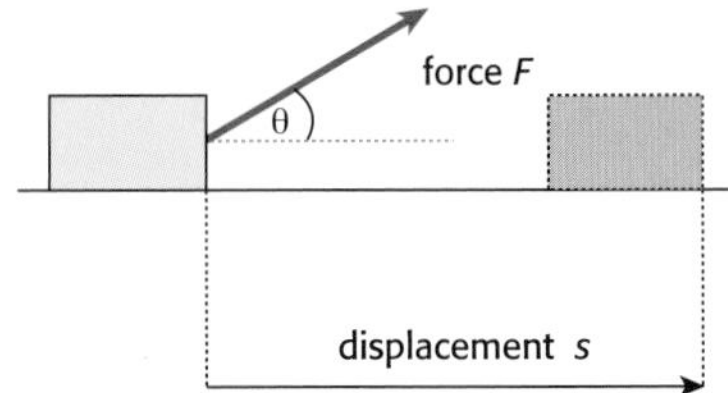

Figure 2 Force and displacement are not in the same direction

We define the work done with the equation

work done, $W = Fs \cos \theta$

By introducing the factor $\cos \theta$, we are effectively considering the component of the resultant force in the direction of the displacement.

Energy

Energy is the ability to do work. Energy and work are both scalar quantities; they are measured in joules. For example, if we give an object 5 kJ of energy, it can do 5000 J of work. How do we give it the energy? The fact that it has gained 5000 J of energy means that something else must have done work on it. The object doing the work lost 5000 J of energy, whereas the object receiving the energy had work done on it. Doing work means that energy is being interchanged.

Many different forms of energy have been identified: see the next section and pages 246–7 for a discussion on the nature of energy. Two extremely important forms that we can analyse quantitatively are **kinetic energy** (KE) and **gravitational potential energy** (gPE). KE is the energy that an object has as a result of its motion, and gPE is the energy that an object has as a result of its position in a gravitational field (i.e. its height).

Kinetic energy

Consider the situation of motion in a straight line when a constant resultant force F causes an object of mass m to accelerate from initial velocity u to final velocity v (Figure 3).

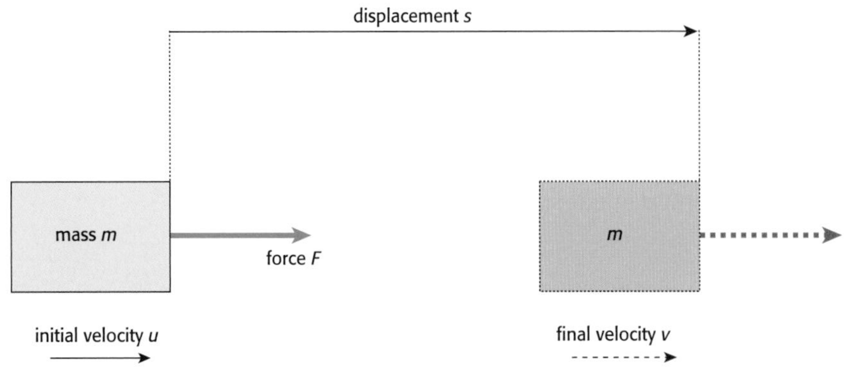

Figure 3 Work done accelerating a mass

1 The diagram below shows the variation with displacement x of the force F acting on an object in the direction of the displacement.

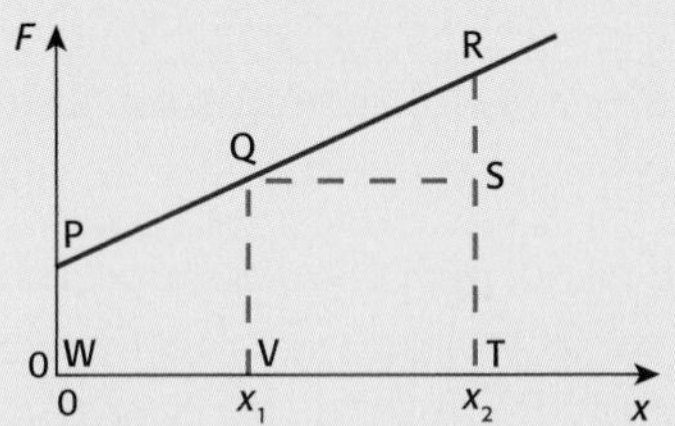

Which area represents the work done by the force when the displacement changes from x_1 to x_2?

A QRS **C** WPQV
B WPRT **D** VQRT

2 A stone of mass m is attached to a string and moves round in a horizontal circle of radius R at constant speed V. The work done by the pull of the string on the stone in one complete revolution is

A zero **C** $\frac{2\pi m V^2}{R}$
B $2\pi m V^2$ **D** $\frac{2\pi m V}{R}$

3 Which **one** of the following is a true statement about energy?

A Energy is destroyed due to frictional forces.
B Energy is a measure of the ability to do work.
C More energy is available when there is a larger power.
D Energy and power both measure the same quantity.

The work, W, done is the product of the force and the displacement:

$$W = Fs$$

The force and the mass are related to the acceleration, a:

$$F = ma \quad \text{so} \quad W = mas \qquad (1)$$

The acceleration and the displacement can be related to the initial and final velocities:

$$v^2 = u^2 + 2as \quad \text{so} \quad 2as = v^2 - u^2 \qquad (2)$$

Combining equations (1) and (2):

$$W = \tfrac{1}{2}mv^2 - \tfrac{1}{2}mu^2$$

The work done equals the change of the quantity

$$\tfrac{1}{2}\text{mass} \times (\text{velocity})^2$$

The change in kinetic energy is the work done on the mass, so we have derived an equation for the kinetic energy of a moving object:

$$\text{KE} = \tfrac{1}{2}mv^2$$

where
KE is the kinetic energy of the object, measured in joules (J)
m is the mass of the object, measured in kilograms (kg)
v is the velocity of the object, measured in metres per second (m s^{-1})

Gravitational potential energy

Consider the situation of an object of mass m being lifted (in a constant gravitational field) a vertical height h above its original position by a force F (Figure 4).

The work done, W, is calculated by

$$W = Fh$$

The force needed to lift the mass is equal to its weight:

$$F = mg$$

so

$$W = mgh$$

This work has increased the object's gravitational potential energy

$$\text{gPE} = mgh$$

where
gPE is the gain in gravitational potential energy of the object, measured in joules (J)
m is the mass of the object, measured in kilograms (kg)
g is the magnitude of the gravitational field strength, measured in newtons per kilogram (≈ 10 N kg^{-1} on the surface of the Earth)
h is the increase in vertical height, measured in metres (m)

Note that this equation only works when the gravitational field is constant. If an object moves a significant distance away from a planet, then the force due to gravity will reduce, and this equation does not apply. In this case, the calculation will involve either graphical work or the use of calculus.

An object of mass m_1 has a kinetic energy K_1. Another object of mass m_2 has a kinetic energy K_2. If the momentum of both objects is the same, the ratio $\frac{K_1}{K_2}$ is equal to

A $\frac{m_2}{m_1}$

B $\frac{m_1}{m_2}$

C $\sqrt{\frac{m_2}{m_1}}$

D $\sqrt{\frac{m_1}{m_2}}$

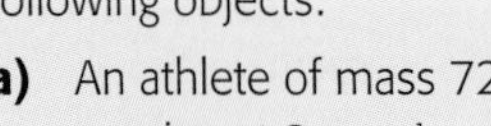

Calculate the KE of the following objects:

a) An athlete of mass 72 kg running at 8 m s^{-1}

b) A tennis ball of mass 60 g moving at 30 m s^{-1}

c) A car of mass 1600 kg moving at 20 m s^{-1}

d) A bullet of mass 150 g moving at 250 m s^{-1}

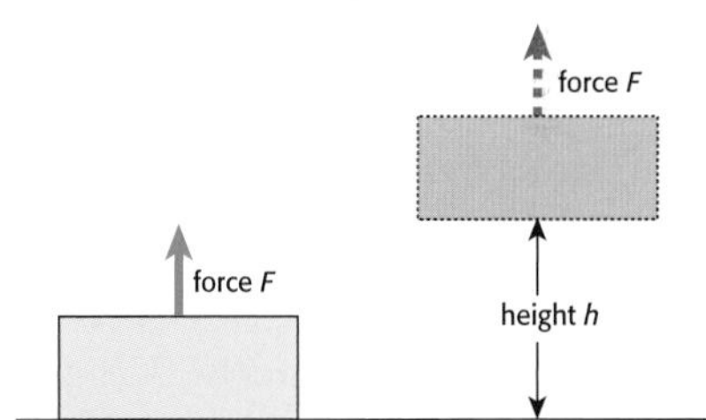

Figure 4 Increasing an object's gPE

Calculate the gPE of the following objects on Earth:

a) A diver of mass 72 kg climbing to a diving board 8 m high

b) A tennis ball of mass 60 g being hit 30 m into the air

c) A car of mass 1600 kg driving up a 20 m hill

d) A bullet of mass 150 g fired 250 m into the air

The conservation of energy

When work is done, energy is interchanged between objects. The amount of energy possessed by any given object can change, but the total amount of energy shared among all the objects involved must remain the same. This principle is known as the **law of conservation of energy:** "energy is neither created nor destroyed, it just changes form" or "the total energy of the universe is constant".

We can use this principle to analyse complex situations without having to predict every physical change that may take place. For example, we began this chapter by considering a mass sliding down a slope (see Figure 1 on p. 47).

If we assume that frictional forces can be neglected, then by analysing the forces involved we can predict the final velocity.

The block starts with gPE, and this is converted into KE. If friction is negligible, then the final KE must equal the initial gPE. This provides a quick method for calculating the final velocity, v.

final KE = initial gPE

$\frac{1}{2}mv^2 = mgh$

$\therefore v = \sqrt{2gh} = \sqrt{2 \times 10 \times 1.8} = \sqrt{36} = 6.0 \text{ m s}^{-1}$

Note that this analysis predicts that the final speed of the block is independent of:

- the mass of the block
- the angle of the slope
- the shape of the final curve.

The same prediction applies whatever route the block takes from its start to the finish. For example, in the absence of friction, the block in Figure 5 would also end up travelling at 6 m s^{-1}.

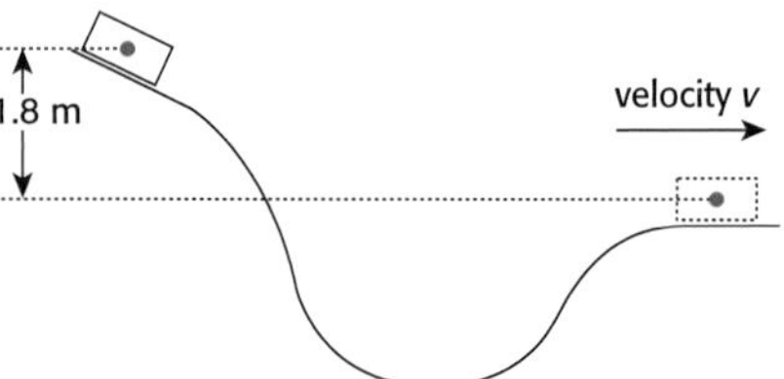

Figure 5 The final velocity only depends on the change in height of the object and not the route

1 A boy is sitting on a swing of length 2 m. The swing is pulled away from its resting position in the middle until it makes an angle of 30° with the vertical. The swing is then released.

   a) How fast will the boy be moving when he passes through the middle?

   b) How far will the boy travel beyond the middle on the other side?

2 The internet can provide lots of examples of simple physics (Newton's laws and energy considerations) being used to analyse complex situations. One very good example is Kung Fu Science, which follows a physics PhD student, Michelle, as she attempts to break three 2 cm thick pieces of pine board using her bare hands (http://www.kungfuscience.org/site.asp accessed July 2009). Do some research for similar sites and create a list of your favourites.

Energy transformations

In the previous section we introduced quantitative equations for two common forms of energy, KE and gPE. There are many different energy forms that we can usefully consider, but ultimately we can classify all forms of energy as either kinetic (if associated with movement) or potential (all other forms).

The various forms of energy include:

- kinetic energy
- potential energy
  - gravitational
  - electrostatic/electric
  - elastic
  - chemical
  - magnetic
- thermal energy
- radiant energy
- light (and the rest of the electromagnetic spectrum) energy
- sound energy
- nuclear energy.

If energy appears to be lost (or gained) in any particular situation, then rather than abandon the law of conservation of energy we need to try and identify a different form of energy. The nature of energy is discussed further on pages 246–7.

Note that in the list above we included "thermal energy" rather than "heat energy". Many scientists use the phrase "heat energy" or even just "heat" to refer to thermal energy. Technically "heat" (or better, "heating") is a verb so using the word "heat" to represent a form of energy can lead to some confusion and it is better to avoid these terms if at all possible.

The quantitative relationships between many of these energy forms are discussed throughout the IB physics course. At this stage, it is sufficient to note the general principle that many everyday practical situations involve much of the initial energy in a system ending up as thermal energy shared among a large number of molecules. This includes all situations in which frictional forces act. The energy is said to be **degraded** (see page 285). If thermal energy is released to the surroundings and cannot be recovered, it is said to be **dissipated.** Thermal energy concepts are discussed in more detail in Chapters 10 and 18.

Collisions and explosions

In any collision or explosion, the total energy involved must be conserved. In addition, the conservation of momentum must also apply to an isolated system. These two principles do not provide enough information for us to predict the final outcome in all situations – this depends on the context. By focusing on one particular type of energy (the kinetic energy) we can categorize different types of collision and explosion, and compare similar situations.

- **Elastic collisions** are those in which the KE is conserved. Everyday collisions are not elastic, because a significant portion of the KE is transferred into thermal energy and sound energy. Perfectly elastic collisions can take place only at the atomic scale.
- **Inelastic collisions** are those in which the total KE is decreased as a result of the collision. Most everyday collisions are inelastic. A **completely inelastic** collision is one in which the objects stick together after the collision.
- **Explosive collisions** are those in which the total KE is increased as a result of the collision. A source of energy (e.g. chemical potential energy or elastic energy) is needed to provide the "extra" KE.

1 An 8 kg truck moving at 5 m s^{-1} collides with a stationary 12 kg truck. The collision is completely inelastic, and the two trucks join together and move off with a common velocity v. Calculate:

a) the common velocity v after the collision

b) the energy "lost" to the surroundings.

2 (mathematically harder) An 8 kg truck moving at 5 m s^{-1} collides with a stationary 12 kg truck. The collision is perfectly elastic. Calculate the velocity of each truck after the collision.

Examples of energy transformations

If we can identify the energy transformations that are taking place in any situation it can often help us to quantify the changes that are taking place. However, remember that changes do not take place as a *result* of the energy changes. A ball does not roll down a hill *because* gPE is transferred into KE. It rolls down the hill because of the gravitational force of attraction on it.

There must always be a physical reason for the changes to take place; there are many things that are energetically possible but could not actually occur. For example, it is straightforward to identify the energy changes that take place when a glass is accidentally dropped. The main energy changes are gPE transforming into KE, which transforms into heat and sound.

From an energy point of view, the reverse process (the fragments coming together to form a whole glass that jumps back on to the table; is not ruled out – but we know it never occurs. See page 157 for an explanation.

?

Outline the principle energy conversions taking place in the following situations. In each case you need to identify the energy *transformations* that are occurring; do not merely list the types of energy involved. Many of the situations involve more than one energy transformation. One possible approach would be for each student in a class to take responsibility for analysing different situations and then report back to the rest of the class to see whether they agree. In addition to the situations listed below, describe a situation of your own choice.

a) A torch that runs on batteries is used to shine light.
b) A doorbell rings when a button is pushed.
c) A burglar alarm sounds when a laser beam is broken.
d) A lift carries passengers to the top floor of a skyscraper.
e) In order to live, you need to eat.
f) A pendulum swings back and forth.
g) Listening to music on a personal hi-fi device of your choice.
i) A firework rocket goes into the air and explodes.
j) Listening to the radio.
k) A voyage to the Moon.
l) A hydroelectric dam used to generate electrical energy.
m) An electromagnet used to pick up and sort metallic waste.
n) An electric hob used for cooking.
o) A microwave cooking a potato.
p) A parachute jump.
q) An aircraft taking off.
r) A pole-vaulter jumping over the bar.
s) A gun being fired.
t) A hair dryer.
u) A mechanical watch or clock.
v) A coal-fired power station.
w) A car being driven at constant velocity.
x) A solar-powered calculator.
y) Playing a piano.
z) A "hole in one" shot at golf.

Power and efficiency

Two useful quantities related to the study of any energy changes are the **power** and the **efficiency.**

Power

Power is the rate at which energy is transferred. This is the same thing as the rate at which work is done. The average power P_{av} over a time interval Δt is defined in terms of the energy transferred ΔW as follows:

$$P_{av} = \frac{\text{energy transferred}}{\text{time taken}} \qquad P_{av} = \frac{\Delta W}{\Delta t}$$

or

$$\text{power} = \text{resultant force} \times \text{velocity}$$

The units of average power are J s^{-1}, and a new unit, the watt (W), is defined:

$$1 \text{ watt} = \frac{1 \text{ joule}}{1 \text{ second}}$$

?

1 A man of mass 75 kg climbs 8 m up a vertical rope in 20 s. Calculate his average power.

2 A car maintains a constant velocity of 25 m s^{-1} against a friction force of 2 kN. How much power is the car's engine developing?

Efficiency

The efficiency of any process is a measure of the proportion of energy that is usefully transferred. What does "usefully" mean? That depends on the context. For example, a light bulb is designed to give out light energy, but it will also dissipate thermal energy to the surroundings at the same time. So in this context the useful energy is the light energy, and the thermal energy is the wasted energy. An electric heater, however, is designed to produce thermal energy and, when it is operating, very little of the energy it produces will be in any other form.

A 60 W light bulb is 15% efficient. Calculate the amount of energy dissipated into heat during 8 hours.

$$\text{efficiency} = \frac{\text{useful work done}}{\text{total energy transferred}} = \frac{\text{useful energy transferred}}{\text{total energy transferred}}$$

We can also express this definition in terms of power:

$$\text{efficiency} = \frac{\text{useful power}}{\text{total power used}}$$

Efficiency is a ratio, and has no units. It is often expressed as a percentage. The efficiency of different fuels is a crucial concept that runs through the whole issue of fuel use. See Chapter 18 for further details.

Work done with non-constant force

The basic defining equation for work done, on page 58, assumed that the force was constant. In many practical situations forces vary significantly with distance. For example, in order to increase the length of a spring, we need to increase the applied force. As a spring is stretched, the varying amount of work done is stored as elastic energy in the spring.

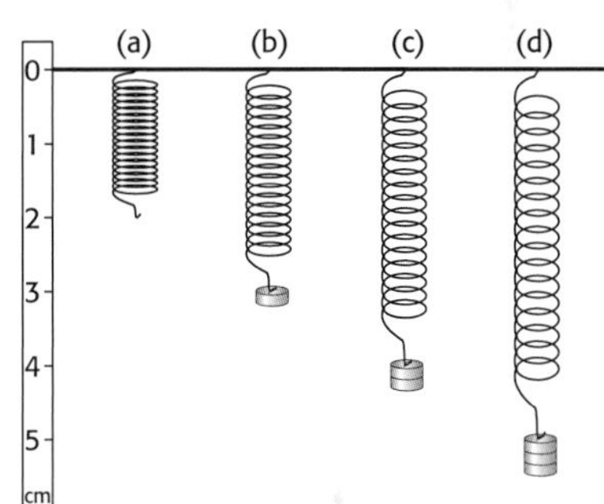

Figure 6 A spring loaded with various masses

The extension, x, is defined as the increase in length compared with the original length. **Hooke's law** applies to springs, and states that the extension is proportional to the force, up to a certain limit. Stretching a spring involves using a varying force.

The equation of the straight line in Figure 7 is given by

$$F = kx$$

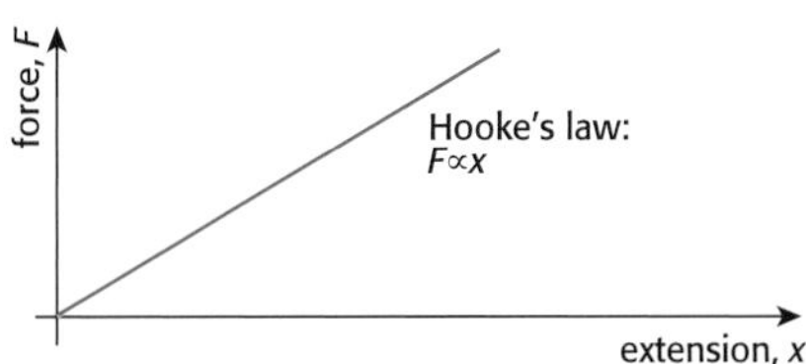

Figure 7 Hooke's Law. Force is proportional to extension

where

F is the force acting on the spring, measured in newtons (N)
x is the extension of the spring, measured in metres (m)
k is a constant, called the **spring constant,** measured in newtons per metre (N m^{-1})

The varying force means that we cannot just multiply the final force and the extension to calculate the work done. Once again we can use graphical techniques (or the mathematics of integration) to analyse the situation. The graph in Figure 8 represents a force that varies with distance. It is reasonable to assume that, over a very small distance δx the value of the force, F, does not change significantly. The work done over this small distance is the product of F and δx, which is represented by the shaded area of the rectangle.

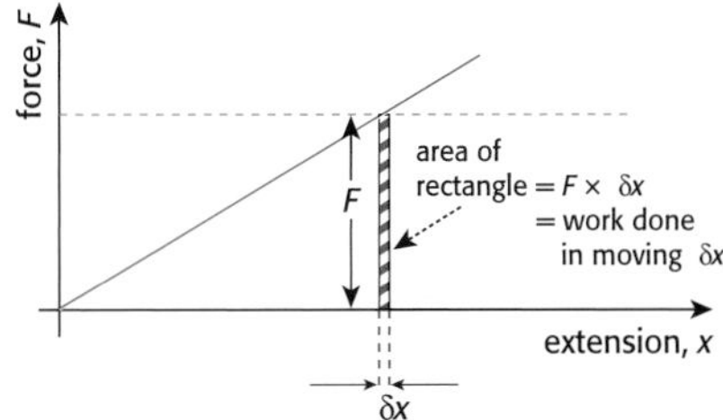

Figure 8 Work done in extending a spring by a small distance

The total work done that results from the varying force is the sum of all the extra small amounts of work done, for the whole distance over which the force acts. This sum is calculated as the area under the graph (Figure 9).

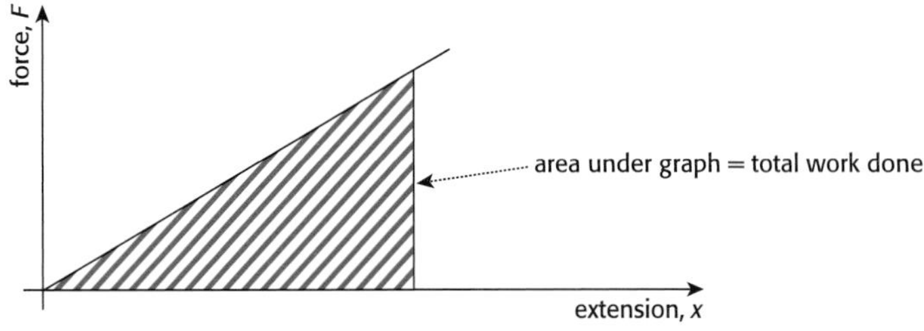

Figure 9 Work done is the area under the graph

$$\text{total work done} = \frac{1}{2}Fx = \frac{1}{2}kx^2$$

Investigating physics: Force–extension graphs

In the previous section we assumed that, for any given value of applied force, there was only one possible value for the extension. This is often the case for springs, but there are situations where this is not the case. Careful measurement can reveal a difference between loading and unloading.

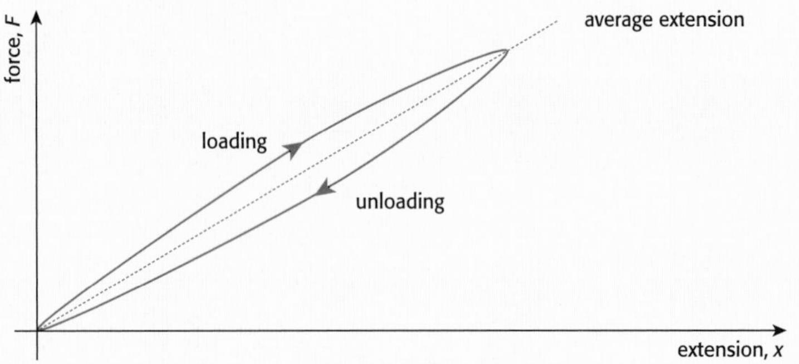

Figure 10 Loading and unloading can give different extensions

The area under the upper curve in Figure 10 represents the work done on the material during loading; the area under the lower curve represents the work done by the material during unloading. The area between the two lines represents the energy stored in the material after one cycle of loading and unloading.

Possible situations to investigate include:

- The extension of a copper wire. What does the shape of the graph of force versus extension tell you about the nature of the forces between copper atoms?
- The extension of a rubber band. What can you deduce about the nature of the forces between molecules of rubber?
- The breaking load for human hair. You could study the effects of different shampoos and/or conditioners.
- The extension of spaghetti and/or noodles to find the ideal cooking time.

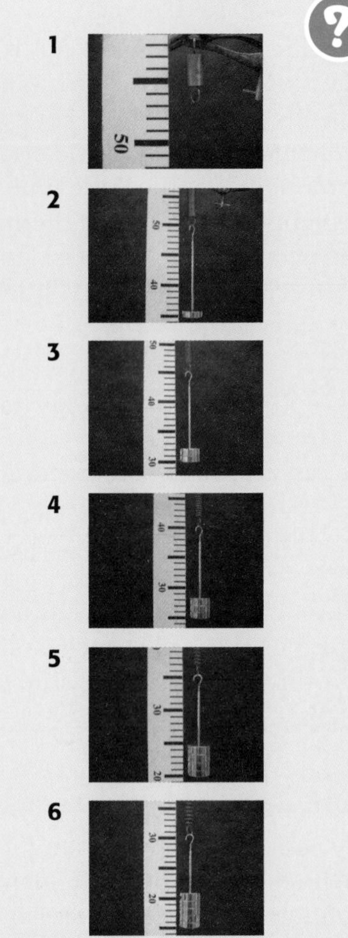

The masses used on the spring shown here are 100 g ± 5 g. Take readings from the figure to see whether Hooke's law is verified within experimental error. Use your graph to estimate:

a) the spring constant

b) the uncertainty range for your value for the spring constant

c) the elastic energy stored in any of the photos showing a stretched spring.

Safety note

There are many possible situations and materials that we can investigate under tension. Any material under tension is potentially dangerous. If the material were to break, some of the energy stored would be transferred into KE of the free end: so safety glasses should be worn at all times.

End of chapter summary

Chapter 5 has three main themes: work, energy and power. The list below summarises the knowledge and skills that you should be able to undertake after having studied this chapter. Further research into more detailed explanations using other resources and/or further practice at solving problems in all these topics is recommended – particularly the items in bold.

Work

- Outline what is meant by work and solve problems involving the work done by a force
- Determine the work done by a non-constant force by interpreting a force – displacement graph.

Energy

- Outline what is meant by kinetic energy and a change in gravitational potential energy.
- State the principle of conservation of energy, list different forms of energy and describe examples of the transformation of energy from one form to another.
- Distinguish between elastic and inelastic collisions.

Power

- Define *power* and *efficiency* and apply the concepts to **solve problems involving momentum, work, energy and power.**

Chapter 5 questions

1 This question is about the kinematics of an elevator (lift).

An elevator (lift) starts from rest on the ground floor and comes to rest at a higher floor. Its motion is controlled by an electric motor. A simplified graph of the variation of the elevator's velocity with time is shown below.

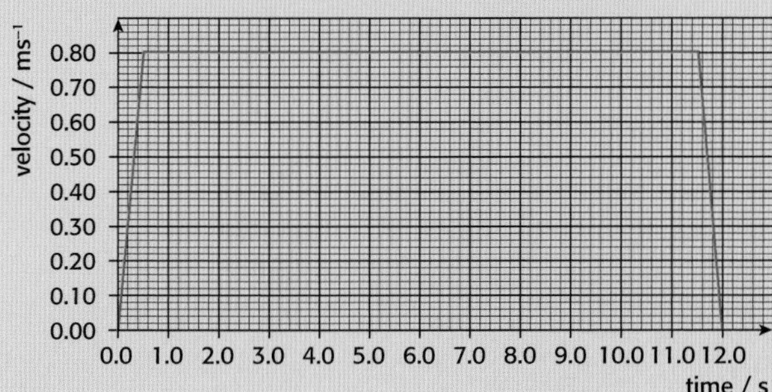

The mass of the elevator is 250 kg. Use this information to calculate

a) the acceleration of the elevator during the first 0.50 s. [2]

b) the total distance travelled by the elevator. [2]

c) the minimum work required to raise the elevator to the higher floor. [2]

d) the minimum average power required to raise the elevator to the higher floor. [2]

e) the efficiency of the electric motor that lifts the elevator, given that the input power to the motor is 5.0 kW. [2]

f) The elevator now returns to the ground floor where it comes to rest. Describe and explain the energy changes that take place during the whole up and down journey. [4]

(Total 14 marks)

2 This question is about driving a metal bar into the ground and the engine used in the process.

Large metal bars can be driven into the ground using a heavy falling object.

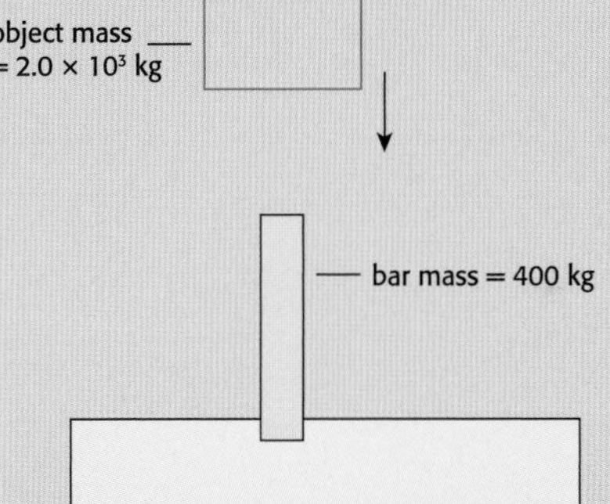

In the situation shown, the object has a mass 2.0×10^3 kg and the metal bar has a mass of 400 kg.

The object strikes the bar at a speed of 6.0 m s^{-1}. It comes to rest on the bar without bouncing. As a result of the collision, the bar is driven into the ground to a depth of 0.75 m.

a) Determine the speed of the bar immediately after the object strikes it. [4]

b) Determine the average frictional force exerted by the ground on the bar. [3]

c) The object is raised by a diesel engine that has a useful power output of 7.2 kW.

In order that the falling object strikes the bar at a speed of 6.0m s^{-1}, it must be raised to a certain height above the bar. Assuming that there are no energy losses due to friction, calculate how long it takes the engine to raise the object to this height. [4]

(Total 11 marks)

3 This question is about the collision between two railway trucks (carts).

a) Define *linear momentum*. [1]

In the diagram below, railway truck A is moving along a horizontal track. It collides with a stationary truck B and on collision, the two join together. Immediately before the collision, truck A is moving with speed 5.0ms^{-1}. Immediately after collision, the speed of the trucks is *v*.

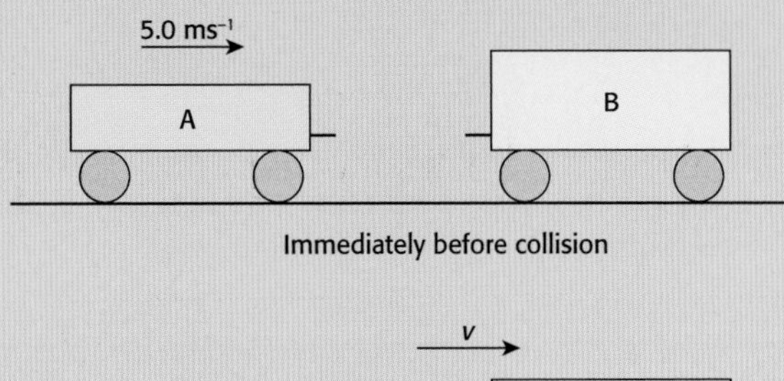

The mass of truck A is 800 kg and the mass of truck B is 1200 kg.

b) i) Calculate the speed v immediately after the collision. [3]

ii) Calculate the total kinetic energy lost during the collision. [2]

c) Suggest what has happened to the lost kinetic energy. [2]

(Total 8 marks)

4 This question is about projectile motion and the use of an energy argument to find the speed with which a thrown stone lands in the sea.

Christina stands close to the edge of a vertical cliff and throws a stone. The diagram below *(not drawn to scale)* shows part of the trajectory of the stone. Air resistance is negligible.

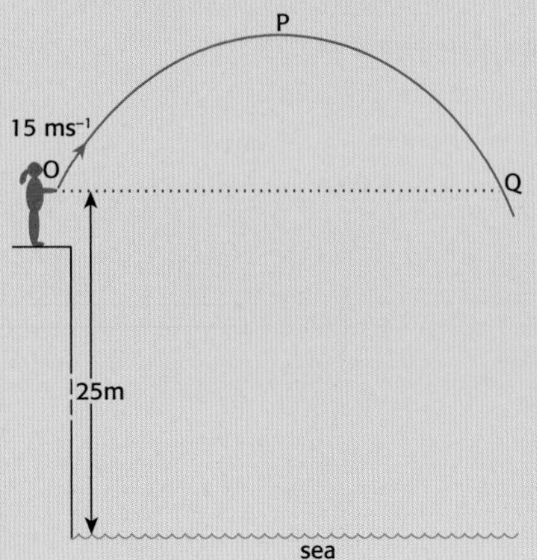

Point P on the diagram is the highest point reached by the stone and point Q is at the same height above sea level as point O.

a) At point P on the diagram above draw arrows to represent

i) the acceleration of the stone (label this A). [1]

ii) the velocity of the stone (label this V). [1]

The stone leaves Christina's hand (point O) at a speed of 15 m s^{-1} in the direction shown. Her hand is at a height of 25 m above sea level. The mass of the stone is 160 g. The acceleration due to gravity $g = 10\text{ m s}^{-2}$.

b) i) Calculate the kinetic energy of the stone immediately after it leaves Christina's hand. [1]

ii) State the value of the kinetic energy at point Q. [1]

iii) Calculate the loss in potential energy of the stone in falling from point Q to hitting the sea. [1]

iv) Determine the speed with which the stone hits the sea. [2]

(Total 7 marks)

5 Linear motion

At a sports event, a skier descends a slope AB. At B there is a dip BC of width 12 m. The slope and dip are shown in the diagram below. The vertical height of the slope is 41 m.

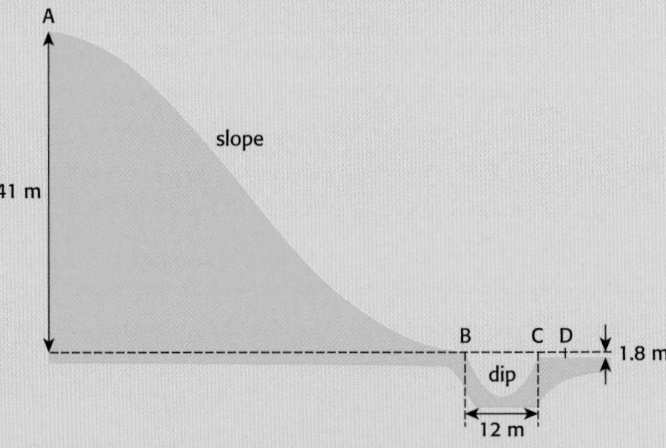

The graph below shows the variation with time *t* of the speed *v* down the slope of the skier.

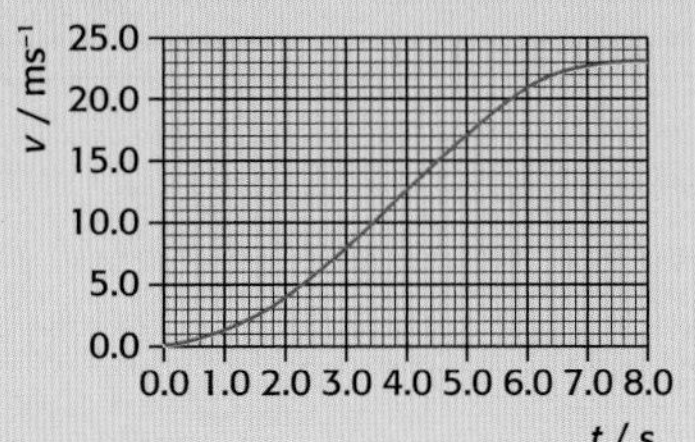

The skier, of mass 72 kg, takes 8.0 s to ski, from rest, down the length AB of the slope.

a) Use the graph to

i) calculate the kinetic energy E_K of the skier at point B. [2]

ii) determine the length of the slope. [4]

b) i) Calculate the change ΔE_p in the gravitational potential energy of the skier between point A and point B. [2]

ii) Use your answers to (a) and (b)(i) to determine the average retarding force on the skier between point A and point B. [3]

iii) Suggest **two** causes of the retarding force calculated in (ii). [2]

c) At point B of the slope, the skier leaves the ground. He "flies" across the dip and lands on the lower side at point D. The lower side C of the dip is 1.8 m below the upper side B.

Determine the distance CD of the point D from the edge C of the dip. Air resistance may be assumed to be negligible. [4]

d) The lower side of the dip is altered so that it is inclined to the horizontal, as shown below.

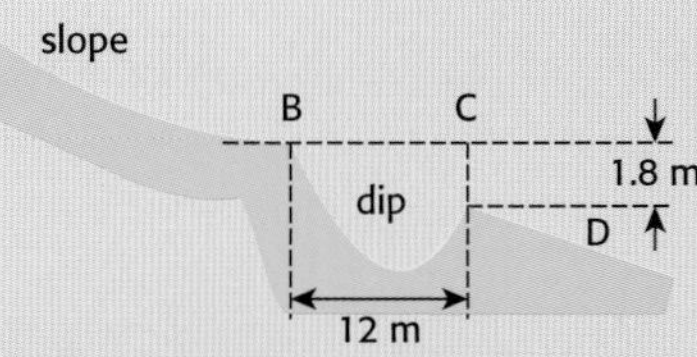

i) State the effect of this change on the landing position D. [1]

ii) Suggest the effect of this change on the impact felt by the skier on landing. [2]

(Total 20 marks)

6 This question is about energy and momentum.

A train carriage A of mass 500 kg is moving horizontally at 6.0 m s^{-1}. It collides with another train carriage B of mass 700 kg that is initially at rest, as shown in the diagram below.

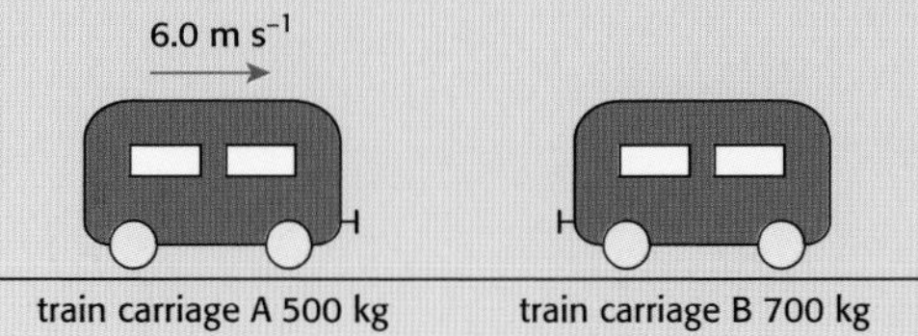

The graph below shows the variation with time t of the velocities of the two train carriages before, during and after the collision.

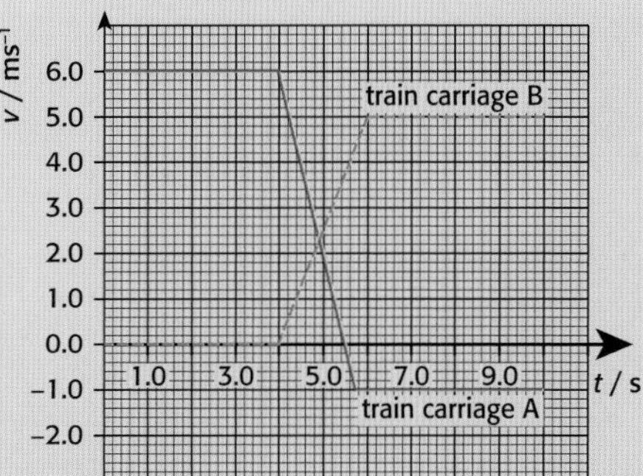

a) Use the graph to deduce that

i) the total momentum of the system is conserved in the collision; [2]

ii) the collision is elastic. [2]

b) Calculate the magnitude of the average force experienced by train carriage B. [3]

(Total 7 marks)

7 This question is about the breaking distance of a car and specific heat capacity.

a) A car of mass 960 kg is free-wheeling down an incline at a constant speed of 9.0 m s^{-1}.

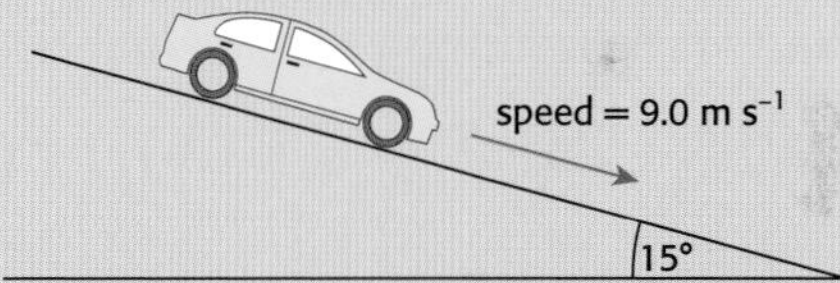

The slope makes an angle of 15° with the horizontal.

i) Deduce that the average resistive force acting on the car is 2.4×10^3N. [2]

ii) Calculate the kinetic energy of the car. [1]

b) The driver now applies the brakes and the car comes to rest in 15 m. Use your answer to (a) (ii) to calculate the average braking force exerted on the car in coming to rest. [2]

c) The same braking force is applied to each rear wheel of the car. The effective mass of each brake is 5.2 kg with a specific heat capacity of 900 J kg^{-1} K^{-1}. Estimate the rise in temperature of a brake as the car comes to rest. State **one** assumption that you make in your estimation. [4]

(Total 9 marks)

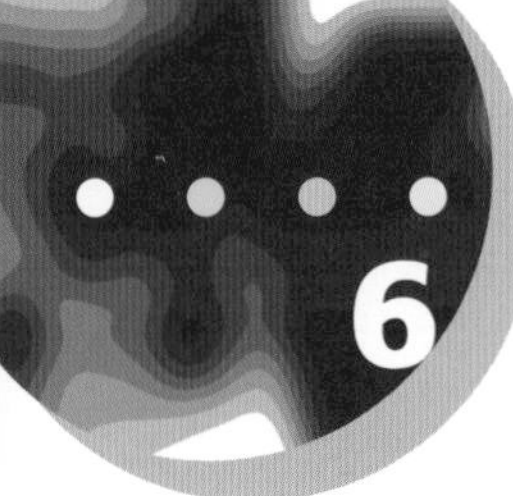

6 Projectiles

HL

This chapter focuses on the motion of projectiles in a gravitational field. Although the chapter has been written for HL candidates, SL candidates could study the material as part of the "wider picture".

Going in more than one direction

The effect that any vector quantity has in a particular direction can be quantified by resolving the vector into components. At an angle of θ, a vector V has an effect $V \cos \theta$ in the horizontal direction, as shown in Figure 1.

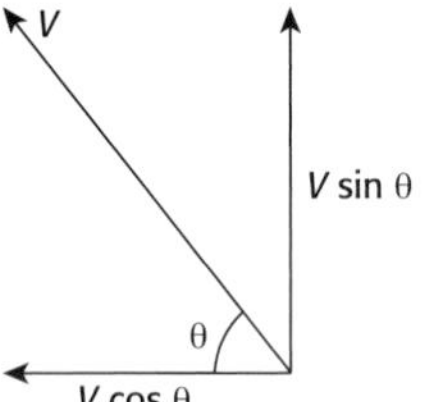

Figure 1 Resolving a vector into components

When θ is 90°, the component at angle θ will be $V \cos(90°) = 0$ (Figure 2). So vectors that are at right angles to one another are completely independent. This means that, for example, horizontal motion and vertical motion can be analysed independently.

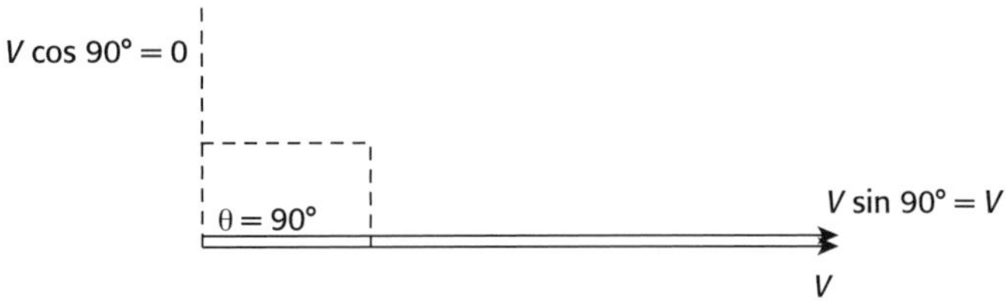

Figure 2 The component at 90° is zero

Which **one** of the following is a true statement concerning the vertical component of the velocity and the acceleration of a projectile when it is at its maximum height? (*The acceleration of free-fall is g.*)

	Vertical component of velocity	Acceleration
A	maximum	zero
B	maximum	g
C	zero	zero
D	zero	g

Consider a ball that rolls at constant velocity along a table and then falls over the edge, as shown in Figure 3. Friction is negligible. Can the shape of the path (called a **parabola**) followed after the ball leaves the table be predicted?

To simplify the analysis, we can first consider the overall motion in the horizontal direction, and then in the vertical direction. When the ball leaves the table, the only thing that changes is that the upward reaction force from the table no longer exists.

When the ball was moving along the table, the resultant horizontal force was zero, and the ball moved equal distances in equal times. When it leaves the table, the resultant horizontal force is still zero, so the ball continues to travel equal horizontal distances in equal times (Figure 4).

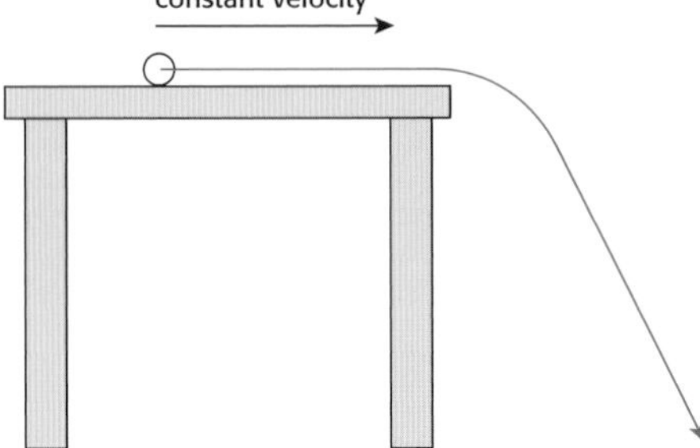

Figure 3 Parabolic motion during fall

A resultant force acts on the ball in the vertical direction when it leaves the table. This means that the ball must accelerate in the vertical direction (Figure 5). The value of the vertical acceleration is g, the free-fall acceleration due to gravity. The vertical distance travelled can be worked out using the constant acceleration formula.

The vertical motion is influenced only by the vertical gravitational force. The time taken for this ball to hit the floor is exactly the same as the time taken for a ball to be dropped from the table vertically down (Figure 6).

It can be helpful to set up the situation and test this statement. If a stationary ball is dropped from the height of the table when a rolling ball leaves the table (this might take a little practice), they will fall together and you will hear only one sound as they both hit the floor together.

Note that the above analysis is valid even if the initial horizontal velocity is extremely large. A bullet fired horizontally from a gun must also follow a parabolic path towards the Earth. The high initial horizontal velocity will mean that the extent to which the bullet falls vertically may be negligible over small distances, but the bullet will still take the same time to fall to the Earth as it takes a bullet dropped from the same height next to the gun.

In the absence of air resistance, the motion of all projectiles must be a constant horizontal velocity and a constant vertical acceleration down. The result of these two independent motions is a parabolic path. The precise shape of the path depends on the initial conditions of the projectile motion.

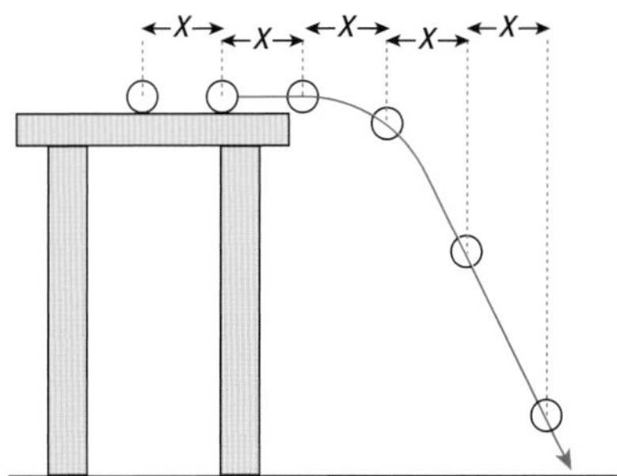

Figure 4 Horizontal motion is constant velocity

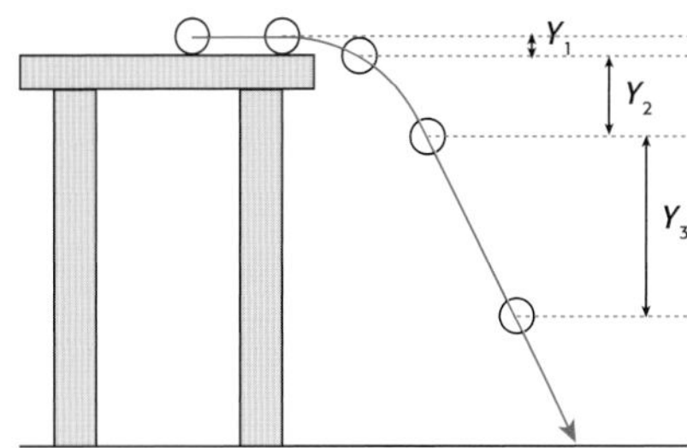

Figure 5 Vertical motion is constant acceleration

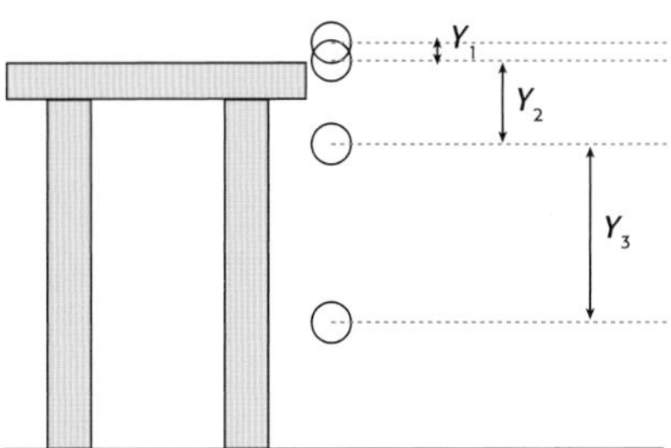

Figure 6 The time taken for vertical motion is independent of the horizontal motion

An object is thrown upwards at an angle of 30° to the horizontal from the top of a 10 m vertical cliff and lands in the sea. The initial velocity of the object is 8 m s^{-1}. See Figure 7.

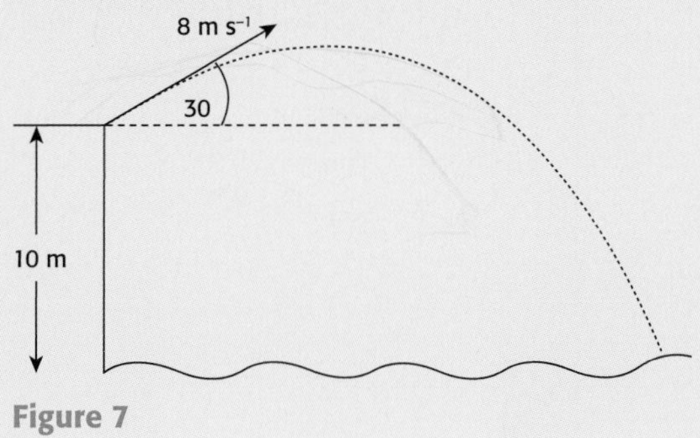

Figure 7

a) Describe how the velocity of the object varies
  i) vertically
  ii) horizontally.

b) Calculate the time taken
  i) before the object gets to the highest point in its flight
  ii) before the object hits the sea.

c) Calculate the velocity of the object
  i) at the highest point in its flight
  ii) when it hits the sea.

d) Calculate the displacement
  i) vertically, at the highest point in its flight
  ii) horizontally, when it hits the sea.

The effect of air resistance

In the above analysis we assumed that frictional effects were negligible. In reality, air resistance will cause any projectile motion to deviate from parabolic motion. Air resistance opposes motion. The magnitude and the direction of the resistive force will vary during an

Describe qualitatively the changes to the answers you have given to the above question that would result if resistive forces were not negligible.

object's flight. The resistive force always opposes the motion, and the magnitude of this force can be taken to be proportional to the speed.

A full analysis is complex, because the varying resistive force has components in both the horizontal and the vertical directions, but these can be analysed independently. The horizontal component of the resistive force will be proportional to the horizontal velocity, and the vertical component of the resistive force will be proportional to the vertical velocity. One way of modelling the situation would be to use spreadsheets.

Overall, the effect of air resistance will be to slow down the object. As the object does work against the resistive force, it will lose energy. See Figure 8.

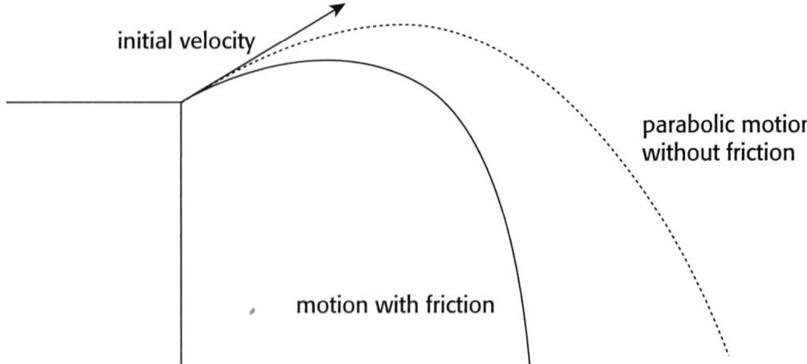

Figure 8 Effect of air resistance (friction) on parabolic motion

Data-based question: Basketball shots

Digital video techniques can be used to capture the motion of any object. A good application is the analysis of basketball shots. Once the data has been captured on video, simple software can be used to trace the motion of the ball. The sketch in Figure 9 of the trajectory of a basketball shot was created using a free piece of software called "vidshell", but many others exist.

The basket shot was recorded using a standard digital video camera, and the software was used to mark the position of the ball in each of the pictures, or frames, taken during its flight. Your task is to analyse Figure 9 in order to calculate a value for the:

a) initial velocity of the basket ball as it leaves the girl's hands

b) acceleration due to gravity.

The following information is available:

- The basket ring is 10 ft (3.05 m) above the ground.
- The time between frames is 0.04 s.

At the time of writing vidshell was available from http://cripe03.rug.ac.be/Vidshell/Vidshell.htm, and useful video clips were available at:

http://www.didaktik.physik.uni-muenchen.de/materialien/inhalt_materialien/videosequenzen/index.html (accessed July 2009).

The same procedure can be used to analyse any type of motion. Many pieces of software allow the user to scale the image, and can automatically calculate the instantaneous displacement, velocity, and acceleration from the captured images. This can be also useful in gaining a good understanding of simple harmonic motion (see Chapter 7).

If you have access to the appropriate equipment, this procedure can allow very detailed investigations to be achieved in a reasonably short amount of time, and would form an excellent basis for an extended essay (see Chapter 25).

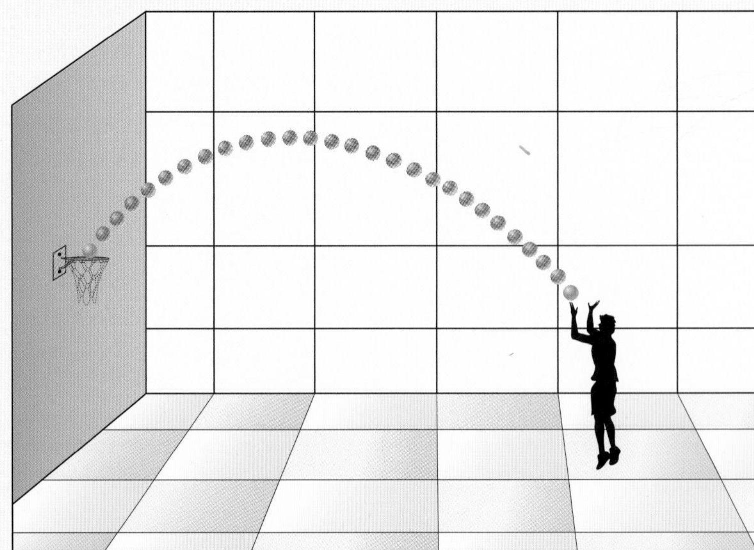

Figure 9 Parabolic path of a basket ball shot

End of chapter summary

Chapter 6 has one main theme: the mathematics of projectile motion. The list below summarises the knowledge and skills that you should be able to undertake after having studied this chapter. Further research into more detailed explanations using other resources and/or further practice at solving problems in all these topics is recommended – particularly the items in bold.

- State the independence of the vertical and the horizontal components of velocity for a projectile in a uniform field.
- Describe and sketch the trajectory of projectile motion as parabolic in the absence of air resistance and describe qualitatively the effect of air resistance on the trajectory of a projectile.
- **Solve problems on projectile motion.**

Chapter 6 questions

1 This question is about projectile motion. A small steel ball is projected horizontally from the edge of a bench. Flash photographs of the ball are taken at 0.10 s intervals. The resulting images are shown against a scale as in the diagram below.

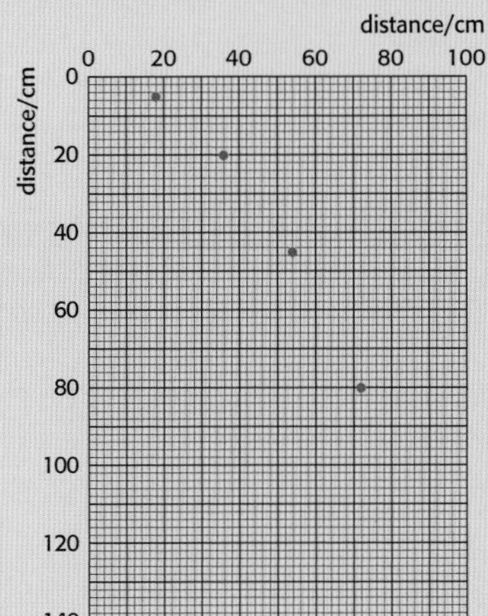

a) Use the diagram to determine

i) the constant horizontal speed of the ball [2]

ii) the acceleration of free fall. [2]

b) Mark on the diagram the position of the ball 0.50 s after projection. [3]

c) A second ball is projected from the table at the same speed as the original ball. The ball has small mass so that air resistance cannot be neglected. Draw on the diagram the approximate shape of the path you would expect the ball to take. [3]

(Total 10 marks)

2 This question is about parabolic motion.

A projectile is launched from the surface of a planet. The initial vertical component of velocity is 40 m s^{-1}. The diagram below shows the positions of the projectile in 0.20 s intervals. Note that no scale has been given on the vertical axis.

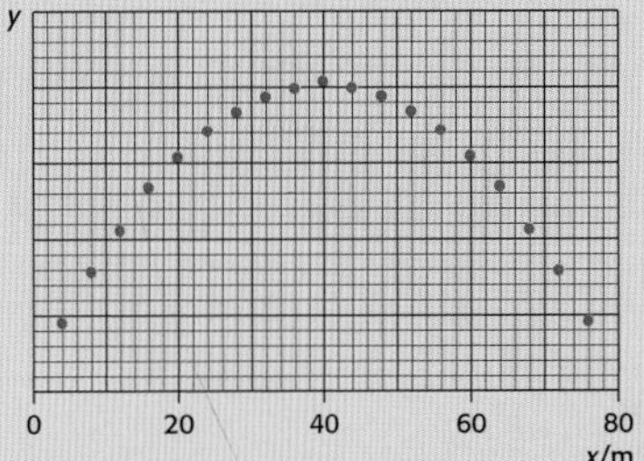

a) Calculate

i) the horizontal velocity of the projectile [2]

ii) the acceleration of free fall at the surface of the planet [3]

iii) the maximum height reached by this projectile. [2]

b) Determine the angle to the horizontal at which the projectile is launched. [2]

c) The projectile is launched with the same velocity from the surface of a planet where the acceleration of free fall is twice that calculated in **a) ii)**. Draw the path of this projectile on the graph above. [3]

(Total 12 marks)

3 This question is about projectile motion.

A projectile is fired at an angle of 45° to horizontal ground such that its speed immediately after leaving the ground is *v*.

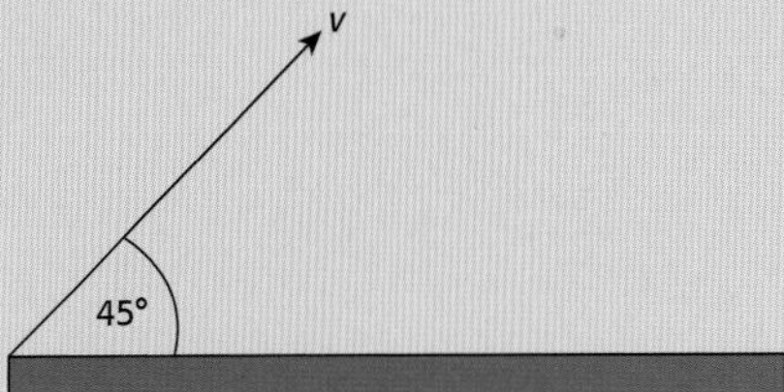

The projectile leaves the ground at time $t = 0$ and returns to the ground at time $t = T$. Air resistance is negligible.

a) Draw a sketch graph to show the variation with time *t* (*x*-axis) of the horizontal velocity v_H and vertical velocity v_V of the projectile from time $t = 0$ to time $t = T$. Label the horizontal velocity with the letter v_H and the vertical velocity with the letter v_V. [4]

b) On your sketch-graph, mark with the letter P the time that corresponds to the projectile at its maximum height. [1]

c) The angle at which the projectile is fired is increased to more than 45°. Explain, in terms of conservation of energy, why the maximum height reached by the projectile increases. [2]

4 This question is about the trajectory of a golf ball.

A golfer hits a golf ball at point A on a golf course. The ball lands at point D as shown on the diagram. Points A and D are on the same horizontal level.

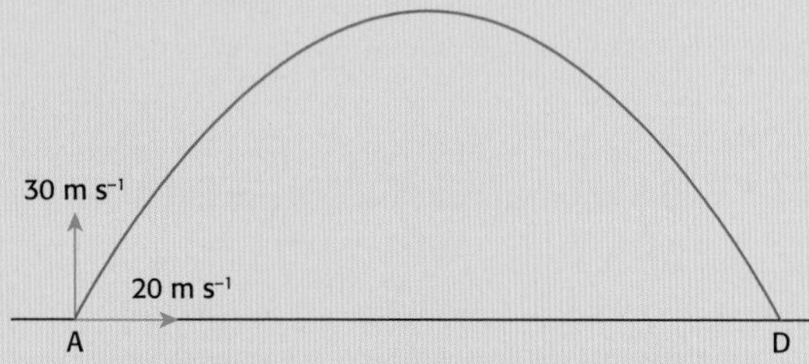

The initial horizontal component of the velocity of the ball is 20 m s^{-1} and the initial vertical component is 30 m s^{-1}. The time of flight of the golf ball between point A and point D is 6.0 s. Air resistance is negligible and the acceleration of free fall $g = 10$ m s^{-2}.

Calculate

a) the maximum height reached by the golf ball. [3]

b) the range of the golf ball. [2]

(Total 5 marks)

5 This question is about projectile motion.

A projectile is fired horizontally from the top of a vertical cliff of height 40 m.

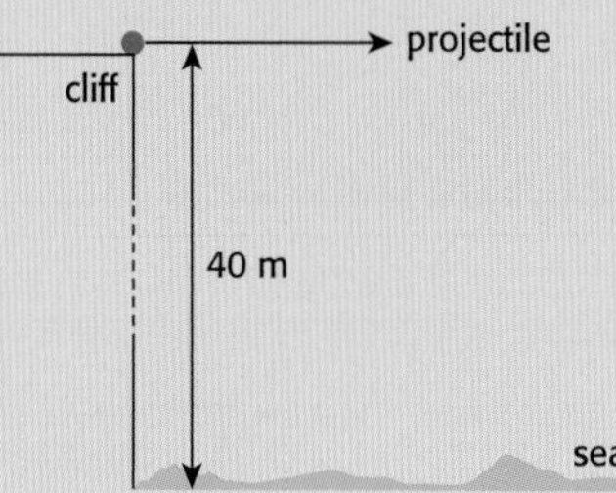

At any instant of time, the vertical distance fallen by the projectile is d. The graph below shows the variation with distance d, of the kinetic energy per unit mass E of the projectile.

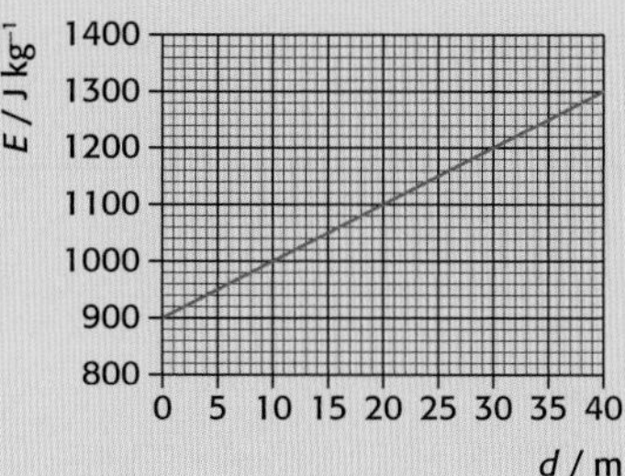

a) Use data from the graph to calculate, for the projectile,

i) the initial horizontal speed. [1]

ii) the speed with which it hits the sea. [1]

b) Use your answers to (a) to calculate the magnitude of the vertical component of velocity with which the projectile hits the sea. [2]

(Total 4 marks)

6 This question is about projectile motion.

A stone is thrown from the top of a cliff of height 28 m above the sea. The stone is thrown at a speed of 14 m s^{-1} at an angle above the horizontal. Air resistance is negligible.

The maximum height reached by the stone measured from the point from which it is thrown is 8.0 m.

a) By considering the energy of the stone, determine the speed with which the stone hits the sea. [3]

b) The stone leaves the cliff at time $t = 0$. It reaches its maximum height at $t = T_H$. Draw a sketch-graph to show the variation with time t of the magnitude of the **vertical** component of velocity of the stone from $t = 0$ to $t = T_S$, the time just before the stone strikes the sea. [4]

(Total 7 marks)

7 Simple harmonic motion

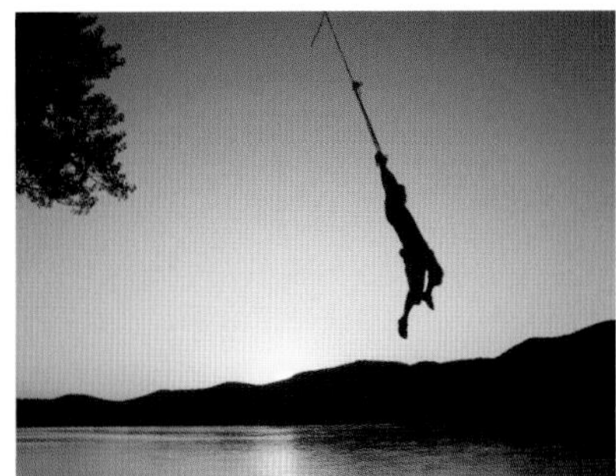

The term "oscillation" describes any vibration that goes "back and forth" without an overall resulting movement. It turns out that many real-world situations approximate to one particular type of oscillation, called Simple Harmonic Motion (or SHM for short). Before starting any of the sections in this chapter it would be useful to:

- review circular motion and how to measure angles in radians
- ensure that you understand and can precisely define what is meant by the basic terminology used when considering oscillations. A good definition stands by itself and does not need further explanation. The units that are used to measure a quantity should be clear from the definition. See if you can come up with correct definitions (and units) for the following terms:
  - *displacement*
  - *amplitude*
  - *frequency and angular frequency*
  - *time period*
  - *angular speed*
  - *phase difference.*
- produce a list of some practical examples of oscillations. As you work through this chapter, check to see if all your examples were appropriate.

Everyday SHM

An object undergoing SHM oscillates about a fixed point. This fixed point is the object's mean position, often called its equilibrium position because it is the point where the object would come to rest when no external forces were acting on it. If displaced away from this point, there must be a **restoring force** on the object. The motion is SHM if this restoring force, and hence the object's acceleration, is always proportional to its displacement from the mean position and always directed towards it.

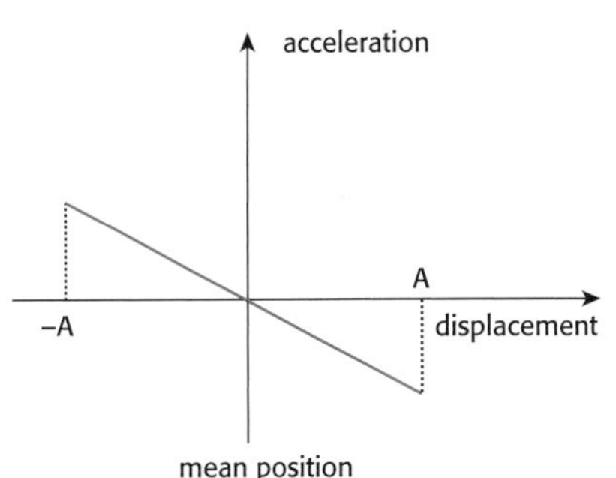

Figure 1 SHM: acceleration is proportional to –displacement

The equation for the resulting displacement oscillation involves a sine function and a graph showing the variation with time of displacement is described as **sinusoidal:**

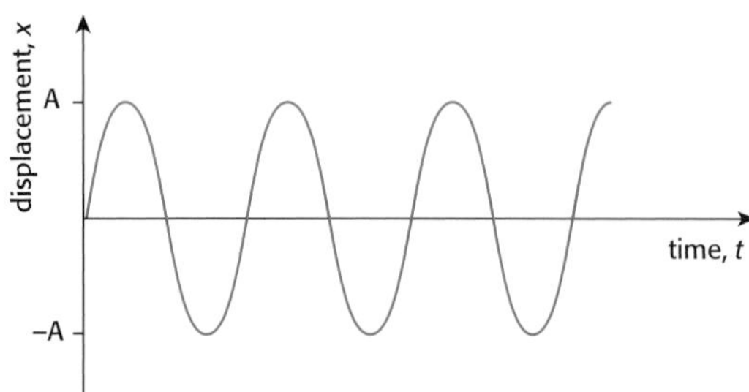

Figure 2 SHM: The variation of displacement with time in sinusoidal

The following situations are some of the many day-to-day examples of SHM that are all around us. Some of these examples can also be considered to be examples of resonance (see page 95). None of these situations are perfect examples of SHM – they are merely approximations to SHM under certain conditions.

Table 1

Situation	Comment
Pendulum	Any suspended object moving from side to side can be considered to be a pendulum. Chandeliers in cathedrals can sometimes be seen to swing back and forth. According to legend, while listening to a Mass in the cathedral of Pisa in 1581, a 17-year-old student noticed something: the chandelier overhead was swaying in the breeze. Sometimes it was barely moving and at other times it was swinging in a wide arc but one complete swing always took the same amount of time (as measured by the student's heart beat). The student was called Galileo Galilei (1564–1642).
Bungee jumping	The fall from the platform is an acceleration due to gravity but once the elastic material has straightened out, the jumper is effectively a mass on the end of a spring and so oscillates vertically. Try to explain why only some parts of the motion can be considered to be SHM. Would it be possible to just bounce back up to the platform?
Diving board	Once the diver has left the board, it will oscillate up and down. The board is one example of a very common situation called a cantilever. It is quite straightforward to build a cantilever in the laboratory to study its motion.
Object bobbing in water	Any object floating in water can bob up and down. If the cross-section going into the water is approximately constant, the oscillations will be SHM. A test tube containing a small mass will float in water and bob up and down with SHM.
Musical instruments	All musical instruments make use of oscillations. For some of these instruments, the cause of the oscillation is obvious, for example, the triangle is hit and vibrates. For some others the musician produces the driving oscillation (e.g. a trumpeter's lips), the amplitude of which is reinforced by the oscillation of the air in the instrument.
Earthquakes	There are different types of waves associated with earthquakes, some of which are SHM waves. Much of the damage (and hence injuries) associated with earthquake disasters are a result of buildings collapsing because they have not been built to withstand the shaking of the ground. See pages 102–4 for further details.

Thinking about science: *Comparison of two different SHM formulae*

Consider three very simple oscillating systems: a pendulum, a mass on a spring and a bouncing ball. A mass on the end of a string (the simple pendulum) and a mass on a spring are common basic examples of Simple Harmonic Motion. The bouncing ball is not an example of SHM.

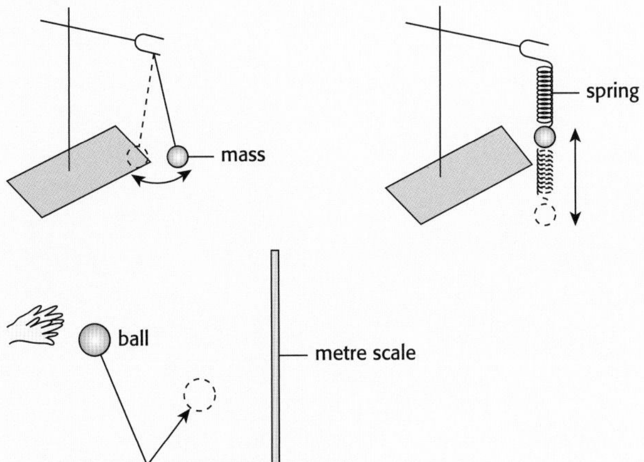

Mathematical analysis is such a powerful tool that these two apparently very different situations can be shown to have aspects of their motion in common. Even though the situations are so different, it is possible to use Newton's laws of motion to derive equations that accurately describe the position,velocity, and acceleration of these objects. Some of the assumptions involved in these derivations will be discussed later.

$$T = 2\pi\sqrt{\frac{m}{k}}$$

Time period of a mass on a spring

$$T = 2\pi\sqrt{\frac{l}{g}}$$

Time period of a pendulum

T time period of the oscillation

m mass of the object on the spring

k spring constant (force per unit extension for the spring)

l length of the pendulum

g gravitational field strength

Perhaps the biggest surprise is that these formulae predict that the time period for each oscillation depends only on two variables. For the spring, m and k fix the time period, T, whereas for the pendulum only l and g are involved. In neither case does the amplitude of the oscillation appear and so we can say that the equations predict that the time period is independent of the amplitude.

The variable g only appears in the equation for the pendulum and is not part of the equation for the mass on the end of the spring. This means that a change of gravitational field strength should not change the frequency of oscillation for a spring. In other words, if a mass on the end of a spring were to be taken to the Moon, the equation predicts that it would oscillate with the same frequency as it did on Earth.

The time period of the pendulum is predicted to be independent of the mass of the pendulum's bob. Do you believe that a 1 m pendulum with a 100 g mass on its end would swing with exactly the same time period as a 1 m pendulum that had a 5 kg mass?

The mass on the end of the spring and the simple pendulum will, for most practical situations, oscillate with a time period that can be correctly predicted by the mathematics of SHM. It is, however, impossible for either of these situations to be anything other than an approximation to SHM as shown below.

For the mass on a spring and the pendulum the SHM equation arises because we assume that the resultant restoring force on the mass is directly proportional to its displacement from the equilibrium position (see page 81). In mathematical notation:

$$F = -kx$$

where k is a constant of proportionality.

Mass on a spring

The derivation of the equation for the time period, T, for one oscillation

$$T = 2\pi\sqrt{\frac{m}{k}}$$

includes the following assumptions (doubtless there are others):

1 the mass of the spring is negligible compared to the mass of the load
2 friction is negligible
3 Hooke's law applies to the spring at all times
4 the gravitational field strength is constant
5 the fixed end of the spring cannot move.

Some of these assumptions (e.g. number 1) may not apply in particular setups but others (e.g. number 4) will always be reasonable assumptions.

Pendulum

The derivation of

$$T = 2\pi\sqrt{\frac{l}{g}}$$

makes the following assumptions:

1 the mass of the string is negligible compared to the mass of the load
2 friction is negligible
3 the angle of swing is small
4 the gravitational field strength is constant
5 the length of the pendulum is constant during the experiment.

Although you would probably agree that the fifth assumption is obvious, a surprisingly large number of students fail to think about how to attach the string to the support when setting up their own simple pendulum.

Physics is often about making simplifications in order to create models that aid understanding of real-world situations. A good model makes enough approximations to turn the situation into one that we can analyse, without making so many that we can no longer relate it to the real world.

If we try to apply a model to a situation where the simplifications no longer apply then, not surprisingly, the model will not make accurate predictions. Often a simple model can be adapted and improved to deal with any over-simplifications – see pages 93–4 on damped SHM.

Inquiry: Pendulum motion

Design and carry out an investigation to measure the effect that the angle of swing has on the time period of a pendulum. In this investigation, the **independent** variable is the angle of swing and the **dependent** variable is the time period of the pendulum. Other aspects such as the length of the pendulum and the gravitational field strength are the **controlled** variables.

Note: as far as the IB Diploma Programme is concerned, this task would **not** be suitable for assessment against the IB Internal Assessment criteria for the **design** skill as the variables have been given. The experiment can, however, be used to assess DCP or CE.

Investigating physics: Using a pendulum with SHM to measure g and assess the uncertainty limits

In calculations involving the gravitational field strength at the surface of the Earth, g, different values can be used: $g = 10\ \text{N kg}^{-1}$ for rough calculations and $g = 9.81\ \text{N kg}^{-1}$ if we want to be more precise. What exactly is the value of g?

If the Earth were a perfectly uniform sphere with constant density, we would expect g to have the same value anywhere on the Earth. At different places

Non-mathematical communication

Mathematical proof is a very powerful technique but is not always useful when discussing ideas. This question asks you to justify, in two different ways, the following statement:

"A bouncing ball's motion is fundamentally different from the motion of a simple pendulum and of a mass oscillating vertically on a spring."

1 By mathematically analysing the forces acting on each of the oscillating masses, show what the final two situations have in common with each other and explain why the first situation is different.
2 Write a brief article justifying the above statement without the use of any mathematical equations or symbols.

a)

b)

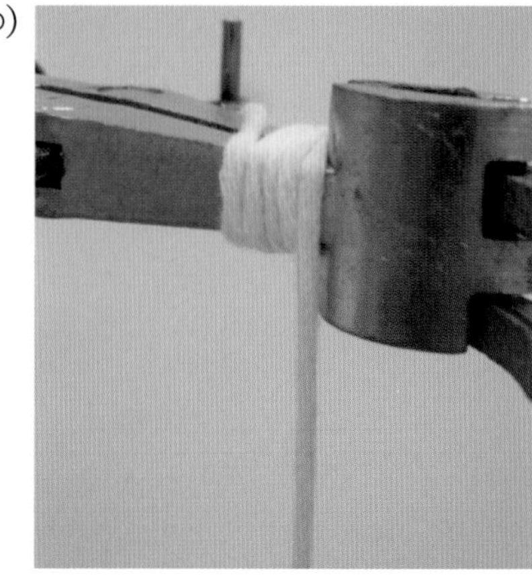

Figure 3 a) Bad way of attaching a pendulum b) good way of attaching a pendulum

around the world, however, we actually observe some variations. Height above sea level is one obvious factor affecting the value of g but local geological conditions and the effects of the Earth's spin at different latitudes can also be important.

The aim of this experiment is to find an experimental value for g and to compare your own value with other values calculated by students both in your school and, if possible, from around the world.

In this experiment you should fix the length of a simple pendulum, l (the **independent** variable) and measure the time period of the pendulum, T, (the **dependent** variable). These values can be substituted into the SHM formula $T = 2\pi\sqrt{\frac{l}{g}}$ in order to calculate g, or by using the SHM equation in the form

$$g = \frac{4\pi^2 l}{T^2}$$

it is possible to find a more accurate value for g.

This approach will only work if all the assumptions made at the start are valid.

Uncertainty in measurements

All experimental measures are subject to some degree of uncertainty. This means that two students working in the same laboratory are unlikely to calculate exactly the same answer. Suppose one student calculated the value $g = 10.22$ N kg^{-1} whereas the other one got $g = 9.73$ N kg^{-1}. Which one is right? A straight comparison between two values is not going to allow you to decide, nor is a comparison with the accepted value. The student with $g = 9.73$ N kg^{-1} is closer to the accepted value than the other student but this does not mean that she is right.

The existence of uncertainties does not mean that you cannot do good science, but it does mean that when drawing conclusions you must calculate and discuss the uncertainty in your result. The aim is to assign a "confidence level" to any value that results from an experiment. Different experimental techniques result in different confidence levels but if two experiments are supposed to be measuring the same value, then they should agree within their confidence levels.

In this context, if two experimenters find different values of g for the same location on the surface of the Earth then, providing they have not made any mistakes, **new physics must have been discovered**. Unfortunately mistakes are possible (even for a physicist!).

Possible sources of mistakes include:

- reading the scales on the instruments wrongly
- the instruments not being correctly set to zero
- making mistakes in the mathematics
- making mistakes when copying data from one place to another.

For this reason, the scientific community would probably not reject a law just because of a single experiment. The experimental evidence needs to be consistently repeatable.

Assessing the uncertainty from first principles

Typical results could be as follows:
Length of pendulum, $l = 42.1$ cm
Time for 10 swings, $10T = 12.75$ s
The calculation gives $g = 10.22$ N kg^{-1}.

The **uncertainty** for the length of the pendulum needs to take into account the readability of the metre rule used (typically ±0.5 mm) as well as the uncertainty resulting from the string not being perfectly flat, say another ±0.5 mm. The uncertainty in the location of the centre of mass of the pendulum bob, approximately ±2 mm, and the uncertainty in the location of fixing should not be ignored. A total figure of ±3 mm or more would not be unreasonable.

Overall the length uncertainty is thus ±0.3 cm in 42.1 cm $= \dfrac{0.3}{42.1}$ or ±0.7%.

Uncertainty in (l) = ± 0.7%.

The uncertainty of the timing due to the readability of the stopwatch is ±0.01 s. If the readability was the only source of uncertainty then the error would be an impressive ±0.01 in 12.75 s or ±0.08%. In reality, assessing when the pendulum bob has completed a full oscillation is not that easy and one should also take into account the reaction time of the experimenter. It is, however, possible to practice taking readings and accurately anticipate the end of a swing. An estimation of ± (a quarter of one time period) or 0.32 s would not seem unreasonable as an upper limit.

The uncertainly over time is thus ±0.32 s in 12.75 s or ±2.5%.

$$\begin{aligned}
\text{Uncertainty in } (T) &= \pm 2.5\% \\
\therefore \text{ Uncertainty in } (T^2) &= 2 \times \text{Uncertainty in } (T) \\
&= \pm 5.0\%
\end{aligned}$$

$$\begin{aligned}
\therefore \text{ Uncertainty in (g)} &= \text{Uncertainty in } (T^2) + \text{Uncertainty in } (l) \\
&= \pm(5.0 + 0.7)\% \\
&= \pm 5.7\% \\
&\approx \pm 6\%
\end{aligned}$$

6% of 10.22 is 0.61 so the appropriate result to quote for this experiment would be:

$g = 10.2 \pm 0.6$ N kg^{-1}

The experimental value is now known to be somewhere between 9.6 N kg^{-1} and 10.8 N kg^{-1}. You could, of course, argue that some of the values given above are pessimistic and that the uncertainty is probably somewhat less than the calculated value.

If the other experimenter assessed the uncertainties to be of the same magnitude then her result would be quoted as $g = 9.7 \pm 0.6$ N kg^{-1} so these two students agree that g is somewhere between 9.6 N kg^{-1} and 10.3 N kg^{-1}.

What is the accepted uncertainty in the values normally quoted for g? The way in which we quote the value implies the uncertainty of the measurement. If we state that $g = 10$ N kg^{-1} this implies an uncertainty of ±1 N kg^{-1}. We are saying that the numerical value for g is 10, not 11 or 9. Writing $g = 9.81$ N kg^{-1} implies an uncertainty of ± 0.01 N kg^{-1} (it is 9.81, not 9.82 or 9.80). When the accepted value is quoted to one **significant digit** we write $g = 10 \pm 1$ N kg^{-1} whereas the latter is quoted to three significant digits and thus implies $g = 9.81 \pm 0.01$ N kg^{-1}.

These experiments both agree with the 10 ± 1 N kg^{-1} and with 9.81 ± 0.01 N kg^{-1} but do not worry if your experiment does not agree with the more accurate figure. This figure is the accepted *average* value of g *at sea level*. There is no particular reason for your local value to be the same as a global average but there is every reason to expect all experiments to get the same value in the same location.

Improving the accuracy

The accuracy of this experiment can be improved by both repeating readings for an individual value and by taking a range of different values of the time period for different lengths of pendulum.

Repetitions allow average values to be calculated but they also provide another quick way of assessing the uncertainties.

If, for example, the time for 10 swings was recorded as 12.75 s, 12.84 s, and 12.63 s, then the average value is 12.74. The maximum value was +0.10 s and the minimum value was –0.12 s, so it would be reasonable to conclude that our readings imply that the correct value is 12.74 ± 0.12 s. It should be noted that this method is not statistically precise and can only be used to provide an estimation of the actual uncertainty of the mean.

One way of analysing these results would be to use each reading to calculate a different value for g and then to average the result. This approach would work but a numerical average treats all results equally and does not take into account any assessment of whether an individual reading is subjected to a large or small amount of error. A better method would be to use each reading to construct a graph that allows you to mathematically identify the most precise link between the variables.

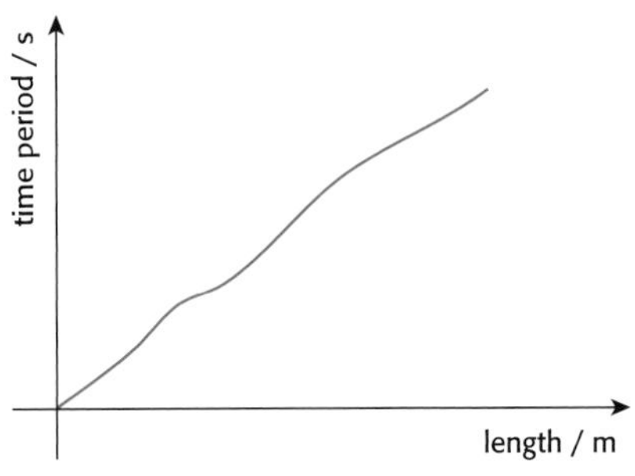

Figure 4 It is hard to see a precise relationship in a graph of this shape

In this situation, many students think it helps to plot a graph showing the variation of the time period with length.

As the best-fit line is a curve, it is not practical to use this plot to assess any formal relationship between time period and length. A better approach is to look at the hypothesised relationship, $T = 2\pi\sqrt{\frac{l}{g}}$, and choose to plot variables (or combinations of variables) in order to produce a straight-line graph.

$$T = 2\pi\sqrt{\frac{l}{g}} \qquad \therefore T^2 = \frac{4\pi^2 l}{g}$$

This equation can be compared with $y = mx + c$.

Table 2

Summary of the comparison with $y = mx + c$		
dependent variable:	y	T^2
independent variable:	x	l
gradient:	m	$4\pi^2/g$
y-intercept:	c	0

Thus if we plot T^2 on the y-axis and l on the x-axis, we will get a straight line going through the origin (intercept c = zero) with gradient, m, given by:

$$m = \frac{4\pi^2}{g} \qquad \therefore g = \frac{4\pi^2}{m}$$

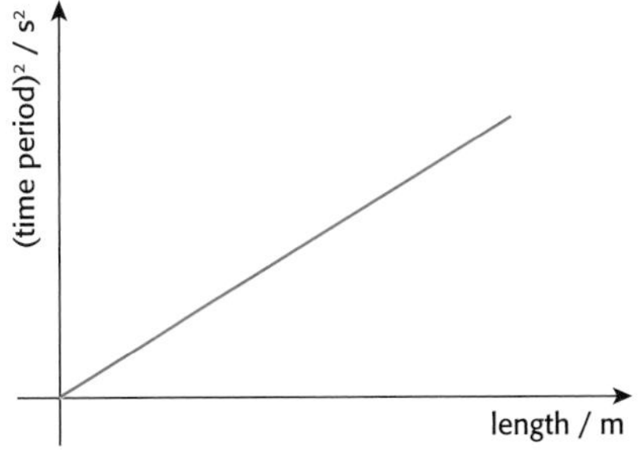

Figure 5 It is clearer from this graph that there is a square root relationship between the two variables

In principle, error bars can be constructed for every point and a best-fit straight line can be added as shown on pages 15–16. In practice it is often only necessary to construct error bars for one or two points (often the largest values and/or the point that seems furthest from the line).

The value for g calculated from the gradient of this line has taken every reading into consideration by creating a visual trend (the straight line) that can be used to identify the actual mathematical link between time period and length.

$$g = \frac{4\pi^2}{\text{gradient of straight line}}$$

This final result is more accurate than any value arising from a single calculation. An assessment of the uncertainty can be achieved by calculating the maximum and minimum values for the gradient of all the straight-line graphs that can be drawn whilst still including every data point.

The time taken for a simple pendulum of varying length to undergo 20 oscillations was recorded and used to calculate the average time period for the pendulum as shown in Table 3. Analyse this data to calculate a value for g. The uncertainty in the time period is negligible.

Table 3 The uncertainty in the length measurement is estimated to be ±1 cm.

Length	Time for 20 oscillations				Period
l	T_1	T_2	T_3	T(average)	T
cm	second				
10	12.84	12.91	12.88	12.88	0.64
20	18.35	18.30	18.21	18.29	0.91
30	22.16	22.16	22.23	22.18	1.11
40	25.36	25.48	25.36	25.40	1.27
50	28.63	28.58	28.62	28.61	1.43
60	31.16	31.09	31.13	31.13	1.56
70	33.85	33.97	33.75	33.86	1.69
80	36.07	35.79	36.07	35.98	1.80
90	38.34	38.31	38.03	38.23	1.91
60	40.33	40.30	40.53	40.39	2.02

Inquiry: The scientific method

It is commonly accepted that it is impossible to prove a law experimentally. An experiment can only produce results that are consistent with the law; if this is the case, we say that the experiment **verifies** the law. However, experiments can disprove (or falsify) a law.

Your experiments and those performed by more experienced scientists are equally valid in this regard. Although highly sophisticated equipment has been designed to measure g with great accuracy, it is important to realize that a simple experiment performed by you in the laboratory is just as valid as a highly technical professional setup as long as the uncertainties are specified.

This aspect of the ***scientific method*** is sometimes referred to as ***"empirical falsifiability"*** and is often used to distinguish scientific theory from non-science. The 20th-century philosopher Karl Popper (1902–1994) is perhaps best known for rejecting the classical ***observationalist-inductivist*** account of scientific method and for advancing the ***principle of falsification*** (Popper 1934).

Research the terms underlined in the paragraph above so that you could explain them to a non-scientist.

Figure 6 Karl Popper (1902–1994)

Mathematical physics: *The "driving" equation for SHM*

If any object is oscillating with SHM, then the most complete and succinct way to describe the variation of its motion over time is to use a branch of mathematics called differential calculus. If you find mathematics challenging, fear not – you do not need to understand calculus to be able to do well at IB Diploma Programme physics. It is true, however, that if you choose to study physics after your IB Diploma, then it is extremely useful to have acquired a high level of mathematical skills. If your mathematics is up to the challenge, you should be using it for IB physics.

At time $t = 0$

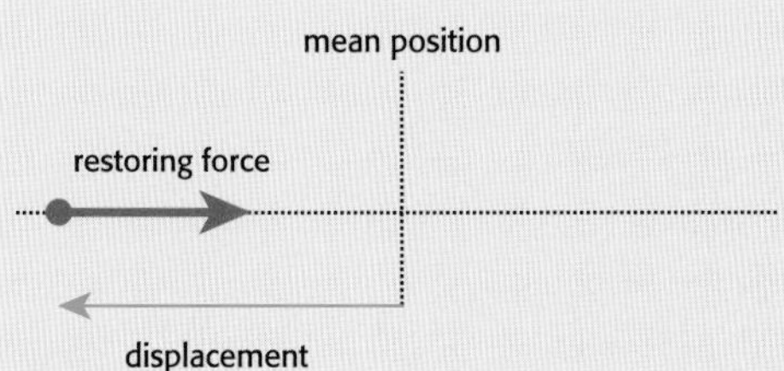

Figure 7

The great benefit of using mathematical analogies is that if a situation can be shown to be an example of SHM, then it is possible to "jump" straight to the known solution without necessarily needing to follow through all of the calculations from first principles.

Perhaps the simplest description of a physical situation that results in SHM is an object oscillating about a mean (average) position. As discussed on page 73, the forces on it are arranged so that the **resultant force** acting on it is always directed towards the mean, or equilibrium position (so it is called a *restoring force*) and is **proportional** to the object's **displacement** from the mean position. Remember: this means that when the displacement from the mean position doubles, the magnitude of the restoring force also doubles.

At a later time, $t = t'$

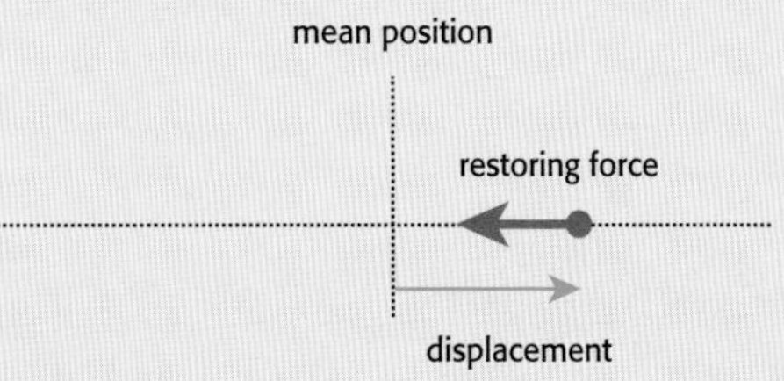

Figure 8

The resulting motion of the object is an oscillation about the mean position.

Mathematically this can be expressed as:

$$\text{resultant force} \propto -\text{displacement}$$

or

$$F \propto -x$$

Therefore

$$F = -kx \qquad (1)$$

Equation (1) defines SHM. The negative sign is important: it is there because the displacement vector and the resultant force acting on the object are always in opposite directions. k is a constant of proportionality and in the case of the mass on a spring, it turns out to be a measure of the force per unit extension.

From Newton's second law of motion we know that an object's acceleration is proportional to the resultant force acting on it:

$$F = ma$$

So we have

$$ma = -kx$$

$$a = -\frac{k}{m}x \qquad (2)$$

The term k/m is a constant and is often written as ω^2, where ω is called the angular frequency. The reason for this is discussed on pages 83–4.

Therefore $a = -\omega^2 x$

The acceleration is the rate of change of velocity and velocity is the rate of change of displacement. This allows equation (2) to be re-written as a second-order differential equation.

Using calculus to solve the "driving" equation for SHM

Acceleration is the rate of change of velocity (or the rate of change of the rate of change of displacement). Using calculus notation this can be expressed as:

$$\frac{d^2x}{dt^2} = -\frac{k}{m}x \text{ this is sometimes written as } \ddot{x} = -\frac{k}{m}x$$

Since k and m are constants for any given situation, we have an equation whose general form is:

$$\frac{d^2x}{dt^2} = -\text{constant} \times x$$

We must now find a solution to this second-order differential equation. Essentially we wish to identify a mathematical function that if differentiated twice, yields the same function. In this situation this means deriving a formula for x in terms of t (and the other constants) that satisfies the condition above, and which can be used to work out the value of x at any given time t. The mathematical way of showing that x is some function of t is by writing $x = f(t)$. f is an unknown function and the variables are inside the brackets.

Solutions to the general equation

Luckily there exists a second-order differential equation of this general form that already has known solutions.

The equation $\frac{d^2x}{dt^2} = -\omega^2 x$ has a solution:

$$x = A\sin(\omega t) \qquad (3)$$

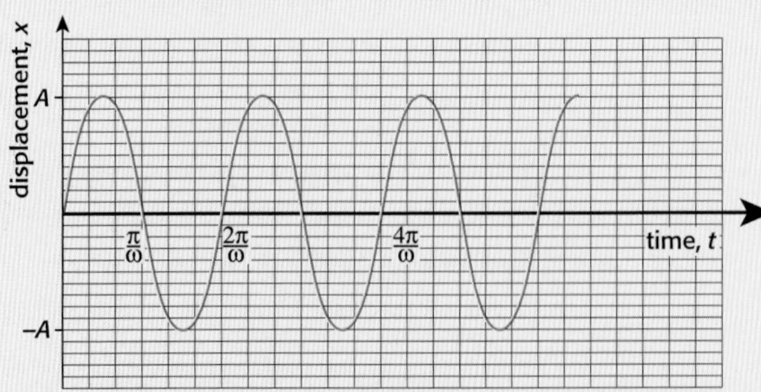

Figure 9 This general shape is called sinusoidal and the pattern repeats with a time period given by $T = \frac{2\pi}{\omega}$.

This equation describes an object that moves back and forth along the x-axis with a maximum displacement A.

Figure 9 shows the motion graphically.

If you have studied the calculus of trigonometric functions, it is reasonably straightforward to prove the solution described by equation (3).

differentiate once to obtain $\frac{dx}{dt} = A\omega\cos(\omega t)$

differentiate again to obtain $\frac{d^2x}{dt^2} = -A\omega^2\sin(\omega t)$

Since $x = A\sin(\omega t)$, this final equation can be written as $\frac{d^2x}{dt^2} = -\omega^2 x$

If you have not studied this level of maths, you should still be able to understand SHM from graphs of the displacement, velocity, and acceleration.

How to apply the general solution

By comparing the equation of the general solution

$$\frac{d^2x}{dt^2} = -\omega^2 x$$

with the equation which describes SHM $\frac{d^2x}{dt^2} = -\frac{k}{m}x$

we can see that they become exactly the same equation if we write $\omega^2 = \frac{k}{m}$ or $\omega = \sqrt{\frac{k}{m}}$. As we found earlier, the first equation has a solution $x = A\sin(\omega t)$, so the SHM equation must have a solution given by:

$$x = A\sin\left(\sqrt{\frac{k}{m}} \times t\right).$$

The time period of SHM must also be given by $T = \frac{2\pi}{\omega}$ or $T = 2\pi\sqrt{\frac{m}{k}}$.

$$\text{Time period of SHM} = 2\pi\sqrt{\frac{\text{oscillating mass}}{\text{restoring force per unit displacement}}}$$

Complete solutions

You may already realize that the general solution does not exactly match the example situation at the start of this section. In that case, the displacement of the object was at its maximum when we started timing the motion (x = maximum when $t = 0$). The solution above has zero displacement when $t = 0$. Has something gone wrong?

The equation that we have found is only one of a general

set of solutions possible. You can see from the way the graph regularly repeats itself that the place where we chose to define $t = 0$ was an arbitrary one. There are an infinite number of other similar solutions possible, providing that they are of the same shape as this graph. These other solutions are, in effect, the same shape graph moved one way or another along the time axis. For example, $x = A\cos(\omega t)$ must also be a solution to the SHM equation because this has the same shape as $x = A\sin(\omega t)$ but is shifted along the time axis by a quarter of the period of the graph.

One way of writing the very general solution is $x = A\cos(\omega t + \phi)$. The variable ϕ is dependent on the initial conditions and is known as the **phase angle**, or in other words, the point where we start timing the motion. So changing its value simply has the effect of moving the graph along the time axis.

Inquiry: SHM

All clocks, both mechanical and electrical, are Simple Harmonic Oscillators (SHOs); as are all transmitters and receivers of electromagnetic waves (radio, microwave etc) and sound waves. Many aspects of engineering design, from the massive to the microscopic, require a detailed knowledge of SHM (e.g. bridges, earthquake protection of buildings, atomic force microscopes for imaging single atoms). Detailed mathematical theories of the behaviour of atoms and molecules (in solids and gases) are also applications of SHM.

Choose one of the situations from the list above, or a completely different situation that you suspect might be SHM, and design an experiment to investigate whether the oscillation involved is an example of SHM (within experimental error).

One possible approach would be to experimentally measure the forces that are causing an object to oscillate. Or you might choose to analyze the displacement of the object over time. If you can devise a way of showing this to be sinusoidal then it must be SHM.

This area of physics can also leads to several possible extended essay research questions. Some examples are given here.

- Is the oscillation of water in a U-tube SHM?
- What factors affect the number of times a bungee jumper oscillates before coming to rest?
- How does the damage caused to a building depend on the frequency of the earthquake waves?

The last two questions may involve some intricate and ingenious modelling in order to collect appropriate data.

Thinking about science: *The links between SHM and circular motion*

Any motion involving oscillations can be analysed from first principles. If the mathematical description of the oscillations can be rearranged and expressed in the form $a = -\text{constant} \times x$ (often written as $a = -\omega^2 x$), then the oscillation under consideration must be SHM.

Once we know this to be the case we know that the general solution must take the form $x = A\cos(\omega t + \phi)$. But what do the variables, A, ω, and ϕ mean, and how do we find their values?

The numerical value for the angular frequency, ω, comes from the equations used in the analysis of the original situation. It is related to the frequency of the oscillation:

$$\omega = 2\pi f$$

The values for A (the amplitude) and ϕ (the phase) also need to be deduced for the situation being considered and it would help to understand what they physically represent.

We can exploit the fact that the mathematics of circular motion is very similar to the mathematics of SHM. This allows us to use our understanding of the physics of circular motion to help build a mathematical interpretation for what ω, A, and ϕ represent in the equations of SHM. As an example, two very different situations are compared on pages 84–5.

In the left-hand column, we are considering an object that is going around in a circle *with constant angular speed*, and in the right-hand column we are considering a mass on the end of a rigid spring. The equations are exactly equivalent but remember that some of the symbols mean different things in the two situations.

Table 4

Circular motion	Symbol	SHM of mass on spring
The time elapsed from the starting conditions	t	The time elapsed from the starting conditions
The component of the displacement in the x-direction	x	The displacement from the mean position
The radius of the circle	A	The maximum displacement (the amplitude)
The angular speed (must be constant)	ω	A constant defined by $\omega = \sqrt{\frac{k}{m}}$
The initial angle when $t = 0$	ϕ	A constant fixed by initial displacement when $t = 0$

Circular motion

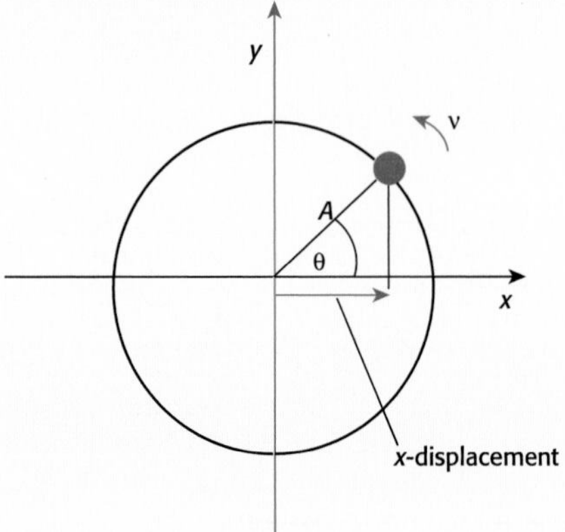

Figure 10 A mass going around in a horizontal circle of radius A at constant speed v

In this situation, the mass is moving in two dimensions. The time period of the motion can be calculated from the time it takes to complete the circle:

$$T = \frac{2\pi A}{v}$$

Another way of expressing this is to use the concept of the angular velocity of the mass, ω. (see page 50).

$$T = \frac{2\pi}{\omega}$$

If we resolve the motion of the mass into the x and y directions, the x-component of the mass's displacement from the centre is given by:

$$x = A\cos(\theta)$$

Another way of expressing this is:

$$x = A\cos(\omega t)$$

SHM of mass on a stiff spring

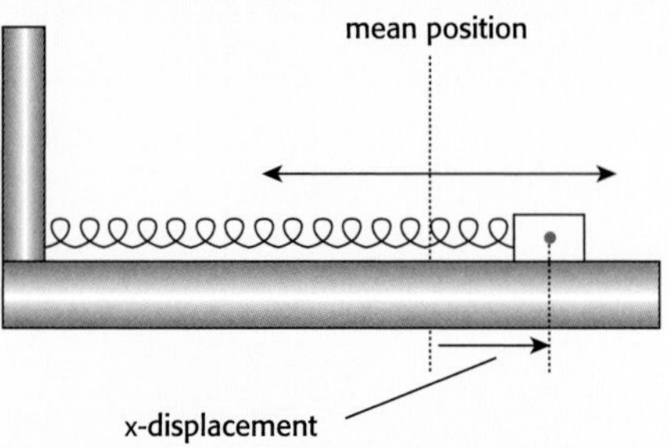

Figure 11 A mass on a frictionless surface attached to a stiff spring undergoing horizontal SHM

In this situation, the mass is moving in one dimension. The time period of the motion can be calculated from the SHM equation:

$$T = 2\pi\sqrt{\frac{m}{k}}$$

Another way of expressing this is to substitute using

$\omega = \sqrt{\frac{k}{m}}$ (see page 81).

$$T = \frac{2\pi}{\omega}$$

Since the motion is known to be SHM, the formula for its displacement is:

$$x = A\cos(\omega t)$$

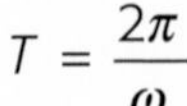

In this example the mass was assumed to have started its motion on the x-axis. We could have started timing from anywhere so the general formula for the x-component of displacement is:

$$x = A\cos(\omega t + \phi)$$

The angle ϕ depends on where the mass is when $t = 0$.

In this example the mass was assumed to have started its motion at the far right-hand side. We could have started timing from anywhere so the general formula for displacement is:

$$x = A\cos(\omega t + \phi)$$

The angle ϕ depends on where the mass is when $t = 0$.

In both situations x represents the displacement in the x-direction and if we concentrate on the motion of the mass moving in a circle projected onto this axis only, it is identical to Simple Harmonic Motion. In other words, there is no difference at all between the two situations as far as the x-direction is concerned. The equation of the mass on the end of the spring is exactly the same as the equation of the mass moving in a circle so long as we only consider the motion in the x-direction and ignore any extra motion in the y-direction. If you look at the shadow of an object moving in a circle, it will look like mass oscillating on a spring.

Since the two motions are exactly equivalent, any algebraic manipulation that applies to circular motion must also apply to SHM. The only difference between the motion of two objects is that the one moving in a circle is also moving in the y-direction at the same time.

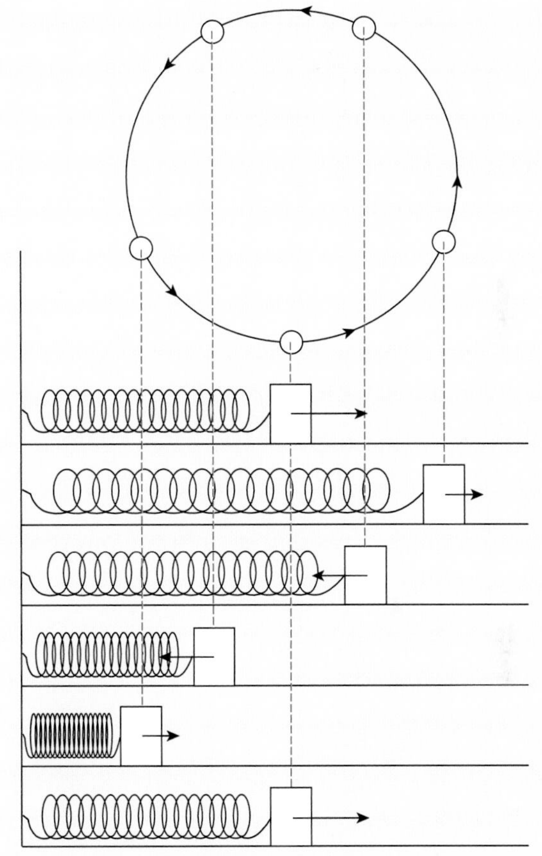

Figure 12 SHM and circular motion

By applying Pythagoras's theorem to the circular motion situation you should be able to show that the x-component of the object's velocity is given by

$$v = \pm\omega\sqrt{A^2 - x^2}.$$

If this works for circular motion, it must also work for SHM.

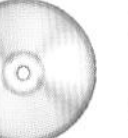

Science tools: Using a graphic display calculator to analyse SHM

A graphic display calculator (GDC) can be used for modelling, data logging or equation analysis.

Modelling

A graphic display calculator, personal computer, or any device with spreadsheets can be used to model SHM using the iterative method introduced on pages 35–7. An outline of the general procedure for modelling the motion of an oscillating mass is given below. In this method we divide the motion into intervals, with each interval representing a fixed period of time (for example 0.1 s or 1 s, depending on the overall time period).

- At any instant of time, the resultant force acting on the oscillating mass is related to the displacement of the mass ($F = -kx$).
- This instantaneous force can be used to calculate the instantaneous acceleration ($F = ma$).
- The change in velocity can be calculated by assuming that the acceleration remains approximately constant over the time interval ($\Delta v = a\Delta t$).
- The initial velocity and the final velocity over the time interval can be used to calculate the average velocity $\left(v_{\text{average}} = \frac{u+v}{2}\right)$.
- The average velocity over the time interval can be used to calculate the change in displacement ($\Delta x = v_{\text{average}}\Delta t$).
- The change in displacement can be used to calculate the new displacement at the end of the time interval.
- The whole process can be repeated with the new displacement.

The procedure for using a GDC (or computer) is outlined here.

1. On the spreadsheet, enter appropriate values for all of the constants:
   - oscillating mass
   - restoring force per unit displacement (spring constant)
   - time interval
   - initial displacement
   - initial velocity.
2. Add headings for the following columns on the spreadsheet:
   - total time elapsed
   - displacement at beginning of time interval
   - instantaneous force
   - instantaneous acceleration
   - initial velocity
   - change in velocity
   - final velocity
   - average velocity
   - change in displacement.
3. Add equations to calculate these values and fill the equations down the sheet.
4. Use the graphing ability of the GDC to display graphs that show the variation with time of:
   - displacement
   - velocity
   - acceleration.
5. The spreadsheet can be modified to answer the following questions:
   - What happens when the restoring force is assumed to be proportional to x^2 rather than x?
   - What happens when frictional forces act on the oscillating mass?

Data logging

If you have position data sensors available to connect to a graphic display calculator, it should be very simple to experimentally log data that records the variation with time of the displacement of a mass undergoing SHM.

A setup such as the one in Figure 13 should not only yield a plot of the variation of displacement with time, but from it you will be able to calculate and plot the variation of velocity and the variation of acceleration with time. There are, however, many adjustments that you will have to make before the graphs plotted by the calculator are useful.

- The position sensor measures the total distance from the sensor to the oscillating mass. Work out the mean position of the mass from the graph.
- Subtract the mean position of the mass from each reading. The displacement graph should then be sinusoidal with a mean value of zero.
- The displacement graph can be used to work out the amplitude, period, and phase shift of the oscillation.
- Use these three pieces of data and the calculator to define your best-fit line through the data. You know the graph should turn out to be sinusoidal so use this option and switch, if you have been working in degrees, to radians.
- The calculator will also be able to work out and plot the variation of the velocity of the oscillating mass. Try to predict the shape of the graph before you look at the answer. You will probably have to rescale the axes in order to view the velocity graph properly.
- At what position do you predict the velocity of the mass will be a maximum? Where is the mass when its velocity is zero?
- Find a mathematical equation for the variation of the velocity with time.
- Finally, the variation in acceleration can also be displayed. What relationship do you expect to see between the displacement graph and the acceleration graph?
- If you know the mass of the oscillating object, you can use the equations that you have discovered to calculate the value for the spring constant.

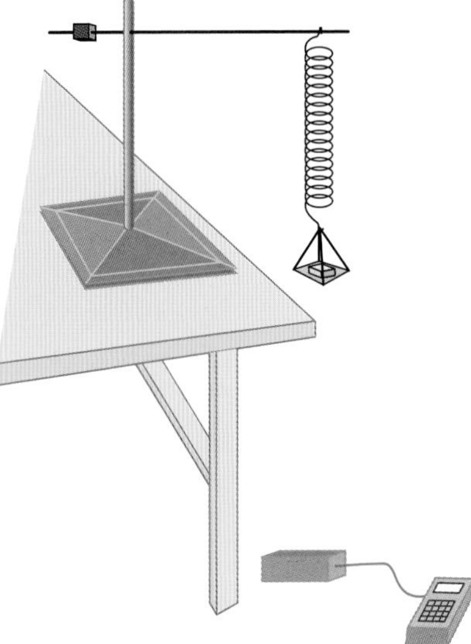

Figure 13 Possible set-up of a position sensor to record SHM

Equation analysis

A graphic display calculator is an extremely powerful tool for analysing experimental data, but it is also possible to gain a good understanding of SHM by using it to manipulate the basic equations. Try the following exercise on your graphic display calculator.

1. Plot a graph of $y = A\sin(\omega x)$ with $A = 2$ and $\omega = 0.05$. The x range can go from 100 to -100 and the calculator needs to be set to work in radians.
2. The GDC should display several complete oscillations of a sinusoidally varying graph. The y value should vary between 2 and -2.
3. Move the curser around the graph and use the values displayed to measure the time period. Does this value agree with the value predicted using $T = \frac{2\pi}{\omega}$?
4. What difference does it make to your graph if you add in a phase angle ϕ to the formula? Try different values of ϕ: π, $\pi/2$, $\pi/4$, and so on.
5. Use the calculator to plot a tangent to your displacement graph for a value of time that you choose. The equation of the tangent can also be displayed.
6. Since it is a straight line, the equation of the tangent will be in the form $y = mx + c$. The value of m is the gradient of the tangent.

7 We are plotting a graph of displacement against time, so the gradient of the tangent to any point on this graph will equal the instantaneous gradient, and therefore the velocity at this time. Select several different positions for plotting the tangent and record the different velocities that these represent.
8 Where is the velocity at its maximum and where is it zero?
9 It should be possible to get the calculator to plot a second graph to represent the variation of the velocity with time. In the mathematical language of calculus, we want to graph the derivative of the first function. The notation for doing this calculation varies among the different makes of calculator. Texas Instruments use the following notation for their TI-83 calculator for graphing the derivative: *nDeriv(function,X,X)*. In this case the second graph to be plotted would be Y2 = *nDeriv*(2 × sin(0.05 × *X*),*X*,*X*).
10 Once the graphs have been rescaled and you can identify which plot represents velocity and which plot represents displacement, you should check to see if you understand the shapes that have been plotted. What is the velocity when the displacement is zero? What is the velocity when the displacement is at its maximum?
11 The derivative function can be used again to display the acceleration. What is the relationship between the acceleration graph and the displacement graph?

Mathematical physics: *Investigating energy in SHM*

The aim of this section is to predict how the energy of the oscillating mass varies during the cycle. The procedure is described in terms of using a GDC, but a similar analysis can be done on a spreadsheet. If you do not have access to a GDC or computer, the graphs can still be plotted manually (it will just take longer). We will consider the variation of the KE and PE for an object moving with a displacement given by:

$$\text{displacement} = A \sin(\omega t)$$

Kinetic energy

The motion of the mass is described by SHM so we know that its velocity can be represented by:

$$\text{velocity} = A\omega \cos(\omega t)$$

A mass, m, moving at a velocity, v, has a kinetic energy given by $\text{KE} = \frac{1}{2}mv^2$. So if we substitute the SHM equation for velocity into this equation we get:

$$\text{KE} = \frac{1}{2}m(A\omega \cos(\omega t))^2$$

or

$$\text{KE} = \frac{1}{2}mA^2\omega^2 \cos^2(\omega t).$$

In order to visualize this variation (a "cos-squared" variation), try plotting the following graphs on your GDC.

1 Plot graphs of the displacement $y = A \sin(\omega x)$ and velocity $\frac{dy}{dt} = A\omega \cos(\omega t)$ with $A = 2$, $\omega = 0.8$. The x range can go from -10 to $+10$ in steps of 0.01 and the calculator must be set to work in radians. You should be able to see several complete oscillations.

2 Now plot the KE equation KE $= \frac{1}{2}mA^2\omega^2 \cos^2(\omega t)$ using 0.7 as the value for mass.

3 Your GDC should display the KE variation shown below:

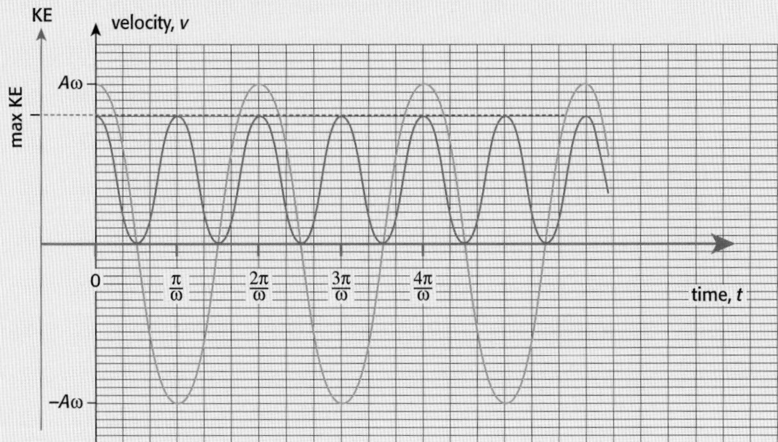

Figure 14 Variation of KE and its relationship with velocity in SHM

Potential energy

In this section we will analyse several different graphs. Remember that as we consider different graphs on the GDC the *y*-axis represents different variables.

As the mass is moving, the change in PE must always be equal to the total work done. We know that if a force *F* is moved through a distance x, then the work done can be calculated from:

work done = force × distance

However, in this situation we cannot simply multiply the value of the force by the value of the displacement because the magnitude of the force on the object is constantly changing. At any given displacement, the magnitude of the force acting on the object depends on x:

$F(x) = k\,x$

This means that the extra little bit of work done in moving a small distance δx is given by:

extra work done = $F(x) \times \delta x$

This is equivalent to using a graph and working out the area of the section of the graph between the line and the *x*-axis.

The total work done between any two points (often called the lower limit and the upper limit) is the addition of all the "extra little bits" of work done during the motion. This work done is equal to the extra potential energy stored when moving between these two points. The way of adding up all the extra little bits of work done is to calculate the area under the variation of force with distance graph. The mathematical technique for working out the area under a graph is called integration and is written like this:

PE stored = total work done = $\int F(x)\,dx$

Since the magnitude of $F(x) = k\,x$ we can deduce that the equation for the PE of a mass moving with SHM is:

PE $= k \int x dx$

Thinking about science

Consider the following points.

- Since the velocity has been squared, the kinetic energy is always positive.
- The axes of Figure 14 have a different scale to the ones that you have plotted on the GDC, but shape of the graphs must be the same.
- Do the maximum and minimum values for the KE agree with the variation of the velocity?

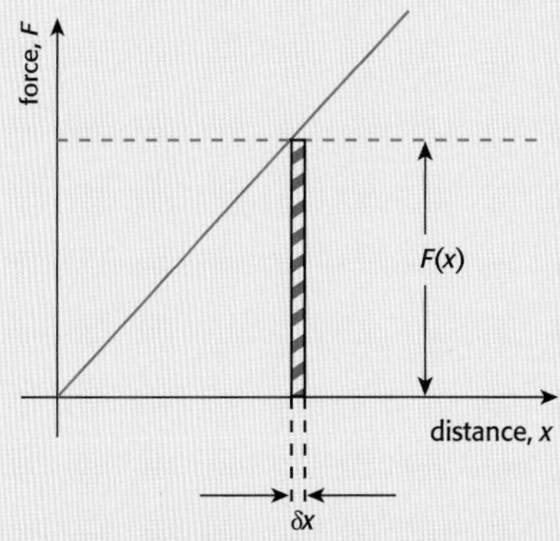

Figure 15 Extra work done is the area under the graph

If you have studied calculus you might recognize the solution to this equation:

$$\text{PE} = \frac{1}{2}kx^2$$

We can now complete this substitution, and you can also verify this formula using the GDC. As before, the calculator must be set to work in radians and we will use the same constants as before: $A = 2$, $\omega = 0.8$, and $m = 0.7$.

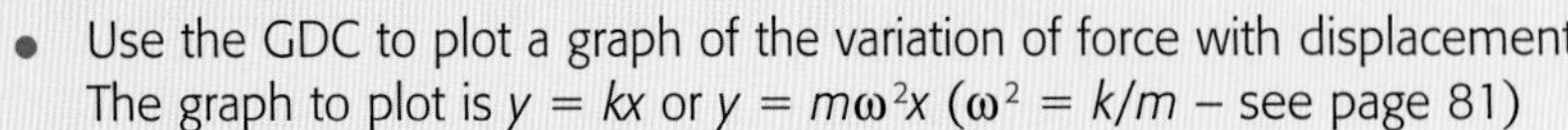

- Use the GDC to plot a graph of the variation of force with displacement. The graph to plot is $y = kx$ or $y = m\omega^2 x$ ($\omega^2 = k/m$ – see page 81)
- Use the "calculate" function to work out the area under the graph between the origin and a range of different values of x.
- The calculator can probably also display a graph of the integral of $y = kx$. The Texas instruments notation for this is *fnInt(function,X,0,X)*, so the equation to enter is Y2 = fnInt($m\omega^2x$, *X,0,X*).

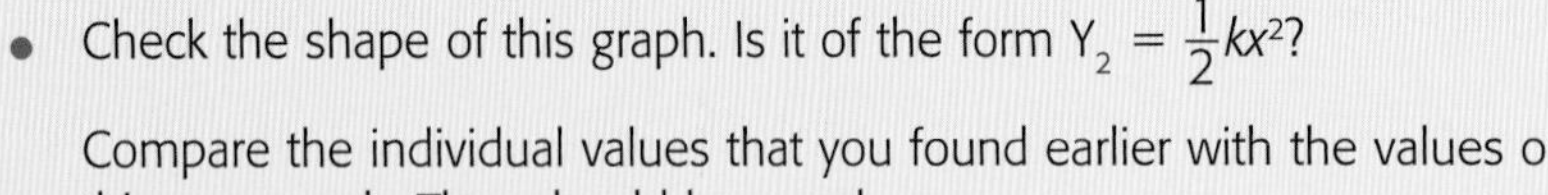

- Check the shape of this graph. Is it of the form $Y_2 = \frac{1}{2}kx^2$?

  Compare the individual values that you found earlier with the values on this new graph. They should be equal.

The two equations that we need to combine are:

$$\text{PE} = \frac{1}{2}kx^2 \text{ and displacement } x = A\sin(\omega t)$$

- Use the GDC to plot the graph $y = \frac{1}{2}m\omega^2(A\sin(\omega t))^2$.
- Is the shape of this graph the one that you expected?

As the mass is moving under SHM we know that its velocity can be represented by:

$$\text{velocity, } \frac{dx}{dt} = A\omega\cos(\omega t)$$

- A mass, m, moving with a velocity, v, has a KE given by $\text{KE} = \frac{1}{2}mv^2$.

  So if we substitute the first equation for the velocity into the second equation for KE, this results in the equation $\text{KE} = \frac{1}{2}mA^2\omega^2\cos^2(\omega t)$. Use the GDC to plot this graph.
- Is the shape of this graph the one that you expected? How does this compare with the PE graph?

Total energy

The above two sections have derived formulae for the KE and PE of an object doing SHM. The formulae are very similar:

$$\text{PE} = \frac{1}{2}mA^2\omega^2\sin^2(\omega t)$$

$$\text{KE} = \frac{1}{2}mA^2\omega^2\cos^2(\omega t)$$

The total energy at any given time is just the addition of PE and KE. Using your graphic display calculator add these two functions together and comment on your answer.

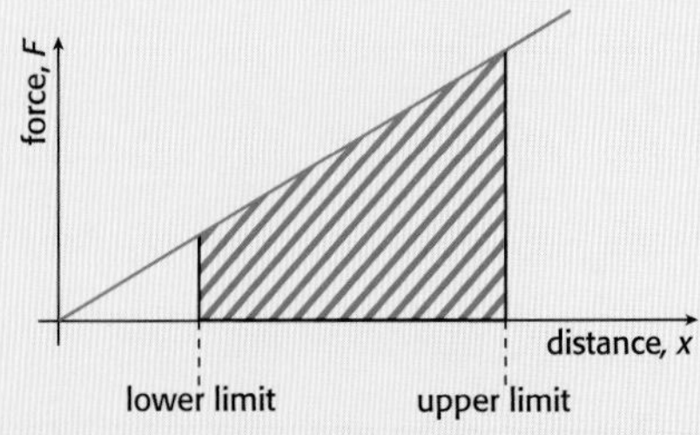

Figure 16 Work done is the area under the graph

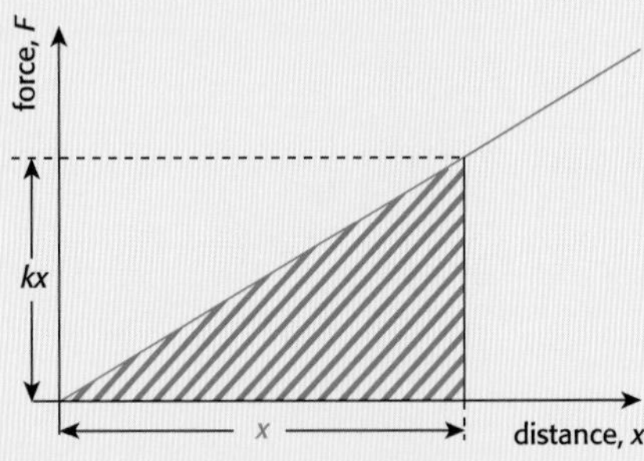

Figure 17 PE gained is the area under the graph

Show algebraically why the result is:

$$\text{Total energy} = \frac{1}{2}mA^2\omega^2$$

Energy in SHM

During SHM energy is converted between kinetic energy and potential energy, but the total energy remains constant. These variations are shown in Figure 18.

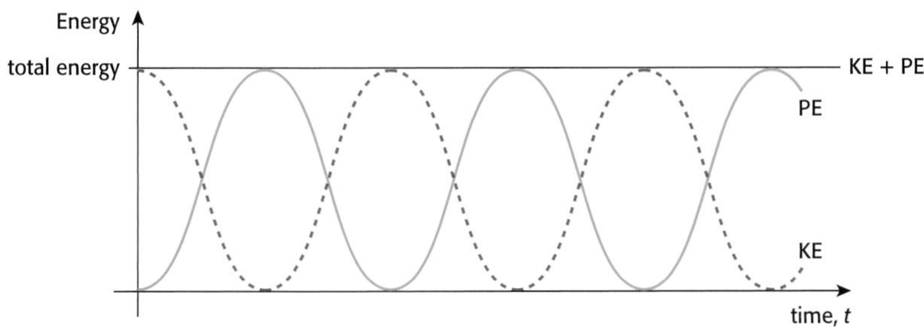

Figure 18 Variation of KE and PE during SHM

The mathematical formulae for these variations are:

$$E_K = \frac{1}{2} mA^2\omega^2 \sin^2(\omega t)$$

$$E_P = \frac{1}{2} mA^2\omega^2 \cos^2(\omega t)$$

$$E_T = E_k + E_P = \frac{1}{2} mA^2\omega^2$$

Since the diplacement $x = A \cos(\omega t)$, the variation in energy can be expressed in terms of the particle's displacement:

$$E_K = \frac{1}{2} m\omega^2 (x_0^{\,2} - x^2)$$

$$E_P = \frac{1}{2} m\omega^2 x^2$$

$$E_T = \frac{1}{2} m\omega^2 x_0^{\,2}$$

where x_0 is the maximum displacement and is therefore equal to A in the equations above. These formulae appear in the data booklet.

The bigger picture: Fourier analysis

The creation and propagation of a sound wave involves oscillations. So far we have only considered SHM but most sound waves (apart from when you whistle) are not sinusoidal – they involve more complicated oscillations. The shape of sound waves made by a musical instrument or everyday speech seem so far away from being perfectly sinusoidal that it is tempting to think that the mathematics of SHM will not be of any use in these situations.

Figure 19 Some different musical instruments

Starting with a simple example does help to understand the more complicated ones. Complicated oscillations can be shown to be the sum of lots of SHM oscillations of different frequencies and different amplitudes. You might like to revise the ideas of superposition of waves before reading more of this section.

In the language of music a single note being played by one instrument is described as having a fundamental frequency (the main note that you hear – also called the first harmonic) along with other frequencies, called harmonics, which give this note its particular characteristic.

If a violin and a flute both played the same note, the two individual sound waves would have the same fundamental frequency but each instrument would also have different harmonic frequencies of different amplitudes. It is the "additional" harmonic frequencies that make a violin sound different to a flute. You can recognise different people's voices from their own unique combination of harmonics. Musicians refer to the sound of a note as its *quality*, or *timbre*.

It is possible to analyse a complicated sound wave and work out what component frequencies (or harmonics) exist. The mathematics of frequency analysis is quite complicated but the result will always be a series of sine and cosine functions of different amplitudes. For many waveforms the (sometimes infinite) series has already been calculated and is called a Fourier series after the French physicist and mathematician Joseph Fourier.

Equipment designed to perform real-time frequency analysis of sounds is generally very expensive and only

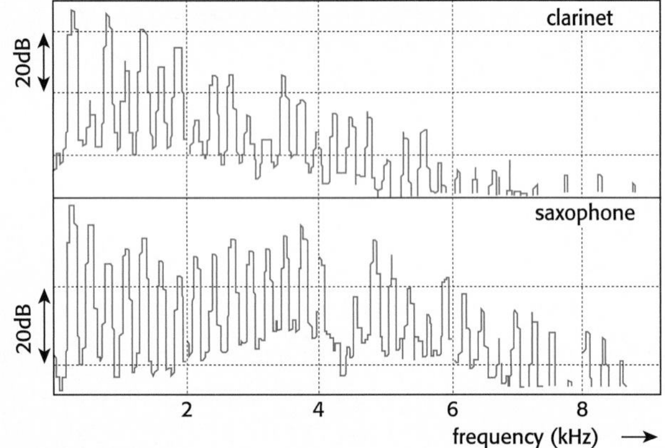

Figure 20 The frequency spectrum of the same note played by a clarinet and by a saxophone

widely used in research laboratories and sound recording studios. These days, however, software is available which allows sound files held on a computer to be analysed and some can perform the analysis directly from a microphone input. There are even versions that are available free on the internet.

If a sound wave can be resolved into its component frequencies, is it possible to create different sounds by adding together different component frequencies? Of course, the answer is yes and this is the basic principle behind the instrument called a synthesizer. The sounds of different instruments are created by combining different sine waves. The internet is, once again, a good place to locate demonstrations of how Fourier-Synthesis can work (see nch.com.au/wave pad/fft.html and also http://www.zeitnitz.de/Christian/Scope/Scope_en.html accessed July 2009)

The first four terms for the Fourier series for a square wave are:

$$y = A\sin(x) + \frac{A}{3}\sin(3x) + \frac{A}{5}\sin(5x) + \frac{A}{7}\sin(7x)$$

Use a GDC to plot this function and to explore to what extent the addition of extra terms alters the shape of the resulting wave.

Damping

Until now we have been considering an SHM system in isolation and we have not considered possible energy losses from the system. But energy loss does happen whenever friction acts on the moving mass (slowing the object down) and is called energy dissipation or **damping**.

When an object is oscillating with SHM its total energy is constant and depends on three things: the object's mass, the initial amplitude of the oscillation, and the angular frequency of the oscillation (which is proportional to its frequency). Note that the total energy is not proportional to its amplitude but proportional to the square of the amplitude. This is an important relationship and it not only applies to mechanical objects undergoing SHM but also to any oscillation that can be described using the SHM formulae and in many other areas of physics (for example, wave–particle duality on page 264). The reasons are shown below.

The total energy of an object moving with SHM can be calculated by adding its potential energy and its kinetic energy together. Mathematically:

$$\text{PE} = \frac{1}{2}mA^2\omega^2\sin^2(\omega t)$$

$$\text{KE} = \frac{1}{2}mA^2\omega^2\cos^2(\omega t)$$

$$\therefore \text{total energy} = \frac{1}{2}mA^2\omega^2 \qquad (\text{since } \sin^2\omega t + \cos^2\omega t = 1)$$

$$\therefore \text{total energy} \propto A^2$$

If we add a resistive (or damping) force to the Simple Harmonic Motion the general "driving" equation for the oscillation will change. The damping (frictional) force usually depends on the velocity with which the object is moving so we can write:

$$\text{Damping force} \propto -\text{velocity}$$

or

$$\text{damping force, } F = -k_{\text{damping}}\frac{dx}{dt}$$

The negative sign is there because the force resists the motion so it must be acting in the opposite direction to the velocity. This situation results in the following differential equation to solve:

$$\frac{d^2x}{dt^2} = \frac{-\left(k_{\text{spring}}\,x + k_{\text{damping}}\frac{dx}{dt}\right)}{m} \qquad (4)$$

where k_{spring} and k_{damping} are different constants and are particular to the practical setup under consideration. Solving this is well beyond the scope of the IB Diploma Programme physics course, However, it is possible to predict the effect of light damping on an object by considering the energy equations again.

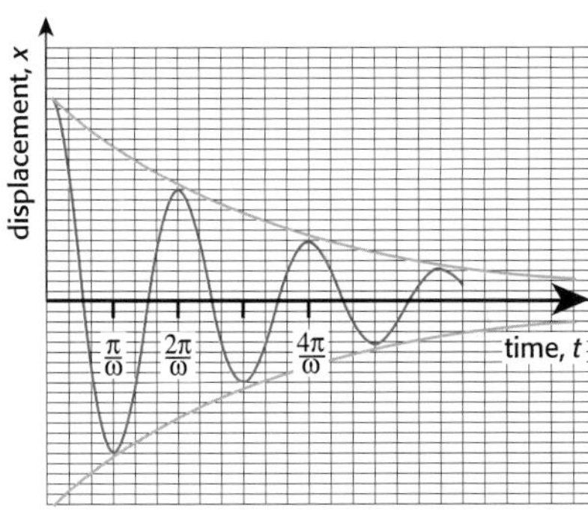

Figure 21 Lightly damped SHM

The energy of an object moving under SHM is proportional to the (amplitude)^2. If work is being done by the system against a resistive force, then the amplitude of the oscillation must fall. Since the time period of the oscillation is independent of the amplitude, the time period, the frequency of the oscillation must be unaffected. Figure 21 shows the variation of displacement with time for **lightly damped SHM**.

Use readings from the graph to show that the amplitude is decreasing exponentially.

The amplitude of the oscillations can be shown to decrease exponentially with time as the system loses energy by doing work against the resistive force.

An exponential decrease of the amplitude only happens if the damping is "light" – that is to say that the frictional force removes only a small portion of the total available energy during one oscillation, e.g. a pendulum moving through air. The air resistance acting on the pendulum bob would cause the amplitude of the oscillations to decrease over time but the time period remains constant.

If, however, the pendulum were to be set up not in air but in viscous oil, the damping force would be so great that oscillations could not take place. In this situation, called **heavy damping** (or even "heavy over-damping"), a pendulum displaced and released would just slowly make its way back to the mean position (Figure 22).

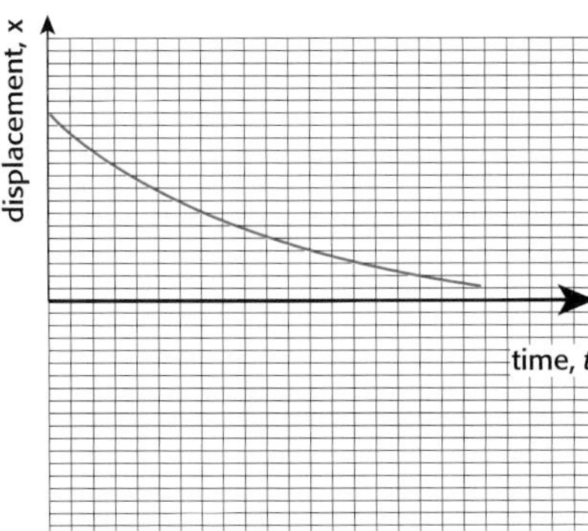

Figure 22 Heavily damped SHM

There are many engineering situations where neither light damping nor heavy damping is desirable. An example of one such situation is the suspension fitted to a car. In all cars, the passengers travel in a compartment that is joined to the wheels of a car by heavy-duty springs.

One of the aims of these springs is to give the driver and passengers a comfortable ride (another is to improve road-holding). If the damping is too light then whenever the car goes over a bump the car will start oscillating up and down with SHM giving a rather unpleasant ride. If the damping is too heavy then the car behaves as though there was no damping at all and the driver feels every bump.

It is possible to arrange for a system to be "critically damped". This describes an arrangement where the damping is somewhere between light and heavy so as to make the oscillations "die out" in the shortest time possible. When this happens, the system does not "overshoot" before returning to zero displacement (see Figure 23).

It is possible to use the iterative method outlined on pages 35–7 to investigate damping. The equation for overall acceleration (equation 4 on page 93) can be inserted into the program, generating new values for velocity and displacement. By changing the coefficient for damping, light, critical, and over-damping can be investigated.

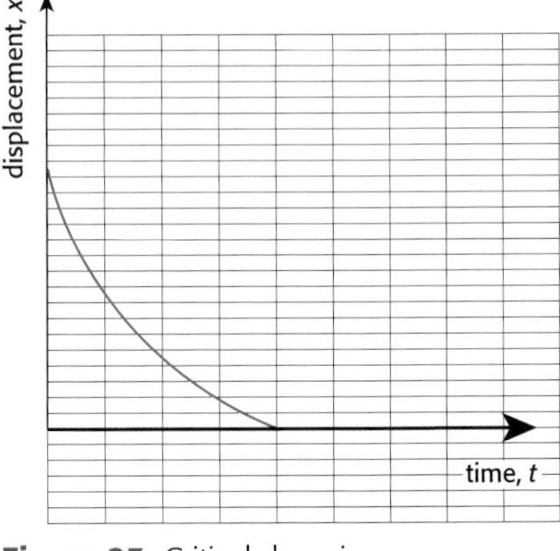

Figure 23 Critical damping

1 Another example of a situation that is critically damped is an electrical meter with a moving pointer. Whenever such a meter is connected to take a reading the needle will, if properly damped, arrive at the correct value in the shortest possible time. What would happen if the meter's needle was too lightly or too heavily damped?

2 Go back and look again at your list of practical examples of SHM. Decide whether no damping, light damping, heavy damping, or critical damping would be most beneficial in each situation. For example, a pendulum will continue to swing for longer if the damping is light.

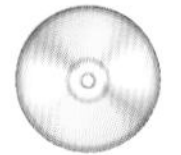

Resonance

Damping removes energy from a system. It is also possible to put energy into the system by having an external force pushing the mass forward, called a **driving force**.

This driving force varies in size and direction so in order to properly derive the solution it is necessary to set up the equation of motion for the object. Once again the full derivation is beyond the scope of the IB Diploma Programme physics course but we can use energy arguments to understand these ideas.

Note that we are now considering two different oscillations and you should ensure that you understand the difference between them. Firstly we consider the oscillations that the system would do if there was no driving force, this is the **natural oscillation** of the system and has a frequency, f_{natural} or f_N, given by the SHM equations. We also consider the driving oscillations with the frequency f_{driving} or f_D.

The driving force can be sinusoidal, defined by the following equation:

$$\text{driving force} = F_D \cos \omega_D t$$

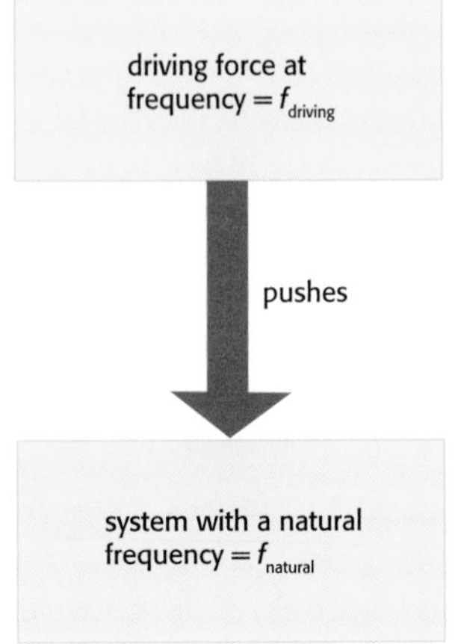

Figure 24 Situation necessary for resonance

This means the equation for the system's motion is:

$$m\frac{d^2x}{dt^2} = -kx + F_D \cos \omega_D t$$

The solution to this equation is still SHM with the system oscillating at the driving frequency, $f_{driving}$. The amplitude of the resulting oscillation depends on how close the driving frequency is to the **natural frequency**. A strange thing happens as the driving frequency approaches the natural frequency. The amplitude of the oscillations increases significantly as the frequencies become nearly equal and the only amplitude-limiting factor is the amount of damping present in the system. This phenomenon is called **resonance**.

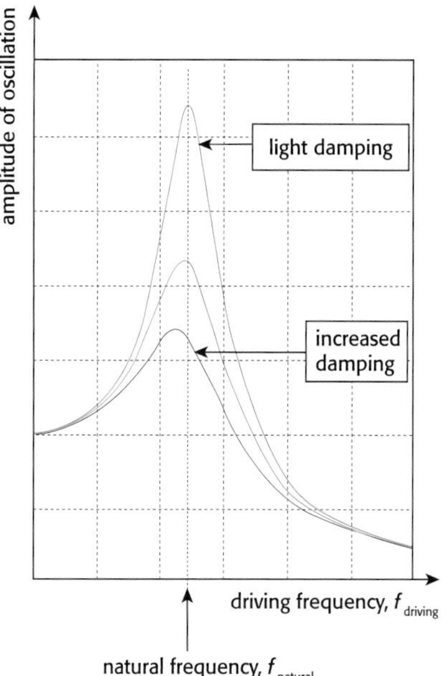

Figure 25 Variation of amplitude with driving frequency

At resonance the driving force is doing work, in other words, putting energy into the oscillating system. As energy goes into the system, its amplitude increases. It stops growing in amplitude when the energy put into the system in one cycle is equal to the energy lost in one cycle due to damping. Resonance is an extremely important phenomenon and occurs in a very wide range of situations. Sometimes resonance is something to be avoided and in extreme cases can cause a lot of damage, whereas in other situations resonance is relied upon to produce a desired effect.

Examples of resonance

The following are all examples of resonance. For each situation, with a small amount of research, you should be able to identify the system providing the driving frequency and the system that has the natural frequency and thus outline the physics of the situation. In addition, you should be able to decide if the resonance condition is something that should be avoided or is something desirable.

1 A removal van's combustion engine is controlled by the accelerator pedal. With the engine out of gear, the engine can be made to idle at different frequencies. At one particular frequency, one of the wing mirrors on the van can be seen to shake violently.
2 A note being played on a clarinet.
3 A microwave oven used to cook a meal.
4 When singing in the shower, an IB Diploma Programme student notices that one particular note sounds louder to him than all the others even though he is not singing that note more loudly.
5 A radio can be tuned so that one particular radio station is heard.
6 An opera singer claims to be able to break a wine glass by loudly singing a note of a particular frequency.
7 If the keys on a piano are pushed down gently enough it is possible to avoid playing any notes. With the keys held down, if any loud noise happens in the room (e.g. somebody shouting), then some of the notes held down will start to sound.

Rediscovering physics: Barton's pendulums

A simple mechanical demonstration of resonance is called Barton's pendulums. The setup can be used to investigate how the system behaves when the amount of damping is changed. In addition, it is possible to explore the relationship between the driving frequency and the phase of the resulting oscillations, as well as the so-called transient oscillations that initially take place before the system settles down in constant motion.

Apparatus:

- heavy pendulum bob e.g. metal or Plasticine, around 0.04 kg is suitable
- several light pendulum bobs e.g. small pieces of Plasticine inside small paper cones
- string
- nylon fishing line or fine string or thread
- clamp stands with G-clamps
- plastic rings, e.g. curtain rings (if you wish to show damping effects)
- slide projector (if available).

Experimental setup:

Make one driver pendulum with a heavy bob. Make several light pendulums of various lengths with one length exactly matching the length of the driver. Suspend all the pendulums from a string as below, and support the ends of the string firmly.

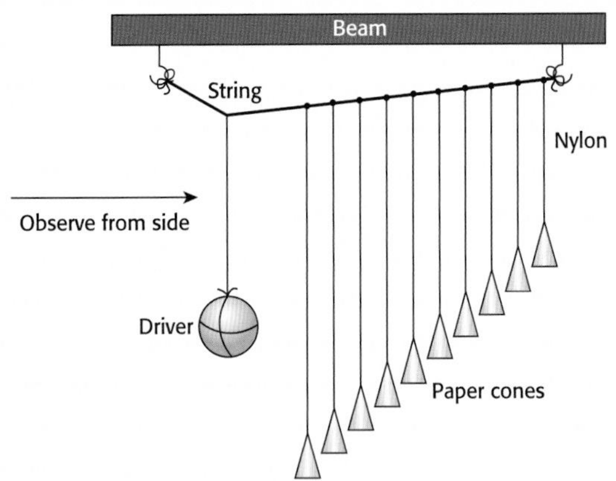

Figure 26 Barton's pendulums

In this situation we have set up 10 different systems each with their own natural frequency, determined by the length of the nylon. The same driving force (from the driver pendulum) is acting on all of these systems at its own frequency. The effective damping may be reduced by slipping plastic rings over the cones (increasing their mass). This is done more easily if the rings have a single cut in them.

The demonstration can be more effective in a darkened room with the cones brightly illuminated by a slide projector.

Notes

- The wooden beam must be firmly clamped and be horizontal. It is very easy for the cones to tangle so take care when setting up the apparatus.
- The lengths of the pendulums can be from a quarter to three quarters of a metre with the driver pendulum a half metre long. Nylon thread supporting the cones may be attached to the string by a half hitch or slipknot; this makes it easier to adjust lengths. The pendulums should be close together. Plasticine can be used successfully to secure the cones though a knot will suffice.

Reference

Popper, K. 1959 (English edn). *The Logic of Scientific Discovery*. London. Hutchinson.

End of chapter summary

Chapter 7 has three main themes: kinematics of simple harmonic motion (SHM), energy changes during SHM and forced oscillations (including resonance). The list below summarises the knowledge and skills that you should be able to undertake after having studied this chapter. Further research into more detailed explanations using other resources and/or further practice at solving problems in all these topics is recommended – particularly the items in bold.

Kinematics of simple harmonic motion (SHM)

- Describe examples of oscillations and be able to define the terms displacement, amplitude, frequency, period and phase difference.
- Define Simple Harmonic Motion (SHM), state the defining equation **and solve problems**.
- Apply the displacement, velocity and acceleration equations for SHM **and be able to solve problems, both graphically and by calculation, for acceleration, velocity and displacement during SHM**.
- Analyse practical data to assess uncertainties and use a graphic display calculator for modelling SHM and equation analysis.

Energy changes during SHM

- Describe the interchange between kinetic energy and potential energy during SHM.
- Use the mathematical equations for the kinetic energy, the potential energy and the total energy of a particle undergoing SHM and be able to solve problems, both graphically and by calculation, involving energy changes during SHM.
- State what is meant by damping and describe examples of damped oscillations

Forced oscillations (including resonance)

- State what is meant by *natural frequency of vibration and forced oscillations*
- Describe graphically the variation of the amplitude of vibration of an object close to its natural frequency of vibration as the driving frequency changes
- State what is meant by resonance and describe examples where the effect is useful and where it should be avoided.

Chapter 7 questions

1 a) A body is displaced from equilibrium. State the **two** conditions necessary for the body to execute simple harmonic motion. [2]

b) In a simple model of a methane molecule, a hydrogen atom and the carbon atom can be regarded as two masses attached by a spring. A hydrogen atom is much less massive than the carbon atom such that any displacement of the carbon atom may be ignored.

The graph below shows the variation with time t of the displacement x from its equilibrium position of a hydrogen atom in a molecule of methane.

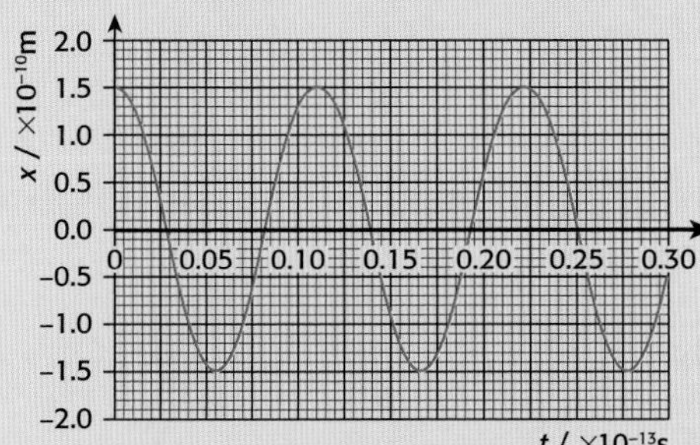

The mass of hydrogen atom is 1.7×10^{-27} kg. Use data from the graph above

i) to determine its amplitude of oscillation. [1]

ii) to show that the frequency of its oscillation is 9.1×10^{13} Hz. [2]

iii) to show that the maximum kinetic energy of the hydrogen atom is 6.2×10^{-18} J. [2]

c) Sketch a graph to show the variation with time t of the speed v of the hydrogen atom for one period of oscillation starting at $t = 0$. (There is no need to add values to the speed axis.) [3]

d) Assuming that the motion of the hydrogen atom is simple harmonic, its frequency of oscillation f is given by the expression

$$f = \frac{1}{2\pi}\sqrt{\frac{k}{m_p}},$$

where k is the force per unit displacement between a hydrogen atom and the carbon atom and m_p is the mass of a proton.

i) Show that the value of k is approximately 560 N m^{-1}. [1]

ii) Estimate, using your answer to (d)(i), the maximum acceleration of the hydrogen atom. [2]

e) Methane is classified as a greenhouse gas.

i) Describe what is meant by a greenhouse gas. [2]

ii) Electromagnetic radiation of frequency 9.1×10^{13} Hz is in the infrared region of the electromagnetic spectrum. Suggest, based on the information given in (b)(ii), why methane is classified as a greenhouse gas. [2]

(Total 17 marks)

2 a) A body is performing undamped simple harmonic motion with amplitude a and time period T. State **two** conditions that must be true for the body's acceleration. [2]

b) i) Sketch a graph to show how the displacement of the object varies with time. Label your line with the letter "d". [1]

ii) On your graph, label T and a. [2]

iii) Using the same scale for the time axis, sketch how the velocity of the object varies with time. Label this line with the letter "v". [2]

iv) Using the same scale for the time axis, sketch how the acceleration of the object varies with time. Label this line with the letter "a". [2]

c) Explain what is meant by damped motion. [2]

d) An object performs lightly damped simple harmonic motion. Sketch graphs to show how its displacement, velocity and acceleration vary with time. [3]

3 A mass is suspended vertically on the end of spring and undergoes simple harmonic motion of frequency 3.5 Hz and amplitude 1.8 cm. Determine

a) the time period of the oscillation [1]

b) the maximum linear speed of the mass [2]

c) the maximum acceleration of the mass [2]

4 a) State two conditions for the phenomenon of resonance to be observable in a mechanical system. [2]

b) Outline one mechanical example of resonance and show how the two conditions identified in (a) apply in this situation. [2]

8 Waves

This chapter builds on the ideas introduced in Chapter 7 and brings together two different, but related, areas of physics: oscillations and waves. Before starting, review the work you have done in Chapter 7 and to try to write down the general definitions/characteristics that are common to all waves.

When imagining a mechanical wave, try to imagine ripples on a pond.

There are some properties that are common to all types of waves, whether they turn out to be mechanical waves, electromagnetic waves, or something else altogether. These properties are that any wave can be shown to be able to undergo **reflection**, **refraction**, **diffraction**, and **interference**. All of these phenomena are discussed later in this chapter. It is the latter two (diffraction and interference) that are often of particular interest to physicists because they are phenomena that are unique to wave motion.
A tennis ball "reflects" off a wall when it bounces and its motion can be described assuming it is a point particle. It is possible to hypothesize an explanation for refraction in terms of particles but there is not yet a successful particle model for diffraction or interference. Sometimes it is not obvious whether a phenomenon is best described by wave motion or a stream of particles.

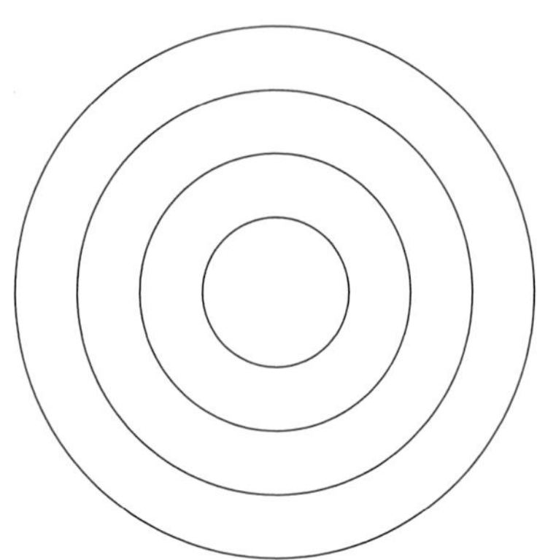

Figure 1 Wavefront diagram of the ripples on the surface of a pond.

Light and sound can be shown to diffract and interfere. This means that waves of some sort must be involved in a proper explanation of how the energy is transmitted from source to receiver. In fact, an absolutely complete description of both light and sound needs to involve wave *and* particle properties (see page 261 for more details).

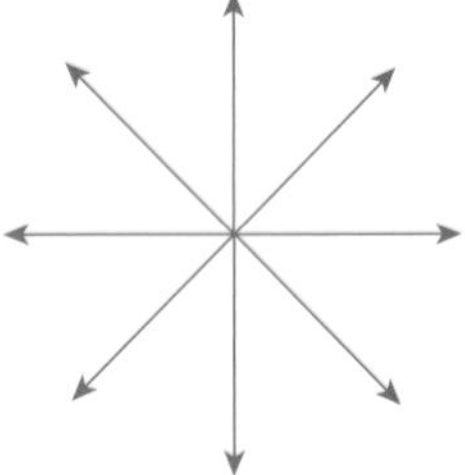

Figure 2 Ray diagram of light spreading out from a bulb.

When representing waves in a diagram one can either draw **wavefronts** or **rays**. Which approach is better depends on the context. A wavefront diagram represents the pattern of the wave at one instant of time. Any given wavefront is a line that joins together different regions that are oscillating together ("in phase"). For example the ripples on the surface of a pond can be represented by concentric circles. Each circle represents the top, or **crest**, of the waves. A ray diagram represents the energy flow that is taking place. Any given ray is a line from the source showing the direction of energy flow. For example light spreading out from a bulb can be represented by radial lines. Each line represents an energy flow direction. Note that in any particular situation, the rays will be at right angles to wavefronts.

Amplitude, intensity, and energy in waves

All waves transmit energy from one place to another. This is achieved *without any overall transfer of mass* using oscillations. In many cases these oscillations can be shown to involve SHM.

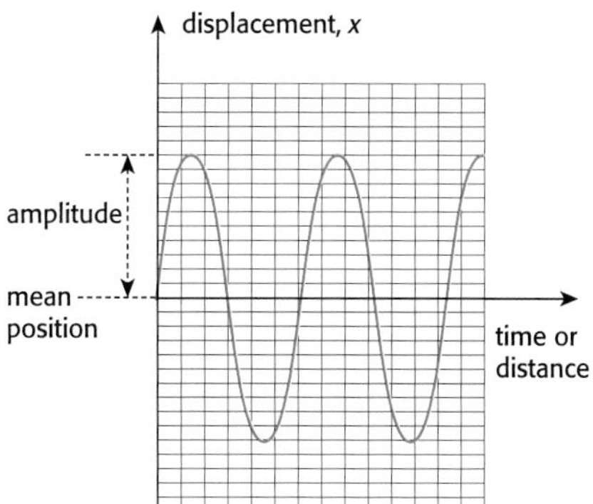

Figure 3 The amplitude of wave motion

Two particular categories of waves can be identified – **longitudinal** and **transverse**. In longitudinal waves, the oscillations are parallel to the direction of energy transfer whereas in transverse waves the oscillations are at right angles to the direction of energy transfer. These two categories are explained further on pages 101–2.

The amplitude of a wave is not the same as its intensity. Amplitude is defined as the maximum variation from the mean value (or rest position) for the oscillation concerned (see Figure 3).

The precise definition of intensity can vary depending on the context in which it is being considered, but in general it is defined as the power received per unit surface area.

$$\text{intensity} = \frac{\text{power received}}{\text{area of receiver}}$$

the unit of intensity is: W m^{-2}

The total energy of an object doing SHM is proportional to $(\text{amplitude})^2$ and the intensity of a wave is also proportional to its $(\text{amplitude})^2$.

When energy is received from a continuous source of waves, the total energy absorbed is proportional to two variables: the area of the receiver and the time for which the wave was received.

Waves always spread out from their sources so the amplitude and the intensity of any wave must decrease with distance from the source. If the source is isotropic (this means that the source emits uniformly in all directions), then the decrease in intensity turns out to be what is called an "inverse square relationship". As you move away from the source, the total wave energy is spread over a greater area so the intensity (and the amplitude) must be reduced. This is illustrated in Figure 4.

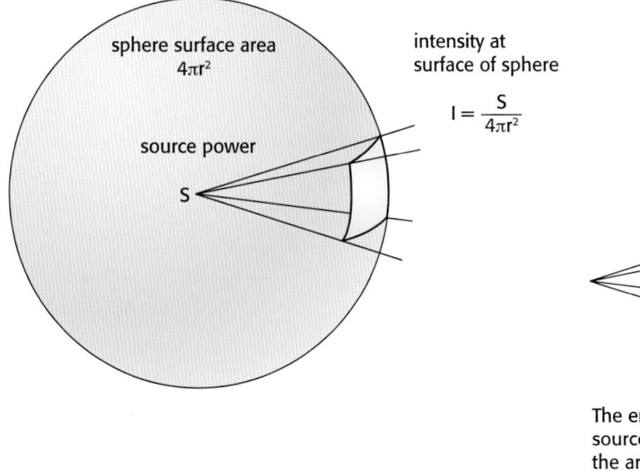

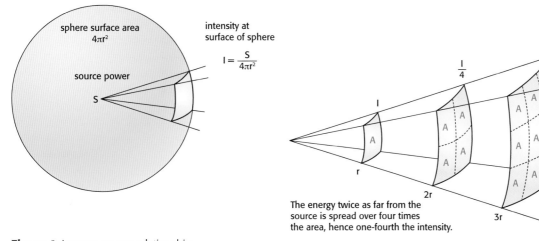

Figure 4 Inverse square relationship

Sound – a longitudinal wave

The human ear is able to detect waves over a very large range of frequencies and amplitudes. The frequency of a sound is related to the pitch that we hear – a low frequency sound is a low pitched note. The range of human hearing is from approximately 20 Hz to 20 kHz. Increasing the amplitude of the sound increases the intensity received by our ears, and hence the loudness that we hear.

Amplitude, intensity, and loudness are all different quantities. Strictly speaking, the amplitude of a sound wave travelling in air would be just the maximum displacement of the air molecules as a result of the sound. This piece of information is hard to measure and not particularly useful. Sound engineers often refer to the amplitude of the electrical signal produced by a microphone or electrical system that is used to process the sound.

The human ear is capable of responding to intensities from as low as 1×10^{-12} W m^{-2} up to about 1×10^{3} W m^{-2}. Different intensities are compared on the decibel scale, which is logarithmic. This means that if a sound intensity is doubled, then doubled again, and so on, the listener experiences, very approximately, the same regular increases in the sensation of loudness.

Table 1 The decibel scale is logarithmic

Source	Intensity/W m^{-2}	Intensity/dB
Threshold of hearing	1×10^{-12}	0
Rustling leaves	1×10^{-11}	10
Whisper	1×10^{-10}	20
Normal conversation	1×10^{-6}	60
Busy street traffic	1×10^{-5}	70
Vacuum cleaner	1×10^{-4}	80
Large orchestra	6×10^{-3}	98
MP3 player at maximum level	1×10^{-2}	100
Front rows of a rock concert	1×10^{-1}	110
Threshold of pain	1×10^{1}	130
Military jet takeoff	1×10^{2}	140
Instant perforation of eardrum	1×10^{4}	160

The apparent loudness of a sound to a given listener depends on many factors and is also very subjective. Increasing the intensity of a sound whilst keeping other variables fixed will make the sound seem louder. The ear does not, however, respond equally to different frequencies. Age is another factor that affects the human ear's response to a sound. Exploring variations in perceptions of loudness could form the basis for an interesting extended essay. See Chapter 25 for more details.

Another possible theme for an extended essay could be an investigation into some aspect of unwanted sound, called noise. Noise pollution is a growing problem in our towns and cities and has been shown to have a significant detrimental effect on both physical and mental health. Technological solutions can prevent some of these effects but much depends on the particular situation being considered. For example, a very low amplitude noise would be disturbing in an examination hall whereas a much louder noise would be ignored in a busy restaurant.

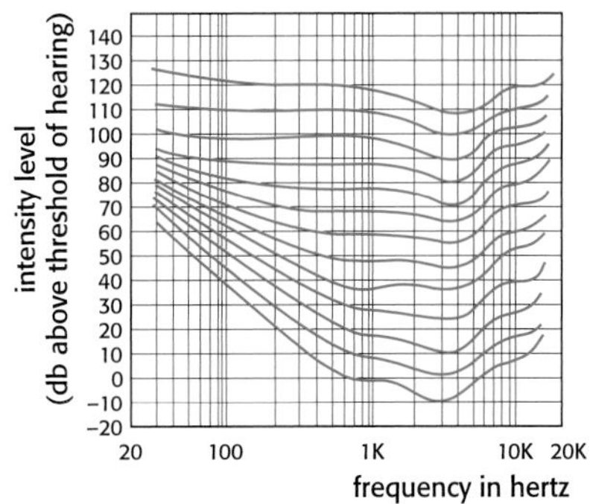

Figure 5 Equal loudness curves (based on Fletcher and Munson)

Light – a transverse wave

Light is an electromagnetic wave (see Figure 12 on page 106). All electromagnetic waves travel at the same speed in a vacuum, $3 \times 10^8\ \mathrm{m\ s^{-1}}$, but they travel at different speeds in different media. This produces a phenomenon called **dispersion**. The speed at which light propagates through glass depends on its frequency and this is the reason why a prism splits white light into its different component frequencies.

We perceive different frequencies of electromagnetic radiation as different colours of light. The human eye is able to respond to frequencies ranging from about 8×10^{14} Hz to 4×10^{14} Hz (corresponding to a wavelength range of 400 nm to 750 nm). The perception of colour is extremely complicated as two frequencies arriving at the same time are not perceived to be a mixture of the two colours: we just see it as one colour. For example, red light and green light added together is seen as yellow light. If all frequencies of visible light arrive together at the same time, this is usually perceived as white light (depending on the amplitudes of the individual frequencies).

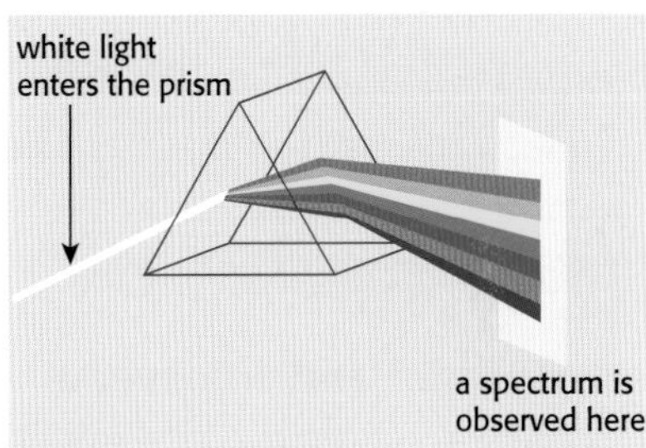

Figure 6 Prism splitting white light into colours

Do not confuse this additive property of light with the mixing of coloured paints. A surface appears coloured because it absorbs most of the frequencies contained in white light. The frequencies that are not absorbed are reflected off the surface and arrive at your eyes. If red paint and green paint are mixed together, we would not get yellow but something more like black.

Light is an electromagnetic wave so if we are precise, the amplitude of light is measured in terms of the maximum electromagnetic field variation. In practical situations the intensity is a much more useful piece of information. Increasing the amplitude of light increases its intensity and this is seen as increasing the brightness – another subjective measurement. Different eyes have different perceptions of brightness.

Earthquakes – longitudinal and transverse waves

The energy involved in earthquake oscillations is very large indeed. As before, the total energy transmitted by the wave is proportional to the square of the amplitude. If the amplitude of one earthquake wave is double that of another, then the energy associated with it is four times more than the energy associated with the smaller earthquake.

The size of earthquakes can be measured on the Richter scale, which was invented by Charles F. Richter in 1934. The Richter magnitude is calculated from a measurement of the amplitude of the largest seismic wave recorded for the earthquake. It is also called the local magnitude scale.

The scale is logarithmic so, for example, a magnitude 5 earthquake would result in ten times the level of "ground shaking" as a magnitude 4 earthquake. The calculation of the total amount of

Figure 7 Earthquake damage

energy transmitted by any given earthquake is complex but we can say that an increase of one point on the Richter scale is roughly equivalent to the release of 32 times as much energy.

Table 2 Describing earthquakes

Descriptor	Richter magnitudes	Earthquake effects	Frequency of occurrence
Micro	Less than 2.0	Microearthquakes, not felt.	About 8,000 per day
Very minor	2.0–2.9	Generally not felt, but recorded.	About 1,000 per day
Minor	3.0–3.9	Often felt, but rarely causes damage.	49,000 per year (estimated)
Light	4.0–4.9	Noticeable shaking of indoor items, rattling noises. Significant damage unlikely.	6,200 per year (estimated)
Moderate	5.0–5.9	Can cause major damage to poorly constructed buildings over small regions. At most slight damage to well-designed buildings.	800 per year
Strong	6.0–6.9	Can be destructive in areas up to about 100 miles across in populated areas.	120 per year
Major	7.0–7.9	Can cause serious damage over larger areas.	18 per year
Great	8.0–8.9	Can cause serious damage in areas several hundred miles across.	1 per year
Rare great	9.0 or greater	Devastating in areas several thousand miles across.	1 per 20 years

Longitudinal and transverse waves: Earthquakes

There are many different types of earthquake waves. The location of the disturbance that causes the earthquake (called the **hypocentre** or **focus**) is below the surface of the Earth. The point on the surface above the hypocentre is called the **epicentre**. The hypocentre is sometimes very deep beneath the epicentre. For example, on 9th June 1994 an 8.2 magnitude earthquake took place beneath Bolivia. The hypocenter was calculated to be 670 km below the surface yet oscillations were recorded in seismic stations around the world. So how was the energy transferred? There are two fundamental types of wave: S waves and P waves.

P waves are an example of **longitudinal waves**. In this type of wave, the oscillations involved are in the same direction as that of energy transfer. Areas of high pressure (**compressions**) and low pressure (**rarefactions**) travel alternately through the medium. Other examples of longitudinal waves include sound waves and one type of wave along a stretched spring.

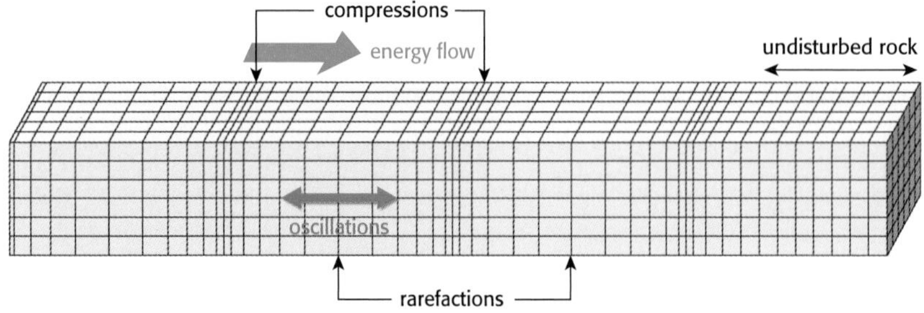

Figure 8 Primary or P waves

Remember that longitudinal waves are NOT just "waves where the oscillations are from left to right". The oscillations are along the direction of energy transfer, whatever that direction might be.

S waves are examples of the other fundamental type of wave: **transverse waves**.

In transverse waves, the oscillations are perpendicular to the direction of transfer of energy. A pattern of **peaks** and **troughs** travel through the medium. There are many other examples of transverse waves: one type of waves on a spring, waves on the surface of a pond, and the whole electromagnetic spectrum of waves: visible light, IR, UV, radio and TV broadcasting, gamma rays, microwaves.

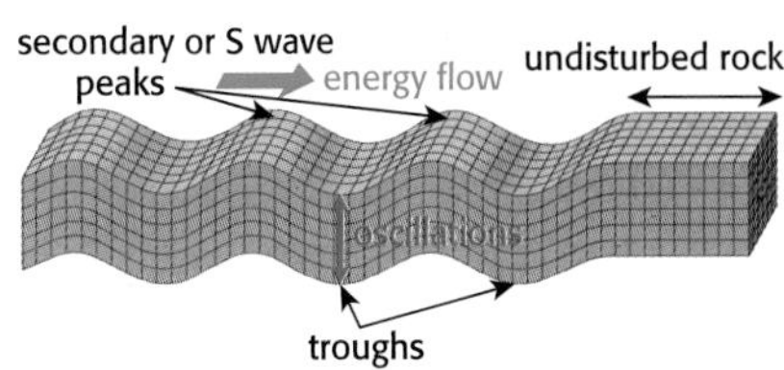

Figure 9 Secondary or S waves

The wave equation

All waves (including, of course, transverse and longitudinal) have a particular wavelength λ, frequency, f and velocity, v. These quantities are related by the **wave equation**:

$$v = f\lambda$$

The wave equation can be derived from the basic wave definitions and is applicable to every wave type, including earthquake waves. It cannot, however, be used to predict the speed of a particular wave from first principles, as this depends on the situation being considered.

Derive the wave equation from the basic wave definitions asked for on page 73.

S and P waves travel through the Earth at different speeds from one another and thus they take different times to arrive at any given point on Earth. Geophysicists use this fact to locate the likely focus of an earthquake from seismic recordings taken from around the world. In general, the P waves travel faster and are thus called **primary waves** because they will arrive at a seismic station first. The S waves, or **secondary waves**, will arrive later. P waves are also sometimes called "pressure (or compression) waves" and S waves are sometimes called "shake waves".

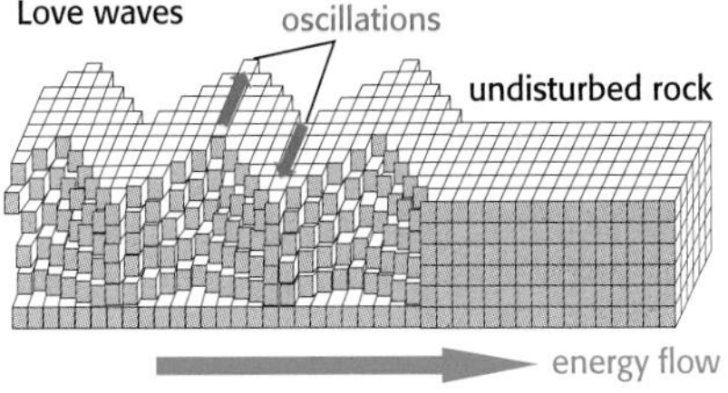

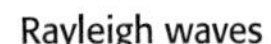

A final class of earthquake waves that travel even slower than S or P waves are called surface waves. As the name suggests, these waves do not travel through the main bulk of the Earth, but they only move over its surface. The speed of these types of wave depends on their frequency and they tend to spread out as they travel away from the epicentre of the earthquake. Two examples of possible surface waves are **Love waves** and **Rayleigh waves**:

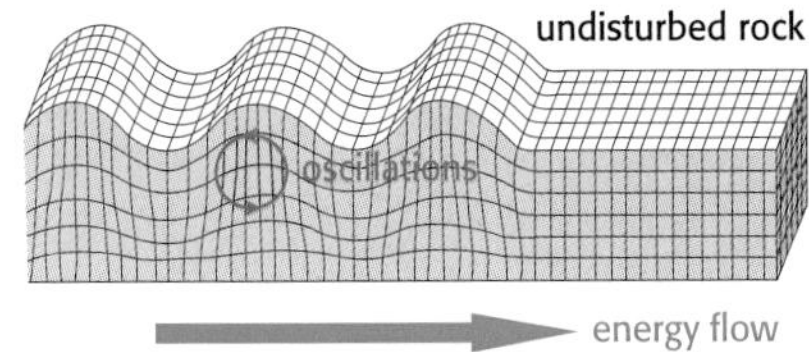

Figure 10 Love waves and Rayleigh waves

Love waves are another example of a transverse wave whereas Rayleigh waves are more complicated and can be considered to be a mixture of transverse and longitudinal.

Thinking about science: An example of the political use of science

In October 2006 the government of North Korea announced that it had successfully detonated a test nuclear bomb. The earthquake wave signature of the explosion was analysed around the world in order to attempt to verify this claim but initially there was widespread disagreement.

Within a few hours, the Russian government announced that it was "100% certain" that a nuclear test had been carried out and measured its size to be equivalent to an explosion of between five to fifteen kilotons of TNT.

South Korea, France, and the US, however, all announced that the measurement suggested the test was less than one kiloton. This could imply that the North Korean test had not been successful or that they had engineered a large conventional explosion and chosen to announce that it was nuclear in nature.

A couple of days later, the US government announced that it had acquired further evidence (traces of radioactive gas in the air near the site of the alleged nuclear test) supporting the claim, but they stressed

that more tests were needed to reach a conclusion. Before this announcement, South Korean and Chinese scientists said that they had detected no evidence of radioactivity in their own air, soil, and rainwater tests. The UN Security Council passed a resolution imposing sanctions on North Korea.

There are, of course, many possible procedures that could be used to check if a nuclear explosion had taken place. Earthquake wave analysis and testing for radioactive substances near the alleged site are just two possible methods. The following questions can be asked:

- Are the two methods of inquiry mentioned above the most reliable methods to use? Can you think of any other possible methods?
- In situations like these, do you think the scientific method can be relied upon to discover the truth behind the political claims or is science being used to make a political point?
- How could a citizen of one of these countries decide what to believe?
- What evidence would you need to convince you either way in this case?

Physics issues: risk assessment and EM waves

Every day we are exposed to risks. Many people would agree that some risks are worth taking, in fact, we sometimes tolerate these "extra" risks because we wish to enjoy all the technological advantages of living in the 21st century. For example, whenever you cross the road there is a risk of being involved in an accident. Most people would not argue for banning all vehicles from our roads in order to save the number of lives that are lost as a result of the occasional accidents that do happen, even though more than 1.2 million people worldwide die in road traffic crashes every year.

Figure 11 Is this an acceptable level of risk?

Perceived risks are often balanced against perceived benefits in this way. For example many people enjoy skiing but there are many accidents associated with this sport. Incredibly, given the often verified direct link between smoking and the increased risk of suffering from many diseases including cancer, there are still thousands of people who choose to ignore the risks associated with smoking.

Often it is not possible to avoid potential risks. As a result of modern communication systems, for example, we are all now exposed to much more electromagnetic radiation than would have occurred only 10 years ago. Some people worry about the potential risks of using mobile phones or living near the associated aerials. So what are the risks and are they risks worth taking? Can the level of risk be quantified or calculated?

The formal study of risk assessment and its management is beyond the scope of the IB Diploma Programme physics course but it is important to understand the sort of checks that companies and governments need to undertake whenever a project is proposed. They need to weigh up the possible consequences of an accident verses the potential benefits that the project brings. In simple terms a risk can be calculated by multiplying the probability of a particular event or accident occurring by some agreed measure of the consequences of the event.

$$\text{risk} = \text{probability} \times \text{consequence}$$

The **probability** is a number between 1 and 0. If, for example, the probability that a new hydroelectric dam would fail was assessed to be 0.0001 year^{-1}, this would mean that on average it would only fail once in every ten thousand years. Most people would accept that this risk is very small. However, if there were several thousand of the same type of dam in existence around the world, then you might

expect an accident every few years. Given the potential consequences, is this an acceptable level of risk?

The **consequence** is a measure of what happens if the event under consideration does take place. In the example above, the failure of a dam would be measured in human lives as well as financial and environmental costs.

Most alleged dangers related to the use of mobile phones are associated with the electromagnetic radiation used. Are all types of electromagnetic radiations as harmful as each other?

All the different types of electromagnetic radiation are the same type of wave – they are a transverse oscillation of electric and magnetic fields and they all propagate at the same speed in a vacuum: 3.0×10^8 m s^{-1}. They do, however, have very different properties as a result of their very different wavelengths.

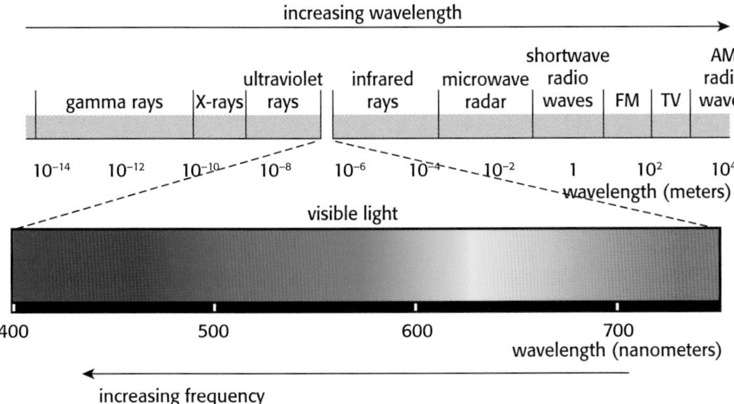

Figure 12 The electromagnetic spectrum

One aspect of the transmission of energy by EM radiation is that it can be modelled as being emitted in "packets" or **photons** of energy. These packets of energy are also called **quanta**. The evidence for this quantum phenomenon is explored more on page 262. The energy of each wave packet is proportional to the frequency of radiation being considered and each photon has an energy given by the following equation:

$$\text{energy of photon, } E = h \times f \qquad (1)$$

The constant h is called Plank's constant and is equal to 6.63×10^{-34} J s

The alleged dangers of mobile phones are largely attributed to the ionizing effects of the radiation they use (microwave radiation) on human tissues. However, equation 1 can be used to show that the photon energies of microwaves are not large enough to be able to cause these ionizations. It is true, however, that some types of electromagnetic radiation, for example gamma waves, can produce these effects.

The experimental evidence available, at the time of writing, does not provide any robust statistical evidence for a definite link between the widespread use of mobile phones and, for example, increased incidence of brain tumours. The World Health Organization coordinates assessments of the health risks. These are in-depth critical reviews conducted through independent, scientific peer-review groups aimed at evaluating the evidence from around the world. Their current conclusions can be downloaded from www.who.int (accessed July 2009).

For a further discussion of the risks associated with a possibly related situation (living near electrical power lines) see pages 226–7.

1 How many photons of visible light are being emitted by a 60W light bulb every second? Use an average value of the wavelength of visible light, say $\lambda = 600$ nm.

2 The ionization energy of hydrogen is 2.18×10^{-18} J. Calculate the frequency of a photon just able to achieve this ionization.

3 Use your answer to question 2 to categorize the different parts of the EM spectrum into those that you would expect to be able to cause ionizations and those that would not be expected to achieve this.

4 Imagine that there is a proposal to install a mobile phone aerial on the roof of your school in return for a small annual fee from the company. Discuss this proposal and create as long a list as possible of the risks associated with this project – what could go wrong and what would be the consequences? What level of risks do you find acceptable? In your opinion what conditions need to be met for the cost–benefit analysis of this project to be favourable?

Data-based question: Graphs of wave motion

Consider the graph (Figure 13) representing the oscillations involved during a particular type of wave motion. Try to write down brief answers to the questions that follow without reading ahead to the solutions.

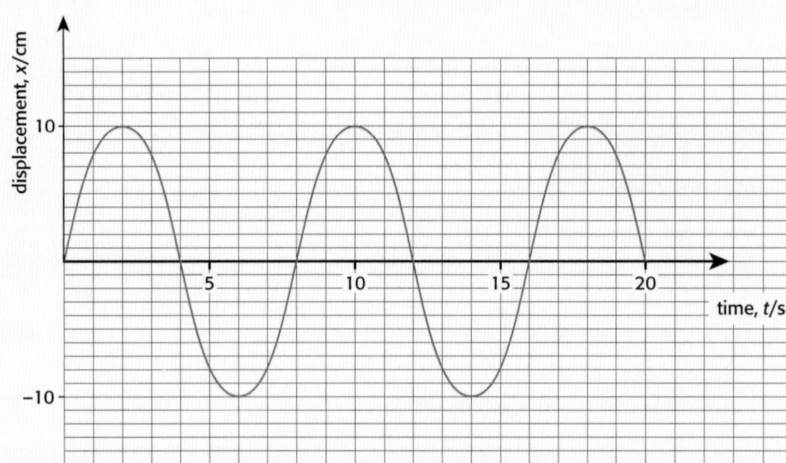

Figure 13 Variation with time of the displacement in a wave

1 Is the wave represented longitudinal or transverse?

2 What is the wavelength of the wave?

3 The time period of the oscillations is 8 seconds. How fast is the wave travelling?

Answers

It is important to understand what exactly is being represented. Under the pressure of an examination situation, many candidates attempt to analyse a graph without reading what is being plotted on each axis!

In Figure 13, the y-axis is labelled displacement and the x-axis, time. Thus the graph represents the variation of displacement with time. In the context of a wave motion, this must be how the displacement **of one particular particle** associated with the wave motion varies as time progresses. It would, of course, be possible to plot other, similar, graphs showing how the different particles oscillate as time progresses but in this graph we have chosen to focus on one particular point.

Now we can discuss the questions posed above.

1 The diagram is a graphical representation and not a physical picture of the waves. You don't have enough information to answer the question; the direction of the displacement is not specified and the graph could equally apply to a transverse wave or a longitudinal wave. Without knowing the direction of energy transfer, you cannot answer this question.

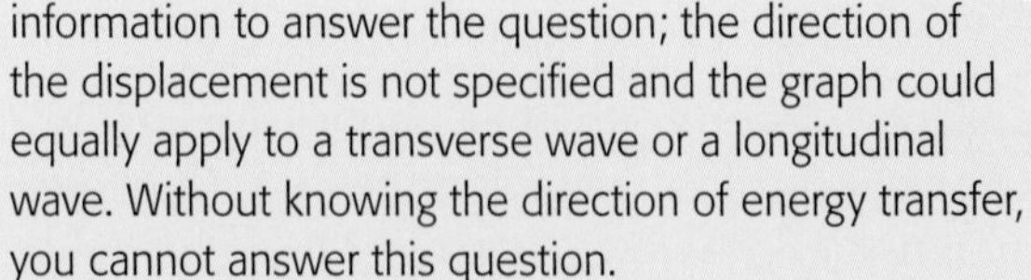

2 Since the graph is only providing information about how one particular particle oscillates with time, we are not being given any information about the separation between different particles along the wave. Another way of thinking about this is that the x-axis is labelled time and thus cannot be used as a measure of wavelength. Any measurements from the y-axis relate to the displacement of an individual particle but this is not the same as the length occupied by one complete oscillation along the wave. The graph cannot be used to provide the information requested.

3 Once again, the graph does not provide enough information to calculate the required quantity. It is possible to calculate the **average speed** of the particle during one quarter (or any complete multiple of a quarter) of an oscillation, as shown below. Remember the speed at which any particle moves during wave motion constantly changes (the graph for speed is a cosine wave).

time taken to complete one complete oscillation = 8 s from the graph,
distance gone in one quarter oscillation = 10 cm
time taken to complete one quarter oscillation = 2 s

$$\text{average speed} = \frac{\text{distance}}{\text{time}} = \frac{10}{2} = 5\,\text{m s}^{-1}$$

During the first quarter oscillation, it started at its maximum speed and slowed down to zero. 5 m s^{-1} thus represents the particle's average speed. This is a completely different quantity to the speed at which the wave pattern is propagated through space so we have not actually answered the question.

The graph provided in Figure 13 is very different to a graph which plots the displacement of **all the points** associated with the wave motion **at one particular time**. An example of this type of graph is shown in Figure 14:

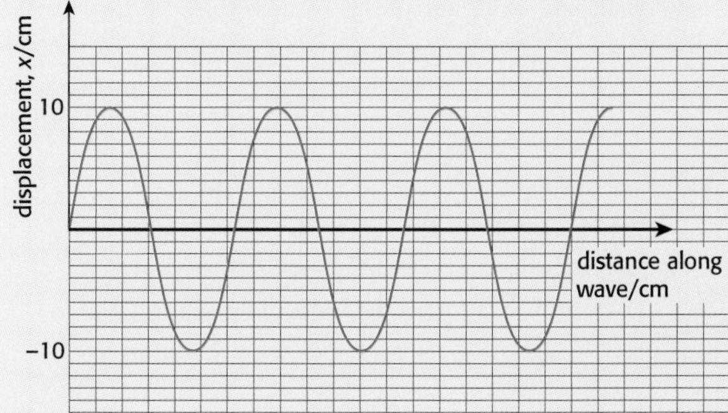

Figure 14 Variation of displacement with distance along a wave

It is potentially very confusing for the two graphs to have very similar shapes, but if the axes are carefully labelled then the difference between the information represented in each should be clear.

To summarize, the first graph (Figure 13) showed the variation **with time** of (the displacement **of one point** along the wave). Different points would need different graphs.

The second graph (Figure 14) shows one "snap shot" of the different variations **with distance along the wave** of (the displacement of all the points along the wave). Different times would need different graphs.

?

1 The paragraph above states that with each approach, further graphs would be needed to fully describe the wave motion. For each approach draw another similar graph to represent either:

i) the variation of displacement for a different point, or

ii) the variation of displacement of all the points at a later time.

2 Annotate your graphs and explain how the following quantities can be calculated from these graphs:

- amplitude
- frequency
- period
- phase difference
- wavelength
- wave speed.

The endoscope: an example of reflection

In this section we will discuss one example of reflection. There are many more and after reading this section you might like to identify and research other examples.

In general, whenever a wave crosses a boundary between two different media it is both partially transmitted and partially reflected. Under particular conditions (see page 109) it is possible for a wave that is travelling in a dense medium to undergo **total internal reflection** (TIR). This means that the entire wave energy is reflected and no wave energy is transmitted into the neighbouring material.

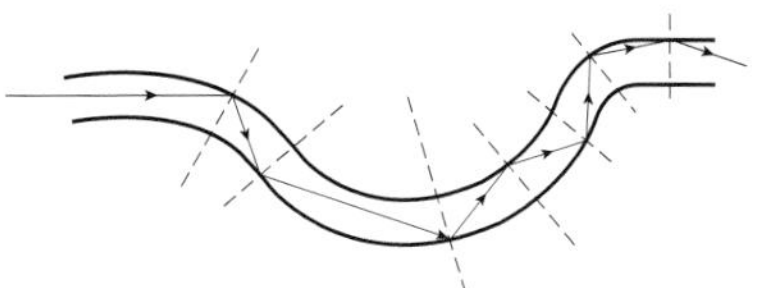

Figure 15 Total internal reflection along a fibre optic

A light ray can be continuously totally internally reflected along a glass fibre, called a fibre optic cable.

Fibre optics are thin strands of optically transparent material that use the principle of total internal reflection to transmit light waves along the path of the fibre with very little loss (see Figure 15). Refer to Chapter 28 for a discussion of some more applications of fibre optic cables. A common application of TIR is in a medical device called an endoscope (Figure 16). This device allows doctors to see inside the body of patient in a way that does not involve surgery.

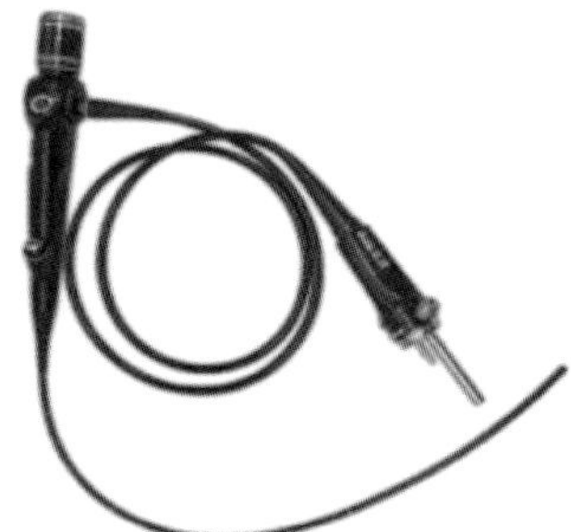

Figure 16 An endoscope

An endoscope consists of two bundles of fibre optic cables which can be introduced into the body. One of the bundles of fibre optics transmits light from an external source to the point of interest and another bundle is used to bring the reflected light back to an external imaging device, effectively providing a view inside the patient. In Figure 17 you can see the image of a polyp, a benign precancerous growth projecting from the inner lining of the colon.

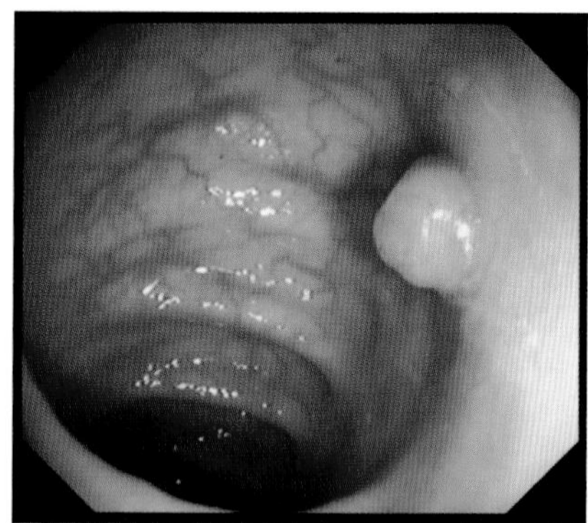

Figure 17 A polyp in the colon

Refraction

As we stated earlier in this chapter, when a wave meets the boundary between two different media, some of the wave is reflected back and some of the wave is transmitted on. The portion that is reflected obeys the law of reflection (angle of incidence = angle of reflection) but the direction of the refracted ray depends on the speeds of the wave in each medium. In the example in Figure 18, the speed of the wave in medium 1 is greater than the speed of the wave in medium 2.

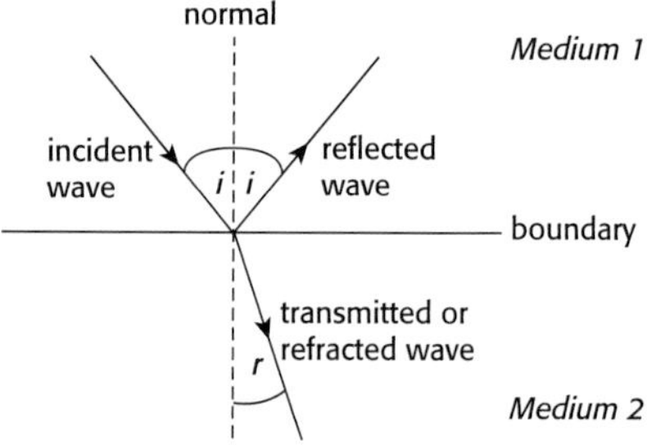

Figure 18 Reflection and refraction

Experiments show that in all situations:

$$\frac{\sin i}{\sin r} = \text{constant} = \frac{\text{speed of wave in medium 1}}{\text{speed of wave in medium 2}} \qquad (2)$$

This equation is known as Snell's law.

The ratio expressed in equation 2 defines a constant called the **refractive index**. Note that the general wave equation ($v = f\lambda$) must still apply as the wave crosses the boundary between the two media. As it crosses the boundary, its wavelength changes in proportion to the change in speed. The frequency of the wave must remain unchanged because the number of waves leaving medium 1 must equal the number of waves entering medium 2, or in other words, the law of conservation of energy must apply.

If the wave is travelling from a medium where its speed is slow into one where its speed is faster, the above situation is essentially reversed. However, there exists a critical angle above which it is impossible for the wave to still be transmitted into the new medium. At exactly the critical angle, the wave is refracted along the boundary, and at any larger angles of incidence all of the wave's energy is reflected (total internal reflection).

This description of refraction only applies whenever a wave makes the abrupt change from one medium to another. If, however, a wave travels through a region where its speed gradually changes then the path of the wave turns out to be a curve. You can imagine the wave passing through thin layers and changing its speed by a small amount each time is passes into a new layer (Figure 19).

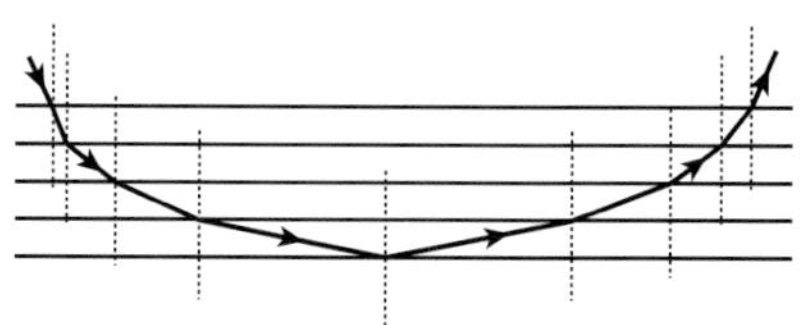

Figure 19 A curved ray path resulting from gradually changing densities

Earthquake waves: an example of refraction

This section outlines one specific example of refraction; after reading it you might like to identify and research some more.

Imagine a huge scale model of the Earth that is 13 m high.

> **Thinking about science: *Discussion***
>
> Given the relatively tiny section of the Earth that the human race is able to explore, how are we able to describe, with a great deal of confidence, the internal structure of the Earth?

Figure 20 A model Earth, 13 m high

On this scale, Mount Everest would be approximately 9 mm high and the deepest diamond mine drilled in South Africa would only be about 3.5 mm deep.

It has been known for a long time that the density of the centre of the Earth is greater than the density at its surface. In the 17th century, Newton was able to use his theory of universal gravity to show that the average density of the Earth was approximately twice the density of the rocks at the surface. If we assume that the density of the material that makes up the Earth increases uniformly with depth, this would result in an increase in the speed of earthquake waves with depth, so the path that they take would be predicted to be curved as shown in Figure 21.

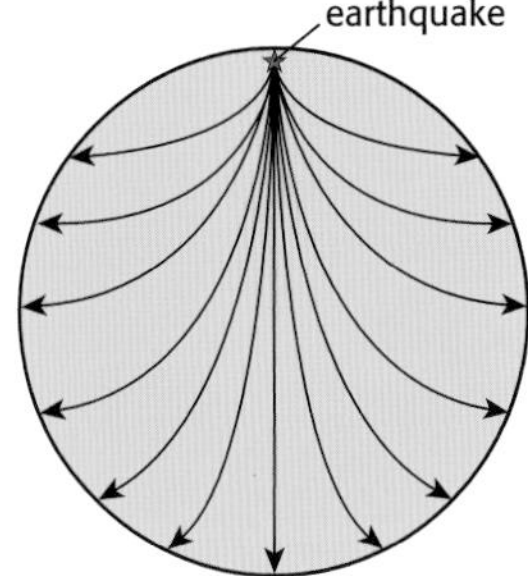

Figure 21 Path of earthquake waves if the density of the Earth varied uniformly

However, what we actually observe depends on whether we are considering P waves or S waves. We will consider these situations separately.

P waves

The P waves do show the expected patterns of refraction but there is a shadow zone where no P waves are observed. This can be explained if we hypothesize a central core region to the Earth where the density (and other properties) of the material suddenly changes.

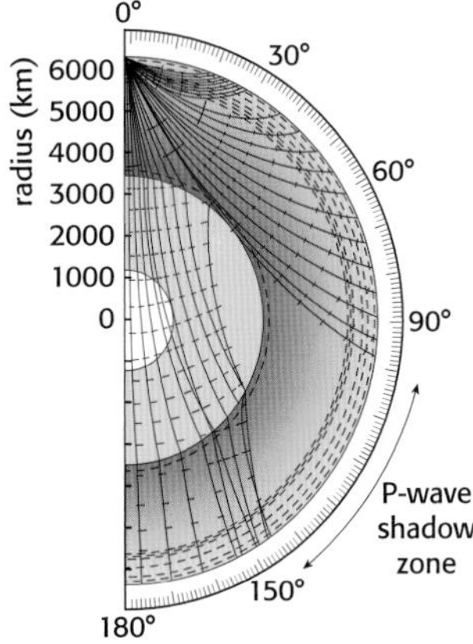

Figure 22 P wave shadow zone

More information about the nature of this core region comes from the study of the behaviour of S waves.

S waves

There is a total S wave shadow zone observed on the opposite side of the Earth to the hypocentre, where no S waves ever arrive.

We can develop an explanation for this shadow zone by considering the way S waves are propagated. They are transverse waves and thus can easily pass through any solid. Transverse mechanical waves are, however, unable to pass through the body of a liquid or a gas. It is possible to move a block of liquid "sideways" without causing a corresponding sideways movement of the adjacent liquid. When trying to visualize the situation, it is important to remember that we are considering the bulk movement of a section of a liquid that is within the body of the liquid. This is very different to the movement of the surface of a liquid when transverse waves will be propagated as a result of the surface tension forces.

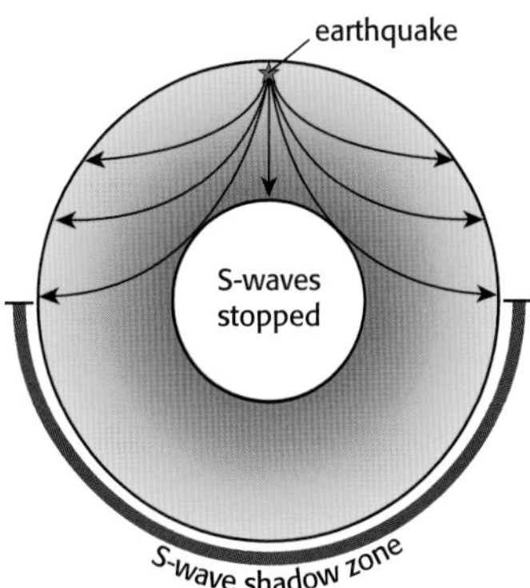

Figure 23 S wave shadow zone

So we can conclude that the S wave shadow zone implies that the core of the Earth contains a liquid zone consisting of molten rocks, or magma. In fact, observations that some P waves arrive at the opposite side of the Earth quicker than expected have been used to show that inside the liquid core, there is a solid inner core.

Why we can't see atoms using light: an example of diffraction

This section outlines one specific example of diffraction; after reading it you might like to identify and research some more.

Why don't scientists build an optical microscope powerful enough to see atoms? The reason is not our inability to make lenses that are powerful enough, but the fact that the wave nature of light makes it impossible. The reason for this is a property that is demonstrated by all waves: **diffraction**, the bending of a wave around an object.

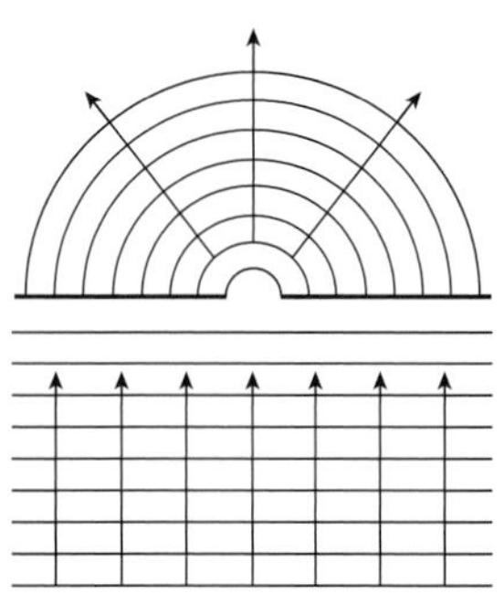

Figure 24 A wave is diffracted by an aperture

In Figure 24 the wavelength of the waves is approximately the same size as the gap and this has resulted in the waves spreading completely around the corners.

In general, if the wavelength and size of the object are approximately the same, the effects of diffraction are maximized. If the object is significantly larger than the wavelength of the wave, almost no diffraction takes place; the wave propagates on unchanged.

We are used to the everyday properties of sound and light. Most people would agree that sound goes around corners whereas light does not. If, for example, you arrived at a concert to find out that your seat was directly behind a pillar, then you would expect to be able to hear the concert but not to see the performers (and you probably now realize why the ticket was so cheap!). If sound and light are both waves, why does light not bend around the pillar in the same way as sound? The reason for this is because of the different wavelengths of sound and of light.

Compare the diameter of an atom with the wavelength of visible light. Can you explain why we will never be able to "see" an atom? Try to write an explanation justifying your answer to a non-scientist.

The human ear can respond to sounds in the wavelength range from approximately 6 cm to 60 m. Everyday objects (including the pillar in the concert hall) are of this sort of size which means sounds diffract around them and we expect to be able to hear the sound even if the source is around a corner or behind a pillar.

Visible light has wavelengths from 400 nm to about 750 nm. This is much smaller than the size of everyday objects so the diffraction of light is negligible. The result is that you can hear the concert but you can't see what is going on. For more details see pages 128–30.

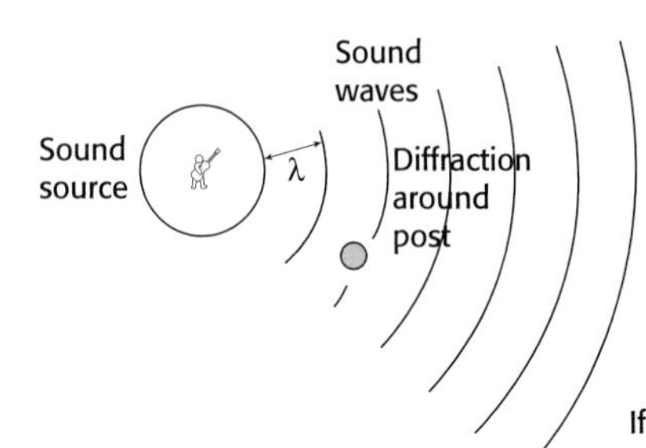
Suppose you bought a concert ticket without looking at the seating chart and ended up sitting behind a large pillar. You would be able to hear the concert quite well because the wavelengths of sound are long enough to bend around the pillar.
Sound source
λ
Sound waves
Diffraction around post
θ
Diffraction past small opening.
If you were outside an open door, you could still hear because the sound would spread out from the small opening as if it were a localized source of sound.
If you were several wavelengths of sound past the post, you would not be able to detect the presence of the post from the nature of the sound.

Figure 25 Sound diffracts around everyday objects but light does not

Answer the following questions and explain to what extent each of the following situations involves the diffraction of waves.

1 When driving a car into a tunnel, why will you be able to carry on listening to a radio programme broadcast on MW (medium wave) for longer than you can if you were tuned into an FM (short wave) station?

2 My friend is unable to pick up terrestrially broadcast TV signals in his house, but can listen to LW radio. The radio and the TV are both broadcast from an aerial that is situated on the other side of a hill from the house. Can you think of an explanation?

3 Imagine you are sitting outside, behind a wall that is two metres high. Are you more likely to hear a man or a woman calling out your name?

4 Figure 26 shows the Shau Kei Wan typhoon shelter on the east side of Hong Kong Island. The entrance for the boats is the gap on the left-hand side of the picture. Explain how the design of the shelter helps to protect boats during a typhoon.

Figure 26 Shau Kei Wan typhoon shelter

Antisound – the sound of silence: an example of interference

This section outlines one specific example of interference; after reading it you might like to identify and research some more.

If you want to listen to music whilst on the move, all the other sounds around you often are distracting. What can be done to block out the background noises? Most mobile music lovers use headphones but the problem with simple earpiece headphones is that you can still hear much of the background sound. Active noise cancellation headphones provide one possible solution.

The basic idea is straightforward; if two waves of the same type (for example, two sound waves) meet at a particular point then the overall displacement at that point is simply the vector sum of the two individual waves. To work out what is happening at any given point in space we just add the individual waves together. This is called the **principle of superposition** (see Figure 28).

Figure 27 Listening to music on a mobile device with headphones

In fact, we have already used the idea of superposition when considering how complex sounds can be viewed as just the addition of a large number of more simple sinusoidal sounds (see Fourier analysis, pages 91–2).

Although waves of any frequency can be superposed (providing they are the same type of wave) the situation becomes particularly interesting if the two waves are:

- of identical frequencies,
- of similar amplitudes, or
- phase linked.

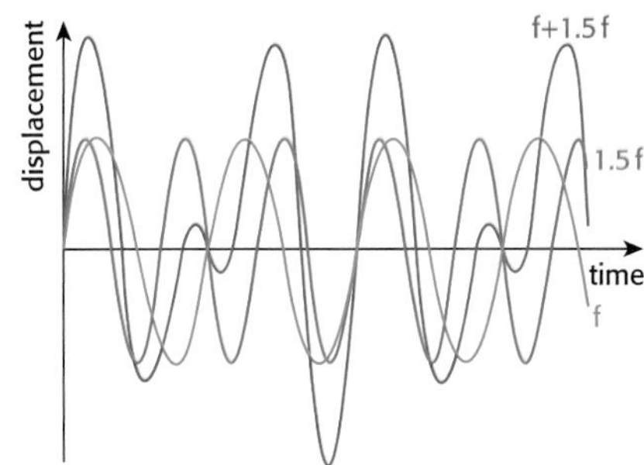

Figure 28 Superposition of two waves of frequency f and 1.5f

Active noise cancellation headphones use two small microphones, one on each earpiece, to detect the ambient noise that is due to arrive at the ears. Complex electronics is then used to create a waveform for sound that is exactly opposite to the arriving noise. This sound is added by the speakers in the earphones and the result is that the noise from outside is cancelled out, leaving just the music playing.

It is also possible to selectively "target" sounds that we would categorize as noise and preferentially allow in the sounds of speech.

Rediscovering physics: Diffraction and interference

Young's double slit experiment provides an example of a situation involving both the diffraction and interference of light. In this experiment, it is essential that the light sources used are **phase linked** (**coherent**) and of a single frequency (**monochromatic**), so light from a laser is usually used.

A laser is shone on a pair of slits. Diffraction takes place at the slits, making each one a separate source of coherent light waves. The light waves from these two sources interfere with each other. Light and dark regions are observed on a screen placed a short distance from the slits.

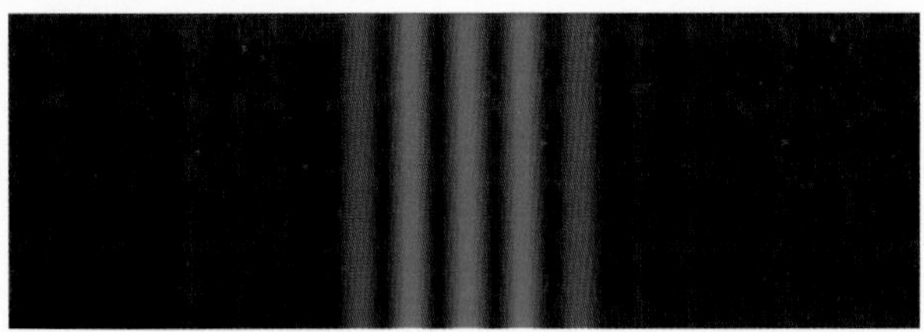

Light regions represent the points where the interfering light waves are both at their maximum displacements (i.e. two crests or two troughs). This is known as **constructive interference**. Constructive interference takes place at any point where the **path difference** is zero, or equal to a whole number of wavelengths.

In these situations the waves arrive in phase. The path difference is the distance travelled by waves from one source to the point in question minus the distance travelled by waves from the other source.

Dark regions appear where a crest and a trough interfere so that the overall intensity is zero. This is known as **destructive interference**. Destructive interference takes place when the waves arrive exactly half a wavelength out of phase. The oscillations from one source exactly cancel the oscillations from the other source.

At all other points the waves are neither perfectly in phase nor completely out of phase, resulting in dim, but not completely dark regions.

There are some very good animations available on the internet to help visualize what happens in two-source interference.

Another example of combined diffraction and interference of light involves an instrument called a **diffraction grating**. If a large number of very narrow slits are arranged very close together they can be treated as a regular array of coherent sources of light. These coherent sources of light then interfere with one another. At certain angles it may be possible for all of the sources to constructively interfere with one another.

Inquiry: Wave properties

Do some research and prepare a brief presentation on one application of your choice that relates to one of the four general wave properties – reflection, refraction, diffraction, and interference. You should present your example in a way that is designed to address the learning needs of the target audience (your classmates). Example topics are given below for reflection but many other possibilities exist.

- Fibre optics for communications/the transfer of digital information.
- Ultrasound for foetal checks.
- Seismic prospecting for oil.
- The virtual image seen in a mirror.
- An echo.
- The "reverberation" of sound in a church.
- Radar location of aircraft.
- The techniques used by trawler fishermen for the location of fish.

Physics issues: seashore waves and tsunami waves

Waves on the seashore are not typical waves. However, this does not mean that we cannot understand them.

Figure 29 A large seashore wave

Sea waves are neither purely transverse nor longitudinal; they are a combination of both types of wave (like Rayleigh waves). When a wave passes through the water, the surface follows an approximately circular pattern (though as one goes deeper in the water it turns out that the oscillations are more elliptical than circular). The velocity of a water wave is dependent on many factors, notably amplitude and depth. As waves head into the shore, they refract. That is, they tend to slow down (if you have ever watched waves you may have noticed that it is possible for a wave that is behind another wave to catch-up and then for the principle of superposition to apply). This effect means that the top of a wave tends to be travelling faster than its base and so in the end the wave breaks.

Another highly complex, atypical, and extremely dangerous wave is the tsunami. "Tsunami" is a Japanese word meaning "harbour wave". A tsunami is virtually unnoticeable in deep oceans (it may only be a few centimetres in height) but it can cause tremendous devastation along a coastline.

Figure 30 Effects of a tsunami

In contrast to traditional water waves (in which there is no overall movement of water – just oscillations near the surface), a tsunami is caused by a major disturbance of water from a significant depth. Typically the source of the wave may be an underwater earthquake, landslide, or volcanic eruption. It is also possible for a meteorite to cause a tsunami.

In this situation, a massive amount of water all moves at the same time. At the surface, the wave tends to have a small amplitude (20 cm), large wavelength (100 km), and very high velocity (200 m s^{-1}). Compare this with a typical surface wave (amplitude 2 m; wavelength 10 m) which moves at about 7 m s^{-1}. The important point is that, unlike normal waves, the movement of the water on the surface is only a tiny fraction of all the water that is actually moving – the total volume involved is huge.

When this huge volume of water arrives at the coast, the abrupt decrease in depth of the ocean means that there will be a sudden increase in amplitude of the wave. The wave only becomes noticeable in the harbour – hence its name. It is this unstoppable wall of water that causes the terrible damage.

On 26th December 2004 a tsunami was caused by an undersea earthquake with a magnitude of 9.0 off the north-western coast of Sumatra, Indonesia. The large tsunami caused tremendous devastation in the region (particularly Sumatra in Indonesia and Phuket in Thailand) and killed approximately 300,000 people. The tsunami even crossed the Indian Ocean and was recorded in Lamu in Kenya. The work to help rebuild the affected communities will go on for a long time and always needs more donations, help, and support. Many IB pupils, teachers, and schools have chosen (via CAS or otherwise) to become involved in some aspect of the IB tsunami appeal launched on 5th January 2005.

Possibilities for exploration and engagement with the IB community theme can be found on the website http://communitytheme.ibo.org. The first theme, sharing our humanity, was officially launched in April 2007 and is intended to serve as a focus for IB World Schools to share the excellent initiatives that are already underway in many schools and encourage and inspire new activities.

Science tools: spider diagrams

A spider diagram is a visual overview, or memory map, of a particular topic. You might find that the process of producing a spider diagram helps you to link ideas when reviewing material and it can be a useful aid to revision. Many different approaches can be used to create a spider diagram. The following task looks at the section of the syllabus dealing with SHM and waves.

The best maps are individually created to highlight all the connections that you have made. One way to get started would be to base your map around the following three main themes:

Simple Harmonic Motion

Waves

Resonance

Use a large piece of paper to highlight the links between these ideas, making sure that each link has a diagram, equation, definition or explanation written by it to remind you of the physics behind the link that you have identified. You will be able to identify many links, including several around each of the following themes:

- situations that involve two themes happening at the same time
- energy considerations
- mathematical similarities

Each of these themes can be explored in many ways including the following:

- terminologies used
- definitions used
- examples/analogies
- assumptions involved

Many of the boxes that result from this exploration can be broken down further or even linked together.

Having created you own spider diagram for this topic, compare your version with others in your school, your country or even from around the world. Interestingly, good spider diagrams can often be understood even if they are written in a language that you cannot understand.

End of chapter summary

Chapter 8 has two main themes: wave characteristics and wave properties. The list below summarises the knowledge and skills that you should be able to undertake after having studied this chapter. Further research into more detailed explanations using other resources and/or further practice at solving problems in all these topics is recommended – particularly the items in bold.

Wave characteristics

- **Describe a wave pulse and a continuous progressive (travelling) wave that transfers energy.**
- Describe waves in two dimensions, including the concepts of wavefronts and of rays and give examples of transverse and of longitudinal waves including the different types of waves associated with earthquakes
- **Describe the terms *crest, trough, compression, rarefaction, displacement, amplitude, frequency, period, wavelength, wave speed* and *intensity*.**
- Draw and explain displacement – time graphs and displacement – position graphs for transverse and for longitudinal waves.

- **Derive** and apply the relationship between wave speed, wavelength and frequency (wave equation).
- State that all electromagnetic waves travel with the same speed in free space, **and recall the orders of magnitude of the wavelengths of the principal radiations in the electromagnetic spectrum.**

Wave properties

- Explain, and give examples of the wave properties: *reflection, refraction, diffraction* and *interference*.
- Describe the reflection and transmission of waves at a boundary between two media and state **and apply** Snell's law.
- Understand the terms: *principle of superposition, path difference, constructive* and *destructive interference*.
- Understand how risks are assessed.

Chapter 8 questions

1 This question is about waves and wave properties.

The diagram below shows three wavefronts incident on a boundary between medium I and medium R. Wavefront CD is shown crossing the boundary. Wavefront EF is incomplete.

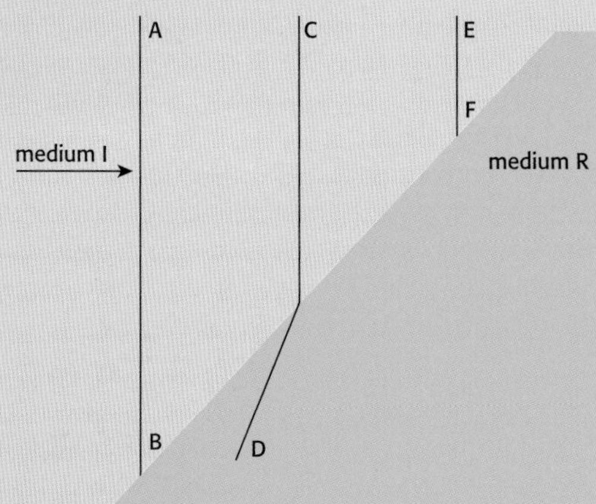

a) i) On the diagram above, draw a line to complete the wavefront EF. [1]

ii) Explain in which medium, I **or** R, the wave has the higher speed. [3]

The graph below shows the variation with time *t* of the velocity *v* of one particle of the medium through which the wave is travelling.

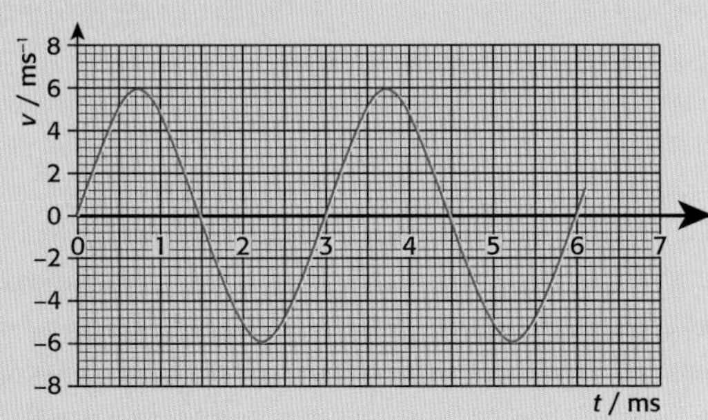

b) i) Explain how it can be deduced from the graph that the particle is oscillating. [2]

ii) Determine the frequency of oscillation of the particle. [2]

iii) Mark on the graph with the letter M one time at which the particle is at maximum displacement. [1]

iv) Estimate the area between the curve and the x-axis from the time $t = 0$ to the time $t = 1.5$ ms. [2]

v) Suggest what the area in b (iv) represents. [1]

c) i) State the principle of superposition. [2]

Two loudspeakers S_1 and S_2 are connected to the same output of a frequency generator and are placed in a large room as shown below.

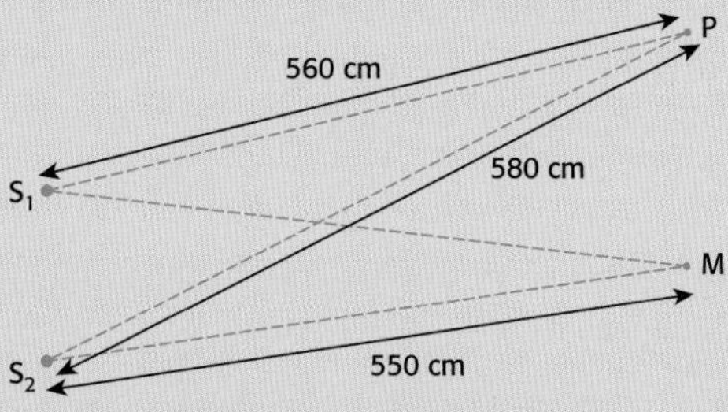

Sound waves of wavelength 40 cm and amplitude *A* are emitted by both loudspeakers.

M is a point distance 550 cm from both S_1 and S_2. Point P is a distance 560 cm from S_1 and 580 cm from S_2.

ii) State and explain what happens to the loudness of the sound detected by a microphone when the microphone is moved from point M to point P. [4]

iii) Referring to the diagram above, the amplitude of the wave emitted by S_1 is now increased to $2A$. The wave emitted by S_2 is unchanged. Deduce what change, if any, occurs in the loudness of the sound at point M and at point P when this change in amplitude is made. [4]

iv) The loudspeakers are now replaced with two monochromatic light sources. State the reason why bright and dark fringes are not observed along the line PM. [1]

Waves of frequency f and speed c are emitted by a stationary source of sound. An observer moves along a straight line towards the source at a constant speed v.

c) State, in terms of f, c and v, an expression for

i) the wavelength of the sound detected by the observer. [1]

ii) the apparent speed of the wave as measured by the observer. [1]

(Total 25 marks)

2 This question is about sound waves.

A sound wave of frequency 660 Hz passes through air. The variation of particle displacement with distance along the wave at one instant of time is shown below.

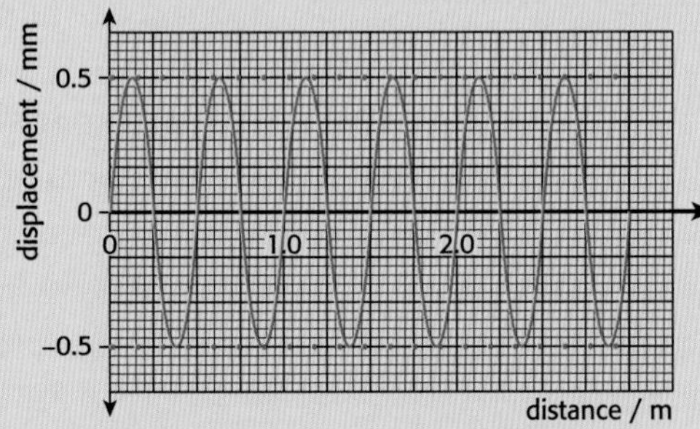

a) State whether this wave is an example of a longitudinal **or** a transverse wave. [1]

b) Using data from the above graph, deduce for this sound wave,

i) the wavelength. [1]

ii) the amplitude. [1]

iii) the speed. [2]

(Total 5 marks)

3 This question is about waves and wave motion.

a) i) Define what is meant by the *speed of a wave*. [2]

ii) Light is emitted from a candle flame. Explain why, in this situation, it is correct to refer to the "speed of the emitted light", rather than its velocity. [2]

b) i) Define, by reference to wave motion, what is meant by *displacement*. [2]

ii) By reference to displacement, describe the difference between a longitudinal wave and a transverse wave. [3]

The centre of an earthquake produces both longitudinal waves (P waves) and transverse waves (S waves). The graph below shows the variation with time t of the distance d moved by the two types of wave.

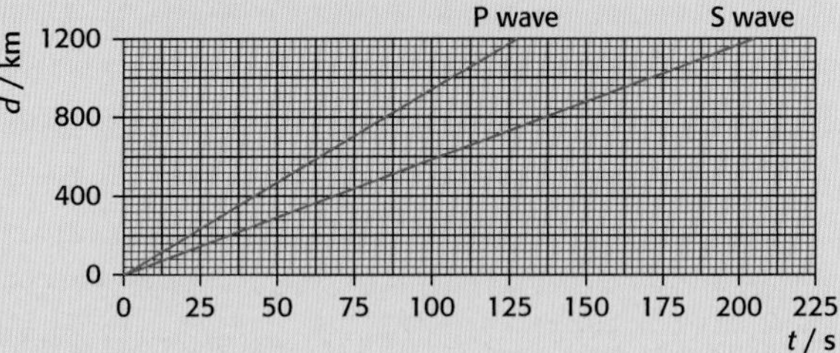

c) Use the graph to determine the speed of

i) the P waves. [1]

ii) the S waves. [1]

The waves from an earthquake close to the Earth's surface are detected at three laboratories L_1, L_2 and L_3. The laboratories are at the corners of a triangle so that each is separated from the others by a distance of 900 km, as shown in the diagram below.

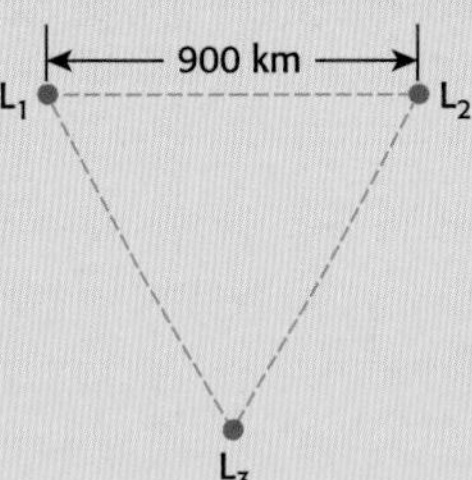

The records of the variation with time of the vibrations produced by the earthquake as detected at the three laboratories are shown below. All three records were started at the same time.

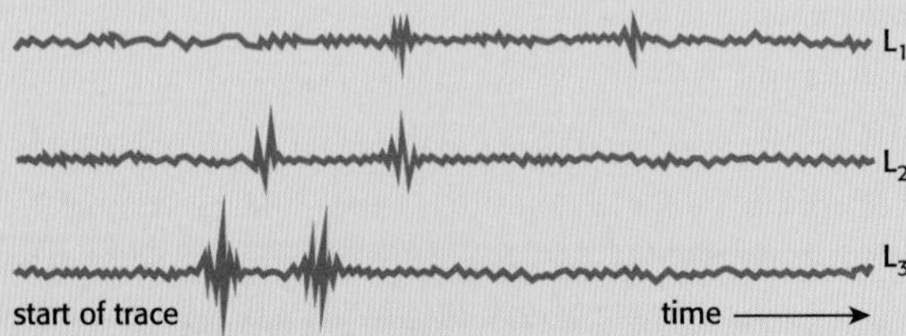

On each record, one pulse is made by the S wave and the other by the P wave. The separation of the two pulses is referred to as the S-P interval.

d) i) On the trace produced by laboratory L_2, identify, by reference to your answers in (c), the pulse due to the P wave (label the pulse P). [1]

ii) Using evidence from the records of the earthquake, state which laboratory was closest to the site of the earthquake. [1]

iii) State **three** separate pieces of evidence for your statement in (d)(ii).

iv) The S-P intervals are 68 s, 42 s and 27 s for laboratories L_1, L_2 and L_3 respectively. Use the graph, or otherwise, to determine the distance of the earthquake from each laboratory. Explain your working. [4]

v) Mark on the diagram a possible site of the earthquake.

There is a tall building near to the site of the earthquake, as illustrated below.

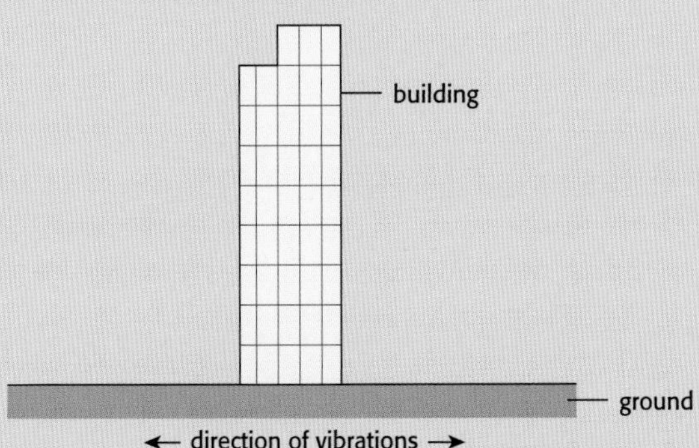

The base of the building vibrates horizontally due to the earthquake.

e) i) On the diagram above, draw the fundamental mode of vibration of the building caused by these vibrations. [1]

The building is of height 280 m and the mean speed of waves in the structure of the building is 3.4×10^3 ms^{-1}.

ii) Explain quantitatively why earthquake waves of frequency about 6 Hz are likely to be very destructive. [3]

(Total 25 marks)

4 Travelling waves

a) Graph 1 below shows the variation with time t of the displacement d of a travelling (progressive) wave. Graph 2 shows the variation with distance x along the same wave of its displacement d.

Graph 1

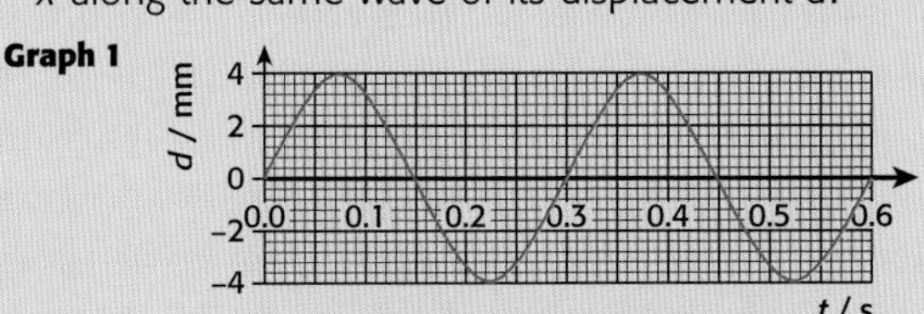

Graph 2

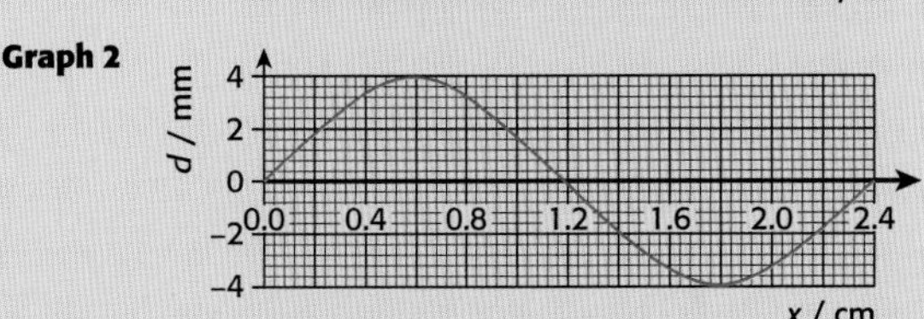

i) State what is meant by a *travelling wave.* [1]

ii) Use the graphs to determine the amplitude, wavelength, frequency and speed of the wave. [4 × 1]

Refraction of waves

b) The diagram below shows plane wavefronts incident on a boundary between two media A and B.

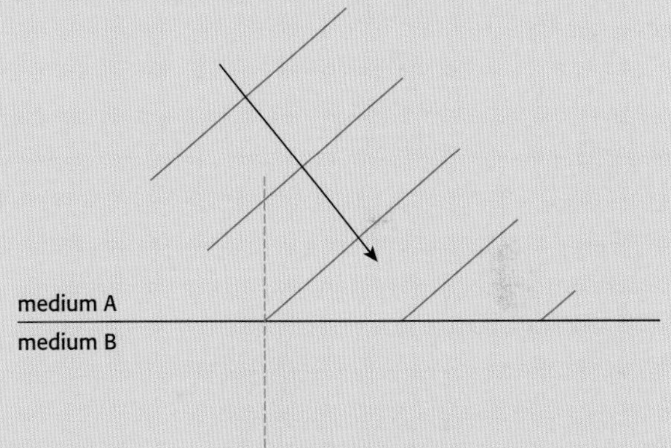

The ratio $\frac{\text{refractive index medium B}}{\text{refractive index medium A}}$ is 1.4.

The angle between an incident wavefront and the normal to the boundary is 50°.

i) Calculate the angle between a refracted wavefront and the normal to the boundary. [3]

ii) On the diagram above, construct **three** wavefronts to show the refraction of the wave at the boundary. [3]

(Total 11 marks)

5 This question is about waves.

a) With reference to the direction of energy transfer through a medium, distinguish between a transverse wave and a longitudinal wave. [3]

b) A wave is travelling along the surface of some shallow water in the *x*-direction. The graph shows the variation with time *t* of the displacement *d* of a particle of water.

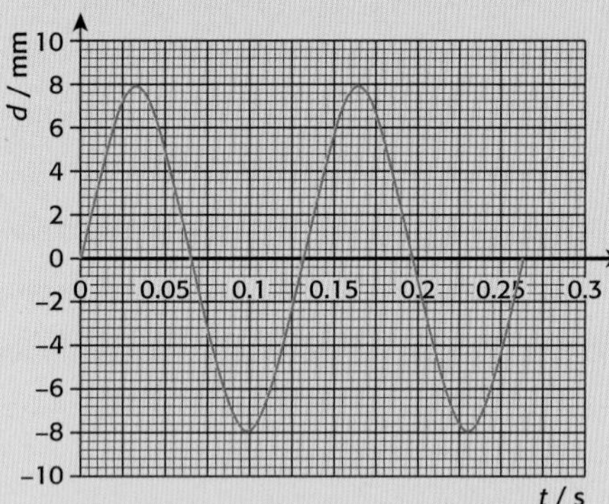

Use the graph to determine for the wave

i) the frequency, [2]

ii) the amplitude. [1]

c) The speed of the wave in (b) is 15 cm s^{-1}. Deduce that the wavelength of this wave is 2.0 cm. [2]

d) The graph in (b) shows the displacement of a particle at the position $x = 0$.

Draw a graph to show the variation with distance *x* along the water surface of the displacement d of the water surface at time $t = 0.070$ s.

e) The wave encounters a shelf that divides the water into two separate depths. The water to the right of the shelf is deeper than that to the left of the shelf.

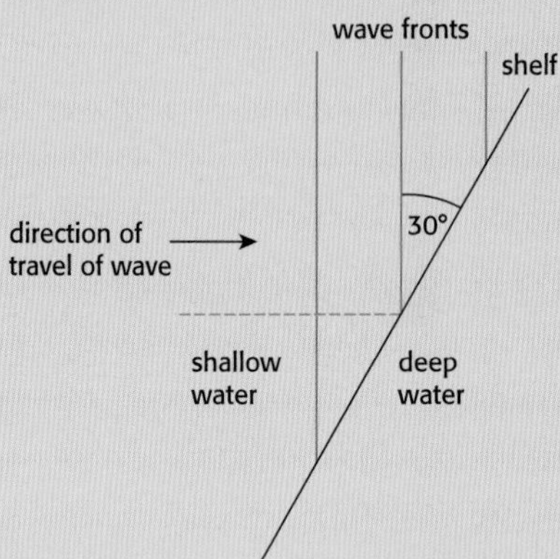

The angle between the wavefronts in the shallow water and the shelf is 30°. The speed of the wave in the shallow water is 15 cm s^{-1} and in the deeper water is 20 cm s^{-1}. For the wave in the deeper water, determine the angle between the normal to the wavefronts and the shelf. [2]

(Total 14 marks)

9 Wave phenomena

In Chapter 8 we introduced the common characteristics and properties of all waves. This chapter builds on these ideas by looking at five different applications:

- standing (stationary) waves
- the Doppler effect
- the mathematics of diffraction
- resolution
- polarization.

Standing (stationary) waves

A transverse or longitudinal wave is sometimes referred to as a travelling wave. Energy is transferred away from the source by a wave pattern or profile that moves through space. For example, Figure 1 represents the variation with distance along a medium of the displacement of a wave (which could be longitudinal or transverse – see pages 101–2).

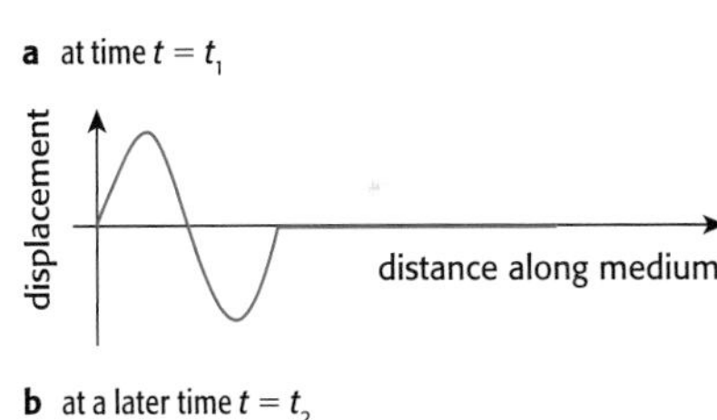

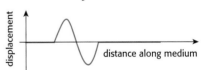

Figure 1 A wave pulse

For a continuous travelling wave in one dimension, the phase of each point changes with time but the amplitude of the wave remains constant.

A standing wave is formed by the superposition (i.e. vector addition) of two waves that are:

- the same type of wave
- of the same frequency and similar amplitudes
- travelling in equal and opposite directions.

Although each of the individual waves is still travelling in their separate directions, the result of the superposition of these waves is a new wave. The phase of each point on the new wave remains fixed with time but the amplitude of the resultant wave varies with time, as represented in Figure 2. We say this wave is fixed in space or, in other words, it is a standing wave.

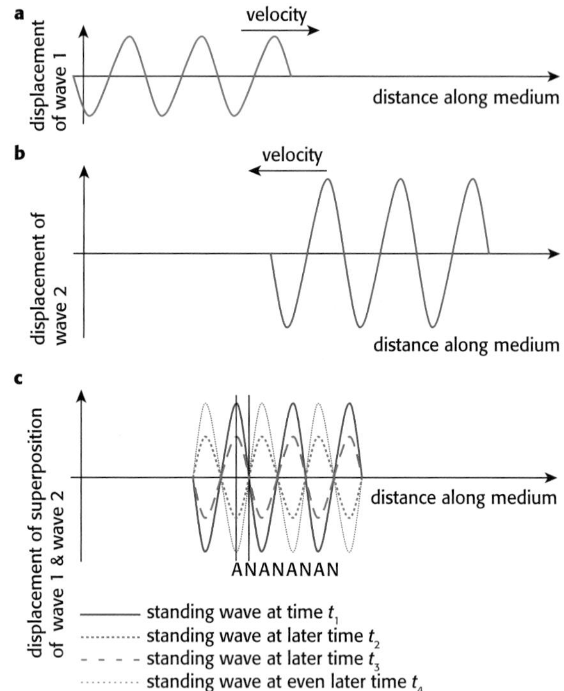

Figure 2 (a) and (b) are component waves that superpose to form (c)

This standing wave does not transfer energy. The points that always show no displacement are called nodes (N). The points that reach the maximum amplitude of displacement are called antinodes (A). See pages 124–5 for a formal derivation of the equation for a standing wave.

A loudspeaker emits sound of frequency f. The sound waves are reflected from a wall. The arrangement is shown in Figure 3.

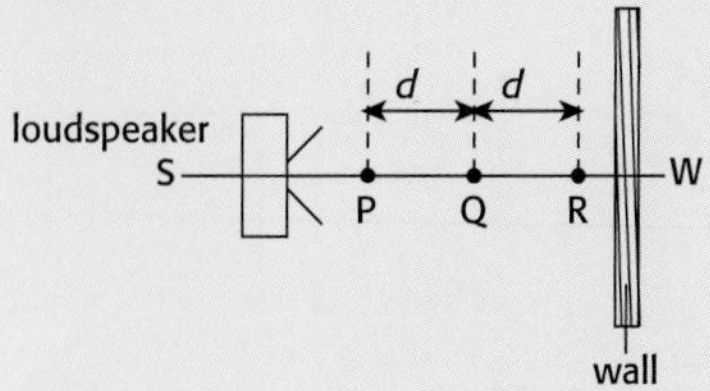

Figure 3

When a microphone is moved along the line SW, minimum loudness of sound is detected at points P, Q, and R. There are no other minima between these points. The separation of the minima is d.

The speed of the sound wave is

A $\frac{1}{2}fd$ **B** $\frac{f}{d}$

C fd **D** $2fd$

The aim of this question is to add together two travelling waves to produce a standing wave as described above. This can be achieved either using a spreadsheet, a GDC or graph paper.

a) Draw the result of the superposition of two sine waves of equal amplitudes that are in phase with one another at a given instant of time as shown in Figure 4.

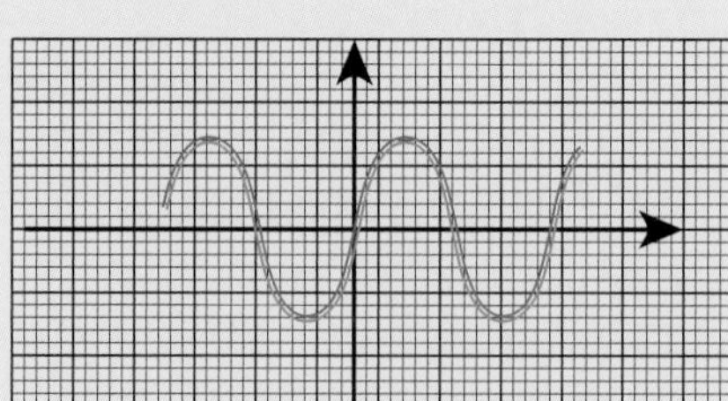

Figure 4

b) i) Move one of the original waves to the left by an eighth of a wavelength and the other wave to the right by the same amount as shown in Figure 5.

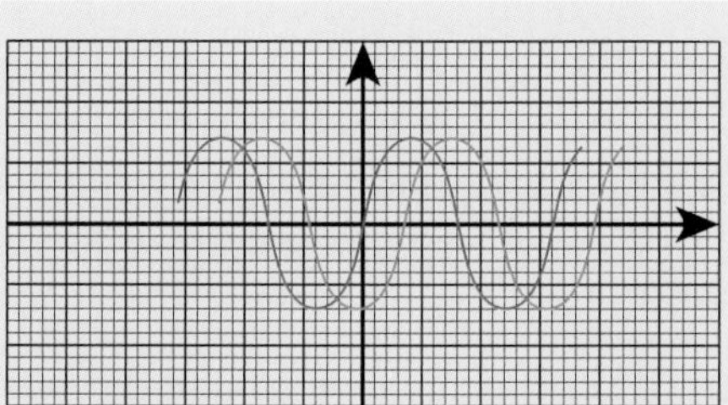

Figure 5

Draw the result of the superposition of the two sine waves in this new position.

c) Repeat this process until the waves are in phase again, moving each wave by an additional eighth of a wavelength each time.

d) Show that any random displacement to the left and right, (i.e. not multiples of an eighth of the wavelength) for example the one shown in Figure 6, fits the pattern you have discovered.

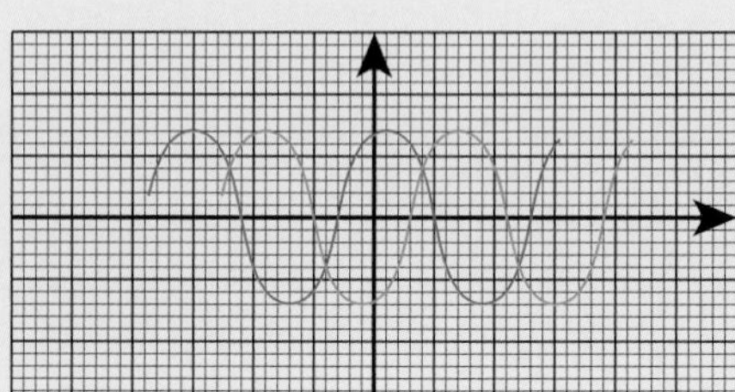

Figure 6

Comparison between standing waves and travelling waves

Table 1 A comparison of standing waves and travelling waves

	Standing	Travelling
Energy	Energy is not transferred but there is an energy associated with the wave.	Energy is transferred from source to receiver at the wave speed.
Amplitude	The amplitude of the wave pattern varies with time. At any particular point on the wave, the maximum amplitude is fixed between zero (at the nodes) to twice the amplitude of the component waves (at the antinodes).	The maximum amplitude of the wave is the same for all points along the wave, providing energy is not dissipated.
Frequency	All vibrations are SHM and at the same frequency as the frequency of the component waves.	All vibrations are SHM.
Wavelength	This is the same as the wavelength of the component waves and is equal to twice the distance between consecutive nodes.	The shortest distance between any two points that are in phase.
Phase	The phase of all the points in a section between two nodes are identical. The points in the next section between nodes are π radians (180°) out of phase with the first section.	All the particles within one wavelength have a different phase.
Wave pattern	Does not move.	Moves.

Boundary conditions

In many situations, one of the waves that is involved in the creation of the standing wave results from a reflection of a travelling wave. The processes involved in causing the reflection at the surface mean that a **boundary condition** is known to apply to the wave. A boundary condition is any principle known to apply to an end point of the wave.

For example, if a transverse wave travelling along a string under tension meets a fixed boundary, the wave will be reflected back along the string. If the end of the string is fixed and cannot move, the standing wave that is created must have a node at the fixed end.

Stringed instruments (piano, guitar, violin, etc.) produce sounds by making strings vibrate under tension. The strings are fixed at either end so the travelling waves produced initially will set up standing waves with nodes at either end. Figure 7 shows some of the possible standing waves on a string that is fixed at both ends.

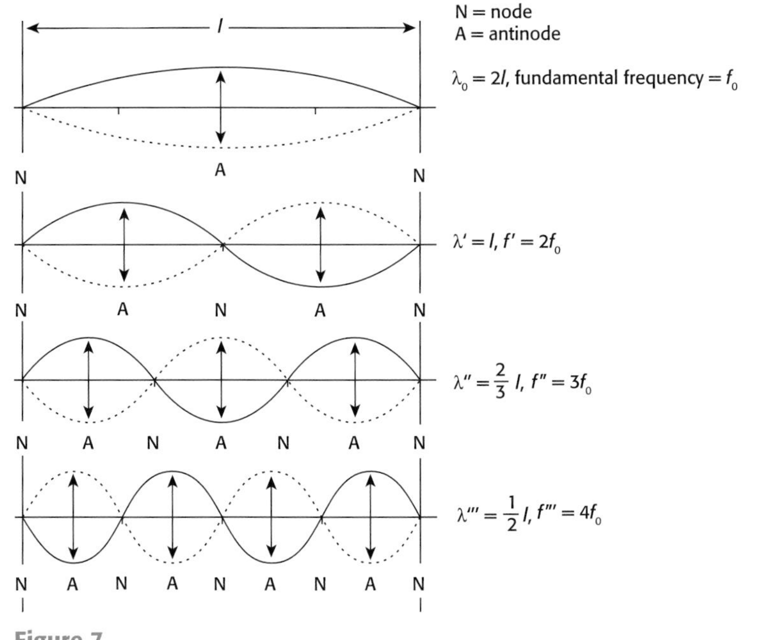

Figure 7

The standing wave with the lowest possible frequency is called the **fundamental** or **first harmonic** and the other possible frequencies are called **harmonics** (also sometimes called **overtones**). The difference between the sound of, for example, a guitar and a violin which are both playing the same note, is a result of the relative amplitude of the different harmonics that are produced by the instrument. This different sound, even though the instruments are playing the same note, is called the **quality** or **timbre**.

Many wind instruments use a similar principle, involving longitudinal standing waves produced in a column of air. The boundary condition at each end of the air column depends on the type of reflection taking place. At a closed end the air is unable to move and there must be a node. At an open end some of the original travelling wave is transmitted out of the instrument and some is reflected back in.

Although we often say that there is an antinode at the open end, in practice, the exact position of the antinode is slightly different to the physical end of the instrument. This can be ignored in many situations but a more accurate approach would be to assign a small end correction c to the length of the instrument l and define the effective length as $(l + c)$.

Note that the nodes referred to above are all displacement node. This means that they mark the position where the displacement does not vary. Similarly, antinodes mark the positions where the displacement shows maximum variation. This is not the same as a pressure node – the point where the pressure does not vary with time.

The fundamental frequency of a pipe closed at one end is f. A pipe of the same length but open at both ends has a fundamental frequency (first harmonic) of

A $\frac{1}{2}f$

B f

C $2f$

D $4f$

Inquiry: Musical instruments from around the world

Investigate a range of different musical instruments from around the world. The different tasks below outline some possible lines of inquiry. Before starting this inquiry, decide which instruments you will investigate, their country/countries of origin and the method you will use to present your findings.

For each instrument chosen you could:

A outline the method of production of the original disturbance that creates the sound

B identify the boundary conditions that apply to the instrument

C use the physical dimensions of the instrument to estimate a typical fundamental wavelength associated within the instrument.

D use your answer to C and the speed of the wave in the instrument to estimate a typical frequency of a note played by the instrument. If the original oscillations are in a column of air then the speed of a sound wave can be taken to be approximately equal to 330 m s^{-1}. If the original oscillations are in a string under tension, then the speed of the sound waves along the string can be calculated from speed, $v = \sqrt{\frac{T}{\mu}}$ where T is tension in the string in N and μ is the mass per unit length of the string in kg m^{-1}

E estimate the range of possible frequencies achievable by the instrument.

Possible practical extensions include:

- the measurement of the fundamental frequency of an individual instrument using a microphone and cathode-ray oscilloscope
- measuring the variation in a string's frequency of vibration with tension. The equation for the speed of a wave in a string under tension (introduced in D, above) can be used to predict the expected frequency
- a digital recording of an instrument can be analysed to discover the main component harmonics involved. This involves the use of a spectrum frequency analyser (see Fourier analysis on pages 91–2).

Rediscovering physics: The speed of sound

This experiment uses a recorder to calculate the speed of sound in air. A recorder is essentially an open pipe, which, when played in the normal way, develops a standing wave between the open end where you blow and the first open hole below the mouthpiece. For this experiment, take the length of the standing wave as between the tip of the mouthpiece and the far end of the first open hole (Figure 8).

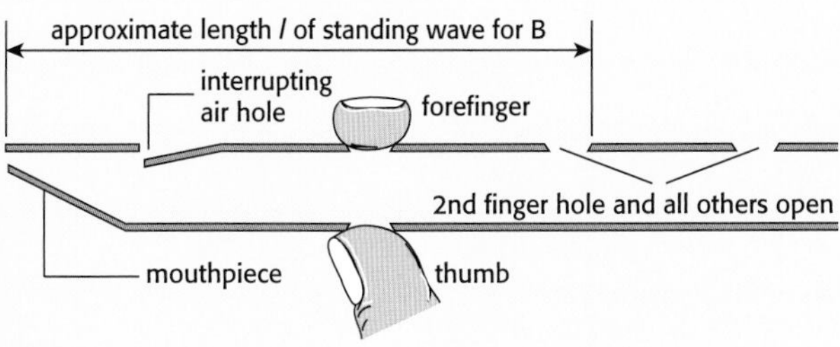

Figure 8 Cutaway side view showing how to play the note B

The wavelength $\lambda = 2l$ and thus the frequency of the note f is
given by:

$$f = \frac{v}{2l}$$

where v is the speed of sound measured in m s^{-1}.

The method involved is to take l as the independent variable, measure f (the dependent variable) and analyse the data obtained to calculate v. Controlled variables include the temperature and pressure of the air in the room.

The method of measurement of frequency will depend on the equipment available. Possibilities include:

- using a microphone and a CRO to display the waveform. An analysis of the waveform against the time-calibrated axis allows the time period and hence the frequency of the fundamental to be assessed
- the sound produced can be recorded. A spectrum analyser (available to download from the internet) can be used to identify all of the frequencies including the fundamental (the lowest frequency present with a relatively large amplitude).

Play the notes C, D, E, F, G, A, and B on a recorder. Figure 9 shows you how to do this.

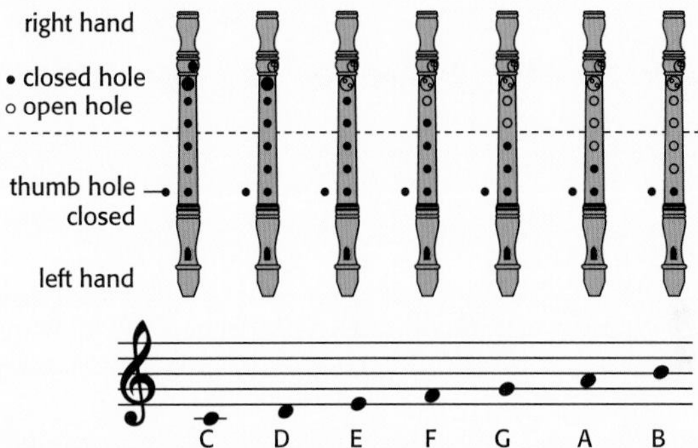

Figure 9 Notes on a recorder

For each note, identify and record the fundamental frequency. Analyse your data to estimate the speed of sound in air, taking care to use a suitable method of graphical analysis. Remember to include your uncertainties.

The bigger picture: Standing and travelling waves

The equation for a standing wave can be derived from the equation for a travelling wave.

The equation for a wave travelling in the positive x direction (to the right) is:

$$y_1 = A\sin(\omega t - kx)$$

Where:

y_1 is the displacement, measured in m

A is the amplitude, measured in m

ω is the angular frequency, measured in radians s^{-1}

t is the time, measured in s
k is a phase constant, called the wave number, measured in radians m^{-1}
x is the distance along the wave, measured in m.

The time period, $T = \frac{2\pi}{\omega}$ and the wavelength, $\lambda = \frac{2\pi}{k}$.

A wave with an equal amplitude, angular frequency, and wave number travelling to the left is represented by:

$$y_2 = A\sin(\omega t + kx)$$

The equation for the standing wave is thus:

$$y = y_1 + y_2$$
$$y = A\sin(\omega t - kx) + A\sin(\omega t + kx)$$
$$y = 2A\cos(kx)\sin(\omega t)$$

This can be analysed as follows:

$$y = (\text{amplitude term})\cdot(\text{time-dependent term})$$

The amplitude term $(2A\cos(kx))$ depends on the position along the wave. The maximum value is $2A$ and the distance between the nodes is $\frac{\lambda}{2}$, or $\frac{\pi}{k}$.

The Doppler effect

The Doppler effect is the change in the measured frequency of a wave that results from the relative motion of a source and/or an observer relative to the medium in which the wave is propagated. An example of the Doppler effect is the change in pitch that is heard as a police car sounding its siren drives past.

Two things should be noted:

- the speed of the wave through the medium is not affected but the *received* frequency is altered
- although an analogous effect can be measured for both sound and light waves, no medium is necessary for the propagation of light and the derivations below only apply to sound waves (see page 127).

In the following discussion,
the velocity of the waves with respect to the medium is v
the velocity of the observer is u_o
the velocity of the source is u_s
the frequency emitted by the source is f_s
the frequency received by the observer is f_o.

Moving observer

An observer moving towards a wave source will receive more waves per unit time when compared with a stationary observer. This is shown in Figure 10.

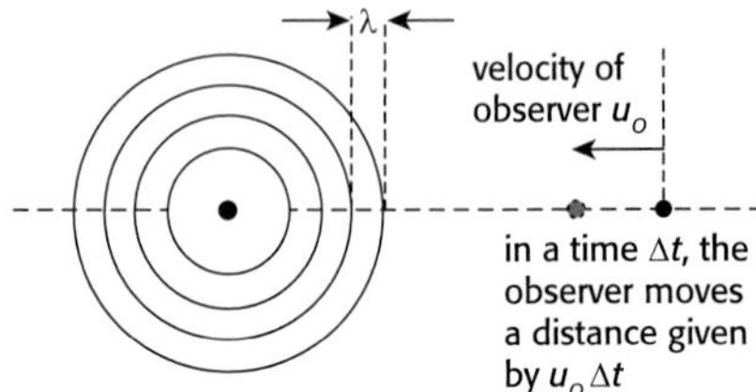

Figure 10 Doppler effect for a moving observer

In a time interval Δt,

The observed frequency $f_o = f_s + \left(\dfrac{\text{number of waves in a length } u_o\Delta t}{\Delta t}\right)$

$$f_o = f_s + \left(\frac{u_o}{\lambda}\right)$$

but $v = f_s\lambda$ so

$$f_o = f_s + \left(\frac{u_o f_s}{v}\right)$$

$$f_o = f_s\left(1 + \frac{u_o}{v}\right) = f_s\left(\frac{v + u_o}{v}\right) \quad \text{[observer moving TOWARDS source]}$$

If the observer is moving away from the source (with a velocity u_o which is less than v) then

$$f_o = f_s\left(1 - \frac{u_o}{v}\right) = f_s\left(\frac{v - u_o}{v}\right) \quad \text{[observer moving AWAY from source]}$$

Moving source

The source movement causes a change in the wave speed relative to the source. This means that the waves will "bunch" together in the forward direction and be spread apart in the reverse direction. This causes a change of wavelength for the observer.

$$f_o = f_s\left(\frac{1}{1 - \left(\frac{u_o}{v}\right)}\right) = f_s\left(\frac{v}{v - u_o}\right) \quad \text{[source moving TOWARDS observer]}$$

$$f_o = f_s\left(\frac{1}{1 + \left(\frac{u_o}{v}\right)}\right) = f_s\left(\frac{v}{v + u_o}\right) \quad \text{[source moving AWAY from observer]}$$

Overall effect

If the velocities of source or observer, v are small compared with the speed of the waves, c (i.e. $v << c$) all four equations can be reduced to one equation:

$$\frac{\Delta f}{f_s} \approx \frac{v}{c} \qquad (1)$$

where

Δf is the difference in frequency between source and observer, measured in Hz

f_s is the frequency of the source, measured in Hz

v is the relative velocity between source and observer, measured in m s^{-1}

c is the velocity of the wave, measured in m s^{-1}

A source of sound emits waves of wavelength λ, period T, and speed v when at rest. The source moving away from a stationary observer at speed V, the wavelength of the sound wave received by the observer is

A $\lambda + vT$ **C** $\lambda + VT$

B $\lambda - vT$ **D** $\lambda - VT$

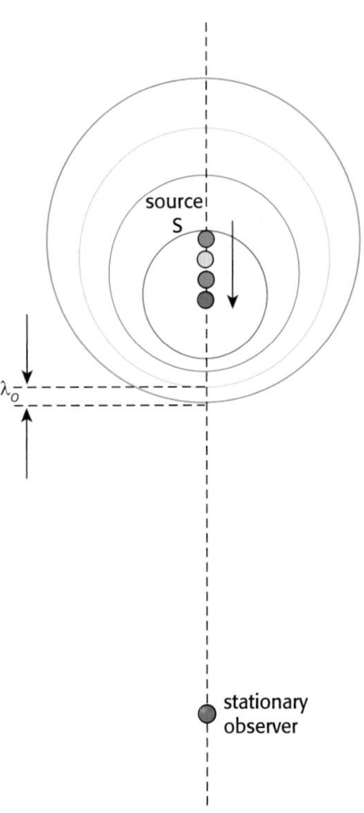

Figure 11 Doppler effect for a moving source

Analyse the situation shown in Figure 11 to derive the frequency relationships stated above for a moving source.

Show that equation 1 applies for each of the above situations providing v is small compared with c.

Thinking about science: *Comparison between sound and EM waves*

The frequency of received sound and EM radiation both depend on the relative motion of source and observer. Both are described by the Doppler effect but the underlying reasons for the change in frequency (and the associated equations) are not completely identical.

The derivations of full equations for the Doppler effect involve using a value for the speed of the wave relative to the speed of the medium *c*. Figure 11, for example, shows waves bunching together because the speed of the wave in the medium is fixed and the source is able to "catch up" with some of the waves that have already been emitted. This means that the relative speed of the waves as measured by the source would be reduced.

This derivation cannot apply to electromagnetic waves, however, as the waves do not travel through a medium. Indeed, a central assumption (**postulate**) of Einstein's theory of special relativity is that all inertial observers will measure exactly the same value for the speed of electromagnetic waves in a vacuum irrespective of their relative motions.

The full analysis for EM radiation results in an equation which predicts the change of frequency when there is a relative motion between source and observer. If the source is moving away from the observer the received radiation will be shifted to a lower frequency, a shift towards the red end of the spectrum, which is called a **redshift**. If the source and observer are moving towards one another, a blueshift is observed. If the relative motion is small compared with the speed of the EM waves in a vacuum, this formula reduces to the same approximation as in the case of sound waves:

$$\frac{\Delta f}{f_s} \approx \frac{v}{c}$$

Einstein's general theory of relativity predicts that gravitational fields will also affect the frequency of electromagnetic waves. When EM waves travel away from a massive object their frequency decreases in an effect called a **gravitational redshift**.

Physics issues: examples of the Doppler effect

We will now highlight some examples and uses of the Doppler effect. It might be useful to research some of them further.

Train going through a station

The sound emitted by a moving train's whistle is of constant frequency, but the sound received by a passenger standing on the platform will change. At any instant of time, it is the resolved component of the train's velocity towards the passenger that should be used to calculate the frequency received.

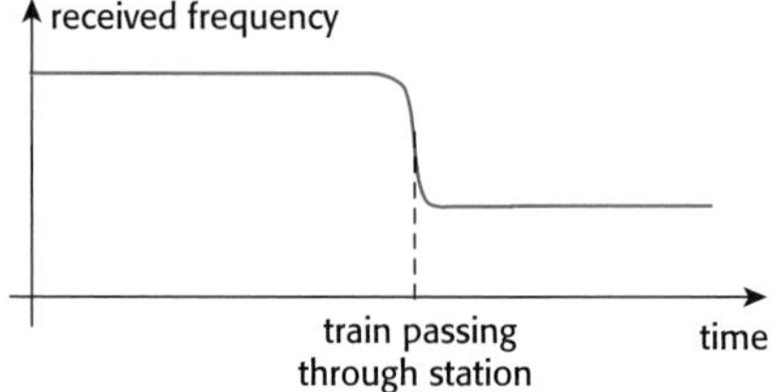

Figure 12 A passenger on a platform hears a change in pitch

Speed measurement for vehicles

In many countries the police use radar to measure the speed of moving vehicles to see if they are breaking the speed limit. A pulse of radio waves of known frequency is emitted, reflected off the moving car and received back at the source. The change between the frequency emitted and the frequency received is used to calculate the speed of the car.

There is a double Doppler effect in this situation. The frequency received by moving car will be higher than the emitted source (if the car is moving towards the radar) and the car also acts as a moving source when the signal is sent back to the receiver.

Figure 13 Policeman using radar to detect vehicle speed

Blood-flow measurements

Doctors use an analogous setup to the speed measurement for vehicles in order to measure the blood flow rate around a body and identify circulation problems.

An ultrasound pulse is arranged to reflect off blood flowing around the body. The frequency that is received can be analysed to calculate the speed of blood flow. Measurements of the dimensions of the vessels are then used to directly calculate the volume flow rate.

Redshift

The light from the vast majority of stars is redshifted. When the light is analysed the frequencies represent the absorption spectrum (see page 238) of the elements contained in the star's outer layers. The measured frequencies are, however, not the same as the frequencies associated with particular elements as measured in the laboratory.

For a given star, all the received frequencies have been shifted by the same amount towards the red end of the visible spectrum, i.e. to lower frequencies. Different stars have different redshifts and the magnitude of the shift can be used to calculate the star's recession velocity. This provides evidence for the big bang model of the creation of the universe.

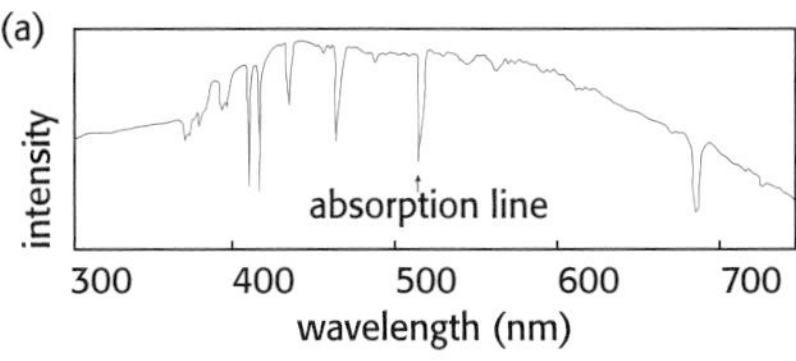

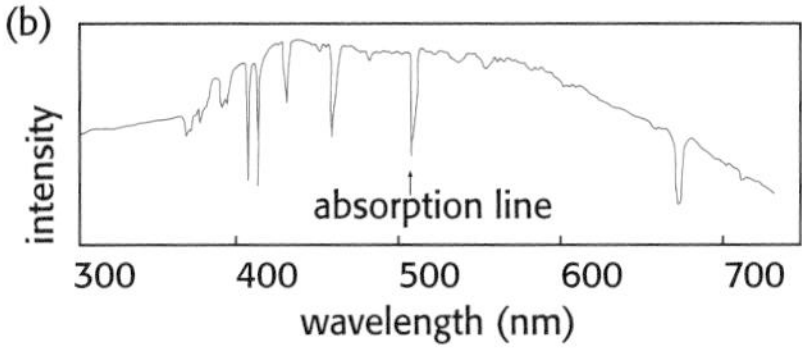

Figure 14 (a) Spectrum of a star moving away from Earth (red shifted); (b) Spectrum of an identical star stationary relative to Earth

Rotating object

When a luminous object rotates, one side will be moving towards the observer and one side will be moving away. For example, the Sun is rotating on its own axis. When an observer looks at the Sun, (it is not safe to do this without special glasses so do not attempt this) the left-hand side is moving towards the observer and the right-hand side is moving away. This means that the absorption spectra for the left-hand side will be blue shifted and the right-hand side will be red shifted. This can be used to find the speed of the rotation of the star.

The same splitting of spectral lines can be used to identify binary stars where only one source of light can be seen. See Chapter 27.

Take measurements from Figure 14 to determine the recessional velocity of the star moving away from Earth.

Broadening of spectral lines

Absorption and emission spectra provide evidence for discrete atomic energy levels. Precise measurements show that each individual level is actually equivalent to a small but defined wavelength range.

The gas molecules that are emitting the light have different velocity components, when resolved along the line of observation. Light from different molecules will be subjected to Doppler shifts so there will be a general "Doppler broadening" of the discrete wavelengths – any given spectral line will broaden into a range of measured wavelengths. In any given spectrum, this amount of broadening must depend on the temperature of the substance emitting the light. A higher temperature means that the molecules have a wider distribution of kinetic energy and hence the spectral line is broader.

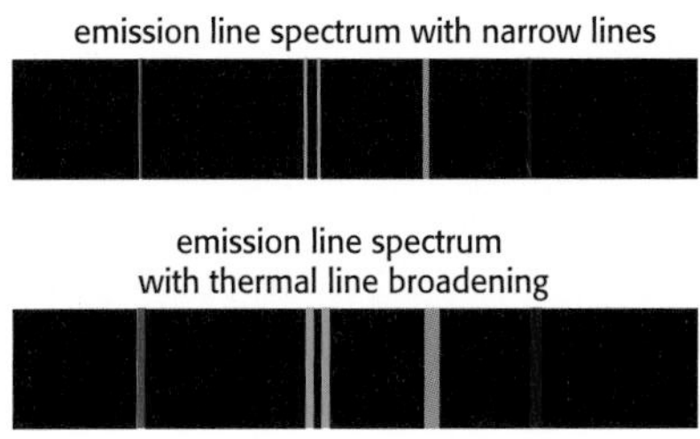

Figure 15 Doppler broadening

Diffraction

In Chapter 8 we introduced the concept of diffraction, that is, any wave that is constrained to pass through a narrow aperture will spread out. The extent of the spreading depends on the size of the aperture compared with the wavelength involved. An extremely narrow aperture behaves like a point source, with equal intensities being transmitted in all directions whereas a wide slit can result in no effective diffraction at all.

The mathematics of diffraction can be very complex and depends on the physical setup used. It is important to distinguish between **Fresnel** and **Fraunhofer** diffraction. Fresnel diffraction involves diverging rays (spherical wavefronts) arriving at the aperture whereas Fraunhofer diffraction involves parallel waves (plane wavefronts). Analysis of the former (Fresnel) is complex whereas the later (Fraunhofer) is more straightforward. Unfortunately, simple experimental setups often involve Fresnel diffraction.

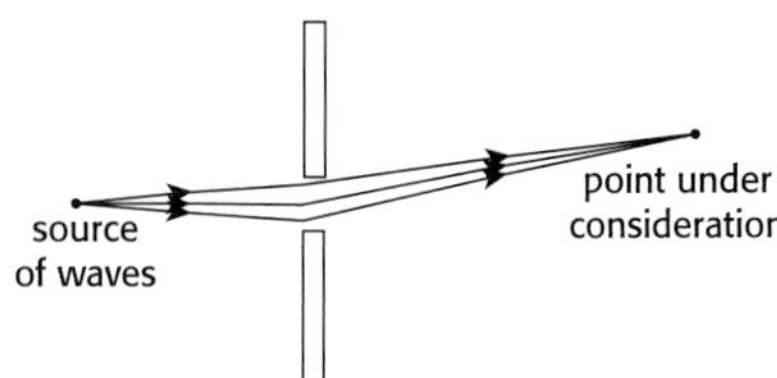

Figure 16a Fresnel diffraction

A wavefront may be considered to be constructed from an infinite number of point sources. The position of the next wavefront can be determined by considering the waves produced by each point source within the previous wavefront. This idea is known as **Huygens principle**. The overall diffraction pattern at an aperture of a given width b is the result of superposition of all the individual point sources in the wavefront across the width of the slit.

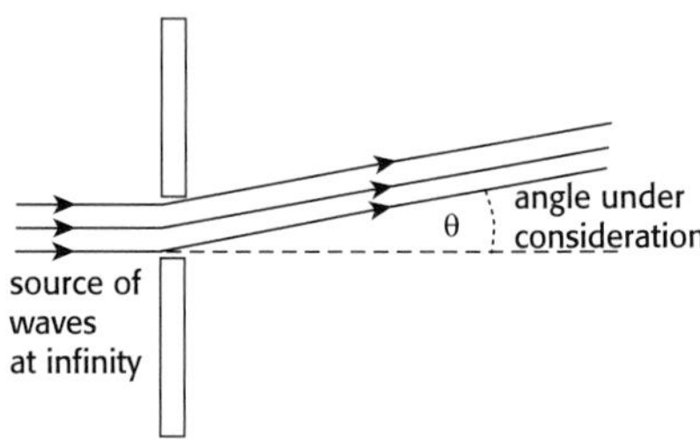

Figure 16b Fraunhofer diffraction

In order to calculate the variation with angle of the intensity of wave energy received after passing through an aperture, it is useful to consider the overall path difference across the slit (Figure 18). As the angle under consideration gets bigger, the path difference between the point sources across the aperture increases.

Consider a slit of width b.

$$\sin\theta = \frac{\text{path difference across aperture}}{\text{slit width, } b}$$

At a certain angle, the path difference across the slit will equal one whole wavelength. Waves from the point sources in one half section of the slit will all cancel with the corresponding point sources in the other half.

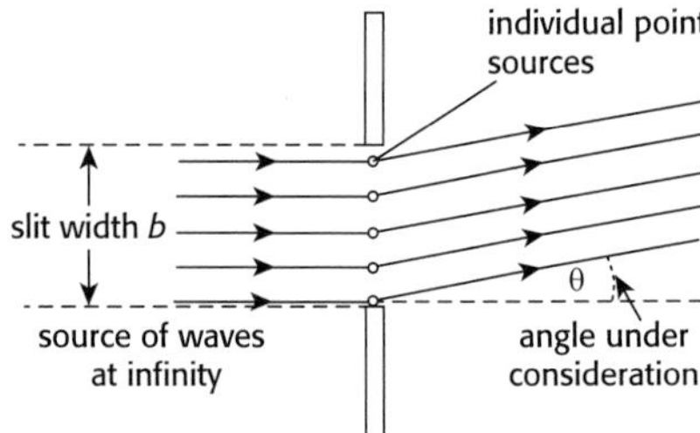

Figure 17 A single slit can be considered to contain many individual point sources

The result will be the first minimum. This takes place when

$$\text{path difference} = b\sin\theta = \lambda$$

$$\sin\theta = \frac{\lambda}{b}$$

Provided θ is small, we can apply the small angle approximation.

$$\sin\theta \approx \theta = \frac{\lambda}{b}$$

θ is the angle between the straight through direction and the first minimum of the diffraction pattern measured in radians
λ is the wavelength, measured in m
b is the slit width, measured in m.

At some larger angles, the path difference across the slit will equal one and a half wavelengths. Waves from the point sources in two of the sections will cancel, leaving the amplitude reduced to a third of the maximum. This is the first maximum. The intensity at this angle (which is proportional to the square of the amplitude) will be one ninth of the initial value.

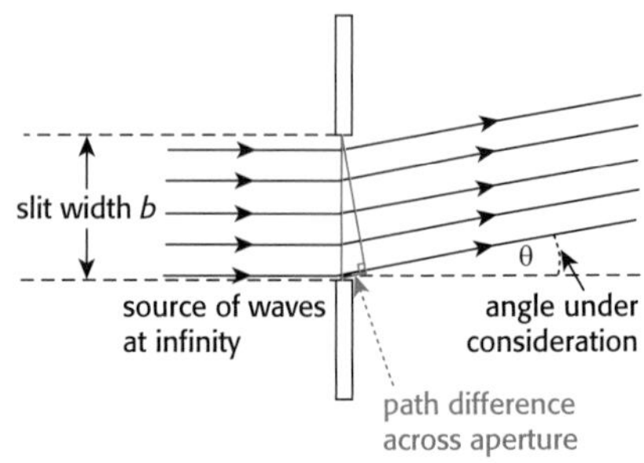

Figure 18 The path difference between individual point sources in a single slit

The next minimum occurs at $\theta = \frac{2\lambda}{b}$ and the next maximum is $\frac{1}{25}$ of the intensity of the initial value. The resulting diffraction pattern is shown in Figure 19:

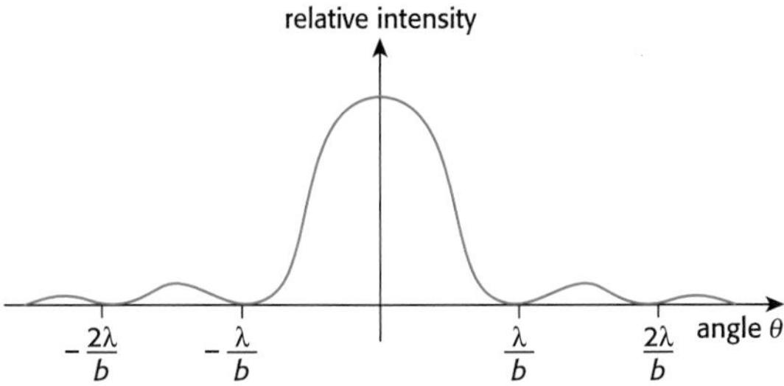

Figure 19 Single slit diffraction

Figure 20 below shows light from a laser (Helium-Neon) that has passed through a single slit of width $\frac{1}{16}$ mm and then projected onto a plastic ruler 3.0 m away from the slit. By taking measurements from the photograph, calculate the wavelength of laser light used. Estimate the uncertainty in your value.

Figure 20

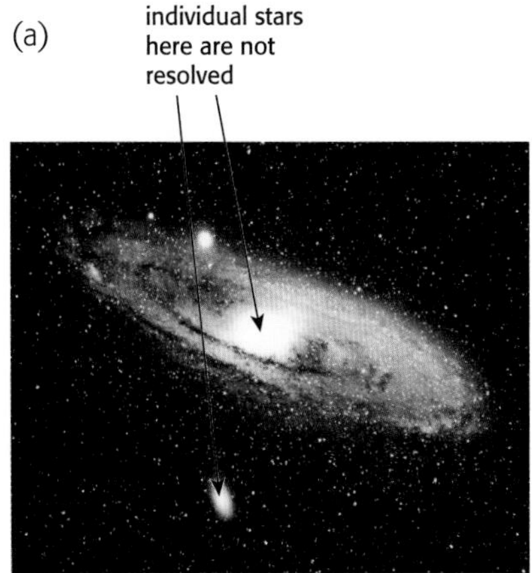

Figure 21a The Andromeda Galaxy

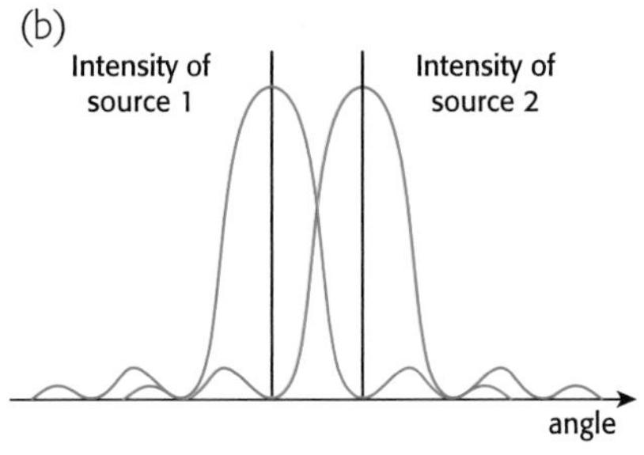

Figure 21b Rayleigh's criterion

Rayleigh criterion

When two sources have a small angular separation, it may be difficult to resolve them as two separate objects, in which case they appear as a single source. For example, many of the "individual" stars that you can see at night actually turn out to be two stars in close orbit around one another (a binary star system). On top of this, a significant number of the "stars" are, in fact, galaxies (large collections of stars). In both cases your eyes are unable to resolve the individual sources because the angular separation between them is too small.

When light from two different sources is brought together, the principle of superposition determines the final image. When viewing two sources through an aperture, such as stars through a telescope, the overall result is the addition of individual diffraction patterns. **Rayleigh's criterion** (which can be practically verified) is the suggestion that:

> *Two sources are just distinguishable when the first minimum of the diffraction pattern of one of the sources falls on the central maximum of the diffraction pattern of the other source.*

The angle at which two sources can just be resolved is fixed by the wavelength being used λ and the size of the aperture involved. If the aperture is a slit of width b then the minimum angle must be $\frac{\lambda}{b}$, as shown on page 129. More typically, the aperture is circular in which case the minimum angle is $1.22\frac{\lambda}{b}$, where 1.22 is a correcting factor for the slightly different geometry. This equation applies equally to all waves. For example, visible light is diffracted when it passes through the circular aperture of a telescope or even the pupil of your eye and radio waves are also diffracted when collected by a radio telescope. Increasing the collecting diameter of a telescope will decrease the angle between two sources that are just resolved.

1 A radio telescope is receiving radio waves of frequency 5.0 GHz and its dish size is 76 m. Two stars that are 100 light years away from Earth are known to be 0.05 lights year apart.

a) Will the radio telescope be able to resolve these two sources?

b) Two separate identical radio telescopes are situated 200 km apart. The signals that they receive can be combined to effectively create a large radio telescope that is 200 km in diameter. Will these be able to resolve the two stars?

2 Figure 23 shows laser light that has passed through a circular hole of width 0.24 mm and then projected onto a plastic ruler 3.0 m away from the slit. By taking measurements from the photograph, calculate the wavelength of laser light used. Estimate the uncertainty in your value.

3 Research one situation or device in which a resolution limit results from the diffraction of waves through an aperture and decide how significant this factor is compared with other factors. The particular example and the method of presenting the results of your research are up to you. In each case:

- outline the situation and explain where diffraction takes place
- state the wavelength or range of wavelengths involved
- state the dimensions of the aperture involved
- calculate the resolution limit according to the Rayleigh criterion
- state the practical resolution limit.

Situations that could be considered include:

a) the human eye

b) an animal's eye

c) a camera

d) a land-based astronomical telescope

e) the Hubble space telescope

f) an array of radio telescopes

g) a visible light microscope

h) an electron microscope.

Figure 22 A radio telescope

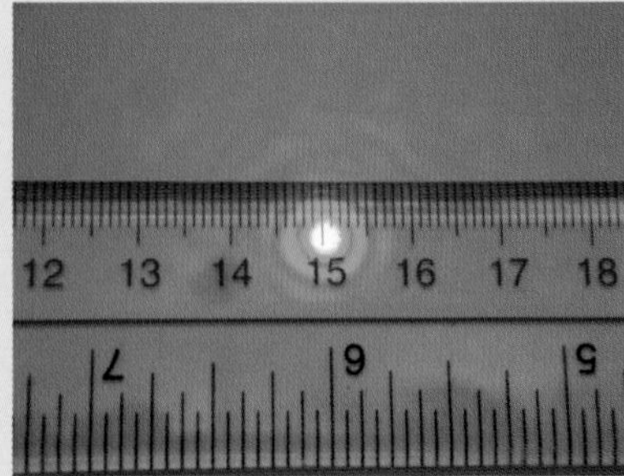

Figure 23

Polarization

Polarization is a property that only applies to transverse waves. Unlike longitudinal waves, where the direction of the oscillations is fixed by the direction of energy transfer, there are an infinite number of possible directions for the oscillations of transverse waves (all at 90° to the energy transfer direction). A wave that is plane polarized is one that only has one direction of oscillation.

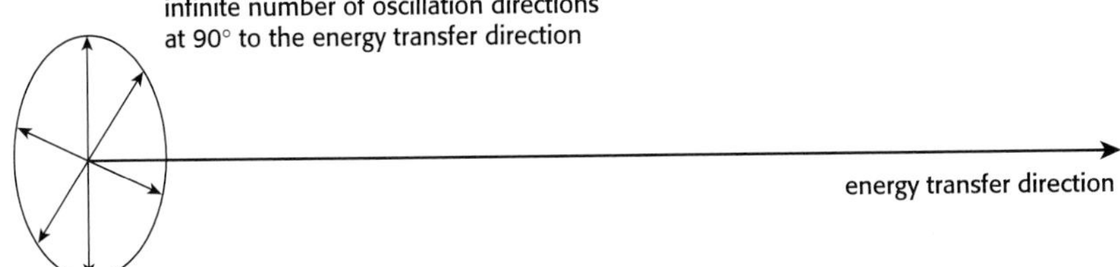

Figure 24 This wave is not polarized

In electromagnetic waves, for example, the plane of vibration is defined as the plane that contains the electric field. Some EM radiation is polarized as a result of its method of production. For example, the radio waves used in the broadcasting of terrestrial television signals are often polarized either horizontally or vertically. The plane of polarization is fixed by the chosen arrangement of the broadcasting aerials.

On the other hand, light from a standard light bulb is not polarized. This means that the oscillations of the electric field associated with the light take place in all directions. It is also possible for light to be partially polarized, which means that the electric field oscillations are more common in one direction than in all other directions. One final possibility is to add together two plane-polarized waves to create a wave whose plane of polarization varies. Depending on the phase of the two waves this is said to be **elliptically** or **circularly polarized**.

The simplest method of polarizing light is to use a sample of a substance called **Polaroid**. This man-made material consists of long-chain molecules that are aligned together. The long chains absorb light that has its electric field in the same direction as the chains and only allows through light with an electrical field at 90° to the chains:

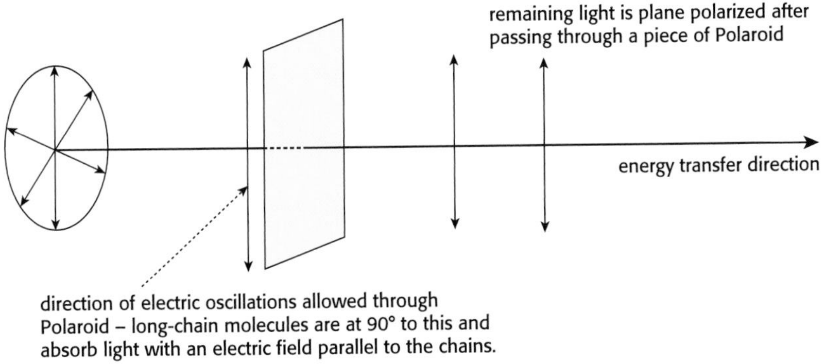

Figure 25 Polarizing a wave

Polaroid absorbs light and can be used to manufacture dark glasses. The advantages of using Polaroid are discussed on pages 134–5.

Explain why "crossed" Polaroids (two samples of Polaroid at 90° to one another) absorb all the light in the region of their overlap whereas they transmit some light when parallel, as represented in Figure 26:

a)

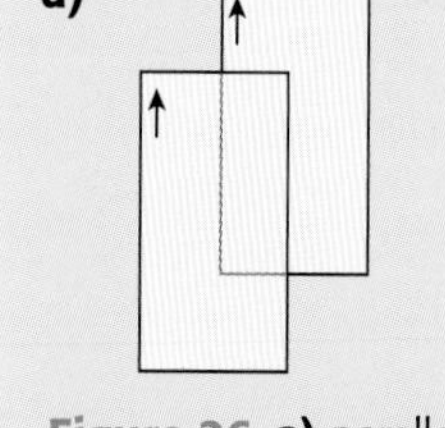

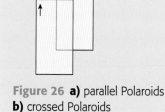

b)

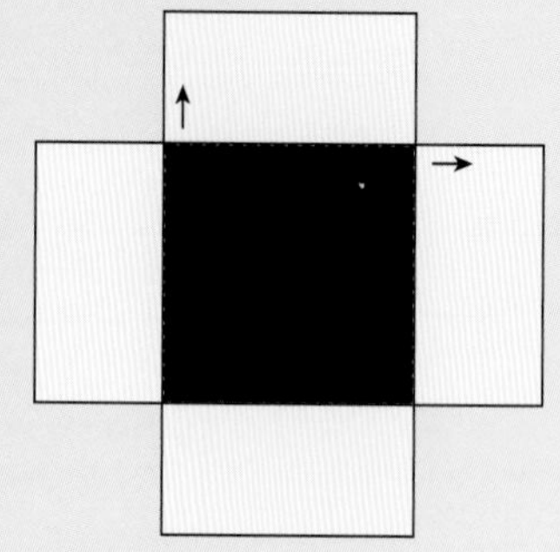

Figure 26 **a)** parallel Polaroids **b)** crossed Polaroids

Some materials (e.g. sugar solutions) are **optically active**. This means that the plane of polarized light is rotated as it passes through the material. This rotation is a result of the interaction between the molecules in the material and the incident light. The degree of rotation for a given substance is measured using two polarizers (one is called an **analyser**) as shown in Figure 27.

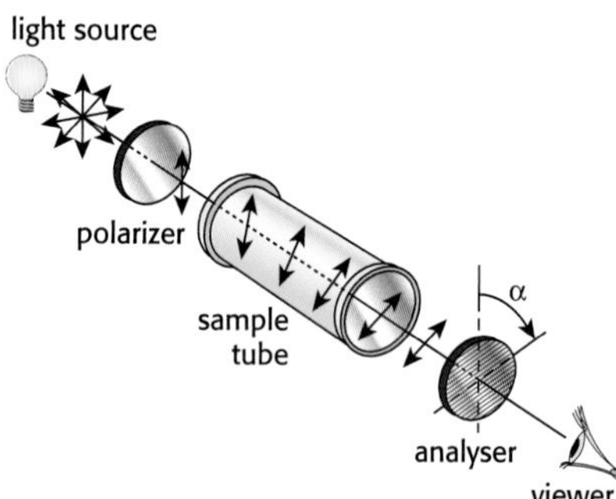

Figure 27 Measuring the optical activity of a sample

The polarizer and analyser are initially aligned without the sample tube being present. The introduction of the optically active sample causes a rotation of the plane of polarization of the light. The analyser is rotated through an angle α to find the maximum intensity of transmitted light. This apparatus is known as a **polarimeter**.

The angle through which the plane of polarization of the light is rotated depends on the length of the sample and its concentration. An example of a polarimeter can be easily constructed using a light source, a beaker, and two pieces of Polaroid. Once built, you can test it using sugar solutions of different concentrations.

1. Different lengths of a particular concentration of sugar solution were analysed in a polarimeter. Table 2 shows the variation with solution length of the angle of rotation of the plane of polarization.
   a) Hypothesize the relationship between angle of rotation and solution length.
   b) Plot a suitable graph to test your hypothesis.
2. Different concentrations of sugar solution were analysed in a polarimeter. Table 3 shows the variation with concentration of the angle of rotation of the plane of polarization. The lengths of each sample were the same.
   a) Hypothesize the relationship between angle of rotation and solution concentration.
   b) Plot a suitable graph to test your hypothesis.

Table 2

Angle of rotation/°	Length of sugar solution/cm
7	2.0
14	4.0
20	6.0
27	8.0
34	10.0

Table 3

Angle of rotation/°	concentration of sugar solution/g cm^{-3}
17	0.25
33	0.50
50	0.75

Mathematical physics: Polarization calculations

There are two mathematical laws associated with polarization – Malus's law and Brewster's law.

Malus's law

The intensity of plane-polarized EM radiation from the analyser depends on the intensity of the plane-polarized light I_0 and the angle between the direction allowed by the analyser and the electric field direction of the polarized light, θ. The analyser only allows one component direction of the electric field through.

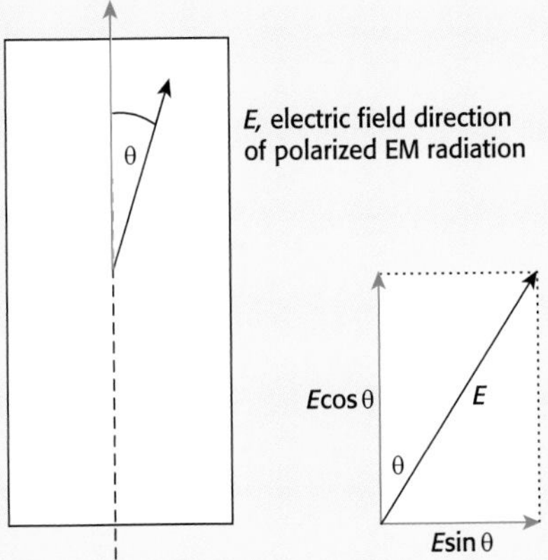

Figure 28 Geometry of transmitted component

The transmitted amplitude of the electric field is $E \cos \theta$. Since the intensity is proportional to the square of the amplitude, the transmitted intensity I is given by:

$$I = I_0 \cos^2 \theta$$

Brewster's law

When light strikes the surface between two media, some light energy is reflected and some is refracted. The reflected light is predominately plane polarized in a direction that is parallel to the surface. If the transmitted ray and reflected ray are at right angles to one another, then the reflected ray is totally plane polarized, as represented below. The angle of incidence for this condition is called **Brewster's angle**.

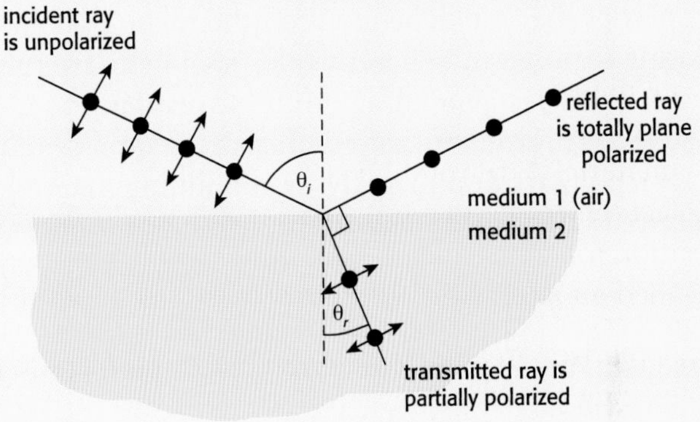

represents electric field oscillation into the paper

represents electric field oscillation in the plane of the paper

Figure 29 Polarization upon reflection

$$\theta_i + \theta_r = 90°$$

refractive index of medium 2 is given by:

$$n = \frac{\sin\theta_i}{\sin\theta_r} = \frac{\sin\theta_i}{\cos\theta_i} = \tan\theta_i$$

A bright source of light is viewed through two sheets of Polaroid whose preferred directions are initially parallel.

a) Calculate the angle through which one sheet needs to be turned in order to reduce the amplitude of the observed electric field vibrations to half its original value.

b) Calculate the effect this has on the transmitted intensity.

c) Calculate the angle through which one sheet needs to be turned to reduce the transmitted intensity to half its original value.

Physics issues: examples of polarization

Polaroid dark glasses

Reflections from all surfaces are predominately polarized in a plane that is parallel to the reflecting surface. In normal use (i.e. a person standing looking at a surface), Polaroid dark glasses are arranged to only transmit light that has a vertical electric field. This means that the intensity of all reflections from horizontal surfaces will be greatly reduced and thus the reflected light, or "glare" will be reduced.

Concentration of certain solutions

An optically active solution such as a sucrose solution will rotate the plane of polarization of light by an angle that is proportional (within limits) to the concentration of the solution. A polarizer and analyser can be used to measure the angle of rotation and, with calibration, the concentration of the solution.

Stress analysis

When subjected to stress, glass (and some plastics) develops optical properties that are dependent on the plane of polarization of the light passing through it. Different colours of light are refracted through different angles depending on the stresses imposed upon the material. This means that when a stressed piece of glass is placed between polarizer and analyser and illuminated by white light, coloured lines will appear in the regions of highest stress.

LCDs

Liquid crystal displays are in extremely common use (e.g. calculator screens). The diagram in Figure 30 represents one possible arrangement.

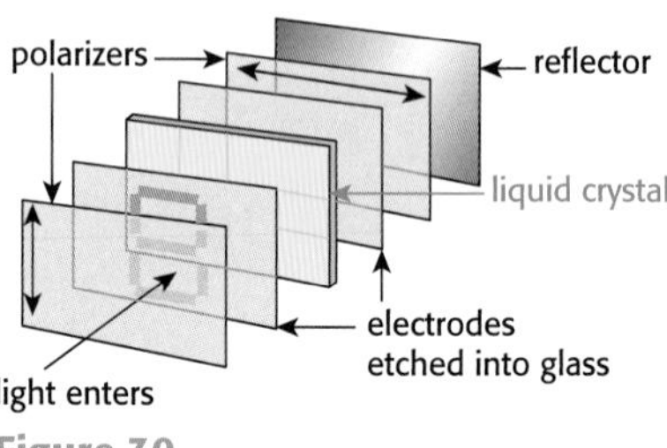

Figure 30

The liquid crystal is sandwiched between two electrodes, which are in turn between two crossed polarizers. Light enters from the front and any light that reaches the reflector at the back is returned to the observer. In the absence of any liquid crystal the whole screen would be dark, but the liquid crystal is optically active and arranged to rotate the plane of polarization of the light by 90°.

When the p.d. between the two electrodes is zero, much of the light entering the front of the display will be returned back and thus the display will appear light. When a p.d. is placed across regions of the liquid crystal, the plane of polarization of the crystal is changed and so the amount of light returned to the observer in these regions will be reduced. It is possible for the plane of polarizations to be such that the region appears black.

> ? Explain what is seen when a calculator display is viewed through a piece of Polaroid that is rotated through a complete circle.

End of chapter summary

Chapter 9 has five main themes: standing (stationary) waves, the Doppler effect, diffraction, resolution and polarization. The list below summarises the knowledge and skills that you should be able to undertake after having studied this chapter. SL candidates studying option A (Sight and Wave phenomena) need study all the themes in this chapter. Further research into more detailed explanations using other resources and/or further practice at solving problems in all these topics is recommended – particularly the items in bold.

Standing (stationary) waves

- Describe the nature of standing (stationary) waves and explain their formation in one dimension.
- **Discuss the modes of vibration of strings and air in open and in closed pipes.**
- Compare standing waves and travelling waves.
- **Solve problems involving standing waves.**

Doppler effect

- Describe what is meant by the Doppler effect and explain it by reference to wavefront diagrams for moving-detector and moving-source situations.

- **Solve problems on the Doppler effect for sound by applying the appropriate equations.**
- Solve problems on the Doppler effect for electromagnetic waves by using the appropriate approximation.
- Outline an example in which the Doppler effect is used to measure speed.

Diffraction

- Sketch the variation with angle of diffraction of the relative intensity of light diffracted at a single slit and derive the formula for the position of the first minimum of the diffraction pattern.
- Solve problems involving single-slit diffraction including sketching the variation with angle of diffraction of the relative intensity of light emitted by two point sources that has been diffracted at a single slit.

Resolution

- State the Rayleigh criterion for images of two sources to be just resolved.
- Describe the significance of resolution in the development of devices such as CDs and DVDs, **the electron microscope** and radio telescopes **and solve problems**.

Polarization

- Describe what is meant by polarized light and **how polarization by reflection takes place**.
- State and apply Brewster's law.
- Explain the terms *polarizer* and *analyser* and calculate the intensity of a transmitted beam of polarized light using Malus' law.
- Describe what is meant by an optically active substance and the use of polarization in the determination of the concentration of certain solutions.
- Outline qualitatively how polarization may be used in stress analysis and the action of liquid-crystal displays (LCDs) and **solve problems**.

Chapter 9 questions

1 This question is about standing waves in pipes. The diagram below shows two pipes of the same length. Pipe A is open at both ends and pipe B is closed at one end.

Figure 31

a) i) On the diagrams above, draw lines to represent the waveforms of the fundamental (first harmonic) resonant note for each pipe. [2]

ii) On each diagram, label the position of the nodes with the letter N and the position of the antinodes with the letter A. [2]

b) The frequency of the fundamental note for pipe A is 512 Hz.

i) Calculate the length of the pipe A. (Speed of sound in air = 325 m s^{-1}) [3]

ii) Suggest why organ pipes designed to emit low frequency fundamental notes (*e.g.* frequency ≈ 32 Hz) are often closed at one end. [2]

(Total 9 marks)

2 This question is about the Doppler effect. Figure 32 shows wavefronts produced by a stationary wave source S. The spacing of the wavefronts is equal to the wavelength of the waves. The wavefronts travel with speed *V*.

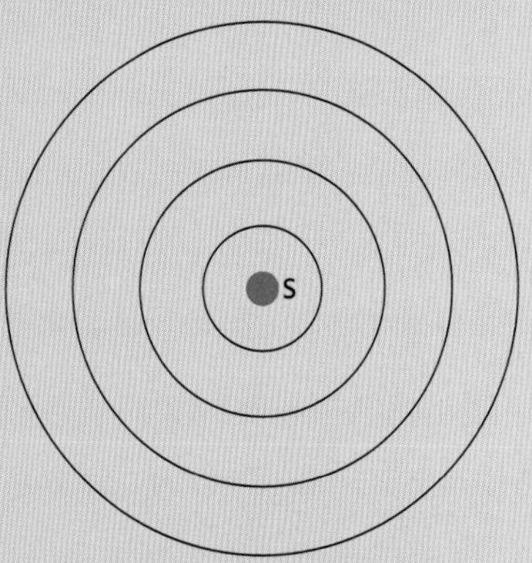

Figure 32 Waves produced by a stationary source, S

a) The source S now moves to the right with speed $\frac{1}{2}V$. Draw **four** successive wavefronts to show the pattern of waves produced by the moving source. [3]

The Sun rotates about its centre. The light from one edge of the Sun, as seen by a stationary observer, shows a Doppler shift of 0.004 nm for light of wavelength 600.000 nm.

b) Assuming that the Doppler formula for sound may be used for light, estimate the linear speed of a point on the surface of the Sun due to its rotation. [3]

(Total 9 marks)

3 This question is about single slit diffraction.

Figure 33 shows an experimental arrangement for observing Fraunhofer diffraction by a single slit. After passing through the convex lens L_1, monochromatic light from a point source P is incident on a narrow, rectangular single slit. After passing through the slit the light is brought to a focus on the screen by the lens L_2. The point source P is at the focal point of the lens L_1.

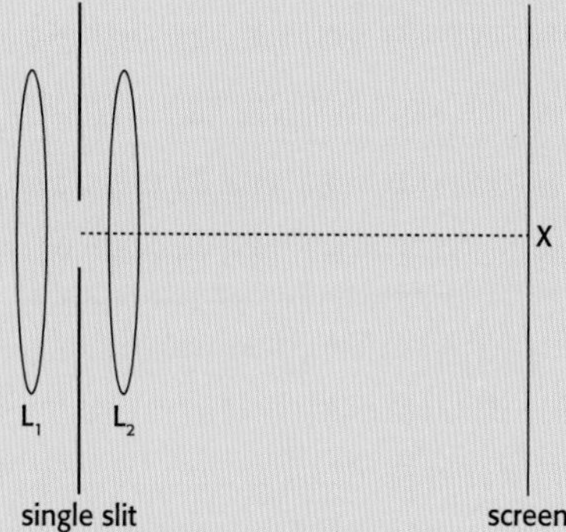

Figure 33 Frauhoffer diffraction

The point X on the screen is directly opposite the central point of the slit.

a) Explain qualitatively how Huygens' principle accounts for the phenomenon of single slit diffraction. [2]

b) Draw a graph to show how the intensity of the pattern varies with distance along the screen. The point X on the screen is shown as a reference point. *(This only needs to be a sketch graph; you do not need to include any numerical values.)*

c) In this experiment the light has a wavelength of 500 nm and the width of the central maximum of intensity on the screen is 10.0 mm. When light of unknown wavelength λ is used, the width of the central maximum of intensity is 13.0 mm. Determine the value of λ. [2]

The lens L_1 is now removed and another point source Q emitting light of the same wavelength as P (500 nm) is placed 5.0 mm from P and the two sources are arranged as shown below.

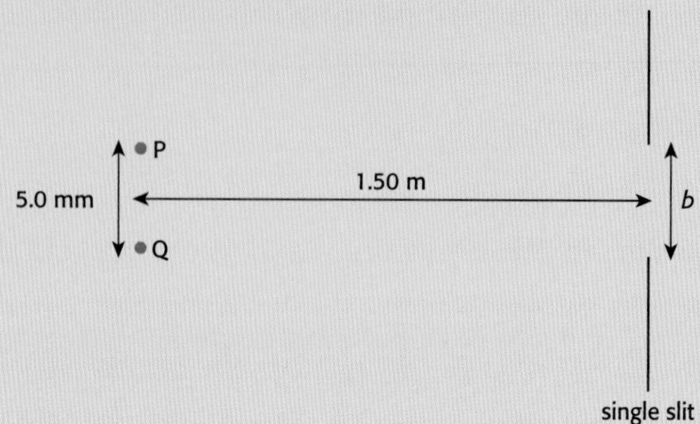

Figure 34

The distance between the sources and the slit is 1.50 m.

d) i) State the condition for the image of P and the image of Q formed on the screen to be just resolved. [1]

ii) Determine the minimum width b of the slit for the two images to be just resolved. [2]

(Total 10 marks)

4 This question is about diffraction at a single slit.

Plane wavefronts of monochromatic light are incident on a narrow, rectangular slit whose width b is comparable to the wavelength λ of the light. After passing through the slit, the light is brought to a focus on a screen.

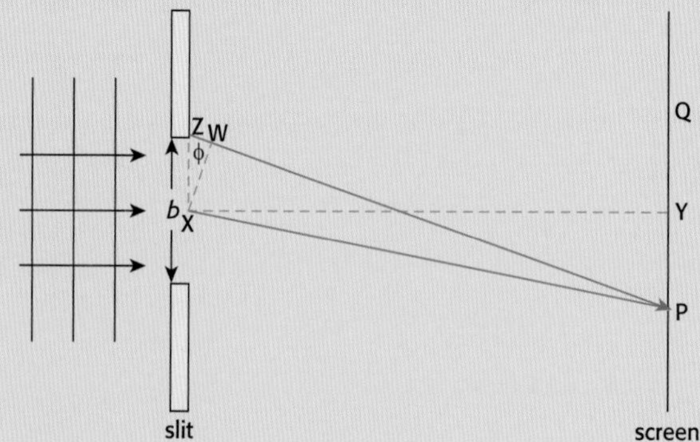

Figure 35

The line XY, normal to the plane of the slit, is drawn from the centre of the slit to the screen and the points P and Q are the first points of minimum intensity as measured from point Y.

Figure 35 also shows two rays of light incident on the screen at point P. Ray ZP leaves one edge of the slit and ray XP leaves the centre of the slit.

The angle ϕ is small.

a) On Figure 35, label the half angular width θ of the central maximum of the diffraction pattern. [1]

b) State and explain an expression, in terms of λ, for the path difference ZW between the rays ZP and XP. [2]

(Total 3 marks)

5 This question is about optical resolution.

The two point sources shown in Figure 36 (not to scale) emit light of the same frequency. The light is incident on a rectangular, narrow slit and after passing through the slit, is brought to a focus on the screen.

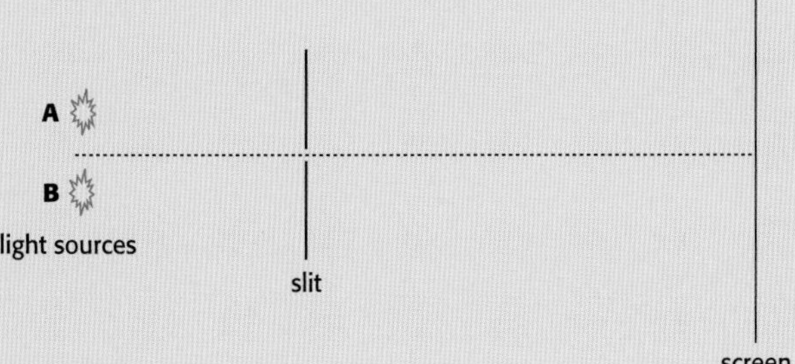

Figure 36

Source B is covered.

a) Draw a sketch graph to show how the intensity I of the light from A varies with distance along the screen. Label the curve you have drawn A.

Source B is now uncovered. The images of A and B on the screen are just resolved.

b) Using the same axes as in **a)**, draw a sketch graph to show how the intensity I of the light from B varies with distance along the screen. Label this curve B. [1]

The bright star Sirius A is accompanied by a much fainter star, Sirius B. The mean distance of the stars from Earth is 8.1×10^{16} m. Under ideal atmospheric conditions, a telescope with an objective lens of diameter 25 cm can just resolve the stars as two separate images.

c) Assuming that the average wavelength emitted by the stars is 500 nm, estimate the apparent, linear separation of the two stars. [3]

(Total 6 marks)

6 An unpolarized parallel beam of monochromatic light is incident on a plane air-glass interface of refractive index 1.50. Calculate the angle of incidence such that the refracted ray is perpendicular to the refected ray.

(Total 2 marks)

7 A ray of plane polarized light of intensity 9.6 W m^{-2} is normally incident on a polarizing filter. The intensity of the transmitted light is 2.4 W m^{-2}.

a) Explain the difference between plane polarized light and unpolarized light. [2]

b) Calculate the angle between the plane of polarization of the light and the preferred plane for the filter. [2]

c) The polarizing filter is rotated around an axis parallel to the ray so that the ray always remains normal to the polarizing filter. State the intensity of the transmitted light when the filter has been rotated from by

i) 90° from its original position. [1]

ii) 180° from its original position. [1]

iii) 270° from its original position. [1]

(Total 7 marks)

10 Thermal physics

Before studying this chapter it would be worth reviewing the different types of energy, and the principles that must apply during any energy conversion (see Chapter 5).

As discussed on page 61, we need to be careful with the word "heat". Heat is best thought of as a process by which energy is transferred, rather than as something physical that flows during the process. Rather than say "heat is lost to the surroundings", it would be better to say that "thermal energy is lost to the surroundings" or "the surroundings have been heated".

If we heat a substance (that is, give it thermal energy) we know that one of two things will happen. Either its temperature will increase (and it might expand), or it will change state (for example, it may melt and go from solid into liquid, which can also be an expansion). Over an extended period of time, both may occur. This chapter begins by considering what is happening when the temperature of an object changes, and then goes on to consider changes of state.

In both cases it is useful to consider the situation from two different points of view: the **macroscopic** (i.e. the changes that can be observed and measured in the laboratory) and the **microscopic** (i.e. the changes that are taking place at the atomic level).

Macroscopic and microscopic: temperature change

If we analyse the energy changes and transfers that take place at the macroscopic scale, some forms of energy can be easily identified (chemical energy or gravitational potential energy, for example); others can be classified as **internal energy** (thermal energy, for example). At the microscopic scale, a substance is composed of particles (atoms and molecules), and its internal energy is the total potential energy and random kinetic energy of these particles. The random kinetic energy of the particles is related to the temperature of the substance.

Macroscopic description

The terms "heat" and "temperature" do not mean the same thing. It is important to use technical language in a precise way. When we heat an object this means that we give it thermal energy (measured in joules). The result of this process is that the object warms up, and a property of the object, called its temperature, increases. Heat and temperature are related, however. The amount by which an object's temperature changes depends on the amount of thermal energy transferred.

Temperature is a scalar quantity that measures how hot something is. The measurement of temperature is done using a predefined scale. Common temperature scales include:

- the Celsius scale, measured in degrees Celsius (°C)
- the Fahrenheit scale, measured in degrees Fahrenheit (°F)
- the thermodynamic scale, measured in kelvin (K).

The zero points for the various scales are set at different temperatures. The sizes of the degrees on the Celsius and Fahrenheit scales are different, but to convert from a temperature T on the thermodynamic scale to one on the Celsius scale (θ), all we need do is subtract 273.15:

$$\theta/°\text{C} = (T - 273.15)/\text{K}$$

An object's temperature is used to predict the direction in which thermal energy naturally flows – that is, from hot to cold. When two objects are put in thermal contact, thermal energy will flow from the object with the higher temperature to the object with the lower temperature.

It can be very difficult to control the experimental uncertainties associated with experiments that involve heat and temperature. This is because any hot object will tend to lose thermal energy to its surroundings and cool down. Conversely, any cold object will tend to receive thermal energy from the surroundings and heat up.

Careful experiments show that, provided a state change does not take place, the temperature change of an object is related to the quantity of thermal energy transferred. For example, a sample of water requires 4200 J for its temperature to change from 22 °C to 23 °C with no loss of thermal energy. It will need the same amount of energy to change from, for example, 57 °C to 58 °C.

The amount of thermal energy Q needed to raise the temperature of a body through 1 K is called its **thermal capacity**, C:

$$Q = C\,\Delta T$$

where
Q is the thermal energy transferred to the object, measured in J
C is the thermal capacity (also called the total heat capacity), measured in J K^{-1}
ΔT is the temperature change of the object, measured in K (or Celsius degrees)

This equation applies so long as there is no change of state.

An object's thermal capacity depends on its material and its mass. The **specific heat capacity** c is the amount of thermal energy required

?

1 Explain the difference between a rate of loss of temperature and a rate of loss of thermal energy. State the units that would be used to measure each quantity.

2 The specific heat capacity of water is 4200 J kg^{-1} K^{-1}, and that of copper is 385 J kg^{-1} K^{-1}. Calculate:

a) the temperature change of 500 g of water given 20 kJ of energy

b) the energy released by 200 g of copper cooling down from 700 °C to 20 °C.

c) the final temperature of a mixture of 200 g of water at 70 °C and 300 g of water at 10 °C, assuming no energy is lost to the surroundings

d) the final temperature when a 200 g block of copper heated to 700 °C is placed in 500 g of water at 10 °C, assuming no energy is lost to the surroundings.

e) Explain why your answer to (d) is likely to be an overestimate of the final temperature.

3 A hole is drilled in an iron block and an electric heater is placed inside. The heater provides thermal energy at a constant rate of 600 W, and its temperature is recorded as it varies with time. The 800 g block increases in temperature and reaches a final maximum temperature. The specific heat capacity of iron is 480 J kg^{-1} K^{-1}. Calculate:

a) the initial rate of increase in temperature

b) final rate of loss of energy of the block.

to raise the temperature of 1 kg of a substance by 1 K.
In this context, the word "specific" means "per unit mass".

$$Q = mc\,\Delta T$$

where
Q is the thermal energy transferred to the object, measured in J
m is the mass of the object, measured in kg
c is the specific heat capacity, measured in J kg^{-1} K^{-1}
ΔT is the temperature change of the object, measured in K (or Celsius degrees)

This equation applies so long as there is no change of state.

The **molar heat capacity** can also be useful; see pages 145–6 for an explanation of the mole.

There is no net flow of thermal energy in or out of an object at constant temperature. Typically the object is probably gaining and losing thermal energy all the time. The rate of energy gained equals the rate of energy lost.

Thermal energy transferred to the surroundings is often described as "lost" energy. This correctly conveys the fact that it would be difficult, from a practical point of view, to harness this energy to do useful work. The conservation of energy means that this energy cannot have been destroyed, but is just in a form (the thermal energy of the surroundings) that makes it hard to utilize. The energy is said to be **degraded** (see page 285).

Microscopic description

Any substance is composed of particles, and these are in random motion whether the substance is a solid, liquid, or gas. Heating an object causes, on average, these particles to move faster. The total thermal energy that is transferred to an object when it warms is shared among all its particles as random kinetic energy (KE).

Note that there is a difference between the organized KE that a moving object's particles must possess (which equals the KE of the object as a whole) and the random thermal kinetic energy that the particles must possess (which is part of the internal energy of the object and is related to its temperature).

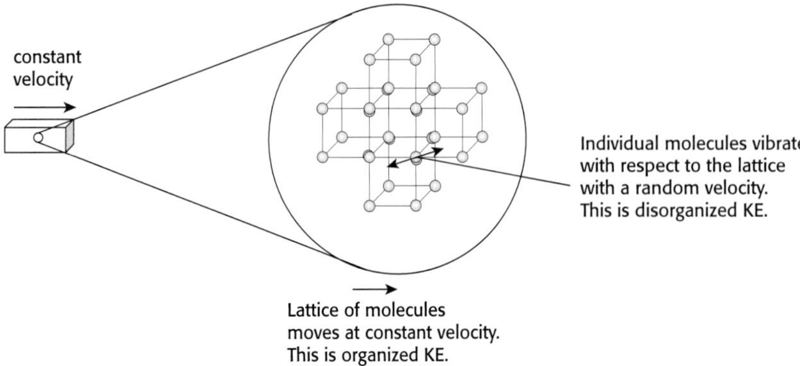

Figure 1 Organized and random KE

1 A substance changes from solid to liquid at its normal melting temperature. What change, if any, occurs in the average kinetic energy and the average potential energy of its molecules?

	Average kinetic energy	Average potential energy
A	constant	constant
B	increases	constant
C	increases	decreases
D	constant	increases

2 The temperature of an ideal gas is reduced. Which **one** of the following statements is true?

A The molecules collide with the walls of the container less frequently.

B The molecules collide with each other more frequently.

C The time of contact between the molecules and the wall is reduced.

D The time of contact between molecules is increased.

Increasing a substance's temperature means that the average random KE of the particles has been increased. The temperature as measured on the thermodynamic scale is proportional to the average KE per particle.

There is no net flow of thermal energy in or out of an object at constant temperature, but individual molecules are interchanging energy all the time. When energy is "lost" to the surroundings this degraded energy is shared amongst all the molecules in the surroundings as disorganized KE. This makes it a difficult practical task to utilize.

A car accelerates from rest up to a constant velocity and then slows down again to rest. When moving forward all the molecules of the car had more KE than when the car was at rest. Does this mean that a moving car has a higher temperature than a stationary one?

Inquiry: Convection, conduction, and radiation

So far we have considered only the effect on an object's temperature of gaining or losing thermal energy. There is another very different property: the ease (or not) with which thermal energy can be transferred between objects or flow through any given object.

There are three important mechanisms by which thermal energy can be transferred from one object to another:

- convection
- conduction
- radiation (particularly in the infrared part of the electromagnetic spectrum).

We shall take the study of radiation further when we consider the electromagnetic spectrum, black body radiation, and the greenhouse effect.

Your task is to research some detail relating to the mechanisms of one of the three processes listed above. You should agree with your teacher the way in which you are going to share your findings with the rest of your group. Possible inquiries include:

1. the application of convection, conduction, and radiation to the design of an astronaut's space suit
2. how gliders are able to stay airborne
3. the design of a hot water system in a house
4. the macroscopic explanation for convection
5. the microscopic explanation for conduction
6. the factors that affect the rate at which infrared radiation is emitted by a surface
7. the factors that affect the rate at which infrared radiation is absorbed by a surface
8. how a vacuum flask works
9. whether the inside of the International Space Station could be kept warm by convection
10. Newton's law of cooling
11. the factors that affect the rate at which thermal energy is conducted along a material
12. the difference between thermal conductors and insulators
13. the coefficient of thermal conduction (also known as the thermal conductivity)
14. how wetsuits/drysuits keep divers warm
15. how the Inuit keep warm in igloos
16. the cooling effect of wind towers.

Investigating physics: Saving Jack

This work can be attempted individually or in small groups. The investigation can be used to assess Design (D), Data collection and processing (DCP), and Conclusion and evaluation (CE).

Cast to your minds back to the tragic end of the film *Titanic*, when Rose is saved but Jack tragically dies of hypothermia because he is totally immersed in almost freezing water. (See youtube clip, http://www.youtube.com/watch?vW1UcB2glx-c accessed July 2009) Your task is to think of ways in which Jack could have been saved, and then try to model them in the laboratory. You need to design an experiment that:

- models Jack
- allows you to vary the conditions suitably
- enables you to take a suitable range of data for analysis.

Produce a plan with an equipment list. Discuss it with your teacher or technician before you carry out the investigation. Try to use some data-logging equipment if possible.

Your final report should be a full laboratory report, including:

- a prediction and outline of the theory involved
- experimental details
- data tables
- graphs
- conclusion and evaluation.

Rediscovering physics: Thermal experiments

This section outlines three experimental situations that can be investigated in order to reinforce your understanding of thermal physics. A safety point to note in all these situations is to take great care near hot objects. Before starting any experiment, you should always discuss an assessment of the risks involved with your teacher.

Calibrating an unmarked thermometer

A mercury-in-glass thermometer (or equivalent) capable of recording temperatures between 0 °C and 100 °C can be used to measure an unknown temperature (room temperature, for example), even if the scale is not clearly marked. Measurements of the length of the column of mercury can be recorded for three different situations: ice/water mix, steam/water mix, and the unknown temperature. The ice/water mix is known to be at 0 °C and the steam/water mix must be at 100 °C. The unknown temperature can be deduced from these measurements. The experiment could provide a good opportunity to practise the treatment of errors.

Measuring a substance's specific heat capacity

A simple electrical heating method can be devised to measure the specific heat capacity of a substance, provided the loss of thermal energy to the surroundings is negligible. The energy dissipated in a time t by an electrical heater with potential difference V across it and current I flowing through it, is calculated using

$$Q = ItV$$

This energy can be assumed to be all transferred to the substance under consideration. The mass of the substance is m and the recorded maximum temperature change is ΔT. The specific heat capacity c is given by

$$c = \frac{Q}{m\Delta T}$$

Again, the experiment could provide a good opportunity to practise the treatment of errors, and to suggest improvements to the technique in the light of the results obtained.

Further research can provide other detailed methods that:

- help correct for any loss of thermal energy
- involve a "method of mixtures" and use a known specific heat capacity to calculate an unknown value
- use a continuous-flow method for a liquid. The apparatus is quite complex to set up, but accurate results can be obtained.

Estimating the temperature of a Bunsen flame

A metal block of mass m_{metal} and specific heat capacity c_{metal} is heated directly in a Bunsen flame and is assumed to reach a steady temperature equal to that of the Bunsen flame, T_{hot}. The block is carefully and quickly transferred to a mass of cold water m_{water} of temperature T_{cold}, and the final combined temperature is T_{final} (Figure 2).

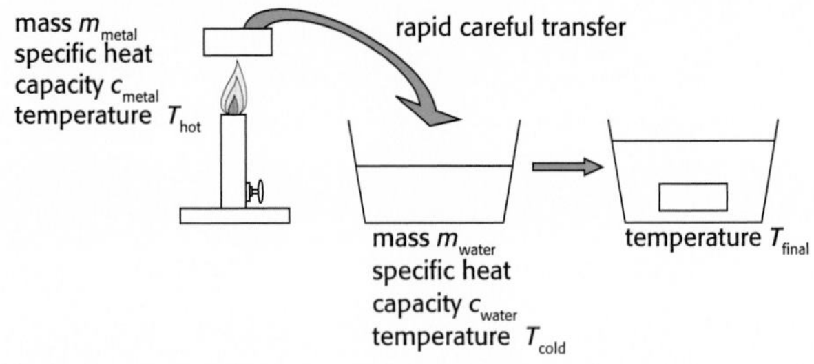

Figure 2 Experiment to estimate the temperature of a Bunsen flame.

Ignoring all heat transfers to the surroundings, the energy lost by the metal cooling down must equal the energy absorbed by the water warming up:

energy lost by metal cooling down = energy gained by water warming up

Therefore

$$m_{metal}c_{metal}(T_{hot} - T_{final}) = m_{water}c_{water}(T_{final} - T_{cold})$$

Substitute in all the values and rearrange for T_{hot}.

A better estimate can be achieved by devising methods to:

- correct for the thermal energy lost while the water is warming up
- allow for the energy involved in changing the temperature of the water container
- correct for the energy involved in the change of state of the small amount of water evaporated during the process (see later work).

Table 1

Phase	Macroscopic	Microscopic	Mean molecular separation	Number per m^3	Volume of molecules relative to that of substance
solid	both volume and shape are approximately constant	molecules are held in position by bonds and vibrate about a fixed point in the lattice	≈ atomic radius	10^{28}	1
liquid	has variable shape but volume is approximately constant	molecules vibrate within ordered clusters but there are no permanent bonds and can move	≈ atomic radius	10^{28}	1
gas	has variable shape and volume	The intermolecular forces are only significant during collisions, molecules are in random motion	≈ 10 × atomic radius	10^{25}	10^{-3}

Source: adapted from P.M. Whelan and M.J. Hodgson, Essential Pre-University Physics, John Murray

Macroscopic and microscopic: phase change

When a substance changes phase, the potential energy of the particles changes as, on average, they move further apart.

Solids, liquids, and gases have different macroscopic properties corresponding to the different structures of the substances in molecular terms. Table 1 summarizes some aspects of these differences, and gives some typical values. A separate fourth phase, **plasma**, exists when a gas is heated to extremely high temperatures (e.g. 10^4 K or above), sufficient to ionize the atoms and molecules. A plasma consists of a mixture of ions and some neutral particles in random motion. The hydrogen and helium in the Sun are in the plasma state.

Macroscopic description

Whenever a substance changes state it does so at a fixed temperature, and there is energy associated with the change: this energy is called the **latent heat**. The amount of thermal energy involved depends on the type of substance, on its mass, and on the phase change involved. It can also be affected by other external factors, such as external air pressure.

The change of state from solid to liquid is called **fusion**, and the change from liquid to gas is called **vaporization**. Each of these changes involves energy being given to the substance. The reverse processes (liquefying and solidifying) must involve energy being released from the substance.

The **specific latent heat** for any phase transformation (at constant temperature) is defined as the latent heat per unit mass:

$$Q = mL$$

where
Q is the thermal energy transferred to the substance, measured in J
m is the mass of the substance, measured in kg
L is the specific latent heat, measured in J kg^{-1}

Microscopic description

The making or breaking of bonds involves energy. When bonds are broken, the potential energy of the molecules is increased, and this requires energy input. In some changes (for example vaporization) the change is also associated with work being done: when a liquid changes into a gas the volume greatly increases, so that during the expansion work must be done (pushing the atmosphere away),

1 The latent heat of vaporization of water is 2300 kJ kg^{-1}. How long would it take a 2 kW electric kettle containing 800 g of boiling water to boil off all the water?

2 In order to maintain a constant body temperature, a sunbather needs to lose about 320 J of thermal energy to the surroundings every second through sweating. Estimate the amount of sweat evaporated from the skin of the sunbather every hour. (The specific latent heat of sweat is about 2300 kJ kg^{-1}.)

Some liquid is contained in a shallow dish that is open to the atmosphere. The rate of evaporation of the liquid does **not** depend on

A the temperature of the liquid.

B the temperature of the atmosphere.

C the depth of the liquid.

D the pressure of the atmosphere.

which adds to the increase in potential energy. When bonds are formed (solidification or liquefaction) then energy must be released.

It is important to note that when intermolecular bonds are made or broken, this happens independently of the KE of the individual molecules. When water boils, the temperature of the liquid and the temperature of the vapour is the same: 100 °C. The bonds between water molecules are being broken, but the average kinetic energy per molecule remains the same, and thus the molecules do not, on average, move faster.

Evaporation and boiling

There are two common processes by which liquids can turn into the vapour state: evaporation and boiling. Boiling takes place at one fixed temperature, and happens throughout the liquid concerned (bubbles appear throughout the body of the liquid). Evaporation, however, is the process by which the faster-moving molecules can escape from the surface of a liquid (Figure 3).

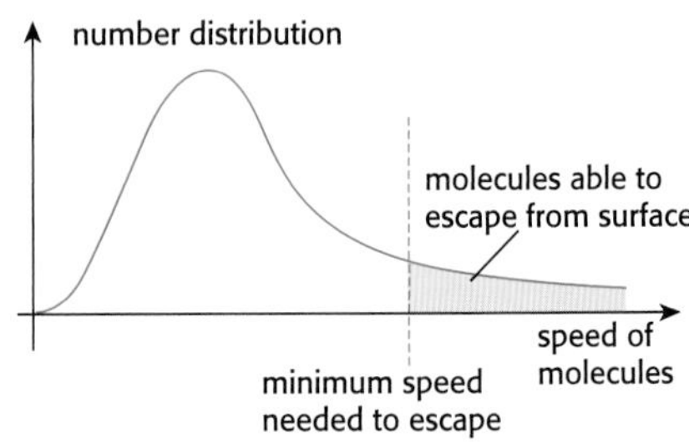

Figure 3 Standard energy distribution curve

As the faster-moving molecules are the only ones that can escape, evaporation causes the average KE per molecule to decrease. In other words, the temperature reduces. Evaporation causes cooling, and is an additional method of transferring thermal energy.

Mathematical physics: The mole

A detailed analysis of the physics (or indeed the chemistry) behind the macroscopic behaviour of materials must involve knowing how many atoms or molecules exist in a given sample of a substance. The numbers involved are extremely large, so it helps to have a unit that deals with large numbers of particles. The amount of substance can then be measured in terms of a fixed number of atoms or molecules. The unit used for the amount of substance is the **mole**.

A mole of any substance will always contain the same number of particles, but the mass will vary depending on the particular substance considered. The mole can be thought of as a very large number. It is defined as the amount of substance that contains the same number of elementary units as there are atoms in 12 g of carbon-12. The number of particles in one mole is called the **Avogadro constant**, and is equal to 6.02×10^{23}.

The word "molar" just means "per mole", so the **molar mass** of any substance is the mass per mole. For example, 1 mole of water molecules (H_2O) has a mass of 18 g, whereas 1 mole of oxygen molecules (O_2) has a mass of 32 g. These masses can be worked out from the individual molar masses for the elements involved (the molar mass of the element hydrogen is 1 g, and for oxygen it is 16 g). Data tables often provide lists of the **relative molecular mass** (rmm) of different elements and molecules. This quantity does not have any units, and is sometimes (wrongly) also called the "molecular weight".

$$\text{relative molecular mass} = \frac{\text{mass of a molecule}}{\text{mass of } {}^{12}_{6}\text{C atom}} \times 12$$

so

$$\text{molar mass} = \frac{\text{relative molecular mass}}{1000}\ \text{kg mol}^{-1}$$

The chemical reaction between hydrogen and oxygen to form water provides an example of how the mole can be used to bring together the microscopic and the macroscopic points of view. A mixture of hydrogen and oxygen is stable at room temperature but will explode if ignited.

The chemical reaction is described by the following equation:

$$2H_2 + O_2 \rightarrow 2H_2O$$

The information represented by this equation is that, at the microscopic scale, two molecules of hydrogen (H_2) and one molecule of oxygen (O_2) combine to produce two molecules of water (H_2O). Since one mole of anything contains the same number of particles, this means that two moles of hydrogen will combine with one mole of oxygen to produce two moles of water. In other words, 4 g of hydrogen (rmm of hydrogen = 2) will react completely with 32 g of oxygen (rmm = 32) and produce 36 g of water.

There are two related pieces of information that can prove useful to those also studying chemistry. First, the charge on a mole of electrons (= 9.6×10^4 C) is called the Faraday constant, and is used in calculations involving electrolysis. Second, a mole of any ideal gas at STP (standard temperature and pressure: 273.15 K and atmospheric pressure) has a volume of 22.4×10^{-3} m^3.

Conversions between the mass of a substance and the number of atoms or molecules involved are relevant when we are considering the behaviour of an ideal gas in detail (see below) and the radioactive decay of nuclei (see page 241).

Ideal gas behaviour

The quantitative analysis of the behaviour of gases is studied in more detail by HL students (see pages 151–4).

Compared with the temperature of solids and liquids, the temperature of a gas is more dependent on external factors. For a given sample of gas, three quantities are interdependent: the pressure p, the volume V, and the temperature T. Macroscopically, many gas experiments show that these quantities are related in approximately the same way, as summarized by the ideal gas equation of state (see page 153). From the macroscopic point of view, an **ideal gas** is one that follows the ideal gas equation for all values of pressure, volume, and temperature.

One of the triumphs of classical physics is that it is possible to predict the macroscopic behaviour of a gas by using Newton's laws to analyse the interactions taking place on the microscopic (atomic) scale. This analysis leads to the identification of temperature as a measure of the average random kinetic energy of the molecules of an ideal gas.

The **kinetic theory** views all matter as consisting of individual atoms and molecules. Certain microscopic assumptions need to be made in order to be able to deduce the macroscopic behaviour of a gas. For an ideal gas, these are as follows:

- The molecules are assumed to behave in an idealized way: that is, Newton's laws of mechanics apply to the individual molecules' motions.
- The intermolecular forces are assumed to be negligible (except during a collision).
- The molecules are assumed to be spherical and their volume negligible (compared with the volume occupied by the whole gas).
- The molecules are assumed to be in random motion.
- The collisions between molecules are assumed to be perfectly elastic.
- The time taken for a collision is assumed to be negligible.

Pressure: macroscopic and microscopic views

From the macroscopic point of view, a general definition of pressure is as follows:

$$P = \frac{F}{A}$$

where

F is the force exerted, measured in N

A is the normal area over which the force acts, measured in m^2

P is the pressure, measured in N m^{-2} (or pascals, Pa)

$$1 \text{ Pa} = 1 \text{ N m}^{-2}$$

In the above equation for pressure, area is at right angles to the direction in which the force acts. If a force does not act at 90° to a surface, then the pressure is calculated using the component of the force that is at 90° to the surface.

Gas pressure can be understood by considering the large number of collisions that take place between the molecules of the gas and the walls of the container:

- When a gas molecule hits the walls of the container, it bounces off.
- The momentum of the gas molecule has changed during this collision, and Newton's second law applied to this situation means that there must have been a force on the molecule from the wall.
- Newton's third law applied to this situation means that there must have been a force on the wall from the molecule.
- Each time a molecule collides with the wall, there will be a small force from the molecule on the wall.
- In a given time, there will be a certain number of collisions.
- The average result of all of these individual molecular collision forces acting for a short time will be a constant force on the wall of the container.
- The value of the constant force on the wall divided by its area is the pressure that the gas exerts on the wall of the container.

A gas exerts a pressure on every surface of its container: this pressure depends on the mass of the gas, its volume, and its temperature. The macroscopic behaviour of an ideal gas can be understood in terms of the motion of the molecules of the gas. In the first three of the following scenarios, one variable has been changed and all but one of the other ones are kept constant.

1 Ideal gases increase in pressure when more gas is introduced into the container. The increase in mass of the gas means that there are more gas molecules in the container, and therefore an increase in the number of collisions that take place in a given time. The force from each molecule remains the same, but an increased number of collisions in a given time means that the pressure increases.
2 Ideal gases increase in pressure when their volume decreases. The decrease in volume means that molecules hit a given area of the walls more often. The force from each molecule remains the same, but an increased number of collisions in a given time means that the pressure increases.
3 Ideal gases increase in pressure when their temperature increases. The increased temperature means the molecules are moving faster, and thus they hit the wall more often. The force from each molecule goes up on average and an increased number of collisions in a given time means that the pressure increases.
4 In this final example, an isolated sample of gas is compressed. Ideal gases increase in temperature when their volume is decreased. As the volume is reduced, the walls of the container move inwards. The molecules are colliding with a moving wall and will, on average, speed up. Faster-moving molecules mean that the average kinetic energy per molecule has increased – that is, the temperature has increased. The smaller volume and higher temperature mean the pressure must have increased.

1 Volume and temperature constant

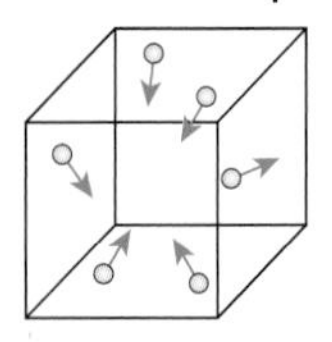

few molecules
low pressure

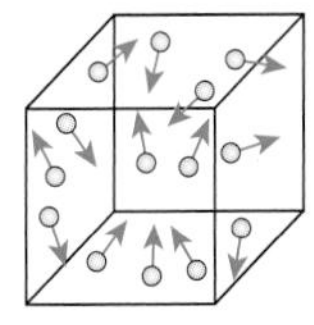

many molecules
high pressure

2 Mass and temperature constant

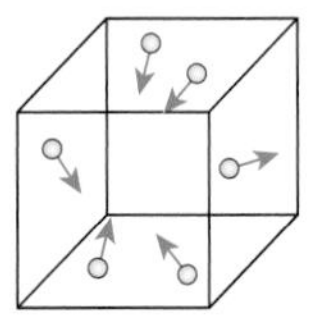

high volume
low pressure

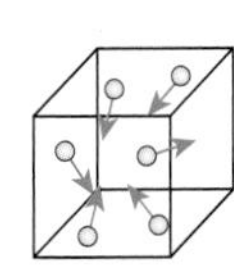

low volume
high pressure

3 Mass and volume constant

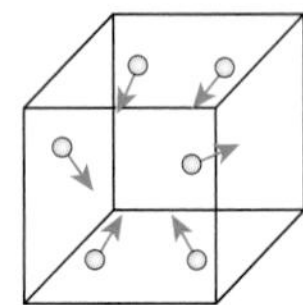

low temperature
low pressure

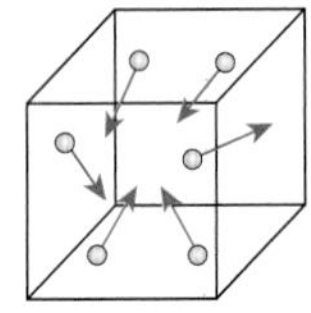

high temperature
high pressure

4 Mass constant

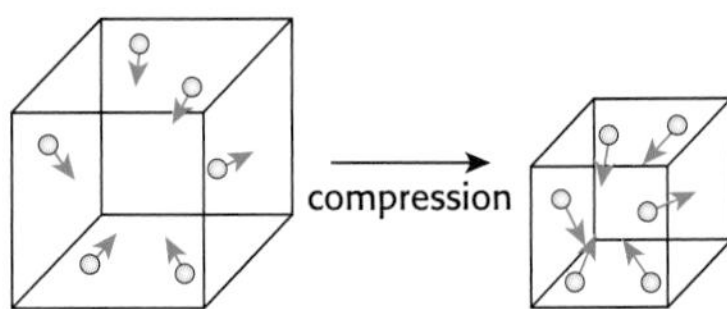

low temperature
high volume
low pressure

high temperature
low volume
high pressure

End of chapter summary

Chapter 10 has two main themes: general thermal concepts and thermal properties of matter. The list below summarises the knowledge and skills that you should be able to undertake after having studied this chapter. Further research into more detailed explanations using other resources and/or further practice at solving problems in all these topics is recommended – particularly the items in bold.

General thermal concepts

- State that temperature determines the direction of thermal energy transfer between two objects, the relation between the thermodynamic and Celsius scales of temperature and that the internal energy of a substance is the total potential energy and random kinetic energy of the molecules of the substance.
- Explain and distinguish between the macroscopic concepts of temperature, internal energy and thermal energy (heat).
- Define the *mole, molar mass* and the *Avogadro constant*.

Thermal properties of matter

(specific heat capacity, phase changes and latent heat and the kinetic model of an ideal gas)

- Define *specific heat capacity* and *thermal capacity* and **solve problems**.
- Explain the physical differences between the solid, liquid and gaseous phases and the process of phase changes (including the reason why temperature does not change during a phase change) in terms of molecular structure, molecular behaviour and particle motion.
- Distinguish be tween evaporation and boiling.
- Define *specific latent heat* and **use the definition to solve problems**.
- Define *pressure* of a gas.
- State the assumptions of the kinetic model of an ideal gas and that temperature is a measure of the average random kinetic energy of the molecules of an ideal gas.
- Explain the macroscopic behaviour of an ideal gas in terms of a molecular model.

Chapter 10 questions

1 Some students were asked to design and carry out an experiment to determine the specific latent heat of vaporization of water. They set up the apparatus shown below

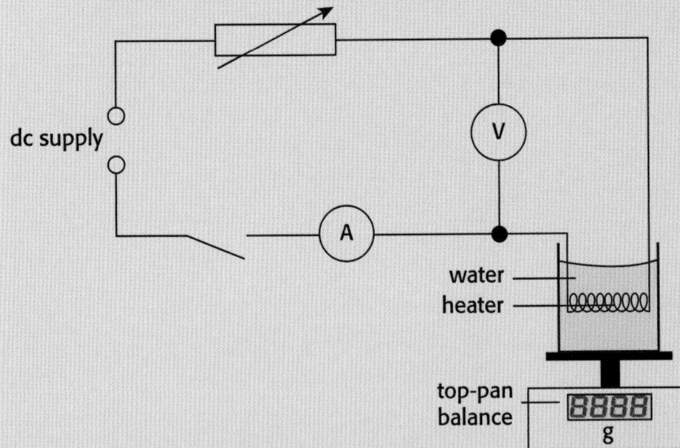

The current was switched on and maintained constant using the variable resistor. The readings of the voltmeter and the ammeter were noted. When the water was boiling steadily, the reading of the top-pan balance was taken and, simultaneously, a stopwatch was started. The reading of the top-pan balance was taken again after 200 seconds and then after a further 200 seconds

The change in reading of the top-pan balance during each 200 second interval was calculated and an average found. The power of the heater was calculated by multiplying together the readings of the voltmeter and the ammeter.

a) Suggest how the students would know when the water was boiling steadily. [1]

b) Explain why a reading of the mass lost in the first 200 seconds and then a reading of the mass lost in the next 200 second interval were taken, rather than one single reading of the mass lost in 400 seconds. [2]

The students repeated the experiment for different powers supplied to the heater. A graph of the power of the heater against the mass of water lost (the change in balance reading) in 200 seconds was plotted. The results are shown below. (*Error bars showing the uncertainties in the measurements are not shown.*)

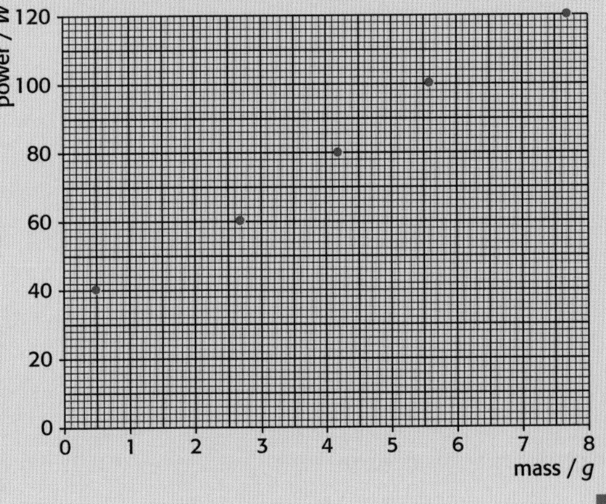

c) **i)** On the graph above, draw the best-fit straight line for the data points. [1]

ii) Determine the gradient of the line you have drawn. [3]

In order to find a value for the specific latent heat of vaporization L, the students used the equation

$P = mL$

where P is the power of the heater and m is the mass of water evaporated **per second**.

d) Use your answer for the gradient of the graph to determine a value for the specific latent heat of vaporization of water. [3]

e) The theory of the experiment would suggest that the graph line should pass through the origin. Explain briefly why the graph does not pass through the origin. [2]

(Total 12 marks)

2 This question is about an experiment to measure the temperature of a flame.

a) Define heat (*thermal*) *capacity*. [1]

A piece of metal is held in the flame of a Bunsen burner for several minutes. The metal is then quickly transferred to a known mass of water contained in a calorimeter.

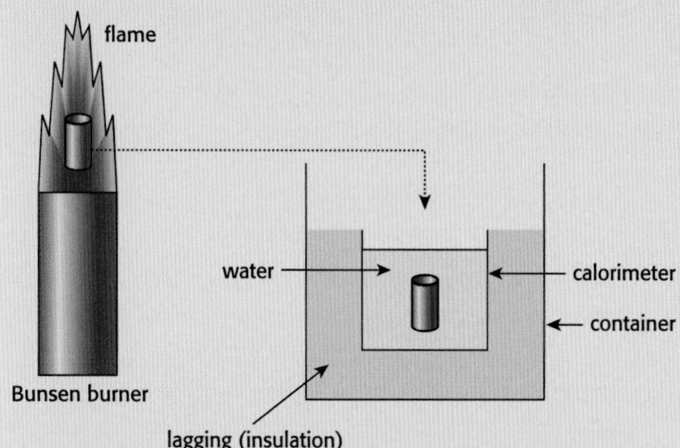

The water into which the metal has been placed is stirred until it reaches a steady temperature.

b) Explain why

i) the metal is transferred as quickly as possible from the flame to the water. [1]

ii) the water is stirred. [1]

The following data are available:

heat capacity of metal = 82.7 J K^{-1}

heat capacity of the water in the calorimeter = 5.46 3 102 J K^{-1}

heat capacity of the calorimeter = 54.6 J K^{-1}

initial temperature of the water = 288 K

final temperature of the water = 353 K

c) Assuming negligible energy losses in the processes involved, use the data to calculate the temperature T of the Bunsen flame. [4]

(Total 7 marks)

3 This question is about the change of phase (state) of ice.

A quantity of crushed ice is removed from a freezer and placed in a calorimeter. Thermal energy is supplied to the ice at a constant rate. To ensure that all the ice is at the same temperature, it is continually stirred. The temperature of the contents of the calorimeter is recorded every 15 seconds.

The graph below shows the variation with time t of the temperature θ of the contents of the calorimeter. (*Uncertainties in the measured quantities are not shown.*)

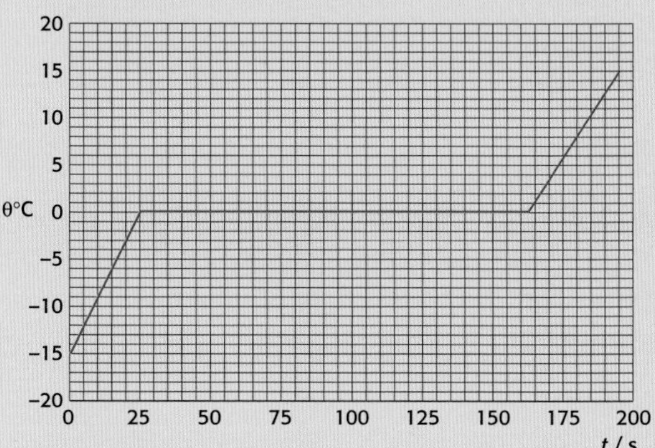

a) On the graph above, mark with an X the data point on the graph at which all the ice has just melted. [1]

b) Explain, with reference to the energy of the molecules, the constant temperature region of the graph. [3]

The mass of the ice is 0.25 kg and the specific heat capacity of water is 4200 J kg^{-1} K^{-1}.

c) Use these data and data from the graph to

i) deduce that energy is supplied to the ice at the rate of about 530 W. [3]

ii) determine the specific heat capacity of ice. [3]

iii) determine the specific latent heat of fusion of ice. [2]

(Total 12 marks)

4 This question is about specific heat capacity and specific latent heat.

a) Define *specific heat capacity*.

b) Explain briefly why the specific heat capacity of different substances such as aluminium and water are not equal in value. [2]

A quantity of water at temperature θ °C is placed in a pan and heated at a constant rate until some of the water has turned into steam. The boiling point of the water is 100 °C.

c) i) Draw a sketch graph to show the variation with time θ of the temperature θ of the water. (*This only needs to be a sketch graph; you do not need to include any numerical values.*) [1]

ii) Describe, in terms of energy changes, the molecular behaviour of water and steam during the heating process. [5]

Thermal energy is supplied to the water in the pan for 10 minutes at a constant rate of 400 W. The thermal capacity of the pan is negligible.

d) i) Deduce that the total energy supplied in 10 minutes is 2.4×10^5 J. [1]

ii) Using the data below, estimate the mass of water turned into steam as a result of this heating process.

initial mass of water = 0.30 kg

initial temperature of the water θ = 20 °C

specific heat capacity of water $= 4.2 \times 10^3$ J kg^{-1} K^{-1}

specific latent heat of vaporization of water $= 2.3 \times 10^6$ J kg^{-1} [3]

iii) Suggest one reason why this mass is an estimate. [1]

(Total 14 marks)

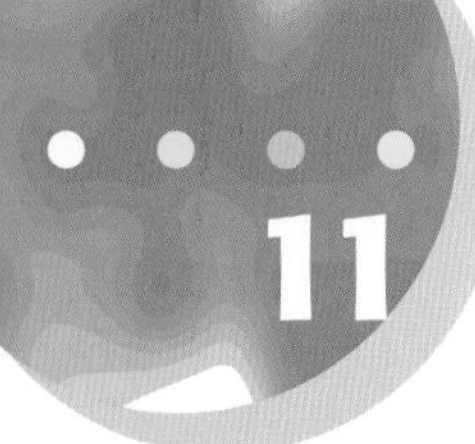

11 Thermodynamics

HL

Before studying this chapter it would be worth reviewing the concepts introduced in Chapter 10. Although the material covered is all higher level material, standard level students would gain useful background information by studying this chapter.

Working with data: Analysis of gases

Three interrelated quantities are needed to describe a fixed mass of gas: its pressure, its volume, and its temperature. In order to investigate how these quantities are related, three separate experiments are needed. In each experiment, one of the variables is kept constant (the controlled variable), one is altered (the independent variable), and the variation in the third variable (the dependent variable) is recorded.

The following three experiments provide outline experimental descriptions and data that should be analysed as practice for the DCP and CE skills.

Experiment 1: Variation of volume with pressure at constant temperature

The apparatus illustrated in Figure 1 was used to subject a fixed mass of trapped dry air to different pressures. As the pressure increases, the oil moves up the graduated tube and the air occupies a smaller volume. The Bourdon gauge records the pressure.

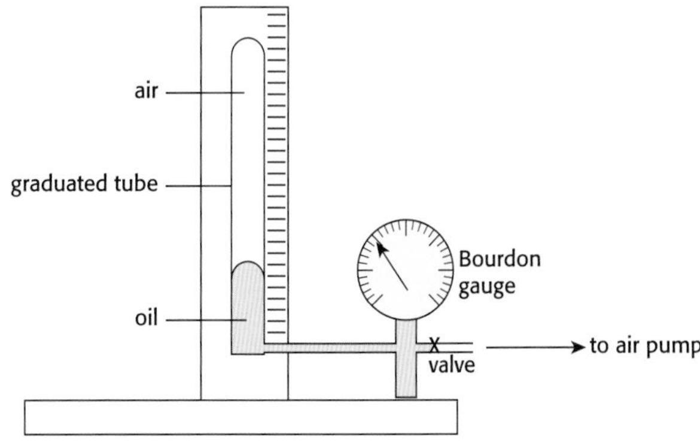

Figure 1 Apparatus to measure the variation of volume with pressure, at constant temperature

The following procedure was followed:

1. Switch on the pump and compress the air.
2. Stop the pump when the Bourdon gauge registers its maximum pressure.
3. Record the pressure on the gauge and the volume of air in the graduated tube.
4. Carefully release the valve to reduce the pressure slightly.
5. Record the new measurements.
6. Repeat steps 4 and 5 until the pressure is back to atmospheric pressure.

Sample data are listed in Table 1.

Table 1 Sample data

Volume of trapped air		Pressure, P / kN m^{-2} (kPa)
V_1 / cm^3	V_2 / cm^3	
34.0	34.0	120
29.0	29.0	140
25.5	25.5	160
22.5	22.5	180
20.0	20.0	200
18.0	18.0	220
16.5	16.5	240
15.5	15.0	260
14.0	14.0	280
13.0	13.0	300
Temp = 17.5 °C	Temp = 18.0 °C	

1 Draw a graph using the data in Table 1 and come to a conclusion about how the volume of a gas is related to the pressure.
2 The actual apparatus used is shown in the photographs on this page. Use your graph to explain the concept of systematic error.

Experiment 2: Variation of volume with temperature at constant pressure

The apparatus shown in Figure 2 was used to subject a fixed mass of trapped dry air to different temperatures.

(a)

(b)

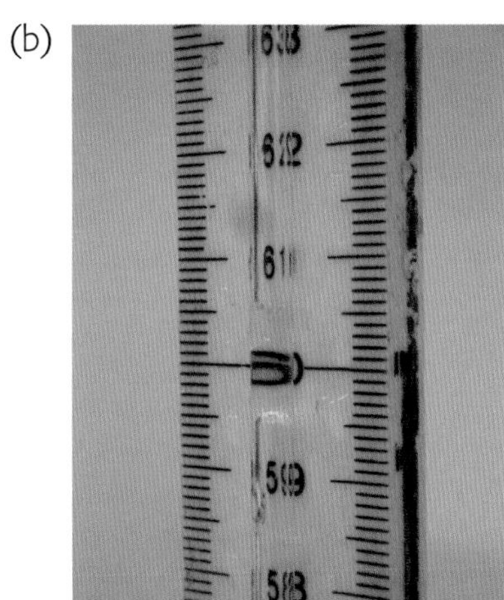

Figure 2 (a) Apparatus to measure the variation of volume with temperature, at constant pressure. (b) Enlargement of submerged bubbles.

Sample data are listed in Table 2.

Table 2 Sample data

Temperature, T / °C	Length of trapped air	
	L_1 / 10^{-3} m	L_2 / 10^{-3} m
80	98.0	98.0
75	96.0	97.0
70	95.0	95.0
65	93.0	94.0
60	92.0	92.0
55	90.0	90.0
50	89.0	89.0
45	88.0	88.0
40	86.0	86.0
35	85.0	85.0
30	83.0	83.0

Estimate the uncertainties associated with the readings and then analyse the data in Table 2 and come to a conclusion about how the volume of a gas is related to the temperature.

Experiment 3: Variation of pressure with temperature at constant volume

The apparatus shown in Figure 3 was used to subject a fixed mass of trapped dry air to different temperatures. Sample data are listed in Table 3.

Figure 3 Apparatus to measure the variation of pressure with temperature, at constant volume

Table 3 Sample data

Temperature, T / °C	Pressure, P / 10^5 Pa
82	1.175
73	1.150
66	1.125
57	1.100
49	1.075
40	1.050
32	1.025
25	1.000

Estimate the uncertainties associated with the readings and then analyse the data in Table 3 and come to a conclusion about how the pressure of a gas is related to the temperature.

Absolute scale of temperature

Note that in experiment 2 (or 3) it would be incorrect to state that volume (or pressure) is proportional to the temperature. It is, however, possible to continue the trend line (this process is known as **extrapolating**) to lower temperatures and redefine a new scale of temperature. For example, the graph in Figure 4 represents the variation with temperature of pressure.

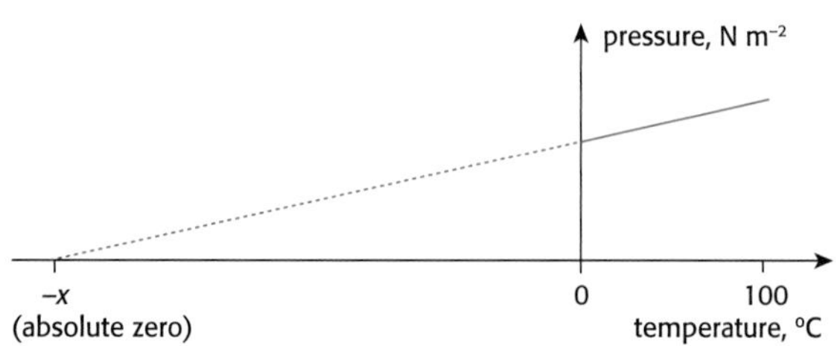

Figure 4 Variation of temperature with pressure

The x-axis can be rescaled so that a new origin is defined to be at the point where the trend line meets the axis (Figure 5). This temperature is called **absolute zero**. For convenience, the size of one degree on the new scale is kept the same as on the Celsius scale. The new temperature scale is called the **absolute scale of temperature**, the **thermodynamic scale of temperature**, or the **kelvin scale**. The unit of this scale is kelvin (K).

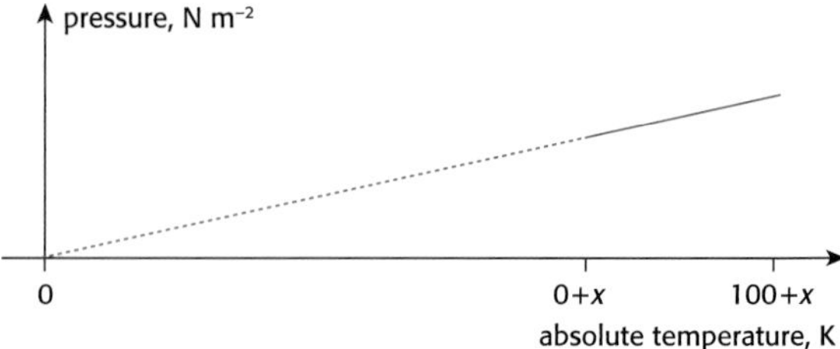

Figure 5 The pressure of a gas is proportional to its absolute temperature

The graph in Figure 5 shows that the pressure of the gas is proportional to its absolute temperature.

Note that by extrapolating the graph we are assuming that the gas continues to vary in the same way at low temperatures as it did at higher temperatures.

Summary

The above three experiments point towards three general laws that can be used for many different gases (particularly at low pressures). For a fixed mass of gas, M, we have:

- *Boyle's law*: If T is constant, pV = constant
- *Charles's law*: If p is constant, $\frac{V}{T}$ = constant
- *Pressure law*: If V is constant, $\frac{P}{T}$ = constant

where
p is the pressure of the gas, measured in N m^{-2}
V is the volume of the gas, measured in m^3
T is the absolute (thermodynamic) temperature of the gas, measured in K

In each case the constant is proportional to the mass of the gas considered. These laws can be brought together into one general law for each gas:

$$\frac{pV}{T} = \text{constant}$$

Many different gases have approximately the same value of constant for one mole, so the equation can be rewritten as the ideal gas equation:

$$\frac{pV}{nT} = R \quad \text{or} \quad pV = nRT$$

where
p, V, and T are as defined above

n is the amount of matter in the gas, measured in moles
R is the universal molar gas constant, which has the value 8.31 J mol^{-1} K^{-1}

An **ideal gas** is one that obeys the above equation for all values of p, V, and T. The **state** of an ideal gas is specified by p, V, and T, and the above equation is sometimes called the **equation of state** for an ideal gas. This use of the word "state" should not be confused with the different states of solid, liquid, and gas. Remember that ideal gases cannot turn into liquids or solids.

The bigger picture: Microscopic explanation for ideal gases

If we apply Newton's laws to the kinetic theory of gases (provided the ideal gas assumptions detailed on page146 apply), it allows us to derive an equation that links the pressure of a gas to the molecular speed distribution. A full statistical analysis is beyond the scope of the IB Diploma Programme, but an understanding of the principal conclusions aids understanding.

Random interactions between the molecules of a gas will result in a distribution of different molecular speeds. A full analysis shows that this distribution, called a **Maxwell–Boltzmann** distribution, has the shape shown in Figure 6.

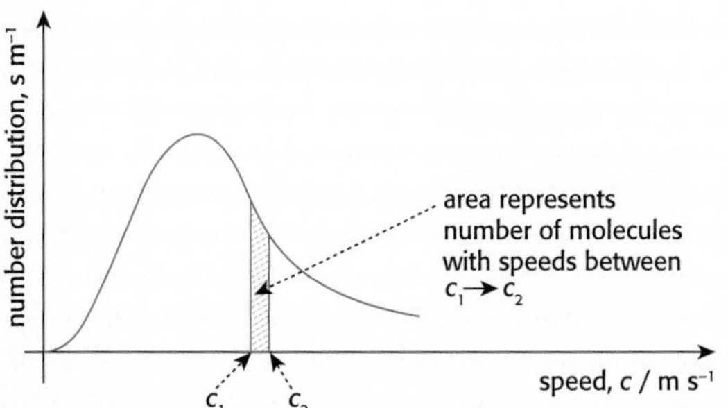

Figure 6 Maxwell–Boltzmann distribution

The area under the graph represents the number of molecules in the chosen speed range (from c_1 to c_2), hence the unit on the y-axis. When a molecule collides with one of the walls of its container, the force from the molecule on the wall will depend on the molecule's velocity. A full analysis shows that the force depends on the square of the velocity, c^2.

The individual forces from each molecule can all be added together and this leads to an expression for the overall pressure in terms of the average value of the squares of the different molecular velocities, $\overline{c^2}$.

Note that:

- The bar above the c^2 represents the average (mean) value.
- This average value is different to the square of the average value of the different molecular speeds, which would be represented by $(\bar{c})^2$.
- The **rms** (root mean squared) value of the molecular speeds is a useful quantity, which is equal to $\sqrt{(\overline{c^2})}$. This is not the same as the average speed, but represents a "typical" speed of the gas molecules. If all the molecules had this speed, then the pressure of the gas would be the same as given by the random distribution of speeds.

Rms values are important in many other areas of physics, notably the analysis of alternating voltages and currents (see page 221).

Ten molecules have the following speeds c in km s^{-1}:
1.0, 2.0, 3.0, 5.0, 5.0, 5.5, 6.0, 7.0, 8.0, 10.0

Calculate:

a) the average speed, $\bar{c}$
b) the square of the average speed, $(\bar{c})^2$
c) the average of speed squared, $\overline{c^2}$
d) the square root of your answer to part **c)**, i.e. the rms speed.
e) Compare and comment on your answers to parts **a)** and **d)**.

A full analysis, applying Newton's laws to the random motion of molecules, predicts that the pressure of an ideal gas is given by

$$p = \frac{1}{3}\rho\overline{c^2}$$

where
p is the pressure of the gas, measured in Nm^{-2}
ρ is the density of the gas, measured in kg m^{-3}
$\overline{c^2}$ is the average of the speed squared for the molecules, measured in m^2 s^{-2}

The gas contains N molecules, each of mass m. The definition of density means that this equation can be rewritten as

$$pV = \frac{1}{3}Nm\overline{c^2}$$

or

$$pV = \frac{2}{3}N\left(\frac{1}{2}m\overline{c^2}\right)$$

This means that the assumptions of kinetic theory predict that

$$pV \propto \frac{1}{2}m\overline{c^2}$$

where $\frac{1}{2}m\overline{c^2}$ is the average kinetic energy of a molecule. This can be compared with

$$pV \propto T$$

which is the ideal gas equation expressed in terms of the absolute scale of temperature. This shows that the absolute temperature of an ideal gas is proportional to the average translational KE of its molecules.

Mathematical physics: Energy changes for ideal gases

There are two different processes by which energy can be transferred to (or from) a gas. Either work is done on a macroscopic scale by or on the gas, or thermal energy is transferred. The latter process is doing work on the microscopic scale.

When a quantity of thermal energy ΔQ is given to an ideal gas, one of three things can happen:

- It can raise the internal energy of the gas by an amount ΔU. The internal energy of the gas is equal to the translation KE of its molecules. The average KE of each molecule is proportional to the absolute temperature of the gas.
- It can enable the gas to do work ΔW. A positive value for ΔW means that the gas has done work. A negative value means that work has been done on the gas.
- Both processes may occur together.

The principle of the conservation of energy tells us that

$$\Delta Q = \Delta U + \Delta W$$

This equation is known as the **first law of thermodynamics**.

The various different ways by which the state of an ideal gas can be altered will have different effects on the internal energy of the gas and on the work done. The energy required per degree change in temperature will depend on the method. For example, the specific heat capacity at constant volume, c_V, will be less than the specific heat capacity at constant pressure, c_p. The gas does work when expanding at constant pressure, but does no work when the volume is constant. Four important changes are considered below.

Calculation of work done

Macroscopically, a gas does work whenever it expands; conversely, work is done on the gas when it is forced to contract. We can calculate the amount of work done by considering a gas expanding and pushing back a frictionless piston, as shown in Figure 7.

Provided the displacement is small, we can assume the pressure remains constant. Since $F = pA$, then

$$\delta W = (pA) \times \delta x$$

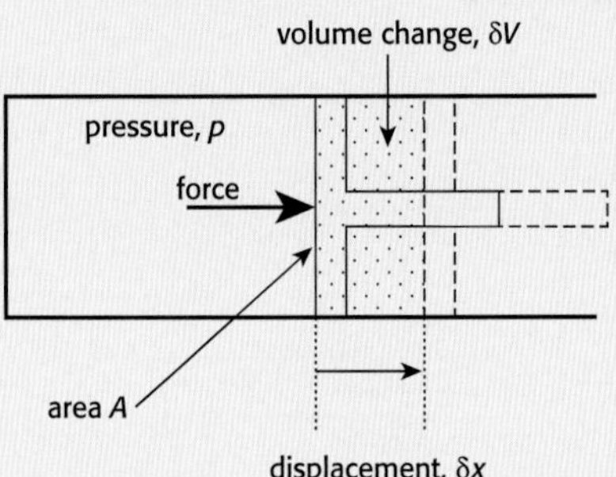

Figure 7 Work done by an expanding gas

Since $A\delta x = \delta V$, then

$$\delta W = p \times \delta V$$

This small amount of work done corresponds to the area under a p–V curve (Figure 8).

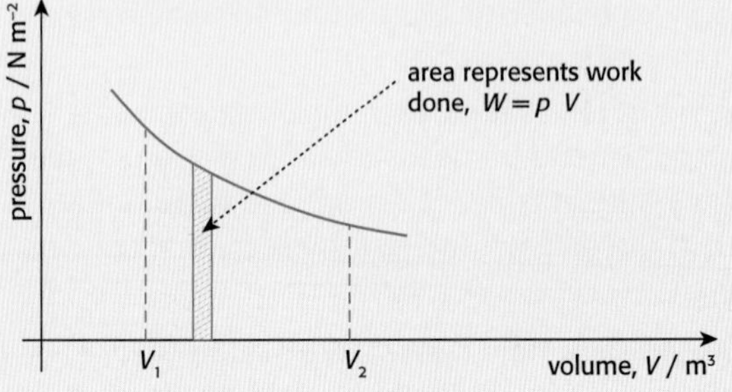

Figure 8 Work done during expansion

In general, the total work done by the gas is the area under the graph if the volume increases (i.e. from V_1 to V_2 in the example in Figure 8). Using calculus notation, $W = \int_{V_1}^{V_2} p \cdot dv$. If the volume decreases, the area represents the work done on the gas.

This principle can be applied to a gas that undergoes a series of state changes and returns back to its original state. The net work done on (or by) the gas is represented by the area contained on the p–V diagram, as shown in Figure 9.

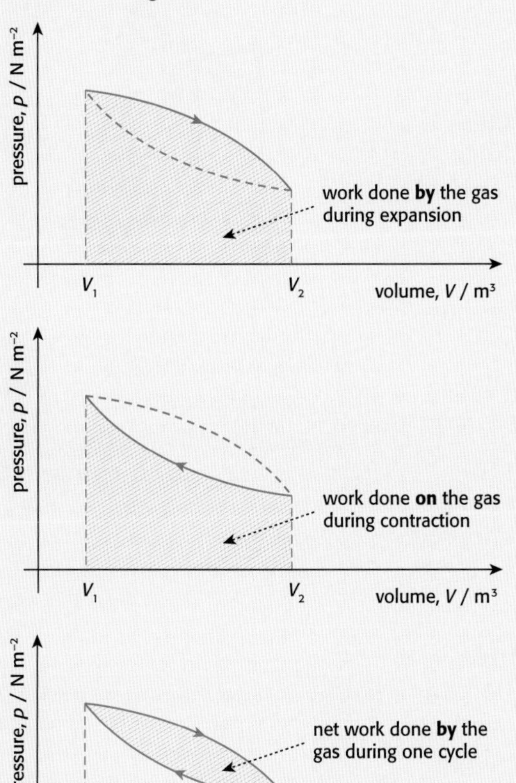

Figure 9 Net work done during one cycle

Isochoric (isovolumetric) change

This state change for a gas takes place at constant volume, so the work done must be zero (Figure 10).

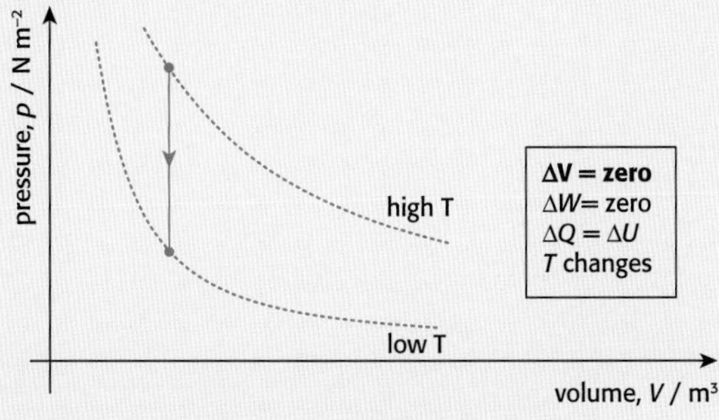

Figure 10 Isochoric (isovolumetric) change

Isobaric change

This state change takes place at constant pressure. The overall work done is easy to calculate (Figure 11).

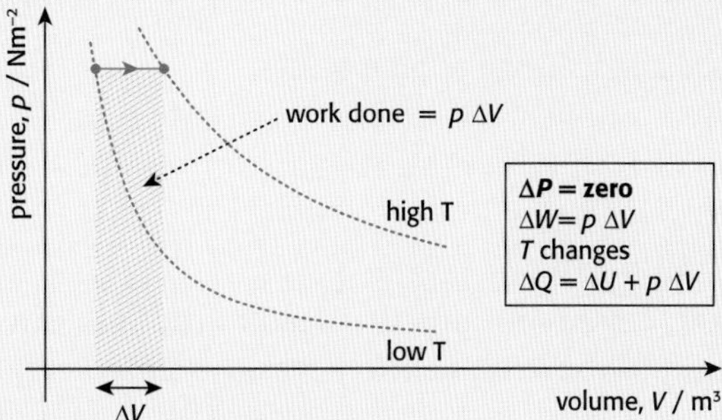

Figure 11 Isobaric change

Isothermal change

This state change takes place at constant temperature. The overall work done is calculated from the area under the curve (Figure 12).

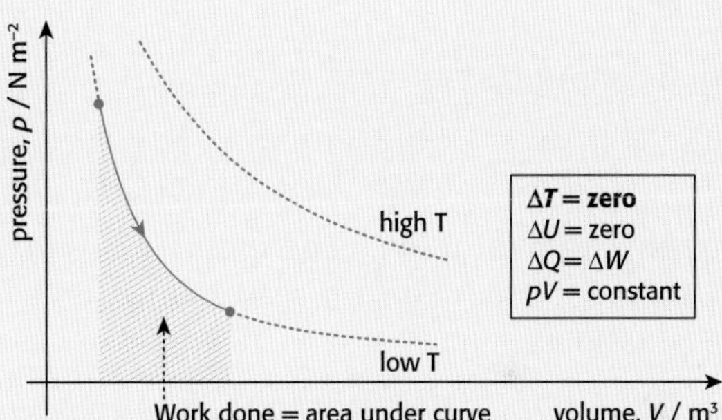

Figure 12 Isothermal change

Adiabatic change

This state change takes place without any transfer of thermal energy to or from the gas. The overall work done is calculated from the area under the curve (Figure 13).

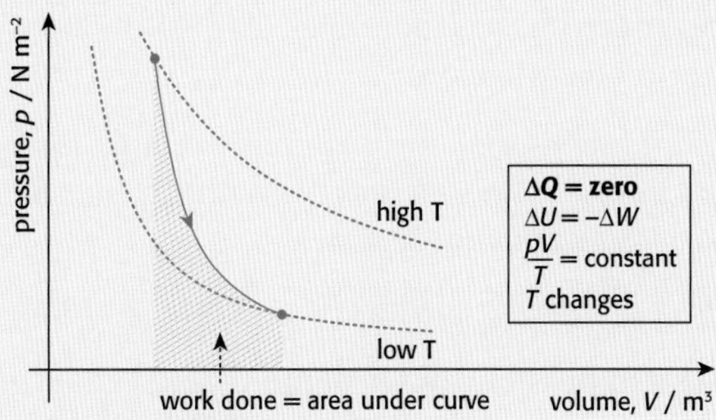

Figure 13 Adiabatic change

One technical point that should be noted is that all the changes that have been considered in this section are classified from a thermodynamic point of view as **reversible**. We have assumed that the universal gas

equation applies all the time. This implies that the gas has never departed from **thermodynamic equilibrium** – that is, there have never been any temperature or pressure gradients *within* the gas that might cause the measured pressure, volume, or temperature to change as time passes.

In order to achieve reversibility in these processes, the practical implications are that all changes need to take place very slowly, so that the gas never significantly departs from equilibrium. In reality, perfectly reversible changes would take an infinite amount of time. Real, non-reversible, changes can be **dissipative** – that is, energy can be lost.

Heat engines to do work

One of the key inventions that lead to the western Industrial Revolution of the 18th century was the steam engine.

Given the importance of the steam engine to industry, any slight improvement in it could be worth a lot of money. Not surprisingly, much intellectual effort was expended on improving the efficiency of the design.

The first law of thermodynamics does not put a limit on the maximum possible efficiency of a heat engine. According to this law, it is possible to design systems that convert thermal energy into useful work completely, but they cannot continue to operate over long periods of time.

For example, an ideal gas being heated under isobaric conditions converts thermal energy into work with 100% efficiency. But the gas cannot freely expand for ever; there must be a physical limit imposed by the equipment used. For this reason an engine must involve a cycle of operations.

1 Explain why an ideal gas cannot be liquefied.

2 For each of the ideal gas assumptions listed on page 146, outline reasons why and how these assumptions may not apply to real gas molecules.

A steam engine is one example of a device called a **heat engine:** other examples include a car engine and a thermal power station. In all of these examples thermal energy is converted into useful work. Figure 14 analyses a generalized heat engine in terms of the energy transfers taking place.

(a)

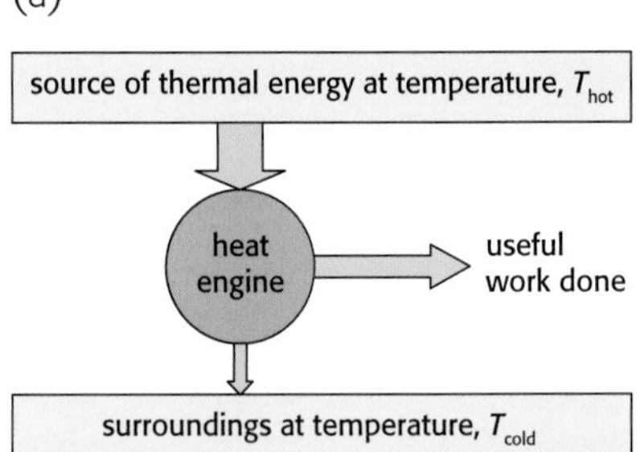

Figure 14a The basic principle of a heat engine

To increase the efficiency of an engine, the amount of energy lost to the surroundings needs to be reduced to a minimum. In the 19th century a general principle called the **second law of thermodynamics** was identified. The following two statements can be shown to be equivalent:

> *"Thermal energy cannot spontaneously transfer from a region of low temperature to a region of high temperature. In order for this transfer to take place, work must be done."*

> *"No heat engine that operates in a cycle can be 100% efficient."*

An analysis of possible cycles demonstrated that the maximum possible efficiency would be achieved if the ideal gas followed the following process:

1 an isothermal expansion
2 an adiabatic expansion
3 an isothermal contraction
4 an adiabatic contraction back to the original state.

(b)

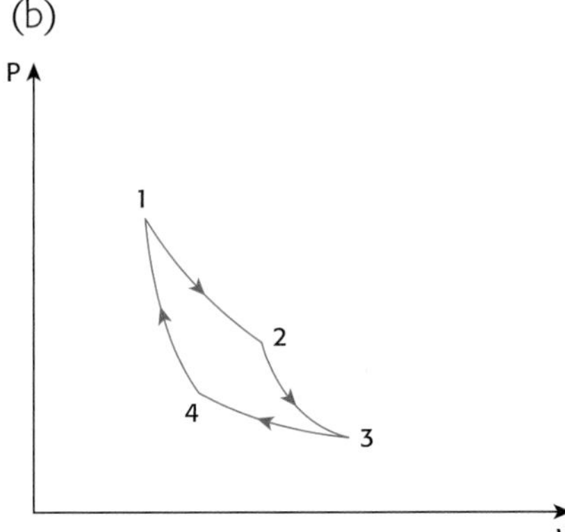

Figure 14b Carnot cycle

The above cycle is called a **Carnot cycle**, and an engine with the maximum possible efficiency is known as a **Carnot engine**. The efficiency of a Carnot engine can be calculated using the following relationship:

$$\text{Efficiency} = 1 - \left(\frac{T_{\text{cold}}}{T_{\text{hot}}}\right)$$

That is, the maximum possible efficiency of any engine is determined by its operating temperature and the temperature of the surroundings. This result has a profound effect on the design of all engines. See page 318 for more information.

Second law of thermodynamics and entropy

The previous section introduced the second law of thermodynamics, as expressed in the 18th century. Nowadays it is much more common to use a concept called **entropy**, S. The entropy of a system is a property that expresses the degree of disorder in the system. For example, the particles in a solid are held in an ordered state compared with the same particles in a gaseous state.

If all the air molecules in a room were collected in one corner of the room, this would be a very ordered state (and so have low entropy) compared with the "normal" arrangement of molecules spread throughout the room in random motion (in which they would have high entropy). In everyday situations, the total entropy of a system is extremely hard to calculate from first principles, but an entropy change can often be quantified.

When thermal energy ΔQ is given to an object at a temperature T the system becomes more disorganized. The increase in entropy can be calculated using:

$$\Delta S = \frac{\Delta Q}{T}$$

where
ΔQ is the thermal energy added to the system, measured in J
T is the absolute temperature of the system, measured in K
ΔS is the increase in the entropy of the system, measured in J K^{-1}

When a thermodynamic process results in a temperature change, we need to use calculus techniques in order to sum up the entropy changes taking place.

The concept of entropy allows another equivalent statement of the second law of thermodynamics:

In any process, the entropy of the universe must either stay the same or increase.

Macroscopically, the second law in this form is a powerful tool to predict thermodynamic changes. It should be noted that on the microscopic scale, there is no guiding control exerted on atoms and molecules making them behave in a particular way. They interact randomly and the changes that take place on the macroscopic scale are simply the ones that are more likely to happen. The macroscopic concept of entropy just quantifies microscopic statistical analysis.

1 The diagram below shows the relation between the pressure and the volume of the air in a diesel engine for one cycle of operation of the engine. During the cycle there are two adiabatic processes, an isochoric process, and an isobaric process.

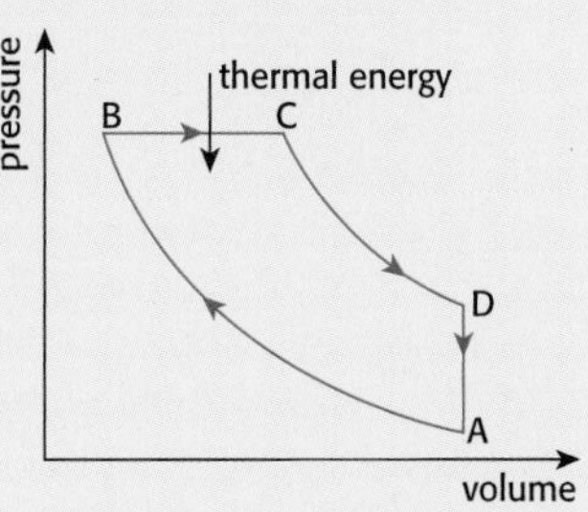

a) Explain what is meant by
  i) an adiabatic process [2]
  ii) an isochoric process [1]
  iii) an isobaric process [1]

b) Identify, from the diagram, the following processes
  i) adiabatic processes [1]
  ii) isochoric process [1]
  iii) isobaric process [1]

During process B → C thermal energy is absorbed.

The diesel engine has a total power output of 8.4 kW and an efficiency of 40%. The cycle of operation is repeated 40 times every second.

c) State what quantity is represented on the diagram by the area ABCD. [1]

d) Determine the value of the quantity that is represented by the area ABCD. [1]

e) Determine the thermal energy absorbed during the process B → C. [2]

[Total 11 marks]

2 When water freezes, the entropy change is −22 J K^{-1} mol^{-1} and the latent heat of fusion is 6 kJ mol^{-1}. Use these data to predict the temperature at which water freezes.

EXAMPLE 1

In Chapter 5 we introduced the example of an impossible situation (the broken bits of a glass spontaneously coming back together and the whole glass jumping back onto the table) that is not excluded by energy considerations. The explanation of this in terms of entropy could be as follows:

- Before falling to the ground, the glass was in an ordered state.
- As a result of falling to the ground, the pieces of glass are randomly spread around the floor, and the original PE of the glass is now shared with the surroundings: this is a more disordered state.
- The process of the glass falling to the ground and breaking into pieces is an entropy increase.
- The reverse process would be an entropy decrease and would violate the second law of thermodynamics.

EXAMPLE 2

The molecules in a solid are arranged in a more ordered state than when the same molecules are a liquid. When a liquid solidifies, the entropy of the substance must go down. The second law of thermodynamics tells us, however, that the entropy of the Universe must increase (or at least stay the same). When a liquid solidifies, bonds are made and thus thermal energy must be released to the surrounds. This energy must increase the entropy of the surroundings.

For a particular substance, the entropy change associated with changing state will depend on the amount of substance.

EXAMPLE 3

When chemists analyse chemical reactions to work out whether a particular process will take place, they need to take the entropy changes into consideration. In order to facilitate the calculations, the concept of Gibbs free energy has been developed.

A chemical reaction is analysed to calculate the overall change in Gibbs free energy, ΔG. The formula includes a term corresponding to the thermal energy released (called the enthalpy), ΔH, and another term corresponding to the entropy change:

$$\Delta G = \Delta H - T\Delta S$$

where ΔH is the thermal energy released as a result of the reaction, measured in J; T is the absolute temperature, measured in K; and ΔG is the Gibbs free energy of the reaction, measured in J.

If the overall change is positive then the reaction could occur spontaneously. A negative change means that the reaction cannot occur spontaneously.

End of chapter summary

Chapter 11 has three main themes: thermodynamic concepts, processes and laws. The list below summarises the knowledge and skills that you should be able to undertake after having studied this chapter. Further research into more detailed explanations using other resources and/or further practice at solving problems in all these topics is recommended – particularly the items in bold.

Thermodynamic concepts

- State the equation of state for an ideal gas **and use it to solve problems**.
- Describe the difference between an ideal gas and a real gas, the concept of the absolute zero of temperature and the Kelvin scale of temperature.
- Deduce an expression for the work involved in a volume change of a gas at constant pressure.

Thermodynamic processes

- Describe the isochoric (isovolumetric), isobaric, isothermal and adiabatic changes of state of an ideal gas.
- Draw and annotate thermodynamic processes and cycles on $p - V$ diagrams and use the diagrams to calculate the work done in a thermodynamic cycle **and to solve problems involving state changes of a gas**.

Thermodynamic laws

- State the first law of thermodynamics and identify it as a statement of the principle of energy conservation.
- State that the second law of thermodynamics implies that thermal energy cannot spontaneously transfer from a region of low temperature to a region of high temperature.
- State that entropy is a system property that expresses the degree of disorder in the system and the second law of thermodynamics in terms of entropy changes.
- Discuss examples of natural processes in terms of entropy changes.

Chapter 11 questions

1 This question is about a heat engine.

A certain heat engine uses a fixed mass of an ideal gas as a working substance. The graph below shows the changes in pressure and volume of the gas during one cycle ABCA of operation of the engine.

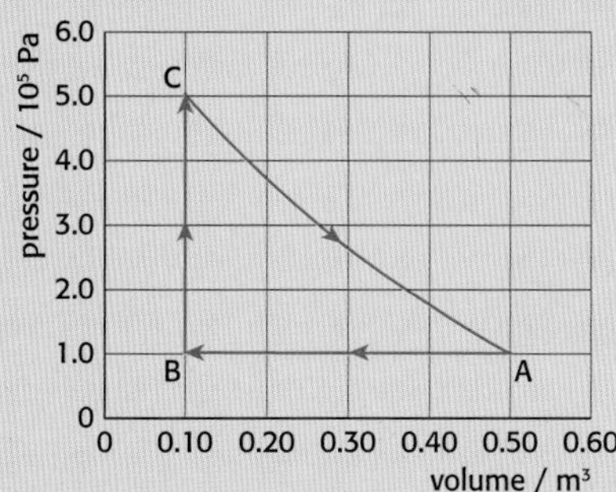

a) For part A → B of the cycle, explain whether

i) work is done by the gas or work is done on the gas. [1]

ii) thermal energy (heat) is absorbed **by** the gas or is ejected **from** the gas to the surroundings. [1]

b) Calculate the work done during the change A → B. [2]

c) Use the graph to estimate the total work done during one cycle. [2]

d) The total thermal energy supplied to the gas during one cycle is 120 kJ. Estimate the efficiency of this heat engine. [2]

(Total 8 marks)

2 A sample of an ideal gas is contained in a cylinder fitted with a piston, as shown below.

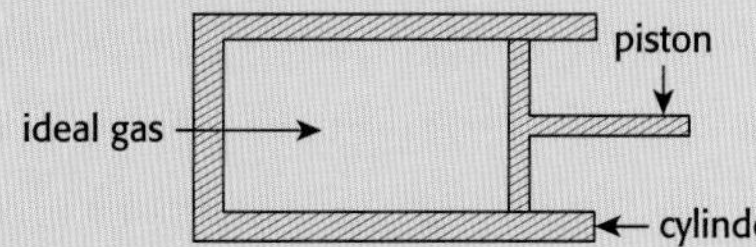

a) i) Explain, in terms of molecules, what is meant by the *internal energy* of the gas. [2]

ii) The piston is suddenly moved inwards, decreasing the volume of the gas. By considering the speeds of molecules, suggest why the temperature of the gas changes. [5]

iii) The gas now expands at constant pressure p so that the volume increases by an amount ΔV. Derive an expression for the work done by the gas. [4]

An engine operates by using an isolated mass of an ideal gas. The gas is compressed adiabatically and then it is heated at constant volume. The gas gains 310 J of energy during the heating process. The gas then expands adiabatically. Finally, the gas is cooled so that it returns to its original state. During the cooling process, 100 J of energy is extracted. The cycle is shown below.

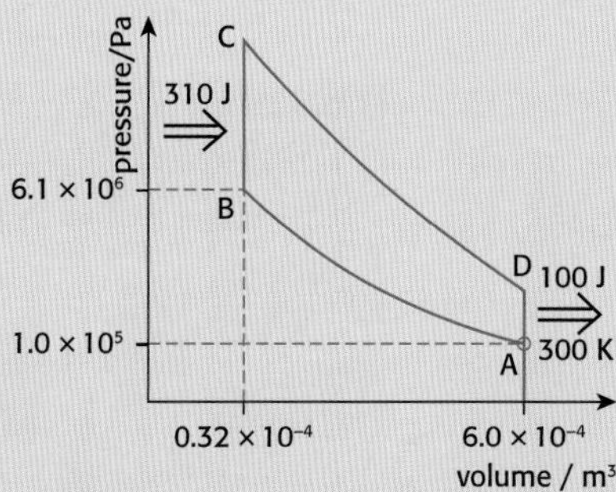

b) i) Mark, on the diagram, arrows to show the direction of operation of the stages of the cycle. [1]

ii) Using data for point A (temperature = 300k), calculate the number of moles of gas. [2]

iii) Determine the temperature of the gas at point B in the cycle. [2]

iv) State what is represented by the area ABCD on the diagram and give the value of this quantity. [2]

v) Calculate the efficiency of the engine. [3]

(Total 21 marks)

3 This question is about an ideal gas and entropy.

a) A fixed mass of an ideal gas is compressed from volume V_1 to volume V_2 at constant temperature. The variation with volume V of the pressure p of the gas is shown below.

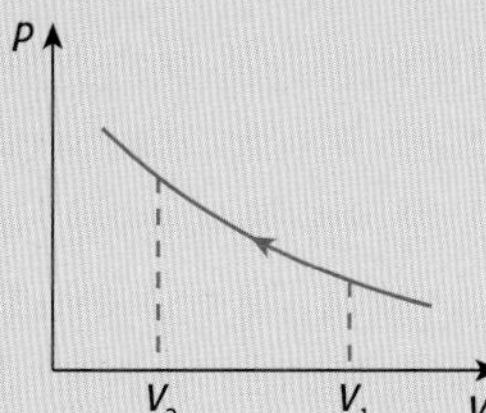

On the diagram above, draw a line to show the variation of pressure p as the volume of the gas is changed from V_1 to V_2 without allowing any thermal energy to enter or leave the gas. [2]

b) On the diagram in (a), identify

i) with the letter G, the line that represents the change that requires the greater amount of work done on the gas. [1]

ii) by shading an area of the diagram, the part of the diagram that represents the difference between the work done in the two changes. [1]

c) For the compression of the gas at constant temperature, deduce what change, if any, occurs in the entropy of the gas and of its surroundings. [3]

(Total 7 marks)

12 Electric currents

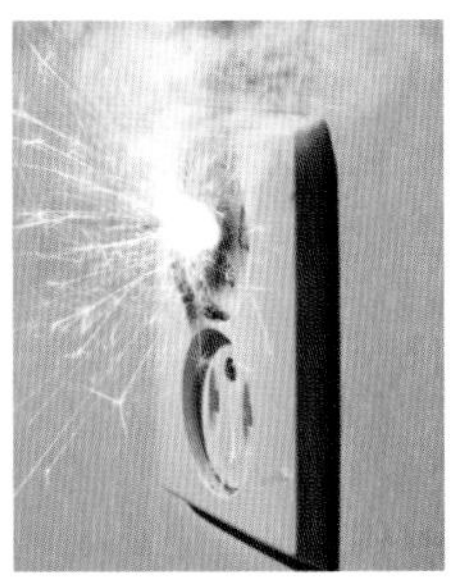

The analysis of an electrical circuit involves many related quantities. This chapter starts by focusing on the differences between the various quantities that are involved, and then introduces the techniques and approaches used when analysing a circuit.

Electrical quantities in brief

The main electrical quantities involved are as follows:

Charge, Q

Charge is a fundamental property of objects. There are two types of charge: **positive** and **negative**. An object with equal amounts of positive and negative charge is **neutral**. Charge is usually measured in terms of the charge on one electron but in electric circuits there are a lot of charge carriers (electrons or ions) so we define a new unit called the **coulomb** (unit symbol C). There are 6.25×10^{18} electrons in one coulomb of negative charge. The letter Q is used to represent charge.

Current, I

Current is the rate at which charge flows. It is measured in amperes (unit symbol A), which are equivalent to coulombs per second. For historical reasons, current is conventionally considered as the flow of positive charge around a circuit. The letter I is used to represent current.

Potential difference (p.d.), voltage, V

Potential difference is the energy difference per charge (i.e. the number of joules per coulomb, or watts per ampere) between two points in a circuit. When measured in volts (unit symbol V) it is usually referred to as **voltage**. The letter V is used to represent potential difference, or voltage.

Resistance, R

Resistance is defined to be the ratio of potential difference to current. The unit is volts per ampere, which is given the name ohm (symbol Ω). As the name suggests, we can think of the resistance as a quantity that measures the resistance to the flow of charge around a circuit. A high resistance needs a high p.d. across it to make a current flow, whereas a low resistance needs a small p.d. across it to achieve the same current. The letter R represents resistance.

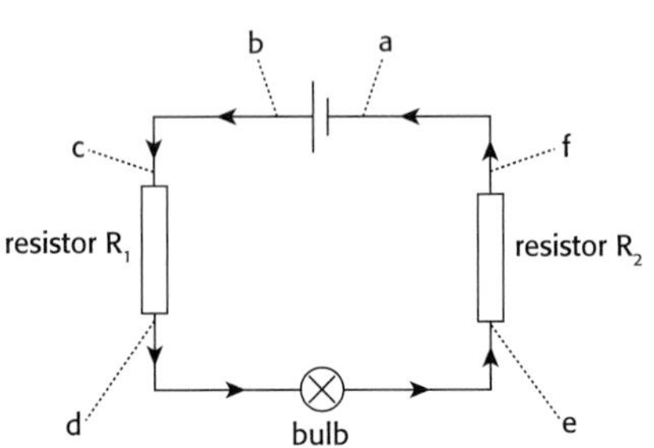

Figure 1

A current involves a charge moving around a circuit and gaining or losing energy. The circuit in Figure 1 illustrates the energy changes that are taking place for each coulomb. When energy is transferred from the moving charge, the energy is said to be **dissipated**.

The result summarized in Table 1 is always true. A complete journey around a circuit exactly uses up the energy per coulomb.

Table 1

a to b	The battery gives some electrical energy to any charge that moves through it. It is a 12 volt battery which means that every coulomb leaving the battery from the left-hand side has 12 joules more energy than when it went in on the right-hand side.
b to c	The wire is a very good conductor, which means that almost none of the energy is used up – a little must be, but it is so small we take it as zero.
c to d	The coulomb must have used up some of its energy moving through this region of higher resistance. Since p.d. measures the energy difference per coulomb we must be able to define a voltage across this component. Let's say it is 6 V in this case.
d to e	Now it goes through the bulb – again it uses up energy, so there must be a potential difference across the bulb. Let's say 2 V.
e to f	Once again more energy is used up. Again there is a p.d. across the resistor, this time 4 V.
f to a	Just like b to c, virtually no energy is used up. We pretend the wire is such a good conductor that no energy is used at all.
	The result is that the coulomb was given 12 joules in order to complete its journey around the circuit, and all of this has been dissipated (6 + 2 + 4 = 12).

You may ask: "What if it hasn't got enough energy to get all the way round? Does it stop after the first resistor?" Remember that the rate of flow of charge (i.e. the current) must be the same before and after the resistor; otherwise more coulombs would go into the resistor than would come out of it. This would mean a build-up of charge, which is impossible.

Where does the energy per coulomb go? To understand this, let's concentrate on one device. We will consider resistor R_2 (Figure 2).

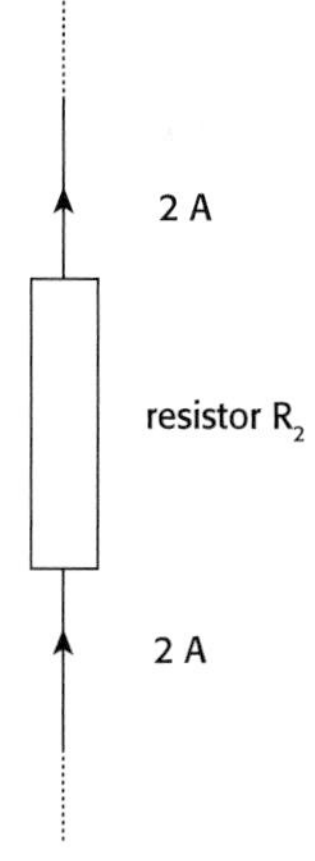

Figure 2

The energy used up per coulomb depends on the value of R_2, and the rate of flow of charge (the current) through it. Assume that in this example the current is 2 A: that is, 2 coulombs move through R_2 every second. We already know that this takes 4 V.

The total energy dissipated is 4 J for every coulomb, and we know that there are 2 C every second. So we use up 8 J every second. This makes R_2 get hotter and hotter.

We now build a completely different circuit incorporating R_2: one that can put 4 A through it. This will take more energy per coulomb. R_2 now needs 8 V across it, because to double the current we need to double the p.d. Here 4 coulombs per second are each dissipating 8 J per coulomb, so the total rate at which energy is used up in this new circuit is 32 J/s.

In summary, when we set up a circuit we fix the *total* energy per coulomb (i.e. the total voltage) by the power supply we choose. The total p.d. it provides is exactly used up in the various devices around the circuit, and the p.d. across each one and the current through it are related.

Resistors are an easy group of devices to work out, because the p.d. across them and the current through them are proportional:

$$\text{p.d. across} = \text{current through} \times \text{resistance} \qquad (1)$$

Example 1: a series circuit

In a series circuit, the components are connected end-to-end (see Figure 3).

In the circuit in Figure 3, 18 V is put across two resistors in series, one of 3 Ω and the other 6 Ω. In this case, 2 A must flow through the circuit for the numbers to work:

The p.d. across the 3 Ω resistor = current through × resistance
= 2 A × 3 Ω
= 6 V

The p.d. across the 6 Ω resistor = current through × resistance
= 2 A × 6 Ω
= 12 V

So the total p.d. across the two resistors = 6 V + 12 V
= 18 V

The numbers balance!

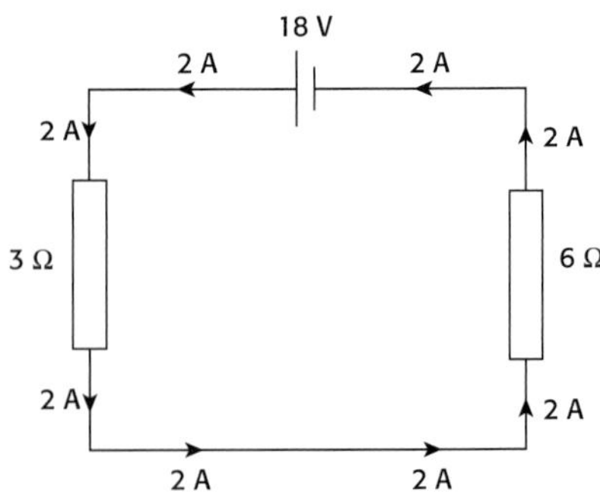

Figure 3 Series circuit

Example 2: a parallel circuit

Components in a circuit are said to be connected in parallel if their ends are connected to the same point in the circuit. In Figure 4, the three components (the battery and the two resistors) are connected in parallel.

In the circuit in Figure 4, the same p.d. (18 V) is put across both resistors by the battery.

The p.d. across the 3 Ω resistor = 6 A × 3 Ω
= 18 V

The p.d. across the 6 Ω resistor = 3 A × 6 Ω
= 18 V

So the total current that flows must equal 6 A + 3 A = 9 A

All you have to do is to apply equation 1 to the resistors. Remember that you must use the p.d. across the resistor, the current through the resistor, and the value of the resistor all in the same equation. You must not use, for example, the p.d. due to the battery, the current through resistor 1, and the value of resistor 2 together, as they don't match.

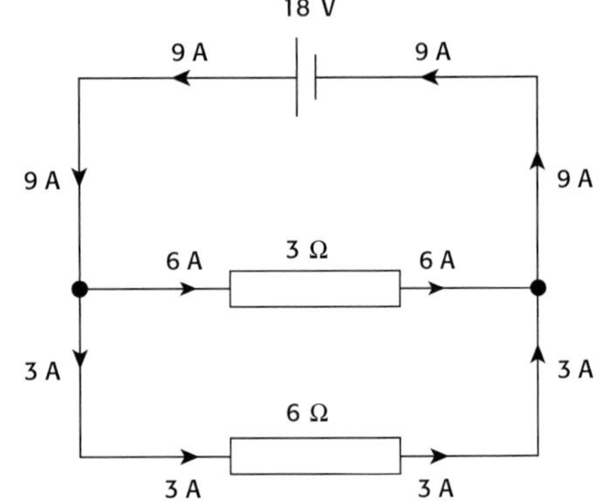

Figure 4 Parallel circuit

***Thinking about science:* What is the real nature of charge?**

Most people are very familiar with everyday electrostatic effects. For example, using a plastic comb may make your hair stand on end. Devices such as photocopiers and air ionizer/purifiers rely on the application of this effect. All electrostatic phenomena are based around charges that are, at least temporarily, at rest.

An electric current always involves charge carriers (usually electrons) moving around a circuit.

Whether we are switching on a light or a complicated piece of electronic equipment, a flow of charge must be involved. But what *is* charge?

As we have already said, there are two types of charge: **positive** and **negative**. They are opposites. If a positive charge and an equal amount of negative charge are in close proximity, the resulting object is uncharged or **neutral**. The SI unit of charge is the coulomb (C).

Some objects can become charged as a result of friction. Study of the charging process shows that the total amount of charge is always conserved. When charge is generated by rubbing two objects together, the positive charge gained by one object is balanced by the negative charge gained by the other object.

For example, if a neutral insulating rod is rubbed with a neutral cloth, and the rod gains a positive charge of $+5 \times 10^{-9}$ C, the cloth will gain an equal but opposite charge (-5×10^{-9} C).

Charge is conserved in electric currents. In any given time, the number of coulombs that flow into a device must equal the number of coulombs that flow out. No charge can be lost; nor can the rate of flow of charge change.

Charge carriers can stay still or be moved about, but an object's charge is something completely different from its mass. Charge can be given to, or flow through, an object. What is happening when an everyday object becomes charged, or when a current flows?

Everyday matter can be thought of as being made up of atoms, or atoms combined into molecules. At this atomic scale we can identify which particles are charged and the type of charge that they carry. In this simple atomic model, everything is made up of fundamental particles – electrons, protons, and neutrons:

- electrons are negatively charged
- protons are positively charged
- neutrons are neutral.

All electrons have the same mass (9.11×10^{-31} kg) and the same amount of negative charge (1.6×10^{-19} C). This is the smallest observable quantity of charge. It is not possible to break an electron down into smaller parts, or remove or change its charge and leave its mass.

A proton has an exactly equal but opposite amount of charge to that of an electron, so a proton and an electron together (as in a hydrogen atom, for example) can be considered neutral. Protons are bound inside nuclei, so the movement of charge is often due to the movement of electrons. In the example in Figure 6 the rod lost electrons, so it has more protons than electrons and, overall, is positive. The cloth now has an excess of electrons and so overall is negative.

When an electric current flows along a copper wire the overall effect is that charge moves around the circuit. In this case the protons are fixed because the atoms are not free to move along the wire: the copper is solid, so the electrons carry the charge in a wire. But current does not always just involve the flow of electrons. Many liquids contain mixtures of atoms or molecules that have gained or lost electrons. These are called **ions**.

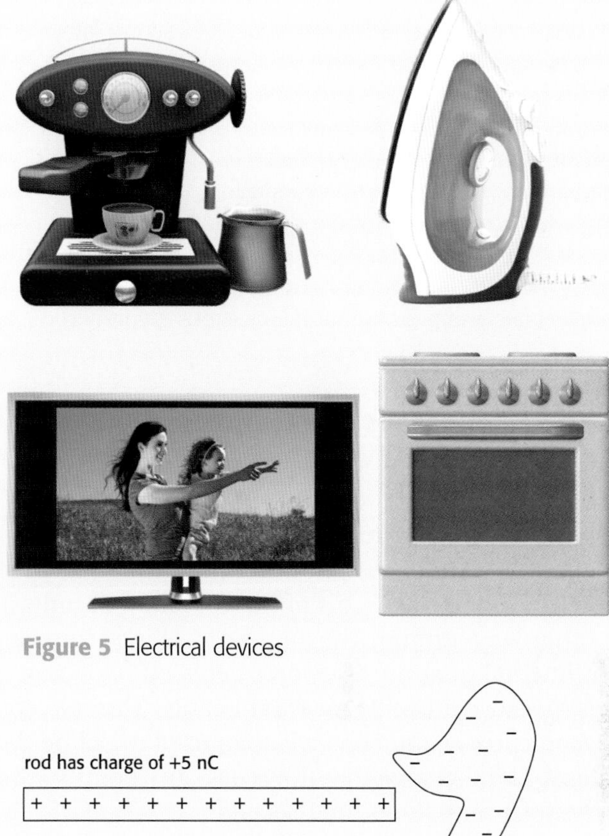

Figure 5 Electrical devices

rod has charge of +5 nC

+ + + + + + + + + + + + +

cloth has charge of −5 nC

Figure 6 Result of a neutral insulating rod being charged by a neutral cloth

For example, a water molecule is a combination of atoms: two hydrogens and one oxygen – H_2O. If a soluble ionic impurity (e.g. table salt) is added to liquid water, the water molecules tend to split into hydroxide ions (OH^-) and hydrogen ions (H^+). When a current flows in this water, it is the ions that are moving, carrying the charge with them.

Semiconductors are another example of conduction in which a current is not just free-moving electrons. In many conducting materials the number of available charge carriers is fairly constant. In semiconductors the number of charge carriers available can be varied by changing the external conditions, for example, the temperature.

Conductors and insulators

A conductor is a substance that allows an electric current to flow through it. This process requires the input of energy but the better the conductor, the less energy that is used up for each coulomb that flows. An insulator is a substance that does not allow the flow of electric current because a great deal of energy would be required for a charge carrier to move.

A substance is classified as a conductor or an insulator according to how free the charge carriers are to move within its structure. In a solid, the only charge carriers that are available to move are the electrons. In insulators, the electrons are tightly bound to their own individual atoms, and are not free to move. In conductors, and particularly in metals, these electrons are **delocalized**: that is, they are free to transfer from atom to atom. At any given time every atom will be neutral, because it has equal numbers of electrons and protons, but the electrons are free to move: when an electron leaves an atom, a replacement is provided from the neighbouring atom.

How fast do the electrons in a normal current move around the circuit? You already know that electrons are moving around randomly inside the metal due to thermal motion. When a p.d. is applied the electrons still move essentially randomly but with an overall or "bulk" motion. Typically, the speed of this "bulk motion" can be measured in mm s^{-1}.

Electrical definitions and approximations

Definition of the units for current and charge

We define currents in terms of the quantity of charge that is moving – current is the rate of flow of charge. It would seem sensible to define the unit of current in terms of the units of charge: that is, to define the ampere in terms of the coulomb. SI units actually do this the other way around. The ampere is defined in terms of an experimental quantity (the force between two long, current-carrying wires) because it can be measured easily. The conceptual link between current and charge is then used to define the coulomb: 1 coulomb is the charge that flows past any point in the circuit when a current of 1 amp flows for 1 second:

$$1 \text{ coulomb (C)} = 1 \text{ ampere second (A s)}$$

Resistance and resistivity, and their units

The relationship between resistance R, potential difference V, and current I is given by

$$R = \frac{V}{I}$$

This equation is the *definition* of the resistance R. Whatever conditions exist in a particular circuit, this relationship must be true, because it is how we define the resistance. This definition leads to the definition of a new unit, the ohm (symbol Ω). It implies that resistance is measured in volts per ampere, and we define

$$1 \text{ ohm } (\Omega) = 1\ \frac{\text{volt (V)}}{\text{amp (A)}}$$

A change in temperature or external pressure can change an object's resistance. Each time we change the p.d. across a device, we should expect a different value for the resistance. In general, the resistance of any object is not just one fixed value but it has a range of values that depend on external circumstances, particularly the circuit to which it is connected and the particular values of current and p.d.

The different components in a circuit can be connected in series, parallel, or a combination of the two. Simple equations can be derived to calculate the total resistance of any given combination.

The derivations of equations 2 and 3 involve applying the laws of conservation of charge and energy.

$$R_{\text{series}} = R_1 + R_2 + R_3 + \ldots \quad (2)$$

$$\frac{1}{R_{\text{parallel}}} = \frac{1}{R_1} + \frac{1}{R_2} + \frac{1}{R_3} + \ldots \quad (3)$$

Resistivity, ρ, is a useful quantity that allows you to find the resistance of a wire experimentally. The definition is given in terms of the resistance of the wire R, its length l, and the area of its cross section, A.

$$R = \rho \frac{l}{A} \quad (4)$$

Using equation 4, show that the units of resistivity must be ohm metres.

This definition means that the units of resistivity must be ohm metres (Ω m). Note that this is the ohm multiplied by the metre, NOT the ohm divided by the metre ("ohms per meter"). Resistivity is related to the microscopic properties of the material.

Perfect electrical meters

The two most common electrical measurements made in circuits are measurements of current and voltage. The general names for the devices that achieve these measurements are ammeters and voltmeters, respectively. When analysing real circuits, the properties of the individual meters need to be taken into consideration.

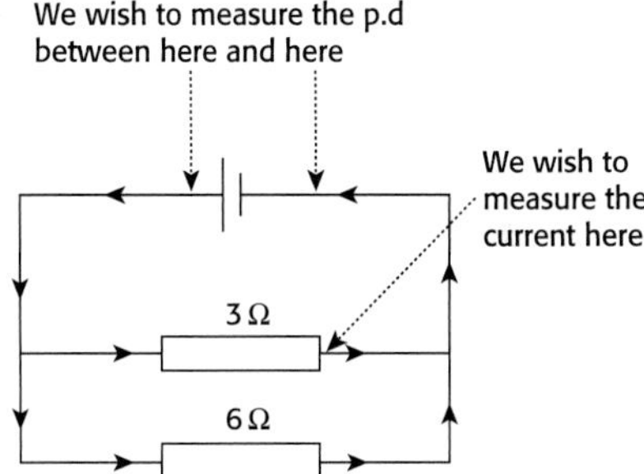

Figure 7 A circuit to be measured

A perfect measuring device should measure and not do anything else. Thus when we imagine connecting a perfect meter into a circuit we assume that it does not affect the circuit in any way. Suppose we wish to measure the current and the potential difference shown in Figure 7.

The ammeter

Ammeters measure the current flowing at a particular point in the circuit and thus the addition of an ammeter requires the meter to be added in series with the rest of the circuit. If we want to measure the current flowing out of the 3 Ω resistor, we would break the circuit at this point and add an ammeter as shown in Figure 8.

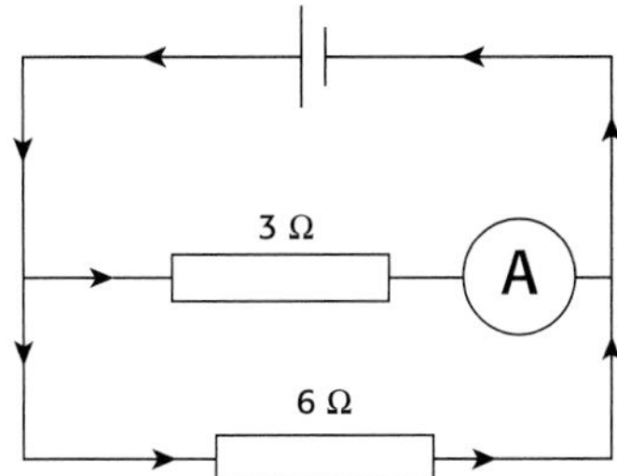

Figure 8 Addition to ammeter to measure current flowing out of 3Ω resister

If the perfect ammeter is not going to affect the circuit with its addition, it should have zero resistance. If the ammeter has a non-zero resistance this means the total resistance of the upper branch would be increased slightly from 3 Ω and the current would also change.

The voltmeter

Voltmeters measure the potential difference between two particular points in the circuit and thus the meter must be added in parallel with the device (or devices) in question.

The perfect voltmeter should have infinite resistance – in other words, no current should flow through it. If the voltmeter does allow a current to flow, this means the total current in the circuit must have changed.

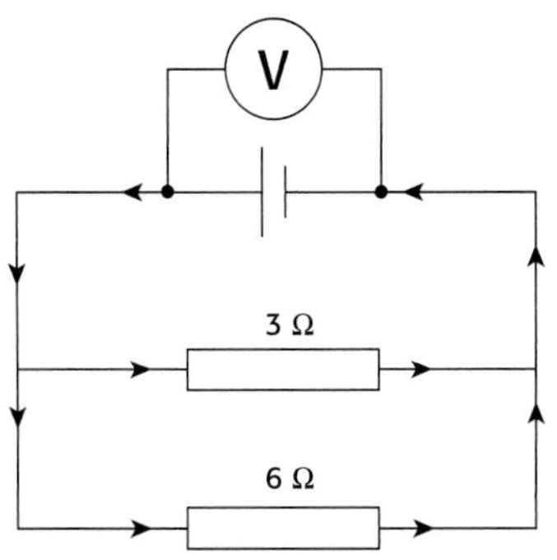

Figure 9 Addition to voltmeter to measure p.d. across the battery

1 The circuit in Figure 10 connects two bulbs, a switch and a 6 V battery together in series:

a) When the switch is closed, which bulb lights up first? Is it A, B, or do they come on together?

b) With the switch open (i.e. no current flowing) what would a voltmeter read if it was connected across

i) bulb A

ii) the battery

iii) the switch?

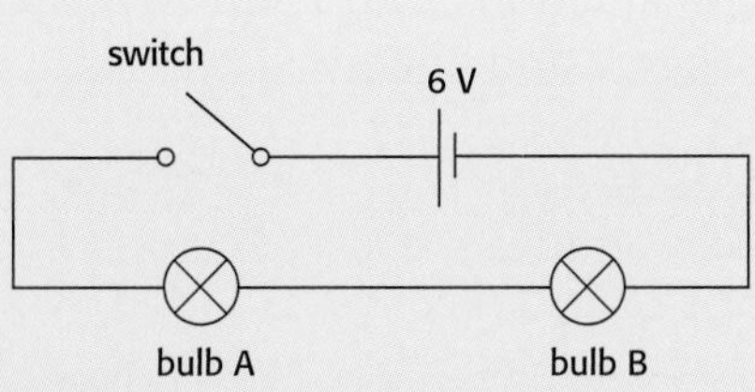

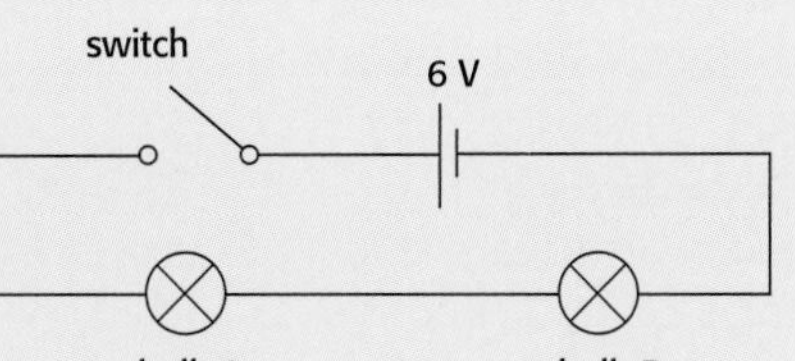

Figure 10

Rediscovering physics: Ohmic and non-ohmic devices

Remember there is no particular reason why the resistance of something should remain constant if other variables are changed. Most of the time the resistance of a device will vary depending on the situation in which it is used. Different devices will have different variations. We need an experiment to measure the variation with p.d. of current for a range of devices. A simple circuit that allows values to be recorded is shown in Figure 11.

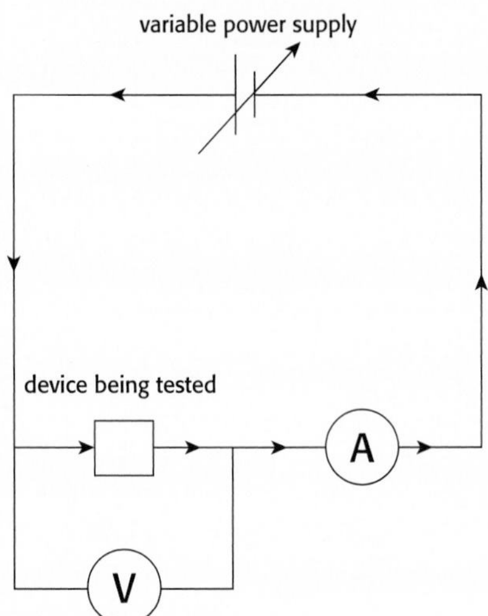

Figure 11 Circuit to measure variation with p.d. of the current for a device

Vary the power supply from zero up to a maximum value (which will depend on the particular device being tested) and record the corresponding values of current that flow through the device. If a smoothly varying power supply is not available then set up a potentiometer circuit (see page 177) to vary the p.d from zero up to the maximum. It is also possible to use a variable resistor in series with the device being tested in order to control the current flowing. The limitation of this latter approach is that the variable resistor will only allow a limited range of values for potential difference to be accessed and ideally we want the full range from zero up to the chosen maximum. The worst effected section of results will probably be at the lower value of p.d. and current. When the variable resistor is set on its maximum value, then the current must be at its minimum value – with this circuit, it is not possible to reduce the current even further. If you have to use this approach, remember that you can still arrange for the current to be zero by unplugging the power supply and thus include the voltmeter and ammeter readings for this situation.

You should use the list below to check that you are following good experimental procedures.

- It is good practice to always check that the whole circuit is connected properly before switching on.
- Always start with the circuit providing the minimum current possible and if the meters offer a choice of possible ranges the they should be set on the least sensitive range. Why should this be the case?
- If the meter gives a reading that is too small to be read accurately then the sensitivity can be increased. Sometimes it may be appropriate to replace the meter with a more sensitive meter.
- It is always possible for devices to overheat. It is advisable to monitor the temperature of components throughout the experiment.
- As a guide, 5 readings over the whole range of voltages is the absolute minimum required to show a relationship, ideally you should try to get readings for around 10 different voltages.
- The data points should ideally be spread out uniformly over the range.
- Individual data points should be repeated / checked and ideally more than once.
- Does the device respond in exactly the same way if its connections are reversed?
- It is good practice to plot the points as they are recorded – this is easily achieved using a graphics display calculator. This allows you to go back and check any points that seem unusual.
- You need to remember to record all of your raw data and include the uncertainties in your measurements.

Possible devices to test could include any or all of the following. See the note below for more details before attempting to build the circuit.

- A tungsten filament lamp
- A resistor
- A diode

If you have time, the following devices can also be tested.

- A different resistor or combinations of resistors
- An LDR
- A thermistor

Note

The characteristics of the components chosen should match the measuring range for the voltmeters, ammeter and milli-ammeters that are available. For safety reasons many schools work with a 12 V supply and ensure that the normal working currents are limited to about 0.1 A, often less. You should bear the following points in mind:

Tungsten filament lamp

Lamps are available with different ratings. These are normally written on the side e.g. "6 V, 0.1A". Sometimes the information is given in terms of the power consumption "6V, 0.6W" in which case you can easily work out the maximum current drawn. This tells you the operating conditions necessary for normal brightness. The bulb will probably be able to cope with a slightly higher p.d.—a bulb should be tested up to and slightly beyond its rated value. Bulbs that are rated identically will not necessarily be absolutely identical.

Resistor

Many resistors have a colour code on them to record the value of the resistor along with the manufacturer's uncertainty limits on this value. e.g red-yellow-orange-gold represents a 24 kΩ ±5%. There is always a risk that resistors can get too hot. The graph that is obtained should be used to calculate the value of the resistance used and this can be compared with the manufacturer's values. If time allows, it is instructive also to compare different results for different resistor combinations. For example, it would be interesting to compare the results for a 500 Ω resistor, to the equivalent combination of 200 Ω and 300 Ω in series and to the combination of 1200 Ω and 820 Ω in parallel.

Diode

The maximum current that semiconducting signal diodes (e.g. silicon diode) should be subjected to can vary. Typically, however, they are unable to withstand currents larger than 10 mA. This means that whenever they are connected in a circuit, they should be protected by a resistor to limit the total current in the diode. You need to remember to measure the p.d. for the diode and not for the combination of the protective resistor and diode.

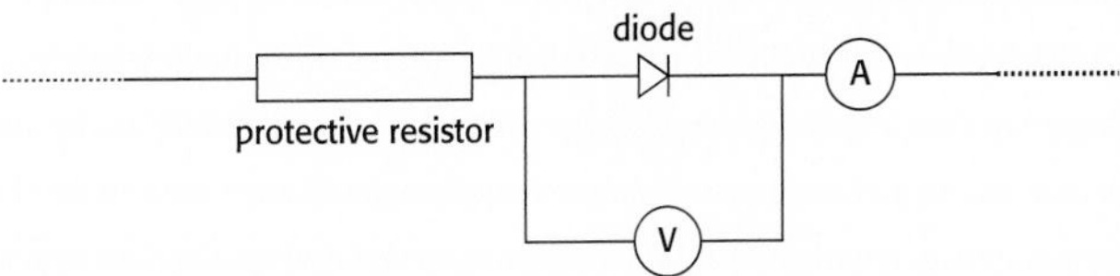

Figure 12 Measuring the p.d. across the diode

The diode is particularly interesting if you investigate it connected in both directions. You will probably need a very sensitive ammeter for some of these measurements.

LDR & thermistor

These are semiconducting devices and there are a very wide range of possible characteristics available. They can initially be treated in the same way as a 5 kΩ or 10 kΩ resistor but you may soon find that they do not always behave in this way. Can you experiment to find out what is going on?

Discussion of results

Idealised results for four different combinations are shown below. Note that the scales of the graphs are not necessarily the same.

Device (a) is a resistor. The straight line through the origin shows that the current flowing through the device is proportional to the p.d. across the device. We have defined resistance as the ratio of p.d to current, so another way of summarizing the results in the graph is to say that the straight line through the origin shows that the resistance is constant over the measured range. The value of the resistance is (gradient of the line)$^{-1}$.

Device (b) is the filament bulb and its resistance is not constant over the measured range. The resistance at any particular value of p.d. is just the value of p.d. divided by the value of current. As the value of the p.d. increases, the resistance of the bulb also increases. Part of the explanation for is that at higher values of p.d., a higher current is flowing causing the temperature of the filament to increase and this causes the bulb's resistance to change.

Device (c) is also a resistor with constant value resistance. If the scales on the axes are the same as in device (a), then device (c) must have the lower resistance because the gradient of its line is higher.

Device (d) is a semiconducting diode. In the forward direction, current can flow with very little resistance whereas in the reverse direction the diode has a very large resistance and so current is virtually reduced to

zero. The reason for using a protective resistor is explained if we look at the nearly vertical section of the graph in the forward direction. It means that effectively the diode allows any forward value of current with a virtually constant p.d. across it (typically 0.7 V for simple diodes). The ability to block current in the reverse direction can also "break down" if the p.d. if high enough. This effect is used in a particular type of diode (Zener diodes) to help regulate voltages within circuits.

Experiments such as this can be used to verify an experimental law, **Ohm's law**. This states that the resistance of a metal at *constant temperature* is constant. Another way of stating this is that for a metal at constant temperature, the potential difference across it will be proportional to the current flowing through it.

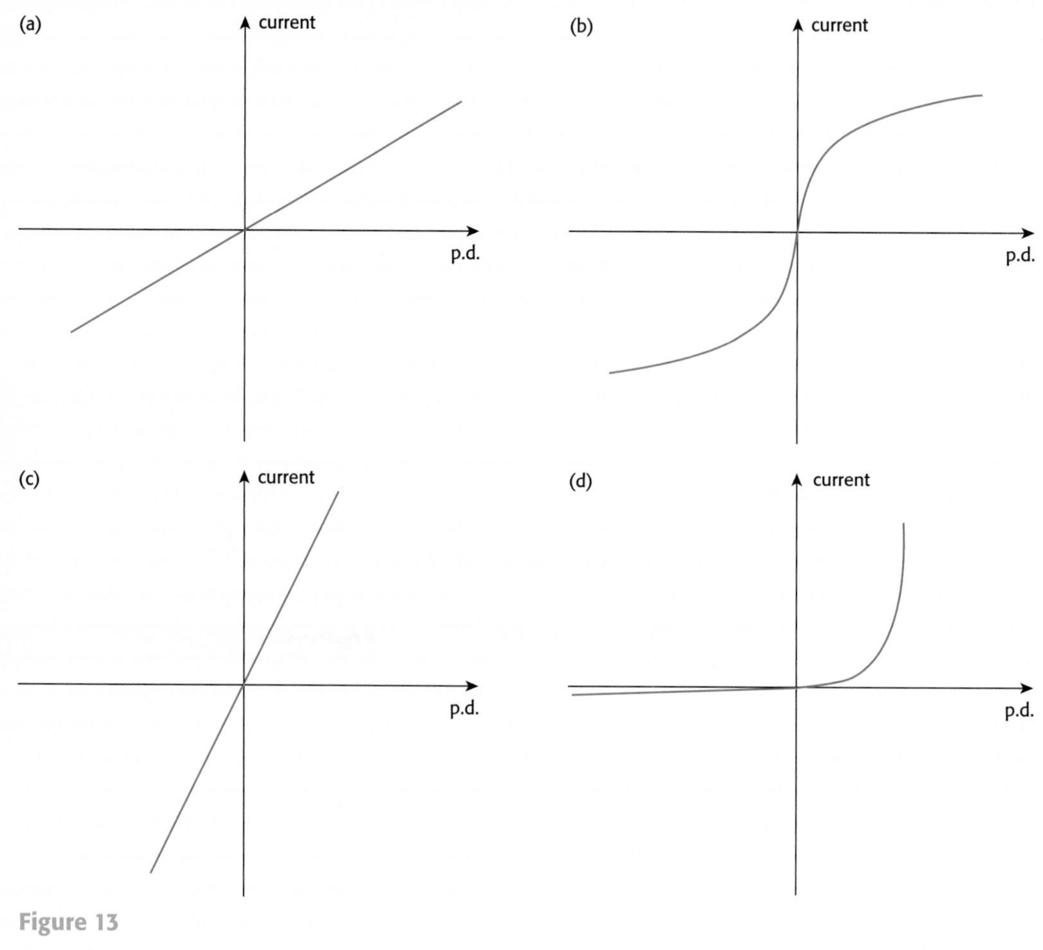

Figure 13

Science tools: Factors that affect the resistance of a material

Your task is to plan an experiment that investigates one factor that affects the resistance of a sample of a material.

The aim of this section is to provide a framework within which you can attempt to plan and carry out an investigation involving electric currents. It has been designed so that as long as you are able to attempt this without further help or assistance, this task could be formally assessed and count towards your final IB Diploma Programme score.

When planning an experiment, a good approach is for you to imagine that you have suddenly turned into a teacher for the day. Unfortunately, nobody left you any instructions and you have to work out the content of the next physics practical lesson. Not only does this mean that you have to invent the experimental procedure that your students will use, but it needs to be reliable enough to produce data that can be usefully analysed. On top of this, you also have to choose

and specify the necessary equipment so that the technician can get it ready for you. Finally, you also have to write out the method so that your class has good instructions to follow. Your instructions need to be sufficiently clear to allow everybody to understand exactly what they have to do. Make sure you leave nothing out.

There are three different aspects to the skill of experimental design. These could be thought of as three stages in the experiment design process:

- Defining the problem and selecting the variables – this involves formulating a focused research question from the general aim stated above and identifying the relevant independent and dependent variables.
- Controlling the variables – this involves designing a method for the effective control of all other possible variables.
- Developing a method for collection of data – the method must allow for the collection of sufficient relevant data.

Power dissipation in circuits

When designing, analysing, or inspecting any electrical circuit, safety considerations should always be the top priority. Two very common sources of electrical fires could be easily prevented if homeowners knew the relationship between power, current, and potential difference.

We want to calculate the power that is **dissipated** in an electrical component. We will start by considering a simple portable heating circuit (see Figure 14). In this circuit there are two resistors (3 Ω and 6 Ω) connected in series to an 18 V battery.

In order to work out the power dissipated in any device, we need to multiply together the p.d. across it and the current going through it.

power (in watts) = potential difference (in volts) × current (in amps)

$P = VI$

In the circuit,

the power dissipated in the 3 Ω resistor = 2 A × 6 V = 12 W

the power dissipated in the 6 Ω resistor = 2 A × 12 V = 24 W

∴ the total power dissipated by the resistors = 12 + 24 = 36 W

Of course we could have calculated the power developed by the battery and we would have found the same answer

the power developed by the battery = 2 A × 18 V = 36 W

This equation comes directly from the definitions of power, p.d., and current.

$$\text{power} = \text{rate of energy consumption} = \frac{\text{energy used}}{\text{time taken}}$$

$$\text{current} = \text{rate of flow of charge} = \frac{\text{charge flowed}}{\text{time taken}}$$

$$\text{p.d.} = \frac{\text{energy used}}{\text{charge flowed}}$$

$$\text{so p.d.} \times \text{current} = \frac{\text{energy used}}{\text{charge flowed}} \times \frac{\text{charge flowed}}{\text{time taken}}$$

$$= \frac{\text{energy used}}{\text{time taken}} = \text{power}$$

Show that the current flowing in the circuit (Figure 14) is 2 A, the p.d. across the 3 Ω resistor is 6 V, and the p.d. across the 6 Ω resistor is 12 V.

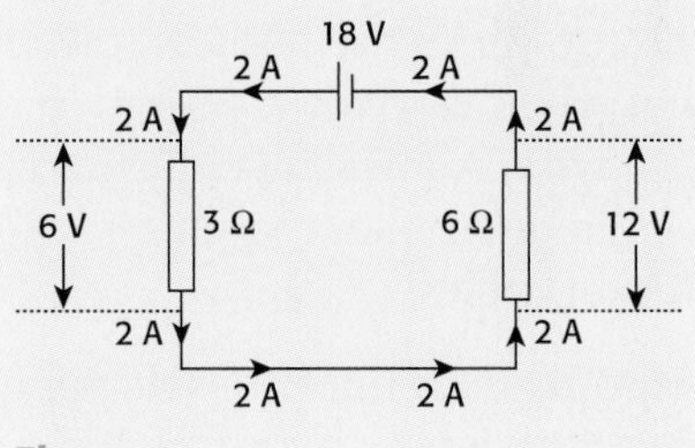

Figure 14

Since p.d., current, and resistance are all interrelated, this equation can be stated in several equivalent ways.

We obtain:

$$P = VI = I^2R = \frac{V^2}{R} \qquad (5)$$

By using equation 1 on page 162, show that the equation $P = VI$ leads directly to $P = I^2R$ and $P = V^2/R$.

We have worked out that the two resistors receive a total of 36 joules every second from the battery. Energy dissipated in a resistor is in the form of thermal energy so the resistors will warm up. When you buy a resistor, it comes with a certain power rating. If you are intending to build the circuit in Figure 18 then you must make sure that the battery is capable of delivering 36 W safely, and the resistors are each capable of safely dissipating the calculated amount of power.

Even if the battery and components were all perfectly safe, other parts of the circuit can go wrong. For example, the wires must be capable of carrying the current that flows in the circuit.

The current and the power calculations were based on the assumption that the wire joining these devices had zero resistance. Even though it probably has a low resistance, it will certainly not be zero. A thick wire has less resistance than a thin wire and this means that when carrying a current, a thick wire will get less hot than a thin wire. A low-power device can be connected using thin wires but a component that requires a lot of power must be connected using thicker cables.

Not only does this principle apply to the connection wires inside an electrical device, it also applies to any power cables or extension leads that are used to connect devices to a source of power. The wiring in your home must be capable of allowing the current to flow without getting too hot. If the wires get too hot, they can cause a fire (see page 179).

Mathematical Physics: ac / dc

In this chapter we have discussed two different types of electric circuits: battery powered circuits and domestic (mains) circuits. The circuit based on the battery involves *direct current* (or dc) whereas the mains electricity is an alternating current (ac). A direct current always flows steadily in the same direction whereas an *alternating current* varies in direction (first one way then the other). It also varies sinusoidally between a maximum current in one direction and the same magnitude maximum current in the other direction. Mains electricity is alternating.

The same equations can be used whether we have a.c or d.c because the numbers used to quantify alternating currents are carefully chosen. Whenever we describe an alternating current we use quantities technically known as the rms (root mean squared) values. These numbers are defined in such as way as to ensure that the same equations can be used for ac and dc. For example, when a bulb is connected to the mains electricity then the current must be constantly changing directions—typically 50 or 60 times each second. If we measure an alternating current of 3 A flowing in a bulb, the value of the instantaneous current at any given time varies sinusoidally from a maximum of about 4.2 A to 4.2 A in the opposite direction.

This can be calculated using the approximation:

$$I_{RMS} = \frac{I_{MAX}}{\sqrt{2}}$$

The overall effect of this varying current is to dissipate **the same amount of power** as if the current was a steady (dc) current of 3A.

In an ac circuit, the idea of resistance is replaced by the analogous idea of impedance.

Real devices: power supplies

Now we will look at the ways in which real batteries and other power supplies depart from their assumed ideal behaviour.

We can use equation 1 on page 162 to predict the current, I, that would be drawn from a 6 V battery by a 2 kΩ resistor.

6 volts across the resistor means that the current is

$$I = V / R = 6 \text{ V} / 2000 \ \Omega = 0.003 \text{ A} = 3 \text{ mA}$$

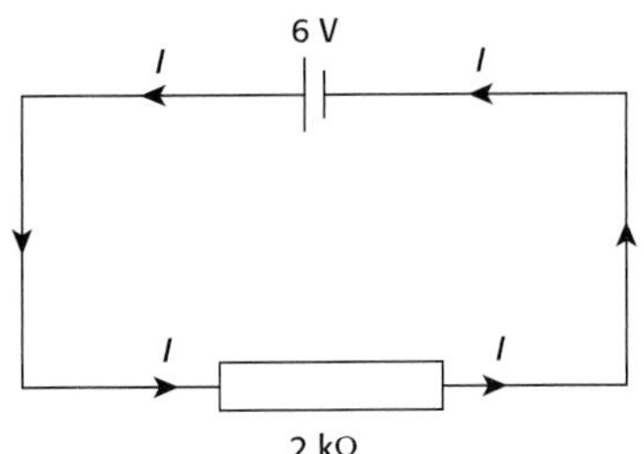

Figure 15

This prediction can be checked experimentally for different values of resistor. Table 2 shows the results of such an experiment.

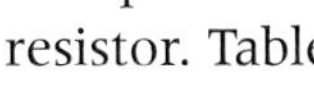

Table 2

Resistance / Ω	Predicted current / A	Actual current / A
2000	0.003	0.003
200	0.03	0.03
20	0.3	0.29
2	3	2.4
0.2	30	8.8
0.02	300	11.5
0.002	3000	12.0
0.0002	30000	12.0

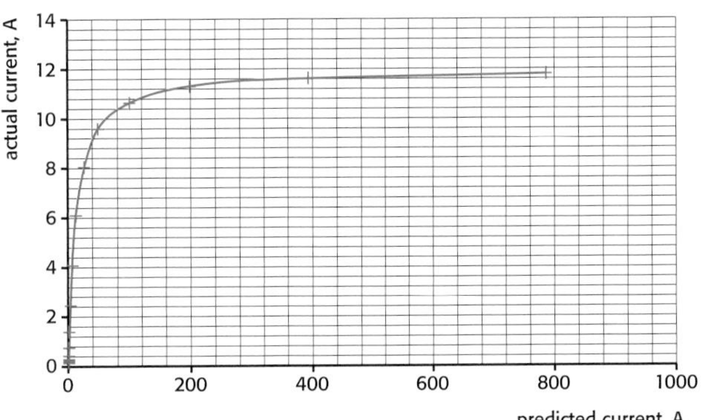

Figure 16 Actual current vs. predicted current

The actual current provided by the battery seemed to be limited to 12.0 A. In order to check that the battery was producing 6.0 V, the experimenter checked to see if the potential difference across the terminals (the **terminal p.d.**) was remaining stable. Table 3 shows the results.

Table 3

Resistance / Ω	Actual current / A	Voltage across battery / V	"Lost" voltage/ V
2000	0.003	6.00	0.00
200	0.03	5.99	0.01
20	0.29	5.85	0.15
2	2.4	4.80	1.20
0.2	8.8	1.71	4.29
0.02	11.5	0.23	5.77
0.002	12.0	0.02	5.98
0.0002	12.0	0.00	6.00

Some of the 6 V seem to have been "lost". Remember that voltage is a measure of the energy per charge so missing voltage means that some energy has apparently gone missing. We know from the law of conservation of energy that this is impossible, so we must find an explanation.

The solution to this problem is that the battery actually has some **internal resistance**. The *total* energy difference per unit charge around the circuit does still remain fixed at 6 V, but some of this energy is used up inside the battery. This is because the energy difference per unit charge from one terminal of the battery to the other is less than the total energy made available by the chemical reaction inside the battery. To distinguish between the terminal p.d. and the total energy difference per unit charge, we introduce a new term—the **electromotive force (emf)**, ε. The emf is equal to the terminal p.d. when no current flows.

Note that the term emf is actually an old phrase, still in use for historical reasons. It is a misleading name because the emf is not actually a force you can measure in newtons. The emf of a battery can be defined as the total energy per unit charge given by the battery. This will be the same as the total energy per unit charge dissipated by a charge during one complete circuit.

An easier way to imagine this situation (and one that works to a very good approximation) is to say that there is an additional resistance in the circuit, an **internal resistance**, which is located between the terminals of the battery. In the example above, the maximum current was limited to 12.0 A, thus the effective minimum resistance of the circuit was given by:

$$\text{internal resistance } r_i = \frac{6.0}{12.0} = 0.5\ \Omega$$

This value is the internal resistance of the battery. All power sources have internal resistance. When a battery becomes "flat" and is unable to provide as high a voltage as it used to provide, its internal resistance must have increased.

The two batteries in Figure 17 are both sold as 12 V batteries – this means they both have an emf of 12 V. The car battery (Figure 17a) has a much lower internal resistance and thus can provide a much larger current.

(a)

(b)

Figure 17 Two 12 V batteries

Figure 18 is similar to Figure 15 on page 172 but shows a way of visualizing the battery as a combined power source and resistor.

Applying $V = IR$ and equation 2 on page 166 to the whole circuit, we get the following equation:

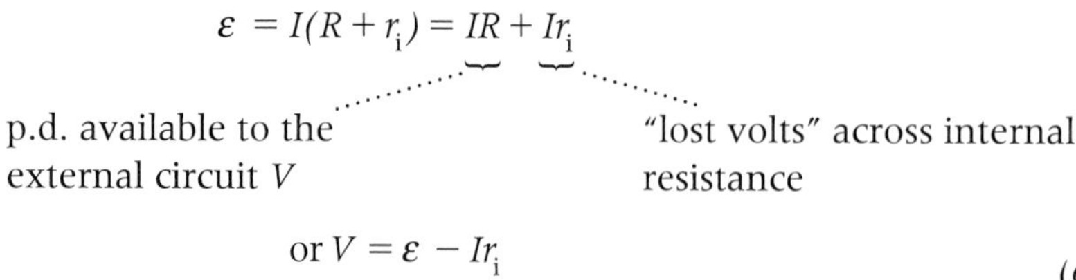

$$\varepsilon = I(R + r_i) = IR + Ir_i$$

p.d. available to the external circuit V (for IR)

"lost volts" across internal resistance (for Ir_i)

$$\text{or } V = \varepsilon - Ir_i \qquad (6)$$

Science tools: data collection and processing

You can use the circuit in Figure 18 to take experimental data that can be processed to calculate the emf and internal resistance of an everyday battery (a dry cell). A simple variable resistor is connected to the dry cell and readings can be taken of terminal p.d., V, and current I. The different values are obtained by varying the external resistance. You should not leave the battery running too long at the higher currents, and you should also take care that nothing overheats. The range of currents that you are able to measure will

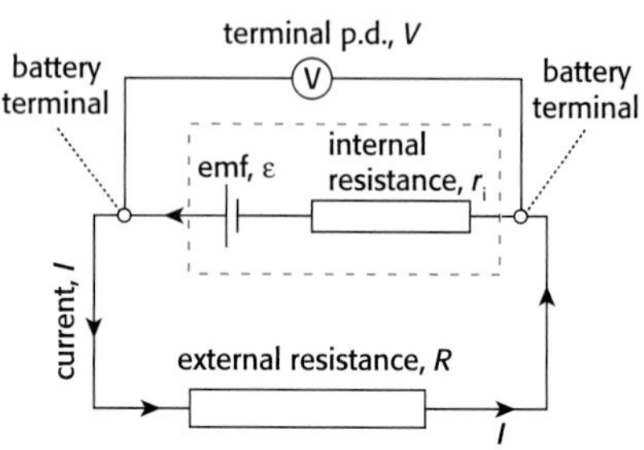

Figure 18

depend on the dry cell that you are using but, as is always the case, you should aim to measure 10 or more different values of the independent variable, current. Ideally, each value should be repeated two or three times. Note that if this experiment is to be used as an assessed practical you should be working on your own.

In general, the three aspects for the internally assessed skill of data collection and processing are

- recording raw data including units and uncertainties where relevant
- processing raw data correctly
- presenting processed data including errors and uncertainties

In this investigation, the raw data consists of an appropriate set of readings for terminal p.d., V, and current, I. You must choose the best method of measuring and recording this data, and you should also use an appropriate number of significant figures for the data that you record. This data then needs to be processed, so you will need to decide on the best way to analyse the numbers to allow you to answer your initial question. Finally, when processing data, you need to take into account the uncertainties that you assessed in the raw data. Often, this is most clearly presented as a separate section in your write-up.

You can also take this experiment on to the conclusion and evaluation stage.

Mathematical physics: *Maximum power*

An interesting mathematical problem arises when considering power supplies with non-negligible internal resistances.

What value of external resistance results in the optimal transfer of energy from the source (the battery) to the sink (the external resistor)?

For any given value of external resistance, the total emf is proportionally shared between the internal resistance and the external resistance but the problem is not as easy as it seems.

If the external resistance is very large then most of the p.d. is recorded across the external resistance and so the external resistor is receiving the greatest proportion of the power. Very little is "wasted" on the internal resistance. The efficiency of the transfer process is high, but unfortunately, the fact that the total resistance is high means that the overall current is low. This means that the actual power that is being transferred from source to sink is small.

If the external resistance is very small then the current is high but this time most of the p.d. appears across the internal resistance and thus, once again, the actual power that is being transferred from source to sink is small.

There must be a value somewhere between these two extremes that results in the maximum possible being transferred to the external resistor. Can you work out what it must be?

Possible ways of attacking this problem include:

- using a spreadsheet or graphics calculator to quickly simulate the result of changing the external resistor on the calculated value of power dissipated in the external resistor
- expressing the power dissipated in the external resistor in terms of the constants of the system and then using calculus to find the condition for the maximum value.

So the answer is that the maximum power transfer takes place when the external resistance and the internal resistance are equal.

The solution to this problem is relevant to many other situations involving electronics. A complicated circuit (e.g. in a recording studio) can be broken down into smaller systems that we can analyse more easily.

There are many stages involved in the high definition recording of music. The process begins with a microphone. The microphone is a device that converts sound energy into electrical energy. The electrical signal from the microphone can then be fed into a preamplifier and then into a series of filters before going into a main amplifier, and so on. In each step we wish to maximize the energy transfer of the electrical signal. The microphone has a certain output resistance and the pre-amplifier section has a certain input resistance so the solution to the problem allows us to say that we must match the resistance values to ensure the maximum possible transfer of signal power. The correct resistance cables must be used from the amplifier to the speakers in a hi-fi system or power will be lost.

Figure 19

Real meters

As we saw on page 166, perfect voltmeters have infinite resistance and perfect ammeters have zero resistance. What happens if this is not the case?

The effect of adding the meter needs to be taken into consideration in the circuit calculations. It is usually necessary to start again with all of your calculations whenever a circuit is altered – it is not always possible to assume that an alteration (for example adding a voltmeter) in one part of the circuit will not effect another part of the circuit. It might help to image a real meter as a combination of a perfect meter and an "additional" resistor. For a real ammeter, the "additional" resistor is in series with the perfect meter, for the voltmeter it is in parallel.

(a) real ammeter:

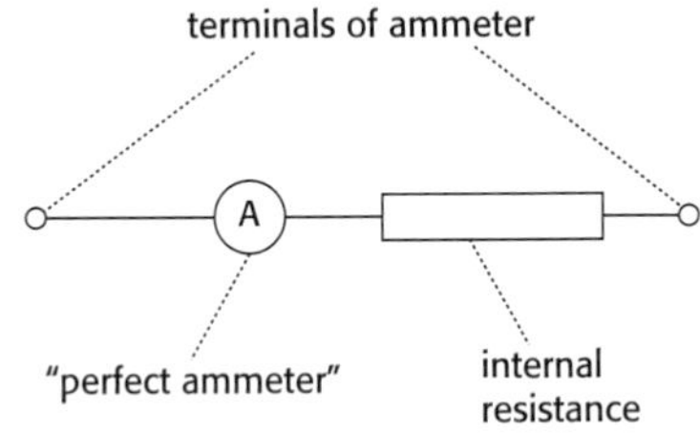

(b) real voltmeter:

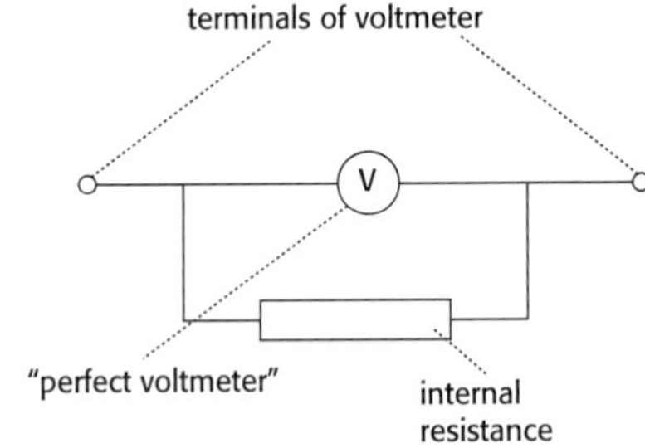

Figure 20

Potential dividers and sensors

For a circuit to be useful, it must interface with the real world; it must have at least one input and one output.

An input device could be as simple as a button or a switch. In a computer circuit, the input devices are things such as the keyboard, the mouse, or a DVD reader. Possible output devices include lights, buzzers, or, in the case of the computer, the screen, the printer, or the loudspeakers. The general name given to any electrical interface that converts energy between different forms is a **transducer**.

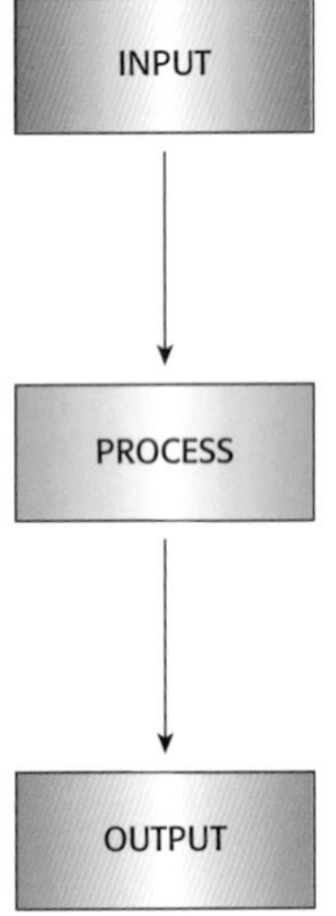

An electrical circuit involves electrical energy and thus input transducers, or **sensors**, are devices that convert some real everyday aspect of the physical world into electrical signals. Examples include microphones (sound into electrical), light sensors (light into electrical), and temperature sensors (thermal energy into electrical).

Sensors can facilitate a whole range of measurements that would be impossible for a lone physicist to observe with a single measuring device. Electronic circuits can be designed in which the measurements are:

- **Automated**: multiple readings of a variable can be recorded over very short response (e.g. air pressure variations during a pulse of sound) or very long time intervals (e.g. the automatic recording of a patient's heart rate over a week).

- **Detailed**: they allow for a large number of precision multiple recordings to be taken at exactly the same time (e.g. the minute variations in the stresses and strains that take place in the body of an aircraft during takeoff).
- **Remote**: they allow for readings to be taken in difficult or even hostile locations (e.g. temperature measurements inside a reactor core or observations on another planet).

So what is happening when, for example, the breaking of a beam of light causes an alarm system to issue a warning?

Many sensors are just a resistor whose resistance varies in a predictable way in response to external factors. Two very important examples of this are *light-dependent resistors* (LDRs) and a type of temperature-dependent resistor called a *negative temperature coefficient (NTC) thermistor*.

In both of these devices, an increase in energy input to the device causes a large decrease in resistance. Thus when the amount of light shining on the LDR increases or the temperature of the thermistor goes up, the resistance goes down. In both cases this is because the devices are acting as semiconductors and the number of charge carriers has been affected.

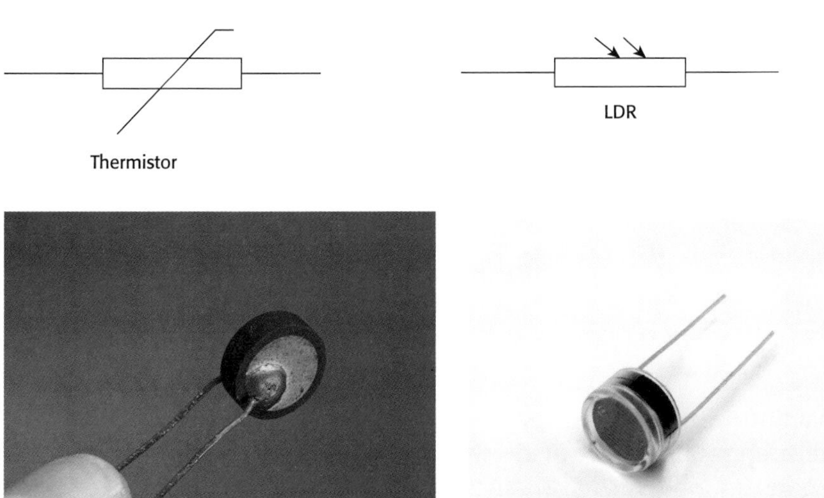

Figure 21 A thermistor and an LDR

A simple way of utilizing this change in resistance is by employing a **potential divider** circuit using two resistors, R_1 and R_2 (see Figure 22).

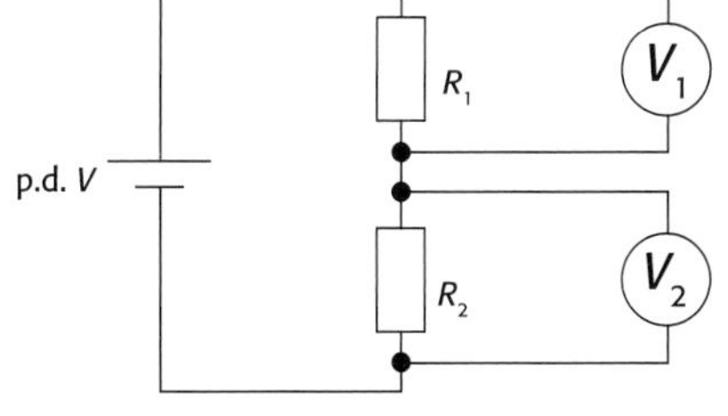

Figure 22

An analysis of the current flowing around the circuit, shows that the total potential difference that is available, V, divides up (hence the name of the circuit) with each resistor taking its "share" in proportion to its resistance.

Since $V_1 = IR_1$, $V_2 = IR_2$ and $I = \dfrac{V}{R_1 + R_2}$

$$V_1 = \left(\frac{R_1}{R_1 + R_2}\right)V \text{ and } V_2 = \left(\frac{R_2}{R_1 + R_2}\right)V \qquad (7)$$

By analysing equation 7 we can see that the larger resistance takes the larger share of the total available p.d.

If either of these resistors was swapped for one of a different value, then the way the potential difference was shared would be different. The share of the p.d. would also vary if one of the resistors varied as a result of external conditions. For example, consider the circuit in Figure 23.

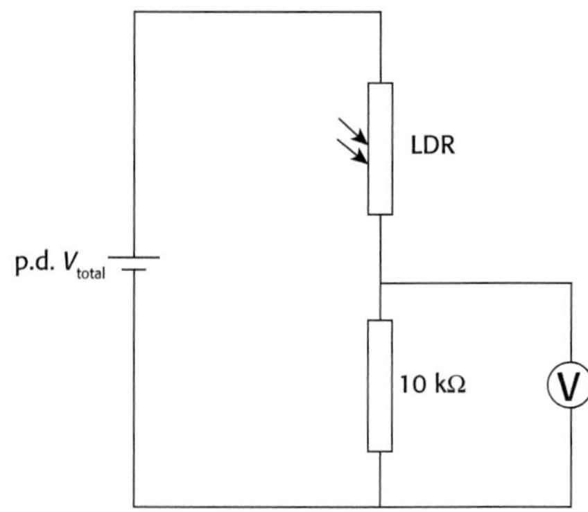

Figure 23

When light shines on the LDR, its resistance is comparatively small, say 200 Ω, but if no light shines then its resistance is greatly increased – perhaps to 20 kΩ.

In the light,

R_1 = resistance of LDR ≈ 200 Ω
R_2 = resistance of fixed resistor = 10 kΩ

∴ the resistance of the fixed resistor >> resistance of LDR
so $R_2/(R_1 + R_2) \approx 1$
and p.d. across the fixed resistor ≈ V_{total}

In the dark,

R_1 = resistance of LDR ≈ 20 kΩ
R_2 = resistance of fixed resistor = 10 kΩ

∴ the resistance of the fixed resistor ≈ $\frac{1}{2}$ × resistance of LDR

$$\text{so p.d. across the fixed resistor} = V_{total} \times R_2/(R_1 + R_2) \approx \frac{V_{total}}{3}$$

So when light stops shining on the LDR there will be a decrease in p.d. across the fixed resistor. An electrical circuit can be designed to compare this p.d. with a known fixed potential difference. We can create a system that will sound an alarm whenever the measured p.d. goes down. The system we are considering will sound an alarm whenever a beam of light that is arranged to shine on an LDR is broken.

To be useful, the circuit we have described still needs further modifications – the simple system would stop sounding the alarm as soon as the light shines on the sensor again. As long as each addition does not fundamentally affect this basic section, we can modify parts of the circuit to make it useful for our purpose.

Remember that it is not possible to assume that the current or potential difference remains constant after a change to the circuit. After a change, the only way to ensure a correct answer is to start the calculations again.

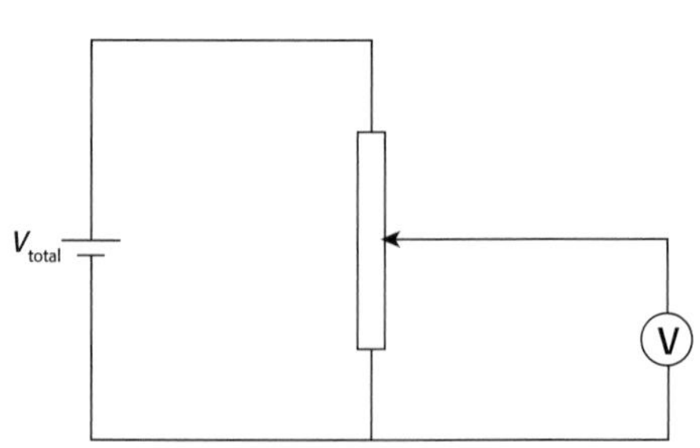

Figure 24 A potentiometer

Sometimes the sensing device can be designed around a **potentiometer** circuit. This is just a potential divider created from a fixed resistor with a movable connection.

As the slider (represented by the arrow in Figure 24) moves from one end to the other, the p.d. measured by the voltmeter varies from the maximum, V_{total}, (when the movable connection is at the top end of the resistor) down to zero (when the movable connection is at the bottom end of the resistor). A simple circuit that uses this approach is the petrol gauge in a car. A device is arranged to float on top of the petrol in the tank. As it moves up and down, it varies the position of the sliding contact on a potentiometer circuit and thus the p.d. recorded is dependent on the amount of petrol in the tank. This reading is calibrated so that it accurately represents the amount of petrol left.

A common sensor in everyday engineering use is the strain gauge. Linear strain is defined in the following way:

$$\text{strain} = \frac{\text{increase in length}}{\text{original length}}$$

A strain gauge measures any small extension or compression that occurs by measuring changes in resistance that take place when a resistor is increased or decreased in length. Often these changes in resistance are very small and so clever techniques need to be incorporated to allow these changes to be measurable.

Real voltmeters have high but not infinite resistance so that when we connect a voltmeter in parallel with a component we change the circuit. The p.d. that we end up measuring is different to the p.d. across the component before we connected the meter. If the resistance of the voltmeter is 10 times greater than the resistance of the device in question, then the maximum error in the reading of p.d. will be 10%. A voltmeter with a higher resistance would further reduce this error.

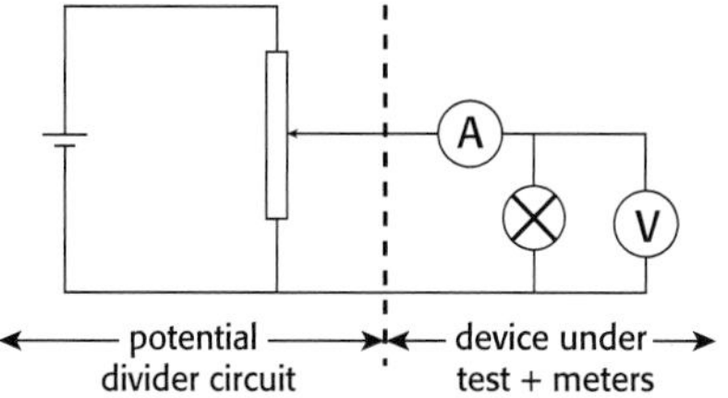

Figure 25 Potentiometer circuit for measuring the electrical characteristics of a bulb

The same consideration needs to be borne in mind when choosing a potentiometer to use in a potential divider circuit. For example, the standard circuit in Figure 25 could be used for measuring the electric characteristics for a bulb.

You can see that the circuit can be split into two halves. The left-hand side of the circuit allows you to fix a p.d. of your choice. When the right-hand side is added, this p.d. is connected to the bulb and meters. We have thus chosen and placed a p.d. across the bulb that will cause a current to flow through the bulb. Even if these meters were perfect, this way of visualizing the circuit is only useful if the resistances of the potentiometer and the bulb were appropriately selected in the first place.

We assume that the p.d. "fixed" by the left-hand side remained constant when the right-hand side of the circuit was added on. This will only be the case if the resistance of the right-hand circuit is significantly greater than the resistance of the potentiometer (like example (e) above). Normal laboratory potentiometers have a resistance across each end of about 1 Ω or 2 Ω and they are constructed to be able to carry a reasonably large current (5 A or so). This means that the resistance of the object under test (the bulb in the above example) should certainly be more than 10 Ω and ideally much larger.

This principle is also used when designing more complicated electric (and electronic) circuits. It is possible, within reason, to add different system "blocks" onto already existing circuits. If the resistance of the extra block is at least 10 times the resistance of the existing block, this addition can be done knowing that the first circuit will be largely unaltered by the addition.

Physics issues: electrical risks

What are the most common causes of cables overheating?

Using inappropriate power cables

If you want to extend the length of a power cable, it is very dangerous to use wire that is too thin. The circuit can appear to work, but the extension wires can heat up very quickly.

Commercial extension leads should always have their maximum power rating clearly written on them and the total power rating of all the devices connected using the extension lead must not exceed this limit. Very long cable extensions (the sort that wind up on a drum) have different maximum ratings depending on whether the cable is wound up or unwound. An unwound cable loses thermal energy more readily that one that is wound up.

Using "splitters" wrongly

Each electrical socket in your home is connected to other electrical sockets. The arrangement of these sockets must be in parallel as each socket is arranged to provide exactly the same p.d. (230 V in Europe and most of Asia, and 110 V in North America). Electronic equipment such as TVs and stereos are designed to operate on this voltage and the more devices you plug into the same socket, the greater the total load and so the socket has to provide a greater total current.

If, for example, you plug in a kettle and this takes 4 A when operating, then two similar kettles must take a total of 8 A. Most domestic wiring is protected by fuses and designed to be able to cope with 15 A safely. If you connected 4 kettles to one socket then the cable connecting the sockets could get too hot and start a fire.

The results of an electrical fire can, of course, be devastating.

A full risk assessment of a domestic electrical circuit might include some of following possibilities:

1 The power leads to most electrical devices are insulated to prevent anybody from accidentally receiving an electric shock. This insulation can degrade with time and with constant movement. It is common for power leads in a plug to be connected to the electrical contacts by trapping the bared copper cables under a screw connection. If the power cable is regularly moved, it is possible for these wires to eventually become frayed or loose.
2 A **short circuit** takes place when power leads are accidentally connected together. If this happens, the current that flows is only limited by the internal resistance of the power supply. Short circuits can be a result of a wiring mistake or poor connections working loose.
3 If you were to touch one of the conductors in an electrical circuit, it would be possible for a current to flow through you to earth. With a low-voltage circuit, the current would be small and you would probably not notice anything. At high voltages, a larger current might flow and this could cause damage, injury, or even death. For this reason, we often carry out risk assessments before working with electrical devices.

?

1 Explain why it is perfectly safe for a mechanic to touch both the terminals of a 12 V car battery, but it would be dangerous if she touched a metal tool that was dropped so it touched both terminals.
2 Why is it important to warn the public of high-voltage electrical installations (e.g. 1000 V) when even higher voltages can be easily generated by friction and these do not cause any harm, e.g. brushing hair or removing nylon clothing.
3 Choose one domestic piece of electrical equipment and make a comprehensive list of all potential risks associated with its use. Do not just include risks that are associated with electric currents as a large number of domestic injuries result from problems such as tripping over the power cables.

Figure 26 A frayed wire

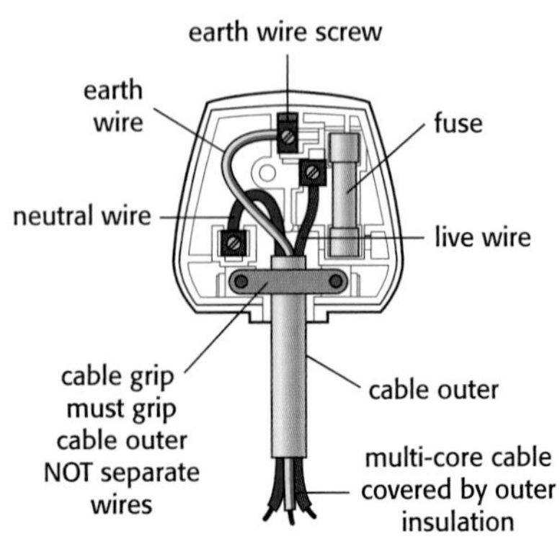

Figure 27 A UK plug

End of chapter summary

Chapter 12 has two main themes: electrical measurements and electric circuits. The list below summarises the knowledge and skills that you should be able to undertake after having studied this chapter. Further research into more detailed explanations using other resources and/or further practice at solving problems in all these topics is recommended – particularly the items in bold.

Electrical measurements

- Define *electric potential difference, electromotive force (emf), electric current, resistance* and *resistivity* and **solve problems involving these quantities** including the determining the change in potential energy when a charge moves through a potential difference.
- State Ohm's law and compare ohmic and non-ohmic behaviour.
- Derive and apply expressions for electrical power dissipation in resistors.
- Describe the concept of internal resistance.

Electric circuits

- **Draw circuit diagrams and apply the equations for resistors in series and in parallel.**
- Describe the use of ideal ammeters and ideal voltmeters.
- Describe a potential divider and explain the use of sensors in such circuits.
- Identify some common electrical risks and **solve problems involving electric circuits**.

Chapter 12 questions

1 This question compares the electrical properties of two 12 V filament lamps.

A lamp is designed to operate at normal brightness with a potential difference of 12 V across its filament. The current in the filament is 0.50 A.

a) For the lamp at normal brightness, calculate

i) the power dissipated in the filament. [1]

ii) the resistance of the filament. [1]

In order to measure the voltage-current (*V-I*) characteristics of a lamp, a student sets up the following electrical circuit.

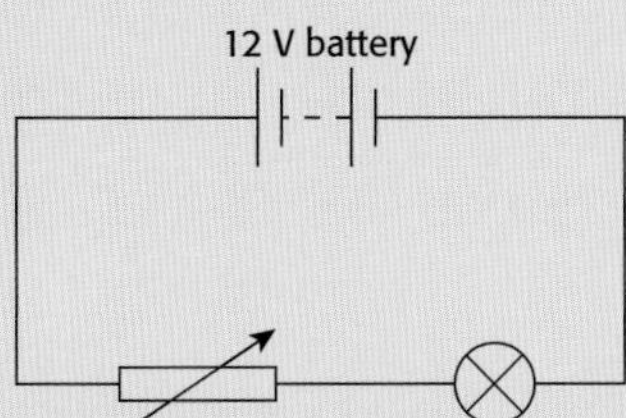

b) Add circuit symbols showing the correct positions of an ideal ammeter **and** an ideal voltmeter that would allow the *V-I* characteristics of this lamp to be measured. [2]

The voltmeter and the ammeter are connected correctly in the previous circuit.

c) Explain why the potential difference across the lamp

i) cannot be increased to 12 V. [2]

ii) cannot be reduced to zero. [2]

An alternative circuit for measuring the *V-I* characteristic uses a *potential divider.*

d) i) Draw a circuit that uses a potential divider to enable the *V-I* characteristics of the filament to be found. [3]

ii) Explain why this circuit enables the potential difference across the lamp to be reduced to zero volts. [2]

The graph below shows the *V-I* characteristic for two 12 V filament lamps A and B.

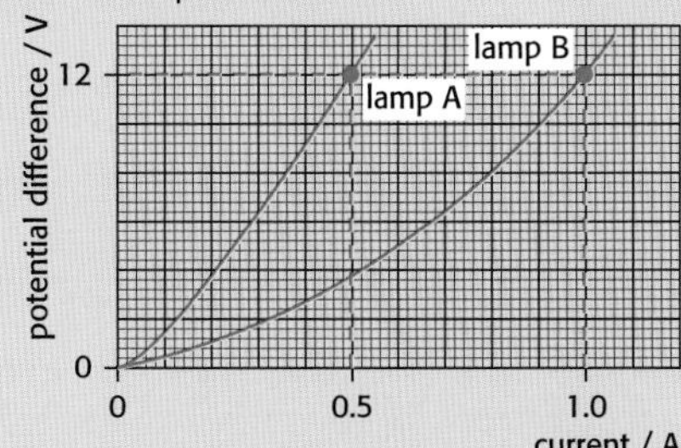

e) i) Explain why these lamps do not obey Ohm's law. [2]

ii) State and explain which lamp has the greater power dissipation for a potential difference of 12 V. [3]

The two lamps are now connected in series with a 12 V battery as shown below.

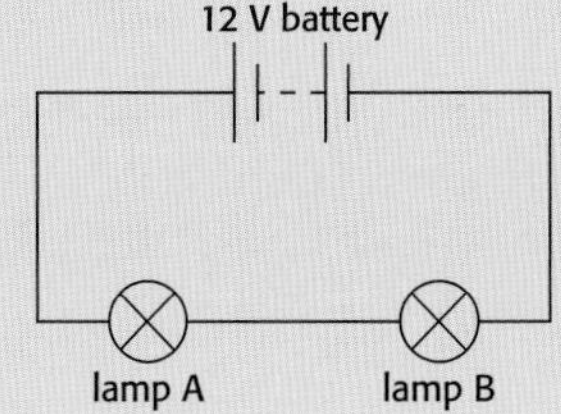

f) **i)** State how the current in lamp A compares with that in lamp B. [1]

ii) Use the *V-I* characteristics of the lamps to deduce the total current from the battery. [4]

iii) Compare the power dissipated by the two lamps. [2]

(Total 25 marks)

2 This question is about emf and internal resistance.

A dry cell has an emf *E* and internal resistance *r* and is connected to an external circuit. There is a current *I* in the circuit when the potential difference across the terminals of the cell is *V*.

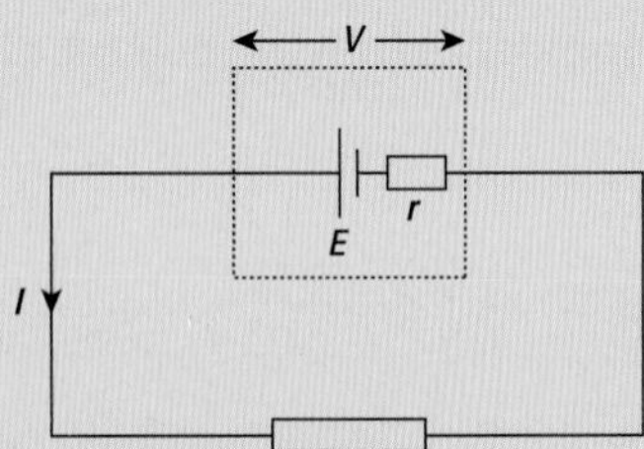

a) State expressions, in terms of *E*, *V*, *r* and *I* where appropriate, for

i) the total power supplied by the cell; [1]

ii) the power dissipated in the cell; [1]

iii) the power dissipated in the external circuit. [1]

b) Use your answers to (a) to derive a relationship between *V*, *E*, *I* and *r*. [2]

The graph below shows the variation of *V* with *I* for the dry cell.

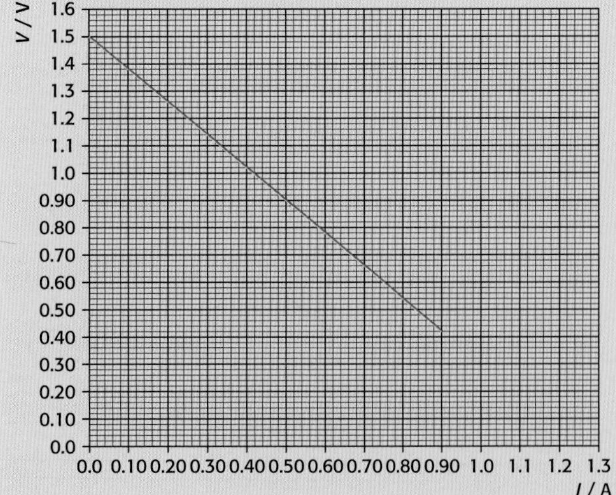

c) Complete the diagram below to show the circuit that could be used to obtain the data from which the graph was plotted.

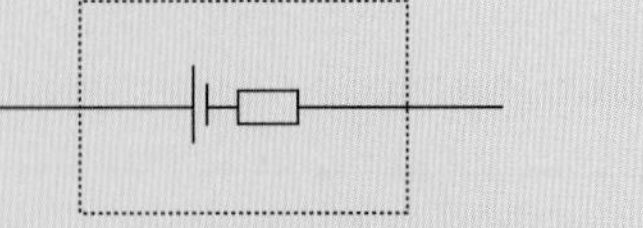

[3]

d) Use the graph, explaining your answers, to

i) determine the emf *E* of the cell; [2]

ii) determine the current in the external circuit when the resistance *R* of the external circuit is very small; [2]

iii) deduce that the internal resistance *r* of the cell is about 1.2 Ω. [3]

(e) The maximum power dissipated in the external circuit occurs when the resistance of the external circuit has the same value as the internal resistance of the cell. Calculate the maximum power dissipation in the external circuit. [3]

(Total 18 marks)

3 This question is about an electric circuit

A particular filament lamp is rated at 12 V, 6.0 mA. It just lights when the potential difference across the filament is 6.0 V.

A student sets up an electric circuit to measure the *I-V* characteristics of the filament lamp.

In the circuit, shown below, the student has connected the voltmeter and the ammeter into the circuit **incorrectly**.

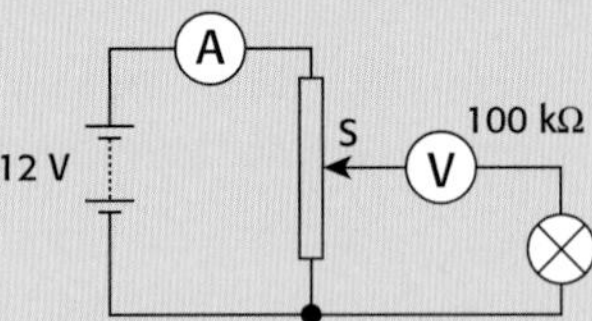

The battery has emf 12 V and negligible internal resistance. The ammeter has negligible resistance and the resistance of the voltmeter is 100 kΩ.

The maximum resistance of the variable resistor is 15 Ω.

a) Explain, without doing any calculations, whether there is a position of the slider S at which the lamp will be lit. [3]

b) Estimate the maximum reading of the ammeter. [2]

c) Draw a circuit diagram showing the correct position of the voltmeter and of the ammeter in order to determine the *I-V* characteristics of the filament lamp. [2]

(Total 7 marks)

4 This question is about electric circuits.

a) Define emf and state Ohm's law. [2]

b) In the circuit below an electrical device (load) is connected in series with a cell of emf 2.5 V and internal resistance *r*. The current *I* in the circuit is 0.10 A.

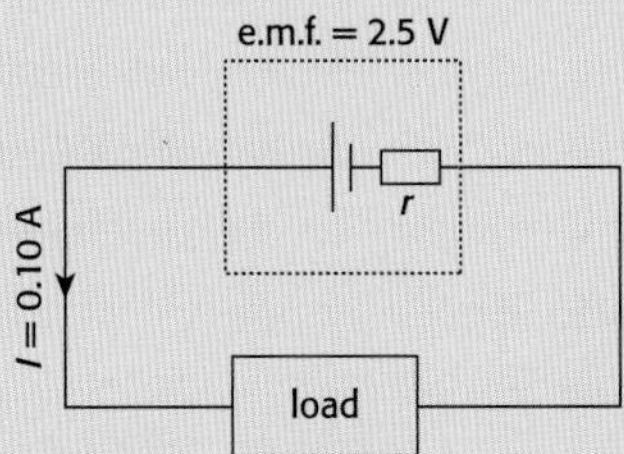

The power dissipated in the load is 0.23 W.

Calculate

i) the total power of the cell; [1]

ii) the resistance of the load; [2]

(iii) the internal resistance *r* of the cell. [2]

c) A second identical cell is connected into the circuit in (b) as shown below.

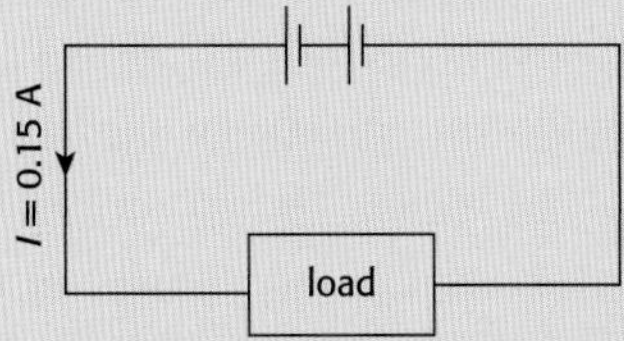

The current in this circuit is 0.15 A. Deduce that the load is a non-ohmic device. [4]

(Total 11 marks)

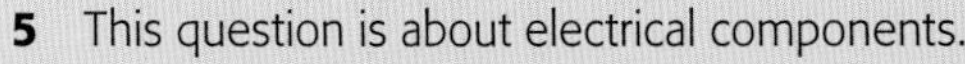

5 This question is about electrical components.

a) Draw a circuit diagram that could be used to determine the current-voltage (*I-V*) characteristics of an electrical component X. [2]

The graph below shows the *I-V* characteristics for the component X.

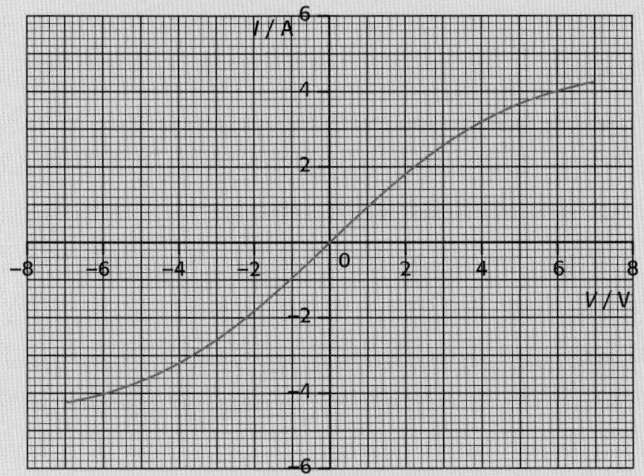

The component X is now connected across the terminals of a battery of emf 6.0 *V* and negligible internal resistance.

b) Use the graph to determine

i) the current in component X; [1]

ii) the resistance of component X. [2]

A resistor R of constant resistance 2.0 Ω is now connected in series with component X as shown below.

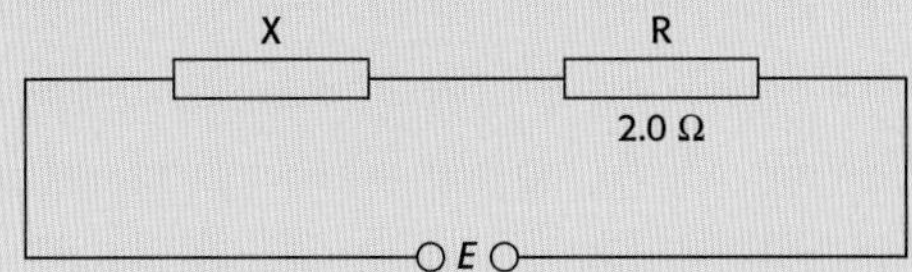

c) i) On the graph above, draw the *I-V* characteristics for the resistor R. [2]

ii) Determine the total potential difference *E* that must be applied across component X and across resistor R such that the current through X and R is 3.0 A. [2]

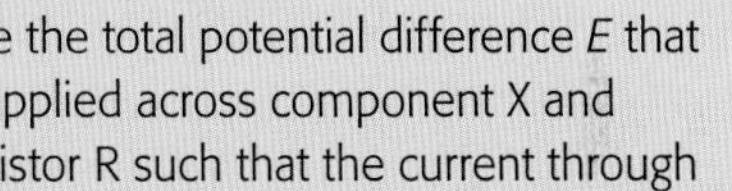

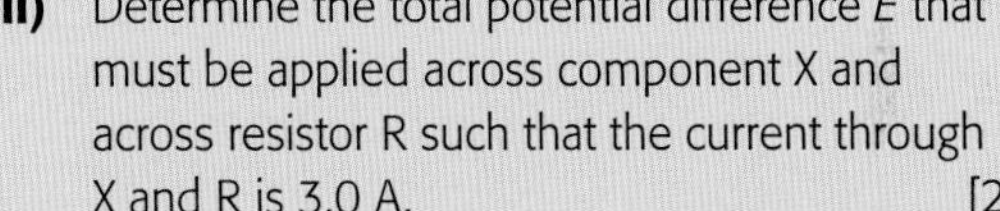

(Total 9 marks)

6 This question is about electrical circuits.

The graph below shows the *I-V* (current-voltage) characteristic of an electrical component T.

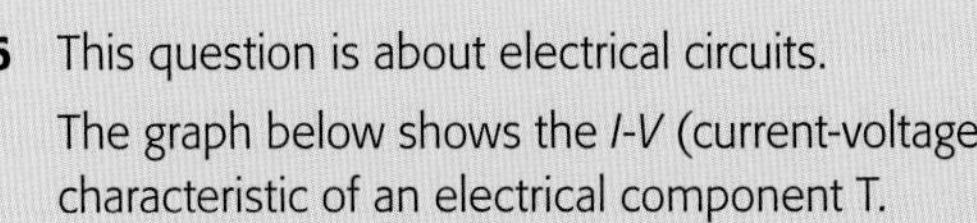

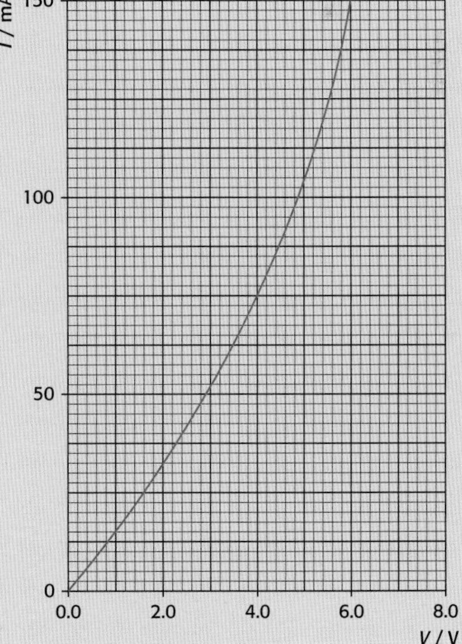

a) On the graph above, draw the *I-V* characteristic in the range *V* = 0 to *V* = 6.0 V for a resistor R having a constant resistance of 40 Ω. [1]

b) The component T and the resistor R are connected in parallel as shown below.

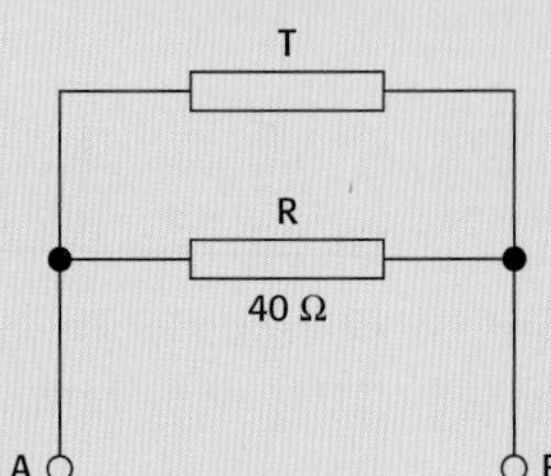

When a battery of constant emf E and negligible internal resistance is connected between the terminals A and B, the current in the resistor R is 100 mA.

i) Calculate the emf E of the battery. [1]

ii) Use the graph to determine the current in T. [1]

iii) Calculate the power dissipation in T. [2]

c) In order to reduce the power dissipation in component T, a second resistor R of resistance 40 Ω is connected in series with T. The circuit is shown below.

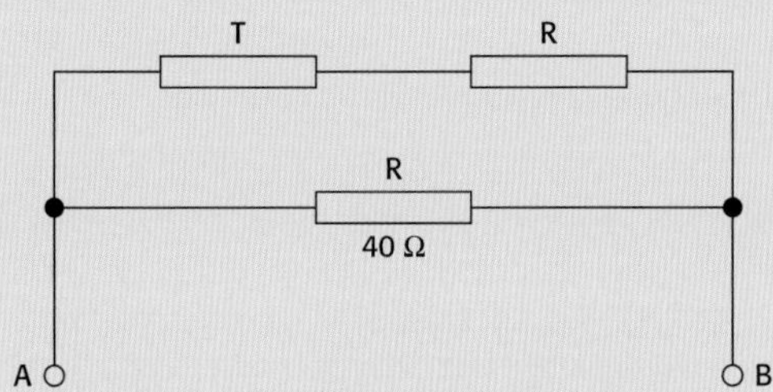

The battery connected between A and B is unchanged.

Use the graph to determine

i) the current in resistor T. [2]

ii) the power dissipation in T. [2]

(Total 9 marks)

7 Heating water electrically

The diagram below shows part of the heating circuit of a domestic shower.

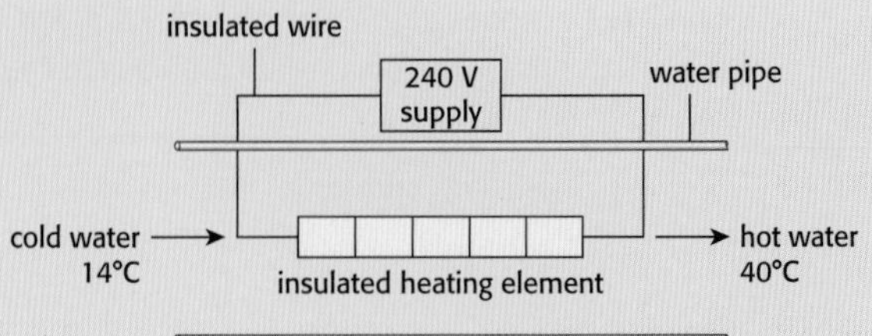

Cold water enters the shower unit and flows over an insulated heating element. The heating element is rated at 7.2 kW, 240 V. The water enters at a temperature of 14°C and leaves at a temperature of 40°C. The specific heat capacity of water is 4.2×10^3 J kg^{-1} K^{-1}.

a) Describe how thermal energy is transferred from the heating element to the water. [3]

b) Estimate the flow rate in kg s^{-1} of the water. [4]

c) Suggest **two** reasons why your answer to (b) is only an estimate. [2]

d) Calculate the current in the heating element when the element is operating at 7.2 kW. [2]

e) Explain why, when the shower unit is switched on, the initial current in the heating element is greater than the current calculated in (d). [2]

f) In some countries, shower units are operated from a 110 V supply. A heating element operating with a 240 V supply has resistance R_{240} and an element operating from a 110 V supply has resistance R_{110}.

i) Deduce, that for heating elements to have identical power outputs

$$\frac{R_{110}}{R_{240}} = 0.21.$$ [3]

ii) Using the ratio in (i), describe and explain **one** disadvantage of using a 110 V supply for domestic purposes. [2]

(Total 18 marks)

13 Three different non-contact forces

In this chapter we will discuss the similarities and differences between three fundamental forces: the force of gravity, the electrostatic force (between charged objects), and the magnetic force (between magnets and/or currents). Examples of these three forces in action can be seen above.

All of these forces have an infinite range (the forces all reduce in magnitude as the objects are moved further apart, but the force never quite gets to zero). It is also worth noting that the objects involved in each of these forces do not have to come in contact for the forces to act between them.

Surprisingly, every force that you observe on a day-to-day basis is an example of one (or more) of these three forces. This is true whether you are considering pushes and pulls (e.g. the push on the tennis ball from your racket when playing a game of tennis, or the friction of your shoes on the ground) or other forces (e.g. the forces that result in you seeing the Sun rising everyday). In fact, the only category of forces that cannot be explained using these three forces are those that exist inside the nucleus of an atom.

Although electric forces and magnetic forces seem quite separate from each other, it turns out that they are related to each other and can both be thought of as different aspects of one fundamental force, called the **electromagnetic force**. It can be helpful, initially, to think of the two effects as entirely separate from one another; after all, the forces each involve different types of object (charges and magnets). Most of the time there is no observable link between the two – a stationary positive charge does not feel a force when next to a magnet.

Before reading any further in this chapter it would be useful to:

- list some examples of everyday forces and categorize them as gravitational, electric, or magnetic (or a combination).
- review the concept of gravitational field strength and recall that $g = 10\ \text{N kg}^{-1}$ on the Earth.
- review the basic force laws between charges (opposite charges attract, like charges repel) and between magnets (opposite poles attract, like poles repel).

- review the shape of the magnetic field around a simple bar magnet.
- review any electromagnetic effects that you have studied.

Force fields

A powerful tool for helping us to understand gravitational, electric, and magnetic forces is the concept of a **force field**. When we use these words, you might think about the sort of invisible force field that you might encounter in cartoons or in science fiction.

Conveniently these fictional force fields can be switched on or off and anybody trying to enter the protected region feels a strong resisting force. As far as we know, unfortunately, no such localized strong resisting force field can exist.

In general terms, a particular force field is defined as the region where a force acts. Gravitational fields, electric fields, and magnetic fields are all infinite in range but the strength of the field varies depending on distance from the mass, charge, or magnet. We can broaden this idea into a mathematical definition.

The actual value of the force depends on the object that is placed into the field. The general idea is to imagine that you could place a small test "object" into the field and measure the value of the resulting force. If you put two test "objects" in the same place, you would expect to measure an equal force acting on each of them so the total force would be doubled. Using ratios, you can calculate the forces acting on any object that you place in the field.

The field strength at a given point in the field is defined as the force per unit test "object" that feels the force at that point:

$$\text{field strength} = \frac{\text{force (N)}}{\text{object feeling the force (kg, C, or A m)}}$$

An important point to note is that the test "object" must be small in magnitude, as we have assumed that it does not change or affect the field that we are testing. It must also be small in physical size, as we are trying to define the field at a particular point in space.

1: Gravitational field strength, g

Gravitational fields act on masses so the unit test "object" in this situation is a point mass. If we place a small point mass m in a gravitational field of magnitude g, it feels a force F acting on it.

$$\text{gravitational field strength } E = \frac{F}{q}$$

As F is a force (in N) and m is a value of mass (in kg), the units of g must be N kg^{-1}. At the surface of the Earth the gravitational field strength is approximately 9.8 N kg^{-1}, which means that a 1 kg mass will feel a force of about 9.8 N and 74 kg will feel a force of about 730 N.

On the surface of the Moon, the gravitational field strength is: g_{moon} = 1.6 N kg^{-1}. This means that a mass of 74 kg will feel a force of about 120 N.

Figure 1 Can you identify the origin of the forces in these pictures?

Use each of the photos in Figure 1 to identify some of the forces that must be acting in each of these situations. State and explain the origin of any of the forces you identify.

Figure 2 Sue Storm from the Fantastic Four protects herself with a force field

Show, by substitution, that the units N kg^{-1} are equivalent to the units m s^{-2}.

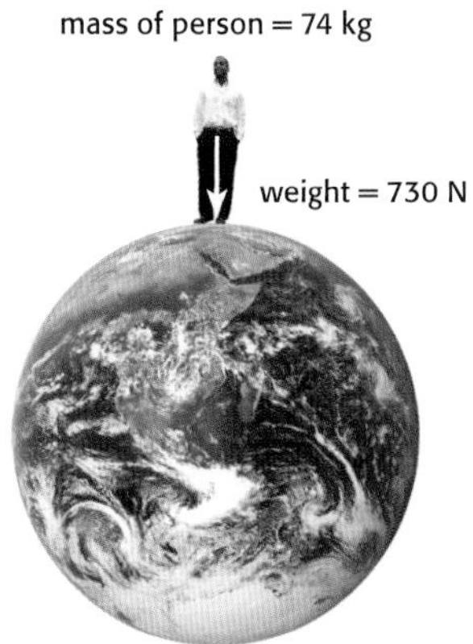

Figure 3 74kg person on the Earth feels a force due to 730 N

mass of person = 74 kg
weight = 120 N

Figure 4 74 kg person on the moon feels a foce due to gravity of 120 N

2: Electric field strength E

Electric fields act on charges so the unit test object is a point charge. If we place a small point charge q in an electric field of magnitude E, it feels a force F. Notice that if q represents a negative charge, this will mean that q is negative and so is E.

What effect does reversing the sign of E have on the physical situation represented?

$$\text{electric field strength } E = \frac{F}{q}$$

As F is a force (in N) and q is a charge (in C) the units of E must be N C^{-1}. We can also define the electric field as a potential gradient – i.e. the change in the voltage per unit distance moved. The unit of E defined in this way is volts per metre (V m^{-1}), which can be shown to be equivalent to N C^{-1}.

For example, there is an electric field between the base of a thundercloud and the ground. If the strength of this electric field is approximately 3.5×10^5 N C^{-1} this means that an electron on the ground whose charge is 1.6×10^{-19} C, will feel a force of about 5.6×10^{-14} N.

Note that even though the unit of electric field strength is N C^{-1}, we cannot define the electric field strength as the force acting on a charge of 1 C. The reason for this is that 1 C is a huge amount of charge. In fact 1 C is so large that it would be impossible to collect this amount of positive or negative charge in one place as the repulsive forces between the individual charges would be too large.

3: Magnetic field strength B

The magnetic field definition is slightly more complex than the other two, as magnetic fields act on several different types of object. There are three basic things that can feel a magnetic force:

- magnets
- currents
- moving charges

A current is, of course, made up of moving charges, but as the situations are very different it is easy to forget that they are, in fact, equivalent. There are some additional complications:

- the direction of the resultant magnetic force is not usually the same as the direction of the magnetic field.

- there can be additional factors that affect the resultant magnetic force for a given field – magnitude of the current, direction of velocity of moving charge, etc.
- there is no such thing as a "unit magnet".

The unit test object in this situation can involve either currents or charges. When considering currents, the length of the piece of wire under consideration will also affect the size of the force.

For a given current, if a certain length of wire feels a force F then double the length of wire will feel double the force, $2F$. The total magnetic force is proportional to the current flowing down the wire, and to the length of the wire.

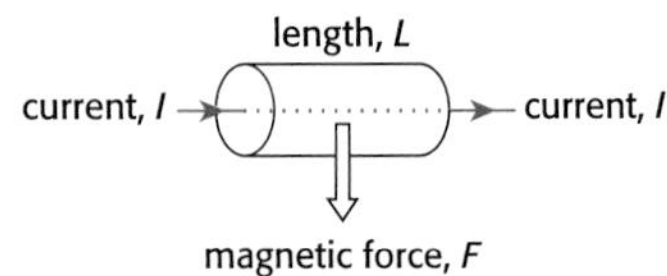

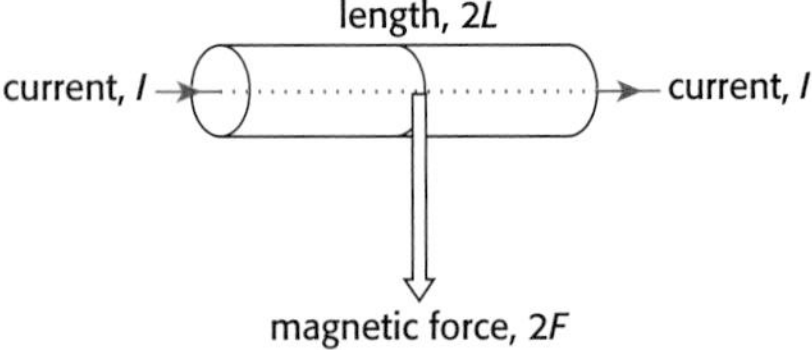

Figure 5 In the example above, the magnetic field must be out of the plane of the paper

In this situation, the unit test object is a small **current-length** which is the value of the current multiplied by the length of wire ($= I \times L$). The units of current-length must be A m (amps × metres). A small current-length IL placed in a magnetic field of magnitude B can experience a range of forces depending on the orientation of the current-length. The maximum force that can act on it, F, is used to define the magnetic field strength, B. Do not assume that the dependency on orientation means that this is a vector equation.

$$\text{magnetic field strength } B = \frac{F}{IL}$$

F is a force (in N) and IL is a value of current-length (in A m) so the units of B must be N A^{-1} m^{-1}. For example, the magnetic field strength close to a particular bar magnet is 7.9×10^{-3} N A^{-1} m^{-1}.
A 5.2 cm wire that has a current of 6.1 A flowing along it placed next to this magnet will feel a maximum force of 2.5 mN.

An alternative, and equally valid, definition of magnetic field strength involves a positive point charge q moving at a velocity v. In this case, the test object is the charge-velocity (in C m s^{-1}) and the maximum force, F, is defined using:

$$\text{magnetic field strength } B = \frac{F}{qv}$$

In this version, the units of magnetic field strength must be N C^{-1} m^{-1} s. These two definitions (and their units) are exactly equivalent, so you can always use SI units. If, for example, an electron (1.6×10^{-19} C) is moving at a speed of 2.3×10^{7} m s^{-1} in a magnetic field of 7.9×10^{-3} NA^{-1} m^{-1}, then it can feel a *maximum* force of up to 2.9×10^{-14} N.

There are two other possible SI units for the magnetic field strength, B – the tesla (T) and the weber per m^2 (Wb m^{-2}). These different possibilities all turn out to be useful in different situations and are absolutely identical – it does not matter which one you use.

$$1\,\text{N}\,\text{A}^{-1}\,\text{m}^{-1} = 1\,\text{N}\,\text{C}^{-1}\,\text{m}^{-1}\,\text{s} = 1\,\text{Wb}\,\text{m}^{-2} = 1\,\text{T}$$

Direction of the force

Force is a vector quantity, which means that we need to specify both its magnitude and its direction. The definitions of the three field strengths discussed so far can be used to calculate the magnitude of a force, but we have not yet considered the direction of the force. All three fields (gravitational, electric, and magnetic) are vectors but the direction of the force field and the direction of the force are not necessarily the same thing.

In gravitational fields, the situation is straightforward. When we put a mass in a gravitational field, it feels a force. The magnitude of the force that is felt is calculated using $g = \frac{F}{m}$. The direction of this force is just the same as the direction of the gravitational field. Objects experiencing a force due to gravity are, as far as we know, always attracted to each other.

Electric fields are also straightforward. The only slight complication is that there are two types of charge that we could put in the field – positive and negative – and the forces that they feel will be in opposite directions. By convention, the direction of the electric field is the same as the direction of the force on a positive test charge. The force on a negative charge must be in the opposite direction to the field.

The magnetic field due to a magnet or charge is the most complicated of the three. We will begin by discussing magnets. In the same way that there are two types of electric charge, there are two types of magnetic pole (north and south). By convention, the direction of magnetic field represents the force that acts on the north pole of a magnet. The direction of force on a south pole must then be in the opposite direction to the field.

We now need to generalize this result to the two other possible situations that involve magnetic forces – a current in a magnetic field and a moving charge in a magnetic field. In both of these situations the maximum force arises when the components are perpendicular to each other. When a conventional current, I, flows at right angles to a magnetic field, B, the force, F, on this current is at right angles to both of these directions. The directions of the three variables B, F, and I define the x, y, and z-axes of a three-dimensional coordinate system.

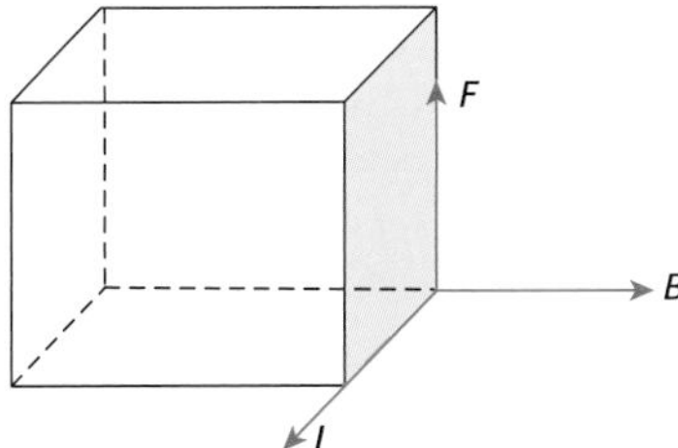

Figure 6 Maximum force, field and current are all at right angles

This rule could not have been predicted – it has only been discovered by experimental measurement. There are several different ways that you can remember this relationship – one possible way is Fleming's left-hand rule, another is the right-hand palm rule (see Figure 7).

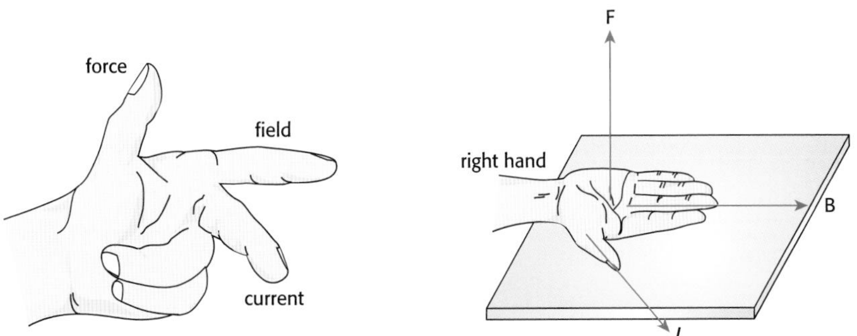

Figure 7 Ways of remembering the orientation of F, B and I

If the current isn't exactly perpendicular to the magnetic field then we must resolve either the current or the magnetic field into two components. In the situation shown in Figure 8, the current has been resolved into a component that is perpendicular to B ($I \sin \theta$) and a component that is parallel to B ($I \cos \theta$).

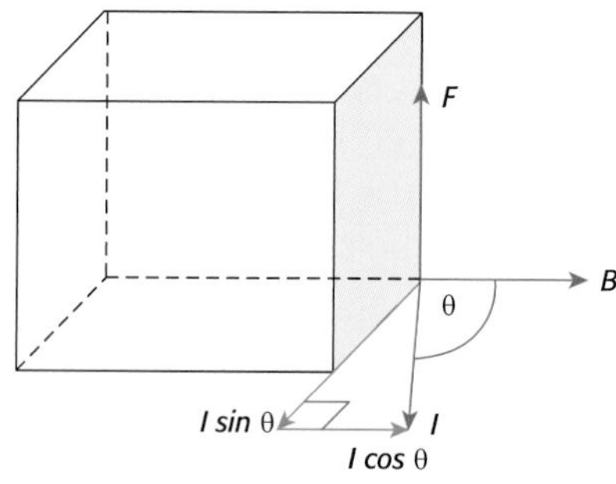

Figure 8 Resolving to calculate force when B and I are not perpendicular

Instead of using $F = BIL$ to calculate the force, we now need to use the appropriate component. Since the component of I parallel to B will not affect the resultant force, F can be calculated using:

$$F = BIL \sin \theta$$

An analogous situation exists when we consider the force on a charge that is moving in a magnetic field. If the charge is moving in the same direction as the magnetic field then it will not feel a force. If the motion of the charge has a component that is perpendicular to the field then it will feel a force given by

$$F = qvB \sin \theta$$

Production and representation of forces

Before we can analyse force fields, we should look at the types of objects that produce them. In each case, the object producing the field is the same as the type of object affected by the resulting force.

- A gravitational field is caused by an object's mass.
- An electric field is caused by an object's charge.
- A magnetic force is caused by a current-length, charge-velocity or a magnet.

Even if they are the same type of object, we must take care to distinguish between the object producing the field and the object feeling the force. However, Newton's third law is, as always, relevant here. For example, two identically charged objects, A and B, repel one another with equal and opposite forces as shown below.

Figure 9 Each charge is +3 mC and neither one is free to move. The forces are equal and opposite.

This can be viewed in several ways:

a) Since A has a charge, it causes an electric field, E_A. Charge B is "sitting" in A's field. Since B is charged, it feels a force, F_B.

b) Since B has a charge, it causes an electric field, E_B. Charge A is "sitting" in B's field. Since A is charged, it feels a force, F_A.

c) Since A and B are both charged, there will be a combined overal field, E_{Total}. If another charged object C, was introduced, it would be "sitting" in the combined field. Since C is charged, it would feel a force, F_c (which would depend on where it was situated). See page 194 for a discussion of combined fields.

Without reading ahead, try to quickly predict what happens to the value of these forces if the charge on B is increased to 6 mC but the charge on A remains the same.

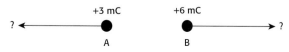

Which of the following statements is true?

A F_A doubles but F_B remains the same

B F_B doubles but F_A remains the same

C F_A and F_B both double

D F_A and F_B both remain the same

Consideration of the fields tells us that:

- Since the charge on A has remained constant we have the same electric field as before, E_A. The charge that is "sitting" in this field has doubled so the force that B feels must have also doubled. Force, F_B will have doubled.
- Since the charge on B has doubled we have doubled the electric field, E_B. The charge that is "sitting" in this field has remained constant but the field has doubled so the force that A feels must have also doubled. Force, F_A will have doubled.

The answer is thus C

Representation of a field: a general approach

In order to pictorially represent the variation in field strength around the source of a field, we need to find a way of representing both the magnitude and the direction of the force on the test object. The general approach is to introduce the idea of **field lines**. These are called **flux lines** in some situations.

At some time in the past, you have probably already used field lines to draw the variation in magnetic field strength around a bar magnet but you might not have thought about what the lines physically represent. The field lines can be drawn in such a way as to represent the field in a mathematically accurate way.

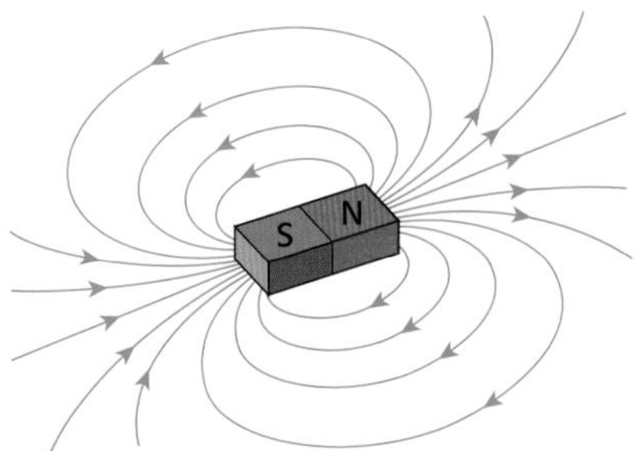

Figure 10 The magnetic force is stronger near the poles of the magnet as shown by the field lines being closer together

Typically, the direction of the force on the test object is shown by the direction of the field lines and the variation in the magnitude of the force on the test object is shown by how close together the lines are drawn. Consider the familiar example of the magnetic field around a bar magnet.

The diagram in Figure 10 is a two-dimensional representation of a three-dimensional field. In general, the number of field lines that pass though a given area that is perpendicular to the field lines represents the strength of the force on the test object.

Figures 11a and 11b both represent a uniform field because the field lines are evenly spaced and parallel to one another. In a field that increases with distance, the lines get closer together and a weakening field is shown by the lines getting further apart.

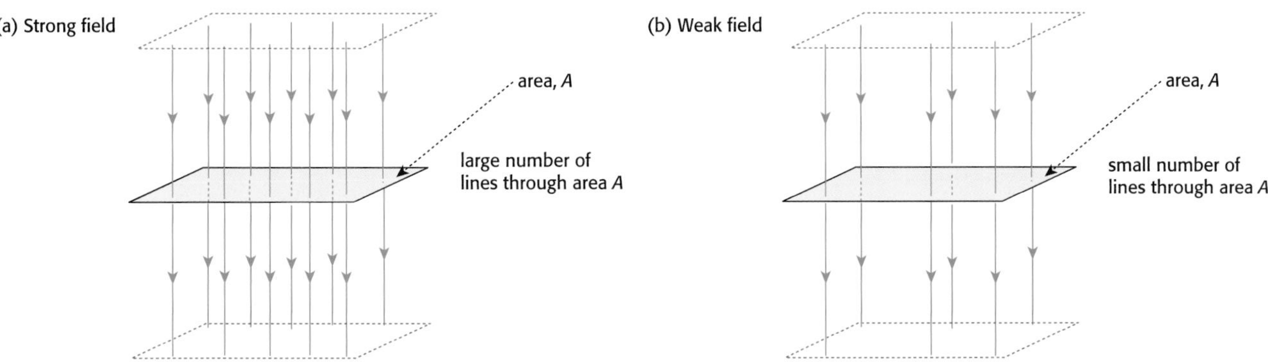

Figure 11

1: Gravitational field lines

A mass attracts all other masses towards it with a force that diminishes with distance of separation. The arrows on the field lines give the direction of the force acting on a small test point mass.

The variation in magnitude of the gravitational field for a point mass is shown in Figure 12.

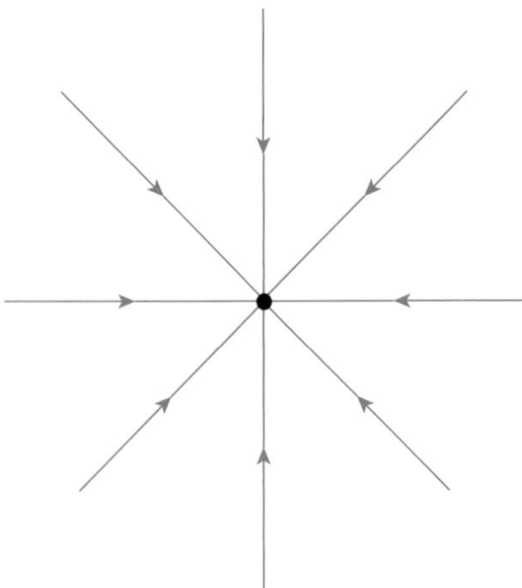

Figure 12 A point mass has a radial gravitational field

This is called a **radial** field. The closer you get to the source of the gravitational field, the stronger the field. The field lines show this by getting closer together towards the source of the field. The field varies according to an inverse square law (see pages 191 and 194) and Figure 13 accurately represents this mathematical relationship because the number of field lines that pass through a given area also decreases according to an inverse square law.

In practical situations, the masses under consideration are not point objects. However, the overall gravitational field outside any uniform spherical object turns out to be the same as that for a point object with the same mass placed at the centre of the sphere. Near the surface of the Earth, we often assume that the gravitational field strength is constant with $g = 9.81\ \text{N kg}^{-1}$. A constant gravitational field would be represented by evenly spaced, parallel field lines (Figure 13).

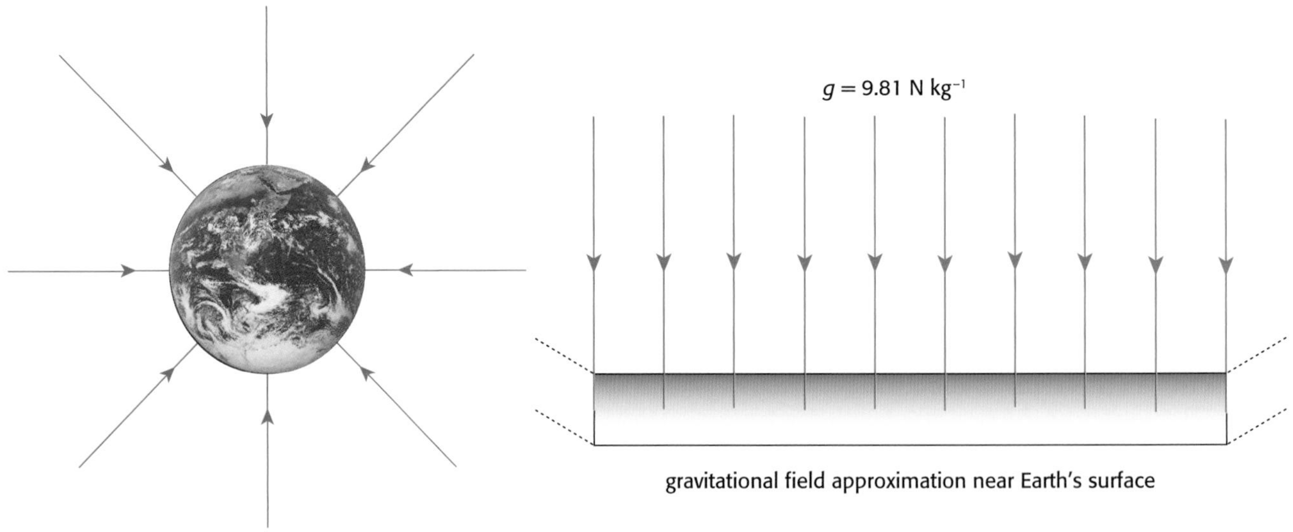

Figure 13

2: Electric field lines

The electric field pattern around a point charge has the same shape as the gravitational field pattern around a point mass, but the direction of the field lines depends on the sign of the charge. By convention, the direction shown indicates which way a positive test charge placed in the field would accelerate, so field lines point towards a negative charge and away from a positive charge.

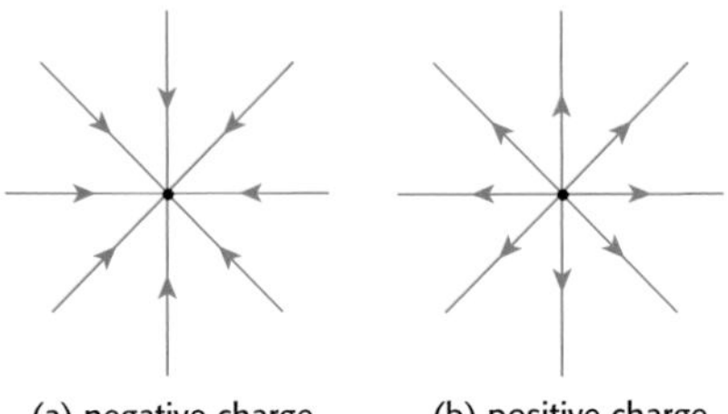

Figure 14 Electric fields around point masses

3: Magnetic field lines

The direction of a magnetic field line represents the direction of the force felt by the north pole of a small test magnet. There is, however, no such thing as an isolated magnetic pole so magnetic field strength is defined in terms of the force on a moving charge, or on a small current-length. Not only does the field depend on the distance to the moving charge, but the geometry of the situation is also important.

Once again, the field that results from a small current-length obeys an inverse square relationship in terms of distance, but the dependence on the geometry of the situation means that the overall magnetic field does not necessarily vary in this way.

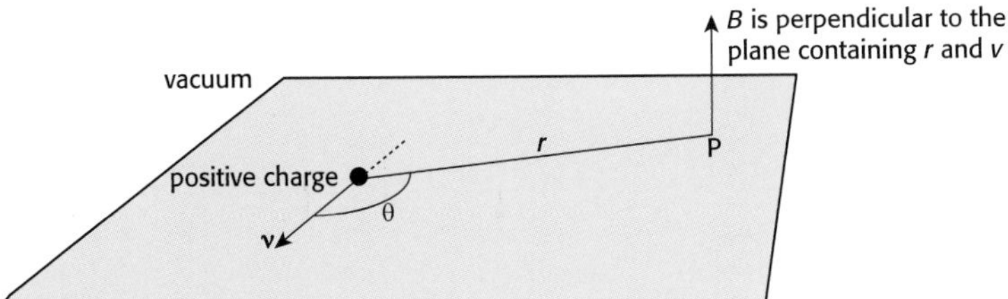

Figure 15 Field at point P due to single moving charge is proportional to the inverse square of the distance away r

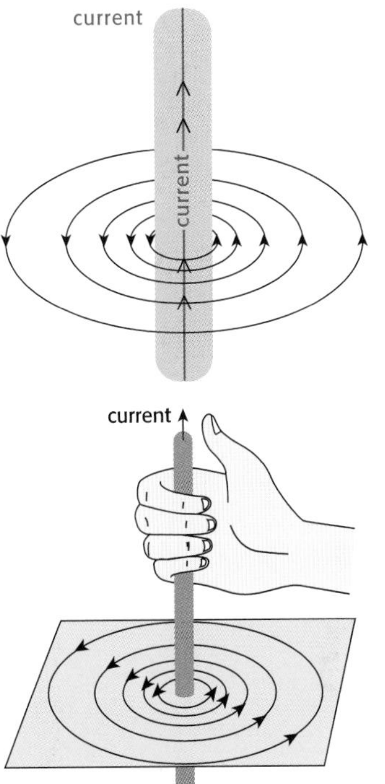

Figure 16 The magnetic field around a current

The overall magnetic field at P is the result of the additions of the magnetic fields due to each small section of the wire. The general method for adding fields together is shown on page 194. In the case of a long straight wire this results in the overall magnetic field pattern shown in Figure 16.

The direction of the overall magnetic field is remembered using the right-hand grip rule.

Three force laws

The following approach is often adopted when trying to predict the effect on an object as a result of a field.

- A single point object can be thought of as the fundamental source of any field.
- Combinations of objects result in a combined field (see page 194 for how to add fields).
- Any other objects placed in this field can be thought of as receivers of a force.
- If the object moves to a different region of the field, then the force on the object is likely to change. The force needs to be recalculated from the field strength.

The first step is to calculate the field strength around a point object (see Figure 18). Whether we are talking about gravitational, electric or magnetic fields, the field strength at a particular point P near the object (*object 1*) depends on only three factors:

- the value of the mass, charge or current-length of the source of the field
- the distance r from the object to the point P
- the material between the object and point P.

Let's consider the simplest case and assume that there is a vacuum between the object and point P. If we put a second object (object 2 – see Figure 19) in this field at point P, then it will feel a force F that only depends on:

- the field due to object 1.
- the value of the mass, charge or current-length of object 2.

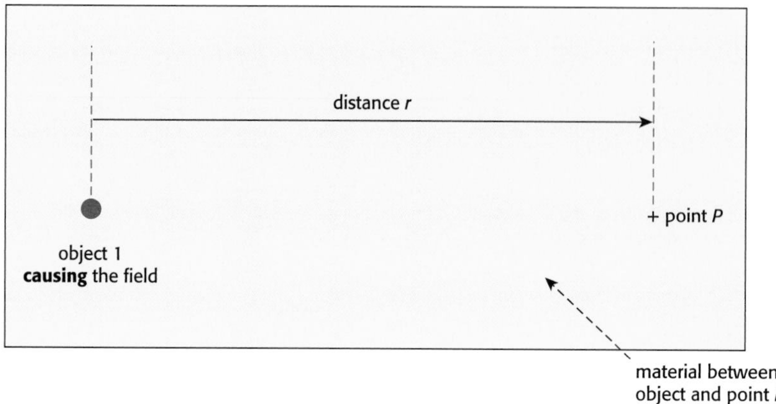

Figure 17 The field strength depends on the value of object 1, *r* and the material

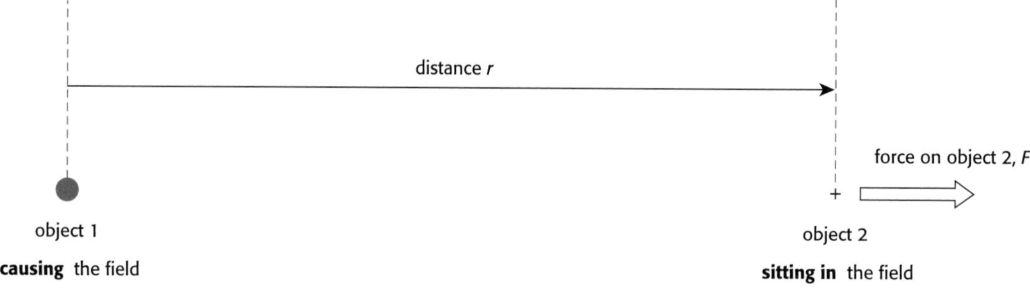

Figure 18 The force on object 2 depends on the field due to object 1 and the value of object 2

1: Calculating gravitational field strength, g

At a point P some distance away from a single point object, experiments show that the gravitational field g is:

- proportional to the mass of the point object causing the field, m_1
- inversely proportional to the square of the distance r from the point object to the point P.

The relationship is described as an **inverse square** relationship. Mathematically this is written as:

$$g \propto \frac{m_1}{r^2} \qquad \text{or} \qquad g = -\text{constant} \times \frac{m_1}{r^2}$$

The negative sign is there because the gravitational force is always attractive, so the direction of the force is towards the source of the field. The constant is given the symbol G and is called the *universal gravitational constant.*

$$g = -G\frac{m_1}{r^2} \qquad G = 6.67 \times 10^{-11}\ \text{N m}^2\ \text{kg}^{-2}$$

So, the force acting on another mass, m_2 that is placed at point P is given by:

$$F = g \times m_2 \qquad \text{or} \qquad F = -G\frac{m_1 m_2}{r^2}$$

This relationship is a mathematical statement of **Newton's law of universal gravitation**. In words, this law states that the gravitational force of attraction between two point masses is proportional to the product of the two masses and inversely proportional to the square of the distance of separation between the masses.

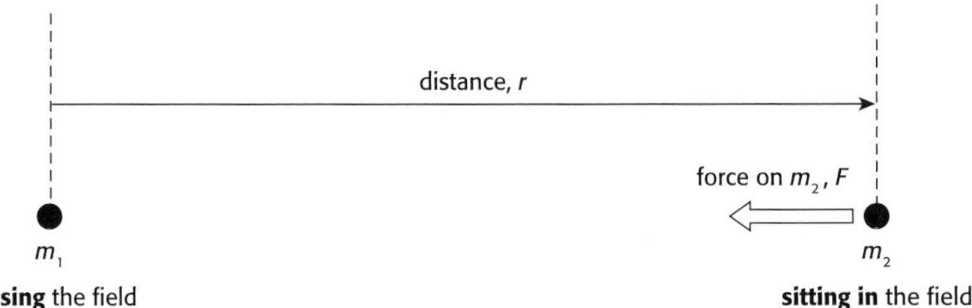

Figure 19 Force on m_2 is calculated using the field due to m_1 at a distance, r

2: Calculating electric field strength, E

At a point P some distance away from a single point object, experiments show that the electric field E is:

- proportional to the charge causing the field, q_1
- inversely proportional to the square of the distance r from the point object to the point P.
- dependent on the material between the object and point P.

Mathematically this is written as:

$$E \propto \frac{q_1}{r^2} \qquad \text{or} \qquad E = \text{constant} \times \frac{q_1}{r^2}$$

By convention, we define the field in relation to a positive charge, and the field is directed away from a positive charge, so this time there is no negative sign. If we once again assume that the charges are in a vacuum, the constant is given the symbol k and is called the Coulomb constant.

$$E = k\frac{q_1}{r^2} \qquad k = 8.99 \times 10^9 \text{ N m}^2 \text{ C}^{-2}$$

This equation is very straightforward, but it sometimes turns out to be useful to write the constant in a different way:

$$k = \frac{1}{4\pi\varepsilon_0} \qquad \varepsilon_0 = 8.85 \times 10^{-12} \text{ C}^2 \text{ N}^{-1} \text{ m}^{-2}$$

The constant ε_0 is called the permittivity of a vacuum (often referred to as the permittivity of free space). If there is some substance between the charges, this will affect the value of k because different materials have different values of permittivity, ε.

The force acting on another charge, q_2, that is placed at point P is thus given by:

$$F = E \times q_2 \quad \text{or} \quad F = k\frac{q_1 q_2}{r^2} = \frac{q_1 q_2}{4\pi\varepsilon_0 r^2}$$

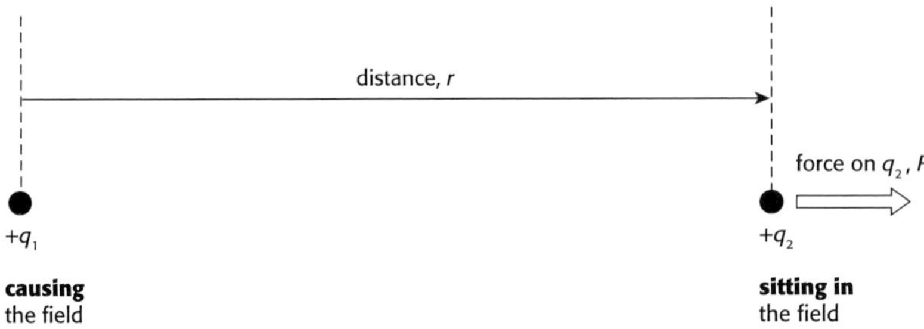

Figure 20 Force on q_2 is calculated using the field due to q_1 at the distance, r

This relationship is a mathematical statement of **Coulomb's law**. In words this law states that the electrostatic force between two point charges is proportional to the product of the two charges and inversely proportional to the square of the distance of separation between the charges.

3: Calculating magnetic field strength *B*

The fundamental relationships for magnetic are beyond the scope of the IB Diploma Programme physics course. The reason for this is that the three dimensional geometries involved with magnetic fields add complications and an analysis from first principles needs to involve a branch of mathematics called calculus.

In general, at a point *P* some distance away from a single moving point charge, experiments show that the magnetic field *B* is:

- proportional t the moving charge that is causing the field, q_1
- inversely proportional to the square of the distance *r* from the point object to the point *P*.
- dependent on the angle θ between the velocity vector and the perpendicular distance to the point *P*.
- dependent on the material between the object and point *P*.

The magnetic field strength is given by:

$$B \propto \frac{q_1 v \sin\theta}{r^2} \text{ or } B = \text{constant} = \frac{q_1 v \sin\theta}{r^2}$$

If the charge is moving in a vacuum, the constant of proportionality is expressed in terms of a constant μ_0 called the **permeability of free space**.

$$B = \frac{\mu_0}{4\pi} \times \frac{q_1 v \sin\theta}{r^2}$$

$$\mu_0 = 4\pi \times 10^{-7}\ \text{T m A}^{-1}$$

A similar result can be derived when considering the magnetic field strength due to a small current-length.

Mathematical physics: How to add fields

All fields are vector quantities (see page 41) so to add two (or more) fields we must correctly add together two (or more) quantities that have both a magnitude and a direction. One key idea is that field lines can never cross one another. It may also be helpful when imagining a combined pattern to remember that field lines repel each other if they are parallel and attract each other if they are anti-parallel.

Another useful property comes from the inverse square nature of all these fields. Although they all have an infinite range, the strength of a field drops rapidy as the separation between source and test object increases. The field due to a nearby object often dominates the resultant field. Figure 21 shows how the gravitational field strength decreases with distance from a point object, starting at a distance of r_0 (the radius of the object).

Whatever the combination of masses, charges, or current-lengths, it should be possible to estimate the overall field felt a large distance from the individual objects. At a large distance we can consider them all together; for example, an object that is a long way from three charges, +3 mC, −5 mC, and +1 mC, will feel an overall field that is equivalent to one charge: +3 − 5 + 1 = 1 mC, no matter how the three individual charges are locally arranged.

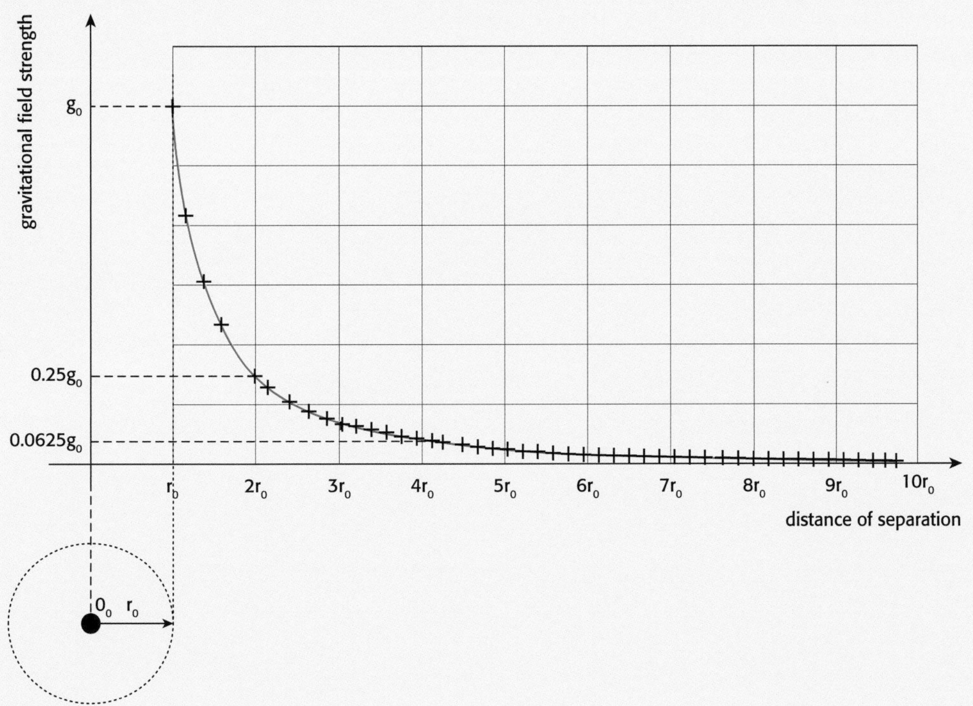

Figure 21 Inverse square relationship

1: Gravitational field patterns

Field due to a planet

Newton's law of universal gravitation gives us a method for calculating the field strength due to a single point mass, but a planet is clearly not a point mass. Theoretically, the overall field of the planet at any point is the addition of the fields for all the individual point masses that make up the planet.

In principle it is possible to work out the distance to every single point and use Newton's law of universal gravitation to calculate the contribution that each point makes to the overall gravitational field. Once we have calculated all the contributions we could then perform a vector sum. As you can imagine this is a complicated mathematical addition.

Fortunately, you do not need to perform this calculation. In fact, the gravitational field anywhere outside a uniform sphere of constant density is exactly the same as if all the mass had been concentrated at its centre. The same idea can be used for an object placed in the field. The force between two spherical masses (whose separation is large compared to their radii) is the same as if the two spheres had their masses concentrated at a point at the centres of the spheres. This simplification means that planets are often treated as point masses (even though they do not have constant density).

For example, the Earth attracts every point on the Moon and vice versa. The situation in Figure 22a can be simplified to the situation illustrated in Figure 22b.

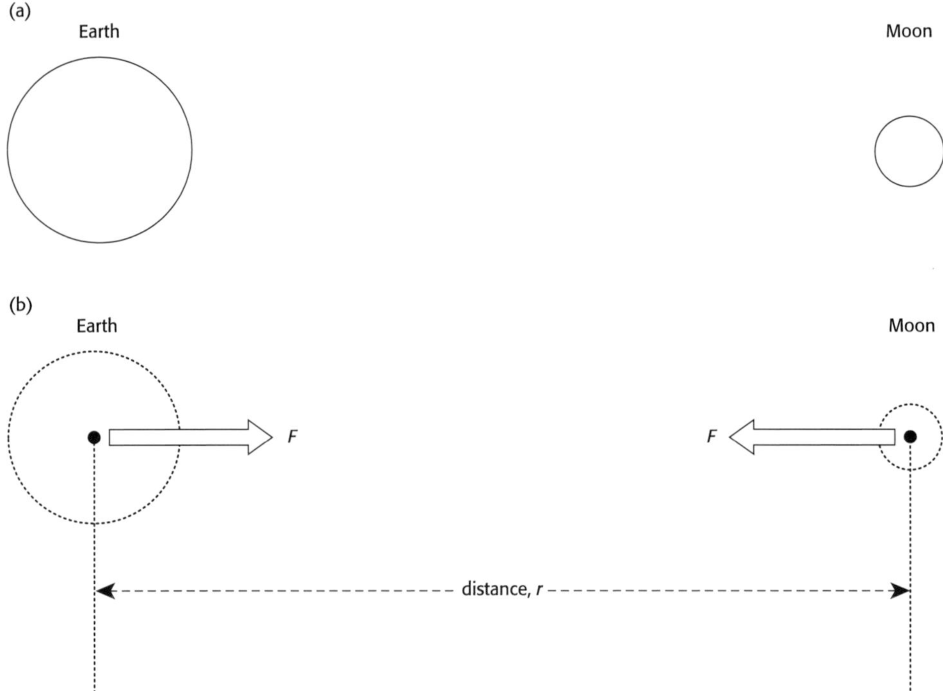

Figure 22 The Earth and the moon can be treated as point objects

Combined fields of two masses

Figure 23 shows the process for calculating the resultant gravitational field between two masses (e.g. the Earth and the Moon) at one particular point P in space.

To find the shape of the overall resultant field, we must repeat this process at every point in space.

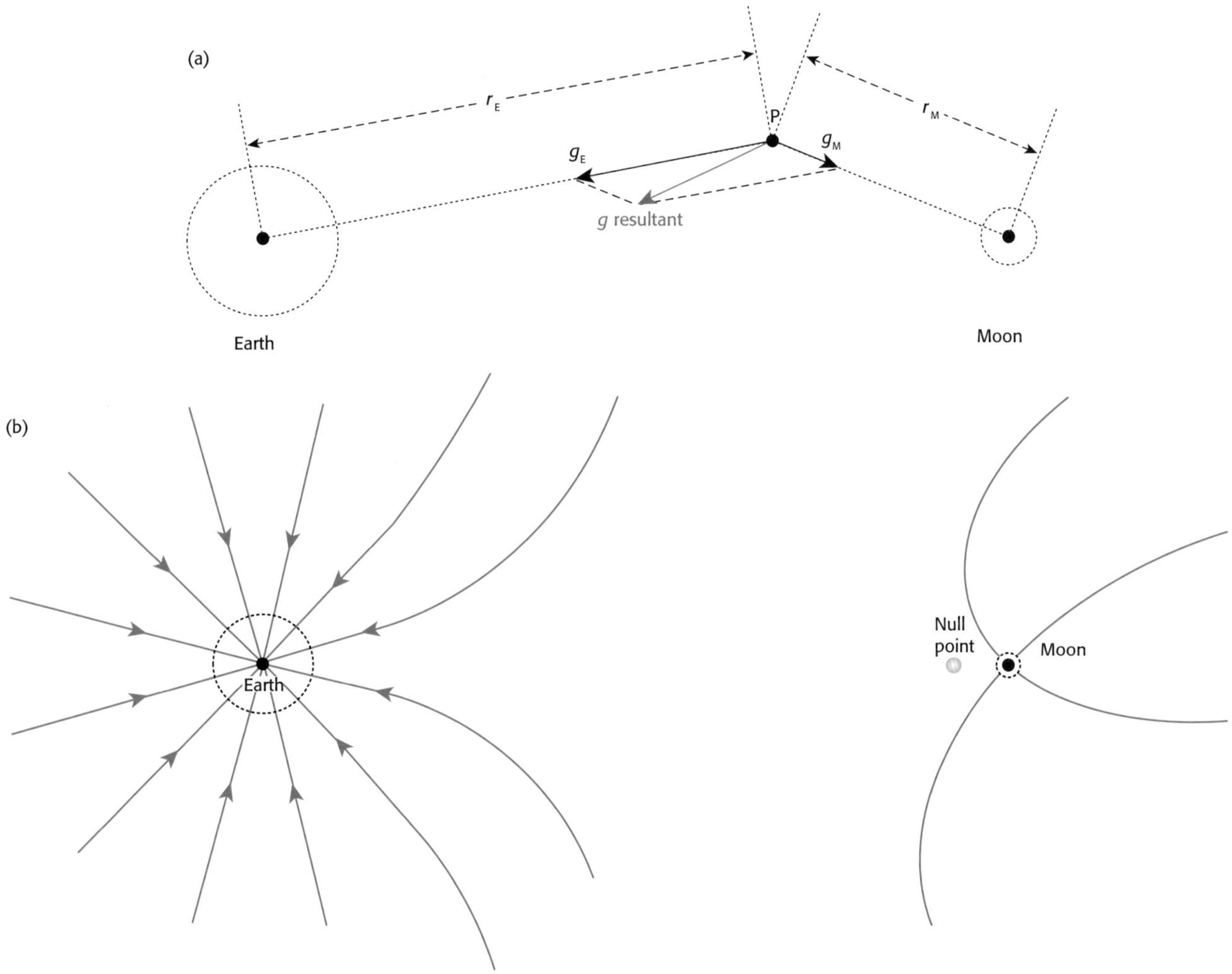

Figure 23 The combined gravitational field due to the Earth and moon: (a) calculation of resultant field at a point (b) overall result

There will be one point somewhere between the Earth and the Moon where the two fields cancel and there will be no resultant gravitational field – the **null point**.

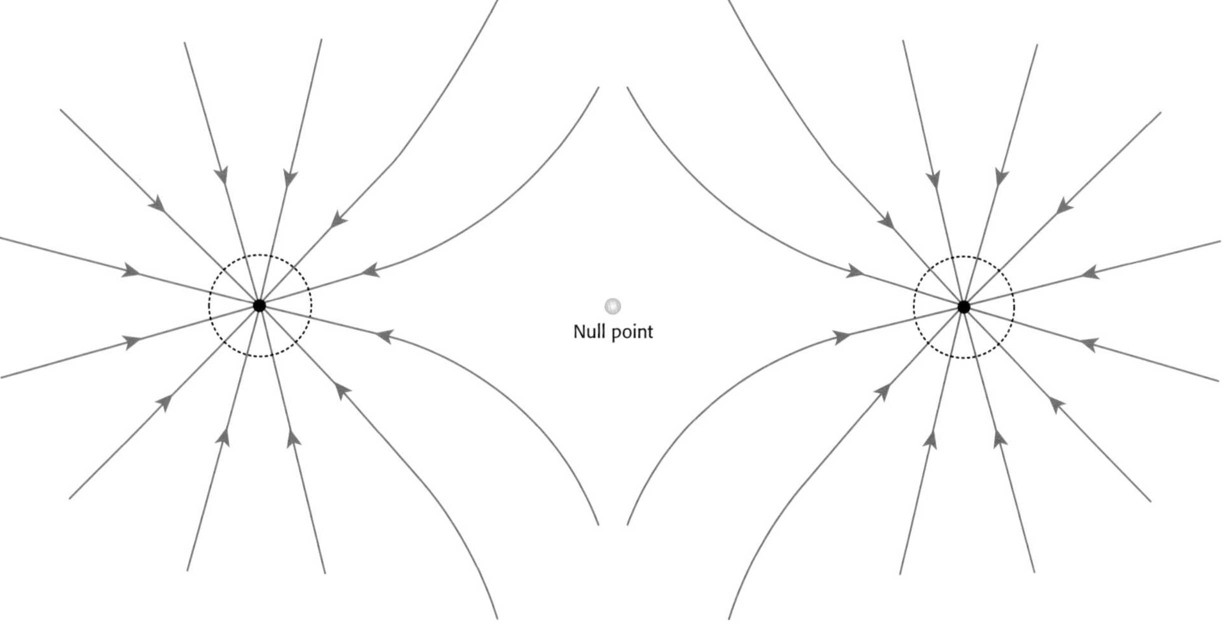

2: Electrostatic field patterns

A charged sphere

Outside a charged conducting sphere, the resultant electric field is identical to the field of an equally charged point charge concentrated at its centre. This statement is analogous to the gravitational field for a spherical mass, but the situations are slightly different. The mass is distributed uniformly throughout the sphere, whereas charge can be shown to always reside on the outside of the sphere. This fact does not affect the pattern of the fields outside the sphere but it does have an effect on the inside. Inside the mass, there is still a gravitational field but inside a hollow conducting object, there is no field at all.

Outside the sphere, the fields are the same as a point charge concentrated at the centre. Both fields are radial.

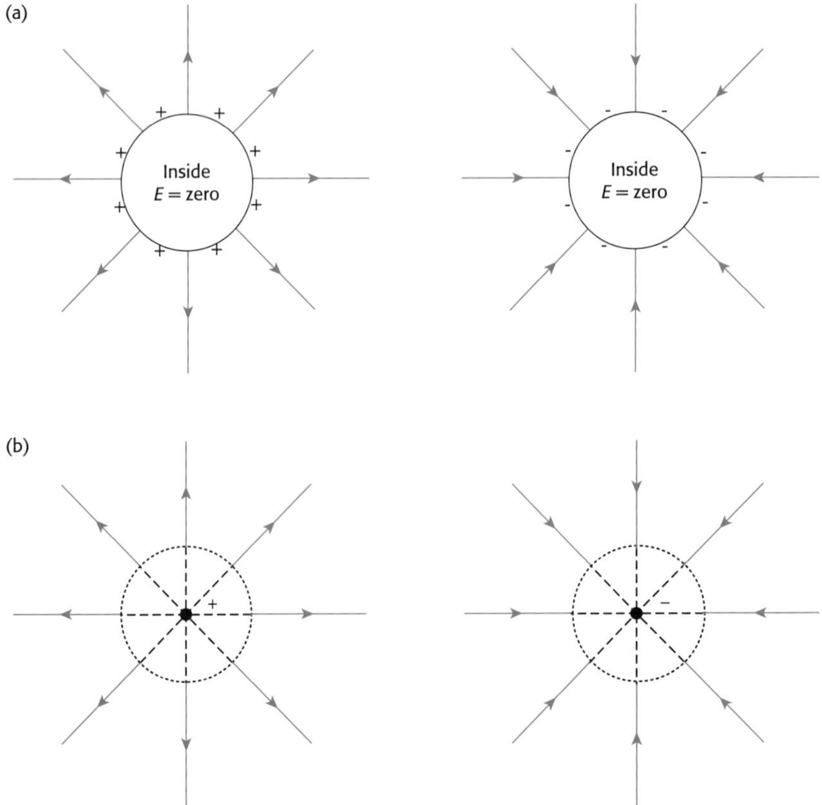

Figure 24 The electric fields around charged conducting spheres are radial outside but zero inside the sphere

Two point charges

If the two point charges are both negative, the overall electric field pattern will be identical in shape to the gravitational field pattern of two masses. A similar shape field exists for two positive charges, but the field directions will be reversed.

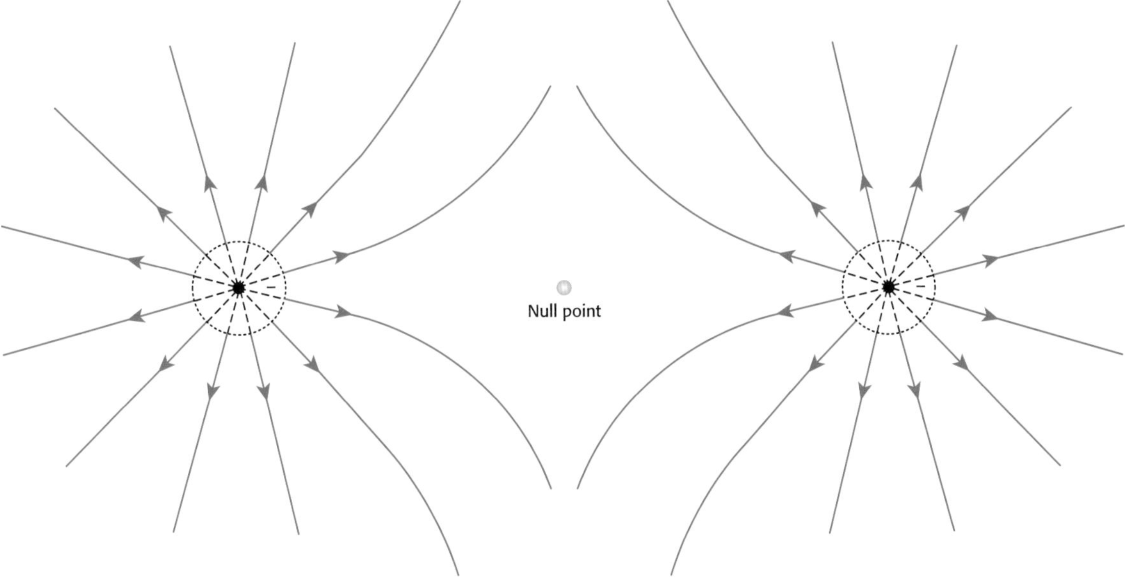

A completely different field pattern is created when the two charges are opposite. As before, the resultant field pattern results from the vector addition of the individual fields.

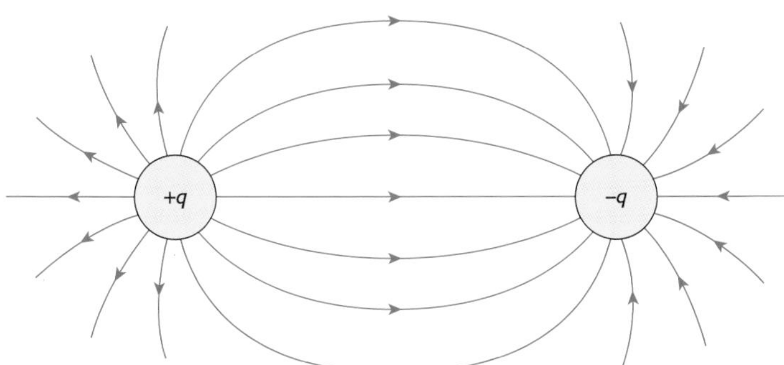

Figure 25 E field between opposite charges

Oppositely charged parallel plates

Notice that the field between the parallel plates is represented by evenly spaced parallel field lines. This means that the electric field between the plates is uniform in the central region. You might think that closer to, say, the positive plate the field would increase. This effect does happen, but it is balanced by an equal decrease in the field resulting from the increase in distance from the other plate.

Near the edge of the parallel plates the field lines spread out. This is known as the **edge effect**. Field lines getting further apart signify that the electric field must be getting weaker.

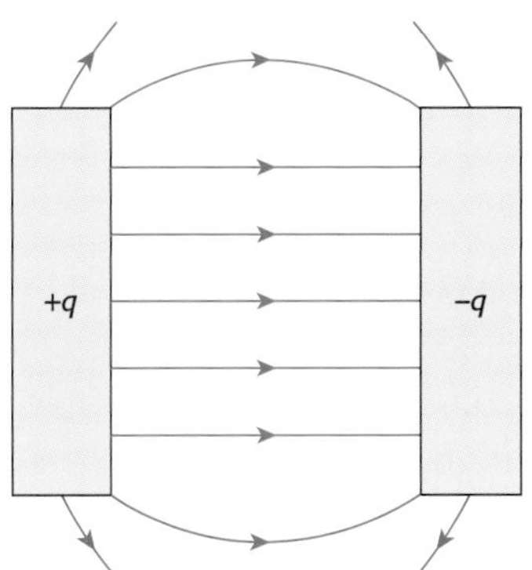

Figure 26 E field between parallel plates

3: Magnetic field patterns

The equations for magnetic field strength in the following sections do not need to be remembered.

A straight wire

The strength of the magnetic field B at a distance r away from an long straight wire which carries a current I is:

$$B = \frac{\mu_0 I}{2\pi r}$$

Where μ_0 is a constant called the permeability of free space.

Note that the magnetic field strength is proportional to the current in the wire. It is also inversely proportional to the distance away from the wire as opposed to being proportional to the inverse square of the distance.

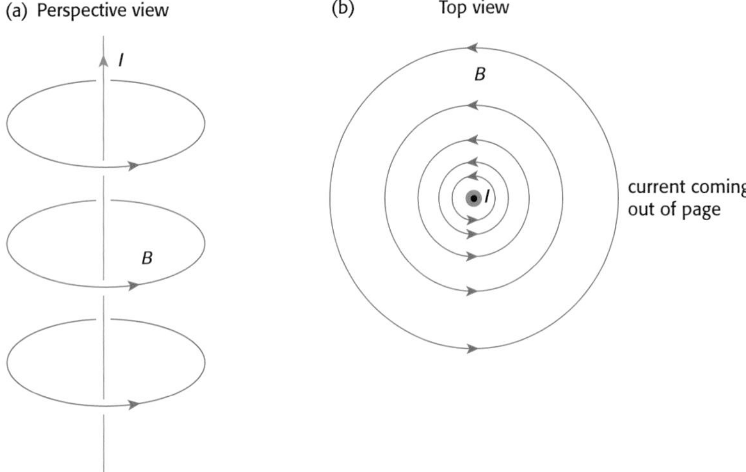

Figure 27 The magnetic field around a current-carrying wire

A flat circular coil

The shape of the magnetic field if a current flows in a flat circular coil is shown in Figure 28.

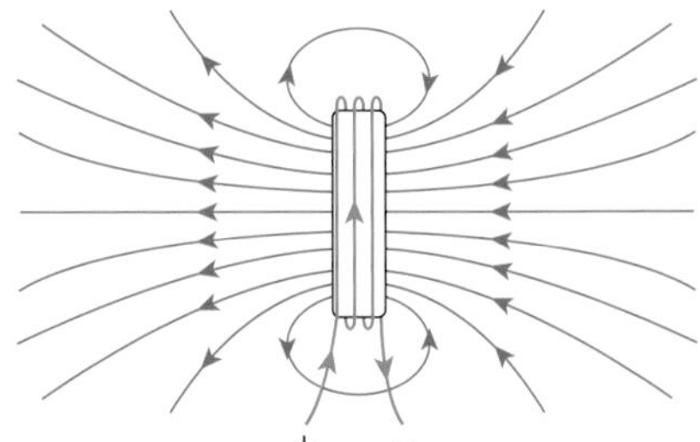

Figure 28

The field in the centre of the coil can be derived from the individual current-lengths around the coil. The result is:

$$B = \frac{\mu_0 NI}{2r}$$

where N is the number of turns of wire, I is the current, and r is the radius of the coil.

A solenoid

A solenoid is just another name for a long coil. A solenoid is an electromagnet, and the field around it is the same as the field around a bar magnet.

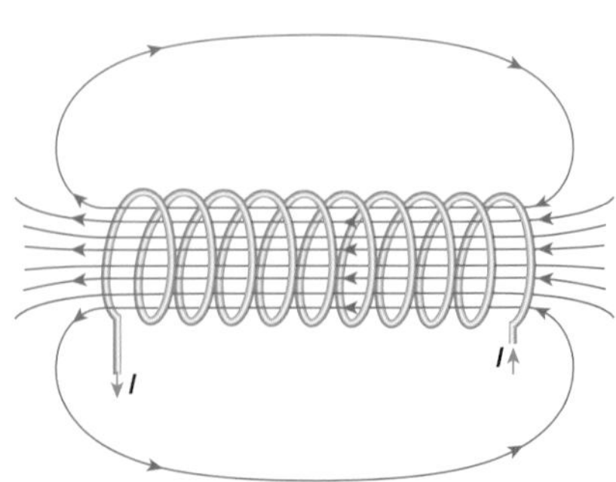

Figure 29 The magnetic field of a solenoid is the same as around a bar magnet

Note that the field in the middle of the solenoid is represented by equally spaced field lines, so we know that the field is constant in the centre of a solenoid. When a current I flows down an infinitely long solenoid with n turns per unit length, the resultant field inside the solenoid is given by:

$$B = \mu_0 nI$$

where n is the number of turns N divided by the length l.

The table below gives some information about the Solar System.

Table 1

Mass of the Moon	7.3×10^{22} kg
Mass of the Earth	6.0×10^{24} kg
Mass of the Sun	2.0×10^{30} kg
Distance from Earth to Sun	1.5×10^{11} m
Distance from Earth to Moon	3.8×10^{8} m
Average surface gravitational field on the Earth	9.8 N kg^{-1}
Average surface gravitational field on the Moon	1.6 N kg^{-1}
Gravitational constant, G	6.7×10^{-11} N m^2 kg^{-2}

Use the information in the table to:

a) calculate the radius of the Earth

b) calculate the distance to the "null point" in space where the gravitational field of the Earth is equal and opposite to the gravitational field of the Moon

c) calculate the value of the gravitational field due to the Earth that would be felt by an astronaut on the surface of the Moon. Would it be possible to measure this field?

d) i) Work out which of the forces due to the Earth, the Moon, and the Sun is strongest at the Earth's surface

ii) If the Earth and the Moon suddenly vanished (not very likely), would you continue to orbit the Sun?

Mathematical physics: *Motion in a field—a general difficulty*

Calculating the precise path of objects moving in fields is often very difficult. Firstly, as the object moves the field is likely to change. Secondly, in the real world every object both acts under the force due to other objects and produces its own field. This means that making mathematical predictions can quickly become extremely complex.

Consider a mass (e.g. an asteroid) that is moving on a path that approaches (but does not collide with) a planet (Figure 30).

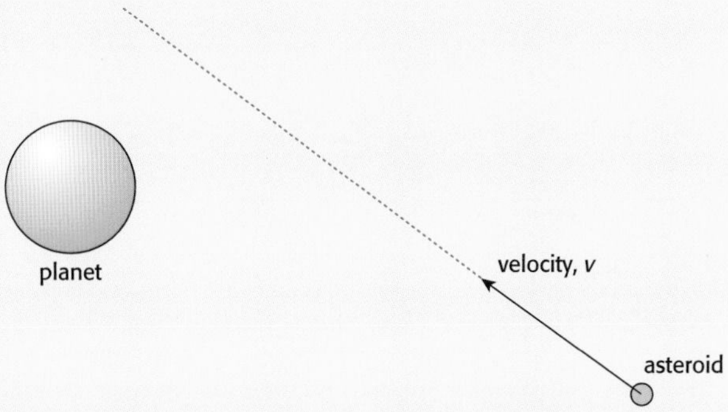

Figure 30 An asteroid is approaching a planet

An approximate description of the resultant motion of the asteroid can be calculated by assuming that the planet's gravitational field at all points in space is fixed.

- From the position of the asteroid, we can calculate what the force on the asteroid will be.
- This force F will cause the asteroid's velocity to change as predicted by Newton's second law.
- Thus we know the position, the velocity and the acceleration of the asteroid and a description of the asteroid's subsequent motion can be found from these starting conditions.
- As soon as it moves to a new position in space, the force on the asteroid will have changed and the subsequent motion will be affected in a different way. We can use a GDC to perform an iteration of the motion of the asteroid, and hence plot its path. To do this, choose initial values for the asteroid r_0 and v_0. The origin is the centre of the planet. In the x-y plane, the component of acceleration in the x direction, a_x, is proportional to $\frac{x}{r^2}$. Similarly a_y is

proportional to $\frac{y}{r^2}$. Using a_x, a_y and kinematics equations it is possible to find the new position after a small amount of time, ΔT. Point by point, the shape of the path of the asteroid can be found.

Newton was able to show that, providing the asteroid does not hit the planet, the path is a hyperbola, a parabola, an ellipse or a circle depending on the initial velocity.

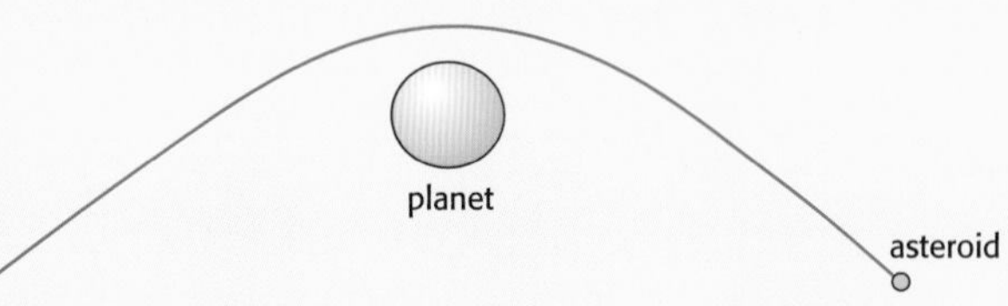

Figure 31 The asteroid path in this case is a parabola

The calculation outlined above has neglected the fact that a force also acts on the planet due to the asteroid. This force will cause the planet to move a little and this would need to be taken into account when calculating the value of the force acting on the asteroid. It is still possible to factor this variation into the equations and arrive at a general solution which predicts the path taken by both the planet and the asteroid, whatever initial conditions of mass and velocity.

Unfortunately, it is not possible to come up with an exact general mathematical solution for more than two objects interacting with each other (leading to the "three body problem"). Most real-world problems involve more than two objects and so solutions must be found by creating models using appropriate approximations.

Rediscovering physics: Magnetic and electric field patterns.

You might have used iron filings and/or plotting compasses to show the overall magnetic field pattern that results from a simple bar magnet. The same procedures can be used to confirm the shape of magnetic fields near current-carrying wires, coils, and solenoids. As the fields can be quite weak, it can be necessary to work with reasonably large currents (e.g. 5 A). For safety, the wires should always be checked to see if they are overheating.

Investigating physics: The strength of an electromagnet

Your task is to plan (and hopefully carry out) an investigation into at least one of the factors that affect the strength of your electromagnet.

Thinking about science: Why do masses attract?

A simple question cannot always be answered properly by a simple answer, e.g. why is the sky blue and why are clouds white? Sometimes answers may seem plausible but actually fail to answer the question. For example, in some parts of the world, some trees lose all their leaves for the winter months whereas others do not.

Asked why some types of tree lose their leaves over winter, some people will say "because they are deciduous". The answer has not really addressed the question at all. A particular tree loses its leaves not **because** it is

deciduous, it is deciduous **because** it loses its leaves. The reason it loses its leaves must be explained by the biochemical reactions happening inside the tree.

In the same way, seemingly simple questions in physics often turn out to be quite complex. It may come as a surprise to you but physicists have only made limited progress, if any, towards explaining **why** the force of gravity exists. Newton came up with an extremely elegant mathematical **description** of the force of gravity, but this mathematical theory only describes the effect; it does not suggest a cause. Even modern theories of gravitation do not necessarily make any attempt to explain the cause – they just describe the effect as accurately as possible and in a way that allows predictions to be tested.

The gravitational force is related to a measurable property that all everyday objects have called **mass**. This property is something with which you are already familiar and you will have built up some understanding of what mass means. There are, however, several possible ways of defining mass and they are not all the same. Perhaps the simplest, but non-mathematical, definition is that mass is *the amount of matter* contained in that object. This definition has the advantage of being very straightforward and aids understanding from the very beginning.

As your studies become more sophisticated, this definition will become more refined. At the same time, simple models can be replaced with ones that provide a more detailed agreement between the predictions and the experimental reality of the universe. In order to improve upon our models, we must be willing to question and reassess our knowledge, assumptions, and understanding.

For example, the law of conservation of mass (mass is neither created nor destroyed) clearly fits well with the idea of mass as the amount of matter, as this does not seem to change. However, it is possible for mass to be converted into energy and vice versa, as summarized in Einstein's famous equation, $E = mc^2$. This means that there are situations where neither mass nor energy is conserved, but the total "mass-energy" of a system is always conserved. Our picture of mass as just the amount of matter now needs to be refined.

The story does not end there. The scientific process is one in which models are continually refined and improved. The following are some of the different descriptions that are associated with mass:

- the amount of matter in an object
- the ratio of resultant force to acceleration of an object (from Newton's equation $F = ma$)
- the property that gives rise to (and is affected by) a gravitational field
- a form of energy (from $E = mc^2$, as discussed)
- a property that results from interactions between objects in the universe and fundamental particles called Higg's bosons.

Thinking about science: Discussion

In the light of the large number of possible descriptions for mass, to what extent do you think that scientific process has been successful in replacing old models with new models. Are all these models compatible with one another?

End of chapter summary

Chapter 13 has considered forces and fields for three different forces: gravitational, electric and magnetic. The list below summarises the knowledge and skills that you should be able to undertake after having studied this chapter. Further research into more detailed explanations using other resources and/or further practice at solving problems in all these topics is recommended – particularly the items in bold.

Gravitational and electrical forces and fields

- **State that there are two types of electric charge and apply the law of conservation of charge.**
- Describe and explain the difference in the electrical properties of conductors and insulators.
- State Newton's universal law of gravitation and Coulomb's law.
- Define *gravitational field strength and electric field strength*.
- Determine the gravitational field due to one or more point masses and the electric field strength due to one or more point charges.
- Derive an expression for the gravitational field strength at the surface of a planet, assuming that all of its mass is concentrated at its centre.

- Draw the electric field patterns for different charge configurations.
- **Solve problems involving gravitational masses, forces and fields and involving electric charges, forces and fields.**

Magnetic forces and fields

- State that moving charges give rise to magnetic fields and draw magnetic field patterns due to currents.
- Determine the direction of the force on a current-carrying conductor in a magnetic field and on a charge moving in a magnetic field.
- Define the magnitude and direction of a magnetic field **and solve problems involving magnetic forces, fields and currents**.

Chapter 13 questions

1 This question is about gravitational field strength near the surface of a planet.

a) i) Define *gravitational field strength.* [2]

ii) State why gravitational field strength at a point is numerically equal to the acceleration of free fall at that point. [1]

b) A certain planet is a uniform sphere of mass M and radius R of 5.1×10^6 m.

i) State an expression, in terms of M and R, for the gravitational field strength at the surface of the planet. State the name of any other symbol you may use. [1]

ii) A mountain on the surface of the planet has a height of 2000 m. Suggest why the value of the gravitational field strength at the base of the mountain and at the top of the mountain are almost equal. [2]

c) A small sphere is projected horizontally near the surface of the planet in (b). Photographs of the sphere are taken at time intervals of 0.20 s. The images of the sphere are placed on a grid and the result is shown below.

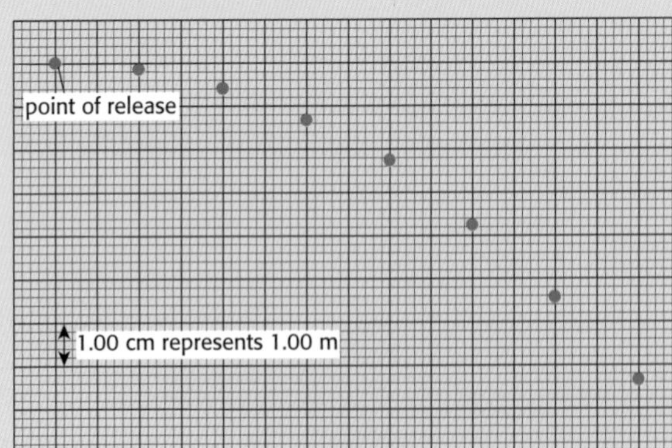

The first photograph is taken at time $t = 0$. Each 1.00 cm on the grid represents a distance of 1.00 m in both the horizontal and the vertical directions.

Use the diagram to

i) explain why air resistance on the planet is negligible; [2]

ii) calculate a value for the acceleration of free fall at the surface of the planet. [3]

(d) Use your answer to (c)(ii) and data from (b) to calculate the mass of the planet. [2]

(Total 13 marks)

2 This question is about fundamental forces.

a) In total, there are approximately 10^{29} electrons in the atoms making up a person. Estimate the electrostatic force of repulsion between two people standing 100 m apart as a result of these electrons. [4]

b) Estimate the gravitational force of attraction between two people standing 100 m apart. [2]

c) Explain why two people standing 100 m apart would not feel either of the forces that you have calculated in parts (a) and (b). [2]

(Total 8 marks)

3 This question is about magnetic fields.

a) Using the diagram below, draw the magnetic field pattern of the Earth.

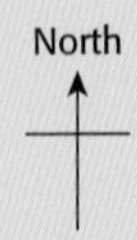

[2]

b) State what other object produces a magnetic field pattern similar to that of the Earth. [1]

c) A long vertical wire passes through a sheet of cardboard that is held horizontal. A small compass

is placed at the point P and the needle points in the direction shown.

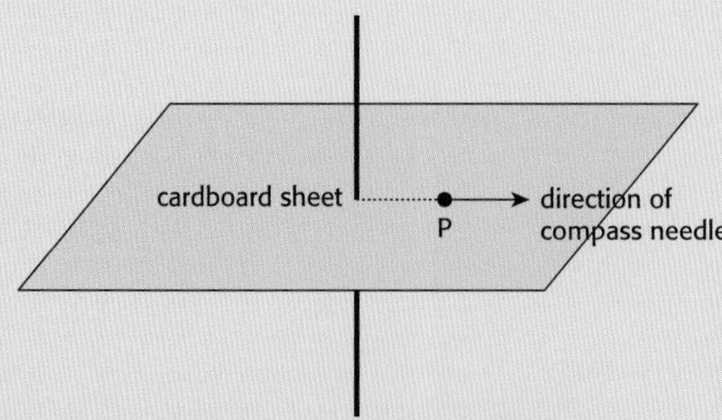

A current is passed through the wire and the compass needle now points in a direction that makes an angle of 30° to its original direction as shown below.

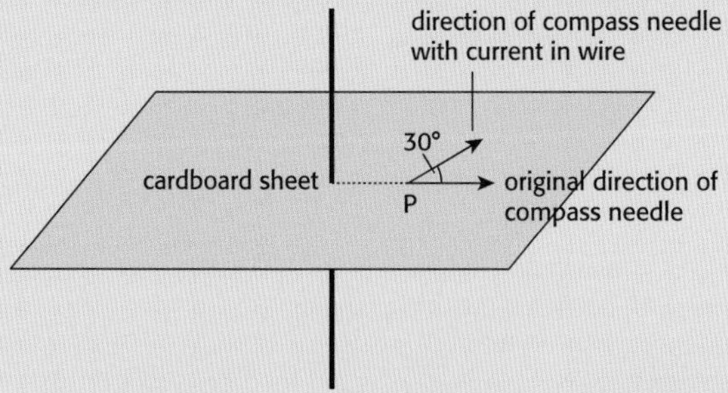

i) Draw an arrow on the wire to show the direction of current in the wire. Explain why it is in the direction that you have drawn. [2]

ii) The magnetic field strength at point P due to the current in the wire is B_W and the strength of the horizontal component of the Earth's magnetic field is B_E.

Deduce, by drawing a suitable vector diagram, that

$B_E = B_W \tan 60°$. [2]

(Total 7 marks)

4 This question is about the electric field due to a charged sphere and the motion of electrons in that field.

The diagram below shows an isolated, metal sphere in a vacuum that carries a negative electric charge of 9.0 nC.

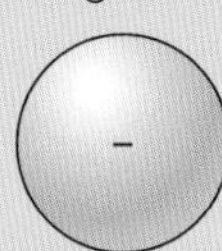

a) Draw arrows to represent the electric field pattern due to the charged sphere. [3]

b) The electric field strength at the surface of the sphere and at points outside the sphere can be determined by assuming that the sphere acts as though a point charge of magnitude 9.0 nC is situated at its centre. The radius of the sphere is 4.5×10^{-2} m. Deduce that the magnitude of the field strength at the surface of the sphere is 4.0×10^{4} Vm^{-1}. [1]

An electron is initially at rest on the surface of the sphere.

c) i) Describe the path followed by the electron as it leaves the surface of the sphere. [1]

ii) Calculate the initial acceleration of the electron. [3]

iii) State and explain whether the acceleration of the electron remains constant, increases or decreases as it moves away from the sphere. [2]

iv) At a certain point P, the speed of the electron is 6.0×10^{6} ms^{-1}. Determine the potential difference between the point P and the surface of the sphere. [3]

(Total 13 marks)

5 This question is about electric charge at rest.

a) Define *electric field strength* at a point in an electric field. [2]

Four point charges of equal magnitude, are held at the corners of a square as shown below.

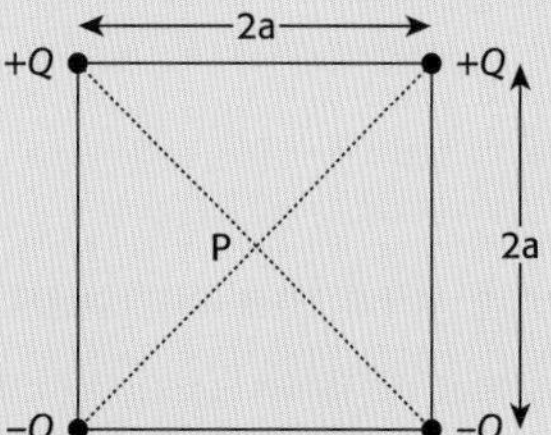

The length of each side of the square is $2a$ and the sign of the charges is as shown. The point P is at the centre of the square.

b) i) Deduce that the magnitude of the electric field strength at point P due to one of the point charges is equal to $\frac{kQ}{2a^2}$. [2]

ii) Draw an arrow to represent the direction of the resultant electric field at point P. [1]

iii) Determine, in terms of Q, a and k, the magnitude of the electric field strength at point P. [3]

(Total 8 marks)

6 This question is about gravitation.

a) State Newton's universal law of gravitation. [3]

b) The average distance of Earth from the Sun is 1.5×10^{11} m. The gravitational field strength due to the Sun at the Earth is 6.0×10^{-3}N kg^{-1}.

Estimate the mass of the Sun. [3]

c) Deduce that the orbital period T of a planet about the Sun is given by the expression

$T^2 = KR^3$

where R is the radius of the orbit and K is a constant. [5]

(Total 11 marks)

14 Field and potential

This chapter builds on the ideas introduced in Chapter 13 and analyses the energy associated with various fields. For the IB Diploma Programme it is sufficient to consider only gravitational and electrostatic fields at higher level.

Potential energy and the concept of potential

A field is the region in which an object feels a force and the magnitude of the field is defined as the force per unit object feeling the force (the test object). As a test object moves in a gravitational, electrostatic, or magnetic field it is likely that work will be done, and the test object will gain or lose potential energy.

The change in the test object's potential energy will depend on the magnitude of a property associated with the field and its change in position. The concept of **potential** at a point is introduced as the energy per unit test object at that point. Note that *potential* is a different quantity from *potential energy*.

As changes in energy are more relevant than absolute values, it is possible to define the zero of potential energy (and thus the zero of potential) wherever we choose. The simplest situation to consider is when we have just two objects – the one causing the field and the test object. In this situation the test object is defined to have zero energy when it is at an infinite distance away from the object causing the field.

The general form for the definition of potential at any point in a field is

$$\text{potential, } V = \frac{\text{energy needed to bring test object from infinity to the point}}{\text{test object feeling the force}}$$

In any field, the principle of conservation of energy means that the energy needed to bring the test object from infinity must be independent of the route taken. There is one value of potential for each point. The technical term describing this property is to say that gravitational and electrostatic fields are **conservative** fields.

Gravitational potential, V_g

Gravitational forces are felt by masses, so the definition of gravitational potential at any point P in the field is

$$V_g = \frac{\Delta E_p}{m}$$

where

ΔE_p is the change in potential energy (or the work done) to bring a small point test mass from infinity to point P, measured in J
m is the mass of the test object, measured in kg
V_g is the gravitational potential at point P, measured in J kg^{-1}

This definition means that all gravitational potentials (and often the total potential energy of a system) are negative. Negative values of energy may seem strange at first, as it is impossible to have a negative value of kinetic energy. But a system with negative gravitational potential energy simply means that energy needs to be put into the system in order to separate the masses involved. This must be the case if the masses are gravitationally attracted to one another.

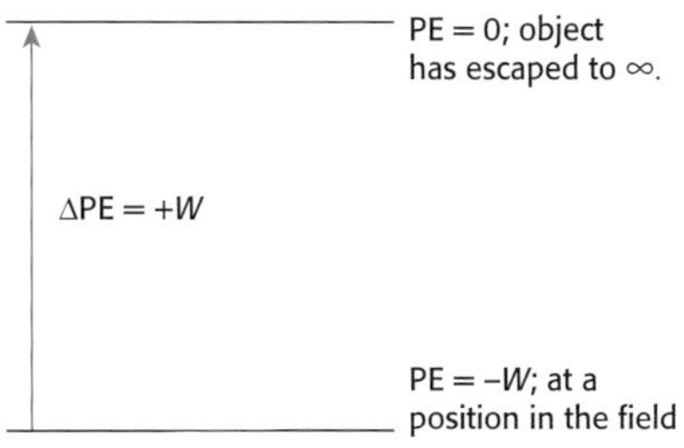

Figure 1 Energy level diagram

In order to give a mass enough energy to escape completely from the gravitational attraction of another object we need to give it energy. Once this mass has escaped it has zero PE, by definition, so the PE within the field must be less than zero. This process is illustrated in the energy level diagram in Figure 1. Note that the horizontal lines represent PEs in the field, not positions in the field.

When a test object is an infinite distance away from another mass, its potential is zero. As it falls in from infinity towards the mass being considered, the force of attraction does work, and the system must lose potential energy (if it falls in freely then the loss of PE is equal to the gain in KE). The mass started with zero PE (by definition) and has lost PE, so its new PE (and the potential at its new position) must be negative(see Figure 2).

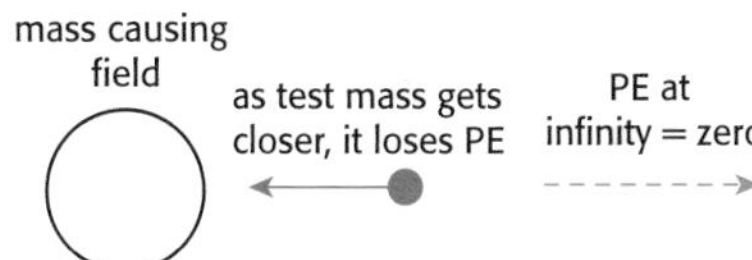

Figure 2 A test object has negative PE within the field

Note also that the definition of zero potential energy being at infinity is different from the one that we have used so far for all practical situations. We are normally interested only in energy *changes*, and so it can be useful to define the surface of the Earth as the zero for PE. This definition helps when we are considering small changes of height near the surface of the Earth, but any significant change needs a fuller analysis.

Figure 3 shows the variation in potential energy of a 1 kg object above the surface of the Earth. The left-hand values use the surface of the

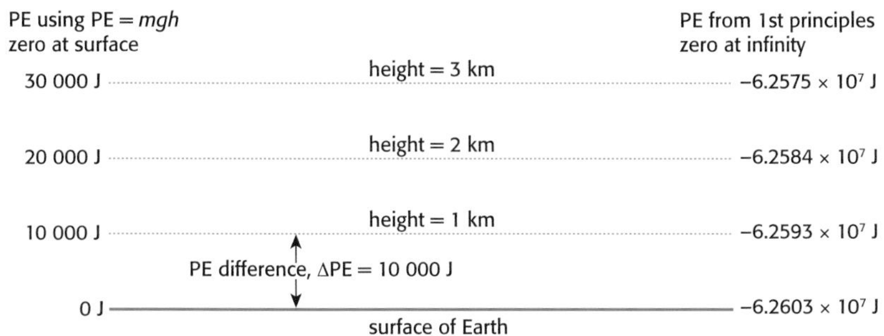

Figure 3 Variation in PE near the surface of the Earth

Earth as the zero for PE and have been calculated using the formula PE $= mg\Delta h$ assuming that the value of gravitational field strength remains constant, $g = 10\ \text{N kg}^{-1}$. The right-hand values use infinity as the zero of potential energy. The numbers have been calculated from first principles using the formula introduced on page 205.

Electrostatic potential, V_e

Electrostatic forces are felt by charges, so the definition of electrostatic potential at any point P in the field is

$$V_e = \frac{\Delta E_p}{q}$$

where

ΔE_p is the potential energy change needed to bring a small point positive test charge from infinity to point P, measured in J

q is the value of the test charge, measured in C

V_e is the electrostatic potential at point P, measured in J C^{-1} (or V)

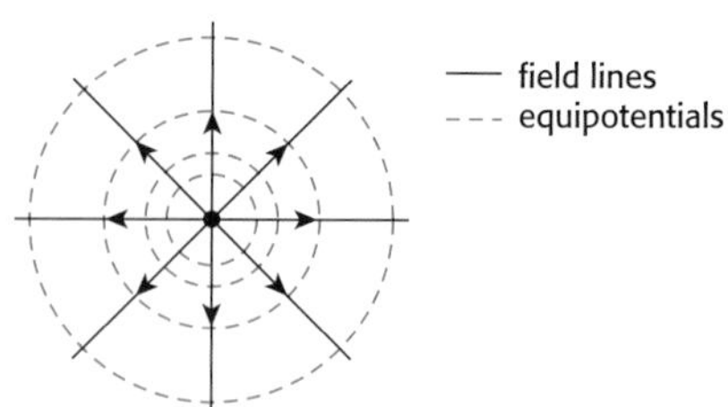

Figure 4 Electric field lines and electric equipotentials around a positive point charge

As charges can be positive or negative, the potential at any point also can be positive or negative. By convention a positive test charge is chosen, so the potentials resulting from the field associated with a positive charge will be positive, and those for a negative charge will be negative.

The concept of difference in potential – potential difference (pd) or voltage – has already been introduced in Chapter 12. An alternative unit for J C^{-1} is V. One possible definition of a perfect conductor is an object for which the electric potential must be constant.

Representation of potential

Potential is a scalar quantity. The variation of potential is most often represented in two dimensions by drawing lines of equipotential and in three dimensions by equipotential surfaces. In either case, the potential is the same at all points on the equipotential.

Contour lines on a map are an example of gravitational equipotential lines, as they join together places that are at the same vertical height.

When a test object is moved along an equipotential, the potential energy must remain the same, and so no work is done. There is a force on the test object, but it is always at right angles to motion along the equipotential. In other words, equipotential surfaces and field lines must be at right angles to one another. For example, Figure 4 represents the electric field lines and electric equipotentials around a positive point charge.

The uniform electric field pattern between two oppositely charged parallel plates was introduced on page 198. In order to calculate, from first principles, the potential at any point between the plates, the individual contributions of each of the point charges on the plates would need to be taken into consideration. This is a complicated mathematical process.

The relationship between field and potential gradient provides a quick way of analysing the situation. See Figure 5. The equipotential lines are always at right angles to the field lines. The constant field in the centre of the plate means that the potential gradient is constant – the equipotential lines are uniformly spaced apart.

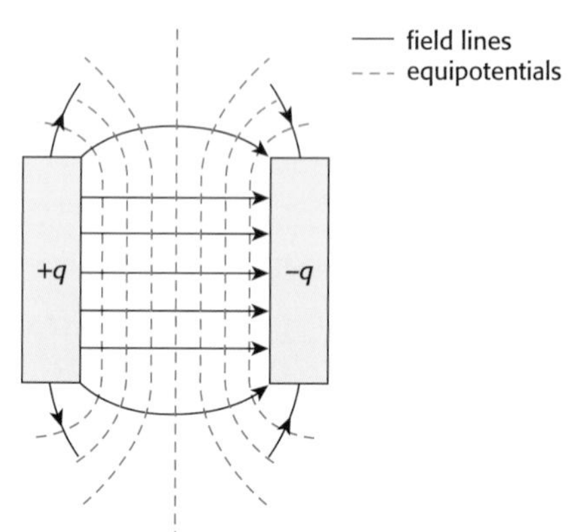

Figure 5 Field lines and equipotentials for parallel plates

Potential gradient

There is a direct mathematical relationship that links the field and the way in which the potential varies. Figure 6 shows a positive test object δq being moved a distance δx between two points A and B in the field due to a positive charge q. In this situation the external force $(= -E\,\delta q)$ does work in moving from A to B and so the potential of the system is increased.

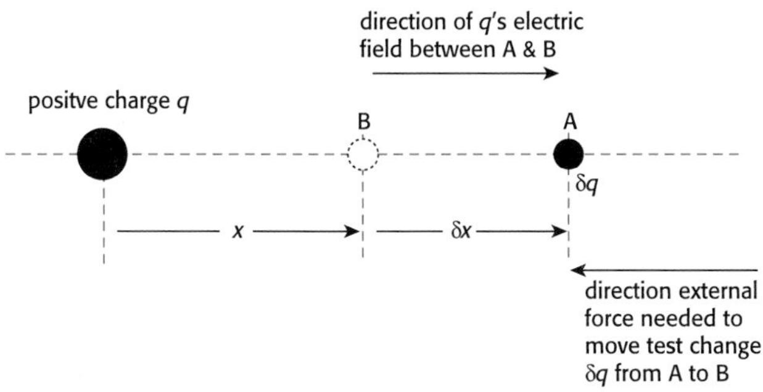

Figure 6

The field between A and B is assumed to be constant and equal to E.

The external work done *on* the system, $\delta W = -E\,\delta q \cdot \delta x$

$$E = -\frac{1}{\delta q}\frac{\delta W}{\delta x} \qquad \text{but} \qquad \delta V = \frac{\delta W}{\delta q}$$

Therefore

$$E = -\frac{\delta V}{\delta x}$$

If δx is very small ($\delta x \to 0$), then this becomes

$$E = -\frac{\mathrm{d}V}{\mathrm{d}x}$$

$\mathrm{d}V/\mathrm{d}x$ represents the **potential gradient**, measured in V m^{-1}
E represents the electric field strength, measured in N C^{-1}

A similar relationship can also be derived for gravitational fields.

In both cases:

field strength = −potential gradient

The negative sign indicates that if potential increases with a displacement, then the direction of the field is in the opposite direction to the displacement.

Calculation of potential

To calculate the potential at a given distance from a source of field (either gravitational or electric), we need to calculate the total work done in bringing a test object from infinity. As the value of the field is constantly changing, this calculation involves a technique called integration. This is not discussed in this course companion, but the results are shown below.

$$\therefore V_{gravitational} = \frac{\frac{Gm_1m_2}{r^2}\times r}{m_2} = \frac{Gm_1}{r}$$

The diagram below shows lines of electric equipotential. The change in potential on moving from one line to the next is always the same. At which point does the electric field strength have its greatest magnitude?

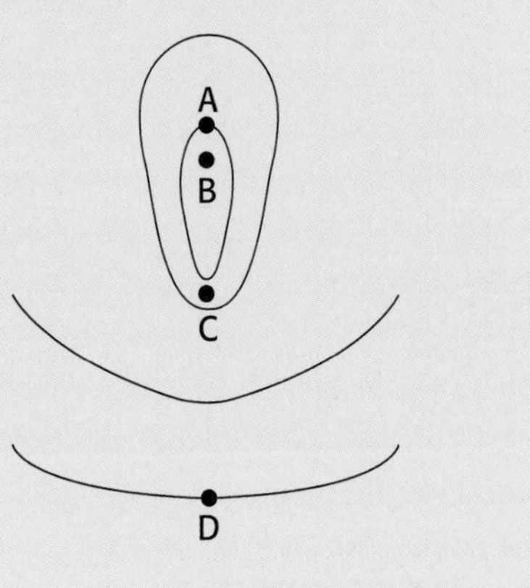

A conducting rod is introduced between oppositely charged parallel plates and is held parallel to the plates. Deduce the changes that will take place in the field and potential pattern as a result of the conductor being placed inside.

1 Show that the units N C^{-1} are identical to V m^{-1}.

2 Using the above derivation as a model, show that a similar relationship applies for gravitational fields.

In both cases the electric and gravitational potential is inversely proportional to the distance away from the point object causing the field.

Mathematical physics: Energy and Potential

The resultant potential at any point in space is just the addition of all the potentials resulting from all the point objects involved. As shown on page 195–6, a spherical object can often be treated as a point object with all the mass or charge concentrated at the centre of the object. In general this approach will work only when considering significant distances away from the spherical object.

Fields and potentials provide equivalent ways of describing the same situation. For example, in order to describe an electrostatic situation fully, we can either

- state values of the electric field strength E, calculate the force on a charge at a point, and draw electric field lines, or
- state values of the potential V, calculate the potential energy of a charge at a point, and draw equipotential lines

?

On a sheet of graph paper, use both of the above approaches to sketch representations of the field and the potential in one of the following situations. Equipotential lines should be labelled with their values. You need to choose the appropriate number of equipotential lines to show in your diagram, and this will depend on the complexity of the situation you choose.

a) An isolated +6 nC point charge (up to 10 cm away).
b) An isolated −6 nC point charge (up to 10 cm away).
c) An isolated 6 kg point mass (up to 10 cm away).
d) Two point charges, both +5 nC, separated by 20 cm.
e) Two point charges, one +5 nC the other −5 nC, separated by 20 cm.
f) Two point charges, one +5 nC the other +10 nC, separated by 15 cm.
g) Two point charges, one +5 nC the other −10 nC, separated by 15 cm.
h) An isolated conducting sphere of radius 5 cm carrying a charge of −8 nC (up to 5 radii away).
i) The Earth, mass 5.98×10^{24} kg and radius 6370 km (up to 10 radii away).
j) The Moon, mass 7.35×10^{22} kg and radius 1740 km (up to 10 radii away).
k) The Earth–Moon system (they are separated by 3.84×10^{8} m).

A good approach would be for each student to work out a different situation, and then present and explain their drawings to their fellow students.

The electronvolt

A potential difference (p.d.) is the difference in potential between two points.

$$\text{potential difference p.d.} = \frac{\text{energy dissipated}}{\text{charge flowed}}$$

This equation was used to define the unit of p.d. to be the volt: hence voltage is an alternative term that means the same as potential difference (or p.d.), and potentials are measured in volts. The same equation is used to create a new unit for measuring energies – the electronvolt. This unit is particularly useful at the atomic level. At this scale, 1 joule is an impossibly large energy for a particle to have, so this new unit is used instead. The definition is as follows:

$$\text{energy difference} = \text{potential difference} \times \text{charge flowed}$$

so 1 electronvolt = 1 volt × charge on one electron

$$1 \text{ eV} = 1.6 \times 10^{-19} \text{ J}$$

Note that this unit is not an "electron per volt" but an "electron × volt". For example, when an electron is accelerated from rest through a potential difference of 1 kV it must gain a kinetic energy of 1.6×10^{-16} J, or 1 keV.

Escape speed

The gravitational potential at a planet's surface provides an easy way of calculating the energy needed by any mass to escape away from the planet's gravitational influence – for example, to escape the Earth's gravitational pull and send a space probe to Mars.

Consider a small mass m at the surface of a planet of mass M and radius r. The small mass has a gravitational potential of $V_{\text{gravitational}}$ and a total energy that is equal to the product of the potential and mass:

$$\text{gravitational potential energy gPE} = V_{\text{gravitational}} \times m$$

All gravitational potentials are negative, because all gravitational forces are attractive, and so the above quantity is negative. If we assume the planet behaves like a point mass, then

$$\text{gPE} = -\frac{GMm}{r}$$

In order to escape the gravitational influence of this planet, the small mass would need enough energy to be able to get an infinite distance away. The minimum amount of positive kinetic energy that is needed is $+\frac{GMm}{r}$. The mass will be able to escape if it has a value of velocity v_{escape} given by:

$$\frac{1}{2}mv_{escape}^2 = \frac{GMm}{r}$$

$$v_{escape} = \sqrt{\frac{2GM}{r}}$$

A mass with less velocity than the above escape velocity must have a negative total energy and will eventually fall back to the planet. A mass with a greater velocity than the escape velocity would theoretically be able to get an infinite distance away from the planet and still be moving.

1 Use the information given in the question on the previous page to compare the escape velocity for the Earth and for the Moon. *Note that the above derivation of escape velocity, and so the answer to question 1, has assumed that there is only one planet exerting a gravitational influence on the mass. In reality, of course, all planets in a system contribute to the total potential energy of a mass on the surface of one of the planets.*

2 Use the information given in the question on the previous page to calculate the escape velocity for the Earth–Moon system from (a) the surface of the Earth and (b) the surface of the Moon.

3 Explain why your answers to the two questions above are different.

Thinking about science: *Kepler's law*

In 1619 a mathematician called Johannes Kepler published *De Harmonices Mundi (Harmony of the Worlds)*. This book contained a statement of his third law: "The square of any planet's orbital period, *T*, is proportional to cube of its mean distance *r* from the Sun".

$$T^2 \propto r^3$$

The table below gives information regarding the masses and distance from the Sun for different planets. The information is given relative to the Earth's mass ($M_\oplus$) and to the Sun–Earth distance (defined as one astronomical unit, 1 AU).

Analyse this information and outline the extent to which the data verify Kepler's third law.

Planet	Orbital time	Distance to Sun / AU
Mercury	87.97 days	0.387
Venus	224.70 days	0.723
Earth	365.26 days	1.000
Mars	687 days	1.523
Jupiter	11.86 years	5.203
Saturn	29.46 years	9.539
Uranus	84.02 years	19.19
Neptune	164.8 years	30.07

Kepler had discovered this mathematical link between the two measured variables, but he could not explain why the relationship was true. Although Kepler's third law made progress towards a deeper understanding of planetary motion, at the time there was no reason to believe that this relationship may not be applicable outside our solar system.

In 1687 Isaac Newton published *Philosophae Naturalis Principia Mathematica (Mathematical Principles of Natural Philosophy)*. This book (often just called the *Principia*) showed how Kepler's third law could be derived from Newton's laws of motion and universal gravitation. The derivation for circular orbits is shown below, but the mathematics works for elliptical orbits as well.

Consider a satellite of mass m in orbit around a planet of mass *M* (Figure 7). The radius of the orbit is *r* and the speed of the satellite in its orbit is *v*.

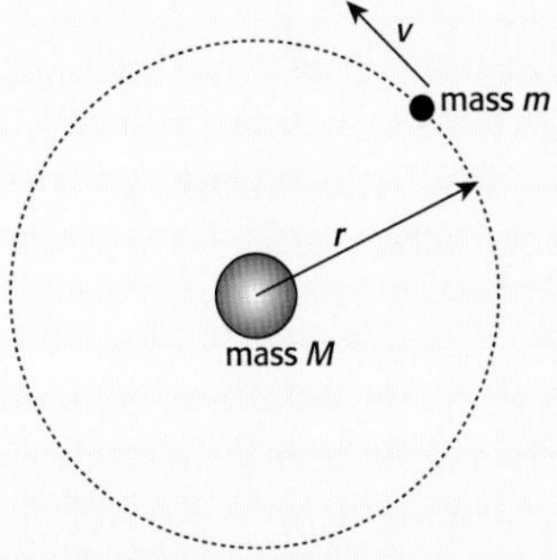

Figure 7 Satellite of mass m orbiting a planet of mass *M*

The gravitational force of attraction on the satellite provides the centripetal force needed to stay in orbit:

gravitational force = CPF

$$\therefore \frac{GMm}{r^2} = \frac{mv^2}{r} \quad \therefore \frac{GM}{r} = v^2$$

The time period for the orbit is T, where

$$v = \frac{2\pi r}{T} \quad v^2 = \frac{4\pi^2 r^2}{T^2} \quad \therefore \frac{GM}{r} = \frac{4\pi^2 r^2}{T^2} \quad \therefore \frac{T^2}{r^3} = \frac{4\pi^2}{GM}$$

Since 4, π, G, and M are all constants (M is the mass of the Sun, which is orbited by all the planets in the solar system),

$$\therefore \frac{T^2}{r^3} = \text{constant} \quad \text{or} \quad T^2 \propto r^3$$

Newton's laws of motion and of universal gravitation are elegant and beautiful mathematical statements. They are extremely powerful and truly universal statements that can still be used to predict accurately the motion of any planets, stars, and galaxies throughout the universe. They do not, however, explain *why* masses gravitationally interact in this mathematical way – see pages 201–4.

Total energy in orbit

An object that is in orbit has both kinetic energy and potential energy, and the total energy is the addition of the two. Using the same symbols as above:

$$\text{KE} = \frac{1}{2}mv^2 = \frac{1}{2}m\frac{GM}{r} = \frac{1}{2}\frac{GMm}{r}$$

$$\text{PE} = V_{\text{gravitational}} \times m = -\frac{GMm}{r}$$

The KE is numerically half the value of the PE, but the KE is positive and the PE is negative. The addition gives the total energy of an object in orbit.

$$\text{Total energy} = \text{KE} + \text{PE} = \frac{1}{2}\frac{GMm}{r} - \frac{GMm}{r} = -\frac{1}{2}\frac{GMm}{r}$$

1. On the same axes, use the above equations to sketch graphs showing the variation with orbital radius of (a) the kinetic energy, (b) the gravitational potential energy and total energy of a satellite.
2. Use the above equations to explain why a satellite in low Earth orbit will *speed up* as a result of the energy it loses doing work against friction with the atmosphere.

Physics issues: the International Space Station

The International Space Station (ISS) is a low-orbit space station that has been continuously inhabited since 2 November 2000, and is a stunning example of true international cooperation. The initial partnership involved 15 different governments: those of the United States, Canada, Japan and the Russian Federation, and 11 member states of the European Space Agency (Belgium, Denmark, France, Germany, Italy, The Netherlands, Norway, Spain, Sweden, Switzerland and the United Kingdom). An international treaty was signed on 28 January 1998 that outlined the purpose of the space station. Article 1 of the treaty states that:

> *This Agreement is a long-term international cooperative framework on the basis of genuine partnership, for the detailed design, development, operation, and utilization of a permanently inhabited civil space station for peaceful purposes, in accordance with international law.*

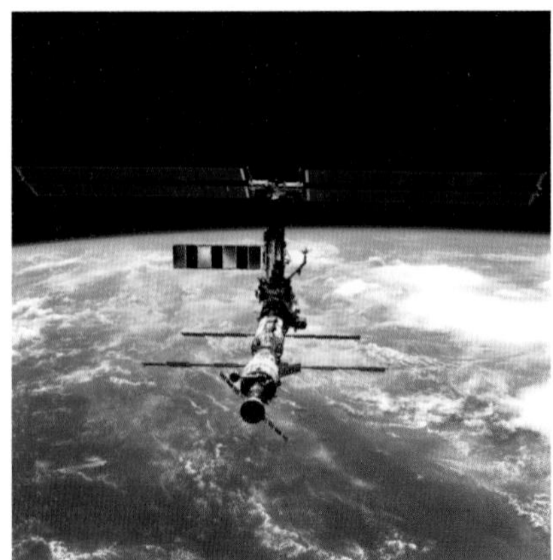

Figure 8 The International Space Station

Some feel that the ISS is an important piece of ongoing international scientific research, and perhaps even the first necessary step taken towards mankind's eventual colonisation of other planets. Others argue that the money spent on its construction and maintenance could be better spent on solving existing problems on Earth.

In your view, is the International Space Station worth the money that is being spent on it? Research some facts and information about the project in order to back up your view.

Thinking about science: *Apparent weightlessness*

Astronauts appear to be weightless. If the spacecraft were a long way away from any massive object, then it would be possible for the force of gravity on the astronaut to be zero, but the spacecraft's orbit is close to the Earth. So the astronaut cannot be far enough away from the Earth to have escaped its gravity.

The explanation for the apparent lack of gravity is that the astronaut and the spacecraft are both in **free fall**. For both objects (the spacecraft and the astronaut), gravity is providing the CPF that is required to constantly accelerate (the direction of motion is constantly changing) and so stay in orbit.

On the surface of the Earth, an astronaut feels the effect of her weight pulling her towards the ground. The spacecraft also has weight, directed towards the ground. As both the astronaut and the spacecraft are accelerating at 9.8 m s^{-2} in the same direction, there is no normal contact force between the astronaut and the "floor" of the spacecraft, and so she feels weightless.

A similar sensation would be felt by a passenger in a lift accelerated downwards at 9.8 m s^{-2}. The two forces acting on the passenger are the force up from the floor, F, and their weight, W.

- When the lift is at rest or moving at constant velocity the passenger is doing the same, and the resultant force on the person must be zero, so $F = W$.
- When the lift is accelerating upwards then the passenger is doing the same and the resultant force on the person must be upwards, so $F > W$.
- When the lift is accelerating downwards, then the passenger is doing the same and the resultant force on the person must be downwards, so $F < W$.

If the lift cable broke, the lift and the passenger would free-fall together towards the Earth, both with an acceleration of 9.8 m s^{-2}. In each case the force of gravity would be doing the accelerating and there would be no additional contact force between the lift and the passenger. The passenger would appear weightless during the fall.

This effect is used to train astronauts to get used to the effects of apparent weightlessness. With careful piloting, an aircraft can be made to execute a series of dives towards the Earth, each with a vertical acceleration of 9.8 m s^{-2}. If the acceleration is too small, the passengers will feel an upward force from the floor. If it is too great, the ceiling will seem to push the passengers down. But during a properly executed dive the passengers will appear weightless and float. When the aircraft comes out of the dive then they will feel the contact force between them and the floor.

It is interesting to note the difference between the perception that trainee astronauts have (that gravity has been temporarily switched off) and the explanation of the effect (the acceleration of the aircraft). Einstein realized that it would be impossible for any observer to tell the difference between a gravitational force of attraction in one direction and an acceleration in the opposite direction. This **principle of equivalence** allowed him to develop a new way of visualizing gravitational attractions – the curvature of space-time. These ideas are studied further in the relativity option.

End of chapter summary

Chapter 14 has considered the three themes of fields, potential and energy for two different forces: gravitational and electric as well as the theme of orbital motion. The list below summarises the knowledge and skills that you should be able to undertake after having studied this chapter. Further research into more detailed explanations using other resources and/or further practice at solving problems in all these topics is recommended – particularly the items in bold.

Field, potential and energy

- Define potential and potential energy for gravitational fields and electric fields.
- Define electric potential difference and the electronvolt.
- Determine the change in potential energy when a charge or a mass moves between two points at different potentials and solve related problems.
- State and apply the expressions for gravitational potential due to a point mass and electric potential due to a point charge.
- State and apply the formulae relating field strength to potential gradient for gravitational and electric fields.
- Determine the potential due to one or more point masses and due to one or more point charges.
- Describe and sketch the pattern of equipotential surfaces due to one and two point masses and due to one and two point charges.
- State the relation between equipotential surfaces and field lines for gravitational and electric fields.
- **Explain the concept of escape speed from a planet and derive an expression for the escape speed of an object from the surface of a planet.**
- **Solve problems involving potential energy and potential for gravitational and electric fields.**

Orbital motion

- State that gravitation provides the centripetal force for satellites in circular orbital motion.
- Derive Kepler's third law.
- **Derive expressions (and sketch graphs showing the variation with orbital radius) for the kinetic energy, gravitational potential energy and total energy of an orbiting satellite.**
- Discuss the concept of "weightlessness" in orbital motion, in free fall and in deep space.
- **Solve problems involving orbital motion.**

Chapter 14 questions

1 This question is about the electric potential due to a charged sphere.

a) Define *electric potential* at a point in an electric field. [3]

The diagram below shows an isolated, metal sphere in a vacuum that carries a negative electric charge of 9.0 nC.

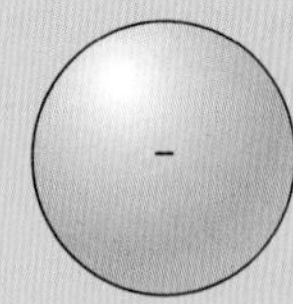

b) On the diagram above draw

i) arrows to represent the electric field pattern in the region outside the charged sphere. [3]

ii) lines to represent three equipotential surfaces in the region outside the sphere. The potential difference between the lines is to be equal in value. [2]

c) Explain how the lines representing the equipotential surfaces that you have sketched indicate that the strength of the electric field is decreasing with distance from the centre of the sphere.

d) The electric field strength at all points inside the conductor is zero. On a copy of the axes below, draw a graph to show the variation with distance r from the centre of the sphere of the potential V. The dotted line is drawn at $r = a$ where a is the radius of the sphere. (*This is a sketch graph; you do not need to insert any numerical values.*)

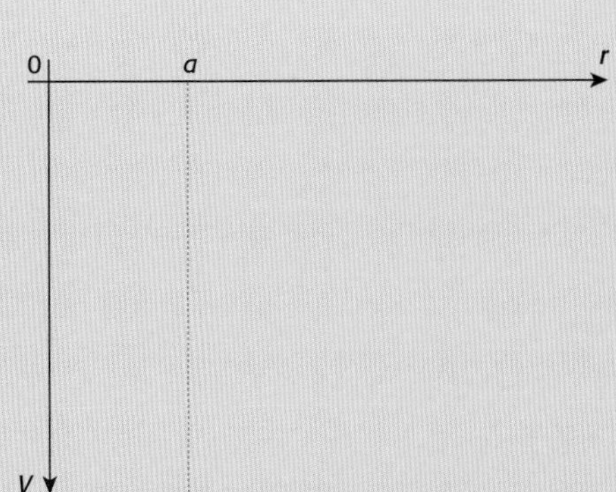

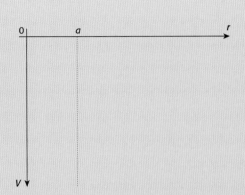

e) The electric field strength at the surface of the sphere and at points outside the sphere may be determined by assuming that the sphere acts as though a point charge of magnitude 9.0 nC is situated at its centre. The radius of the sphere is 4.5×10^{-2} m. Deduce that the potential at the surface of the sphere is -1800 V. [1]

An electron is initially at rest at the surface of the sphere.

f) i) Describe the path followed by the electron as it leaves the surface of the sphere. [1]

ii) Determine the speed of the electron when it reaches a point a distance 0.30 m from the centre of the sphere. [4]

(Total 18 marks)

2 This question is about gravitational potential energy.

The graph below shows the variation of gravitational potential V due to the Earth with distance R from the centre of the Earth. The radius of the Earth is 6.4×10^6 m. The graph does not show the variation of potential V within the Earth.

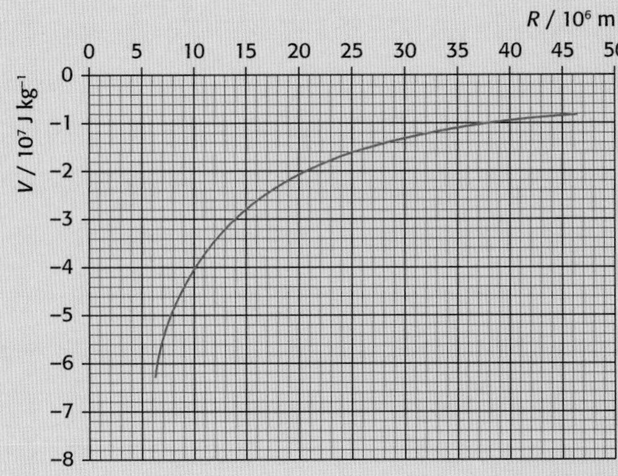

a) Use the graph to find the gravitational potential

i) at the surface of the Earth [1]

ii) at a height of 3.6×10^7 m above the surface of the Earth [1]

b) Use the values you have found in part **a)** to determine the minimum energy required to put a satellite of mass 1.0×10^4 kg into an orbit at a height of 3.6×10^7 m above the surface of the Earth. [3]

c) Give **two** reasons why more energy is required to put this satellite into orbit than that calculated in **b)** above. [2]

(Total 8 marks)

3 This question is about escape speed and Kepler's third law.

Jupiter and Earth are two planets that orbit the Sun. The Earth has mass M_e and diameter D_e. The escape speed from Earth is 11.2 km s^{-1}.

Data for Jupiter are given below.

Mass: 1.90×10^{27} kg (318 M_e)

Mean diameter: 1.38×10^5 km (10.8 D_e)

a) i) State what is meant by *escape speed*. [1]

ii) Escape speed v is given by the expression

$$v = \sqrt{\left(\frac{2GM}{R}\right)}$$

Determine the escape speed from Jupiter. [2]

b) i) State Kepler's third law. [1]

ii) In 1610, the moon Ganymede was discovered orbiting Jupiter. Its orbit was found to have a radius of 15.0 R and period 7.15 days, where R is the radius of Jupiter. Another moon of Jupiter, Lysithea, was discovered in 1938 and itsorbit was found to have a radius of 164 R and a period of 260 days. Show that these data are consistent with Kepler's third law. [2]

(Total 6 marks)

4 This question is about forces on charged particles.

a) A charged particle is situated in a field of force. Deduce the nature of the force-field (magnetic, electric or gravitational) when the force on the particle

i) is along the direction of the field regardless of its charge and velocity.

ii) is independent of the velocity of the particle but depends on its charge.

iii) depends on the velocity of the particle and its charge. [5]

b) An electron is accelerated from rest in a vacuum through a potential difference of 2.1 kV. Deduce that the final speed of the electron is 2.7×10^7 m s^{-1}. [3]

The electron in **b)** then enters a region of uniform electric field between two conducting horizontal metal plates as shown below.

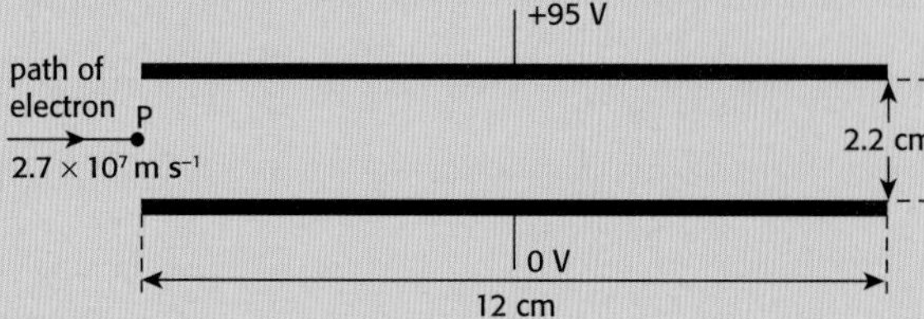

The electric field outside the region of the plates may be assumed to be zero. The potential difference between the plates is 95 V and their separation is 2.2 cm.

As the electron enters the region of the electric field, it is travelling parallel to the plates.

c) i) On the diagram above, draw an arrow at P to show the direction of the force due to the electric field acting on the electron. [1]

ii) Calculate the force on the electron due to the electric field. [3]

d) The plates in the diagram opposite are of length 12 cm. Determine

i) the time of flight between the plates. [1]

ii) the vertical distance moved by the electron during its passage between the plates. [3]

e) Suggest why gravitational effects were not considered when calculating the deflection of the electron. [2]

f) In a mass spectrometer, electric and magnetic fields are used to select charged particles of one particular speed. A uniform magnetic field is applied in the region between the plates, such that the electron passes between the plates without being deviated.

For this magnetic field,

i) state and explain its direction [3]

ii) determine its magnitude [2]

g) The electric and magnetic fields in **f)** remain unchanged. Giving a brief explanation in each case, compare qualitatively the deflection of the electron in **f)** with that of

i) an electron travelling at a greater initial speed

ii) a proton having the same speed

iii) an alpha particle (2 protons and 2 neutrons) having the same speed [7]

(Total 30 marks)

15 Electromagnetic induction

The generation of electrical energy and the transformation of alternating currents between different voltages both rely on electromagnetic induction.

Chapter 13 introduced the motor effect, in which a current placed in a magnetic field feels a force. This force can produce motion if the current is free to move (e.g. in an electric motor), and to get the maximum force the directions of the current, field, and force all have to be at right angles to one another.

Electromagnetic induction is the process by which an emf is generated. If a complete circuit is available then the emf can cause a current to flow.

Generating emf using motion

When a conductor (e.g. a length of wire) is moved through a magnetic field, a small emf (i.e. a p.d.) is induced across the ends of the conductor, and can be measured with a sensitive voltmeter (Figure 1). The magnitude of the emf is maximized when the direction of motion, the magnetic field, and the conductor are all at right angles to one another.

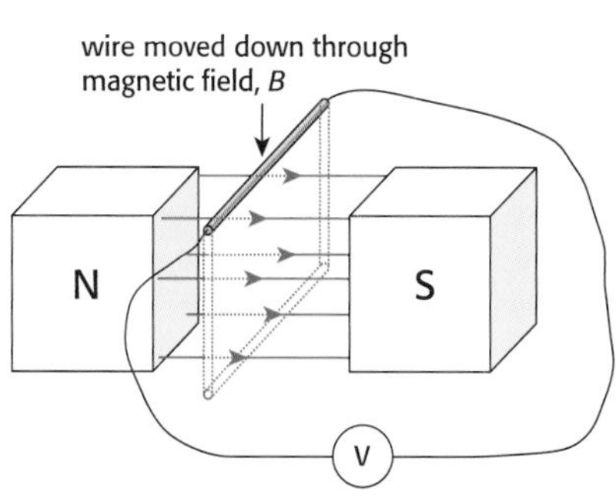

Figure 1 Motion of a conductor through a magnetic field generates a voltage

In the situation in Figure 1, a voltage will be recorded on the voltmeter when the wire is moving through the magnetic field. Effectively, the wire behaves like a small cell as it moves through the magnetic field. The value of the voltage depends on three different factors:

- the strength of the magnetic field, B
- the speed of the wire, v
- the length of the wire in the field, l.

The generation of emf as a result of the motion of a conductor through a magnetic field needs to be explained from both the microscopic and macroscopic points of view.

Microscopic point of view

The moving wire must contain equal numbers of positive and negative charges. Although the resulting movement of charge is not along the wire it may be considered as a virtual current, as shown in Figure 2.

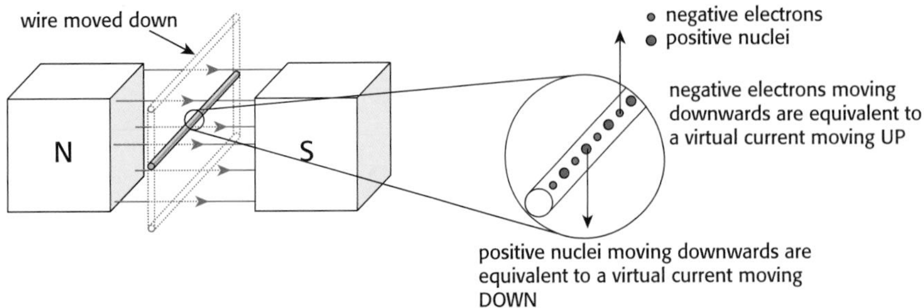

Figure 2 Motion of the wire can be considered to be equivalent to virtual currents

Note that these virtual currents are not *along* the wire, but are in the same direction that the wire is moving. These virtual currents are in the magnetic field, so they feel a force, and the direction of the force can be determined. In this example you should confirm that, as the wire moves down, the electrons feel a force *into* the page (the plane of the paper), whereas the nuclei feel a force out of the page (Figure 3).

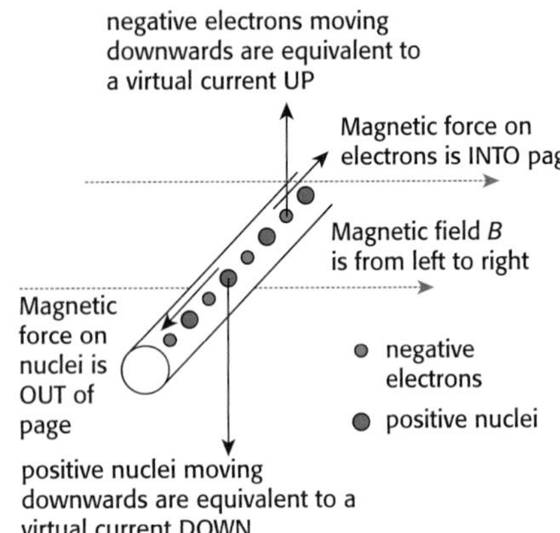

Figure 3 Force on electrons and nuclei in the moving wire

The nuclei cannot move, so the magnetic force on a nucleus has no effect. The force on the electrons, however, will cause them to move: in this example they move into the page, making the far end negative and leaving the near end positive.

As the electrons move, a p.d. is generated along the wire. Equilibrium is reached when the value of this p.d., V, produces an electric force on electrons in the wire that exactly balances the magnetic force (Figure 4).

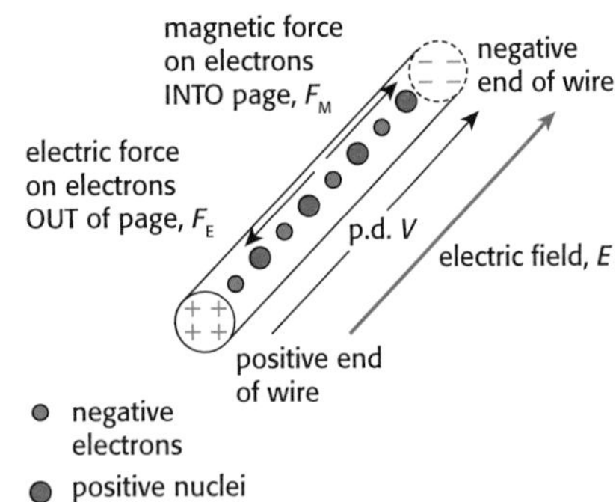

Figure 4 Forces on an electron in the wire are balanced

At equilibrium, the forces on an electron in the wire are balanced.

$$F_E = F_M$$

The length of the wire is l, so (using the relationship introduced on page 184) the electric force F_E on an electron can be calculated from the electric field, E, and the charge on the electron, e:

$$F_E = Ee = \frac{V}{l}e \qquad \text{since } E = \frac{V}{l}$$

The magnetic force can be calculated from the equation given on page 185:

$$F_M = Bev$$

$$\therefore V = Blv$$

where
V is the p.d. (emf) generated, measured in V
B is the magnetic field strength, measured in teslas (T)
l is the length of the wire, measured in m
v is the velocity of the wire, measured in m s^{-1}

Note that this p.d. has been induced with a specific orientation. If the two ends of the wire were joined to complete a circuit outside the magnetic field then a conventional current would flow, as shown in Figure 5.

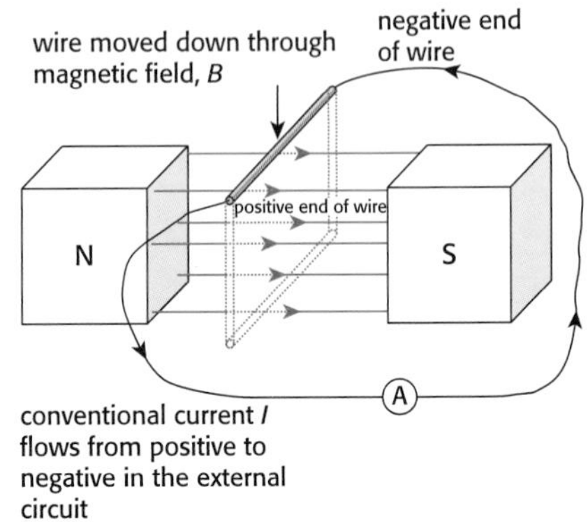

Figure 5 Direction of induced current

The current that flows is in a magnetic field, so it will feel a force. The direction of this force can be predicted using the motor rule; it always opposes the motion that set it up. This is known as **Lenz's law**:

> *The direction of the induced emf is such that, if a complete circuit were available, the current direction would oppose the change that caused the induced emf*

In the above example, the motion is down, which induces a current that flows out of the page. This current feels an opposing force up. If the motion had been up, then the current would have been out of the page and the force would be down – still opposing the change that caused the induced emf

Macroscopic point of view

Energy considerations can be used to deduce the value of the induced emf of a moving wire. Provided there is a completed circuit, the induced emf will cause a current to flow and energy to be dissipated. This energy must come from the work done by the external force that is needed to move the wire. The diagram in Figure 6 represents a rectangular coil that is being pulled at a steady velocity v through a region of magnetic field B.

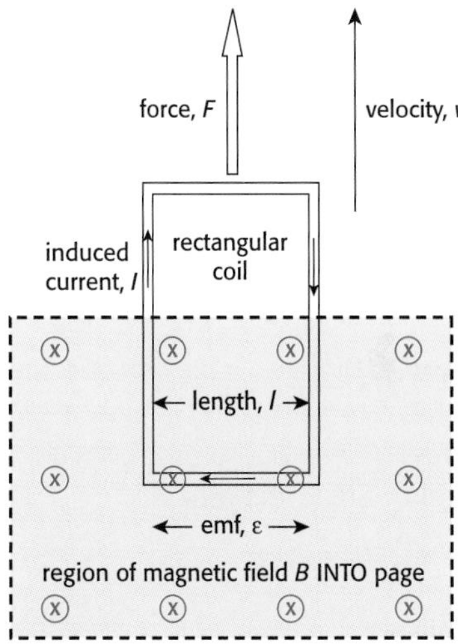

Figure 6 A coil pulled at a steady velocity

The force pulling the coil, F, is opposed by a force of equal magnitude in the opposite direction. This force arises because the induced current in the bottom section of the coil is in a magnetic field and must therefore feel an opposing force, as predicted by Lenz's law. On this macroscopic scale, Lenz's law is just an application of the law of conservation of energy.

Using the motor rule, the magnitude of the force opposing the motion must be

$$F = BIl$$

Since the wire is moving at constant velocity, the resultant force is zero, so the force pulling up the page must have magnitude $F = B\,I\,l$.

The rate at which work is done by the external force $= F \times v$

The rate at which electric energy is dissipated in the wire $= \varepsilon \times I$

From the conservation of energy,

$$\varepsilon \times I = BIl \times v$$

$$\therefore\ \varepsilon = B\,l\,v$$

where
ε is the induced emf, measured in V
B is the magnetic field strength, measured in T
l is the length of the vertical wire, measured in m
v is the velocity of the wire, measured in m s^{-1}

Note that, in the above derivation, current flows around the coil because the top side of the coil is outside the field. If the magnetic field covered the whole circuit, then a similar emf would be generated in the top side coil. Since the two emfs are exactly the same, they would cancel each other, and no current would be induced in the coil (see Figure 7).

Use the law of conservation of energy to deduce that it is impossible for the induced current direction to be in the opposite direction in Figure 6.

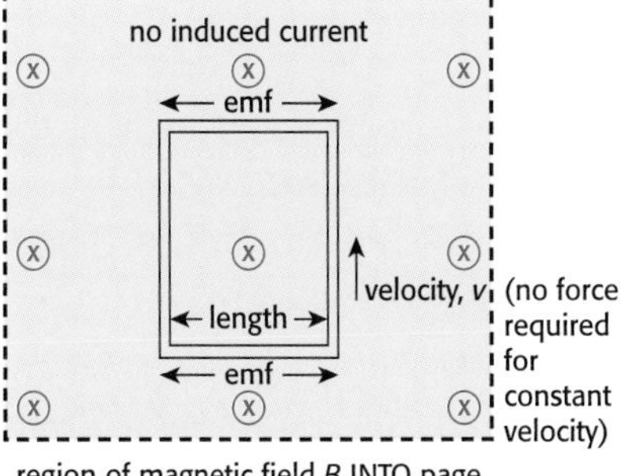

Figure 7 A moving coil completely inside a magnetic field does not induce a current

Mathematical physics: Concept of flux

When a conductor moves at right angles to a magnetic field, an emf is induced. A useful method for calculating the value of the emf involves the definition of a new mathematical concept associated with magnetic fields –the **flux** ϕ. Chapter 13 introduced the idea of the strength of field being represented using flux lines. Flux lines that are close together represent a strong field (Figure 8).

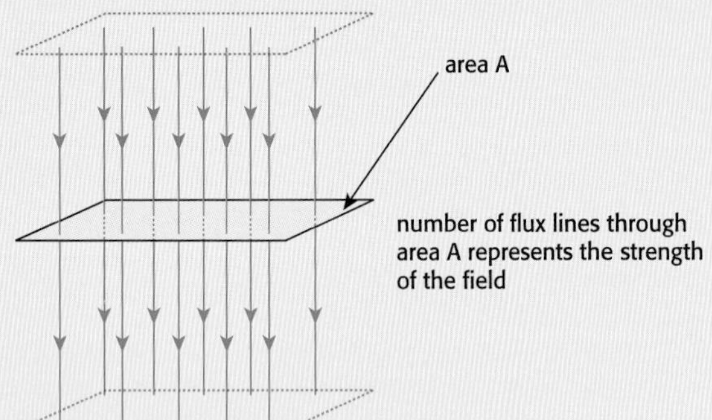

Figure 8 Flux lines represent field strength

We define flux ϕ to be the product of the magnetic field strength B and the area A:

$$\phi = B \times \mathbf{A} \times \cos\theta$$

where
B is the magnetic field strength, measured in T
A is the cross-sectional area, measured in m2
θ is the angle between the field lines and the normal to area, measured in degrees or radians (Figure 9)
ϕ is the flux, measured in T m^2

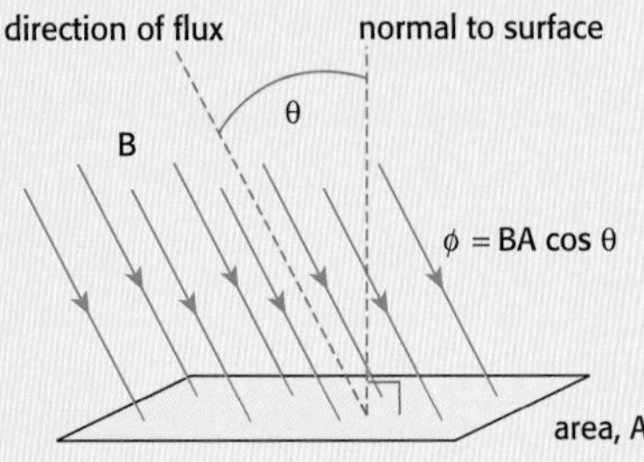

Figure 9 Definition of flux

If the area is normal to the field, then $\cos\theta = 1$ and so the $\cos\theta$ term disappears.

A new unit is defined for flux – the **weber** (Wb).

$$1 \text{ Wb} = 1 \text{ T} \times \text{m}^2$$

This means that an alternative unit for magnetic field strength is the Wb m^{-2}, which corresponds closely to the concept of representing field strength by the number of flux lines going through an area. The number of lines drawn in a field pattern represents the flux. An alternative name for magnetic field strength B is the **magnetic flux density** – that is, the flux per unit area.

This definition can be used to express the equations that we have already derived for the induced emf in a useful new way. In a time Δt, the wire generating the emf's velocity v moves a distance Δx, which sweeps out an area of flux (Figure 10).

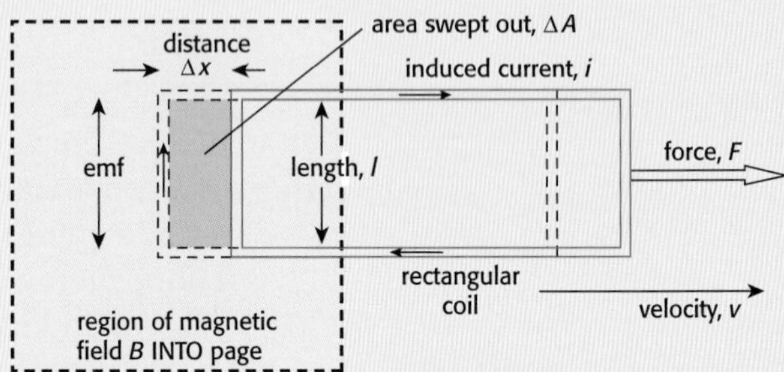

Figure 10 Change of flux through a coil

In a time Δt,

area swept out, $\Delta A = l \times \Delta x = l\, v\, \Delta t$
flux swept out, $\Delta\phi = B\,\Delta A = B\, l\, v\, \Delta t$
rate of change of flux through wire $= \dfrac{\Delta\phi}{\Delta t} = B\, l\, v$
$=$ magnitude of induced emf ε

Flux can be imagined as the number of field lines. An emf is generated whenever flux lines are "cut", and the above relationship shows that the induced emf in a wire is equal to the rate of cutting of flux.

Flux linkage

It is possible to increase the induced emf in a given situation by using a moving coil of wire (with several loops) instead of one single length of wire. The concept of **flux linkage** is introduced to take this into account. When the plane of a coil is right angles to the field then

flux linked with a coil = flux going through the coil × number of turns

The emf induced in a coil is equal to the rate of change of flux linkage with the coil. This is known as **Faraday's law** of electromagnetic induction. Faraday's law and Lenz's law can be combined into one statement (sometimes known as **Neumann's equation**):

$$\varepsilon = -N\frac{\Delta\phi}{\Delta t}$$

where
ε is the induced emf, measured in V
N is the number of turns on the coil

$\Delta\phi$ is the change in flux through the coil, measured in Wb
Δt is the time taken for the change of flux, measured in s

The full calculus version also applies:

$$\varepsilon = -N\frac{d\phi}{dt}$$

where $d\phi/dt$ is the rate of change of flux through the coil, measured in Wb s–1.

As explained above, adding extra turns to the loop increases the emf induced only if part of the coil remains outside the field. The identification of flux with the number of field lines make this easy to visualize. Consider a coil of wire moving out of a region where a magnetic field acts, as shown in Figure 11.

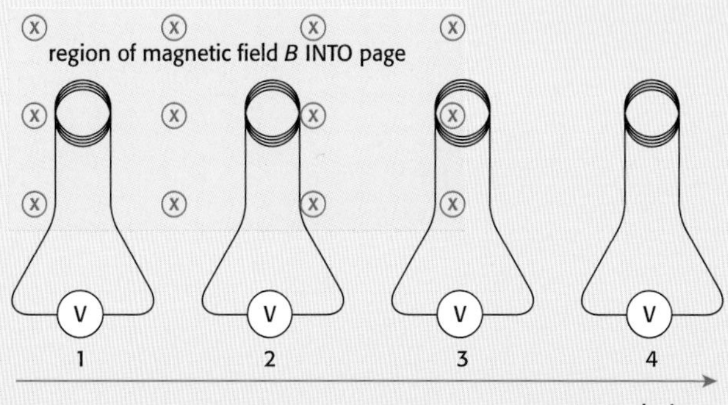

Figure 11 A coil moves out of a magnetic field

As the coil moves between positions 1 and 2, no p.d. will be registered on the voltmeter. The flux linked with the coil does not change, so overall no emf is generated. An alternative explanation is that identical emfs are induced in the left-hand and right-hand sides, making the overall induced emf equal to zero. The only time when emf is generated is when there is a change in flux in the coil at position 3.

Generators

The rotation of a coil in a magnetic field must also generate an emf, as the total amount of flux linked with the coil changes over time. Figure 12 represents a flat coil rotating at a constant angular velocity ω in a uniform magnetic field B.

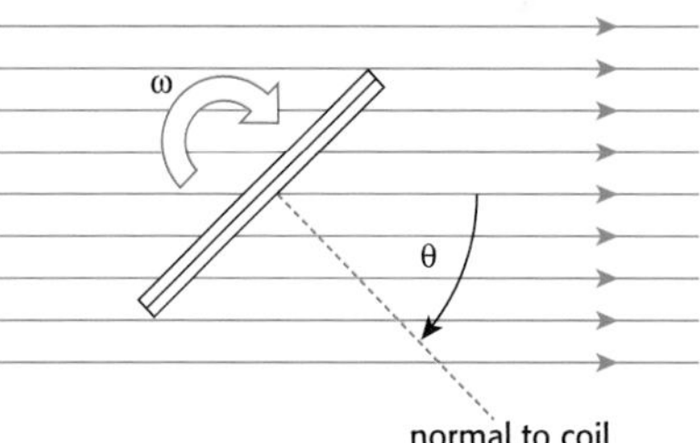

Figure 12 A coil rotating in a magnetic field changes the flux linked

The flux linking with each turn in the coil in the position shown is

$$\phi = BA\cos\theta \qquad \text{where } \theta = \omega t$$

The emf ε is worked out using Faraday's law:

$$\varepsilon = -N\frac{d\phi}{dt}$$

$$\therefore \varepsilon = N\omega AB\sin(\omega t) = \varepsilon_0\sin(\omega t) \qquad (\text{where } \varepsilon_0 = N\omega AB) \qquad (1)$$

A coil rotating at a uniform angular velocity in a constant field generates an emf that is sinusoidal. The maximum emf is proportional to four factors:

- the number of turns, N
- the angular frequency of rotation, ω (which is related to the frequency of rotation, f)
- the cross-sectional area of the coil, A
- the magnetic field strength, B.

Transformer emf

So far, we have considered induced only emfs caused by the physical motion of a conductor through a magnetic field. Emfs are generated whenever there is a change of flux linkage, even if there is no physical motion. The following examples all give rise to an induced emf In the final two examples, the objects are both stationary and yet an emf is still induced. This is knows as a **transformer-induced emf**

a) A coil of wire is moved towards the north pole of a bar magnet.

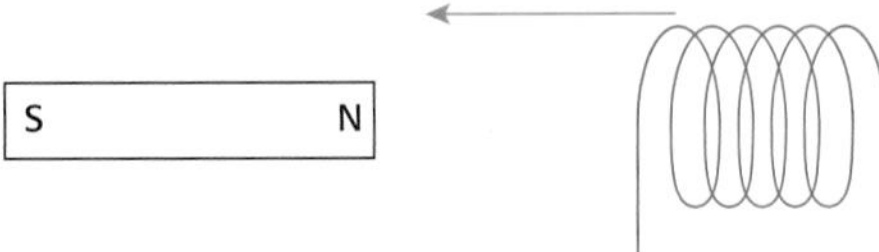

b) A coil of wire is moved away from the north pole of a bar magnet.

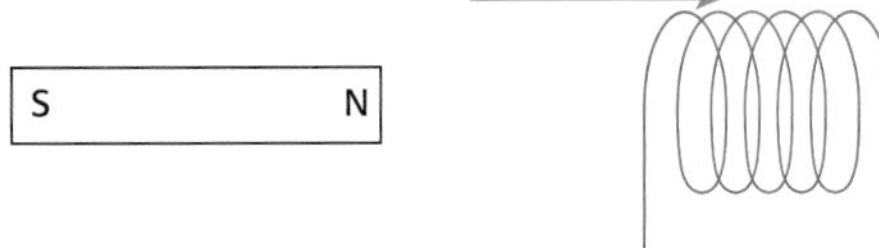

c) The north pole of a bar magnet is moved towards a coil of wire.

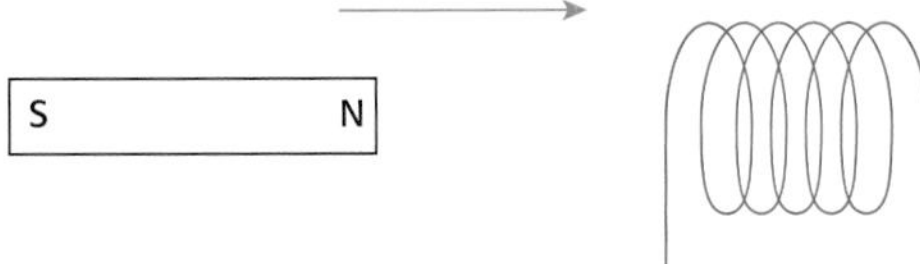

d) The north pole of a bar magnet is moved away from a coil of wire.

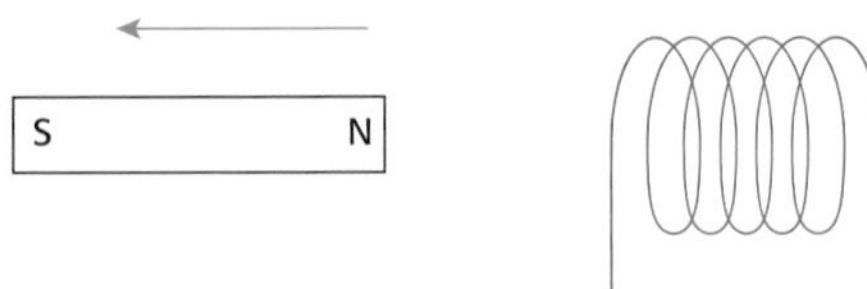

e) The current in an electromagnet is increased. There is a coil of wire nearby.

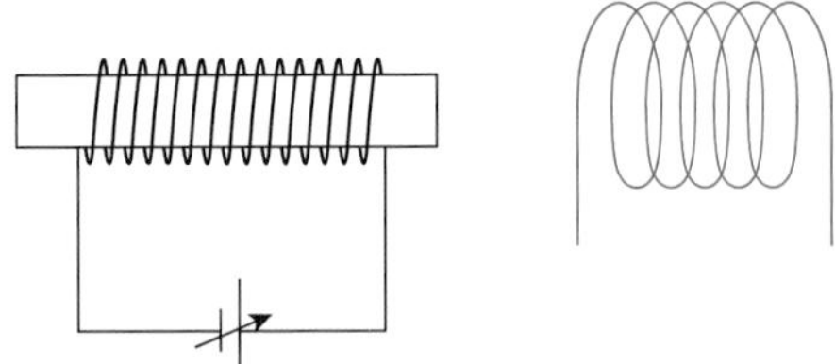

f) The current in the above electromagnet is decreased.

Although it seems very reasonable to assume that any change of flux must induce an emf, the mechanism by which we explained the generation of the emf relied upon the actual physical motion of a conductor through a magnetic field. The final two situations appear directly related, but we have not identified a mechanism by which a changing current in one coil can induce a current in a nearby coil.

The analysis of the interaction between magnetic and electric effects is highly complex, and is beyond the scope of IB Diploma Programme physics. The full link can be explained using Maxwell's equations and relativity.

For each of the above situations **a)** to **f)** use Lenz's law to predict the direction of the induced current.

The north pole of a permanent bar magnet is pushed along the axis of a coil as shown below.

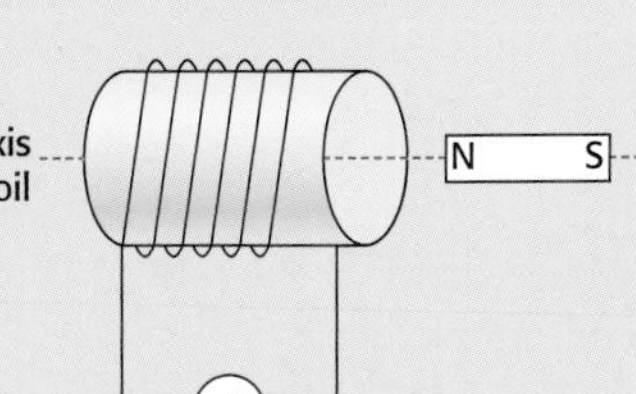

The pointer of the sensitive voltmeter connected to the coil moves to the right and gives a maximum reading of 8 units. The experiment is repeated, but on this occasion the south pole of the magnet enters the coil at twice the previous speed.

Which of the following gives the maximum deflection of the pointer of the voltmeter?

A 8 units to the right

B 8 units to the left

C 16 units to the right

D 16 units to the left

1 State the units for each of the variables on the left-hand side and the right-hand side of equation 1 on the previous page. Show that the units for the left-hand side are the same as for the right-hand side.

2 The axes in Figure 13 represent the output of a simple ac generator with a coil rotating 50 times a second. On the same axes, sketch the output you would expect if:

a) the magnetic field is doubled

b) the angular speed is doubled

c) the angular speed is halved

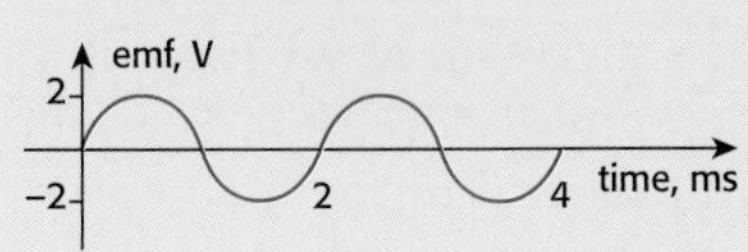

Figure 13 Output of an ac generator

Mathematical physics: rms values

Mains electricity provides alternating current (ac), whereas a battery supplies direct current (dc). These two types of sources are treated the same, however, when analysing the energy transfers that take place in circuits. A statement such as "The mains electricity provides 110 V ac" may cause confusion, as the voltage value must be changing all the time, so it is not clear what value is being quoted.

A cell's voltage does not vary greatly with time (Figure 14).

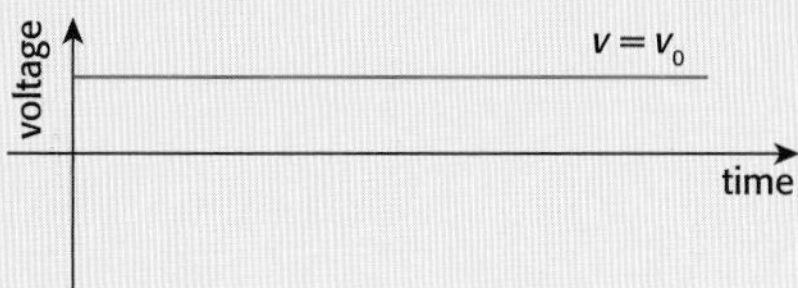

Figure 14 Output of a cell

A power pack output will typically vary quite a lot, but the polarity is always the same (Figure 15).

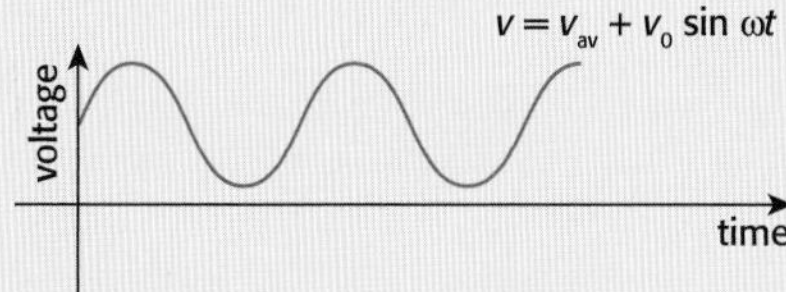

Figure 15 Output of a power pack

In mains electricity the variation with time of the voltage is sinusoidal, and it alternates in polarity (Figure 16).

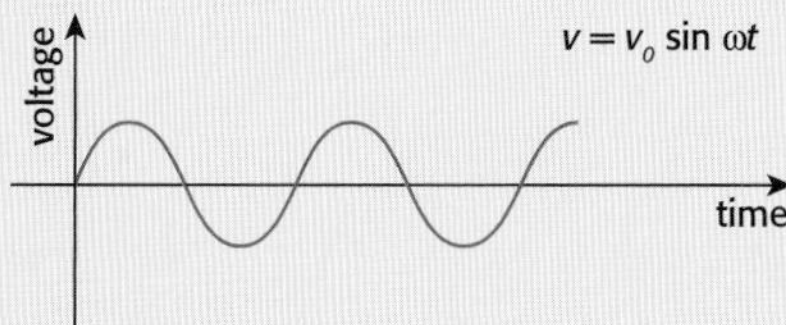

Figure 16 Mains electricity is sinusoidal

The quoted value for mains electricity is not the peak value, nor the mean (this is zero, as the voltage is equally positive and negative), nor the average over half a cycle. It is a value called the **root mean square** or **rms** value.

The definition of rms involves three different stages in the calculation:

1. Determine the variation with time of the square of the voltage. This is the square value.
2. Determine the average (mean) value of the squared voltage. This is the mean square value.
3. Determine the square root of this average value. This is the root mean square value.

This process may seem over-complicated, but there is a reason for it: the power P dissipated by a resistor R is proportional not to the voltage V put across it, nor to the current through it, I, but to the product of the two.

The rms value of any alternating electrical quantity is equal to the constant dc value that would give the same average power dissipation. In other words, when we are calculating the energy or power for an ac circuit, we can continue to use the dc equations so long as we substitute the rms values.

At any given instant, the instantaneous power can be calculated using

$$P = IV = I^2R = \frac{V^2}{R}$$

The resistance R is constant, but all other quantities vary with time. Since $V = I\,R$, the instantaneous current is given by $I = I_0 \sin\omega t$, where $I_0 = V_0/R$.

The variation with time of power dissipated is

$$P = \frac{(V_0 \sin \omega t)^2}{R} = \frac{V_0^{\ 2}}{R}\sin^2 \omega t$$

The power varies all the time, because the sin2(ωt) term introduces a factor that varies between 0 and 1.

Over one cycle,

the average value of power dissipation

$$= \frac{V_0^{\ 2}}{R} \times \text{average value of sin2}(\omega \text{t}).$$

A full calculus analysis shows that the average value of sin2(ωt) = ½. This means that

$$P_{average} = \frac{1}{2} \times \frac{V_0^{\ 2}}{R}$$

The rms value of an ac voltage V_{rms} is the constant dc voltage that would give the same average power dissipation as the sinusoidal ac voltage.

$$\therefore \frac{V_{rms}^2}{R} = \frac{1}{2} \times \frac{V_0^{\ 2}}{R}$$

$$\therefore V_{rms} = \frac{V_0}{\sqrt{2}}$$

A similar relationship applies for sinusoidal currents:

$$I_{rms} = \frac{I_0}{\sqrt{2}}$$

1 An ac voltage of frequency 50 Hz and peak value 325 V is connected across a 500 Ω resistor. Use a computer spreadsheet, a graphics display calculator, or graph paper to:

a) plot a graph (over at least one complete cycle) of the variation with time of

i) the voltage across the resistor

ii) the current flowing through the resistor

iii) the power dissipated in the resistor.

b) Use your answer to **a) iii)** to estimate

i) the energy dissipated over a whole cycle

ii) the average power dissipated over a whole cycle.

c) Use your answer to **b) ii)** to calculate the equivalent constant dc voltage that would give the same power dissipation as **b) ii)**.

d) Compare your answer to **c)** with the rms value calculated using the relationship between the peak value and rms value. Comment on your answer.

2 Using a GDC, or otherwise, estimate the rms value for the following ac voltages:

a) a square wave of amplitude 2 V:

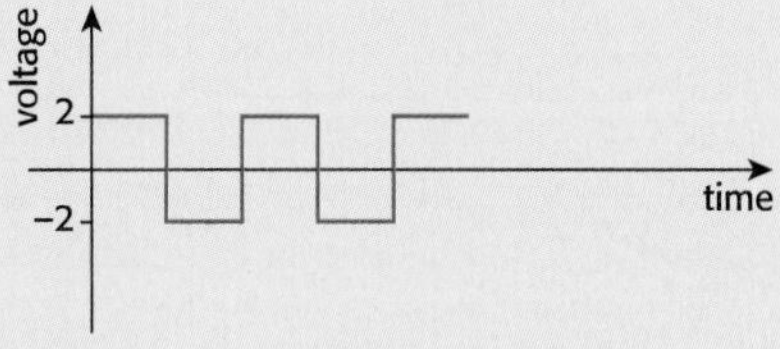

b) a triangular wave of amplitude 2 V:

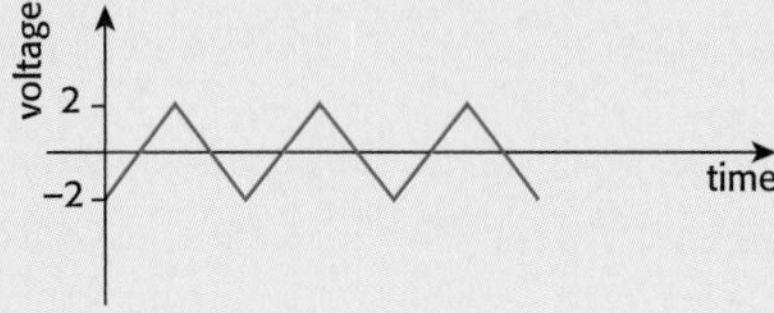

c) a "sawtooth" wave of amplitude 2 V:

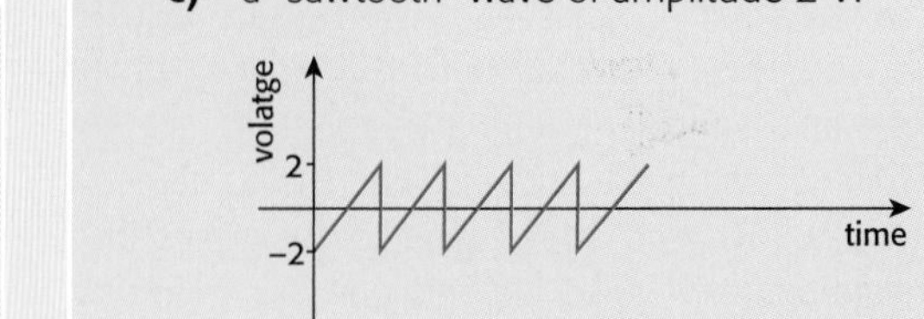

Phase of *V* and *I*

At any given time, the potential difference across a resistor, V, and the current through it, I, must be in phase, because they are linked by

$V = IR$.

There are, however, devices in which the current and the p.d. are not in phase – **capacitors** and **inductors**.

The full study of these devices (and thus the response of a general electrical circuit to ac) is beyond the IB Diploma Programme syllabus. In general, however, there is a phase difference between the alternating potential difference across a group of components and the current flowing through the components. The concept of resistance was introduced for dc The ac equivalent definition is called the **impedance** Z of a circuit, and is defined as the ratio of the rms voltage to the rms current.

HL

Transformers

We saw on page 220 how the magnetic variation that results from a change of current in one circuit can be used to induce a variation of emf in another unconnected circuit. The transformer is a device that uses this effect to alter voltages.

Figure 17 shows the principal components in a transformer. Two coils of wire – the **primary** and the **secondary** – are wrapped around the same iron core. An alternating voltage is placed across the primary coil, and this induces an alternating voltage across the secondary coil. Adjusting the number of turns in either coil can change the rms value of the induced voltage that is generated on the secondary.

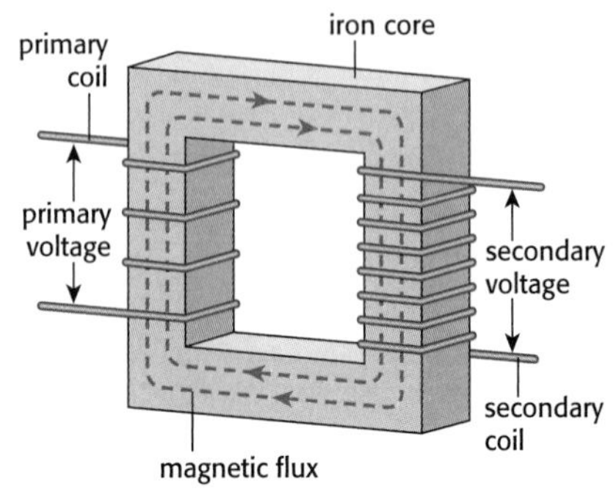

Figure 17 A transformer

The transformer circuit symbol is shown on the right:

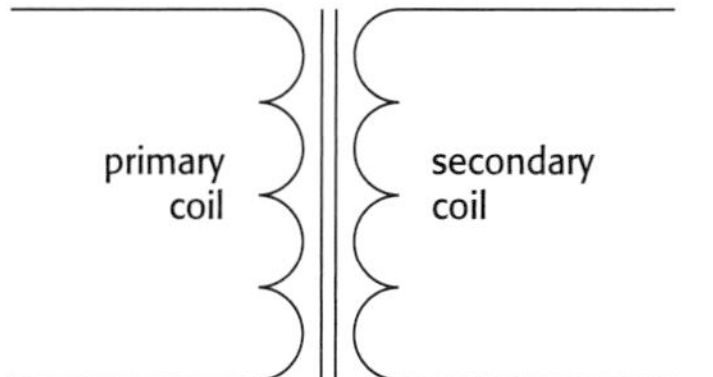

Figure 18 The electric circuit symbol for a transformer

The process by which an induced voltage is created across the secondary is as follows.

1 The alternating voltage input on the primary creates an alternating magnetic field.
2 The iron core ensures that almost all of the flux created by the primary links with the secondary coil.
3 The changing magnetic field means that the flux linkage with the secondary coil is constantly changing.
4 The rate of change of flux in the secondary induces an alternating potential difference across the secondary.
5 Increasing the number of turns on the secondary increases the flux linked with the secondary, and so the value of the alternating p.d. across the secondary.

The net effect is that an alternating voltage placed across the primary coil can be transformed into any chosen value of alternating p.d. out at the secondary, depending on the turns ratio chosen.

In a **step-up transformer**, the output voltage is greater than the input voltage – for example, an adapter that allows devices designed for operation in Europe, and thus requiring 230 V, to operate normally using the mains in the USA, which is at 110 V. In a **step-down transformer**, the output voltage is less than the input voltage – for example, a mobile phone (cellphone) charger that plugs into the mains.

The following relationship applies to an ideal transformer:

$$\frac{V_s}{V_p} = \frac{N_s}{N_p}$$

where
V_s is the rms value of the p.d. induced across the secondary, measured in V
V_p is the rms value of the p.d. across the primary, measured in V
N_s is the number of turns in the secondary
N_p is the number of turns in the primary

At first sight it can seem as though the ability to alter the p.d. contradicts the law of conservation of energy, but the currents that flow in the primary and the secondary are also fixed by the turns ratio. Assuming the transformer is 100% efficient:

power out = power in

$$V_s I_s = V_p I_p$$

$$\therefore \frac{V_s}{V_p} = \frac{I_p}{I_s}$$

where
V_s is the rms value of the p.d. induced across the secondary, measured in V
V_p is the rms value of the p.d. across the primary, measured in V
I_s is the current flowing in the secondary, measured in A
I_p is the current flowing in the primary, measured in A

In order to calculate the voltages and currents that are involved with the use of ideal transformers, the following procedure should be used:

- The output voltage is fixed by the input voltage and the turns ratio.
- The load that is connected to the secondary device fixes the output current.
- The turns ratio can be used to calculate the input current that is drawn from its supply.

Physics issues: power considerations in real situations

When power is being transmitted, for example from a power station to a customer, the aim is to avoid as much energy loss as possible. Transformers play a significant role in reducing the losses to an acceptable level.

Loss in power lines

An ideal power line would have zero resistance. Lines with non-zero resistance would dissipate energy by warming up, and real power lines will inevitably have some resistance. For a wire of given material and length, a decrease in resistance can be achieved by increasing the thickness of the wire. In practice there has to be a trade-off between the extra material costs associated with lower-resistance power lines and the cost of the energy lost to the environment.

1 **a)** Estimate the diameter of an overhead power cable.
b) Assuming the wire is made of aluminium (resistivity $\rho = 5 \times 10^{-8}\ \Omega$ m), calculate the resistance of a 100 km section of the power line.

2 A small factory is directly connected to a power station by power cables of total resistance 50 Ω. The output voltage of a power station is 25 kV, and the current drawn by the factory is 100 A. Calculate:
a) the power output of the station
b) the power lost in the power cables
c) the efficiency of the power transmission.

Figure 19

Reasons for using transformers

The main reason for using ac and transformers in the transmission of electrical energy is the reduction of power loss. The power lost, P, in transmission lines is calculated from the current flowing, I, and the resistance of the lines, R, using $P = I^2R$. When designing the transmission system used, the operating voltage has to be chosen.

A higher operating voltage would mean that the current necessary for a given overall power would be less, and so the power loss would be reduced. A halving of current would result in the power loss dropping to a quarter of its original value.

Unfortunately, high voltages are potentially dangerous, and so are unsafe to use in the home. Transformers are used to step up the output just before transmission to a high voltage in order to reduce power lost in the cables. The voltage can then be stepped down for safe consumer use.

A practical transmission energy grid around a country will involve many different users being supplied at different voltages (Figure 20).

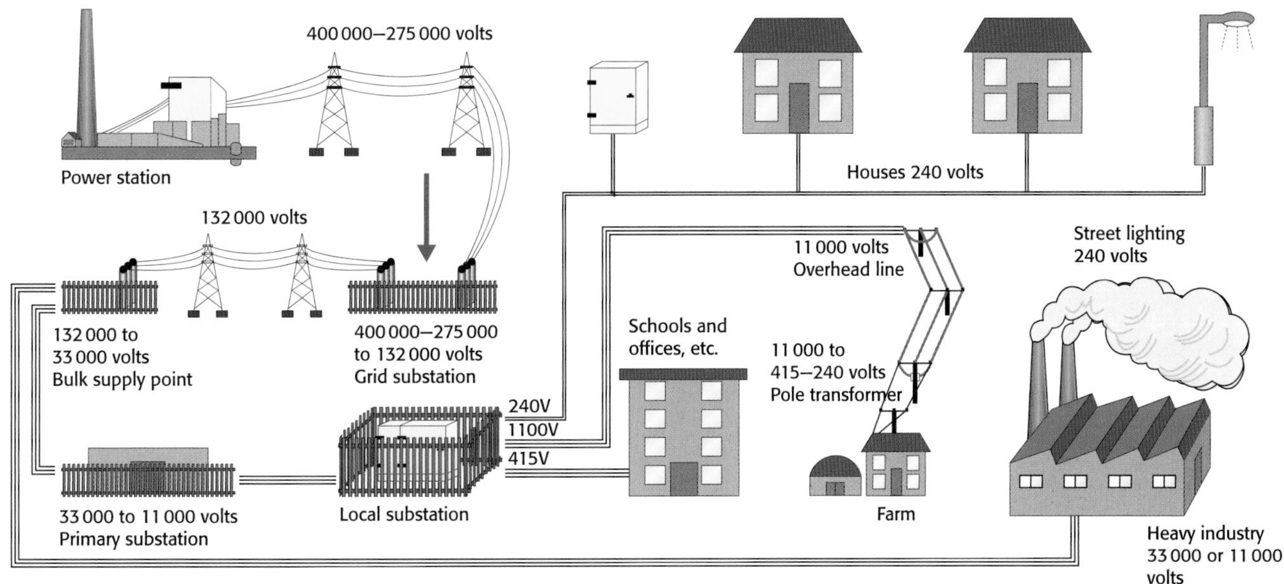

Figure 20 Typical energy distribution grid

Transformers are also used to change the mains voltage into lower, even safer voltages for individual devices such as battery chargers for mobile phones, radios, and cameras. Occasionally the need exists to step the voltage back to higher voltages (e.g. within the older models of television).

1 The efficiency of the energy transfer system described in question 2 (page 224) is improved by adding 2 transformers of turns ratios 5 and 0.2 appropriately into the system.

a) Explain how the circuit should be modified to include the transformers.

b) Calculate

i) the new power loss in the power cable

ii) the improved efficiency of the system.

Losses also happen in transformers

Transformers prevent a great deal of energy loss in power cables, but unfortunately they are not ideal, and some energy is lost in the transformer itself. The possible sources of loss include:

- heating in the wires of the primary and secondary coils resulting from the resistance of the wires
- heating in the iron core resulting from eddy currents (see below)
- **hysteresis** losses in the core. The core warms up as a result of the cycle of magnetic changes taking place. An iron core will result in less energy lost compared with, say, a steel core.
- losses associated with magnetic "leakage". Any magnetic flux that is produced by the primary coil but does not link with the secondary coil represents energy being lost. The iron core is there to ensure that most of the magnetic field is concentrated within the core.

An emf will be induced in any conductor that links with the magnetic field produced by the primary. The core itself is a conductor, and thus an emf will be induced in the iron. Unless the design of the transformer can prevent a current from flowing in the core, this voltage would cause large currents to flow around the core, as shown in the cross-sectional diagram in Figure 21.

These unwanted induced currents are known as **eddy currents.** The complete iron core has a low resistance, so the induced eddy currents would be large, and a great deal of the input energy would be wasted in heating the core. The solution is to prevent the current from flowing by splitting the core into very thin independent sections that are electrically insulated from one another. Each thin section is known as a **lamina**, and the core is said to be **laminated** (Figure 22).

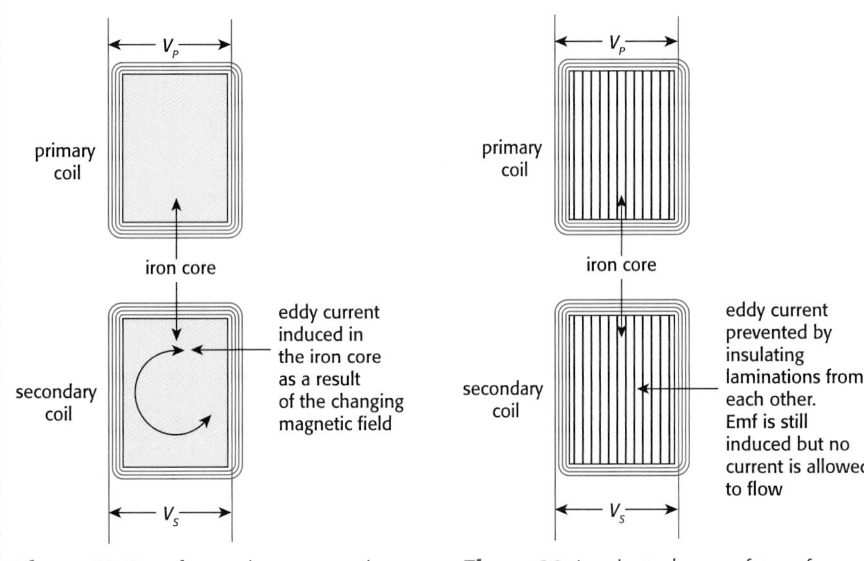

Figure 21 Transformer in cross section

Figure 22 Laminated core of transformer

Inquiry: Dangers of power lines

Many people have strong feelings about whether (or not) it is safe to live very close to an overhead power line. The large alternating currents that flow in power lines must have EM radiation associated with them, and this invisible radiation is a worry to many.

As discussed on page 105, any individual photon associated with the field will not have enough energy to cause ionization. The frequency of the ac is extremely low (50 or 60 Hz), so the EM frequencies are also low. The effects of ionizing radiations are discussed further in the next chapter (page 254), but the extremely low frequency means that these fields are non-ionizing.

The extra-low-frequency EM radiation is still capable of inducing small alternating currents in the body, and some argue that these could cause a detrimental health effect. If this effect does exist, it could be hypothesized that the risks would be dependent on the frequency, the length of any exposure-induced current, and the degree to which the induced current is localized within the body, as measured by the **current density**, defined by the equation below:

$$\text{current density} = \frac{\text{induced current}}{\text{cross-sectional area through which the current flows}}$$

If this effect does exist it should be identifiable with research, but so far no evidence of a clear link has been established between low-frequency fields and any harm caused to genetic material in the laboratory nor detrimental health effects in a population.

Some statistical studies, however, have located regions where more children are diagnosed with leukaemia – a cancer of the blood – than would be expected if the causes were random. The reason for these **leukaemia clusters** has not yet been discovered, but some have suggested a link with overhead power cables.

In 1994 the Royal Society of Edinburgh held a Symposium on Leukaemia Clusters. The report of this meeting is available on the Leukaemia Research Foundation's web page (www.lrf.org.uk accessed July 2009). It includes the following statement:

Ionizing radiation is responsible for less than 10% of childhood leukaemias. It is generally agreed that neither waste from nuclear installations, nor the preconceived exposure of fathers to radiation, is a likely cause of leukaemia in children. The latter theory is biologically implausible on the basis of what is known about the frequency of human gene mutations. At present, the evidence for the implication of other causes is very weak. The possibility that non-ionizing radiation might be responsible is unlikely, but worthy of further exploration. There is increasing evidence that extremely low-energy electromagnetic fields can induce metabolic changes in cells, notably a change in calcium permeability across cell membranes. To suggest that this increases the risk of cancer is, as yet, without evidence, but it is possible that such electromagnetic fields may have biological effects.

More than 10 years later, the debate is continuing. The following article appeared in a UK paper in April 2007:

Power lines link to cancer in new alert

By Nicholas Cecil, *Evening Standard* 20 April 2007

Homes and schools could be banned from being built near power lines.

A secret report has raised fresh fears of a link between power lines and cancer. The confidential study, obtained by the Evening Standard, *urges ministers to consider banning the building of homes and schools close to overhead high-voltage power cables because of possible health risks.*

It says a ban is the best way to reduce significantly exposure to electromagnetic fields from the electricity grid system.

The report was drawn up by scientists, electricity company bosses, the National Grid, government officials and campaigners over two years after the Health Protection Agency accepted there was a weak statistical "association" between prolonged exposure to power fields and childhood leukaemia.

But the 40 members of the panel have clashed over the final details and conclusions. It stops short of specifically recommending a ban on new homes and schools within 60 metres of power lines, or vice versa, which could wipe a total of £2 billion off property prices across Britain and limit land for housing developments.

But the report concludes that the Government should consider such a move, stating: "We urge government to make a clear decision on whether to implement this option or not."

The report, to be signed off by panel members next week, has sparked conflict at a series of hearings, according to a Whitehall source.

Two members of the panel, regulator Ofgem and Scottish & Southern Energy, are understood to have quit.

Some members of the panel took the view – adopted by the Government's health advisers and the World Health Organization – that childhood leukaemia is the only adverse health effect where evidence is strong enough for precautionary measures to be considered.

According to this view, if there is a link, the building ban would cut just one case of childhood leukaemia every year or two, and the costs would outweigh the benefits by a factor of at least 20.

The second group generally backed views highlighted by the California Department of Health Services which suggested electromagnetic fields are "possibly carcinogenic" in terms of childhood leukaemia, and placed four other health effects in this risk category. They were adult leukaemia, adult brain tumours, miscarriages and a form of motor neurone disease, although some scientists believe there are links with more diseases.

"The advice to government from following this 'California' view would therefore be to tend to favour implementing the 'corridors for new build' option," SAGE added, stressing that in this scenario the costs and benefits would be at least comparable.

The panel is set to recommend that the Health Protection Agency should issue more information about how to reduce the impact of exposure to electromagnetic fields. It will also call for a change to the working of overhead lines to reduce the radius of intense electromagnetic fields.

In a response to this article, the Leukaemia Research Foundation released a press statement:

Statement concerning the connection between power lines and childhood leukaemia

"As the UK's leading funder of research into childhood leukaemia we would welcome any new research that may help save the lives of the 500 children a year diagnosed with leukaemia.

"However, it is important to note that the report referred to in the Evening Standard *is not new research; nor does it offer any new evidence as to the cause of childhood leukaemia.*

"We have been funding research for almost 50 years, including one of the largest studies ever carried out into the causes of childhood cancers (UKCCS study). Scientists strongly believe the cause of childhood leukaemia to be an abnormal immune response to an infection.

"There is no biological evidence anywhere in the world that proves a clear link between power lines and

childhood leukaemia; nor is there any compelling evidence to suggest a link between any form of cancer and power lines.

"50 years ago, very few children with leukaemia survived. Thanks to decades of research that have resulted in improved treatments, childhood leukaemia survival rates are now at an all time high. Leukaemia Research remains committed to investing in research that will benefit the 110,000 people living in the UK with a blood cancer."

Cathy Gilman, Chief Executive of Leukaemia Research.

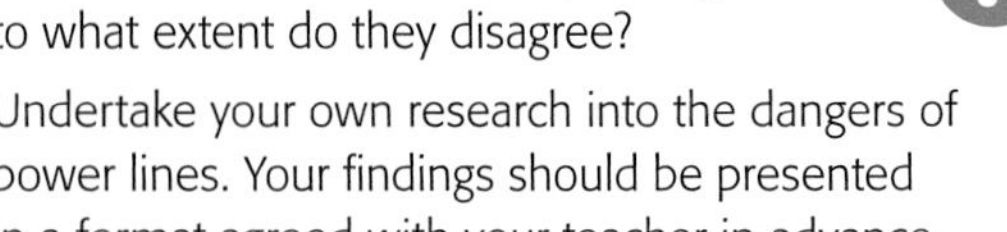

1 To what extent do the two reports agree, and to what extent do they disagree?

2 Undertake your own research into the dangers of power lines. Your findings should be presented in a format agreed with your teacher in advance (e.g. poster/article for school magazine/web page/podcast/essay/presentation). There are many web pages available on both sides of the argument. You should ensure that any evidence you present is properly cited.

Thinking about science: Correlation and cause

Powerful statistical techniques have been devised that allow data to be analysed. They can provide a quantitative measure of the degree to which two (or more) variables are correlated together, but great care needs to be taken when interpreting statistical evidence. It is important to keep in mind the difference between a **cause** and a **correlation**.

Even if two variables are highly linked together, there is no guarantee at all that there is a causal relationship between them. The two factors could be linked by a third factor, or it is possible for the factors to move together by chance. It is important to understand the difference.

For example, an analysis of children's ability to read and their foot size shows a very strong correlation. On average, the bigger the size of a child's feet, the better the child is at reading. This does not mean that large feet cause a child to be better at reading. In this case the children's ages cause the correlation – the older the child, the more likely he or she is to be a good reader (they have learned more), and as the child gets older their feet will also grow.

Although the lack of cause should be obvious in the above example, it is all too easy to fall into the trap of assuming that a link is there even if this is not the case.

Sometimes, however, the lack of a mechanism does not mean that two apparently unrelated events do not have a causal link. In 2001 a Danish study of 7000 women found that those who worked predominantly at night were 50% more likely to develop breast cancer. The figures were adjusted to consider other possible factors, including alcohol consumption and age at birth of first and last children (see http://news.bbc.co.uk/1/hi/health/1108590.stm accessed July 2009). One possible theory is that the exposure to light during the night could trigger hormonal changes, which then cause breast cancer.

There is a correlation between owning a washing machine and the risk of dying in a car accident. It would be poor logic to assume that buying a washing machine increases the chances of an accident. Suggest a reason for this correlation.

End of chapter summary

Chapter 15 has three main themes: induced electromotive force (emf), alternating current and the transmission of electrical power. The list below summarises the knowledge and skills that you should be able to undertake after having studied this chapter. Further research into more detailed explanations using other resources and/or further practice at solving problems in all these topics is recommended – particularly the items in bold.

Induced electromotive force (emf)

- Describe the induction of an emf by relative motion between a conductor and a magnetic field and by a time-changing magnetic flux.
- Derive the formula for the emf induced in a straight conductor moving in a magnetic field.

- Define *magnetic flux* and *magnetic flux linkage*.
- State Faraday's law and Lenz's law **and use them to solve electromagnetic induction problems**.

Alternating current

- Describe the emf induced in a coil rotating within a uniform magnetic field.
- Explain the operation of a basic alternating current (ac) generator and describe the effect on the induced emf of changing the generator frequency.
- Discuss what is meant by the root mean square (rms) value of an alternating current or voltage and state the relation between peak and rms values.
- **Solve ac circuit problems for ohmic resistors using peak and rms values.**
- Describe (**and solve problems on**) the operation of an ideal transformer.

Transmission of electrical power

- Outline the reasons for power losses in transmission lines and real transformers and explain the use of high-voltage step-up and step-down transformers in the transmission of electric power.
- **Solve problems on the operation of real transformers and power transmission.**
- Suggest how extra-low-frequency electromagnetic fields, such as those created by electrical appliances and power lines, induce currents within a human body and discuss some of the possible risks involved in living and working near high-voltage power lines.

Chapter 15 questions

1 A bar magnet is suspended above a coil of wire by means of a spring, as shown below.

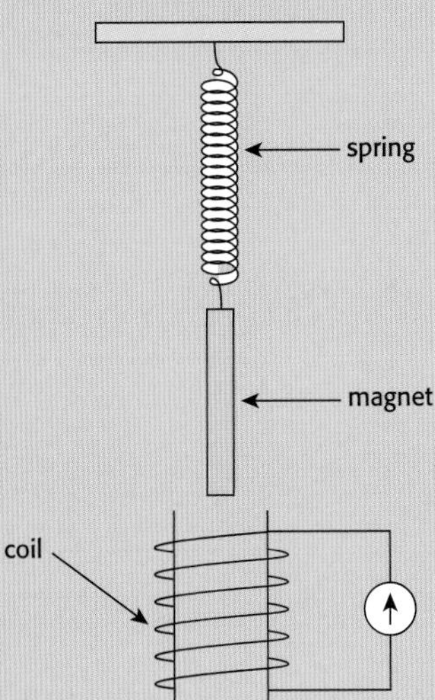

The ends of the coil are connected to a sensitive high resistance voltmeter. The bar magnet is pulled down so that its north pole is level with the top of the coil. The magnet is released, and the variation with time t of the velocity v of the magnet is shown below.

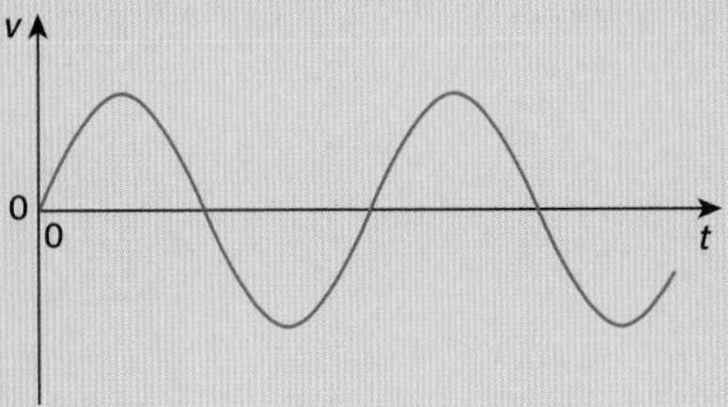

a) On the diagram above,

i) mark, with the letter M, one point in the motion where the reading of the voltmeter is a maximum.

ii) mark, with the letter Z, one point where the reading on the voltmeter is zero. [2]

b) Explain, in terms of changes in flux linkage, why the reading on the voltmeter is alternating. [2]

2 This question is about electromagnetic induction.

In 1831 Michael Faraday demonstrated three ways of inducing an electric current in a ring of copper. One way is to move a bar magnet through the stationary copper ring.

a) Describe briefly a way that a current may be induced in the copper ring using a **stationary** bar magnet. [1]

You are given the following apparatus: copper ring, battery, variable resistor, lengths of insulated copper wire with connecting terminals at each end.

b) Describe how you would use all of this apparatus to induce a current in the copper ring.

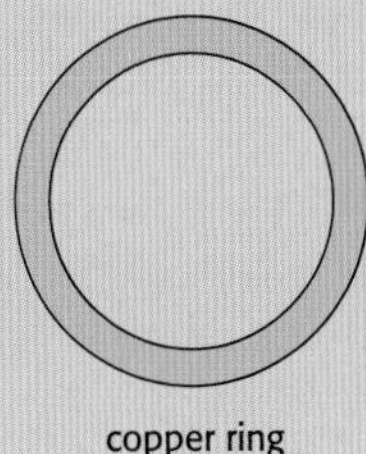

In the diagram below, a magnetic field links a circular copper ring. The fi eld is uniform over the area of the ring, and its strength is increasing in magnitude at a steady rate

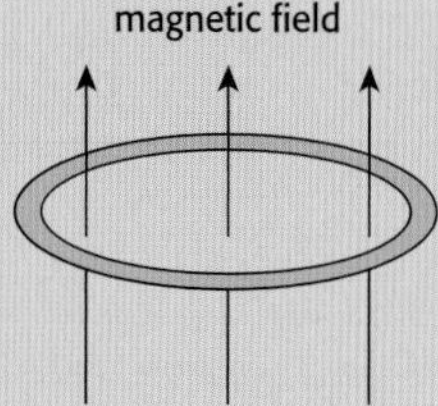

c) i) State Faraday's law of electromagnetic induction as it applies to this situation. [2]

ii) Draw, on the diagram, an arrow to show the direction of the induced current in the copper ring. Explain how you determined the direction of the induced current. [3]

iii) The radius of the copper ring is 0.12 m and its resistance is 1.5×10^{-2}. The field strength is increasing at rate of 1.8×10^{-3} T s^{-1}. Calculate the value of the induced current in the copper ring. [3]

(Total 13 marks)

3 This question is about an ideal transformer.

a) State Faraday's law of electromagnetic induction. [2]

b) The diagram below shows an ideal transformer.

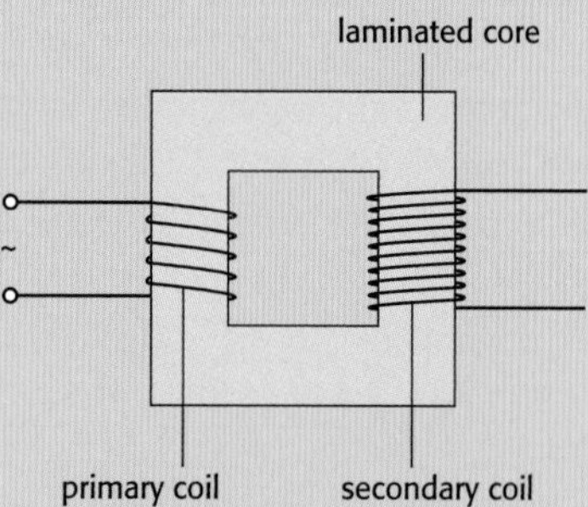

i) Use Faraday's law to explain why, for normal operation of the transformer, the current in the primary coil must vary continuously. [2]

ii) Outline why the core is laminated. [2]

iii) The primary coil of an ideal transformer is connected to an alternating supply rated at 230V. The transformer is designed to provide power for a lamp rated as 12V, 42W and has 450 turns of wire on its secondary coil. Determine the number of turns of wire on the primary coil and the current from the supply for the lamp to operate at normal brightness. [3]

(Total 9 marks)

4 Electromagnetic induction

A small circular coil of area of cross-section 1.7×10^{-4} m^2 contains 250 turns of wire. The plane of the coil is placed parallel to, and a distance *x* from, the pole-piece of a magnet, as shown below.

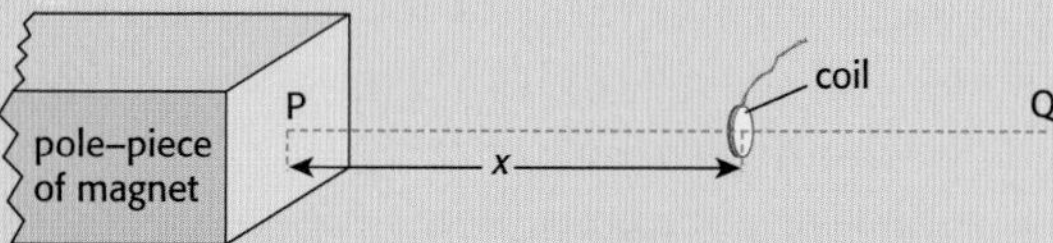

PQ is a line that is normal to the pole-piece. The variation with distance *x* along line PQ of the mean magnetic field strength B in the coil is shown below.

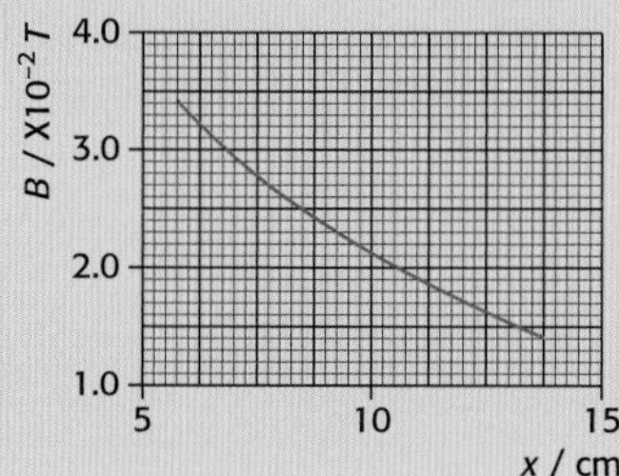

a) For the coil situated a distance 6.0 cm from the pole-piece of the magnet,

i) state the average magnetic field strength in the coil; [1]

ii) calculate the flux linkage through the coil. [2]

b) The coil is moved along PQ so that the distance *x* changes from 6.0 cm to 12.0 cm in a time of 0.35 s.

i) Deduce that the **change** in magnetic flux linkage through the coil is approximately 7×10^{-4} Wb. [2]

ii) State Faraday's law of electromagnetic induction and hence calculate the mean emf induced in the coil. [2]

c) i) State Lenz's law. [1]

ii) Use Lenz's law to explain why work has to be done to move the coil along the line PQ. [4]

(Total 11 marks)

5 This question is about electromagnetic induction.

A small coil is placed with its plane parallel to a long straight current-carrying wire, as shown below.

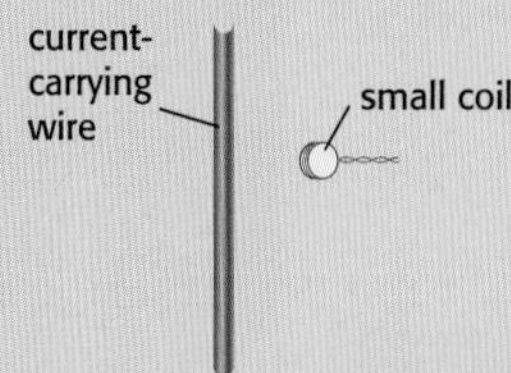

a) i) State Faraday's law of electromagnetic induction. [2]

ii) Use the law to explain why, when the current in the wire changes, an emf is induced in the coil. [1]

The diagram below shows the variation with time t of the current in the wire.

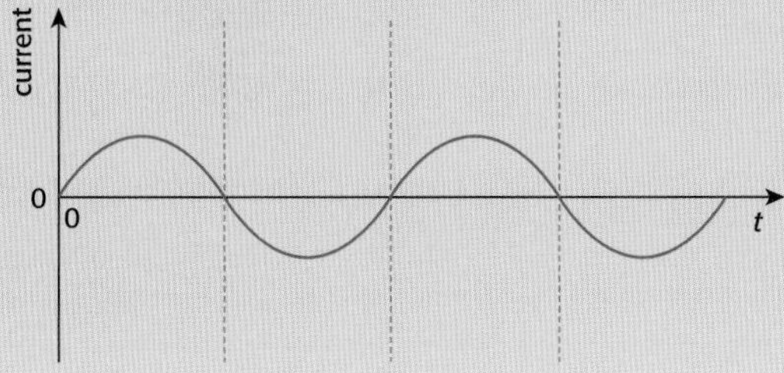

b) i) Draw a sketch graph to show the variation with time t of the magnetic flux in the coil. [1]

ii) Construct a sketch graph to show the variation with time t of the emf induced in the coil. [2]

iii) State and explain the effect on the maximum emf induced in the coil when the coil is further away from the wire. [2]

c) Such a coil may be used to measure large alternating currents in a high-voltage cable. Identify **one** advantage and **one** disadvantage of this method. [2]

(Total 10 marks)

16 Atomic and nuclear physics

This chapter considers different models for the basic structure of all matter. The discovery of radioactive decay led to a major revision in the models used, but the story does is not over. The most commonly used simple atomic model (involving protons, neutrons, and electrons) can be shown to have major shortcomings, and thus a better, more sophisticated, model is required.

Before starting any of the sections in this chapter it would be useful to answer the questions on the right.

1 Estimate, in metres, typical sizes of the following objects:
 a) the width of a human hair
 b) a strand of DNA
 c) a molecule of water
 d) an atom of carbon
 e) a nucleus of iron.

 Now research the accuracy of your estimates.

2 Explain the difference between an atom and a molecule.

3 There are approximately 100 different elements; explain how they differ from one another.

4 Summarize your knowledge of the atomic model. Where are the protons, neutrons, and electrons located in this simple model? Do you already know any experimental evidence that would back up this model?

Thinking about science: Models and experimental evidence

Question 4 asks you to recall a simple model of the atom. In order to explain where the electrons, neutrons, and protons are found, many people draw diagrams like the ones shown in Figure 1.

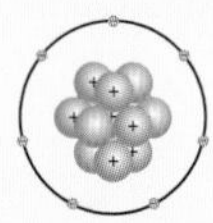
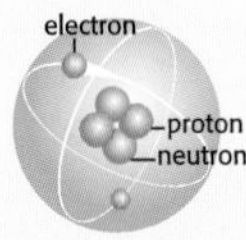

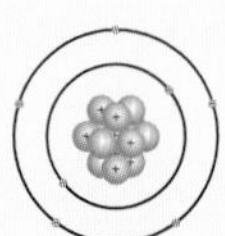

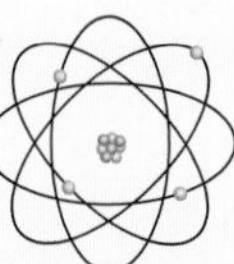
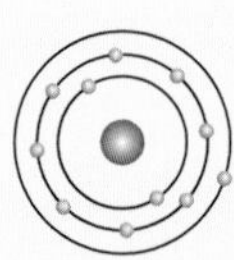

Figure 1 Simple models of the atom

These diagrams are attempts to represent two atoms of nitrogen, one of beryllium, one of helium, one of neon, and one of sodium. Although they have some ideas in common, they are very different: so which, if any, is right?

From a theory of knowledge perspective, understanding why particular atomic models have been developed allows you to critically question the model and develop a real understanding of the nature of the physical universe. As an IB learner, you should already be familiar with the IB learner profile (see page 3), which summarizes the skills you need to develop. All IB learners should strive to become "thinkers" – especially critical thinkers:

> *"They exercise initiative in applying thinking skills critically and creatively to recognize and approach complex problems, and make reasoned, ethical decisions."*

In the context of this chapter, thinking critically means investigating the experimental evidence that we have relating to atomic structure and questioning how well any models fit with that data.

Historical models of matter

Ancient models of the fundamental structure of all matter can seem strange today, but more up-to-date ideas are not always easier to believe. Many people accept that all matter is made out of a combination of just three things – electrons, neutrons, and protons. The common structure that is used to put these three things together – the atom – is mostly empty space: the central nucleus and the tiny electrons occupy a small percentage of the overall size of the atom. This is true whatever the object, be it living, dead, or never alive (e.g. you, the earth on which you walk, the air that you breath, the food that you eat, this book, the computer on which it is being written, a star). Are you comfortable with the idea that most of the volume you occupy is empty space?

Ancient philosophers, whose ideas were not formally based on experimental evidence, were free to propose and justify any model that seemed appropriate. The modern scientific method (which you study as part of your theory of knowledge course) is based on the objective use of measurable experimental data to critically test theories and hypotheses.

The Greek philosopher Democritus (*c.* 460–370 BC) was one of the earliest philosophers to introduce the concept of an **atom** – the smallest particle into which anything can be broken. In this model, each of the different types of atom (water atoms, fire atoms, earth atoms, and so on) had different shapes, and these could be used to explain matter's different properties. For example, fire atoms might be covered in thorns, which is why you can get burned; water atoms would be smooth, which is why water flows easily. This model was not, however, widely accepted at the time.

Empedocles (*c.* 490–430 BC) proposed that all matter was made up of a combination of four **elements** or basic principles: earth, fire, air, and water. For example, when something living dies and decays, it seems to change into earth, so this makes sense.

A fifth element, aether, was later added by Aristotle (384–322 BC) as the basic building block used in "the heavens". This philosophical model was generally accepted up until the late 17th century.

An interesting aspect of this theory was that it might be possible to change one type of object into another. **Alchemy** was the attempt to find a way of changing base metal into gold; often this involved the search for something called the philosopher's stone. This transmutation was envisaged happening at high temperatures.

Figure 2 *The Alchemist*, by David Rychaert (1612–1661)

In the 17th century, the playwright Ben Jonson (1572–1637) wrote the play *The Alchemist*, in which a character called Subtle pretends to be an alchemist and cheats gullible people out of their money on the pretext of being able to achieve this transmutation. In the following extract Subtle is trying to persuade a suspicious character, Pertinax Surly, that his ability to change lead into gold is less surprising than nature's ability to change an egg into a chicken:

SUBTLE Why, what have you observ'd, sir, in our art,
Seems so impossible?

Surly But your whole work, no more.
That you should hatch gold in a furnace, sir,
As they do eggs in Egypt!

Subtle Sir, do you
Believe that eggs are hatch'd so?

Surly If I should?

Subtle Why, I think that the greater miracle.
No egg but differs from a chicken more
Than metals in themselves.

Surly That cannot be.
The egg's ordain'd by nature to that end,
And is a chicken in potentia.

Subtle The same we say of lead and other metals,
Which would be gold, if they had time.

Isaac Newton (1643–1727) is remembered these days as one of the truly great scientists of all time, and as a founder of the scientific method, but he was also interested in alchemy. He conducted many secret experiments in the hope that this theory might be valid. To quote John Maynard Keynes, Newton was "not the first of the Age of Reason; he was the last of the magicians…".

It was the study of gas pressures that led the English chemist John Dalton (1766–1844) to revive Democritus's idea and propose that all matter consisted of small particles called **atoms**. The French chemist Joseph Proust (1754–1826) experimentally verified that some substances were compounds of different elements, and that the proportions of these elements were fixed in precise ratios. This supported Dalton's idea of individual atoms. However, at that time it was not clear how many different types of atom existed. The Russian chemist Dmitri Ivanovich Mendeleyev (1834–1907) arranged the known elements into a table (the **Periodic Table**) based on their masses and on how the elements combined with other elements – that is, their chemical properties.

We can use very simple experiments to try to set upper and lower limits to the volume of space that atoms and molecules occupy. One famous experiment involves carefully measuring how far an oil drop spreads out when it floats on water (Figure 3).

If we know the volume of the oil drop before we place it on the water, and then measure the area of the oil patch after it spreads out, it is a simple calculation to work out the thickness of the patch. This thickness is an upper limit for the size of atoms and/or molecules.

(a)

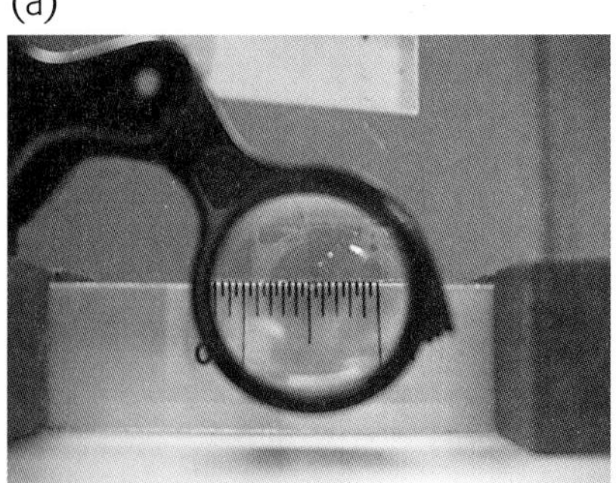

(b)

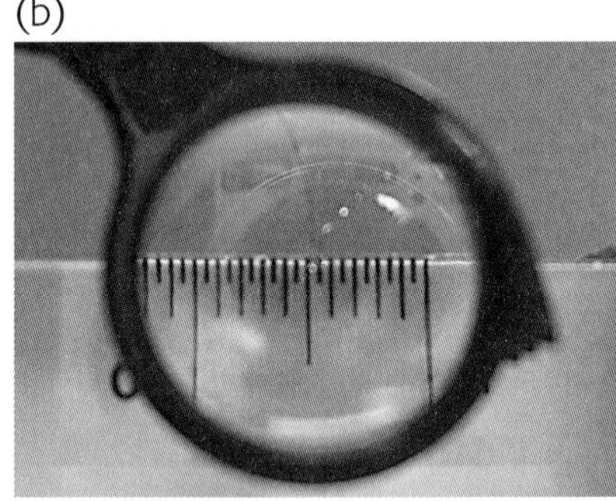

(c)

60 50 40 30 20 10

(d)

(e)

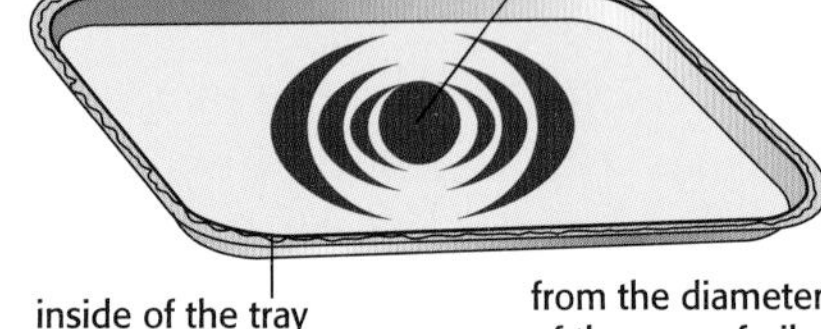
shallow tray filled to the brim with water, coated with lycopodium powder
small drop of oil placed in the centre
inside of the tray is coated to the rim with paraffin wax
from the diameter of the area of oil, and its volume, the thickness of the oil layer can be calculated.

Figure 3 (a) and (b) Oil drop
(c) Tray of water (d) Oil drop on water
(e) Experimental setup

a) Use readings from the pictures in Figure 3 to estimate
- **i)** the volume of oil used
- **ii)** the surface area of the oil patch.

b) Use your answers to part **a)** to calculate the thickness of the oil patch.

Another simple experiment is to consider a stone step that has worn down over time, see Figure 4. If we assume that each time a shoe slides on a stone step, it can rub off one atom, we can estimate the size of an atom. The calculation starts by finding a way of estimating the number of times that the steps in Wells Cathedral in the UK have been walked upon since the steps were built (1319). If we then use the photo to estimate the volume that has been worn away, we can estimate the size of an atom.

Figure 4 Worn out steps in Wells cathedral provide evidence for the existence of atoms

Towards the end of the 19th century, the atomic model was based around the number of elements that were known to exist, and all matter was considered to be made of different combinations of these elements. It was only during the 1890s that some progress was made with modelling the *internal structure* of the atom.

The electron

Professor J.J. Thomson (1856–1940) was awarded the Nobel Prize in Physics in 1906 for the discovery of the **electron**. He was investigating the recently discovered phenomena of "cathode rays" – invisible rays that were emitted from the negative electrode when an electrical potential difference was placed across two metal plates in a vacuum. The rays caused the glass or a zinc sulfide screen to glow. By making the anode out of specific shapes (Figure 5), it was possible to show that the cathode rays must travel in straight lines.

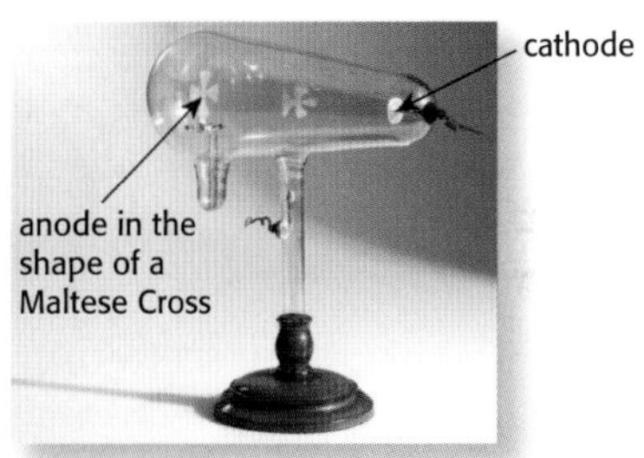

Figure 5 Maltese cross tubes

Thomson demonstrated that the cathode rays were small negative charges because they were deflected in an electric field. By adding a suitably arranged magnetic field at right angles to the electric field he could bring the cathode rays back to zero deflection. He used this result to calculate the velocity of the negative particles and the deflections measured in the electric field experiment to calculate the ratio of charge to mass for the particles. This value was always the same, whatever substance he used for the cathode. Deflections had been achieved with the charged ions of different elements; the cathode ray particle had a much smaller mass for a similar charge.

1. Draw the arrangement of electric field and magnetic field that would allow for the magnetic deflection in Thomson's experiment to balance the electric deflection.
2. Derive an expression for the speed of electrons if they show no overall deflection when subjected to an electric field E crossed with a magnetic field B in Thomson's experiment.
3. When cathode rays were first discovered, many physicists thought that they must be a form of wave motion similar to light. Outline the observations this model does explain and those that it does not.
4. Thomson's experiments allowed him to calculate the charge-to-mass ratio for the electron, but not the electron's actual charge or its mass. The first accurate measurement of the charge on an electron was achieved by the American physicist R.A. Millikan. Research this experiment, and summarize both the experiment and Millikan's calculations.
5. All atoms were known to be neutral. The discovery that atoms contained negative electrons meant that some positive matter must also exist. Thomson's atomic model of the time envisaged the electrons held together in an atom, like plums in a pudding – the "plum pudding" model. The "pudding" in the atoms would be some sort of positive "glue" that normally kept the electrons trapped inside the atom. Research this model, and create a presentation showing how this model could explain the experiments of the time.

When doing these experiments Thomson also discovered that different forms of the same element existed. These had exactly the same chemical properties, but differed slightly in terms of their overall mass. He had discovered **isotopes**.

Thomson obtained the same value of charge-to-mass ratio for cathode rays as for the negative particles that were emitted by some metal surfaces when ultraviolet light was shone on them (the photoelectric effect – see pages 263–4). He suggested that every atom contained these tiny particles, now called electrons.

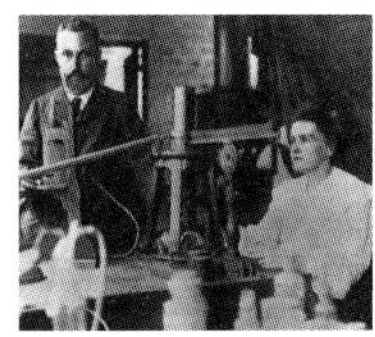

Figure 6 Henri Becquerel (left); Marie and Pierre Curie (right)

The nucleus

On Sunday 1 March 1896 the discovery of radioactivity was made by Henri Becquerel. This phenomenon was subsequently investigated by Pierre and Marie Curie, and three types of radioactive emission were identified – alpha, beta, and gamma.

Ernest Rutherford, who was a student of J.J. Thomson, worked in collaboration with Thomas Royds and showed that alpha particles were helium nuclei. He also calculated the charge-to-mass ratios for alpha and beta particles using a similar technique to J.J. Thomson. When doing this experiment, he noticed that a small piece of mica (a type of mineral) placed in the path of the alpha particles caused the particles to be scattered, and he designed an experiment to measure the degree of scattering that took place when alpha particles passed through a thin piece of gold foil (Figure 7).

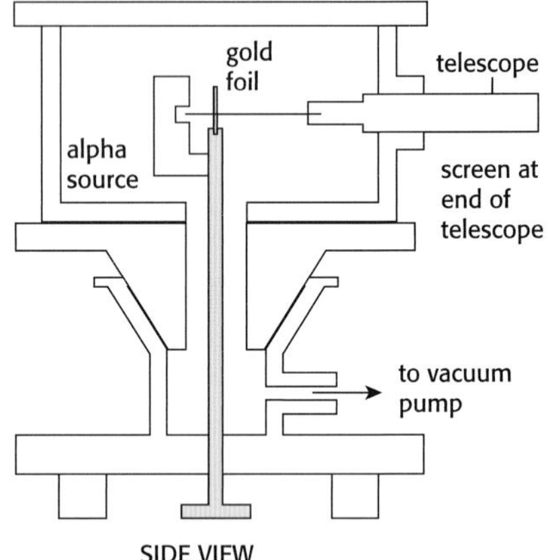

Figure 7 Rutherford's experiment on scattering of alpha particles

The vast majority of the alpha particles went through the foil virtually undeflected. A member of Rutherford's research group, Hans Geiger (who later developed the Geiger counter), was looking for a project for one of his PhD students called Ernest Marsden. Marsden was given the task of using the apparatus to check whether there were any deflections through large angles.

Figure 8 Left to right: Ernest Rutherford (1871–1937); Hans Geiger (1882–1945); Ernest Marsden (1889–1970)

The amazing discovery was that a small number of the alpha particles were deflected through very large angles indeed: in effect, some of them "bounced back". Rutherford described this discovery as astonishing: "It was almost as incredible as if you fired a 15-inch shell at a piece of tissue paper and it came back and hit you." The mathematics of the situation was analysed in detail, and the evidence could only be explained if all the positive charge in the atoms of the gold foil was concentrated at the centre, along with all the effective mass: i.e. a nucleus. Rutherford calculated that the diameter of the nucleus was 100 000 times smaller than the diameter of the atom, and the positive nucleus would repel the incoming alpha particle according to Coulomb's inverse square law. He was able to calculate the theoretical percentage of alpha particles that would be deflected through any particular angle range according to this model, and the predictions agreed exactly with the experimental measurements.

The proton

Rutherford's experiments continued to advance understanding of nuclear physics. In 1919 he was able to show that a fast-moving alpha particle could interact with a nitrogen nucleus. The nitrogen nucleus was transmuted into oxygen, and the result was the emission of a hydrogen nucleus.

?

1. Do some research into Rutherford and Royds' experiment. Explain how alpha particles were shown to be helium nuclei.
2. Outline how you could experimentally prove that beta particles are, in fact, electrons.
3. Draw a careful and accurate diagram to show the paths taken by alpha particles as they are deflected by gold nuclei.

In Rutherford's own words: "We must conclude that the nitrogen atom is disintegrated under the intense forces developed in a close collision with a swift alpha particle, and that the hydrogen atom which is liberated formed a constituent part of the nitrogen nucleus."

$${}^{14}_{7}\mathrm{N} + {}^{4}_{2}\mathrm{He} \rightarrow {}^{17}_{8}\mathrm{O} + {}^{1}_{1}\mathrm{H}$$

This transmutation meant that Rutherford had succeeded in doing something that the alchemists had unsuccessfully attempted to do – change one element into another. He went on to show that other elements could also be transmuted, but the hydrogen nucleus was always emitted. He therefore proposed that the hydrogen nucleus must be contained inside all nuclei, and it was renamed the **proton**. However, it was clear from the beginning that something else must be inside the nucleus for the calculations of total mass to make sense. Although a proton + electron combination would be neutral, and was originally proposed as the missing mass, detailed calculations showed that a nucleus formed of just protons and electrons could not be possible. An additional neutral particle inside the nucleus was proposed – the **neutron**.

Figure 9 Sir James Chadwick (1891–1974)

The neutron

James Chadwick was awarded the 1935 Nobel Prize in Physics for his discovery of the neutron in 1932. His research built on an observation by two German physicists and a subsequent French experiment.

In 1930 the German physicists Walther Bothe and Herbert Becker found that a very penetrating radiation was emitted when they bombarded beryllium with alpha particles. This radiation was non-ionizing. In 1932 Irène Joliot-Curie (the daughter of Pierre and Marie Curie) and her husband Frédéric let the radiation hit a block of paraffin wax, and found it caused the wax to emit protons.

The debate at the time was to decide whether this neutral radiation was a sort of high-energy radiation (like gamma rays) or whether it could be a stream of neutrons. Chadwick and Rutherford were able to show that gamma rays would not have sufficient energy to cause protons to be emitted, but that neutrons would be able to do this. Chadwick's elegant analysis involves applying the conservation of momentum to the collisions. The equation representing the creation of the neutrons is

$${}^{4}_{2}\alpha + {}^{9}_{4}\mathrm{Be} \rightarrow {}^{1}_{0}\mathrm{n} + {}^{11}_{6}\mathrm{C}$$

Do some research into Chadwick's experiment and his analysis. Explain how he used the law of conservation of momentum in his calculations.

The Bohr model of the hydrogen atom

Before the neutron had been discovered, a Danish physicist, Niels Bohr, had proposed an important mathematical model of the hydrogen atom. This brought together Rutherford's proposed nucleus with the ideas of electric fields and forces into a mechanism by which the phenomena of atomic spectra could be understood.

Figure 10 Niels Bohr (1885–1962)

Successes of the model

It had been known for some time that, when different elements became hot enough, they gave off light. If this light was analysed in detail it was seen to contain specific frequencies of light, which

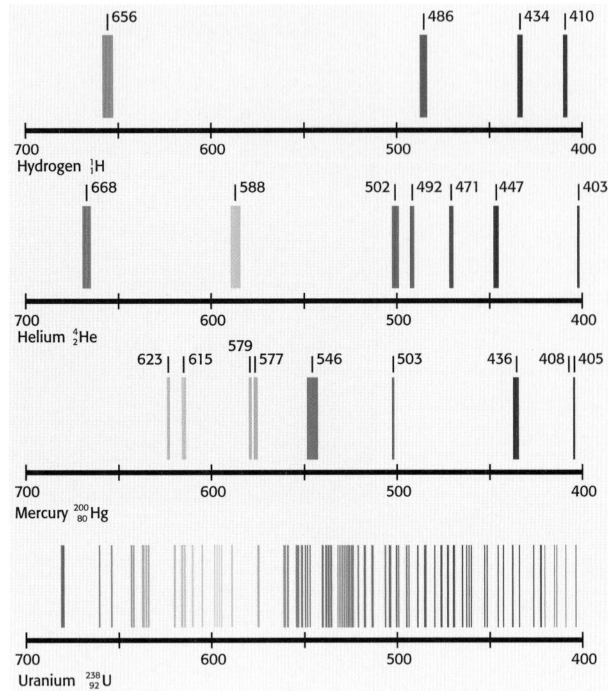

Figure 11 Emission spectra. Top to bottom: hydrogen, $^1_1$H; helium, $^4_2$He; mercury, $^{200}_{80}$Hg; uranium, $^{238}_{92}$U.

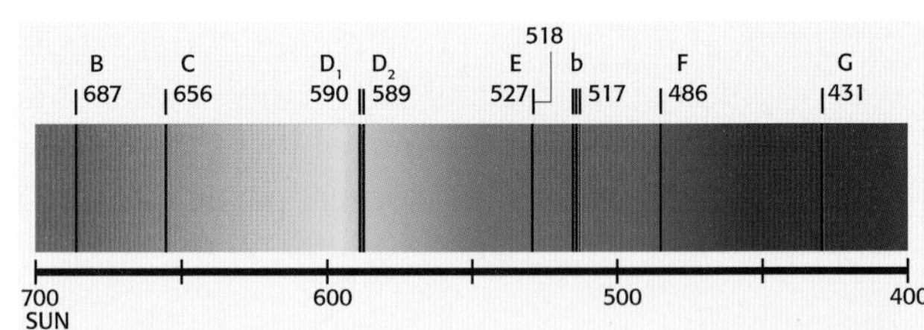

Figure 12 Absorption spectrum from the Sun

corresponded to particular colours – the **emission spectrum** for each element (Figure 11). Different elements had different frequencies.

Light from some other sources (e.g. the Sun) showed most frequencies but with a few missing – an **absorption spectrum** (Figure 12).

Emission and absorption spectra are discussed further on pages 268–9. The different colours of light can be explained using an aspect of quantum theory that was just beginning to be developed at the same time. The relevant idea that needs to be understood here is that light is not a continuous wave but is emitted as "packets" of energy. These "packets" are called **photons**. The energy of a photon, E, is related to its frequency f by the equation $E = hf$, where h is a constant called **Planck's constant** ($h = 6.63 \times 10^{-34}$ J s). The different colours of light correspond to different photon energies.

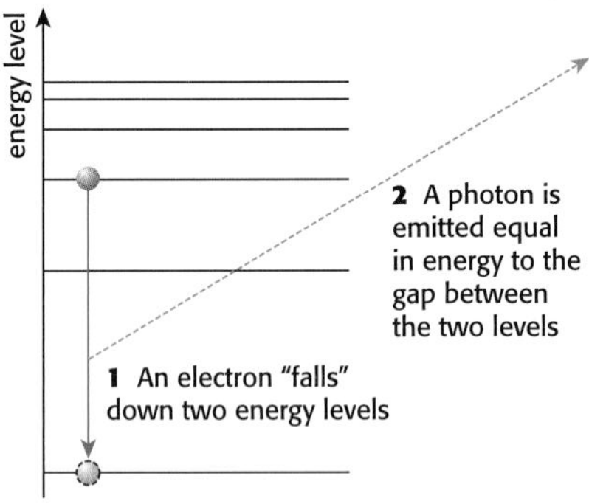

Figure 13 The emission of a photon when an electron drops to a lower energy state

The fact that specific colours exist in an emission spectrum implies that the electrons in atoms can only have specific energies. Not all energies are possible; there are only a limited number of energy levels available.

Bohr proposed that it was the electron that moved between different energy levels. In order to move up an energy level, an electron needs to absorb a photon. When an electron moves down energy levels it emits a photon (Figure 13).

Using Coulomb's law and the mathematics of circular motion, Bohr was able to show that different radii for the electron's orbit in the hydrogen atom would correspond to different energies. He discovered that, provided he made a few simple assumptions about which orbits were available and which were denied, he could calculate all the energy levels that existed in the hydrogen atom. His calculated available orbit energies were in close agreement with many of the energies calculated from analysis of the hydrogen emission spectrum.

With the discovery of the neutron, the final piece of a jigsaw seemed to have fallen into place. The model for an atom involved tiny electrons (containing all the negative charge and hardly any mass) continuously in orbit around a tiny nucleus (containing all the positive charge and all the mass). The force that provided the centripetal force necessary to retain these electrons was the electrostatic attraction between the electrons and the nucleus.

To summarize, over a period of less than 40 years at the beginning of the 20th century the whole of atomic theory had been proposed, experimentally researched, and refined. Teams of physicists from around the world, working together and sharing results with one another, were able to significantly improve our understanding of the basic building blocks of the universe. This incredibly rich period of scientific research needs to be put in perspective against the backdrop of the First World War, which took place between 1914 and 1918.

At the same time as new atomic models were being created, two fundamentally important branches of physics were also being developed: **quantum theory** (which was associated with the atomic models being created, including the Bohr model), and **relativity**. Before this, physics had not undergone such a radical new approach to the problem of understanding the Universe since the time of Isaac Newton in the 17th century.

Known problems of the model

Unfortunately, things are not as simple as they seem. Although the above simple model can explain many experimental observations, it is unable to explain others, and there are some serious problems. The following issues need to be addressed if a proper understanding of atomic structure is going to be developed based upon this model.

- There is no available mechanism by which the protons and neutrons would stay inside a small nucleus. The only known force that might help balance the force of repulsion between the protons is the force of gravitational attraction. At the atomic and subatomic level the gravitational force is so small that it is often ignored. There must be at least one other type of force that isn't electromagnetic or gravitational.
- Even if another force or interaction is used to explain why nuclei are kept together, we still don't have a mechanism for radioactive decay.
- Even if we ignore the stability of the nucleus for a short time, there is another problem with the model. Accelerating charges radiate energy. An electron that is moving in constant circular orbit around a proton must be accelerating (it is changing direction all the time), and so it should radiate energy all the time. As it loses energy it should spiral into the nucleus. This predicts that atoms can't exist! Clearly they do exist, so something is very wrong with the model.
- The mathematical ability of the Bohr model to be able to predict the hydrogen emission spectrum frequencies is impressive, but if we try to use the same approach with other elements, the predictions do

?

1. Do some research into how emission spectra and absorption spectra are observed. You do not need to memorize the details of any particular experimental setup, but if you understand how the spectra are observed then you can better understand the physics behind their creation.
2. Use the mathematics of circular motion and of electrostatic fields to calculate the speed v of an electron moving in a circular orbit around a proton at a radius r.
3. The proton, the neutron, and the electron are needed to explain the structure of an atom. This chapter started with some diagrams representing how they are arranged in different atoms. For each diagram try to list ways in which the picture helps understanding, and ways in which the picture might cause misunderstandings.

not agree with experiment. Moreover, the Bohr model fails to predict some of detail of the lines: the so-called "fine structure" and the relative amplitudes of different frequencies.

- Finally, although some of the assumptions used in the Bohr model make sense, there is no underlying theory to justify the mathematical rule that allows some orbit energies and disallows others.

Energies on the atomic scale: the electronvolt

When considering atomic changes, the joule is an impossibly large energy for a particle to have. In its place is a derived unit called the **electronvolt**, and the related units of megaelectronvolts (MeV) and gigaelectronvolts (GeV). The definition of potential difference (p.d.) comes from a consideration of the energy changes when a charge is accelerated by a potential difference:

$$\text{potential difference p.d.} = \frac{\text{energy change}}{\text{charge transferred}}$$

so

$$\text{energy change} = \text{potential difference} \times \text{change transferred}$$

One electronvolt is the energy gained by an electron when it moved through a p.d. of one volt:

$$1 \text{ electron volt} = 1 \text{ volt} \times \text{charge on one electron}$$

$$1 \text{ eV} = 1.6 \times 10^{-19} \text{ J}$$

$$1 \text{ MeV} = 1.6 \times 10^{-13} \text{ J}$$

$$1 \text{ GeV} = 1.6 \times 10^{-10} \text{ J}$$

The need for new forces

As explained above, there must be an additional force (not gravitational or electromagnetic) keeping the protons and the neutrons bound together. What information do we already know about this new force? First, it has not been observed to act anywhere other than inside the nucleus, and so it must have a very short range, and cannot vary with distance in the same way that gravitational or electromagnetic forces do; it is certainly not an "inverse square" type of force. Second, when it *does* act, it must be strong in order to keep the protons bound inside the nucleus. The unimaginative name given to this new force is the **strong nuclear force** or the **strong interaction**.

The mathematics of the strong interaction is not straightforward, because the resultant force varies considerably, depending on the situation. For example, an energy of 1.7 MeV is needed to separate the single neutron from the proton that is inside the nucleus of the isotope of hydrogen called deuterium, whereas 8.0 MeV are needed to separate the two neutrons and the proton that are inside hydrogen's other isotope, tritium. Ever-increasing energy is required as more and more **nucleons** (protons or neutrons) are involved. For example, 27.3 MeV are needed to separate the two protons and two neutrons inside a helium nucleus.

The introduction of a third force, the strong interaction, to our model is not enough if we wish to be able to explain all aspects of the nucleus. Currently we have no mechanism by which, for example, beta decay can take place inside a particular nucleus, so a fourth interaction must be involved. The associated forces are not as large as in the strong interaction, and so the name chosen for this fourth force is the **weak interaction**.

Thinking about science: The nature of forces or interactions

In everyday physics the concept of a force is easy to understand: it is a push or a pull. Newton's laws predict that a resultant or unbalanced force on an object must cause acceleration: thus a force is something that can change an object's velocity. This definition is not quite complete, because forces can also cause an object to distort in shape. In many situations a change in shape is the result of two forces acting in opposite directions (when, for example, you squeeze an eraser), but a single force can also cause an object to change its shape: for example, when a golf ball is hit it temporarily changes shape (as we can see if we use high-speed photography).

In particle physics a force is seen as the cause of *every* change: sometimes this can be a change in motion, but it could also be the creation of new particles or even the disintegration of one particle into new particles. To emphasize this difference it is usual in particle physics to talk not about forces but about **interactions**. The four known interactions, which can act anywhere in the universe, are thus gravitational, electromagnetic, strong, and weak.

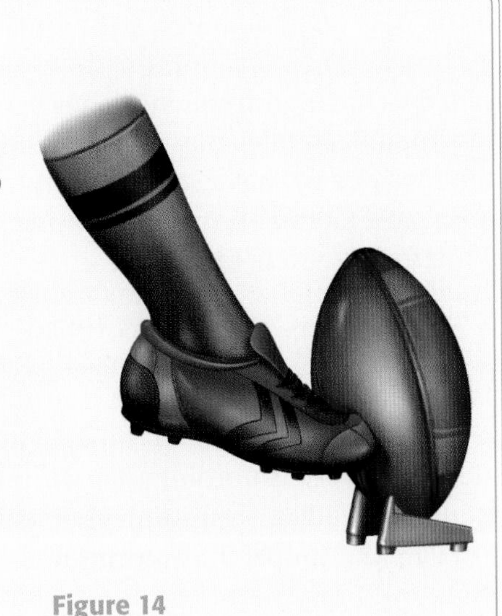

Figure 14

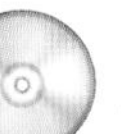

Radioactive decay – nothing lasts for ever

The phenomenon of radioactive decay is a *nuclear* reaction. This means that when it occurs, alpha, beta, or gamma radiation originates in the nucleus. Alpha radiation is just a part of the nucleus (the same as a helium nucleus), and gamma radiation is just electromagnetic radiation. Beta particles, however, are electrons, and electrons do not exist in the nucleus.

They are created when the weak interaction makes a neutron change into a proton and an electron.

Properties of Alpha, Beta and Gamma Radiations

Property	Alpha, α	Beta, β	Gamma, γ
Effect on photographic film	Yes	Yes	Yes
Approximate number of ion pairs produced in air	10^4 per mm travelled	10^2 per mm travelled	1 per mm travelled
Typical material needed to absorb it	10^{-2} mm aluminium; piece of paper	A few mm aluminium	10 cm lead
Penetration ability	Low	Medium	High
Typical path length in air	A few cm	Less than one m	Effectively infinite
Deflection by E and B fields	Behaves like a positive charge	Behaves like a negative charge	Not deflected
Speed	About 10^7 m s^{-1}	About 10^8 m s^{-1}, very variable	3×10^8 m s^{-1}

The result of any radioactive decay is that the nucleus changes. In the case of alpha or beta decay, the new nucleus must belong to a new element. We can predict the product, or **daughter nuclide**, that is created by considering the conservation laws for mass and charge. Daughters are not always stable; it is common for the nucleus that has been created to go on and decay further. Each stage of the decay will have its own half-life, and thus the different stages will progress at different rates. Figure 17 represents the uranium-238 decay chain. The final end product will be a stable nucleus of lead.

In the case of gamma decay the nuclide remains the same type, but energy has been lost because a photon of electromagnetic radiation (of a specific high frequency) has been emitted. This last piece of information implies that there must be nuclear energy levels in the same way as we have seen that atomic energy levels exist. Very often gamma radiation accompanies (or follows very soon after) an alpha decay or a beta decay.

The processes involved are completely random. There is a fixed chance that any particular nuclide will decay in a particular way in a given amount of time, but we cannot know whether or not this decay will actually happen for any nucleus that we choose to consider. There is nothing that we can do to alter this probability; for example, heating a substance or putting it under great pressure does not alter the probabilities associated with radioactive decay. Very often, rather than quoting the probability of a particular decay taking place, radioactive half-lives are used to compare different nuclear stabilities. It is true that a nucleus can be bombarded with other particles in order to make a decay take place, but these **stimulated emissions** involve the capture of the incoming particle, which creates a new but unstable nuclide, and then its own decay.

The random nature of spontaneous radioactive decay results in an **exponential decrease** in two quantities with time: the number of nuclei available to decay and the rate of decay. This exponential decrease has a very specific mathematical property – the time it takes for either quantity to reduce to half its initial value is a fixed length time – the **half-life**. The more likely the decay is to take place, the shorter the half-life. The more stable the nucleus, the longer the half-life. Half-lives can vary from fractions of a second to millions of years. See page 276 for more mathematical details.

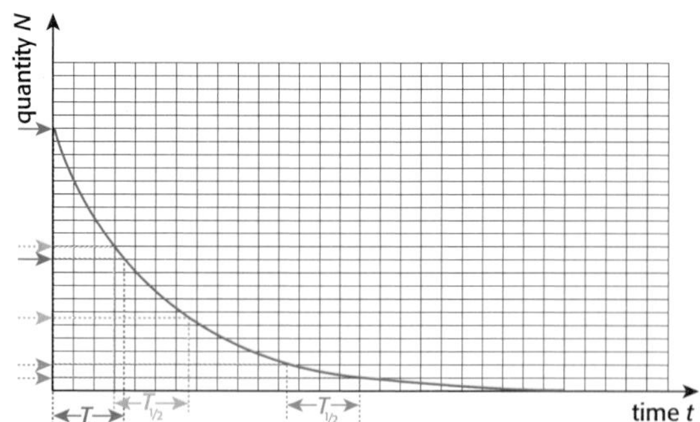

Figure 16 The half life of exponential decay is constant

The definition of half-life is the time taken for the number of nuclei (that are available to decay) to halve. It is also the time taken for the rate of decay to halve. The table below shows how the numbers of nuclei that are available to decay continues to get smaller with time.

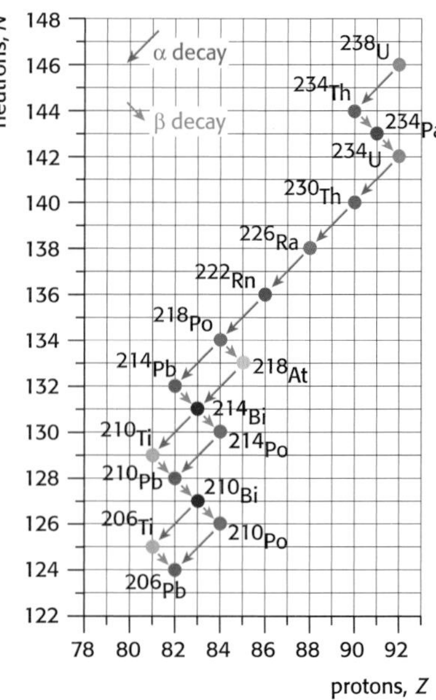

Figure 15 Nuclear decay reactions for uranium-238

type of radiation	nuclide	half-life
	uranium-238	4.5×10^9 years
α	thorium-234	24.5 days
β	protactinium-234	1.14 minutes
β	uranium-234	2.33×10^5 years
α	thorium-230	8.3×10^4 years
α	radium-226	1590 years
α	radon-222	3.825 days
α	polonium-218	3.05 minutes
α	lead-214	26.8 minutes
β	bismuth-214	19.7 minutes
β	polonium-214	1.5×10^{-4} seconds
α	lead-210	22 years
β	bismuth-210	5 days
β	polonium-210	140 days
α	lead-206	stable

Figure 17 The uranium-238 decay chain

Time taken / half lives	Number of nuclei available to decay
0	$N = 100\%$
1	$\frac{N}{2} = 50\%$
2	$\frac{N}{4} = 25\%$
3	$\frac{N}{8} = 12.5\%$
4	$\frac{N}{16} = 6.25\%$
5	$\frac{N}{32} = 3.125\%$
etc	etc

There are many good simulations that show how random processes can result in an exponential relationship. For example, the planetqhe site (www.planetqhe.com) provides a simulation of radioactive decay (see link under "compound events I"). Analyse the data provided by the spreadsheet to see how well it fits exponential decay.

Of the 1600 or so different nuclides that either exist or have been created, only about 370 can be called stable, and of these there are thought to be about 100 that can still decay but have half-lives greater than 10^{10} years. No element with an atomic number greater than 82 (lead) has any naturally occurring stable isotopes.

An important consideration is the natural relative abundance of different nuclides. For example, there are eight possible nuclides of carbon ranging from carbon-9 to carbon-16. The vast majority of naturally occurring carbon is one of the two isotopes carbon-12 (98.9%) or carbon-13 (1.1%): both of these are stable. A tiny fraction (approximately $10^{-10}\%$) also exists as carbon-14, which undergoes beta-decay with a half-life of 5700 years. Analysis of this tiny percentage that is radioactive can be used in the process of **carbon dating**, which can be used to estimate the date of some historical items.

A neutron is only relatively stable inside a nucleus; outside this situation, it decays into a proton with a half-life of 889 seconds. Electrons, protons, and photons seem to be stable, though some theories predict mechanisms by which the proton might decay. Recent experiments in Japan seem to have put a lower limit of 10^{35} years on any half-life for the proton.

Predicting the relative stability of a particular nuclide depends on the interactions that are taking place within the nucleus: these are complex, and involve interactions between the protons and neutrons. However, we can generalize the types of decay that are likely to take place by comparing the numbers of protons and neutrons in stable nuclides.

Small nuclei seem to be stable with equal numbers of protons and neutrons. For example, oxygen-16 is stable, and has 8 protons and 8 neutrons. As the proton number increases, stable nuclei tend to have a proportionately larger number of neutrons. For example, lead-208, which is stable, contains 82 protons and 126 neutrons. Graphically this can be seen by plotting the number of neutrons against the number of protons for all the stable nuclides that exist (Figure 18).

Define the following terms as precisely as possible. If there are any terms that you do not know, do some quick research to find out what they mean:

a) nucleus
b) nuclide
c) atomic number Z
d) mass number A
e) nucleon number
f) neutron number
g) alpha decay
h) beta decay
i) gamma decay
j) half-life.

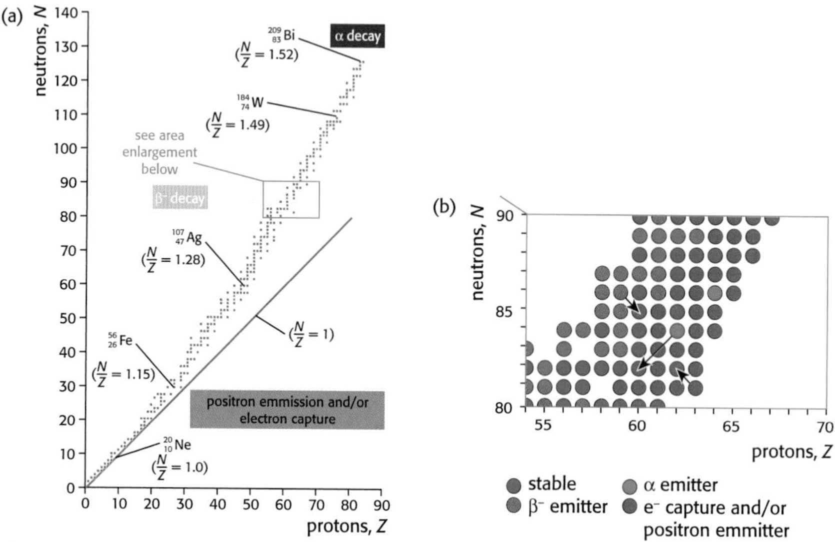

Figure 18 (a) Graph of neutrons against protons for non-radioactive isotopes, showing the band of stability. (b) Close-up of a section of the band of stability, showing the types of decay a radioactive isotope might undergo to become stable.

Nuclides that have a higher number of neutrons than those in the "band of stability" will tend to decay emitting a negative beta particle. Only large nuclei tend to undergo alpha decay.

Is energy conserved in radioactive decay? $E = mc^2$

Perhaps the most famous equation in all of physics is Einstein's mass–energy equivalence.

This most famous of formulae was originally proposed using different symbols: "If a body releases the energy L in the form of radiation, its mass is decreased by L/V^2" (Einstein 1905). In its current form, so long as you know the speed of light c (3×10^8 m s^{-1}), this equation allows you to calculate the amount of energy E in joules needed to create a given amount of mass m in kilograms, or the amount of energy released if matter is converted into energy.

Whatever form the equation is quoted in, it is simple enough to apply, but the concepts it introduces should be extremely worrying. The laws of conversation of mass and conservation of energy seem no longer to apply.

Conservation of mass–energy

Rather than think in terms of the conservation of mass, or the conservation of energy, we can now think in terms of the conservation of mass–energy. There is a **binding energy** associated with every nucleus. Whenever bonds are formed, be they chemical or nuclear, energy is released, and this energy must come from somewhere. The mass of any nucleus is less than the sum of the masses of the protons and neutrons that are contained within the nucleus. The difference between these two masses is called the **mass defect**. Whenever a nuclear reaction takes place (fission, fusion, or artificial transmutation), any energy change that takes place must be understandable in terms of the conservation of mass–energy.

The following list highlights the orders of magnitude of some of the possible mass changes that take place:

- The mass of a single carbon nucleus is 1.992×10^{-26} kg, whereas the total mass of six individual protons and six individual neutrons is 2.009×10^{-26} kg.
- Uranium-238 radioactively decays into thorium-234 with the release of an alpha particle. The mass of the uranium nucleus is 7.614×10^{-30} kg greater than the combined mass of the thorium nucleus and the alpha particle.
- In order to create an electron–positron pair, a minimum of 1.635×10^{-13} J are required.
- The Sun radiates energy from its surface at a rate of 3.8×10^{26} W. Its temperature is constant, so this is the rate of energy release as a result of hydrogen fusion, which equates to a mass loss of 4.2×10^{9} kg s^{-1}.

1 For each of the first two bullet points, calculate the energy associated with the quoted mass difference.

2 For each of the last two bullet points, deduce the mass associated with the quoted energy change.

Mass units

The SI unit of mass is the kilogram, but this is a very large unit when considering particle interactions. In order to simplify comparisons on the atomic scale, the unified atomic mass unit (u) is often used. One unified atomic mass unit is defined as one-twelfth of the mass of a carbon-12 atom. A proton and a neutron at rest both have an approximate mass of 1 u. An object's overall mass increases whenever it gains energy and decreases whenever it loses energy. In order to break a carbon-12 nucleus up into its constituent parts energy would need to be added, and so the proton and the neutron have masses slightly larger than 1 u even after the masses of the electrons have been taken into consideration:

Proton rest mass = 1.007 276 u

Neutron rest mass = 1.008 665 u

Electron rest mass = 0.000 549 u

It is very common to find atomic mass units used in tables comparing different isotopic masses. Remember that most tables quote the overall *atomic* mass (i.e. the mass of the atom including the electrons). In order to work out individual nuclear masses we need to subtract the appropriate number of electron masses from the total given in the table.

All physics equations must balance in terms of units as well as numbers. Einstein's mass–energy equivalence equation is no exception, and it can be used to define new units for mass that are often more suitable to use when considering particle interactions.

$$\text{units of } m = \text{units of } \frac{E}{c^2}$$

A common unit of energy at the atomic scale is the electronvolt (eV) or its SI multiples keV or MeV. 1 MeV = 1.6×10^{-16} J. This unit, the mega-electronvolt, is used to define a unit for mass: MeVc^{-2}, which includes the speed of light squared.

$$\text{units of } m = \frac{\text{units of } E}{c^2} = \text{MeV}\, c^{-2}$$

Using these units, an electron has a mass of 0.511 MeVc^{-2}, and thus an electron–positron pair has a total mass of 1.022 MeVc^{-2}, which means that an energy of 1.022 MeV is needed to create an electron–positron pair. In general, an energy of x MeV is needed to create a particle with rest mass = x MeVc^{-2}.

Data-based question: Graph of binding energy per nucleon

In order to break up a nucleus, a specific amount of energy needs to be provided. There is a binding energy associated with different nuclides. The larger the nucleus, the bigger the total binding energy. The relative stability of different nuclides can be compared by calculating the binding energy per nucleon for each nuclide. The resulting graph is shown in Figure 19. The nucleus of iron-56 turns out to have the most binding energy per nucleon, is radioactively stable, and does not decay. Remember that this graph is comparing different nuclides; it cannot be used to compare the chemical stability of different elements.

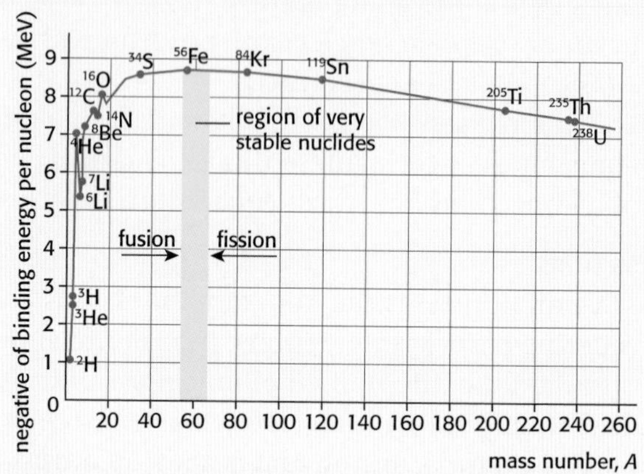

Figure 19 The stability of different nuclides

For each of the following situations, take readings from the graph in Figure 19 to quantitatively explain the energy changes involved. You should start by writing the appropriate nuclear equation in each case.

a) In the Sun, the fusion of different isotopes of hydrogen to form helium releases energy. The three-stage process involves

- the fusion of two protons to form deuterium (hydrogen-2)
- the fusion of deuterium and a proton to form helium-3
- the fusion of two helium-3 nuclei to form a helium-4 nucleus with the emission of two protons.

b) In a red giant star other possible reactions include the fusion of:

- two helium-4 nuclei to form beryllium-8
- beryllium-8 with another helium-4 nucleus to form carbon-12
- carbon-12 with another helium-4 nucleus to form oxygen-16.

c) When a red giant star stops shinning, its core is iron.

d) Uranium-238 decays naturally into thorium-235 with the emission of an alpha particle.

e) If bombarded with a neutron, uranium-235 can break up to form barium-141, krypton-92, and three other neutrons. This reaction can produce a chain reaction.

f) If bombarded with a neutron, uranium-235 does not completely break up to form helium nuclei and/or other neutrons.

Thinking about science: *What is energy?*

Energy is a fundamental concept in physics. It is not something physical that can be studied, but rather an abstract idea that arises from the mathematical principles that seem to govern all physical changes. There are many situations in which we can use a formula for calculating the energy before and after a physical change, and to date no exception has ever been found in which the total energy before a change has not equalled the total energy after the change. Energy considerations allow us to analyse and predict the outcome of physical changes, but they do not necessarily help us understand the mechanisms by which these changes take place.

Over time, energy formulae have been developed to acknowledge that energy can take different forms (gravitational potential energy, kinetic energy, thermal energy, elastic energy, electrical energy, chemical energy, radiant energy, nuclear energy, etc.) but the principle of energy conservation has always held.

For example, when an electrical heater is placed in liquid water, connected to an appropriate battery and the current is switched on, the temperature of the water increases. The processes involved at the microscopic level are complex. From the energy perspective, chemical energy in the battery has been converted (via electrical energy) into thermal energy in the water. Equations have been developed that relate the increase in thermal energy in the water to its temperature change, and which relate the electrical energy to the current, potential difference, and time taken.

In a simple experiment to verify the law of conservation of energy, any "missing" energy can be explained in terms of energy being used to heat the wires in the electrical circuit, or energy that has been dissipated to the surroundings.

If the heater remains connected to its power source, the water's temperature will continue to increase until it gets to the boiling point. While the water is boiling, electrical energy is still being provided by the power source, but no temperature change is taking place. This does not mean that the law of conservation of energy has been broken, but rather that we need to modify the equations to include another form that is not immediately obvious. This form of energy is called latent heat ("latent" means hidden). The breaking of the intermolecular bonds in water uses energy, and the making of bonds will release the same amount of energy if the gaseous water turns back into liquid water later on. The law of conservation of energy applies to this situation so long as we include all the possible forms.

One of the very best descriptions of energy was given by the Nobel Prize winning physicist Richard Feynman (1918–1988), in his famous published series of undergraduate lectures given at the California Institute of Technology in the 1960s. (Feynman *et al.* 1963).

Figure 20 Richard Feynman (1918–1988)

In his analogy he imagines a child playing with toy building blocks. The mother of this naughty child develops an experimental law that helps her work out where the child has placed the blocks – whether they are locked inside his toy box (she can measure the change in the weight of the box) or thrown into the dirty water in the bath (she can measure the change in the height of the water). Her abstract mathematical formula will accurately predict where the blocks (representing the energy) have gone, but says nothing about the nature of the blocks themselves.

Einstein's formula does not contradict the law of conservation of energy; it just provides us with another factor that we need to include when considering possible energy changes – an object's mass. The formula represents the idea that mass and energy are equivalent: if the energy of an object changes, then so does its mass. In everyday chemical changes the energy changes involved are equivalent to only tiny changes in mass, so no such change is noticed. For example, 42 kJ of energy is required to increase the temperature of 1 kg of water by 10 °C. The mass change associated with this amount of energy is calculated as follows:

$$m = \frac{E}{c^2} = \frac{42000}{\left(3 \times 10^8\right)^2} = 4.7 \times 10^{-13} \text{ kg}$$

This is a tiny mass increase compared with the 1 kg, but it is very large compared with the mass of an individual water molecule (approximately 3×10^{-26} kg). In this context water molecules are not being created or destroyed, but the mass of each water molecule has increased slightly. Mass can no longer be imagined as just the amount of matter present in an object, but a mathematical quantity that is related to the total energy concentrated in an object (see page 201).

In all nuclear changes, the energies involved can be understood in terms of the mass changes that are taking place. In extreme situations (e.g. high-energy particle collisions) it is possible for new particles to be created. Other conservation laws (e.g. conservation of linear momentum, conservation of charge) can help explain the creation or annihilation that is observed.

Thinking about science: *The neutrino*

When a particular nuclide decays via α-decay, the kinetic energy of the α-particle that is emitted is always the same. Similarly, γ-decays always have fixed frequencies for any particular decay under consideration. In each case, the measured energy release associated with the decay is in agreement with the measured mass changes. β-decay is problematic, however, because for any particular decay, experimental measurements demonstrate that the emitted β-particle comes out with a range of

possible energies. The technical language describing this is that the β-spectrum is continuous.

For example, chlorine-36 decays into argon-36, emitting a β-particle. The β-particles have a range of kinetic energies from almost zero up to a maximum value of about 0.71 MeV.

Analysis of the masses of the particles involved in this decay predicts that all the β-particles should have the maximum energy of about 0.71 MeV. What is happening to the "missing" energy?

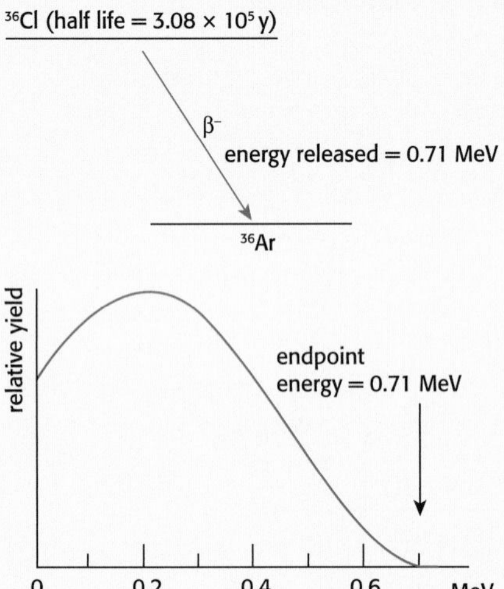

Figure 21 Decay of chlorine-36 and the resulting beta particle energy distribution

It is tempting to answer this question in the same way as most everyday energy calculations and state that the missing energy must have gone into thermal energy ("heat") and/or sound, but this answer is not appropriate here.

Remember that we are talking about a nuclear decay. On this scale, thermal heat is just the random kinetic energy of the atoms involved, and the only particles that concern us are a chorine nucleus, an argon nucleus, and an electron. We are considering the kinetic energy of each of these particles; there is no such thing as extra "heat". Sound is also the movement of atoms or molecules and thus is not an appropriate answer either. No gamma rays are emitted (nor is there any other electromagnetic radiation), so there does not appear to be any mechanism by which this energy could have become "lost".

Data books rarely give values for atomic nuclei as the original data come from measurements with ions. Atomic masses can be used to calculate nuclear masses by subtracting the mass of the appropriate number of electrons. The following data apply to the β-decay of $^{36}_{17}$Cl.

Atomic mass of $^{36}_{17}$Cl = 35.968 307 u

Atomic mass of $^{36}_{18}$Ar = 35.967 545 u

Mass of $^{0}_{-1}\beta$ = 0.000 549 u

Mass defect

= mass of $^{36}_{17}$Cl nucleus – mass of ($^{36}_{18}$Ar nucleus and β-particle)

= (35.968 307 – 17 × 0.000 549) – (35.967 545 – 18 × 0.000 549) – 0.000 549 u

= 0.000 762 u

Energy released = 0.000 762 × 931.5 MeV

= 0.7098 MeV = 0.71 MeV

This energy will be shared between the argon atom and the β-particle. The share can be calculated by applying the law of conservation of momentum to the decay. Given the much larger mass of the daughter product, the argon will essentially be at rest if the chlorine particle was at rest when the decay took place. The energy of the decay should effectively all go into the kinetic energy of the β-particle.

Thus the mass defect predicts the kinetic energy of the emitted β-particle, but most have less energy than this. Some of the energy available cannot be accounted for. Does this mean that the conservation of energy – a firm tenet of all science for more than 400 years – is false? If so, a lot of physics will have to be rewritten. To quote from a lecture that Niels Bohr delivered in 1930:

> *At the present stage of atomic theory we have no argument, either empirical or theoretical, for upholding the energy principle in β-ray disintegrations, and are even led to complications and difficulties in trying to do so.*

In a now famous letter to a meeting of physicists at Tübingen University the Austrian-Swiss physicist Wolfgang Pauli wrote (in German) suggesting a desperate remedy to allow physics to hang on to the law of conservation of energy:

> *Dear Radioactive Ladies and Gentlemen,*
>
> *I beg you to listen most favourably to the carrier of this letter ... I have hit upon a desperate remedy to save ... the law of conservation of energy. Namely, the possibility that there could exist in the nuclei electrically neutral particles, that I wish to call*

neutrons, which … do not travel with the velocity of light. The mass of the neutrons should be of the same order of magnitude as the electron mass and in any event not larger than 0.01 proton masses. The continuous beta spectrum would then become understandable by the assumption that in beta decay a neutron is emitted in addition to the electron such that the sum of the energies of the neutron and the electron is constant …

I agree that my remedy could seem incredible because one should have seen those neutrons long ago if they really exist. But only he who dares wins and the serious situation, due to the continuous structure of the beta spectrum, is lightened by a remark of my honoured predecessor, Mr Debye, who recently said to me in Bruxelles: "Oh, it's much better not to think about this at all, just like the new taxes."

… Thus, dear radioactive people, look and judge. Unfortunately, I cannot appear in Tübingen personally since I am indispensable here in Zurich because of a ball on the night of 6/7 December.

With my best regards to you …

Your humble servant

W. Pauli

Pauli is suggesting is that there is another particle that is emitted in this decay. If a neutral particle of very low mass were to be emitted at the same time as the β-particle, then the two particles would share the energy that is available. The only problem is that this particle had not been detected by anybody at the time. The choice was between the particle not existing and the law of conservation of energy being wrong, or the particle existing but being virtually undetectable. At the time Pauli proposed calling it a neutron (the particle we now know as a neutron had not yet been discovered), but this name was changed a few years later to the **neutrino**. To be absolutely precise, the particle that is emitted at the same time as a β-particle is called an **anti-neutrino** (see pages 273–4 for a discussion of antimatter in general).

$${}^{36}_{17}\mathrm{Cl} \rightarrow {}^{36}_{18}\mathrm{Cl} + {}^{0}_{-1}\beta + \bar{\nu}$$

anti-neutrino

In order for this hypothesis to be plausible, the properties of the anti-neutrino read more like science fiction than agreed science fact. To be undetected it must not feel the strong nuclear force at all (electrons also do not feel the strong force), and the fact that it is electrically neutral means that it is also unaffected by an electromagnetic field. Zero, or close to zero, mass means that it is also unaffected by a gravitational field. With these three properties it can therefore pass through the whole Earth, and the chance of any interaction taking place is very small indeed.

The nuclear reactions happening in the Sun mean that millions upon millions of these particles arrive at every square metre of the surface of the Earth every second. They pass straight through all the buildings, everything else (including you), go all the way through our planet, come out on the other side and head off into space. In 1963 John Updike (an American novelist and poet) published his poem "Cosmic Gall" in his collection *Telephone Poles and Other Poems*:

Neutrinos, they are very small.
They have no charge and have no mass
And do not interact at all.
The earth is just a silly ball
To them, through which they simply pass,
Like dustmaids down a drafty hall
Or photons through a sheet of glass.
They snub the most exquisite gas,
Ignore the most substantial wall,
Cold shoulder steel and sounding brass,
Insult the stallion in his stall,
And, scorning barriers of class,
Infiltrate you and me! Like tall
And painless guillotines, they fall
Down through our heads into the grass.
At night, they enter at Nepal
And pierce the lover and his lass
From underneath the bed—you call
It wonderful; I call it crass.

Without any supporting experimental evidence, the neutrino hypothesis would have just remained an interesting or crass proposal (depending on your point of view). Indeed if neutrinos "do not interact at all", then their existence could be only an unverifiable hypothesis. Luckily, Updike's third line is not scientifically accurate, and neutrinos do interact, but only rarely. Nowadays it is possible for these chargeless and virtually massless particles to be detected occasionally.

In 1987 an extraordinary piece of luck along with careful measurements convinced many scientists of the existence of neutrinos. On 23 February that year a supernova (SN1987a) was observed in the Large Magellanic Cloud.

One hundred and seventy thousand years ago a blue giant star (50 times larger than our Sun), Sanduleak-69 202, exploded. The light had been

travelling through space since the explosion and started to arrive on Earth in February 1987.

Supernovae that are visible to the naked eye are extremely rare events in our galaxy, and this was the first such event to take place since 1604. Galileo built and made observations with his new instrument, the telescope, in 1609, so this supernova was the first "local" one to be observed since stars had been observed with telescopes.

The explosion also released a huge number of neutrinos (10^{56}) in a couple of seconds, and these had also been travelling through space for thousands of years. Two neutrino detectors (neither had originally been designed to look for cosmic neutrinos) had both recently undergone refurbishments that allowed them to be able to detect the extra burst of activity.

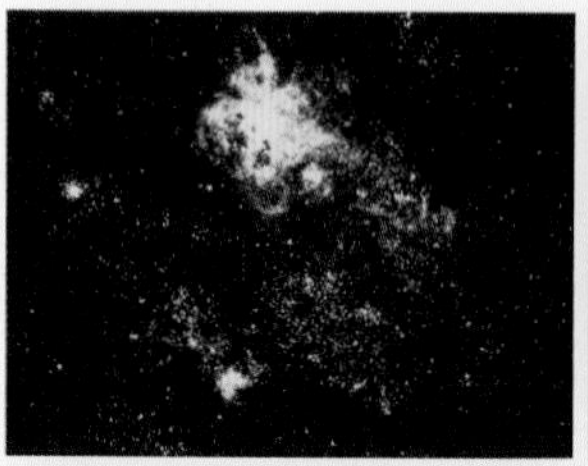

Figure 22 SN1987a

One detector was the Kamioka Underground Observatory, located in Kamioka-cho, Gifu, Japan. In 1985 the detector (Kamiokande) was upgraded so as to be able to observe neutrinos of cosmic origin. The other neutrino detector, the IMB (Irvine–Michigan–Brookhaven) experiment is located in Mentor, Ohio, USA. Its refurbishment was completed in September 1986.

At 07:35:41 GMT on 23 February 1987 each detector received a burst of approximately 10^{16} neutrons passing through it. As a result, a grand total of 19 neutrino interactions were recorded over 13 seconds. Eight were recorded in the IMB experiment and 11 in Kamiokande. The linking of these observations to the detailed theories of the nuclear reactions taking place inside a supernova provides strong evidence that these ephemeral particles really do exist.

According to Richard Feynman (see page 249):

> *The principle of science, the definition, almost, is the following:* The test of all knowledge is experiment. *Experiment is the* sole judge *of scientific "truth".*
>
> Feynman et al. 1963, p. 1-1.

The process of how scientific ideas are modified and developed was analysed in an important work, *The Structure of Scientific Revolutions*, which was published by Thomas Kuhn in 1962. The book aims to analyse the history of science. Its publication was a landmark event in the sociology of knowledge.

1. Have you been persuaded that neutrinos exist? Explain your beliefs as succinctly as possible to somebody who does not study physics.
2. How would you answer John Updike?
3. Why is a total of 19 events out of 10^{16} seen as statistically significant?
4. Research what is meant by the term "paradigm shift" as used by Thomas Kuhn. Is the discovery of neutrinos such a shift?
5. The fictional character Sherlock Holmes remarked to his assistant Dr Watson: "How often have I said to you that when you have eliminated the impossible, whatever remains, *however improbable*, must be the truth" (Conan Doyle 1890). How far does this sentence apply to the discovery of the neutrino? Is it possible that the law of conservation of energy will be shown to be incorrect at some time in the future?

Physics issues: tracking particles

Subatomic particles are too small to be seen with an optical microscope, but there are many techniques that can be used to work out where they used to be. The most common technique for visualizing a particle track is to use the fact that many of these particles cause ionizations. As a moving particle passes through a medium, it can cause a neutral atom to lose an electron and thus become a positive ion. All the ionizations caused by a subatomic particle can be used to find out where the particle has been.

For example, one technique arranges for a volatile liquid to boil. The bubbles that form do so around the ions that the particle has created, and so the ionization tracks become visible as a trail of bubbles.

Unfortunately, neutral particles rarely cause ionizations, so their paths cannot be seen in many detectors, but their existence can often be inferred from the other tracks that *do* exist.

A common technique used to identify the various particles is to ensure that a constant magnetic field is acting all the time while the particles are being tracked. As a result of the magnetic force that acts on moving charged particles, a positive particle will curve in one direction and a negative particle will curve in the opposite direction. The angle of curvature depends on the particle's charge, mass, and velocity. Detailed measurements of the tracks can be used to calculate the ratio of the particle's charge to its mass, which often allows the particle to be identified.

Figure 23 Fermilab bubble chamber: 4.6 m in diameter in a 3 T magnetic field

Physics issues: background radiation, biological effects, and risk

Radioactive decay is a natural phenomenon, and is going on around you all the time. The activity of any given source can be measured in terms of the number of individual nuclear decays that take place in a unit of time. This information is quoted in **becquerels** (Bq).

1 Bq = 1 nuclear decay per second

Experimentally, this would be measured using a **Geiger counter**, which detects and counts the number of ionizations taking place inside the **GM tube** (Figure 24). Some Geiger counters display only the total count in a given time, whereas others can display the measured count rate directly.

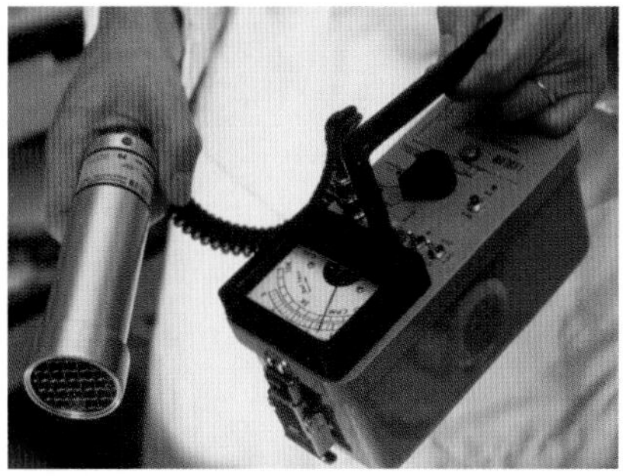

Figure 24 Geiger counter and GM tube

A working Geiger counter will always detect some radioactive ionizations taking place even when there is no identified radioactive source: there is a **background count** as a result of the **background radiation**. A reading of 30 counts per minute, which corresponds to the detector registering 30 ionizing events, would not be unusual. Some cosmic gamma rays will be responsible, but there will also be alpha, beta, and gamma radiation received by the detector as a result of radioactive decays that are taking place in the surrounding materials. The pie chart in Figure 25 identifies typical sources of background radiation, but remember that the value of background radiation will vary from country to country and from place to place.

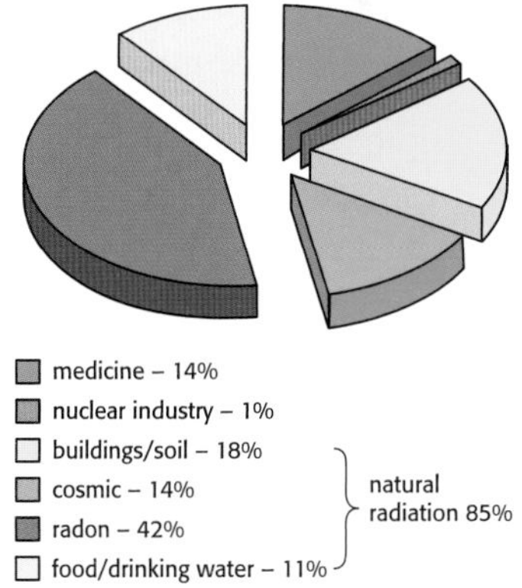

Figure 25 Typical sources of background radiation

Most people are concerned about any exposure and want to know whether the background radiation where they live is safe. Unfortunately, the process of estimating the potential risk that results from exposure to ionizing radiation is complex, and full of uncertainties.

One aspect that contributes to the difficulty in making predictions is that many of the effects of radiation are **stochastic**. This means that a greater dose will, in general, be more likely to have a detrimental effect, but there is an underlying randomness in the process. Given this randomness, it is impossible to assume that there is a minimum dose that is safe. Typically effects (such as cancer) can develop many years after the exposure that caused them. There is an analogy here with smoking. Some of the effects of smoking are stochastic, and although some smokers manage to avoid developing lung cancer, there is a direct correlation between the number of cigarettes that have been smoked and the likelihood that cancer will develop.

One aspect that potentially adds to the fear of ionizing radiation is its invisible nature, and thus the possibility of individuals being harmed without realizing anything has happened. The total energy associated with a harmful radioactive dose is very small: for example, a lethal dose might transfer only about 750 J to an individual. This same energy, if given to a mug of water, would cause a temperature rise of much less than one degree Celsius. How can this tiny amount of energy be so harmful?

Much of the biological damage caused by radiation can be related to DNA damage as a result of the ionizations. DNA is a long and complex molecule, and the biochemistry of its interactions with other chemicals will be affected by the breaking or making of bonds, which can result from the ionizations caused. The viability of any cell in our bodies depends on the DNA that it contains. A typical cell has the structure shown in Figure 26.

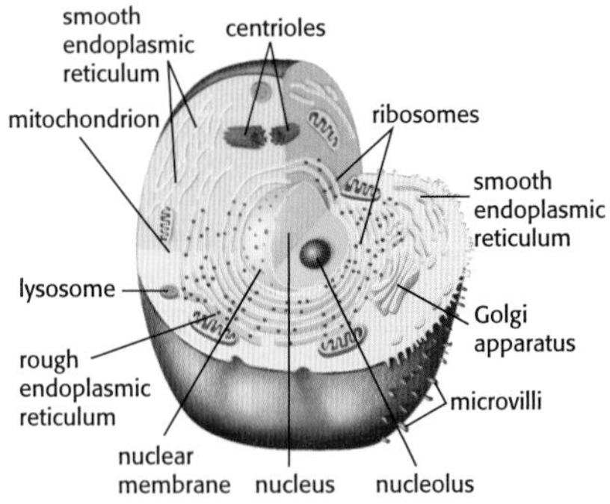

Figure 26 A typical human cell

When incoming radiation causes an ionization in the cell nucleus it may cause damage to the DNA, called DNA mutation (Figure 27).

To put this into perspective, there are approximately 10^{14} cells in a human. Naturally occurring background radiation means that ionizations are occurring all the time. On average there will be one ionization in the DNA molecule in every cell in your body every year. The human body has impressive processes for dealing with this damage. In general, if a mutation occurs, there are three possible outcomes, depending on the amount of damage: repair, cell death, or cell mutation.

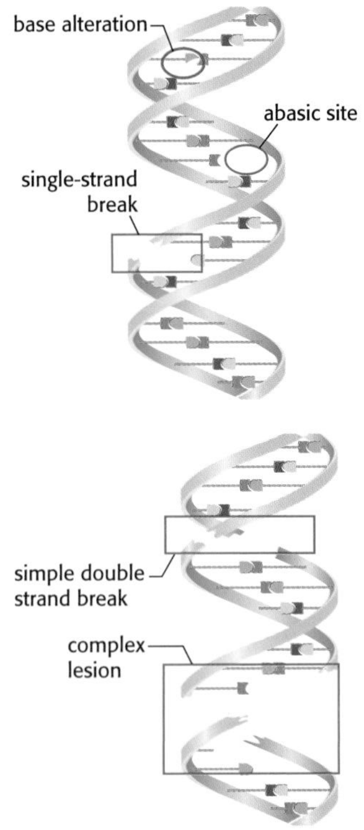

Figure 27 Possible DNA damage

A simple break, or **lesion**, is not usually a problem; the vast majority of these lesions are repaired by the body's own biochemical processes, and the cell can survive. Many cells cannot be repaired, and die, but this is not a problem, because millions of cells die every day in every human. The third possibility is that the cell continues to survive, but the damage is not repaired, or is incorrectly repaired. Some cells continue to survive in their damaged state. If damaged cells continue to multiply in an uncontrolled way then it is called cancer.

At very high doses a large number of cells would be damaged and die, and this would affect the ability of an organ or of the whole body to be able to survive. Some types of cell are more susceptible to damage than others. At lower doses it may take some time for effects to become noticeable. The list below indicates the likely effects of a range of whole-body radiation doses and dose rates on individuals.

The unit used to compare radioactive doses is the **sievert** (Sv). Technically a quantity called the **dose equivalent** is being measured. This takes into account the fact that different amounts of damage will result from different types of radiation, even if the total energy received per unit mass is the same, see Biomedical physics option, pages 455–6.

- **10 000 mSv** (10 sieverts) as a short-term and whole-body dose would cause immediate illness, such as nausea and decreased white blood cell count, and subsequent death within a few weeks. Between 2 and 10 sieverts in a short-term dose would cause severe radiation sickness, with increasing likelihood that this would be fatal.
- **1000 mSv** (1 sievert) in a short-term dose is about the threshold for causing immediate radiation sickness in a person of average physical attributes, but would be unlikely to cause death. Above 1000 mSv, severity of illness increases with dose.
  Doses greater than 1000 mSv occurring over a long period are less likely to have early health effects, but they create a definite risk that cancer will develop many years later. The estimated risk of fatal cancer is 5 of every 100 persons exposed to a dose of 1000 mSv (i.e. if the normal incidence of fatal cancer were 25%, this dose would increase it to 30%).
- Above about **100 mSv** the probability of cancer (rather than the severity of illness) increases with dose.
- **50 mSv** is the lowest dose at which there is any evidence of cancer being caused in adults. It is also the highest dose that is allowed by regulation in any one year of occupational exposure. Dose rates greater than 50 mSv/yr arise from natural background levels in several parts of the world but do not cause any discernible harm to local populations.
- **20 mSv/yr** averaged over 5 years is the limit for radiological personnel such as employees in the nuclear industry, uranium or mineral sands miners, and hospital workers (who are all closely monitored).
- **10 mSv/yr** is the maximum actual dose rate received by any Australian uranium miner.
- **3–5 mSv/yr** is the typical dose rate (above background) received by uranium miners in Australia and Canada.
- **3 mSv/yr** (approx.) is the typical background radiation from natural sources in North America, including an average of almost 2 mSv/yr from radon in air.
- **2 mSv/yr** (approx.) is the typical background radiation from natural sources, including an average of 0.7 mSv/yr from radon in air. This is close to the minimum dose received by all humans anywhere on Earth.
- **0.3–0.6 mSv/yr** is a typical range of dose rates from artificial sources of radiation, mostly medical.
- **0.05 mSv/yr**, a very small fraction of natural background radiation, is the design target for maximum radiation at the perimeter fence of a nuclear electricity generating station. In practice the actual dose is less.

The chance of developing a fatal cancer from a dose of 1 Sv is rated at about 5% or 0.05 (5 people in 100 would die as a result of this dose). The risk associated with smaller doses is assumed to be proportional. A typical dose from a lung investigation involving technetium-99 might be about 1 mSv: thus the associated risk is 0.005% or 5×10^{-5}. This means that 5 fatalities would be expected in 100 000 people exposed to this dose.

Many treatments involve much lower doses and the calculated risk may well be lower, e.g. 1×10^{-6} (one in a million). It is extremely hard to put this level of risk into context and assess how concerned one should be. Mathematically, the following activities all have a fatality risk associated with them (not, of course, as a result of radioactive decay) of approximately one in a million:

- smoking 1.4 cigarettes
- living 2 days in a polluted city
- travelling 6 min in a canoe
- 1.5 min mountaineering
- travelling 480 km in a car
- travelling 1600 km in an airplane
- living 2 months together with a smoker.

The technetium-99 lung investigation mentioned above is thus 50 times more dangerous than each of the above.

There is no such thing as a safe dose. Radioisotopes have many medical applications, both in terms of diagnosis and in terms of treatment, and the risk of not treating a particular condition needs to be weighed against any extra risk that involves ionizing radiation. In general, all additional exposure needs to be as low as can be reasonably achieved, and needs to show a positive overall benefit to the patient.

In order to make informed decisions, it is important to understand the risks and the benefits of any proposed procedure, and to compare these with everyday risks. Sometimes our own perceptions of the risks are widely different from the statistical analysis of the data. For example, Professor Gerd Gigerenzer of the Max Planck Institute in Berlin has analysed the road accident statistics from the US Department of Transportation following the appalling events of 11 September 2001. In the months that followed, many people chose to drive rather than fly, and thus exposed themselves to a greater risk of being involved in a car accident. Professor Gigerenzer's analysis (Gigerenzer 2004) suggests that the number of Americans who lost their lives on the roads by avoiding the risk of flying was higher than the total number of passengers killed on the four fatal flights.

1 Radioactive decay is a *random* process. This means that

A a radioactive sample will decay continuously.
B some nuclei will decay faster than others.
C it cannot be predicted how much energy will be released.
D it cannot be predicted when a particular nucleus will decay.

2 An isotope of radium has a half-life of 4 days. A freshly prepared sample of this isotope contains N atoms. The time taken for $7N/8$ of the atoms of this isotope to decay is

A 32 days. **C** 12 days.
B 16 days. **D** 8 days.

3 A nucleus of the isotope xenon, Xe-131, is produced when a nucleus of the radioactive isotope iodine I-131 decays.

a) Fill in the boxes below in order to complete the nuclear reaction equation for this decay.

$$^{131}_{\square}\text{I} \rightarrow {}^{131}_{54}\text{Xe} + \beta^- + \square$$ [2]

The activity A of a freshly prepared sample of the iodine isotope is 6.4×10^5 Bq and its half-life is 8.0 days.

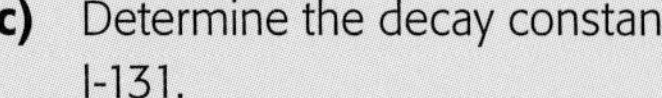

b) Draw a graph to illustrate the decay of this sample. [3]

c) Determine the decay constant of the isotope I-131. [2]

HL

The sample is to be used to treat a growth in the thyroid of a patient. The isotope should not be used until its activity is equal to 0.5×10^5 Bq.

d) Estimate the time it takes for the activity of a freshly prepared sample to be reduced to an activity of 0.5×10^5 Bq. [2]

(Total 11 marks)

Artificial transmutations

Is it possible to make a particular atom become radioactive? Strawberries are sometimes exposed to gamma radiation so they stay fresh for longer, but this does not make them become radioactive. It is possible to contaminate a substance with radioactive material, but this just means that atoms that are already radioactive have been mixed with those that were originally present.

For example, as a result of the 1986 accident at Chernobyl in the Ukraine, restrictions were placed on the movement and sale of sheep farmed in Wales (2700 km away). The aim of the restrictions was to prevent sheep with high levels of radioactive caesium-137 (caused by eating contaminated grass) to enter the human food chain. The European agreed safe maximum for meat is 1000 Bq kg^{-1}. Individual sheep within defined restricted areas were monitored, and the resulting measured count rate converted to an estimated count rate per mass. As radioactive decay is a random process, a working limit of 645 Bq kg^{-1} was applied. In practice this means that the chance of a sheep giving a "false low" one-off reading is only 1 in 40. Any sheep that exceeded this limit had to remain within the restricted area, and could not be used for human consumption. Immediately after the accident the restrictions and monitoring affected two million sheep in Wales. By 2006, because of the reduction in background (caesium-137 has a half-life of 28 years) the number of sheep at risk had fallen to 180 000.

Note that no atoms in the sheep had been turned radioactive; it is just that the sheep's diet means that they had potentially ingested radioactive material, which then became part of its body mass. Your own body contains some radioactive elements, which contribute to the average background count.

Under certain circumstances it *is* possible for nuclear reactions to take place that alter a previously stable nuclide into an unstable one. Typically the reaction involves the collision between an individual nucleus and something else – e.g. a neutron, an alpha particle, or even (in the case of nuclear fission) another nucleus. In this situation a successful reaction can be thought of as one in which the incoming particle interacts with the nucleus to form an intermediate state, which then decays into two or more fragments. Overall, the atom of one element has been changed into another element, and the reaction is known as an **artificial transmutation**. One model of how this reaction takes place is to imagine the nucleus behaving like a drop of liquid (Figure 28).

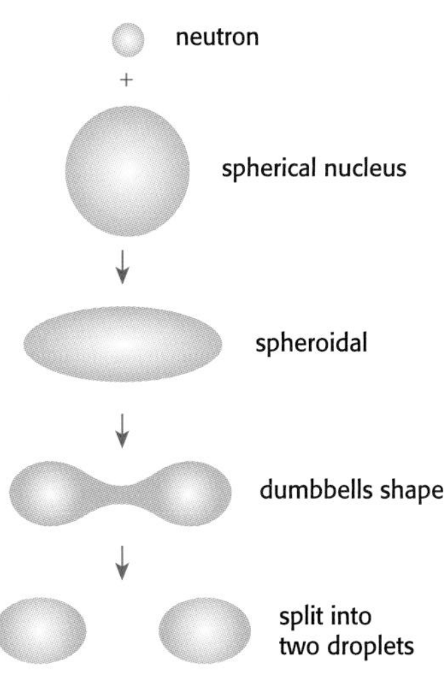

Figure 28 Liquid drop model of the fission mechanism

The first artificial transmutation reaction was achieved by Rutherford in 1919. Nitrogen nuclei that are bombarded by alpha particles form oxygen nuclei:

$$^{14}_{7}N + ^{4}_{2}\alpha \rightarrow ^{17}_{8}O + ^{1}_{1}H$$

The discovery of the neutron is another example, as is the physics of nuclear fission reactors (page 333), and the nuclear fusion that is taking place in our Sun.

References

Doyle, A.C. 1890. *The Sign of Four*. London. Spencer Blackett.

Einstein, A. 1905. "Is the inertia of a body dependent on its energy content?" *Annalen Der Physik*. Vol 18, number 13. pp 639–641.

Feynman, R.P., Leighton, R.B. and Sands, M. 1963. *The Feynman Lectures on Physics*. Vol 1, section 4–1. Reading, MA. Addison-Wesley.

Gigerenzer, G. 2004. "Dread risk, September 11, and fatal traffic accidents". *Psychological Science*. Vol 15, number 4. pp 286–287.

Kuhn, T. *The Structure of Scientific Revolutions*. Chicago. University of Chicago Press.

Updike, J. 1963. *Telephone Poles and Other Poems*. New York. Knopf.

End of chapter summary

Chapter 16 has three main themes: the atom, radioactive decay and nuclear reactions. The list below summarises the knowledge and skills that you should be able to undertake after having studied this chapter. Further research into more detailed explanations using other resources and/or further practice at solving problems in all these topics is recommended – particularly the items in bold.

The atom (atomic structure, nuclear structure)

- Describe a model of the atom that features a small nucleus surrounded by electrons and outline one limitation of the simple nuclear model.
- Outline the evidence that supports a nuclear model of the atom and for the existence of atomic energy levels.

Radioactive decay (radioactivity, half-life)

- Describe the phenomenon of natural radioactive decay and explain why some nuclei are stable whereas others are unstable.
- Describe the properties of α and β particles and γ radiation and outline the biological effects of ionizing radiation.
- State that radioactive decay is a random and spontaneous process, and that the rate of decay decreases exponentially with time.
- Define the term *radioactive half – life* and determine its value from a decay curve.
- **Solve radioactive decay problems** involving integral numbers of half-lives.

Nuclear reactions (fission and fusion)

- Describe and give an example of an artificial (induced) transmutation **by constructing and completing nuclear equations**.
- Apply the Einstein mass – energy equivalence relationship.
- Define the terms *unified atomic mass unit, mass defect, binding energy, electronvolt and binding energy per nucleon* **and solve problems involving mass defect and binding energy**.
- Draw and annotate a graph showing the variation with nucleon number of the binding energy per nucleon, and apply the graph to account for the energy release in the processes of fission and fusion.
- State that nuclear fusion is the main source of the Sun's energy.
- Describe and **solve problems** involving the processes of nuclear fission and nuclear fusion.

Chapter 16 questions

1 This question is about nuclear binding energy.

a) i) Define *nucleon.* [1]

ii) Define *nuclear binding energy of a nucleus.* [1]

Draw axes to show values of nucleon number A on the horizontal axis (from 0 to 250) and average binding energy per nucleon E on the vertical axis (from 0 to 9 MeV). Binding energy is taken to be a positive quantity.

b) Mark on the E axis, the approximate position of

i) the isotope $^{56}_{26}$Fe (label this F). [1]

ii) the isotope $^{2}_{1}$H (label this H). [1]

iii) the isotope $^{238}_{92}$U (label this U). [1]

c) Using the grid in part (a), draw a graph to show the variation with nucleon number A of the average binding energy per nucleon E. [2]

d) Use the following data to deduce that the binding energy per nucleon of the isotope $^{3}_{2}$He is 2.2 MeV.

nuclear mass of $^{3}_{2}$He = 3.01603 u

mass of proton = 1.00728 u

mass of neutron = 1.00867 u [3]

In the nuclear reaction $^{2}_{1}\text{H} + ^{2}_{1}\text{H} \rightarrow ^{3}_{2}\text{He} + ^{1}_{0}\text{n}$ energy is released.

e) i) State the name of this type of reaction. [1]

ii) Use your graph in (c) to explain why energy is released in this reaction. [2]

(Total 13 marks)

2 This question is about radioactivity and nuclear energy.

a) Define the following terms,

i) *Isotope* [1]

ii) *Radioactive half-life* [1]

Thorium-227 (Th-227) results from the decay of the isotope actinium-227.

b) i) Complete the following reaction equation.

$^{227}_{89}\text{Ac} \rightarrow ^{227}_{90}\text{Th} +$ [1]

Th-227 has a half-life of 18 days and undergoes α-decay to the isotope Ra-223 (Ra-223). A sample of Th-227 has an initial activity of 32 arbitrary units.

ii) Draw a graph to show the variation with time t (for t = 0 to t = 72 days) of the activity A of Th-227. [2]

iii) Determine from your graph, the activity of thorium after 50 days. [1]

iv) Outline the experimental procedure to measure the activity of Th-227. [2]

In the decay of a Th-227 nucleus, a γ-ray photon is also emitted.

c) Use the following data to deduce that the energy of the γ-ray photon is 0.667 MeV.

mass of Th-227 nucleus = 227.0278 u

mass of Ra-223 nucleus = 223.0186 u

mass of helium nucleus = 4.0026 u

energy of α-particle emitted = 5.481 MeV

unified atomic mass unit (u) = 931.5 MeV c^{-2}

You may assume that the Th-227 nucleus is stationary before decay and that the Ra-223 nucleus has negligible kinetic energy. [3]

(Total 11 marks)

3 This question is about radioactive decay.

A nucleus of the isotope xenon, Xe-131, is produced when a nucleus of the radioactive isotope iodine I-131 decays.

a) Explain the term *isotopes*. [2]

b) Fill in the boxes below in order to complete the nuclear reaction equation for this decay. [2]

$^{131}_{\square}\text{I} \rightarrow ^{131}_{54}\text{Xe} + \beta^{-} + \square$

c) The activity A of a freshly prepared sample of the iodine isotope is 3.2×10^{5} Bq. The variation of the activity A with time t is shown below.

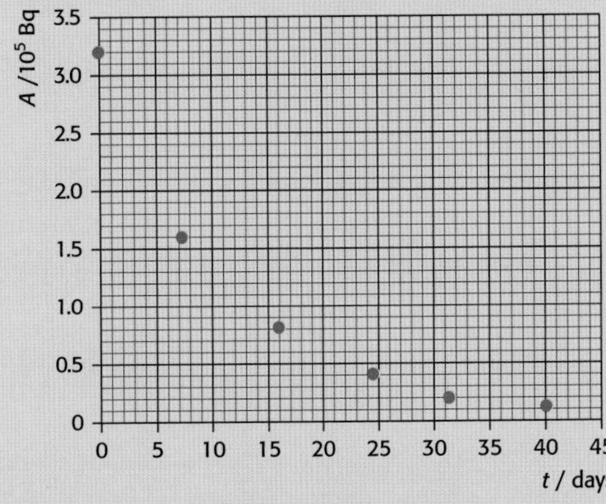

Draw a best-fit line for the data points. [1]

d) Use the graph to estimate the half-life of I-131. [1]

(Total 6 marks)

4 This question is about nuclear binding energy.

The table below gives the mass defect per nucleon of deuterium (^2_1H) and helium-4 (^4_2He).

	Mass defect per nucleon / u
(^2_1H)	0.00120
(^4_2He)	0.00760

a) Explain the term *mass defect.* [2]

b) Calculate the energy, in joule, that is released when two deuterium nuclei fuse to form a helium-4 nucleus. [4]

(Total 6 marks)

5 This question is about nuclear reactions.

a) State the meaning of the terms

i) nuclide [2]

ii) isotope [1]

b) The isotope sodium-24 undergoes radioactive decay to the stable isotope magnesium-24.

i) Complete the nuclear reaction equation for this decay. [2]

$$^{24}_{11}Na \rightarrow ^{24}_{12}Mg$$

ii) One of the particles emitted in the decay has zero rest-mass. Use the data below to estimate the rest mass, in atomic mass units, of the other particle emitted in the decay of $^{24}_{11}Na$

rest mass of $^{24}_{11}Na = 23.99096u$

rest mass of $^{24}_{12}Mg = 23.98504u$

energy released in decay = 5.002160 MeV [3]

c) The isotope sodium-24 is radioactive but the isotope sodium-23 is stable. Suggest which of these isotopes has the greater nuclear binding energy. [2]

(Total 10 marks)

6 Radioactive decay

a) Carbon-14 is a radioactive isotope with a half-life of 5500 years. It is produced in the atmosphere by neutron bombardment of nitrogen. The equation for this reaction is

$$^{14}_{7}N + ^{1}_{0}n \rightarrow ^{14}_{6}C + X.$$

i) Explain what are meant by *isotopes.* [1]

ii) Define the term *radioactive half-life.* [1]

iii) Identify the particle X. [1]

b) Living trees contain atoms of carbon-14. The activity per gram of carbon from a living tree is higher than that per gram of carbon-14 from burnt wood (charcoal) found at an ancient campsite.

i) A living tree continuously takes in carbon dioxide from the atmosphere. Suggest why the activity of the carbon from the charcoal is less than that of the living wood. [3]

ii) Each gram of a living tree contains approximately 1×10^{-12} g of the isotope carbon-14. Deduce that each gram of carbon in living wood contains approximately 4×10^{10} atoms of carbon-14. [2]

c) Draw a graph to show the variation with time of the number of carbon-14 atoms in one gram of wood from a tree. Your graph should indicate the number of atoms for a period of 1.8×10^4 years after the tree has died. (Half-life of carbon-14 = 5500 years) [3]

d) The activity of a radioactive sample is proportional to the number of atoms in the sample. The activity per gram of carbon from a living tree is 9.6 disintegrations per minute. The activity per gram of carbon in burnt wood found at the ancient campsite is 1.9 disintegrations per minute.

i) Estimate the number of atoms of carbon-14 in the burnt wood. [1]

ii) From the graph you have drawn in (c), estimate the age of the burnt wood. [1]

(Total 13 marks)

7 Radioactivity

One isotope of potassium is potassium-42 ($^{42}_{19}K$). Nuclei of this isotope undergo radioactive decay with a half-life of 12.5 hours to form nuclei of calcium.

a) State what is meant by the term *isotopes.* [2]

b) Complete the nuclear reaction equation for this decay process.

$$^{42}_{19}K \rightarrow {}_{20}Ca +$$ [2]

c) The graph below shows the variation with time of the number N of potassium-42 nuclei in a particular sample.

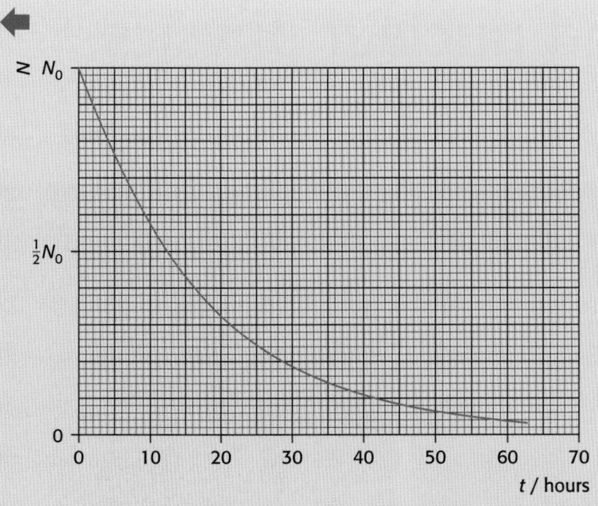

The isotope of calcium formed in this decay is stable.

Draw a line to show the variation with time t of the number of calcium nuclei in the sample. [1]

d) Use the graph in (c), or otherwise, to determine the time at which the ratio

$$\frac{\text{number of calcium nuclei in sample}}{\text{number of potassium-42 nuclei in sample}}$$

is equal to 7.0. [2]

(Total 7 marks)

17 Quantum physics and nuclear physics

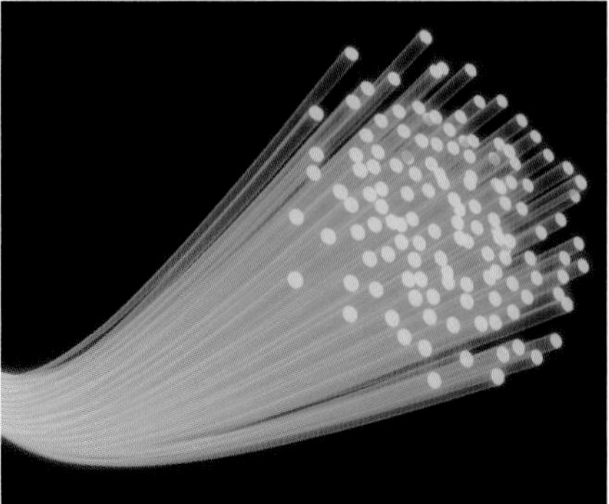

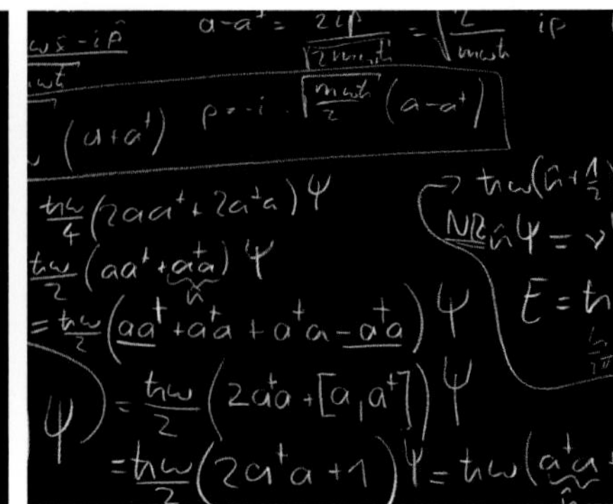

HL

At the beginning of the 20th century, there was a crisis in physics. The latest experimental observations on the properties of light and atoms could not be explained using the "classical" ideas of the time, such as those dealing with motion and waves. A new approach was needed, and it led to the creation of a completely new branch of physics: **quantum theory**, or **quantum mechanics**. The last time that such a complete scientific revolution had taken place was when Newton had put forward his theories on motion and gravity some 300 years earlier. In this chapter, you will discover more about this new revolution and why it became necessary.

Thinking about science: The nature of light

For centuries, the nature of light had been a mystery. By Newton's time, two opposing models had emerged. One treated light as a stream of **particles**, the other as a set of **waves**. To be successful, a model should be able to explain all the observed properties of light including the energy transfer, shadows, reflection, refraction (the change in direction when it enters a different medium), diffraction (the spreading of light around an object), and interference (the addition of light from two sources to form bright and dark "fringes"). This is how the models compare.

Light as a stream of particles

In the early 1700s, in his book *Opticks*, Newton proposed that light is a stream of tiny *corpuscles* or particles. In this model, a typical light beam consists of a large number of particles, each travelling in a straight line and each carrying energy. During reflection, the particles bounce off a surface – all in the same direction if the surface is smooth like a mirror. Newton also proposed a mechanism to explain refraction but, unfortunately, this was based on the assumption that the particles speed up when entering a medium such as glass or water, which we now know to be wrong. Newton was unable to explain the diffraction of light without using some wave concepts.

Light as a wave motion

Huygens first put forward a wave model of light in the late 1600s. In this model, energy is carried, not by a stream of particles, but by waves travelling out from their source. Over the years, the wave model superseded Newton's particle model because it was able to explain effects such as diffraction and interference. In the late 1800s, James Clerk Maxwell, doing theoretical work on electric and magnetic fields, predicted the existence of electromagnetic (EM) waves which would have the same speed as that measured experimentally for light. As a result, by the beginning of the 20th century, scientists felt confident that all the properties of light could be explained by treating it as an electromagnetic wave. However, their confidence was about to be undermined, as you will find out below.

Quantum explanations to replace classical ideas

This section considers three problems: the photoelectric effect, black body radiation, and electron diffraction. With these, classical models do not work. New ideas are needed to explain them. These are the **particle properties of light,** the **wave properties of matter,** and the concept of **quanta**.

Photoelectric effect

This is the emission of electrons from a surface, such as polished metal, because light or ultraviolet (UV) is shining on it. The effect was discovered by Hertz in 1887. The emitted electrons are sometimes called **photoelectrons** and, in a suitable setup, they can flow as a current in a circuit.

The electrons need energy to leave the surface. In the wave model of classical physics, the energy is supplied as a constant flow by the incoming light waves. However, this fails to explain the following experimental observations:

- In general, there is a minimum frequency, called the **threshold frequency,** below which no photoelectrons are emitted, whatever the intensity of the light.
- Different metals have different values for this threshold frequency.
- When photoemission is taking place, the magnitude of the photoelectric current depends on the intensity of the light source.
- No time delay is observed between the arrival of the light and the emission of the electrons, even at very low intensities.
- The energy of the photoelectrons is independent of the intensity of the light but varies linearly with the frequency of the incident light, as in Figure 1.

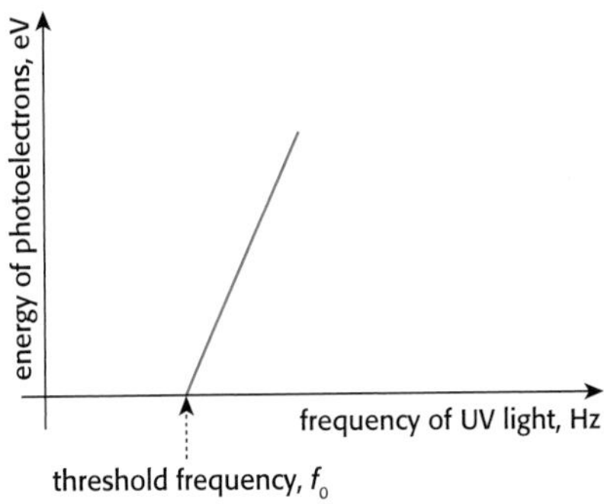

Figure 1

In 1905, Einstein came up with a radical model to explain the features of the photoelectric effect. He considered the incoming light not as a constant flow, but broken up into "packets" of energy. Each packet is called a **quantum**. Its energy E is given by this equation:

$$E = hf$$

h is the **Planck constant:** 6.63×10^{-34} J s

f is the frequency in Hz

Einstein's explanation can be used to predict how the maximum energy of the electrons should vary and this prediction is experimentally verified. In the apparatus below, photoelectrons are emitted by the cathode and accelerated by a potential difference.

If the potential of the variable power supply is reversed, then the photo electrons are not accelerated but they are repelled. As this negative potential is increased, the photocurrent gets smaller and smaller. Eventually at a certain reverse potential called the **stopping potential**, V_s, the photocurrent stops. This is shown in Figure 2.

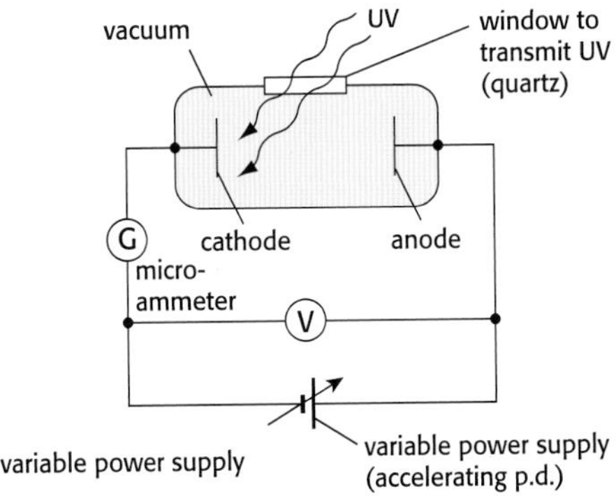

Figure 2 Millikan's stopping potential experiment

In Figure 1, the maximum energy of the photoelectrons can be calculated from the stopping potential using the charge on an electron, e.

$$\text{Maximum KE of electrons} = V_s \times e$$

Use Einstein's photoelectric equation to show that (a) the gradient of the graph in figure 1 equals the Plank constant, h. (b) the negative intercept on the y-axis would be equal in magnitude to the work function, ϕ.

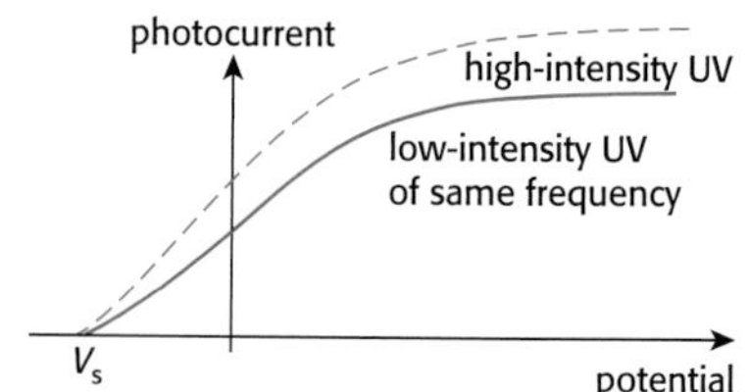

Figure 3 Variation of photocurrent with accelerating potential for the photoelectric effect.

In Einstein's model, the light travels in the form of particles called **photons**, each carrying one quantum of energy. For any given frequency, the more photons arriving per second, then the greater the intensity of the beam. A 60 watt light bulb emits about 1.8×10^{20} photons every second.

If an electron absorbs a photon, the energy it gains is hf. But for any particular metal, there is a minimum amount of energy needed to make an electron break free of the surface. This amount is called the **work function**, ϕ. If the photon provides *less* than this minimum, no electron is emitted. If it provides *more*, then the excess is transferred to the electron as kinetic energy, as given by this equation:

$$\frac{1}{2}mv^2 = hf - \phi$$

This equation has the same form as the data in Figure 1. In 1921, Einstein was awarded the Nobel Prize for Physics for his explanation of the photoelectric effect.

Black body radiation

Hot objects radiate electromagnetic (EM) energy at a range of frequencies. More accurately, *all* objects do. But if the temperature is increased, two changes happen. First, the total energy radiated per second increases. Second, there is an increase in the proportion of energy emitted at the higher frequencies – in other words, shorter wavelengths. At room temperature, an object mainly radiates in the infrared part of the EM spectrum. A hotter object may emit some visible radiation, becoming "red hot" or, if the temperature is high enough, "white hot".

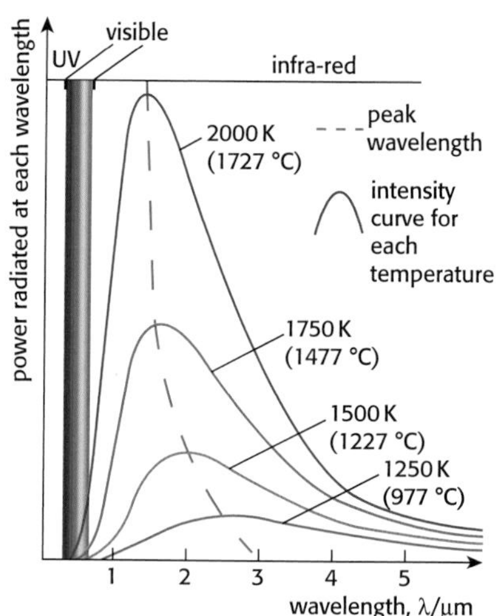

Figure 4 Black body radiation curves for a range of temperatures

Good emitters of radiation are also good absorbers. A **black body** is a theoretical object that will absorb all frequencies falling on it. So, if light is falling on it, it will absorb all frequencies, reflect none, and therefore appear black. As a black body is the best possible absorber, it is also the best possible emitter. Odd though it may sound for something so bright, the Sun is a black body emitter.

Figure 4 shows how, at different temperatures, energy is distributed within the spectrum of radiation from a black body emitter. The Rayleigh-Jeans law used a classical model to correctly predict the distribution of wavelengths for the low energy part of the spectrum, but the mathematics broke down at high energies, shown in Figure 5. This failure, known at the time as the **ultraviolet catastrophe** results from the assumption that radiation is emitted as a continuous flow. If the atoms are treated as emitters of energy in the form of photons, the observed spectum exactly matches the theory.

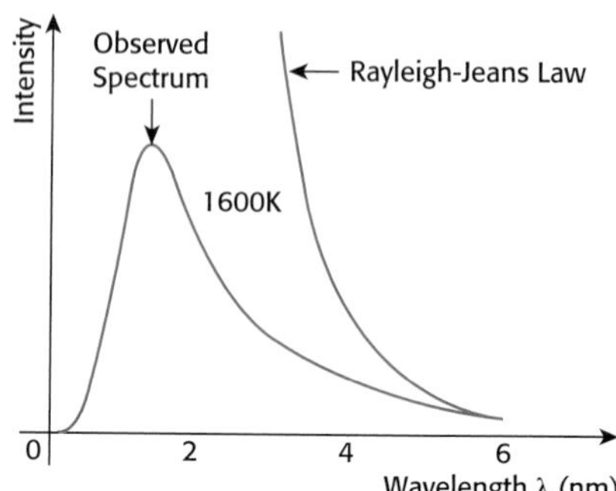

Figure 5 The classical Rayleigh-Jeans Law fails to correctlypredict the observed spectrum

Electron diffraction

The effects described above suggest that EM waves have particle-like properties. This raises an important question: can particles have wave-like properties? In 1927, Davisson and Germer scattered a stream of electrons from the surface of a crystal of nickel and found that, at some angles, the amplitude of the scattering was increased. This was equivalent to the diffraction patterns that can occur with light waves. In other words, electrons can behave like waves.

Thinking about science: Wave–particle duality

Refraction, interference, and diffraction can be explained by assuming that EM radiation travels as waves. On the other hand, the photoelectric effect and nature of black body radiation, require explanations based on quanta and particles called photons. Is this a contradiction? No. It just means that each wave or particle model is incomplete. A complete model must include both aspects. Similarly, some properties of electrons require that they be treated as particles. But electron diffraction shows that they can have wave-like properties. Overall, the evidence suggests that waves can behave like particles, and particles like waves. This is called **wave–particle duality**.

Probabilities

When light is diffracted by a slit, the interference between different waves causes a pattern of bright and dark fringes whose positions be calculated using the traditional wave model. Quantum mechanics should predict the same pattern. However, its mathematics does not give the precise locations and paths of particles or waves. Instead, it calculates probabilities. Each photon can take many possible paths. Each path has a probability associated with it, and it is the combination of all the probabilities that produces the pattern.

Matter waves

According to the de **Bröglie hypothesis**, all matter – for example, an electron or a proton – has a **probability wave** (a **matter wave**) associated with it. The wavelength of this probability wave can be calculated from a particle's momentum, p.

$$p = \frac{h}{\lambda}$$

This matter wave provides a way of calculating the probabilities of any future positions or interactions. The mathematics is complicated, but the information is described by an equation called the particle's **wave function**. The probability of a particle being found in any particular region depends on the amplitude of this wave function (to be more precise, it is proportional to the square of the amplitude). With many particles, the wave functions combine and can interfere.

When electrons are scattered by the surface of a crystal, they are more concentrated at certain angles. These correspond to the regions of constructive interference that result from the electrons' wave functions combining. Once again, the path of an individual electron cannot be predicted, but the relative probabilities of its arrival at different angles can be calculated using quantum mechanics. The experimental measurements are in close agreement with the theoretical predictions.

Working with data: Davisson and Germer experiment

Figure 6 shows the principle behind the Davisson and Germer electron diffraction experiment.

A beam of electrons strikes a target nickel crystal. The electrons are scattered from the surface. The intensity of these scattered electrons depends on the speed of the electrons (as determined by their accelerating potential difference) and the angle.

When the electron beam had an energy of 54 eV a maximum scattered intensity was recorded at an angle of 50°. The experiment data can be represented in one of two ways. Figure 7 is a traditional representation of the data.

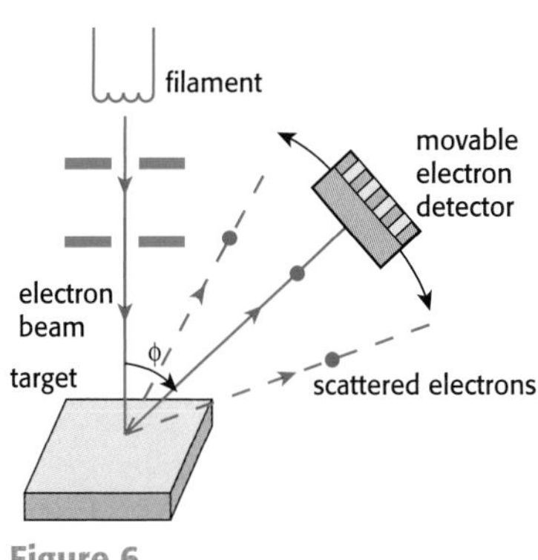

Figure 6

Figure 8 is known as a polar plot. In this graph, the scattered intensity is represented by the distance (the **radial distance**) from the origin to the line (i.e. along the dashed line) at any given angle.

The peak amplitude of scattering takes place at a given energy of electron beam

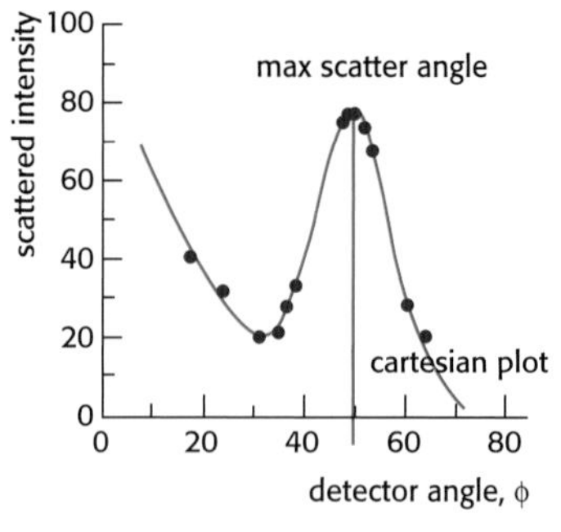

Figure 7

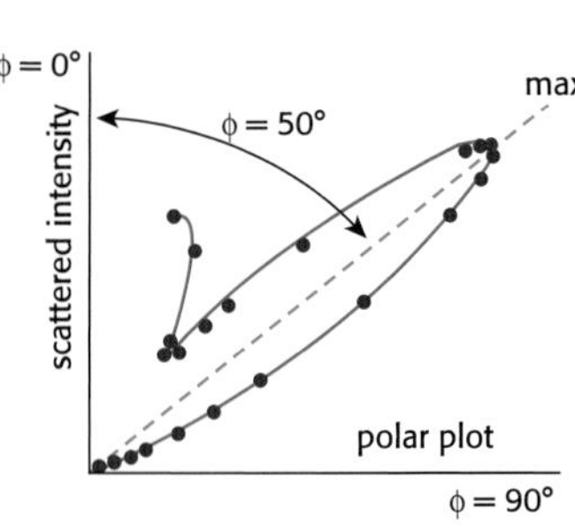

Figure 8

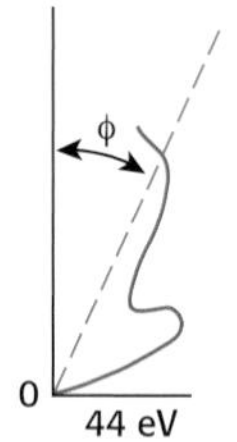

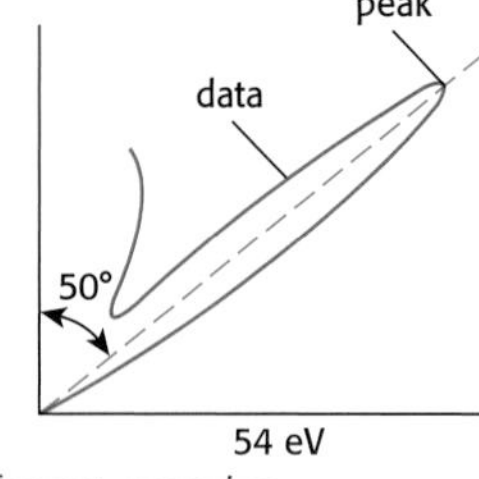

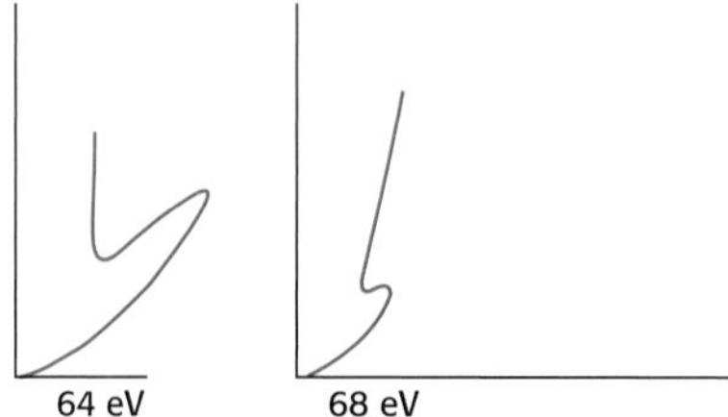

Figure 9 Polar plots for different energies

(see Figure 9). The reason for this increased amplitude is constructive interference of the electrons' matter waves at that particular angle. For an atomic spacing of d, the path difference AB between two adjacent atoms is $d \sin \phi$ (see Figure 10).

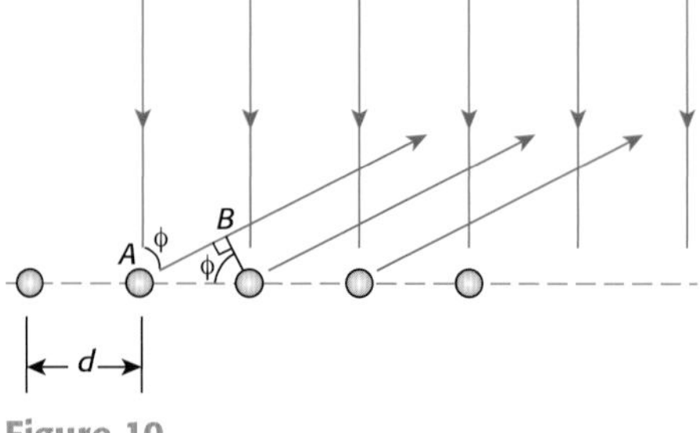

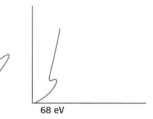

Figure 10

For the electrons that have been accelerated through a p.d. of 54 V:

1. Summarize the information represented by the radial graph plots in Figure 8.
2. The mass on an electron is 9.11×10^{-31} kg, calculate the momentum of an electron.
3. Use the de Bröglie relationship to calculate the wavelength of the electrons.
4. The separation of the nickel atoms is known to be 2.15×10^{-10} m.
   a) Calculate the path difference between two nickel atoms at the maximum intensity.
   b) Comment on your answer.

The experiment was repeated for many different accelerating potential differences V. Figure 10 shows how the effective electron wavelength varies with the reciprocal of the square root of accelerating PD $\frac{1}{\sqrt{V}}$.

5. Use the de Broglie relationship to deduce a general expression for the wavelength of the electron in terms of its mass m, its charge e, and the accelerating pd V.
6. Use the graph in Figure 11 to estimate a value for Planck's constant h.

Further investigation work on the Davisson–Germer experiment is readily available on the internet. One good example is the simulation published by the Physics Education Technology Project, which is available at http://phet.colorado.edu/simulations/sims.php?sim=DavissonGermer_Electron_Diffraction accessed July 2009.

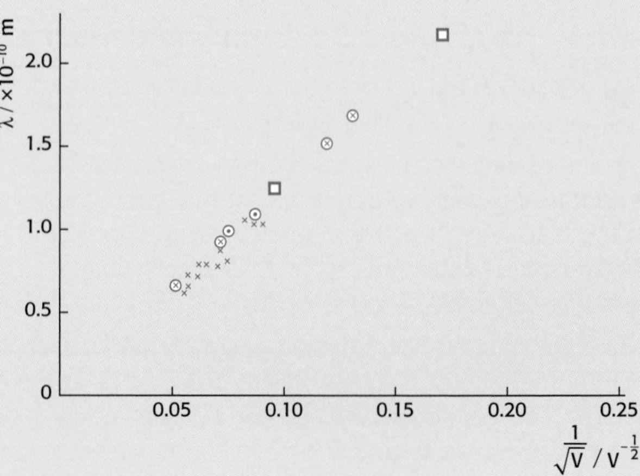

Figure 11

Thinking about science: *Probabilities and quantum states*

The objects around us exist on a macroscopic scale: they are large enough to be seen. On this scale, we can use classical mechanics (Newton's laws) to predict the paths taken by things as they move around. One characteristic of classical mechanics is that its laws are **deterministic**. This means that the future is absolutely determined by the past. For example, if you know the mass, velocity, and position of a ball and the forces on it at any instant – in other words, the starting conditions – you could, in principle, work out exactly where the ball was going to be at all times in the future.

At the atomic level, classical mechanics must be replaced by a model using the quantum theory. This has two key features:

The first feature is that, in quantum theory, certain states – for example, the amount of energy a particle has – are only "allowed" to have specific values. For more about this, see the section on orbitals below.

The second feature has already been described: the model is not deterministic but **probabilistic**. This means that whatever information we have about the starting conditions of a system, the best we can do is to calculate the relative probabilities of various outcomes. We can never predict with certainty what will happen.

Both features are apparent in the quantum mechanics model of the atom, as in the following example:

In a hydrogen atom, the nucleus has just one electron in orbit around it. However, you should not think of this as a point object, fixed in a particular orbit. There are different possible quantum states for the electron, each of which corresponds to a different total energy. In each state, or **orbital**, the electron has a range of different possible locations. One can picture the electron as being "smeared-out", with its wave function being used to calculate the probability distribution of the different locations.

The use of wave functions for electrons may seem complicated, and can only be used to give the probability (rather than the certainty) of finding the electron located in a region of space. But, to date, all experiments have confirmed that this is the best method of predicting what takes place. The measurements agree with the calculated probabilities. Fundamental particles really do seem to behave in this strange and mysterious way, and interactions between individual atoms can only be accurately described by using quantum theory.

The probabilistic rules of quantum physics do not mean that classical physics is wrong. The effects we see in everyday objects are the result of many billions of quantum events. Classical physics is a very good, and useful, approximation for the averaged-out effects, on a larger scale, of many quantum events.

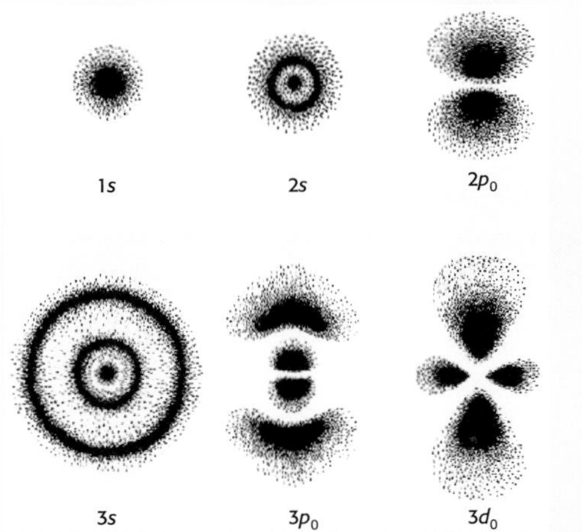

Figure 12 Probability density plots of some hydrogen atomic orbitals. The density of the dots represents the probability of finding the electron in that region.

Thinking about science: *The uncertainty principle*

Quantum theory suggests that one cannot determine outcomes with any certainty. It also imposes a limit on the accuracy with which any experimental measurement can be achieved, no matter how hard one tries. This limit is given by **Heisenberg's uncertainty principle**.

In quantum mechanics, there is a wave function associated with each electron, proton, photon, or other particle. These wave functions are used to calculate the relative probabilities of various future outcomes. Using this idea, Heisenberg was able to show that certain variables are linked in a way that limits how

much one can know about both of them at the same time. For example, for a moving electron, position and momentum are two such variables. To quote Heisenberg: "The more precisely the position is determined, the less precisely the momentum is known in this instant, and vice versa" (*Zeitschrift für Physik*, *43*:1927). In symbolic form the relationship is:

$$\Delta x \Delta p \geq \frac{h}{4\pi}$$

Δx and Δp are the uncertainties in position and momentum (respectively), and h is the Planck constant (6.63×10^{-34} J s).

Two other related measurements are energy E and time t. For these, the uncertainty relationship has the same format as the previous one:

$$\Delta E \Delta t \geq \frac{h}{4\pi}$$

There are three additional points to make here:

- For each variable, the uncertainty refers to a statistical quantity called the **standard deviation**.
- The value of the Planck constant is so small that, in everyday events, the effects of quantum uncertainty are not usually observed.
- The uncertainties in the above equations come from the mathematics of quantum mechanics. They are associated with the particles themselves, and *not* due to limits imposed by experimental techniques.

Practical implications

The practical implications of both of the above relationships can be seen in particle physics experiments, and in some modern electronic devices. For example, some radioactive materials emit alpha particles. According to the rules of classical physics, these particles should remain trapped in the nucleus because they do not have enough energy to escape. However, the uncertainty principle "allows" them to borrow energy, provided they can "pay it back" in the time limit imposed by the equation. As a result, the particle can escape its trap. This is an example of a process called **quantum tunnelling**. Quantum tunnelling by electrons is used in **scanning tunnelling microscopy,** a procedure that allows individual atoms to be imaged.

Inquiry: Electron diffraction tube

When a stream of electrons hits a sample of graphite powder, the electrons form a pattern of rings:

Figure 13

The rings correspond to regions of constructive interference. The path difference between different layers of atomic spacing corresponds to a whole number of electron wavelengths.

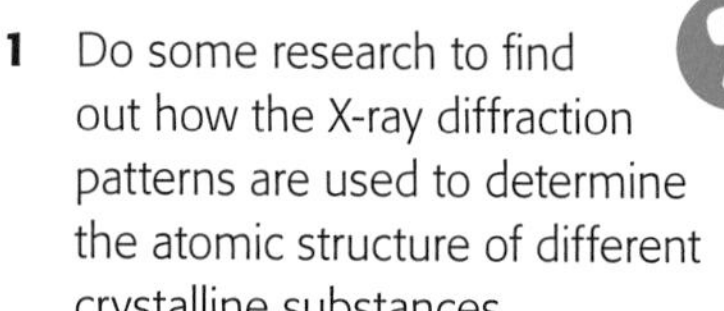

1 Do some research to find out how the X-ray diffraction patterns are used to determine the atomic structure of different crystalline substances.

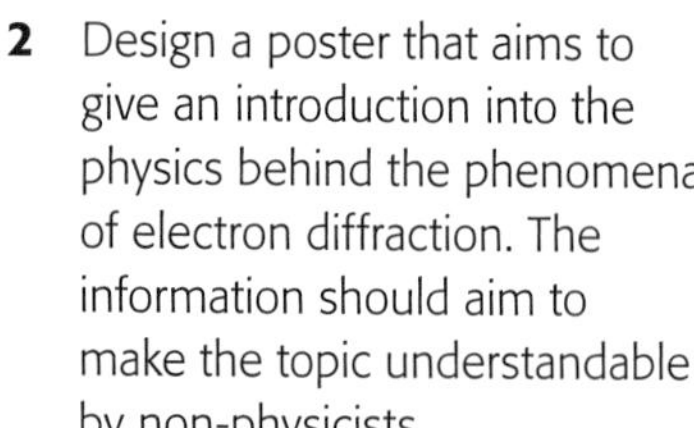

2 Design a poster that aims to give an introduction into the physics behind the phenomena of electron diffraction. The information should aim to make the topic understandable by non-physicists.

3 Predict what will happen to the size and brightness of the circles in Figure 13 if:

a) the accelerating pd is decreased

b) the accelerating pd is increased.

Emission and absorption spectra

If chemists want to identify the metals present in a compound, they can use a flame test. A sample of the metal salt is placed in a flame, and the colour observed helps identify the metal involved. An example is available here www.youtube.com/watch?v = jJvS4uc4TbU (accessed July 2009).

One common type of street light gives off an orange colour because it uses low-pressure sodium vapour to produce the light (see Figure 14).

Emission spectra

In a spectrum, the different colours are associated with different frequencies of light. Each element has its own **emission spectrum:** the particular frequencies produced when the element is hot enough to radiate light.

To analyse a spectrum, the emitted light needs to be split into its component frequencies. The instrument used is called a **spectrometer**. It uses either a prism to refract and disperse the light or a **diffraction grating**.

Different frequencies of light are recorded at different angles. The measurement of each angle is used to calculate the observed wavelengths, from which the frequency can be calculated. Typically, the practical arrangement involves shining light through narrow slits so that the resulting spectrum is displayed as a series of lines at different wavelengths. This is sometimes called a **line spectrum**.

Experimentally, as the temperature of the element is increased, the bands become broader. An extremely hot source of light will tend to give off a **continuous spectrum** of light – white light.

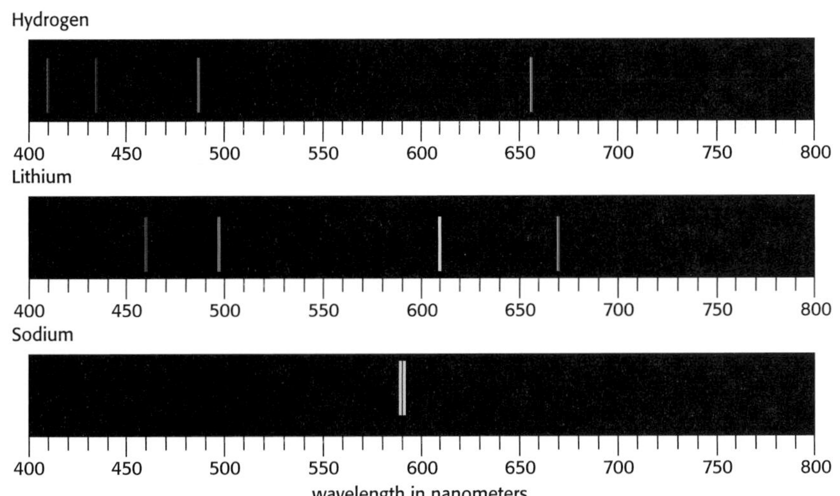

Figure 16 Atomic emission spectra for H, Li, and Na

Absorption spectra

Newton was one of the first scientists to analyse the light from the Sun. He observed that the spectrum is not, strictly speaking, continuous. It is nearly a continuous spectrum of white light but there are some frequencies missing (or of reduced intensity). This is

Table 1

Metal	Flame colour
barium	light green
calcium	brick red
copper	blue/green
lead	blue/white
potassium	lilac
sodium	bright orange

Figure 14 Flame tests showing different elements emitting different spectra when heated

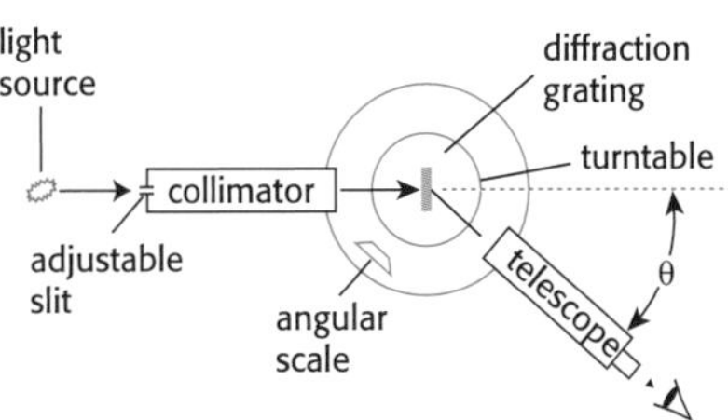

Figure 15 Spectrometer

typical of an **absorption spectrum.** The missing frequencies correspond to those found in emission spectra.

There is a link between emission and absorption spectra. An element emitting particular frequencies of light is also able to absorb the same frequencies. If a continuous spectrum (white light) shines through a gas, then it will preferentially absorb certain frequencies. This absorbed energy will be re-radiated, but if that happens, it will take place in all directions. So the intensity of the characteristic frequency in the forward direction will be reduced.

The light from the Sun passes through its outer, colder, layers where the particular frequencies are absorbed. These are characteristic of the elements that exist in the outer layers.

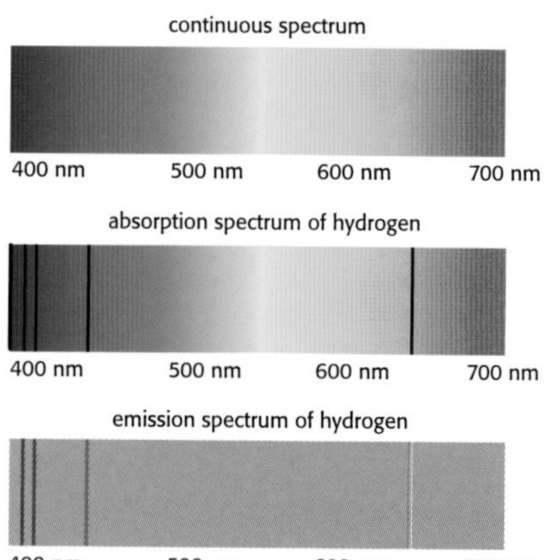

Figure 17 Absorption and emission spectra of hydrogen

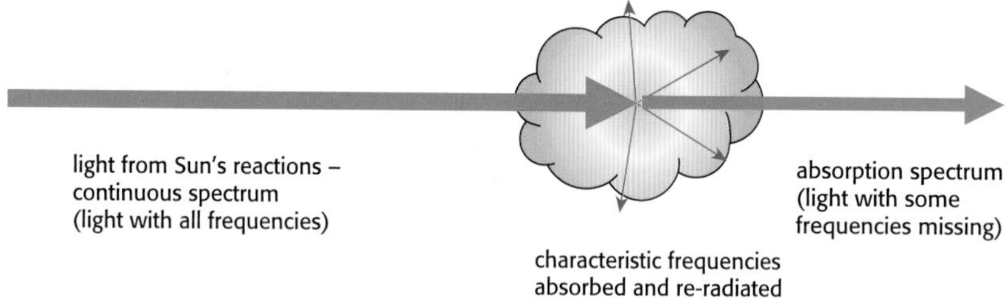

Formation of a line spectrum

One key question is "how is a line spectrum formed?". In other words, why are only certain specific frequencies present? To understand that, one needs to go back to the link between the energy of a photon and its frequency:

$$E = hf$$

If only certain frequencies are present in a spectrum, then it follows from the equation that only certain photon energies are possible. Quantum theory explains this as follows. For the electrons in an atom, there are only a limited number of discrete (separate) energy quantum states. Electrons are "allowed" to occupy these, but not have any energy in between. If an electron loses energy by making a transition to a lower level, then a photon of equivalent energy (and of one particular frequency) is emitted. Similarly, an incoming photon can lift an electron to a higher state, provided the energy change is the correct match.

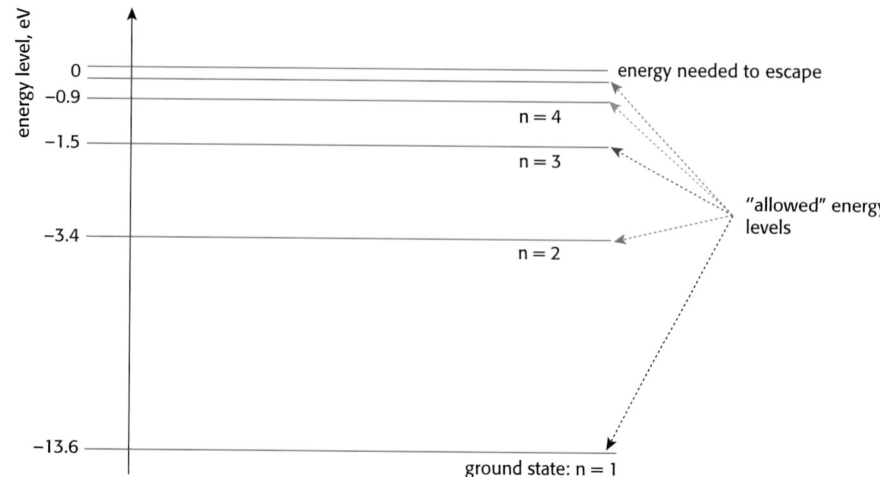

Figure 18 Electron energy levels in hydrogen

The electron energy levels in an atom are often represented by lines on an energy plot. The chart in Figure 18 is for a hydrogen atom. The lowest energy level represents the **ground state** of the electron. The top level (defined as zero) is for an electron that has enough energy to escape from the nucleus. It is the transitions between the different energy levels that correspond to particular frequencies in the hydrogen spectrum. For example, if an electron moves between the n = 4 and the n = 2 energy levels, this must involve an energy loss:

energy change $\approx -0.9 - (-3.4)$ eV = 2.5 eV

This "lost" energy must go into a photon of energy 2.5 eV

$$\text{So: frequency of photon} \approx 2.5\ \text{eV} / h = \frac{2.5 \times 1.6 \times 10^{-19}}{6.6 \times 10^{-34}}$$

$$= 6.0 \times 10^{14}\ \text{Hz}$$

So: wavelength of photon $\approx 5 \times 10^{-7}$ m $\approx$ 500 nm

The experimentally measured wavelength turns out to be 486 nm.

In the hydrogen spectrum, the lines can be arranged into sets, each representing a different series of possible transitions. The series are named after the scientists who first discovered them.

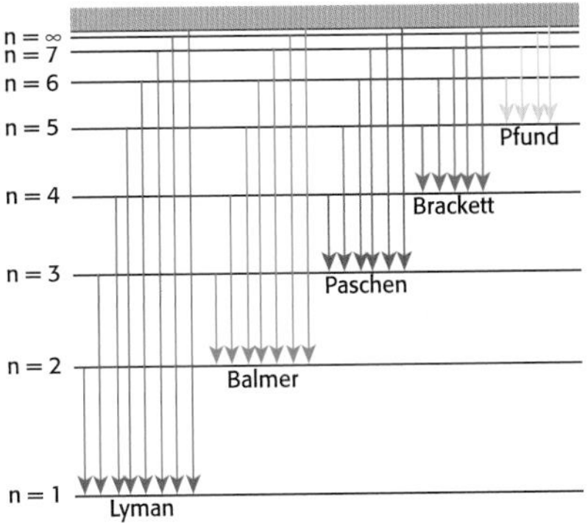

Figure 19 Electron transitions in the Hydrogen atom

The above model is much simplified. In a more detailed analysis using the quantum theory, different energy levels are associated with different orbitals (see earlier section) and the electrons are assumed to behave as standing waves. This method of modelling electrons is known as **wave mechanics**.

Modelling electron orbits

In the most simple model of an atom, electrons orbit a nucleus made up of protons and neutrons. This offers no explanation for the existence of atomic energy levels. However, by making some additional assumptions about quantization, Bohr was able to accurately predict many aspects of the hydrogen emission spectrum.

Bohr started by quantizing a quantity called the angular momentum of an electron, in units of $\frac{h}{2\pi}$. In other words, the allowed values for the angular momentum are $\frac{h}{2\pi}$, $2\frac{h}{2\pi}$, $3\frac{h}{2\pi}$ ... and so on.

Bohr argued that an electron with one of these amounts of angular momentum must be in a stable orbit. Using Newtonian mechanics, and assuming that the orbits were circular, he calculated their energies. He found that these matched the values needed to produce the emission spectrum of hydrogen. This was an impressive result but, unfortunately, it did not work well for more complicated atoms with more protons and electrons. Also, it failed to explain some of the fine detail of the observed spectrum and the relative intensities of the various lines. It took a wave model of electron orbits to overcome these difficulties.

The mathematics of an accurate wave model is very complicated, but an insight into the principles involved can be gained by starting with a simple "electron in a box" model. We know that the electron is trapped in the atom because it is attracted by the nucleus. This model imagines that the electron exists somewhere within a box whose sides are of length L. For simplicity, we imagine an atom to be cubic (rather than spherical), and start by considering how a probability wave (a matter wave) might fix in the box. From an earlier section, you will remember that the probability of finding an electron at any particular location depends on the amplitude of this wave.

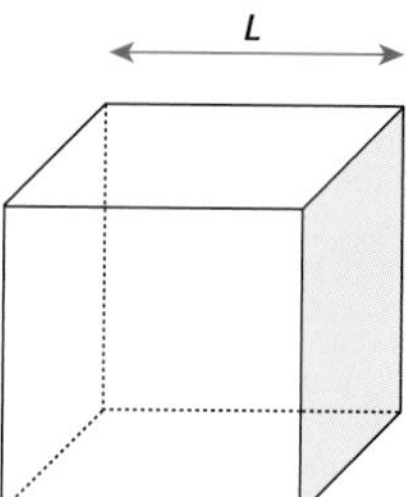

Figure 20 Electron in a box model

In this model, the electron is trapped inside the box, so the probability of finding it outside the box must be zero. This means that the wave must have zero amplitude outside the box, but can have a value inside. Only a standing wave will be able to fit these conditions, and there must be a node at each wall. Possible standing waves are shown in Figure 21.

In a "real" atom, an electron is not, of course, trapped inside a box. However, the concept of the wave function can be modified to calculate the chances of finding the electron located at any given distance from the nucleus. **Schrödinger's model** of the hydrogen atom used wave mechanics to develop a more accurate wave function for the electron. It took various other properties into account, including angular momentum, spin, and the variation in electrical potential energy associated with the charges on the electron and the proton. In this model, the wavelength of the electron's wave function is not fixed, but varies with distance.

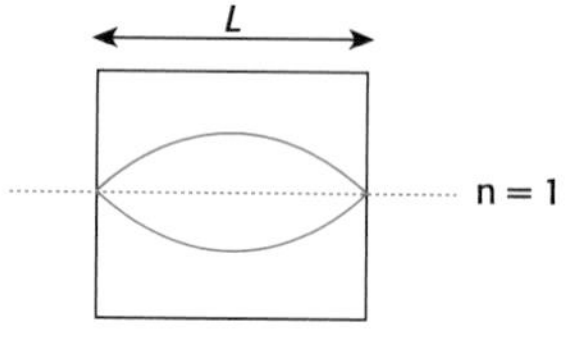

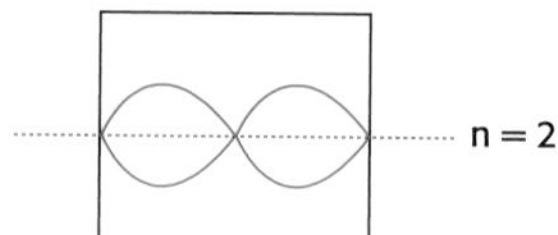

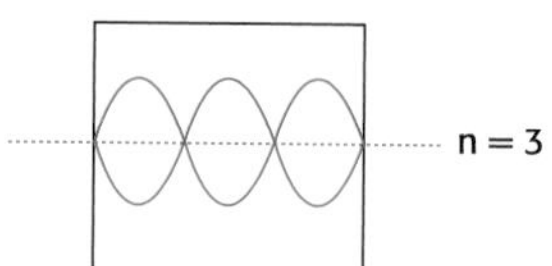

Figure 21 Possible wave functions (standing waves) for an electron in a box

For the wave function, **boundary conditions** apply. The amplitude must be zero at infinity otherwise the electron will not be bound to the nucleus. A wave function has no physical meaning, but the square of its amplitude is proportional to the probability of finding the electron in the region being considered.

For some problems, you also need to understand the link between an electron's kinetic energy (KE) and electrical potential energy (PE). The total energy associated with its wave function is the sum of these two. When an electron is an infinite distance from the nucleus, its PE is zero. When it is close to the nucleus, its PE is negative, so its KE must be positive. However, for a bound electron, its total energy is negative.

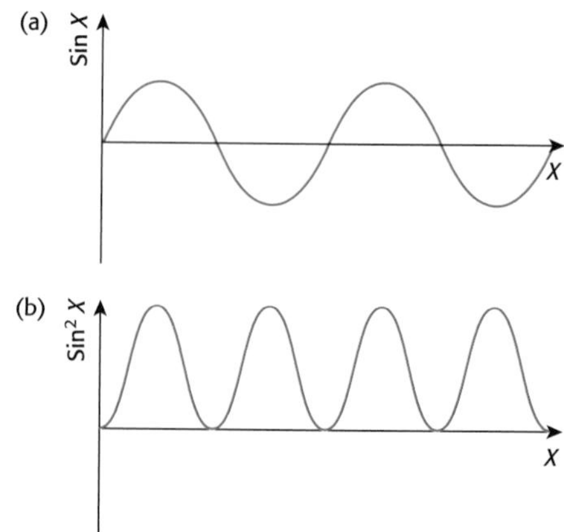

Figure 22 A sinusoidal wave function (a) will result in the probability varying as shown by (b)

1. Use the "wave in a box" diagrams to calculate the first three possible values for the wavelength, λ, of the electron in this model in terms of L.
2. Use the de Bröglie relationship to predict the first three possible values for the momentum p of the electron in this model in terms of λ.
3. The mass of the electron is m_e. Use the relationship between momentum and kinetic energy to calculate the first three possible values for the KE of the electron in this model.
4. Derive a relationship for the possible values for the KE of the electron in this model in terms of n, m_e, h, and L.

The nucleus

Data booklets give specific information about nuclear radii and masses. This section explains how these two quantities can be estimated. The existence of the nucleus was deduced as a result of Geiger and Marsden's alpha particle scattering experiment (see page 236). If the original energies of the alpha particles are known, then the same experiment can be used to estimate an upper limit for the size of a nucleus. The different deflections are a result of the different original paths for the alpha particles.

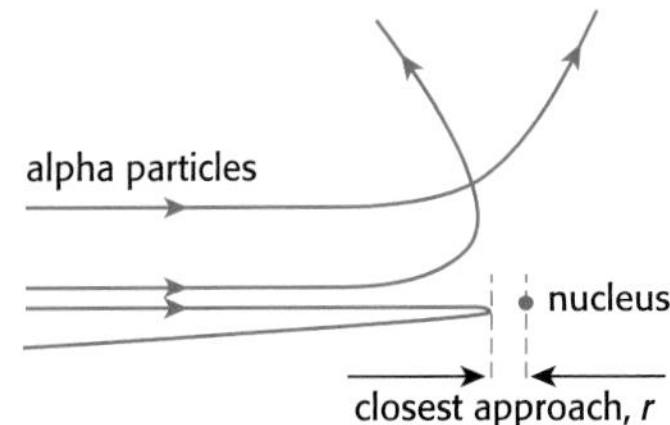

Figure 23 The closest approach of an alpha particle is a result of heading directly towards a nucleus

None of the alpha particles collide with the target nuclei, but ones that get closest are the ones whose original paths head directly towards a particular nucleus. As an alpha particle gets closer and closer to the nucleus, the force of repulsion slows down it down. Energetically, the alpha particle is losing kinetic energy and gaining electrostatic potential energy. At the point of closes approach, the alpha particle is temporarily stationary and all its energy has been converted into electrostatic potential energy. Since we know the charge on the nucleus, the charge on the alpha particle and the original energy of the alpha particle, we can calculate the distance of closest approach, *r* using the electrostatic potential energy equation:

$$\textit{Energy of alpha particle} = \frac{(\textit{charge on alpha}) \times (\textit{charge on nucleus})}{4\pi\varepsilon_0 \times (\textit{closest approach})}$$

The distance of closest approach is an upper limit on the size of the target nucleus.

Different nuclei have different masses and these can be determined using a Bainbridge mass spectrometer.

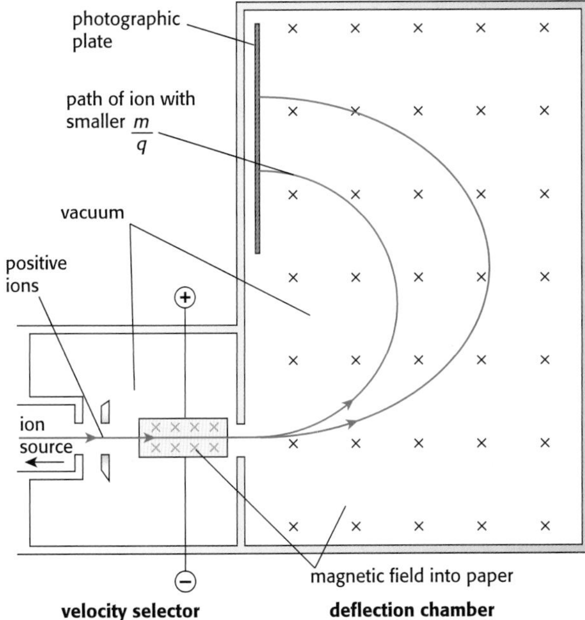

Figure 24 A Bainbridge mass spectrometer

The process starts with the atoms being analysed being ionised. The resulting ions are then accelerated by a potential difference and then enter a region where a constant magnetic field *B* is applied. The field is perpendicular to their velocities and this results in the ions following circular paths with the centripetal force being provided by the magnetic force on the moving charge. The radius of the path depends on many factors

- The charge on the ion, q
- The speed of the ion, v
- The mass of the ion, m
- The magnetic field strength, B

In order to identify the particular masses involved it is important to select ions that have the same charge and the same speed. In Figure 24 this is achieved in the region which is labelled as the **velocity selector**. There are different technique used to achieve this, but in the example shown crossed electric and magnetic fields are used to select ions moving at the same speed – those that are moving at a different speed will be deflected and not enter into the deflection chamber. Since they were all accelerated from rest by the same accelerating potential difference – these will also have the same charge.

Clearly different elements will be expected have different masses. Larger mass ions will travel in larger circles. Analysis of a single element showed different paths taken and thus provide evidence for the existence of isotopes.

Nuclear energy levels

Discrete energy levels also exist for nuclei so the nucleus, like the atom, is a quantum system. When a nucleus falls from a high-energy state (the "excited" state) down to a lower energy state this must correspond to the emission of a photon. Nuclear energy levels correspond to much larger energies than atomic energy levels and the gaps between them are also larger. When a nucleus emits a photon in this way the photon has a very large energy – it is in the gamma ray section of the electromagnetic spectrum. Most of the time gamma rays are associated with the prior emission of an alpha or a beta particle. The first emission creates a nucleus in a high-energy state, which then drops into a lower energy state.

For a given decay mode, the excited nuclear state and the ground nuclear state have a fixed quantum level and thus the energy of the emitted alpha particle and the subsequent photon are predictable.

For example, when Magnesium-27 decays into Aluminium-27, it has two possible modes of β-decay. Both states of aluminium-27 are excited states and gamma rays of three possible discrete energies can be emitted as the aluminium-27 returns to the ground state. Figure 25 represents the nuclear energy levels:

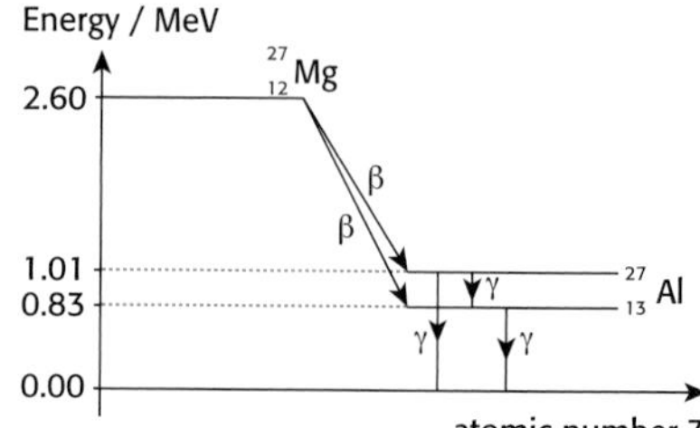

Figure 25

Radioactive decay

Antimatter

In August 1932 a photograph was published that resulted in an important realization (Figure 26). The photograph showed a **particle track** (the path taken by an atomic particle) that demonstrated that, in addition to the matter that we observe on a day-to-day basis around us all the time, there also exists a "parallel" type of particle called antimatter.

Figure 26 The first piece of evidence for the existence of a positron

A magnetic field was arranged at right angles to the plane of the paper, going into the page. The track in Figure 26 shows a particle passing through a 6 mm lead plate. We know that the particle must lose energy as a result of going through the lead plate, and thus it slows down. The upper portion of the track is more curved, so the particle must have entered from the bottom and be moving up the page. It curves to its left, so we can deduce that the particle must be positive. It can be shown (by more detailed analysis of the track) that it has exactly the same charge-to-mass ratio as an electron, but it is positive. The track shows the path of a previously unknown particle – the positive electron or **positron**.

Every form of matter that exists has an equivalent "opposite" form of antimatter. Any property that the particle has (e.g. charge) will have the same magnitude but the opposite sign in the antiparticle. Although the properties of antimatter are not easy to verify, most theories predict that an antiparticle would still gravitationally attract a normal matter particle. The gravitational force is, however, negligibly small on the atomic and nuclear scale.

Antimatter is not a regular everyday observation, because, if matter and antimatter come into contact with one another, they will annihilate one another. All their combined mass–energy will be converted into another form – often electromagnetic radiation. But antimatter is definitely a real part of the composition of the Universe around us. Many naturally occurring nuclides radioactively decay with the emission of a positron, or β^+ decay. In this case a neutrino is emitted at the same time (as opposed to the antineutrino that is emitted during β^- decay). For example, chlorine-36 undergoes β^- decay, whereas the isotope chlorine-34 undergoes β^+ decay:

$$^{36}_{17}\mathrm{Cl} \rightarrow {}^{36}_{18}\mathrm{Ar} + {}^{0}_{-1}\beta + \bar{\nu}$$

$$^{34}_{17}\mathrm{Cl} \rightarrow {}^{34}_{16}\mathrm{S} + {}^{0}_{+1}\beta + \nu$$

All the antimatter that has been observed to date exists only in minute quantities, but there is no reason to believe that antimatter atoms or antimatter molecules would not be stable. Antihydrogen (consisting of a positron in orbit around an antiproton) has already been produced. It was first produced at CERN (the European Centre for Nuclear Research) towards the end of 1995. Experimenters are hoping to be able to design experiments to measure its emission/absorption spectrum, which is predicted to be identical to that of normal hydrogen.

If the spectral lines of matter and antimatter were the same, there would be no way in which we could tell whether some of the stars and galaxies that we observe are made of matter or of antimatter! Current theories of the evolution of the universe propose that all the matter we currently observe is just the small amount "left over" after an initial early period in which much larger (and very nearly equal) amounts of matter and antimatter were annihilated.

Mathematical physics: exponential decay

The driving equation for exponential increase is

$$\frac{d}{dt}(N) = kN$$

which has the solution

$$N = N_0 e^{kt}$$

So

$$\frac{dN}{dt} = kN_0 e^{kt} = kN$$

The driving equation for exponential decrease is

$$\frac{d}{dt}(N) = -kN$$

which has the solution

$$N = N_0 e^{-kt}$$

So

$$\frac{dN}{dt} = kN_0 e^{-kt} = -kN$$

You can use these general exponential equations to prove that the common ratio property always applies. For radioactive decay it is particularly useful to note that this means there is a fixed relationship between the half-life and the constant of proportionality called the **decay constant** λ. For a given mode of decay λ is defined as the probability of decay of a given nucleus per unit time:

$$\frac{dN}{dt}(N) = -\lambda N$$

So the equation for the number of atoms available for radioactive decay is

$$N = N_0 e^{-\lambda t}$$

This can be rearranged to show that the half-life is linked to the decay constant by the following relationship:

$$T_{\frac{1}{2}} = \frac{\ln 2}{\lambda}$$

Thus if you are given (or can calculate) the half-life then it is a simple substitution to work out the decay constant (and vice versa).

Working with data: How do you tell if it's exponential?

Many natural or everyday phenomena (population growth, radioactive decay, the charging of a capacitor, the rate of loss of heat from a hot object, the decrease in amplitude of a pendulum, compound interest—even the response of our eyes and ears to light and sound) are, or approximate to, exponential, and thus an understanding of the nature of exponential growth or decay is hugely important.

The underlying "driving" function behind all exponentials associated with time is the idea that the rate of change of any quantity depends on the amount that is available to change. For example, more babies are born when there are more adults around to be parents: that is, the rate at which an animal population increases depends on the overall size of the population (and therefore the number of adults in the population). Similarly, the overall rate at which a sample undergoes radioactive decay depends on the number of nuclei that are left and available to decay.

The radioactive process is random. We cannot predict when an individual nucleus is going to decay, but we do know that the likelihood of a decay increases if more nuclei are available to decay.

The rate of change means the change that takes place per unit time. Using calculus notation, the rate of change of a quantity R is can be related to the number available N as follows:

$$R = \frac{d}{dt}(N)$$

So for exponential growth or decay the following relationships are all equivalent:

Exponential growth:

$$R \propto N$$

$$\frac{d}{dt}(N) \propto N$$

$$\frac{d}{dt}(N) = kN$$

where k is a constant.

Exponential decay:

$$R \propto -N$$

$$\frac{d}{dt}(N) \propto -N \frac{d}{dt}(N) = -kN$$

where k is a constant.

Both the exponential increase and the exponential decrease share a surprising property associated with how the quantity changes with time—the **common ratio** property. For exponential decay, the time taken to get from any value to half that value is always the same and is called the **half-life**, $T_{½}$. See Figure 27.

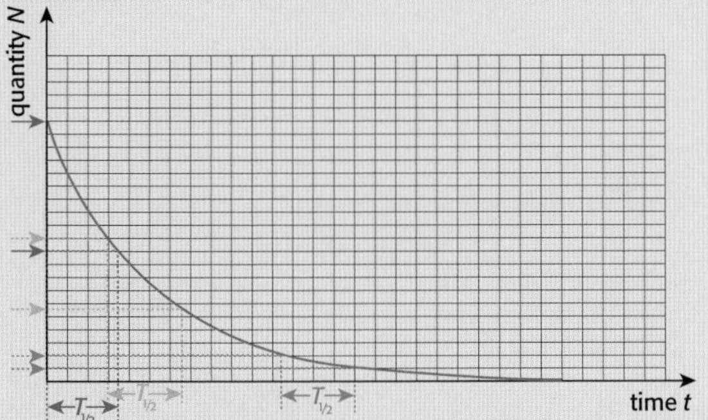

Figure 27 The half life of exponential decay is constant

There is nothing special about the ratio of a half. The time taken to go to one third, a quarter or any other fraction is also always fixed (Figure 28).

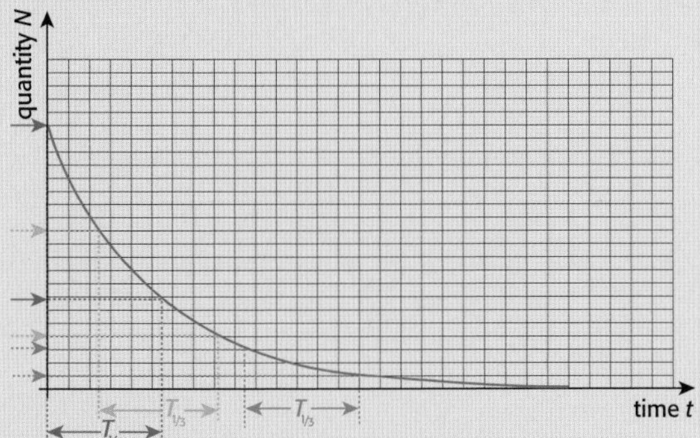

Figure 28 The time taken for any ratio is fixed in exponential decay

Exponential rise also shows the same constant ratio property for any value that you choose (Figure 29).

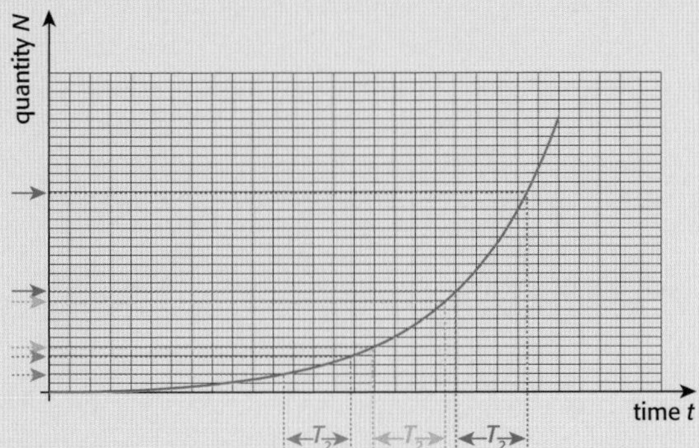

Figure 29 In exponential growth, the time taken for the quantity to double is constant

It is all too common for students to start thinking that all graphs that curve downwards towards the *x*-axis must be further examples of exponential decay, and all graphs that curve upwards must be exponential growth. This is not true, so if you are given some data that relate two quantities, how can you tell whether the relationship is an exponential one? There are essentially three different techniques:

1 *Find/verify the constant ratio property.*

All exponential functions must show the constant ratio property. A plot can be analysed to see whether this is the case as follows:

- Chose a ratio: halving, doubling, or anything appropriate.
- Use the graph to calculate at least three values (preferably more) of increase in the *x*-value that equates to this change in the *y*-value.
- Given the uncertainties involved, it is important to use all sections of the graph for this calculation.
- The values should be equal within an appropriate uncertainty range.

2 *Plot a graph that has one axis logarithmic and the other linear.*

The mathematics of this technique do not need to be understood by SL students, but it can always be used. Suppose we have an exponential decrease governed by the following relationship:

$$N = N_0 e^{-\lambda t}$$

$$\ln N = \ln\left(N_0 e^{-\lambda t}\right)$$

$$= \ln\left(N_0\right) + \ln\left(e^{-\lambda t}\right)$$

$$= \ln\left(N_0\right) - \lambda t$$

Thus a plot of ln *N* on the *y*-axis against *t* on the *x*-axis will give a straight line going through the origin with *y*-intercept = $\ln(N_0)$ and gradient = $-\lambda$.

This plot can be achieved in two ways. Using normal graph paper it is simply a matter of calculating the natural logarithm of each value, using the ln function key on a calculator, and then plotting this directly on a graph. Small numbers (less than e) will work out to be negative. Alternatively, special **log-linear graph paper** can be used to plot the numbers directly.

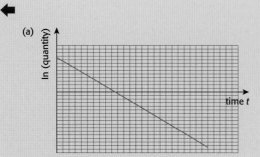

(a) ln (quantity) time t

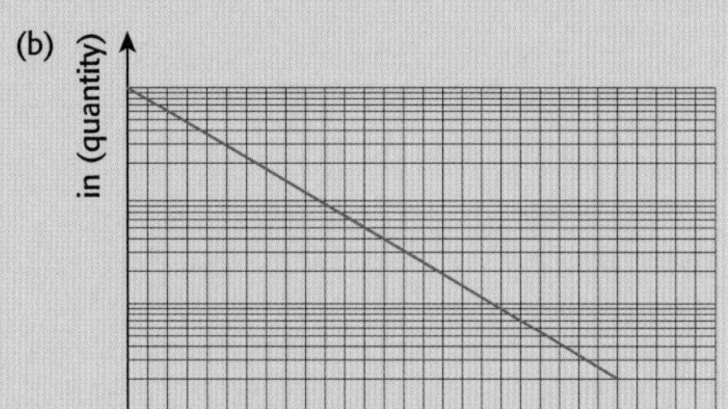

Figure 30 **(a)** Using normal graph paper to plot a log-linear graph. **(b)** Using special log-linear graph paper to produce the same plot.

If the line is straight to within an appropriate amount of uncertainty, then the original data must be exponential. Calculating the gradient is straightforward on normal graph paper, but log-linear graph paper can prove slightly more complex.

Note that logs to base e (also called natural logs or ln(**x**)) and logs to base 10 (log(x)) will both produce straight-line graphs. The only difference between the two is a scaling factor:

$$\ln(N) = 2.3026 \times \log(N)$$

Table 1

Number N	log (N)	ln (N)
1	0.0	0.0000
10	1.0	2.3026
100	2.0	4.6052
1000	3.0	6.9078

3 *Show proportionality between the rate of change and the quantity itself.*

In some situations it may be possible to demonstrate an exponential relationship by demonstrating that the rate of change is proportional to the quantity that is changing.

The radioactive decay equation

$$\frac{dN}{dt} = -\lambda N$$

shows that the *rate of decay* is proportional to the number of atoms available to decay (Figure 31).

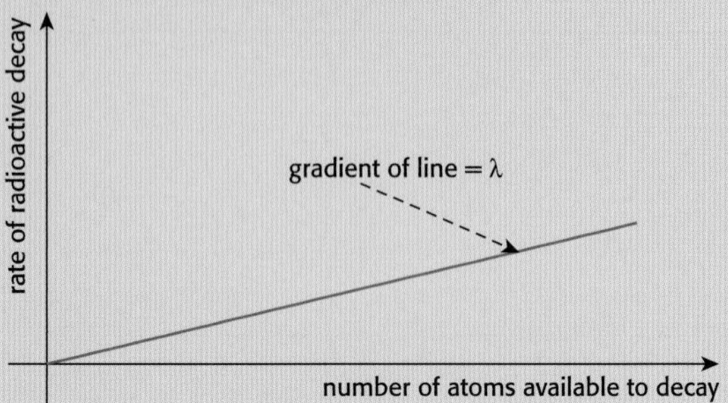

Figure 31 Graph of the radioactive decay equation

Sometimes merely proving the exponential nature of a change is not enough; you may need to analyse the data further. For radioactive decay the aim of data collection may well be to find the half-life of a particular radioactive substance. If the rate of decay varies significantly over a reasonable period of time, then methods 1 or 2 can be used to show its exponential nature. With method 1 the half-life can be estimated directly from the graph, but method 2 would provide a better way of utilizing all the data, and thus a more precise method for estimating the half-life. The gradient of the log-linear graph would equal the decay constant λ, and there is a direct mathematical relationship between the decay constant and the half-life that allows the half-life to be determined.

In many situations, however, the rate of decay is effectively constant, because the overall decay rate is low (the decay constant is a very small number). For example, the half-life of uranium-238 is 4.5 billion years. Because this is such a long time period, no laboratory experiment will be able to collect data to allow the half-life to be determined using methods 1 or 2. In this situation the half-life can be calculated using method 3. The number of atoms that are available to decay comes from a measurement of the mass of the sample, and a measured activity can be used to calculate the rate of decay.

End of chapter summary

Chapter 17 has two main themes: Quantum physics and nuclear physics. SL candidates studying option B (Quantum and Nuclear Physics) need to study both themes. The list below summarises the knowledge and skills that you should be able to undertake after having studied this chapter. Further research into more detailed explanations using other resources and/or further practice at solving problems in all these topics is recommended – particularly the items in bold.

Quantum physics (the quantum nature of radiation, the wave nature of matter, atomic spectra and atomic energy states)

- Describe the photoelectric effect, the concept of the photon and an experiment to test the Einstein model of the photoelectric effect.
- Describe the de Broglie hypothesis, the concept of matter waves and an experiment to verify the de Broglie hypothesis.
- **Solve problems involving photoelectric effect and matter waves.**
- Outline a laboratory procedure for producing and observing atomic spectra and explain how they provide evidence for the quantization of energy in atoms and how to calculate wavelengths of spectral lines from energy level differences and vice versa.
- Describe one piece of evidence for the existence of nuclear energy levels and explain the origin of atomic energy levels in terms of the 'electron in a box' model.
- Outline the Schrödinger model of the hydrogen atom and the Heisenberg uncertainty principle with regard to position – momentum and time – energy.

Nuclear physics (nuclear physics, radioactive decay)

- Explain how the radii of nuclei may be estimated from charged particle scattering experiments and how the masses of nuclei may be determined using a Bainbridge mass spectrometer.
- Describe β^- decay, including the existence of the neutrino.
- State the radioactive decay law as an exponential function, define the decay constant and derive the relationship between decay constant and half-life.
- Outline methods for measuring the half-life of an isotope and **solve problems involving radioactive half-life**.

Chapter 17 questions

1 This question considers some aspects of the atomic and nuclear physics associated with isotopes of the element helium.

Atomic aspects

a) The element helium was first identified from the *absorption spectrum* of the Sun.

i) Explain what is meant by the term *absorption spectrum*. [2]

ii) Outline how this spectrum may be experimentally observed. [2]

b) One of the wavelengths in the absorption spectrum of helium occurs at 588 nm.

i) Show that the energy of a photon of wavelength 588 nm is 3.38×10^{-19} J. [2]

ii) The diagram below represents some of the energy levels of the helium atom. Use the information in the diagram to explain how absorption at 588 nm arises.

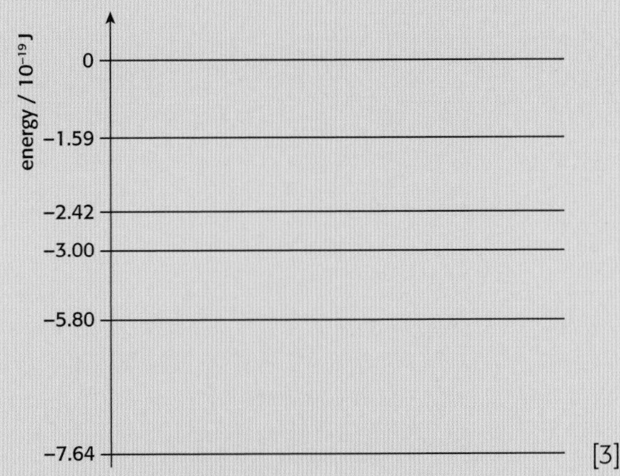

[3]

Two different models have been developed to explain the existence of **atomic** energy levels. The **Bohr model** and the **Schrödinger model** are both able to predict

the principal wavelengths present in the spectrum of atomic hydrogen.

c) Outline

i) the Bohr model, and

ii) the Schrödinger model. [6]

Nuclear aspects

d) The helium in the Sun is produced as a result of a nuclear reaction. Explain whether this reaction is burning, fission, or fusion. [2]

At a later stage in the development of the Sun, other nuclear reactions are expected to take place. One such overall reaction is given below.

$${}^{4}_{2}\text{He} + {}^{4}_{2}\text{He} + {}^{4}_{2}\text{He} \rightarrow \text{C} + \gamma + \gamma$$

e) i) Identify the atomic number **and** the mass number of the isotope of carbon C that has been formed.

Atomic number:

Mass number: [2]

ii) Use the information below to calculate the energy released in the reaction.

Atomic mass of helium = $6.648\,325 \times 10^{-27}$ kg

Atomic mass of carbon = $1.993\,200 \times 10^{-26}$ kg [3]

Another isotope of helium ${}^{6}_{2}\text{He}$ decays by emitting a β^{-}-particle.

f) i) State the name of the other particle that is emitted during this decay. [1]

ii) Explain why a sample of ${}^{6}_{2}\text{He}$ emits β^{-}-particles with a **range of energies**. [2]

iii) The half-life for this decay is 0.82 s. Determine the percentage of a sample of ${}^{6}_{2}\text{He}$ that remains after a time of 10 s. [2]

(Total 27 marks)

2 This question is about the photoelectric effect.

The following are two observations relating to the emission of electrons from a metal surface when light of different frequencies and different intensities is incident on the surface.

i) There exists a frequency of light (the threshold frequency) below which no electrons are emitted whatever the intensity of the light.

ii) For light above the threshold frequency, the emission of the electrons is instantaneous whatever the intensity of the light.

Explain why the wave model of light is unable to account for these observations.

(Total 6 marks)

3 This question is about the wave nature of matter.

a) Describe the concept of matter waves and state the de Broglie hypothesis [3]

b) An electron is accelerated from rest through a potential difference of 850 V. For this electron

i) calculate the gain in kinetic energy. [1]

ii) deduce that the final momentum is 1.6×10^{-23} Ns. [2]

iii) determine the associated de Broglie wavelength. (Electron charge $\boldsymbol{e} = 1.6 \times 10^{-19}$ C, Planck constant $\boldsymbol{h} = 6.6 \times 10^{-34}$ J s) [2]

(Total 8 marks)

4 This question is about atomic spectra.

An electron undergoes a transition from an atomic energy level of 3.20×10^{-15} J to an energy level of 0.32×10^{-15} J. Determine the wavelength of the emitted photon.

(Total 3 marks)

18 Global energy sources

This chapter covers a crucial area of physics, which impacts on all of us every day, no matter what part of the world we live in. Our societies are completely dependent on energy resources and power generation in order to function.

Before starting this chapter, make sure you:

- understand the difference between renewable and non-renewable energy sources
- know how electrical energy is generated from fossil fuels
- understand the principle of conservation of energy and how to apply it in a range of examples
- understand how thermal energy is transferred from place to place.

This chapter will compare energy use, electricity generation and the pollution they cause, between the richer and poorer nations of the world. You will need to understand the topics covered from the perspective of different nations, and show some empathy for the varying economic and social conditions around the world.

As a useful starting point we shall take the member countries of the OECD (Organization for Economic Cooperation and Development) as examples of richer, more developed countries, and non-OECD countries as the less developed or developing nations.

Use the OECD website, www.oecd.org (accessed July 2009), and other sites to answer the following questions.

1 Which countries are currently members of the OECD?
2 What are the requirements for membership of the OECD? List some common features of the countries that are OECD members.
3 Brazil, Argentina, India, and China are not members of the OECD, but Mexico, South Korea, and Japan are. Explain why this could be, giving suitable reasons.

World energy consumption

All aspects of the global economy, including industrial production, financial and other services, agriculture and transport, require an adequate supply of **primary energy**, predominantly in the form of fossil fuels, and of **secondary energy**, in the form of electricity and consumable fuels. A primary energy source such as coal, crude oil, or gas is fuel in its initial form. Secondary energy sources such as electricity and petrol (gasoline) are produced from primary sources. The thirst for economic growth and technological development, particularly in the more developed countries, has led to an **unsustainable** growth in the demand for energy, with the unwelcome side effect of pollution and its possible impact on the world's climate.

In 1987 the UN World Commission on Environment and Development (the Brundtland Commission) defined the term **sustainable development** as:

> *"Development which meets the needs of the present without compromising the ability of future generations to meet their own needs".*

Although this definition is 20 years old, it is still regarded as applicable today.

World economic growth has been uneven. The rate of growth in the richer areas has far outpaced that in the poorer, less developed countries. In the race for development, partly to alleviate poverty, the less developed countries have often been forced to make choices about their energy generation that have resulted in accelerating pollution with few controls on output through legislation.

Globally, society has to balance the ever-increasing demand for energy with the need to make changes sustainable so that future generations can enjoy the benefits of new technologies. The world map in Figure 1 illustrates some of the differences between the main regions of the world. It shows total annual electrical energy consumption per head of population in 2006.

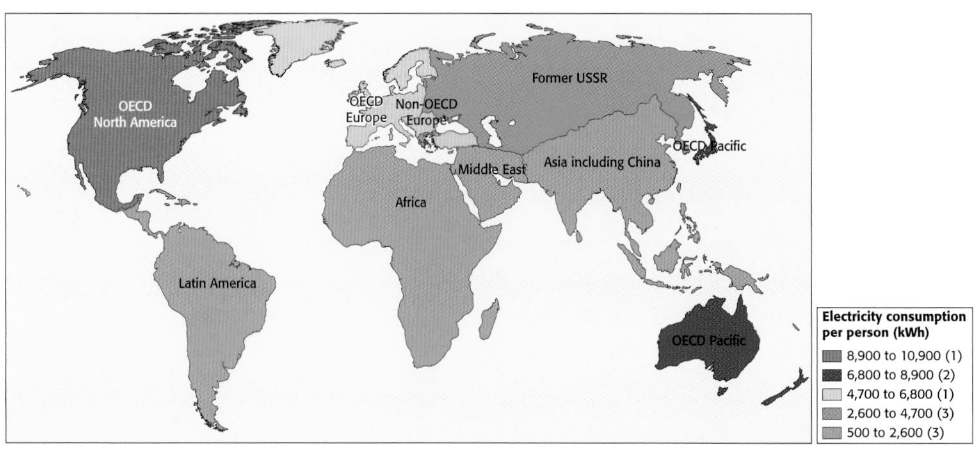

Figure 1 World map illustrating annual electricity consumption per capita in 2006

Inquiry

Data such as that contained in Figure 1 can be misleading. For example, it does not show us which areas of the world use the most electricity. To know this you will need information on human populations. A good website to start researching this would be the UN site, www.un.org (accessed July 2009).

1 Use the data in the map in Figure 1 to outline differences between OECD and non-OECD countries in terms of electricity consumption.

2 How do you think a map showing total electricity consumption would differ from Figure 1?

See www.iea.org/Textbase/country/maps/world/ele.htm (accessed July 2009).

Trends in energy use: The view from an oil company

Governments need to predict energy demand many years in advance to ensure that the necessary power-generating and refinery capacity are built and ready to meet the expected usage. In making these projections they need to consider factors such as:

- predicted population growth
- projected economic growth over several years
- costs, efficiencies and trends in the use of different fuels
- the link between carbon dioxide emissions (produced in the combustion of fossil fuels) and climate change.

The passage below is adapted from the Shell Petroleum Company website, http://www.shell.com.

The International Energy Agency (IEA) and Shell's own scenarios expect energy use to grow by more than half over the next quarter century. Demand could double by 2050—(see the graph in Figure 2).

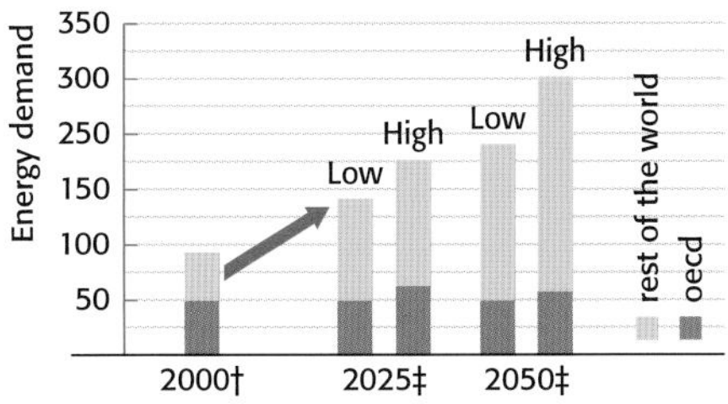

† Source: Shell Global Scenarios 2004
‡ Source: Shell Long-Term Energy Scenarios 2001

Figure 2 Rising global energy demand. (100 = Global primary energy demand in year 2000)

The increase by 2025 represents more than the current energy consumption of North America and the European Union combined. Almost all of this growth looks set to come from developing countries, in particular China and India, as they continue to industrialize and lift billions of people from poverty. Reducing poverty in the developing world and maintaining prosperity in today's industrialized economies depends on expanding the supply of convenient and secure modern energy.

At the same time, supplies need to be kept safe from interruptions. A wide range of energy options will be needed to avoid over-dependence on any one region or energy source. Energy conservation can provide part of the answer. Substantial improvements in efficiency can be made, quickly and cost-effectively. But conservation alone cannot meet the challenge of supplying the vast quantities of energy needed for development.

Alternative energy such as wind, solar power and biofuels can provide some of the energy required. Today these sources meet less than 1% of the world's energy needs, but with government support and the cost reductions we and others are working to achieve, their use could expand quickly. Our scenarios expect them to grow several times faster than fossil fuels and to become a larger part of the energy mix. Even so, with so much extra energy needed, these alternatives would still be supplying less than 10% of energy demand by 2025.

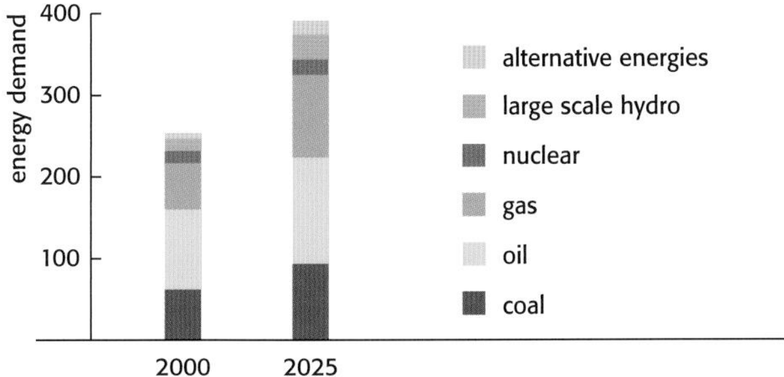

Figure 3 Changing energy mix (million barrels oil equivalent per day).

The greatest part of the energy needed will continue to come from fossil fuels. Oil and natural gas supply more than 50% of the world's energy today. Yet finding new sources is getting tougher. This is not because the world is running out. There are still enough resources left to be discovered and developed, but these are in increasingly remote and challenging locations and in "unconventional" oil deposits, such as oil sands and oil shales. More investment, more advanced technology and new operating practices will be required to develop them cost-effectively and responsibly.

Managing the environmental and social impacts of this growing demand for fossil fuels is a crucial and complex task. Climate change poses the biggest challenge. Despite more fossil fuel use, the world's greenhouse gas (GHG) emissions will need to be no higher than today's level (and falling) by 2050, if GHG levels in the atmosphere are to stabilise by the end of this century. Technology, government policy and changes in behaviour will all be needed to manage GHG emissions. Technology and operational changes must also be found to address impacts on biodiversity, air quality and water, and to generate benefits for local communities.

1. Use the bar chart in Figure 2 to estimate the percentage growth in global energy demands by 2025 and 2050 assuming (a) a low-growth scenario and (b) a highgrowth scenario.
2. Describe some of the conditions that may ensure that the high-growth scenario is achieved.
3. Should governments should plan for high growth, low growth, or somewhere between the two? Explain.
4. Explain why countries such as China and India need to plan for much higher growth in energy demand than most OECD countries, despite concerns about pollution and climate change.
5. Outline why OECD countries need to plan for energy growth, even though their populations are relatively stable.
6. Explain why it is important that countries do not become dependent on only one type of energy source.
7. Use the bar chart in Figure 3 to outline how the use of different energy sources is predicted to change.
8. Oil companies operate to sell energy for profit. Explain whether you think this is a responsibly written article or an example of bias from an oil company. Give clear reasons to support your arguments.

Some of the reasons for your answers to question 7 will be discussed later in Chapter 22 as we consider global development in the light of possible global warming and climate change.

Thinking about science: *Ethics and morality*

The IB Diploma Programme TOK guide describes ethics as involving,

> *"a discussion of the way we ought to live our lives, the distinction between right and wrong, the justification of moral judgments, and the implications of moral actions for the individual and the wider group."*

This chapter gives you plenty of opportunity to discuss the ethical issues raised by energy generation and its consequential contributions to pollution and climate change. Ask yourself: is there such a thing as absolute right and wrong? Your answer may depend on your personal moral values, which will be influenced by your cultural background, education, and conscience. In a situation when two "moral" people disagree on an ethical judgment, how do we decide who is right or wrong?

Consider the two statements below, concerning a coal-fired power station and a plan to increase its output.

A Increasing the output of power station X will provide cheaper energy because of economies of scale, and will enable more people to have access to an affordable and reliable supply of electricity.

B Increasing the output of power station X will increase carbon dioxide and sulfur dioxide levels locally, which will contribute to global climate change.

Each statement identifies winners and losers, whether the output is increased or whether it stays the same. There are different winners and losers depending on the point of view. If moral judgments are based on values, and values differ from one person to another, there is an ethical dilemma here. How can we resolve this dilemma?

One simple method is to apply the **principle of utility**, which states that the supreme good is that which tends towards the greatest good of the greatest number of people. In this sense "good" refers to happiness, satisfaction, or pleasure. If an action tends to achieve this, it is deemed good; if it does not, it is considered wrong. This can lead to problems, though: for example, cutting power output in order to reduce climate change globally may lead to unemployment, hardship, and eventual poverty for power workers.

1. Apply the *principle of utility* to the power station dilemma mentioned above and suggest whether scenario A or B is preferable.

2 Explain whether you think the *principle of utility* gives the "best" outcome in this case.

The Three Gorges Dam on the Yangtze River in China should produce 22 000 MW of electrical power when at full capacity in 2011. This will replace at least 20 conventional coal-fired power stations. The reservoir formed behind the dam will have an area greater than 1000 km^2, and will displace over 1.4 million people from their cities, towns, and farms (see Chapter 22).

3 Use the principle of utility to decide whether the construction of the Three Gorges Dam can be justified. Can the displacement of so many people be considered an ethically correct decision?

The conservation of energy

A foundation stone of much of physics is the law of conservation of energy:

> *Energy may not be created or destroyed; but it can be converted from one form into another.*

The common phrase "the generation of energy" is misleading, because it implies energy can be created out of nothing. We should instead be focusing on energy changes. Rather than say that we generate electrical energy when a fuel is burned, it would be better to say:

> *"We release the stored chemical energy from a fuel in the form of thermal energy, which may then be converted into kinetic energy that can turn a generator, which converts this into electrical energy."*

Figure 4 shows this process pictorially.

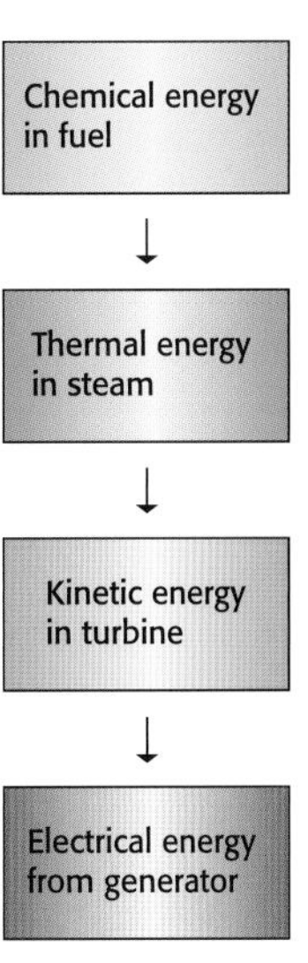

Figure 4 Energy flow from fuel to electrical energy in a thermal power station

We are not creating energy; we are simply converting it from one form into another. Inevitably there are losses of useful energy along the way (often in the form of thermal energy or sound). This leads to **energy degradation**, in which thermal energy is transferred to the surroundings, as increased particle kinetic energies that cannot be recovered. The law of conservation of energy can always be strictly applied in any physical process, but there will be some losses that cannot be prevented: some energy will be degraded into thermal energy that can no longer usefully be used.

Engineers work hard to reduce the amount of energy degraded. From a theoretical perspective, it is possible to design single processes where thermal energy is completely converted into useful work. If, however, we wish to design a machine for the continuous conversion of thermal energy into work then we need to think in terms of a cyclical process. In this case, further theoretical analysis shows that energy must always be degraded.

For example, the expansion of an ideal gas at constant pressure as a result of thermal energy input could convert thermal energy into useful work and avoid all degradation of energy but it would be impossible to design any machine around the continual expansion of a gas. The reason is we cannot allow the physical expansion of the gas to continue forever. In order to limit the size our machine we need to consider a cycle of changes that returns the machine back to its starting position after one cycle. This has to involve the degradation of energy.

From your previous studies you will know that the efficiency of any energy transfer process is defined as follows:

$$\text{efficiency} = \frac{\text{useful energy obtained}}{\text{total energy in}}$$

The efficiencies of most coal-fired power stations can be as low as 30%. Reductions in energy degradation can improve efficiencies significantly (see Chapter 21).

Renewable and non-renewable sources

Renewable sources of energy are those which are produced at a quicker rate than they are consumed (Table 1).

Table 1 Renewable and non-renewable sources of energy

Renewable sources	Non-renewable sources
Hydro • water stored in dams • tidal • pumped storage	Fossil fuels • coal • oil • gas
Wind	Nuclear (uranium)
Solar	Waste
Wave	
Geothermal	
Biofuels, e.g. ethanol	

Some biofuels such as wood could be renewable or non-renewable.

Figure 3 illustrates the world's reliance on fossil fuels. Most of the category "other" in the figure is hydroelectricity, but it also includes the combustion of wastes and biofuels, as well as solar, geothermal, wave, and wind energy.

The primary source

Nearly all the sources of energy mentioned above derive their energy from the Sun. Electromagnetic **radiation** arrives at the Earth's surface. Some of this is then re-emitted back into space, but a small amount is absorbed by plants through the process of **photosynthesis**, in which water and carbon dioxide in the atmosphere combine to form glucose, a chemical energy store. The overall reaction can be summarized as

carbon dioxide + water → glucose + oxygen

or in chemical notation

$$6CO_2 + 6H_2O \rightarrow C_6H_{12}O_6 + 6O_2$$

The key idea here is that carbon dioxide is removed from the atmosphere by plants, which act as **carbon stores**.

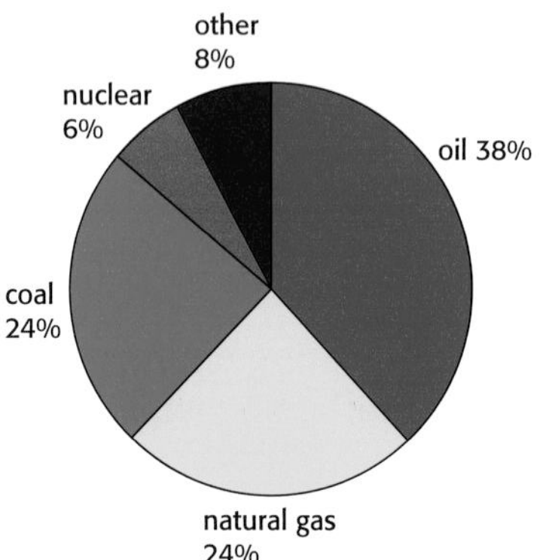

Figure 5 World's proportional energy consumption by fuel in 2003

The fuel crisis

A frequent cause for concern over the last 40 years or so has been the so-called "energy crisis". As we mentioned earlier, energy is conserved, and so does not run out. However, it *does* become degraded, and is therefore no longer conveniently usable.

A fuel is a store of energy in a convenient form. When we talk about fuels we usually think about fossil fuels – combustible sources that may be burnt in the presence of oxygen to produce thermal energy – and nuclear fuels. The **fuel crisis** indicates that we are running out of fossil fuels. Figure 6 illustrates how quickly we have exploited oil resources, for example.

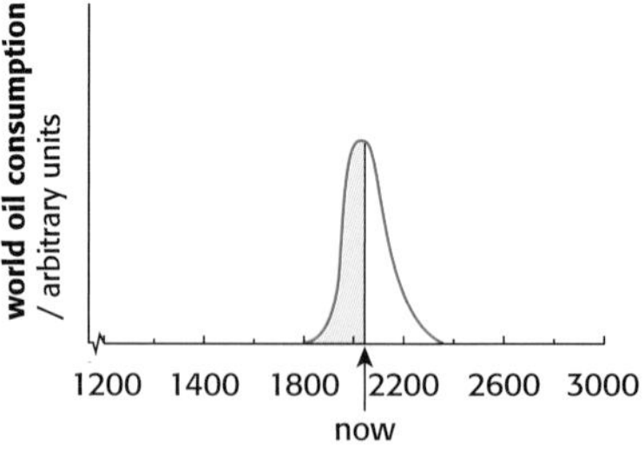

Figure 6 Historical illustration of world oil consumption

It seems that all oil will be exhausted within a few centuries at current estimates – assuming that, as reserves start to run out, so our consumption falls. At current rates of consumption oil reserves will be exhausted significantly sooner.

Oil companies and governments invest significant amounts of capital in predicting the amount of available crude oil. These predictions are complicated, and expensive geophysical surveys are needed to determine how much oil is actually stored under the ground and how much is physically recoverable. Figure 7 illustrates how the estimates of known oil reserves have changed over time.

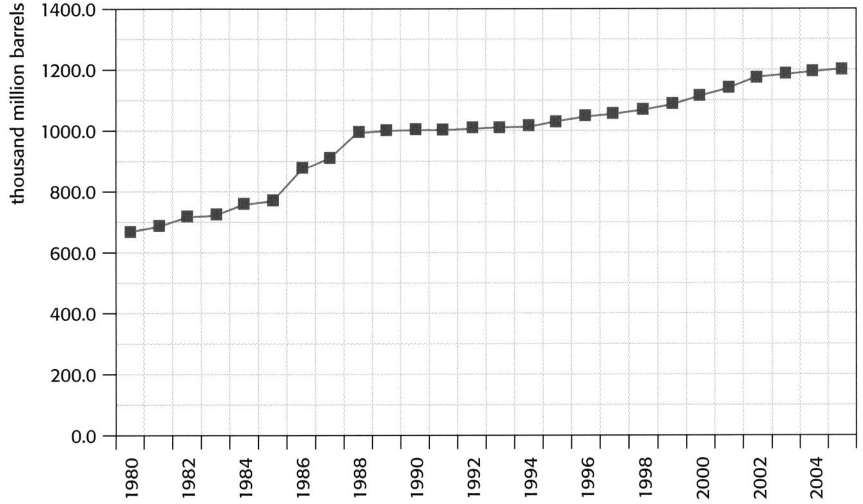

Figure 7 World oil reserves over the last few decades

There are several reasons why the estimate of known reserves has actually increased.

- More oil has been discovered, in new deposits.
- Thanks to advances in technology, known as enhanced oil recovery, more oil can now be extracted from reservoirs that were previously thought to be exhausted.
- The rising price of oil has meant that some previously discovered sources are now economically worth exploiting.

Oil companies use the term **"ultimately recoverable resource"** as an estimate of the total amount of oil that will ever be recovered and produced. This resource can be broken down into three main categories:

- cumulative production
- discovered reserves, and
- undiscovered resources.

Cumulative production is an estimate of all of the oil produced up to now, **discovered reserves** are an estimate of known oil deposits with a good probability of recovery, and **undiscovered resources** are based on geological knowledge together with some estimation.

Despite these predictions, it is widely projected that the world's fossil fuel dependence will continue to increase over the next few decades, as shown in Table 3. All values are in quadrillion British thermal units (10^{15} Btu), or quads. (The quad is an internationally used unit of energy; 1 quad = 1.1×10^{18} J.)

1. Briefly outline how fossil fuels are formed, and suggest a timescale for this formation.
2. Describe how the chemical energy stored in fossil fuels originated from the Sun.
3. Outline how the following energy sources derive their energy from the Sun
   a) hydroelectricity – water stored behind dams
   b) wind power
   c) ethanol.
4. Nuclear fuels do not derive their energy from the Sun. Describe the origin of the energy stored in uranium.
5. Outline the origins of geothermal energy.

1. Explain the difference between the terms resources and reserves when applied to fossil fuels.
2. Use Table 2 to estimate how long fossil fuels are expected to last at current rates of production.

Table 2 Fossil fuel reserves and production in 2005

Fuel (and unit used)	Proved reserves in 2005	Production in 2005
Coal (10^9 tonnes)	909.1	5.85
Oil (10^9 barrels)	1200.7	29.6
Gas (10^{12} m^3)	179.8	2.76

Source: BP Statistical Review of World Energy June 2006

3. Give reasons why the fuels may last for either (a) a longer time or (b) a shorter time than your estimate in question 2.

Table 3 Projection of fossil fuel dependence / quads

	1990	2002	2003	2010	2015	2020	2025	2030
Total non-OECD								
Oil	52.7	63.3	65.1	82.3	91.6	100.4	109.8	120.0
Natural gas	38.0	44.7	47.2	63.1	75.1	87.0	99.6	113.1
Coal	45.9	51.8	54.8	79.1	92.9	105.7	118.4	132.4
Nuclear	3.1	4.0	4.2	5.5	7.0	8.7	10.0	10.6
Other	10.3	14.6	15.2	23.6	26.8	29.6	33.1	36.7
Total	150.0	178.4	186.4	253.6	293.5	331.5	371.0	412.8
Total OECD								
Oil	83.4	95.4	97.0	103.3	107.5	110.3	114.5	119.1
Natural gas	37.2	51.2	51.9	58.0	64.7	69.1	72.9	76.8
Coal	43.5	44.9	45.6	49.7	51.4	54.3	58.3	63.1
Nuclear	17.3	22.7	22.3	23.4	24.0	24.3	24.1	24.1
Other	5.9	17.6	17.4	21.7	22.3	23.5	24.7	25.7
Total	197.4	231.9	234.3	256.1	269.9	281.6	294.5	308.8

Source: History, Energy Information Administration (EIA), International Energy Annual report 2003 (May-July 2005), website www.eia.doe.gov/iea/. Projections: EIA, *Annual Energy Outlook 2006.*

Inquiry

The data in Table 3 would be much easier to analyse in graphical or visual form. Use suitable graphs to prepare presentations on the following topics.

1 Demonstrate how the world total use of fossil fuels is projected to increase over this timescale, showing clearly the change in share of each fossil fuel.

2 Compare the total energy dependence of OECD countries with that of non-OECD countries.

3 Compare the growth in the use of coal with the use of nuclear energy in both OECD and non-OECD countries.

4 The "other" category refers to renewable sources of energy. Compare the projected growth of these renewable sources with that of non-renewable sources.

Describe the trends in detail, and use your knowledge and suitable websites to explain your findings. Think carefully about the best type of graph to use to display your data, and remember that appropriate use of colour can have a strong visual impact on your audience. Some good websites to start with are:

www.eia.doe.gov/iea/

http://www.bp.com/statisticalreview (both accessed July 2009)

The suggested topics above are merely ideas, and should not limit your research. The data can be used to suggest many interesting hypotheses and conclusions.

During your research you may have encountered several different energy units.

5 Find out the value of the following energy units in joules:

a) megatonne of coal equivalent (Mtce)

b) megatonne of oil equivalent (Mtoe)

c) cubic metre of natural gas (m^3)

d) barrel of oil (bbl)

e) kilowatt-hour (kWh).

Fossil fuel dependence

Fossil fuels can be used directly in the home to provided heating but historically their use tended to grow as a result of industries being established that relied upon their use. Initially these industries tended to be located near natural deposits but soon infrastructure networks were established to transport the fuels around and between countries. The high energy density of fossil fuels (see pages 286–7) means that transportation and storage can be achieved relatively easily. Their use is, however, not without some disadvantages. Not only are fossil fuels non-renewable but their use has been linked with global warming (see chapter 20), their combustion products are polluting (notably producing acid rain) and the extraction of fossil fuels can damage the environment.

Despite all this, most forecasts of global energy consumption over the next few decades predict that:

- Global energy demand is going to increase.
- Fossil fuels will continue to provide most energy.

This may seem surprising: fossil fuels are running out, and there is a long-term need for any future development to be sustainable. Are governments being irresponsible in planning to use up valuable resources? Are they planning with due consideration for the environment?

Sustainable development is built on three key ideas or pillars: economic aspects, social aspects, and environmental aspects. There is a tendency to focus only on environmental issues when considering sustainable development, but the social and economic aspects are just as important. Growth in energy production, through increased use of coal for example, can contribute significantly to a country's economic growth, generating much-needed revenue. Although increased fossil fuel use has a detrimental effect on the environment, electrification can do much to reduce poverty in developing countries, improve healthcare, and provide employment. Improving efficiencies of power production and modern technologies may significantly reduce the impact of fossil fuel use on the environment.

1 Making predictions such as those in Figure 8 is very challenging. Outline the information a forecaster might need in order to make these predictions.

2 Outline the circumstances that may cause the predictions to be:

   a) an overestimate

   b) an underestimate.

3 Calculate the predicted dependence on fossil fuels in 2010 and 2030 as a percentage.

4 Outline some reasons why increased fossil fuel use may be considered:

   a) sustainable

   b) unsustainable.

5 Explain, with reference to economic, social, and environmental developments, why governments planning for increased use of fossil fuels can still be considered to be responsible.

6 One way to reduce greenhouse gas emissions and still plan for increased energy demand would be to increase the share of renewable fuels and nuclear power. However, if you compare the projections for 2010 and 2030, the nuclear and renewable share (shown as "other" in the graphs above) remains virtually constant. Outline some reasons why the proportions of

   a) nuclear power, and

   b) renewable energy generation

   are not expected to increase despite the pressure for development to be more sustainable.

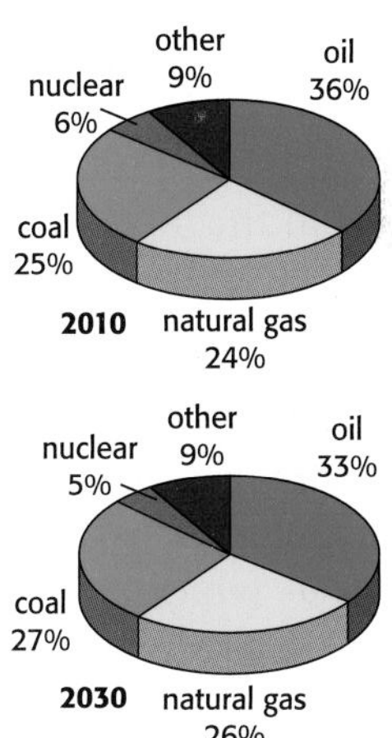

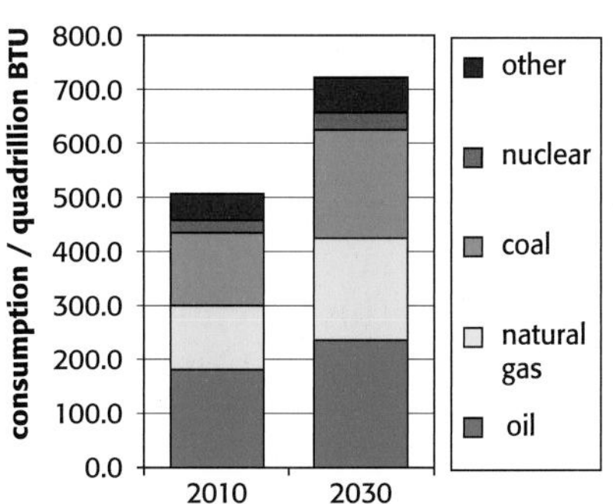

Figure 8 Predicted world energy consumption by fuel, in 2010 and 2030

Inquiry: Historical use of fossil fuels

The use of coal has a long history. The first commercial use is thought to have been in China around 1000 bc, where it was used to smelt copper and cast coins. Coal proved to be the main fuel in the Industrial Revolution in the West in the 18th century, when it was used in the iron-making process. The 1960s saw oil overtake coal, with the growth in the transportation sector, and the use of gas is likely to increase rapidly in the future as gas turbines to generate electricity are proving cleaner and more efficient than coal or oil (see page 319), although in 2006 coal is still the major fuel used for electricity production.

Research the historical use of fossil fuels and produce a timeline of your findings – this may help you understand why we rely so much on these sources today.

Thermal generation of electricity

About 20% of primary energy is converted into electrical energy. For many people, electricity provides a relatively cheap and reliable source of energy. However, more than 1.6 billion people in 2006 had no access to electricity – about 25% of the world population – and this is not predicted to fall significantly (1.4 billion by 2030).

Figure 9 illustrates the proportions of different energy sources used to generate electrical energy in 2006.

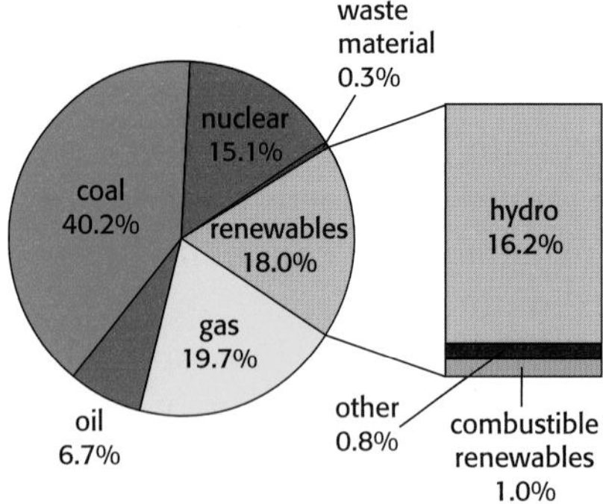

Figure 9 The proportions of different energy sources used to generate electrical energy in 2006

The "other" category under renewables includes wind, solar, and geothermal together. Although only 0.8% at present, it is likely that the use of wind turbines will increase dramatically in the future (see page 327).

In Chapter 15 you learnt about the physics of electricity generation. The basic principle is to rotate a magnet inside a coil. In a typical thermal power station coal is used to produce steam at high pressure, which then turns a turbine connected to the generator. The process is illustrated in the block diagram shown in Figure 10.

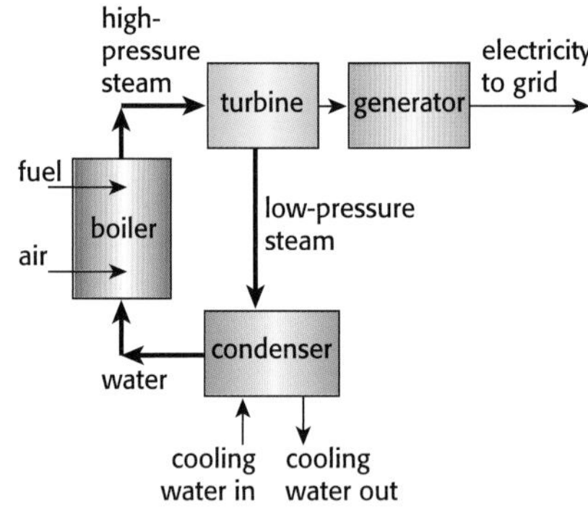

Figure 10 Block diagram of a thermal power station

Burning coal in power stations contributes significantly to global carbon dioxide emissions. In addition, coal usually contains impurities, often sulfur. Nitrogen from the air also reacts in the high temperatures used for combustion. Additional waste products include sulfur dioxide (SO_2) and various nitrogen oxides indicated by the formula NO_x, which produce acid rain. Despite this, coal is – in 2007 – used more than any other fuel in the generation of electrical energy, and this is likely to remain true for some time.

Technological improvements in the use of coal and design of power stations mean that coal has the potential to be cleaner and more efficient in the future: see Chapter 21.

Estimating the energy density of fuels

Fossil fuels have a relatively high **energy density** compared with other fuels. This is defined as the total energy released per kg burnt, and is measured in J kg^{-1}.

Energy is released from fossil fuels by combustion, as summarized in highly simplified form by the following equation:

$$\text{fuel} + \text{oxygen} \rightarrow \text{carbon dioxide} + \text{water}$$

Figure 11 Heating water to find the energy density of methylated spirits

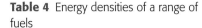

It is quite straightforward to design and carry out an experiment that gives a reasonable estimate of the energy density of various fuels using apparatus available in your school science laboratory. One way to estimate energy density would be to set up an arrangement so that a fuel is used to heat a known mass of water (Figure 11).

The formula $Q = mc\Delta T$ enables you to estimate the thermal energy required to raise the temperature of the water by a measured amount.

If the mass of fuel is measured before and after heating, then you can calculate the mass of fuel used to produce the temperature rise in the water, and hence the amount of thermal energy needed.

Any experiments involving heating will involve the loss of a significant amount of energy to the surroundings – the air, the vessel containing the water, the thermometer, and so on. Great care is needed to ensure that the water is heated in the same way by each of the fuels selected and, to ensure some reliability, tests have to be repeated.

This investigation would be easily assessible for the three skill areas of:

- data processing and presentation (DCP)
- concluding and evaluating (CE)
- manipulative skills (MS).

Any experimental work involving the burning of fuels will require careful attention to safety, and this is assessed in MS. Suitable precautions must be taken to remove the fumes produced, either by ensuring adequate ventilation in the laboratory with extractor fans or, preferably, by doing the whole experiment in a fume cupboard.

Table 4 Energy densities of a range of fuels

Simple compounds	Energy density / MJ kg^{-1}
carbon (to CO_2)	32.8
ethanol	29.7
hydrogen	142
propane	50.3

Food	Energy density / MJ kg^{-1}
carbohydrates	17.2
fats	38.9
proteins	17.2

Biomass fuels	Energy density / MJ kg^{-1}
charcoal	29.0
coconut husks	9.8
dung, air-dry	12.0
rice husks	15.5
maize cobs	18.9
maize stalks	18.2
peat	14.6
wood, green	10.9
wood, air-dry	15.5
wood, oven-dry	20.0

Fossil fuels	Energy density / MJ kg^{-1}
gas (methane)	55.5
coal, anthracite	31.4
coal, bituminous	$>$ 23.9
coal, sub-bituminous	17.4–23.8
coal, lignite	$<$ 17.43
diesel	45.3
gasoline, automotive	45.8
gasoline, aviation	43.1
jet fuel	43.4
kerosene	46.3
oil, crude (petroleum)	41.9
oil, heating	42.5

Source: http://hypertextbook.com/physics/matter/energy-chemical/ (accessed July 2009)

Use the data in Table 4 to answer the following questions.

1. Calculate the mass of (a) anthracite coal, (b) automotive gasoline (petrol), and (c) methane that must be burnt to raise the temperature of 500 g of water from 15 °C to 85 °C. (The specific heat capacity of water is 4200 J kg^{-1} K^{-1}.)
2. Outline any assumptions you have made in answering question 1.
3. Without considering any other factors, explain which fuel would be the fuel of choice for a power-generating company.
4. What other factors would need to be considered when selecting a suitable fuel for the generation of electrical energy?

The Colbert fossil plant in Alabama, USA, uses anthracite coal to generate up to 7.8 billion kWh of electrical energy annually, supplying over half a million homes.

5. Assuming a maximum efficiency of 30%, estimate the mass of coal consumed in: (a) one day; (b) one year.
6. Estimate: (a) the rate of electrical energy production; (b) the mass of coal used per household.

The German power company RWE is planning to build a more efficient coal-fired power station in Germany, using clean coal technologies, possibly by 2014. It is likely to use lignite as the fuel, and generate electrical energy at 450 MW with a planned efficiency as high as 45%. Carbon dioxide produced will be captured and stored deep underground – see pages 321–2.

7. Estimate the minimum mass of lignite used: (a) in one day and (b) over 1 year to operate this power station.

8 Explain why the mass of lignite calculated is a minimum value.

The Pratum Rice Mill, located in Pathumthani Province, Thailand, generates electrical energy in an innovative way. The primary fuel is rice husks, which are used to run a generator rated at 9.2 MW. The electricity is used to power the rice plant itself, and a surplus of up to 5.5 MW will be used in the Thailand national grid.

9 Assuming an overall efficiency of 28%, estimate the mass of rice husks needed to operate the plant annually.

10 Compare the energy density of biofuels and fossil fuels. Discuss whether large-scale or small-scale use of biofuels in electrical energy generation is most appropriate. Give reasons for your answer.

Inquiry: Energy end use

It is convenient to categorize the end use of energy into four sectors:

- industry
- commercial
- residential
- transportation.

The US-based Energy Information Authority (EIA) website presents a huge amount of world data for public use: www.eia.doe.gov (accessed July 2009). It includes several convenient spreadsheets that you can download for your own inquiry work. The world data for 2003 (Figure 12) is quite useful when considering the end use of primary energy sources.

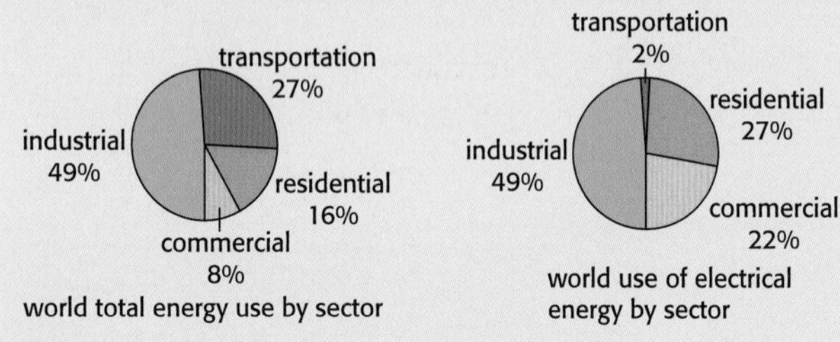

Figure 12 World end use of energy in 2003

1 Use the EIA website or other sources to define clearly what each category of end use means.

2 Into which category would educational establishments – schools, universities, and colleges – fit?

3 Find out the proportion of use of each fossil fuel in each sector, and explain why a certain fuel is preferred in each case – it may be for historical, social, or economic reasons.

4 Currently, transportation uses up more than a quarter of our primary energy. Find out: (a) what is predicted to happen in the future to the amount of energy used for transport; (b) what is projected for the proportion of electricity used in transport. Back up your findings with suitable data.

5 A challenging task would be to try and calculate, from the data available, the proportion of electrical energy used in our total end use of energy – one difficulty being the need to convert all energy units into a common unit before making any calculations. (An answer of between 15% and 20% would be reasonable, depending on which data source you use.)

Table 4 can be used to gain an insight into the energy content of some fuels, particularly those available to many people living in developing countries.

Physics issues: energy and poverty

The alleviation of poverty is a major global challenge. One aim of sustainable development is to provide accessible supplies of affordable energy to the rural poor in developing countries, to help to lift them out of poverty. However, simply providing a supply of electricity will not necessarily reduce poverty. The most successful development projects in rural areas involve local people at all stages, from initial planning to final completion. Such "ownership" is crucial for the long-term viability of such projects.

The Institute of Energy (IEA) states some key observations and predictions:

- Currently, some 1.6 billion people – one-quarter of the world's population – have no access to electricity. Without new policies, 1.4 billion people will still be without any supply of electricity in 2030.
- Four out of five people without electricity currently live in rural areas of the developing world, mainly in South Asia and sub-Saharan Africa, but the pattern of electricity deprivation is set to change: 95% of the increase in population in the next three decades will occur in urban areas.
- Some 2.4 billion people currently rely on traditional biofuels – wood, agricultural residues, and dung – for cooking and heating. That number will *increase* to 2.6 billion by 2030. In developing countries, biomass use will still represent over half of residential energy consumption in 2030.
- Lack of electricity and heavy reliance on traditional biomass are hallmarks of poverty in developing countries. Lack of access to electrical supplies exacerbates poverty and contributes to its perpetuation, as it precludes most industrial activities and the jobs they create.
- Investment will need to focus on various energy sources, including biomass, for thermal and mechanical applications to bring productive, income-generating activities to developing countries. Electrification and access to modern energy services do not necessarily guarantee poverty alleviation.
- Renewable energy technologies such as solar, wind, and biomass may be cost-effective options for specific small-scale, local applications, while conventional fuels and established technologies are likely to be preferred to generate electrical energy for the large-scale national grid.

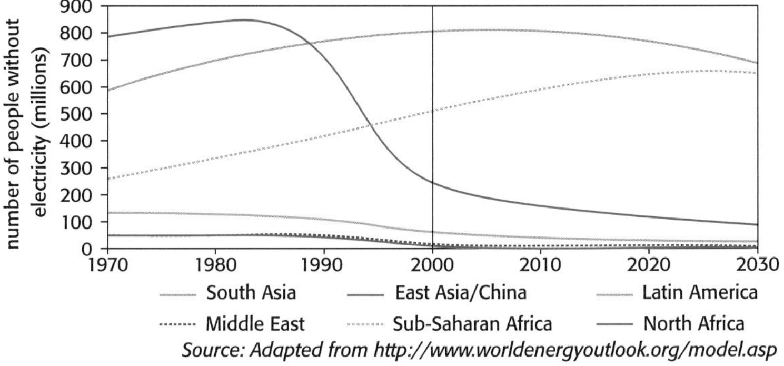

Figure 13 Number of people without electricity, 1970–2030

If a convenient supply of electricity is an indicator of development, then many countries still have some way to go to help reduce some of the constraints of the poverty. Poor people have limited access to bank loans or other capital to develop small-scale income-earning schemes. They may be forced to borrow money from friends, family, or unlicensed "loan sharks", often ending up in debt, which impoverishes them further. Lack of access to clean water and sanitation spread disease, thus making poor people less effective at work and so less likely to earn an income. Such poverty traps are often referred to as the poverty cycle.

This question is about the sustainable generation of electricity.

a) State and explain whether the energy sources used in the following power stations are renewable or non-renewable:

i) coal fired

ii) hydroelectric. [4]

b) Biogas can be produced from cow dung. This may be used as a fuel for an electricity generator. State and explain whether cow dung may be considered a renewable or non-renewable source of energy. [2]

c) One cubic metre of biogas produces 6.36 kWh when burnt in oxygen, and is used to generate up to 2 kWh of electrical energy.

i) Calculate the number of joules in 1 kWh. [1]

ii) Calculate the efficiency of the generator. [1]

d) Small-scale biogas generators can be used in the rural areas of developing countries to provide a cheap and sustainable source of electrical energy. Explain what the term sustainable means in this context. [2]

e) Assume that 100 cows produce enough dung to operate a 10 kW generator. Use the data in part (c), and your answers to it, to calculate the volume of gas burnt each second in the generator. [3]

Two illustrations of how the lack of a reliable supply of electricity can contribute to the poverty cycle are outlined below.

- Dried dung is commonly used in rural South Asia for heating and cooking. It is a cheap and reliable, though inefficient, energy source. The fumes produced are highly polluting, which causes illness, particularly amongst women, who are the major care givers, educators, and providers in the family.
- In many parts of rural Kenya it takes several hours every day to fetch and carry enough water for cooking, drinking, washing, and sanitation, so time for any income-earning activities is limited. Water cannot be pumped to many villages unless there is a reliable supply of electrical energy.

Electrification in China: A case study

China secured electricity access for almost 700 million people in two decades, enabling it to achieve an electrification rate of more than 98% in 2000.

From 1985 to 2000, electricity generation in China increased by nearly 1000 TWh, 84% of it was coal-fired, and most of the rest hydroelectric. The electrification goal was part of China's poverty alleviation campaign in the mid-1980s. The plan focused on building basic infrastructure and on creating local enterprises. China's economy grew by an average annual 9.1% from 1985 to 2000.

A key factor in China's successful electrification programme was the central government's determination and its ability to mobilize contributions at the local level. The electrification programme was backed by subsidies and low-interest loans. The programme also benefited from the very cheap domestic production of elements, ranging from hydro generators down to light bulbs. China has also avoided a trap into which many other nations have fallen: most Chinese customers pay their bills on time. If they do not, their connections are cut off.

China's transformation and distribution networks still need very large investment to match standards in more developed countries. Electricity services can be unreliable and wiring and meters in homes and offices are undependable, sometimes even unsafe. Usage is low, especially in rural areas, where consumers tend to restrict their electricity use to lighting their homes because of the cost and inconsistency of supply; power cuts are common in such areas.

1. Plot a suitable graph of the data in Table 5, and determine the relationship between population and time.
2. Calculate the proportion of the total population, based on the 2000 value, who gained access to a supply of electrical energy during the previous two decades.
3. Estimate the proportion of the population without access to electricity in 1980.
4. The expression "an electrification rate of more than 98% in 2000" used above is ambiguous: it could be interpreted in more than one way. Assuming it means that 98% of the total population have access to a supply of electricity, calculate how many people still do not have such access.
5. If the expression in question 4 refers to the growth in supply of electrical energy from 1980 to 2000, estimate the proportion of the population without access in 1980.

Figure 14 Village in Ethiopia

Figure 15 Fetching and carrying water in Kenya

Table 5 China's population growth

Year	Population
1950	562 000 000
1960	648 000 000
197 0	820 000 000
1980	984 000 000
1990	1 147 000 000
2000	1 264 587 054
2010	1 347 000 000*

**This figure is a projection, made in 2000.*

6 Compare your answers to questions 3 and 5, and comment on the consistency of the data within the passage.

7 Such rapid growth in the provision of electrical energy can have both costs and benefits to the country overall. Outline: (a) some of the costs; (b) some of the benefits.

8 Reflect on your answer to question 7 and explain whether, on balance, you think such rapid development has been beneficial to China. Your answer should include references to the three pillars of sustainable development – social, economic, and environmental.

The project planning cycle

Many IB world schools in developed countries have formed relationships with schools in less developed countries with the aim of working together to help contribute, in a small way, to the alleviation of poverty. Creativity, Action, Service (CAS) is an integral part of the IB Diploma Programme, and all students are expected to become involved with the service aspect. Initiating and developing a long-term development project with the aim of alleviating poverty in a sustainable way can be extremely rewarding and worthwhile for everybody involved.

Project planning is remarkably similar to the planning required for a full experimental investigation incorporating all group 4 practical skills, the group 4 project, or even the extended essay.

Identifying the main aims of the project is crucial to the ultimate success. The aims must be sufficiently focused to enable specific goals to be achieved, yet flexible enough to allow for changes in circumstances. The project must have positive impacts on the economic, social, and environmental needs of the communities targeted for intervention. A key aspect of this is that the local community you have chosen to work with must be fully involved in all aspects of the project planning cycle. Partnerships with schools, charities, and NGOs (non-governmental organizations) in the host country could be of great assistance, providing local knowledge and raising awareness of cultural issues on both sides.

Raising funds is one challenge, but the next is to use the funds to complete the project, e.g. building a rural school, drilling boreholes, or building suitable pit latrines to ensure healthy sanitation. It is also vital to ensure that funding for the project can be maintained, so that the capital resources can be kept functioning over several years. Sadly, some projects fail after only a few years. Often this is because funding for maintenance runs out. A visit to the country and getting actively involved in the implementation part of the project can be very rewarding. Many students find the experience of working with people who are significantly disadvantaged compared with themselves very positive, and sometimes even life-changing. This is illustrated in some quotes by students from Sha Tin College, Hong Kong, after working in an orphanage for street children in Cebu in the Philippines.

Inquiry

Use a variety of sources, including, the UN Food and Agriculture Organization (FAO) and IEA websites, to complete the research outlined below.

1 Briefly outline the predicted trends for electrification in each region illustrated in Figure 11 and try to explain the trend for: (a) China; (b) South Asia (India); (c) Africa.

2 Outline what is meant by a poverty trap, using some specific examples from your own research.

3 Discuss how access to cheap electricity may help rural communities to escape from the poverty trap. You should refer to economic and social development in your answer.

4 Describe any possible negative impacts of electrification, using a range of global examples.

"I realized that something so little can mean so much to someone, and if you didn't do that small thing it would not be the same."

"After this trip I cannot just ignore the problem of poverty, I cannot possibly see things the way I did before ... I was there with poverty – I hugged it and I changed forever."

"I truly feel that this trip has changed my life and the way I see things."

"I have learned to be truly thankful for everything I have been blessed with."

Perhaps some of these quotes will motivate you to start considering how you can help, in a small way, to alleviate poverty in the world.

References

Thomas, A. et al. 1994. *Third World Atlas*. Buckingham: Open University Press. p 62; quoting from World Commission on Environment and Development (1987), *Our Common Future* (The Brundtland Report). Oxford University Press, Oxford.

Tomkinson, J.L. 2004. *The Enterprise of Knowledge: A Resource Book for Theory of Knowledge*. 2nd edn. Athens. Anagnosis Books. p 348.

World Commission on Environment and Development. 1987. Our Common Future (The Brundtland Report). Oxford. Oxford University Press.

End of chapter summary

Chapter 18 has three main themes: energy degradation and power generation, world energy sources and fossil fuel power production. The list below summarises the knowledge and skills that you should be able to undertake after having studied this chapter. Further research into more detailed explanations using other resources and/or further practice at solving problems in all these topics is recommended – particularly the items in bold.

World energy sources

- Identify different world energy sources (both primary and secondary), compare trends in energy use between developed and developing countries and state the relative proportions of world use of the different energy sources that are available.
- Outline and distinguish between renewable and non-renewable energy sources, give examples of each energy source and discuss the relative advantages and disadvantages of various energy sources
- Describe what is meant by sustainable development and the fuel crisis.
- Define the energy density of a fuel and discuss how this influences the choice of fuel in given situations.

Fossil fuel power production

- **Outline the historical and geographical reasons for the widespread use of fossil fuels.**
- Discuss the energy density of different fossil fuels with respect to the demands of power stations and their final overall efficiency.
- Discuss the relative advantages and disadvantages associated with the transportation and storage of fossil fuels and in particular the environmental problems associated with the use of fossil fuels in power stations.
- Outline some examples of the links between energy demands and poverty.

Energy degradation and power generation

- Realise why the continuous conversion of this energy into work requires a cyclical process and the transfer of some energy from the system even though thermal energy may be completely converted to work in a single process.
- Explain what is meant by degraded energy and use Sankey diagrams to identify where the energy is degraded.
- **Outline the principal mechanisms involved in the production of electrical power.**
- Describe the stages involved in development project planning.

Chapter 18 questions

Sankey Diagrams

Energy transfers are shown in Sankey diagrams. They are useful to work out efficiencies in a system. The energy flow is represented by an arrow. The thickness of each arrow is drawn to scale to show the amount of energy.

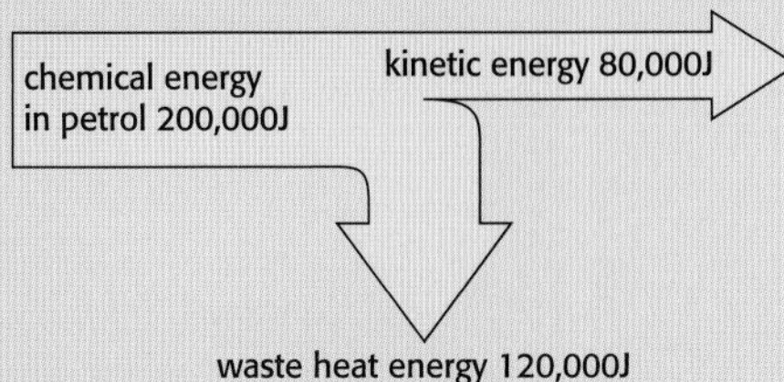

1 This diagram shows the simplistic conversion of chemical energy in petrol to kinetic energy and waste heat in a car. Calculate the efficiency of this car.

2 When 500J of coal are burnt in a power station the following energy losses occur;

a) 75J lost in the boiler

b) 225J lost to the cooling water

c) 25J is lost in the generator.

i) Calculate the amount of energy converted into useful energy.

ii) Draw a Sankey diagram, to scale, to illustrate this process.

iii) Calculate the efficiency of the power station.

3 The Sankey diagram below illustrates the energy transfer processes involved when using a steam engine to light an electric lamp. Assume it is drawn to scale.

The energy transfer processes involved in using a steam engine to light an electric lamp (source: Nuffield Physics)

Take appropriate measurements to calculate the

i) overall efficiency of the system and the proportion (as a % of the input energy) of energy lost as heat

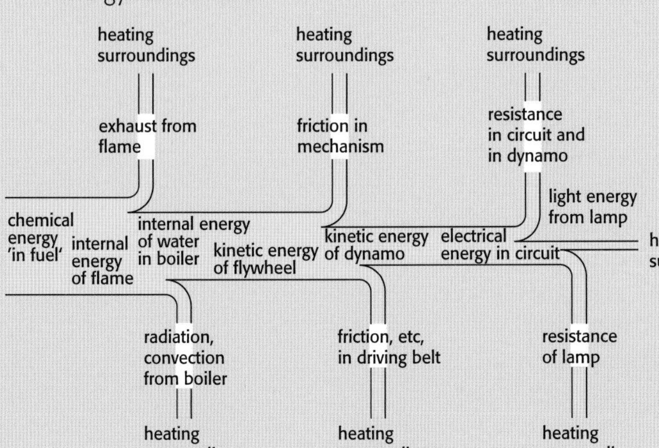

ii) due to frictional processes,

iii) resistive heating in circuits and

iv) thermal energy transfer.

4 **i)** Construct first a non-quantitative Sankey diagram for a car travelling at a constant speed.

ii) Then make it as quantitative as you can using these data: At 90 kmh^{-1} the car will use 7 litres of petrol per 100km; in order to keep the speed steady the power output at the flywheel is 15 kW and at the wheels is 12 kW. The energy produced when 1 litre of petrol is completely burnt is 10 kWh.

5 This question is about the sustainable generation of electricity.

a) State and explain whether the energy sources used in the following power stations are renewable or non-renewable.

i) Coal fired

ii) hydro electric. [4]

b) Bio gas can be produced from cow dung. This may be used as a fuel for an electricity generator. State and explain whether cow dung may be considered a renewable or non-renewable source of energy. [2]

c) One cubic meter of biogas produces 6.36 kWh when burnt in oxygen and can be used to generate up to 2 kWh of electrical energy.

i) Calculate the number of Joules in 1 kWh [2]

ii) Calculate the efficiency of the generator. [1]

d) Small scale biogas generators can be used in the rural areas of developing countries to provide a cheap and sustainable source of electrical energy. Explain what the term sustainable means in this context. [2]

e) Assume that 100 cows produce enough dung to operate a 10 kW generator. Use the data in and your answers to part (c) to calculate the volume of gas burnt each second in the generator [3]

19 Solar radiation and the greenhouse effect

Radiation from the Sun is called **solar radiation**. This chapter considers how its energy is trapped by the Earth's atmosphere, and how certain human activities are adding to this effect. Before starting any of the sections in this chapter it would be useful to:

- know that the combustion of fossil fuels releases carbon dioxide into the atmosphere
- know that our demands for energy are projected to increase, leading to increased carbon dioxide production
- explain how conduction, convection, and radiation transfer thermal energy
- understand that there is evidence of a link between atmospheric carbon dioxide concentrations and global temperatures
- appreciate how scientists use data to support their arguments, but remember that there may be different possible interpretations of the same set of data.

The greenhouse effect

Some of the gases in the atmosphere (notably water vapour, carbon dioxide, and methane) effectively form an insulating blanket around the Earth. Without this, the Earth would be a rather cold and miserable place, with an average temperature of about –20 °C. The extra warming is called the **greenhouse effect**, because greenhouses also trap the Sun's thermal energy. The gases that cause it are called **greenhouse gases**.

The greenhouse effect is necessary for life on Earth. It keeps global temperatures high enough for biodiversity to flourish. However there is evidence to suggest that human activity, mainly burning fossil fuels for energy production, has put significant amounts of extra carbon dioxide into the atmosphere over the last century or so. At present global temperatures are rising and there are predictions that this may cause dramatic climate changes and rising sea levels. Most (but not all) climate scientists believe that this **global warming** is being caused by the human, or **anthropogenic**, addition of greenhouse gases to the atmosphere. They call it the **enhanced greenhouse effect**.

Greenhouses

These buildings, made of glass or clear plastic, are used in many parts of the world to grow plants and food crops where the outside temperature is too low. For example, mangoes, a tropical fruit, can be grown successfully in Japan inside greenhouses.

Figure 1 A greenhouse

So how does a greenhouse work? A common explanation suggests that the glass allows shorter wavelength radiation from the Sun to pass through to heat the ground, but reflects back the longer wavelength infrared from the warmed ground. However, if polythene is used for a greenhouse, it performs just as well as glass, even though it is almost as transparent to longer wavelengths as it is to shorter ones. A more likely explanation is that the temperature increase occurs because the warmed air inside a greenhouse is trapped and cannot rise and flow away. In other words, convection is prevented.

The atmospheric greenhouse effect does not work like a greenhouse. So a different model is needed. By the end of this chapter, you should be able to explain the greenhouse effect using the correct physics.

1 Briefly outline how thermal energy is transferred by conduction, convection, and radiation.

2 State which of the three mechanisms is normally responsible for transferring thermal energy from soil, plants, etc.

i) to the air, and

ii) within the air.

3 A small 50 W electrical immersion heater is used to heat some water for 10 minutes (specific heat capacity of water = 4200 J kg^{-1} K^{-1}). Calculate the temperature rise in

i) 250 cm^3 of water (a beaker), and in

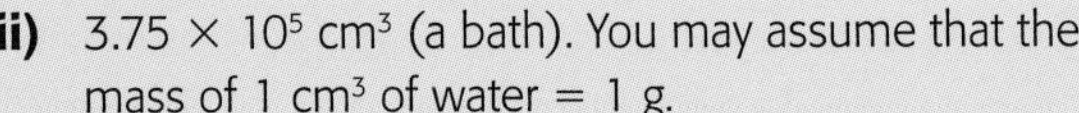

ii) 3.75×10^5 cm^3 (a bath). You may assume that the mass of 1 cm^3 of water = 1 g.

4 Use the ideas in question 3 to suggest why a greenhouse is warmer inside than outside. To do this it might be useful to consider two identical sets of tropical plants and soil, one set inside a greenhouse and the other outside.

Black body radiation

The concept of a black body is covered in detail on page 263. Stars are considered to be black bodies, but many cooler objects also approximate to black body radiators. The Earth is not a perfect black body, and this must be considered when developing a climate model to explain its observed average temperature.

Figure 2 shows some features of the radiation emitted by a black body. All wavelengths are radiated, but at different intensities, depending on the surface temperature of the object. The wavelength of the peak radiation is given by the **Wien displacement law**:

$$\lambda_{max} T = \text{constant}$$

The constant has a value of 2.90×10^{-3} m K.

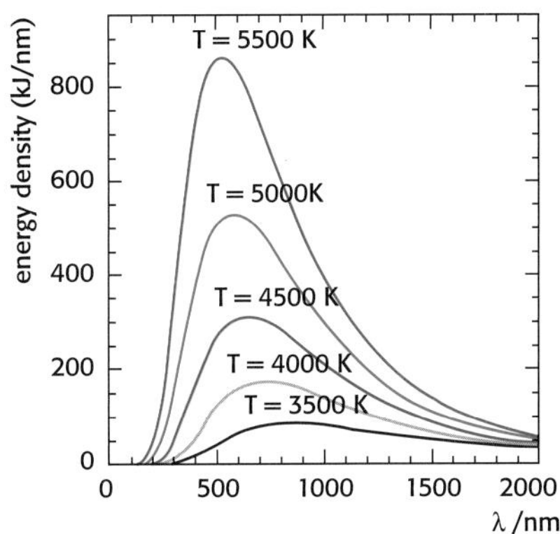

Figure 2 Black body radiation spectrum for a range of temperatures

By analysing the radiation curve of a hot, distant object, such as a star, its surface temperature can be estimated using the equation $\lambda_{max} T$ = constant. Doing the calculation for the Sun, for example, gives a surface temperature of just under 6000 K (see question 2).

1 Use the graph in Figure 2 to find the wavelength of peak radiation for

i) T = 5500 K

ii) T = 3500 K.

2 Calculate the peak wavelength radiated by the Sun assuming the surface temperature is 5800 K. State which part of the electromagnetic spectrum this is found in.

3 Compare the curve for 3500 K, shown in Figure 2, with a curve for 280 K, which is similar to the Earth's mean temperature. State how the maximum energy density and peak wavelength positions differ.

4 Suggest a meaning for the physical quantity represented by the area under the black body curve.

It is very difficult to determine the Earth's mean temperature, averaged over one year. Temperatures vary significantly at different latitudes and altitudes, and from season to season. For the next question assume a value of 15 °C.

5 Write down the Earth's average temperature in kelvin (K) and calculate the peak wavelength radiated, assuming black body behaviour. In which part of the electromagnetic spectrum is this?

Solar radiation

To investigate this more closely, one needs to start with some definitions:

Power is the rate at which energy is transferred. A power of one watt (1 W) means a rate of energy transfer of one joule per second (1 J s^{-1}).

The **solar constant**, *S*, is the average solar power per m^2 arriving at the top of the Earth's atmosphere. Its value is 1.37 kW m^{-2}. In other words, 1.37 kilojoules of solar energy reach each perpendicular square metre of the top of the Earth's atmosphere every second.

The Sun's **luminosity** is the *total* power radiated by the Sun. Although it fluctuates very slightly, the solar constant is normally assumed to be reasonably stable at 1.37 kW m^{-2}. A knowledge of the solar constant enables one to estimate the luminosity of the Sun and hence its surface temperature.

The Sun is approximately 149×10^6 km from the Earth. Astrophysicists call this distance one **astronomical unit** (1 AU). At the distance of the Earth, all the power emitted by the Sun is spread out over a sphere of radius 1 AU.

The solar power flux at the Earth's surface is significantly less than the solar constant.

There is a significant amount of absorption and reflection by the atmosphere. Also, the actual value of the solar power flux in any one region varies depending on the latitude, the season, and the time of day (at night it is, of course, zero). Typically, the average solar power flux at the Earth's surface at mid-European latitudes $\approx$ 150 W m^{-2}.

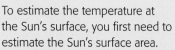

1 Use the value of the solar constant *S*, to calculate the Sun's luminosity.

To estimate the temperature at the Sun's surface, you first need to estimate the Sun's surface area.

2 The angular diameter of the Sun measured from the Earth is 0.52°. Calculate

i) the angular diameter in radians,

ii) the Sun's radius in metres,

iii) the surface area of the Sun, assuming that it is a sphere.

The **Stefan-Boltzmann** law gives the relationship between the luminosity L of a star, the surface area A, and the surface temperature T:

$$L = \sigma A T^4$$

where σ is the **Stefan-Boltzmann constant:** 5.67×10^{-8} W m^{-2} K^{-4}

iv) Estimate the temperature of the Sun's surface.

3 Calculate the total power and the power per m^2 radiated by the Earth at a temperature of 288 K. You can assume the Earth is a sphere of radius 6400 km.

Estimating the solar power flux

Power received per m² is called the **power flux**. Using measurements taken in the following experiment, you can estimate the solar power flux at the Earth's surface in your region.

- Pour some water into a regular-shaped glass vessel. A boiling tube or a measuring cylinder may be appropriate, but you should carry out pilot tests to see which one works best. Stain the water black using some ink.
- Record the mass of water and the mass of the glass vessel.
- Place in direct sunlight, making sure that the glass vessel is normal ('square on') to the Sun. The experiment will work best on a hot, dry day with little wind.
- Record the initial temperature of the water. Measure the temperature rise over a known period of time. Continue until a rise of 3 K to 5 K is obtained.

1. Calculate the cross-sectional area of the vessel facing the Sun.
2. Calculate the thermal energy gained by the water and the glass using $\Delta Q = mc\Delta T$. (Specific heat capacity of water = 4190 J kg^{-1} K^{-1}, specific heat capacity of Pyrex glass = 750 J kg^{-1} K^{-1})
3. Use your values in parts 1 and 2 to estimate the value of solar power received per m².

You could do some research to find out how your value for the solar power flux (part 3 above) compares with published values from more sophisticated experiments.

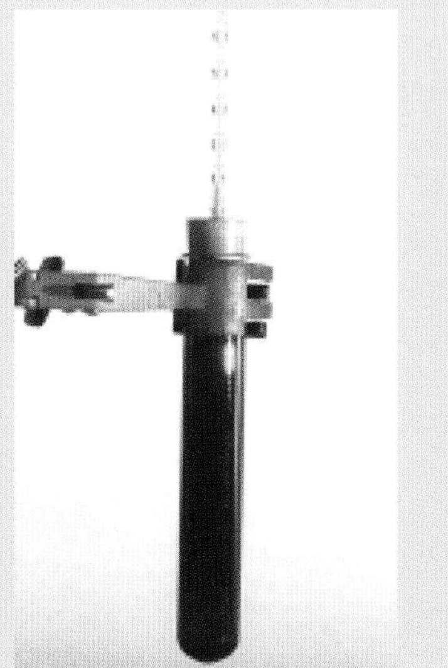

Figure 3 Measuring the solar constant

Mathematical physics: A simple energy balance model

If the Earth is at thermal equilibrium, this means that it is emitting energy at the same rate as it is absorbing it. So:

power radiated by the Earth = power received from the Sun

Assuming that the Earth is equivalent to a disc of radius r, normal to the incoming solar radiation:

total incoming power = $S \times \pi r^2$
(S is the solar constant)

Some of this radiation is reflected straight back into space. This fraction is called the **albedo**, α.

So: total power absorbed by the Earth = $S(1-\alpha)\pi r^2$ (1)

The next stage is to calculate the power radiated by the Earth, assuming that this is emitted over a sphere of area $4\pi r^2$, and that this behaves as a black body radiator. For this, we apply the Stefan-Boltzmann law for an object at temperature T:

So: power radiated by the Earth = $\sigma \times 4\pi r^2 \times T^4$ (2)

So: $4\sigma\pi r^2 T^4 = S(1-\alpha)\pi r^2$ (3)

So:

$$T = \sqrt[4]{\frac{S(1-\alpha)}{4\sigma}}$$

Using S = 1370 W m^{-2}, α = 0.3, and $\sigma = 5.67 \times 10^{-8}$ W m^{-2} K^{-4} gives

$$T = \sqrt[4]{\frac{1370 \times (1-0.3)}{4 \times 5.67 \times 10^{-8}}}$$

So: T = 255 K

This simple analysis predicts the temperature of the Earth to be lower than it actually is. So the model needs to be modified. The atmosphere (and the greenhouse gases in it) have an important role in keeping the planet at the much more acceptable global annual mean temperature of ≈288 K.

The Earth with its atmosphere is not a perfect black body. **Emissivity**, ε, is defined as the ratio of power radiated per unit area by an object to the power radiated per unit area by a black body at the same temperature. Allowing for this modifies equation 2 to this:

$$\text{power radiated by the Earth} = 4\varepsilon\sigma\pi r^2 T^4$$

You should be able to show that the Earth's temperature will be given by:

$$T = \sqrt[4]{\frac{S(1-\alpha)}{4\varepsilon\sigma}}$$

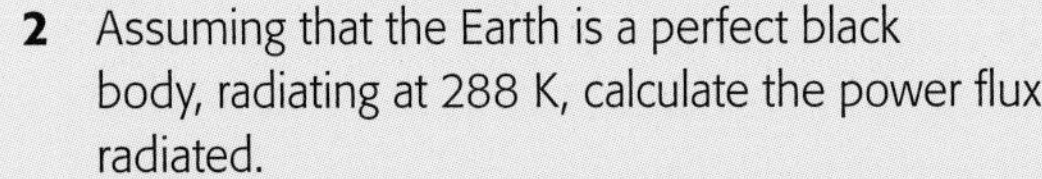

1. Calculate the emissivity of the Earth assuming the mean temperature is 288 K.
2. Assuming that the Earth is a perfect black body, radiating at 288 K, calculate the power flux radiated.
3. Compare your answer to the previous question with the lower flux value radiated for a black body at 255 K.

Modelling energy absorption

In this section, the aim is to develop a model to explain how the atmosphere helps to keep the Earth warm. Before studying this section, make sure that you have read Chapter 7, dealing with the mathematical models used for simple harmonic oscillations.

Certain molecules in the atmosphere absorb particular wavelengths radiated by the Earth and the Sun. However, the Earth is significantly cooler than the Sun, so its black body radiation spectrum is shifted to regions of longer wavelength compared to that of the incoming solar radiation. Because of the shift in peak wavelength, molecules in the atmosphere absorb more of the outgoing (terrestrial) radiation than they do incoming (solar) radiation.

Several gases in the atmosphere absorb energy in these wavelengths but the most important ones are water vapour, carbon dioxide, and methane. All occur naturally, although human activities have significantly increased the concentrations of the first two.

To see how energy is absorbed, carbon dioxide will be used as an example (Figure 4).

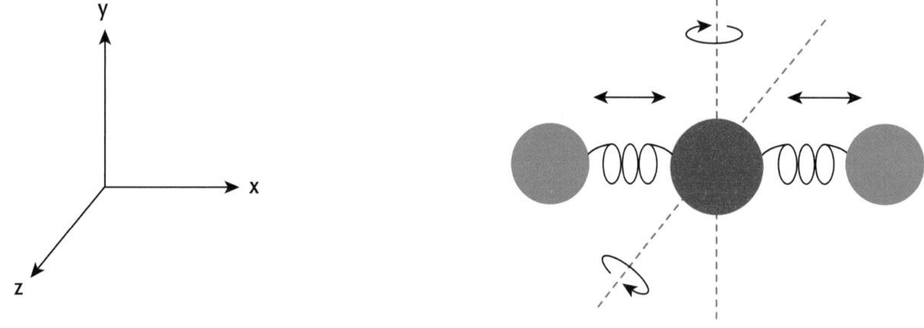

Figure 4 Carbon dioxide molecule and possible modes of translation, vibration, and rotation

Carbon dioxide has a linear molecule with a shape as shown in Figure 4. Each oxygen atom (blue) is joined to the carbon atom (red) by a double bond. Essentially, this bond is formed by clouds of electrons: see the chemistry course companion for further details. In the model in Figure 4, it is convenient to think of the bonds as a spring. When this spring is stretched or compressed, forces try to return the atoms to their equilibrium position.

The above molecule has kinetic energy which may be considered in separate modes:

i) three linear modes due to movements in the x, y, and z directions;
ii) two rotational modes due to rotation around the two dotted axes illustrated;
iii) four vibrational modes (Figure 5), each one excited by a different wavelength.

Inquiry: Temperatures throughout the Solar System

The surface temperature of a planet depends on several factors, including:

- the distance from the Sun
- the composition of the atmosphere
- the albedo (the proportion of incoming solar energy reflected back into space)
- the emissivity.

Choose a planet or moon in the Solar System and carry out some research to find out about the temperature range you would expect to find there. Present your findings to your group in a suitable way.

You should include the following:

- An estimate of the average solar power flux for your chosen planet or moon. Show how you calculated this.
- Details of the atmosphere, including its composition

A good approach would be to research the key data for your object then use a spreadsheet or graphical calculator to develop your model.

1 Describe carefully the movement of each atom in the four modes of vibration shown in Figure 5.

In a linear molecule such as carbon dioxide, the number of vibrational modes is given by $3N - 5$, where N is the number of atoms in the molecule. In a non-linear molecule such as water or methane, the number of modes is given by $3N - 6$.

2 Calculate the number of vibrational modes in

i) a water molecule (non-linear)

ii) a methane molecule (also non-linear).

3 Draw a simple diagram to illustrate these modes in a water molecule.

From Chapter 7, you will know that the frequency of vibration of a spring is given by this equation:

$$f = \frac{1}{2\pi}\sqrt{\frac{k}{m}} \quad \text{where } k \text{ is the spring constant.}$$

4 Compare modes A and C in Figure 5. Explain whether it is the stretching mode or the bending mode that

i) requires more energy

ii) has a higher frequency. (Hint: compare the value of k in each case.)

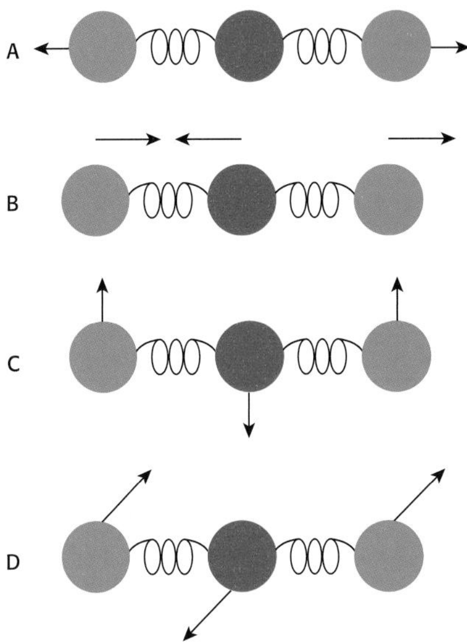

Figure 5 The four vibrational modes of carbon dioxide

The rotational and vibrational modes have a **natural frequency** which corresponds to some of the wavelengths radiated by the Earth, particularly those in the infrared part of the spectrum. The carbon dioxide molecules absorb energy from precise wavelengths, and the oscillatory modes are excited. This is an example of **resonance** (see page 95). The internal energy of the molecules increases and they re-radiate this energy in random directions: this is called **scattering**. The net flux of outward energy is therefore reduced, as some of this scattered energy is re-absorbed by the Earth. In this way, the planet warms up until a new equilibrium temperature is reached.

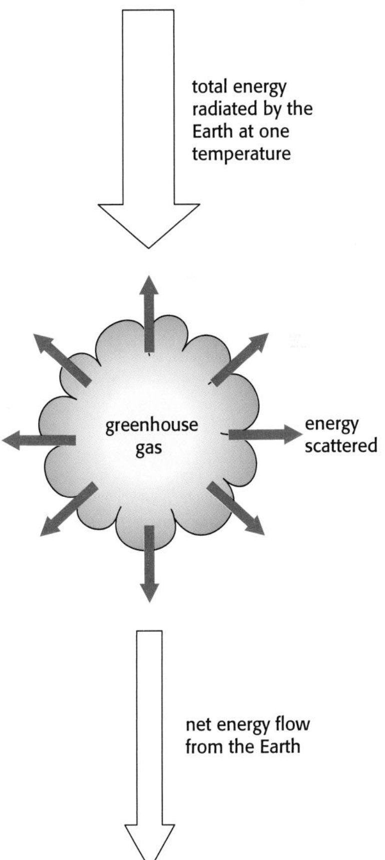

Figure 6 Absorption of energy by greenhouse gases leads to scattering

Data-based question: Greenhouse gas absorption spectrum

The graph below shows the transmission of different wavelengths from a particular waveband through carbon dioxide gas.

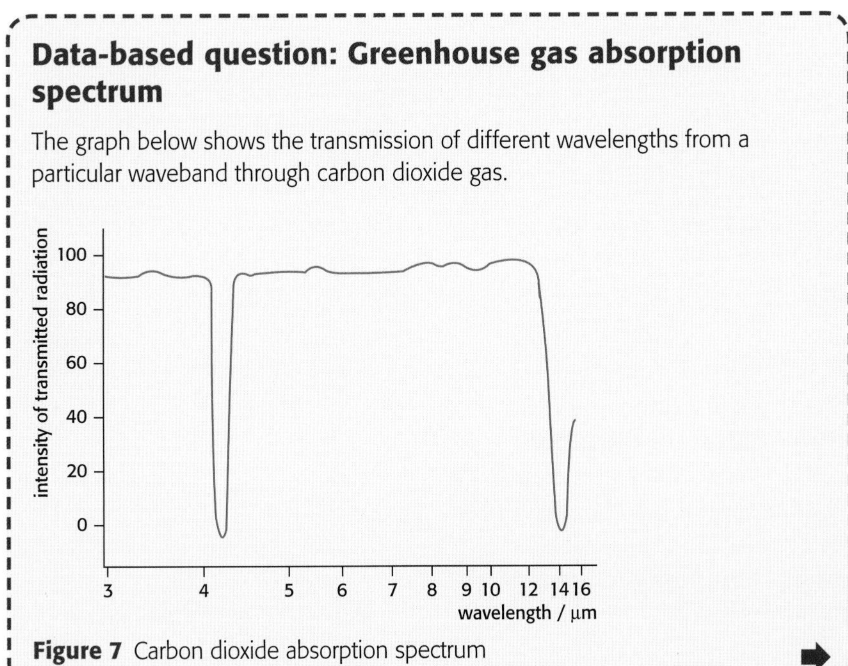

Figure 7 Carbon dioxide absorption spectrum

1 Write down how many μm there are in 1 m.
2 State the range of wavelengths illustrated in the graph and to which waveband in the electromagnetic spectrum they belong.
3 Write down the wavelengths of radiation that are absorbed most efficiently by carbon dioxide gas.
4 Outline the different molecular mechanisms possibly responsible for absorption in each wavelength.
5 Before going any further, refer back to the rather simplistic explanation of the greenhouse effect at the start of this chapter.
You should now be able to write a much more detailed and physically accurate description of the process.

Evidence for an enhanced greenhouse effect

The enhanced greenhouse effect is another name for the global warming caused by the activities of humans.

The Inter-Governmental Panel on Climate Change (IPCC), a body within the UN, published its fourth assessment report in February 2007. It stated the following:

> *"Warming of the climate system is unequivocal, as is now evident from observations of increases in global average air and ocean temperatures, widespread melting of snow and ice, and rising global mean sea level".*

The report backed up this statement with evidence from temperature records going back over nearly 200 years. Some of this data is included below. The range of uncertainty in any value is estimated and included in square brackets – [ ].

1 *Eleven of the last twelve years (1995–2006) rank among the 12 warmest years in the instrumental record of global surface temperature (since 1850).*
2 *The updated 100-year linear trend (1906–2005) of 0.74 [0.56 to 0.92] °C is therefore larger than the corresponding trend for 1901–2000 given in the TAR (Third Assessment Report) of 0.6 [0.4 to 0.8] °C.*
3 *The linear warming trend over the last 50 years (0.13 [0.10 to 0.16] °C per decade) is nearly twice that for the last 100 years.*
4 *The total temperature increase from 1850–1899 to 2001–2005 is 0.76 [0.57 to 0.95] °C.*
5 *Urban heat island effects are real but local, and have a negligible influence (less than 0.006 °C per decade over land and zero over the oceans) on these values." (IPCC, 2007, p4)*

?

1 State three of the observed consequences of the enhanced greenhouse effect reported in the IPCC assessment on this page.
2 Use the data in point 2 to calculate the mean temperature increase per decade from 1906 to 2005.
3 Use the data in points 2 and 3 to plot
   i) the global mean temperature deviation from the 1956 value on a graph over the period 1906–2005
   ii) on the same axes, plot the global mean temperature deviation from the 1956 value averaged from 1956 to 2005.
4 Use your graph from question 3 to outline any changes in the rate of temperature rise in recent decades.
5 Explain
   i) the meaning of the term "urban heat island effect" used in point 5, and
   ii) why this point was included with the other data.

Modelling an enhanced greenhouse effect

Developing an accurate model to predict the consequences of the enhanced greenhouse effect is very difficult. Many of the models produced so far have led to different conclusions. A sensible approach is to look at the predictions of as many of the models as possible, then try to reach a consensus. However, this is not quite the same as reaching an average, because the scientific community may judge some models to be more accurate than others.

In the simple model discussed earlier in the chapter, this formula was used to predict the average global temperature:

$$T = \sqrt[4]{\frac{S(1-\alpha)}{4\varepsilon\sigma}}$$

S, α, and ε are all variables which can change over time.

The value predicted by this equation is too low by about 35 K. We now know that the atmosphere and its composition have a significant effect on the temperature. Also, climate is influenced by other mechanisms, such as the fluid dynamics of the atmosphere and oceans, combined with the effects of the Earth's spin. There is another process that is in action: positive feedback.

Positive feedback

To understand this, consider the following examples:

- Snow and ice reflect significant amounts of solar radiation straight back into space. Global warming has led to melting of the polar ice caps, which reduces the Earth's albedo. This in turn increases the absorption of solar radiation, which increases global warming.
- The oceans store huge quantities of dissolved carbon dioxide. Oceanic warming reduces the solubility of carbon dioxide, so gas is released into the atmosphere. This in turn causes more global warming.

In both cases, a change in one factor causes a change in another which, in turn, "feeds back" to magnify the original change.

Global warming may well lead to desertification (the formation of deserts) in certain parts of the planet. Outline an argument of how this may lead to:

a) positive feedback, increasing the effects of global warming, or

b) negative feedback, reducing the effects.

Assumptions for a simple model

This model is based on an idea from the climateprediction.net website. A spreadsheet or graphical calculator is used to model changes in the Earth's temperature over time.

The model assumes the following:

1 The initial power flux of incoming solar radiation at the Earth's surface is 390 W m^{-2}.

2 The outgoing power flux depends only on the temperature of the Earth, initially 288 K (the 2007 value), and is given by the Stefan-Boltzmann law:

$$\text{outgoing power flux} = \sigma T^4$$

3 The change in temperature of the Earth over time depends on:
   - the energy difference between the incoming and outgoing radiation
   - the surface heat capacity, C, of the Earth. (C is the thermal energy needed to raise the temperature of 1 m^2 of the Earth's surface by 1 K. In the model, $C = 4.0 \times 10^8$ J m^{-2} K^{-1}.)

Over a period of time, Δt, the Earth absorbs energy which causes a temperature rise ΔT.

The energy absorbed = incoming energy − outgoing energy

But: energy = power × time

So: $\text{temperature rise} = \dfrac{(\text{incoming power} - \text{outgoing power}) \times \text{time}}{\text{heat capacity}}$

The new temperature is calculated like this:

new temperature = old temperature + temperature change

The idea is to model what would happen if the effective incoming solar radiation changes. This could happen if:

- the solar luminosity changes due to solar flares or sunspot activity
- the Earth's albedo changes
- the Earth's emissivity changes due to increased greenhouse gas concentrations in the atmosphere.

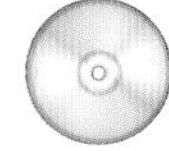

Setting up the spreadsheet

If designed well, a spreadsheet will quickly re-calculate all your dependent variables when you change just one of your independent variables. It is worth taking a little time to plan your spreadsheet design before inputting your variables. An Excel spreadsheet for this simple climate model is illustrated:

	A	B	C	D	E	F
1	Modelingclimate change.					
2						
3	σ	c	Δt (years)	Δt (s)	T_0	% change in energy
4	5.67E-08	400000000	0.5	15768000	288	5
5						
6						
7	time (years)	Incoming radiation	Outgoing radiation	ΔT	T	
8	0	390.07939	390.07939	0	288	
9	0.5	409.58336	390.07939	0.7688465	288.769	
10	1	409.58336	394.26154	0.6039864	289.373	
11	1.5	409.58336	397.57044	0.4735496	289.846	
12	2	409.58336	400.17928	0.3707091	290.217	
13	2.5	409.58336	402.2305	0.2898498	290.507	
14	3	409.58336	403.8398	0.2264114	290.733	
15	3.5	409.58336	405.10023	0.1767253	290.91	
16	4	409.58336	406.0861	0.1378621	291.048	
17	4.5	409.58336	406.85643	0.1074959	291.155	
18	5	409.58336	407.45783	0.0837884	291.239	
19	5.5	409.58336	407.92707	0.0652912	291.305	
20	6	409.58336	408.29299	0.0508664	291.355	
21	6.5	409.58336	408.57825	0.0396218	291.395	
22	7	409.58336	408.80054	0.0308588	291.426	
23	7.5	409.58336	408.97374	0.0240314	291.45	
24	8	409.58336	409.10865	0.0187131	291.469	
25	8.5	409.58336	409.21373	0.0145708	291.483	
26	9	409.58336	409.29557	0.0113449	291.495	
27	9.5	409.58336	409.35929	0.0088329	291.503	
28	10	409.58336	409.40891	0.0068768	291.51	
29						

Figure 8 Spreadsheet to model climate change

1 The constants, σ and *C*, are entered in cells A4 and B4. Note the format for the value of σ.

2 The time increment, Δ*t*, is entered in cell C4 in years for ease of use. As calculations use seconds, a formula is entered in cell D4 to covert years to seconds = C4*365*24*3600. The term C4

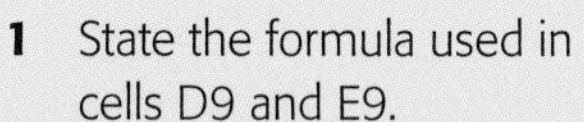

1 State the formula used in cells D9 and E9.

2 Use the graph to determine

  i) the new steady state temperature

  ii) the time to reach this new steady state.

Change the time interval to (i) 0.25 years and (ii) 1 year.

3 Describe any similarities and differences between your new graphs and the one illustrated in Figure 9.

4 State, for each new time interval

  i) the final steady state temperature

  ii) the time to reach this steady state.

The formula in cell C9 is = A4*E8^4

5 Explain what would happen when filling down if the formula = A4*E8^4 was used. (You can test this out by trying it).

6 Explain the use of the "$" notation throughout the spreadsheet – it is very important to understand this when designing spreadsheets.

7 Explain whether you think this kind of model is useful for scientists to attempt predictions of future climate change.

refers to the number in cell C4. The "$" s are used in Excel to "fix" the reference – see below.

3 The initial temperature is entered into cell E4.
4 The number in cell F4 is the % increase or decrease in the incoming solar radiation.
5 Row 8 is used to set up the initial conditions.
6 The formula = E4 is used in cell E8.
7 Cell C8 is calculated using the Stefan-Boltzmann law; = A4*E8^4.
8 As the incoming radiation must initially be equal to the outgoing radiation, the formula = C4 is entered into cell B4.
9 Row 9 makes the first set of calculations following the increase (or decrease) in the incoming solar radiation.
10 The time interval chosen in cell C4 is generated in cell A9 using the formula = A8 + C4.
11 The increased solar power in cell B9 is generated using the formula = B8*(1 + F4/100)
12 The outgoing radiation in cell C9 is assumed to be constant in the model over the selected time interval. It is calculated using the formula = A4*E8^4. The temperature in cell E8 is the previous temperature.
13 The temperature change and new temperature are generated in cells D9 and E9.
14 Highlight the cells A9 to E9, grab the small black square in the bottom right-hand corner of cell E9 and fill down for 20 rows.
15 Plot a graph of temperature (y-axis) against time (x-axis). To do this highlight the cells A7 to A28, press and hold the CTRL key and highlight cells E7 to E28. Click the chart wizard button, and select the various options to complete the graph.

The graph for this model is shown below.

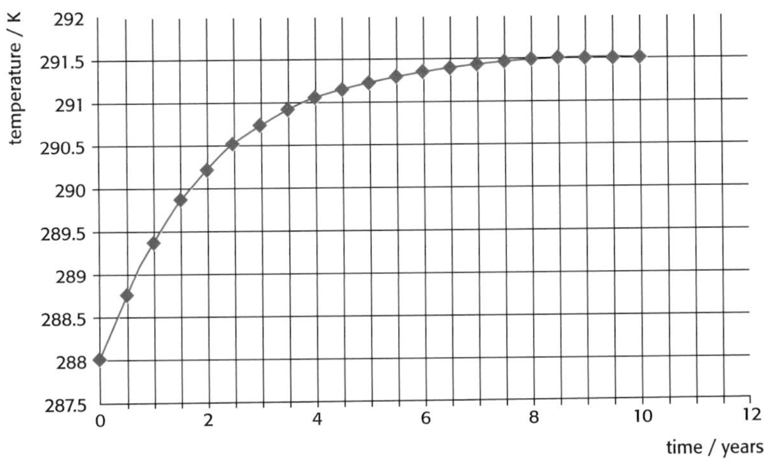

Figure 9 Graph of temperature against time for a simple climate model

This model is very flexible. Try different scenarios and find out what the model predicts. For example:

- Change the % change in solar energy to +8% and –5%.
- Change the initial temperature.
- Change the time interval.

- Try to improve the sophistication of the model by including albedo and emissivity in the calculations.
- Try to find out if there is a relationship between atmospheric carbon dioxide concentration and emissivity. If so, this could be included in the model.

An interesting feature of this model can be explored if you use a large time interval. For example, a time interval of 5 years gives the graph below.

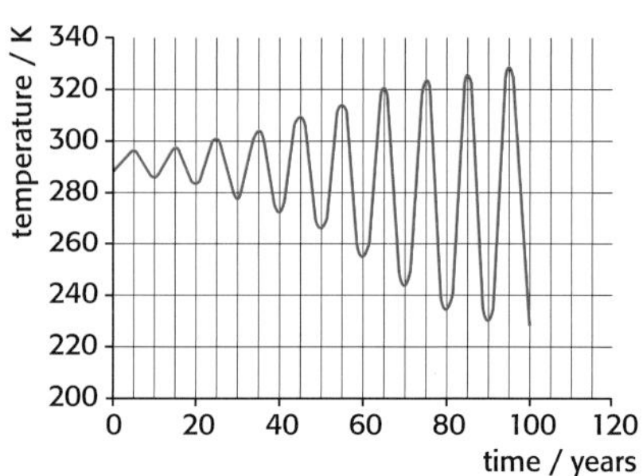

Figure 10 An unstable model of climate change

This suggests an unstable model, with an oscillating temperature. Think about why the model predicts this result. Does temperature change in discrete time steps, or is it a continuous change? You might want to investigate this feature of the model further.

End of chapter summary

Chapter 19 has three main themes: solar radiation, the greenhouse effect and its enhancement. The list below summarises the knowledge and skills that you should be able to undertake after having studied this chapter. Further research into more detailed explanations using other resources and/or further practice at solving problems in all these topics is recommended – part icularly the items in bold.

Solar radiation

- Calculate the intensity of the Sun's radiation incident on a planet.
- Define a planet's *albedo* and state factors that determine it.

The greenhouse effect

- Describe the greenhouse effect.
- Identify the main greenhouse gases, their sources and the molecular mechanisms by which greenhouse gases absorb infrared radiation.
- **Analyze absorption graphs to compare the relative effects of different greenhouse gases.**
- Outline the nature of black body radiation and be able to draw and annotate a graph of the emission spectra of black bodies at different temperatures.
- State the Stefan – Boltzmann law and apply the concept of emissivity to compare emission rates from different surfaces.
- Define *surface* heat capacity and solve problems on the greenhouse effect and the heating of planets using a simple energy balance climate model (in direct calculations and using spreadsheets).

The enhanced greenhouse effect

- State what is meant by the enhanced greenhouse effect and identify the increased combustion of fossil fuels as the likely major cause of the enhanced greenhouse effect.

Chapter 19 questions

1 This question is about the enhanced greenhouse effect and climate change.

Here is some data:
solar constant = 1.37 kWm^{-2}
Earth's radius = 6.4×10^3 km
Stefan-Boltzmann constant = 5.67×10^{-8} $W\ m^{-2}k^{-4}$
Surface heat capacity of Earth = 4.0×10^8 $J\ m^{-2}K^{-1}$

a) i) Show that, using a simple energy balance model, the Earth re-radiates about 340 $W\ m^{-2}$ of power per unit area [3]

ii) State two assumptions you have made. [2]

b) i) Use the Stefan-Boltzmann law to estimate the mean temperature of the Earth using this data. [2]

ii) Briefly explain why the Earth is warmer than your answer above suggests. [2]

A change in the atmosphere causes the effective power flux absorbed by the Earth to increase by 10%, causing the planet to become hotter.

c) Calculate:

i) the new power flux absorbed by the Earth. [1]

ii) the total energy gained per m^2 in 1 year. [1]

iii) the total energy radiated per m^2 in 1 year, assuming that the Earth continues to radiate at 340 W m^{-2} during this time. [1]

d) i) Determine the temperature rise of the planet over 1 year [2]

ii) State one additional assumption of this model. [1]

2 This question is about black body radiation and the greenhouse effect

The power absorbed from the Sun at the surface of the Earth is 240 W m^{-2}.

i) Calculate the mean temperature of the Earth. [4]

ii) State any assumptions you made in your calculation [2]

The solar constant is measured to be 1.37 kW m^{-2}.

iii) Estimate the mean albedo of the Earth and state any assumptions you have made [6]

iv) Describe the *greenhouse effect* and compare it with the *enhanced greenhouse effect*. [6]

3 a) A student sets up an experiment to find the solar power flux using a boiling tube containing blackened water. She adjusts the apparatus to make sure the boiling tube is normal to the sun and measures the temperature rise over 30 minutes. The temperature of the water is observed to rise by 3.5 K over this time. Suggest why adding black ink to the water improves the experiment.

b) Explain why the experiment would potentially give a more accurate result on a still, dry (low humidity) day.

c) The boiling tube has a diameter of 2.0 cm and a length of 10.2 cm and is 0.2 cm thick. Estimate (i) the mass of the boiling tube and (ii) the maximum mass of water inside. (density Pyrex glass = 2.23 gcm^{-3}, density water = 1gcm^{-3})

d) Use your answers to 19.10 and the data above, assuming that the boiling tube is full of water, to calculate the thermal energy rise in the boiling tube and the water.

e) Estimate the solar power flux required to cause this temperature rise.

20 Climate change

This chapter will explore the evidence for the causes of global warming, and then consider some of the possible effects of climate change. The consequences of climate change are so serious on a global scale that tackling global warming is one of the most important challenges facing us all. Failure to do so could drastically affect the entire planetary ecosystem.

Some scientists think that the possible effects of the enhanced greenhouse effect have been underestimated and that climate change may cause even more drastic effects than predicted.

The ice age is coming, the sun is zooming in
Engines stop running and the wheat is growing thin
A nuclear error, but I have no fear
London is drowning—and I live by the river"
(The Clash, Joe Strummer / Mick Jones, "London's Calling", 1979)

Thinking about science: Interpretation of climate data

Two things are happening:

- global temperatures are rising, and
- global concentrations of carbon dioxide are increasing.

Figure 1 shows the estimated mass of atmospheric carbon dioxide since 1750, and Figure 2 shows the variations in global annual mean temperatures for the period 1881–2005 from the mean value averaged over 1950–1975.

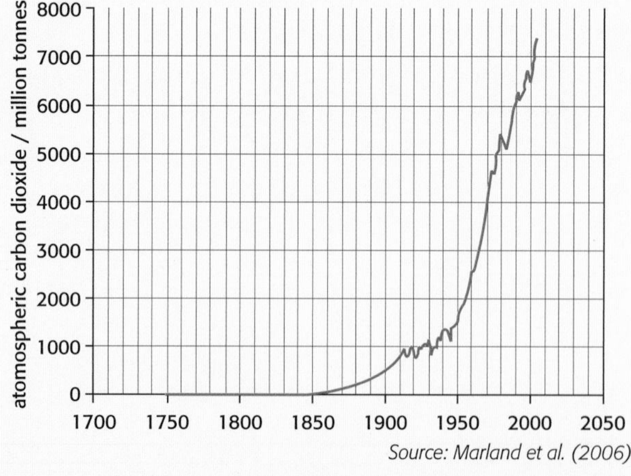

Figure 1 Atmospheric carbon dioxide levels since 1750

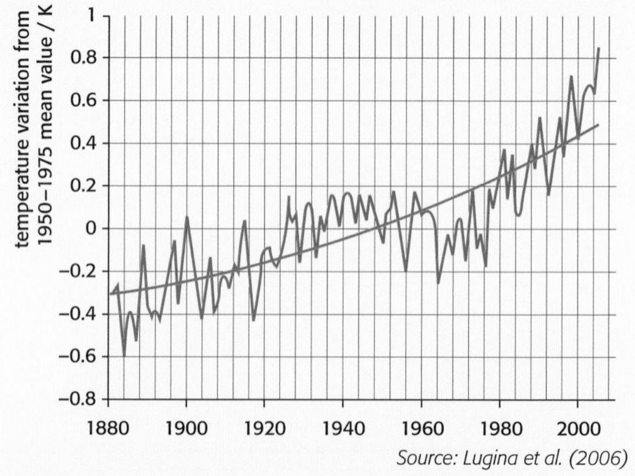

Figure 2 Monthly surface air temperature time series, area-averaged over the 30° latitudinal belts of the globe, 1881–2005

One of the important questions asked in the TOK course is "How do we know what we know?" The answer – from the scientific point of view – is found in data.

1 Describe the trends shown in Figures 1 and 2.

2 If taken at face value, outline any evidence from the graphs that suggests there is a correlation between the two sets of data.

Question 2 is quite difficult to answer without knowing the specific data values. One way to see if there is indeed a correlation between the two data sets would be to plot the temperature variation against the carbon dioxide levels and see whether there is any meaningful relationship (Figure 3).

A problem is that the mean temperature may well be chaotic. We can predict values using the deterministic, mechanical rules applied to the fluid dynamics of the atmosphere, but when we apply these rules to such a complex system involving so many variables (atmospheric pressure, wind speed, and so on), the outcome appears chaotic. **Chaos theory** is a relatively new branch of science, and it has produced much fascinating research.

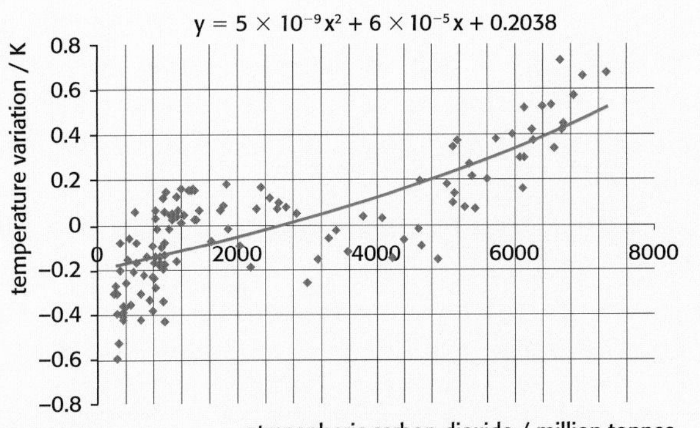

Figure 3 Plot of mean temperature variation against atmospheric carbon dioxide levels from 1881 to 2003

To see whether there is any correlation between temperature and carbon dioxide levels, and so decide whether recent changes may be **anthropogenic** (that is, originating in human activity) we need to go much further back in time.

Long-term trends

The data illustrated in Figures 1–3 was determined from ice core records drilled at the Russian Antarctic base Vostok (Figure 4). In an example of international cooperation between Russia, France, and the USA, several ice cores were drilled between 1987 and 1998, the deepest going more than 3600 m below the surface.

Ice core data are unique: every year the ice thaws and then freezes again, forming a new layer. Each layer traps a small quantity of the ambient air, and sophisticated analysis of this trapped air can determine mean temperature variations from the current mean value and carbon dioxide concentrations. The depth of the cores obtained at Vostok means that a data record going back more than 420 000 years has been built up through much painstaking and careful analysis. In Figure 5 the time axis goes back in time as you move to the right, and is measured in kiloyears (kyr) BP. The term BP means *before present* and refers to the arbitrary date of 1950 when radioisotope records were first established.

The two graphs in Figure 5 show many similarities, which suggests that there is a correlation between the two sets of data. The data have been analysed precisely, and the trends are summed up on the Carbon Dioxide Information Analysis Centre website, as quoted below.

> *There is a close correlation between Antarctic temperature and atmospheric concentrations of CO_2.... The extension of the Vostok CO_2 record shows that the main trends of CO_2 are similar for each glacial cycle. Major transitions from the lowest to the highest values are associated with glacial–interglacial transitions. During these transitions, the atmospheric concentrations of CO_2 rise from 180 to 280–300 ppmv.... The extension of the Vostok CO_2 record shows the present-day levels of CO_2 are unprecedented during the past*

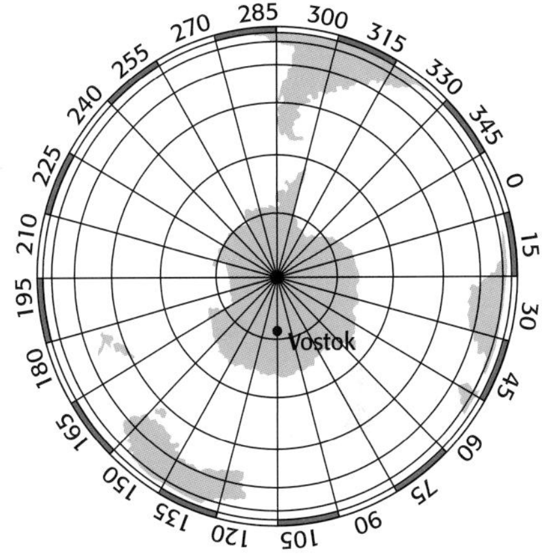

Figure 4 Map showing the location of the Vostok Antarctic base

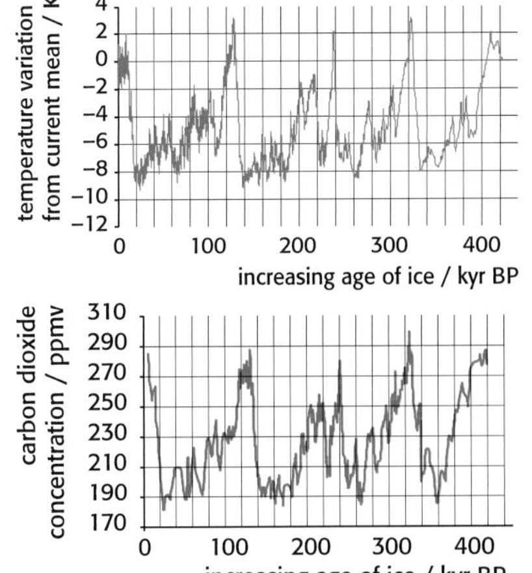

Figure 5 Temperature and carbon dioxide data from the Vostock ice cores. Source: (top) Petit *et al.* (2000); (bottom) Barnola *et al.* (2003)

420 kyr. Pre-industrial Holocene levels (~280 ppmv) are found during all interglacials, with the highest values (~300 ppmv) found approximately 323 kyr BP. *When the Vostok ice core data were compared with other ice core data … for the past 30 000–40 000 years, good agreement was found between the records: all show low CO_2 values [~200 parts per million by volume (ppmv)] during the Last Glacial Maximum and increased atmospheric CO_2 concentrations associated with the glacial–Holocene transition…. these measurements indicate that, at the beginning of the deglaciations, the CO_2 increase either was in phase or lagged by less than ~1000 years with respect to the Antarctic temperature, whereas it clearly lagged behind the temperature at the onset of the glaciations.*

Source: Barnola et al. (2003)

Use the temperature and carbon dioxide graphs determined from the Vostock Ice cores, and the passage above, to answer the questions in the margin.

1. Explain whether the graphs as shown can be used to determine which comes first: changes in carbon dioxide levels or changes in temperature.
2. From the passage:
   a) Determine which comes first: the change in temperature or the change in carbon dioxide levels.
   b) Explain how a change in this variable causes the other variable to change.
3. What evidence from the passage or in the graphs suggests that the current interglacial warmer period is different from the previous natural cycles?

The balanced view

The IPCC report (mentioned on page 304) suggests not only that global warming is leading to climate change, but that human activities are directly to blame. It concludes that:

Most of the observed increase in globally averaged temperatures since the mid-20th century is very likely due to the observed increase in anthropogenic greenhouse gas concentrations. Discernible human influences now extend to other aspects of climate, including ocean warming, continental-average temperatures, temperature extremes and wind patterns. (IPCC, 2007, p.8)

Some scientists argue against this, but in comparison with the work of the IPCC there is very little supporting evidence for their point of view. Professor Richard Lindzen of MIT acknowledges that:

- global mean temperatures are 0.5 K higher than a century ago,
- carbon dioxide levels have risen over the past two centuries, and
- carbon dioxide is a greenhouse gas.

However, he goes on to say that there is much uncertainty in making predictions, because:

…the climate is always changing; change is the norm. Two centuries ago, much of the Northern Hemisphere was emerging from a little ice age. A millennium ago, during the Middle Ages, the same region was in a warm period. Thirty years ago, we were concerned with global cooling.

Distinguishing the small recent changes in global mean temperature from the natural variability, which is unknown, is not a trivial task. All attempts so far make the assumption that existing computer climate models simulate natural variability, but I doubt that anyone really believes this assumption.

We simply do not know what relation, if any, exists between global climate changes and water vapour, clouds, storms, hurricanes, and other factors, including regional climate changes, which are generally much larger than global changes and not correlated with them. Nor do we know how to predict changes in greenhouse gases. This is because we cannot forecast economic and technological change over the next century, and also because there are many man-made substances whose properties and levels are not well known, but which could be comparable in importance to carbon dioxide.

Inquiry

Fossil and geological records over many thousand of years suggest there has been a natural cycle of warming and cooling, leading to ice ages and corresponding warmer periods in between. Some possible causes of this oscillation in global temperatures are:

- periodic wobbling of the Earth's axis of rotation
- increased or decreased solar luminosity due to solar flare or sunspot activity
- changing volcanic activity.

Choose one of these ideas and research it thoroughly. Describe the physical mechanisms involved, and explain how these explain the historical cooling–warming cycles deduced.

What we do know is that a doubling of carbon dioxide by itself would produce only a modest temperature increase of one degree Celsius. Larger projected increases depend on "amplification" of the carbon dioxide by more important, but poorly modelled, greenhouse gases, clouds and water vapour. (Lindzen, The press gets it wrong, our report doesn't support the Kyoto treaty, 2001)

Data-based question: The causes of global warming

It would seem that the Earth's climate is naturally cyclical, with a time period of approximately 100 000 years or so. There have been cooler, glaciated periods followed by much warmer periods, as established by the Vostok base and verified by other ice core records. What is not so easy to establish, however, is whether we are currently in a naturally warm period or whether this warm period is somehow different. The evidence would suggest that the properties of this warm period are fundamentally different from those in the past. Levels of carbon dioxide—the major greenhouse gas—are unprecedented, and it is no coincidence that power production and other processes developed in the last 200 years produce large quantities of this gas.

The IPCC 2007 report concluded that:

> *Global atmospheric concentrations of carbon dioxide, methane and nitrous oxide have increased markedly as a result of human activities since 1750 and now far exceed pre-industrial values determined from ice cores spanning many thousands of years. The global increases in carbon dioxide concentration are due primarily to fossil fuel use and land-use change, while those of methane and nitrous oxide are primarily due to agriculture.* (IPCC, 2007, p2)

The two graphs in Figure 6, taken from the IPCC report, indicate changes in atmospheric concentrations of carbon dioxide and methane over the last 10 000 years. The atmospheric gas levels are measured in ppm (parts per million) or ppb (parts per billion).

Radiative forcing is a measure of how each greenhouse gas changes the effective balance of incoming and outgoing radiation. Power flux is also measured in W m^{-2}. If a gas has a positive radiative forcing effect, this is seen as equivalent to an increase in the solar constant. An increase in the albedo due to an increase in volcanic particles in the atmosphere, for example, would have a negative radiative forcing effect.

1. Describe and compare the trends shown in the two graphs in Figure 6. In your description, point out any differences in pre-industrial (before ~1750) and industrial concentrations of the gases, and specifically any changes after 1950.
2. Explain what the "thickness" of each trend line might indicate.
3. Use the two graphs to compare the radiative forcing effect of carbon dioxide compared with that of methane.

The IPCC report states that

> The global atmospheric concentration of carbon dioxide has increased from a pre-industrial value of about 280 ppm to 379 ppm in 2005. The annual carbon dioxide concentration growth rate was larger during the last 10 years (1995–2005 average: 1.9 ppm per year) than it has been since the beginning of continuous direct atmospheric measurements (1960–2005 average: 1.4 ppm per year), although there is year-to-year variability in growth rates. (IPCC, 2007, p2)

4. Estimate the atmospheric carbon dioxide concentration in 2050, assuming an annual growth rate of (a) 1.4 ppm and (b) 1.9 ppm.
5. Outline the major causes of recent anthropogenic carbon dioxide emissions.
6. From the data provided, and any other sources, write your own conclusion for the causes of global warming.

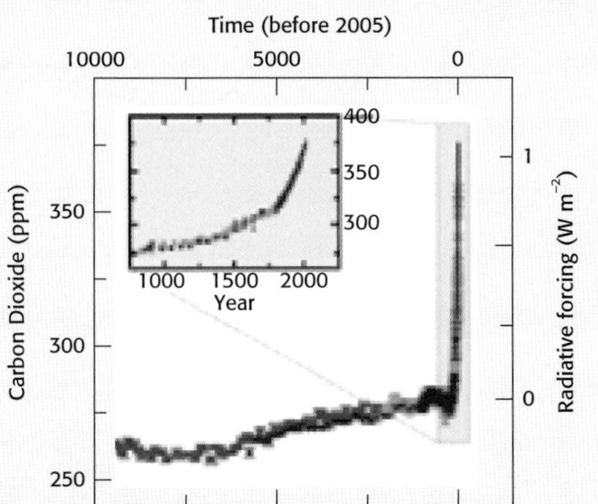

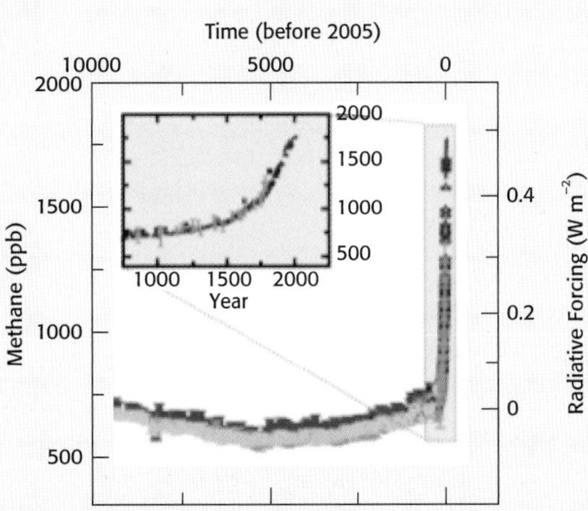

Figure 6 Changes in carbon dioxide and methane levels from ice core and modern data.

Physics issues: chaos and the effects of climate change

Our climate is driven by solar radiation and the Earth's spin. The changes in the oceans and the atmosphere are complex, and difficult to predict. In principle, all these changes are deterministic: that is, they obey well-known mechanical and thermodynamic rules, but the interactions between all possible climate "drivers" can only be modelled to some extent on very large computers.

Inquiry

The following are some predictions of climate change:

- temperature change
- erratic precipitation
- sea level rise
- increased number and strength of tropical cyclones.

There is a tendency to blame all unusual climate phenomena on global warming, when there is no evidence to do so. Extreme weather has occurred throughout the Earth's history, but can any recent events be blamed on climate change?

Consider some of the predictions of climate change made above, and carry out some research on a particular region to see whether there is any evidence of climate change. Some examples of specific events are given below, but you may want to choose something relevant to where you live.

- Hurricane Katrina caused great destruction in New Orleans in the Southern USA in August 2005.
- The summer of 2003 was the hottest ever on record in Europe – possibly the hottest for 500 years. In addition, the ten hottest years have occurred since 1990.
- Much of Queensland and New South Wales in Australia have received significantly lower rainfall over the last ten years than would be expected from long-term climate records, leading to severe water shortages and an increased risk of bush fires.
- Early 2007 saw excessive flooding in Ethiopia and Somalia in North-East Africa, displacing over a quarter of a million people. This followed a period of drought affecting around 11 million people in Kenya, Ethiopia, Eritrea, Somalia, and Djibouti.
- The Chinese government estimates that floods in Fujian province in June 2006 were the worst for 50 years.
- Many residents in Mumbai, India, reported in July 2005 that the flooding was the worst they could remember.
- In July 2001, more than a million peasant farmers in Honduras, Nicaragua, El Salvador, and Guatemala were severely affected by crop failure due to a lack of rain. At the same time around 9000 Miskito Indians living on the Atlantic coast of Nicaragua had their rice crops destroyed by flooding.

Make sure your research is thorough, and that you avoid looking at just individual events, though the list above provides some possible starting points. You should consider climatic patterns and trends over the longer term. The IPCC website is an excellent source of climate data – see www.ipcc.ch (accessed July 2009).

Data-based question: Rising sea levels

Sea level rises provide one of the most immediate and direct consequences of global warming. Two islands of the Pacific nation of Kiribati have already been covered by rising seas, and several other nations are under threat. Most of the Maldives, an island nation in the Indian Ocean, are less than 2 m above sea level, and a sea level rise of 1 m would cause severe difficulties. However, it is not only small island nations that are under threat. More than 70% of the world's population live on low-lying coastal plains, which include 11 of the 15 biggest cities.

You can find out about the threat posed by rising sea levels where you live with the website http://flood.firetree.net/ (accessed July 2009).

Source www.greenpeace.org.uk

Figure 7 Pita Meanke, of Betio village, stands beside a tree as he watches the "king tides" crash through the sea wall into his family's property, on the South Pacific island of Kiribati

The data in Table 1 are from the IPCC fourth assessment report.

Table 1 Estimated rate of sea level rise due to different sources

Source of sea level rise (Observed rate of sea level rise and estimated contributions from different sources)	Rate of sea level rise (mm per year) 1961–2003	1993–2003
1. Thermal expansion	0.42 ± 0.12	1.6 ± 0.5
2. Glaciers and ice caps	0.50 ± 0.18	0.77 ± 0.22
3. Greenland ice sheets	0.50 ± 1.2	2.1 ± 0.7
4. Antarctic ice sheets	1.4 ± 4.1	2.1 ± 3.5
Sum of individual climate contributions (above) to sea level rise	1.1 ± 0.5	2.8 ± 0.7
Observed total sea level rise	1.8 ± 0.5	3.1 ± 0.7
Difference (observed minus sum of estimated climate contributions)	0.7 ± 0.7	0.3 ± 1.0

Source: IPCC (2007, p5)

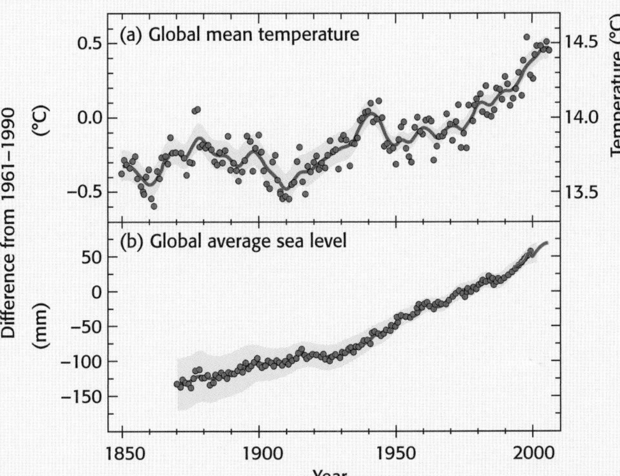

Figure 8 Sea level data from the IPCC report

1 Compare the observed mean sea level rise during 1961–2003 with that for 1993–2003. Suggest a reason (or reasons) for the difference.

2 Describe the trends shown in the graphs in Figure 7 and suggest whether there is a correlation between the global mean temperature and sea level rises.

Some of the sea level rise is due to melting ice sheets in Greenland and Antarctica.

3 Calculate the percentage uncertainties from both time periods in the sea level rise due to (a) the Greenland Ice sheet and (b) the Antarctic ice sheet.

4 Comment on the value of these uncertainties, compared with data you might normally measure in the lab.

5 Suggest a reason for the reduction in these uncertainties over the two time periods.

6 Explain the apparent discrepancies between the sum of the individual values in rows 1–4 in Table 1 with the sum of the individual climatic contributions and the observed sea level rise.

A major contribution to sea level rise is thermal expansion. Volume expansivity, γ, is defined in the equation $\Delta V = \gamma V \Delta T$, where ΔV is the change in original volume V caused by a temperature change ΔT.

7 Deduce the units of volume expansivity.

Water has a rather unusual expansivity in that it contracts between 0 °C and 4 °C before expanding. Use $+210 \times 10^{-6}$ (in SI units) as the mean value of γ for water in the following questions.

8 A swimming pool of length 25 m and width 10 m is filled to an average depth of 1.5 m. Calculate the increase in depth due to volume expansion when the temperature rises by 10 K.

The Indian Ocean is estimated to have a total area of 73 556 000 km^2 and a volume of 292 131 000 km^3.

9 a) Calculate the average depth of the Indian Ocean.

b) Estimate the mean sea level rise in the next 50 years, assuming a temperature rise of 0.16 K per decade.

10 List some of the assumptions you made in your answer to question 9.

You can decide for yourself whether the people of the Maldives should be concerned. It is quite clear that decisive, international action is needed in the short term in order to reduce the effects of global warming.

Physics issues: global efforts to reduce climate change

Reaching the conclusion that the current phase of global warming is most likely anthropogenic is a big step forward. The unanswered question is: Can anything be done in time to slow down and reverse any harmful climate change? If changing practices are to have any effect, the nations of the world need to unite and work together in planning and initiating concrete action.

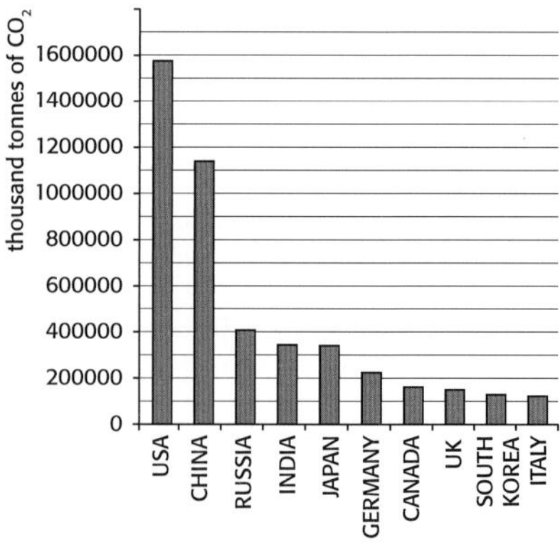

Figure 9 The world's top ten carbon dioxide emitters in 2003

The graph in Figure 9 illustrates the top ten carbon dioxide emitters in the world. It is no surprise that these countries have some of the largest economies and energy needs. Many are OECD countries, but China and India are developing countries with rapidly growing economies. Any reductions in carbon dioxide output from all these countries will clearly have the most significant effect on the global total.

A similar graph of carbon dioxide emissions per person (Figure 10) shows quite a different story. The second set of countries forms a different group. Only the USA is found in both. These countries are arguably the most inefficient in their energy production.

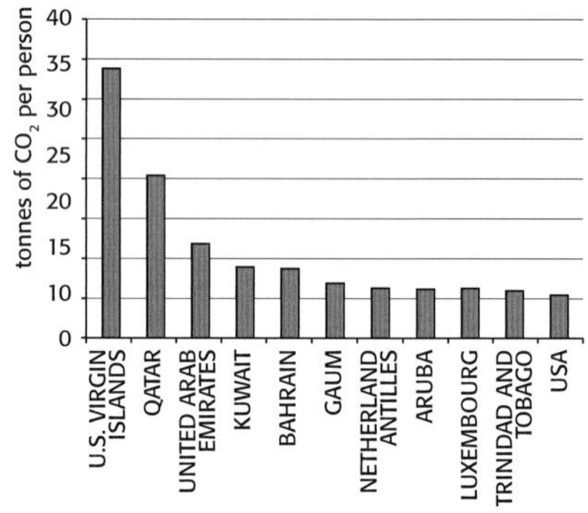

Figure 10 Per capita carbon dioxide production

The Kyoto Protocol was first approved in December 1997 by more than 160 of the world's countries, in Kyoto, Japan. The basic principle is for developed countries to reduce their greenhouse gas emissions back to 1990 levels during the period 2008–2012. The protocol allows developing countries still to increase output, as they are not responsible for the current situation, but it is hoped that China and India will voluntarily cut their emissions, though they have no legal obligation to do so under the current treaty.

It is still to be seen how effective this agreement will be, but it is widely believed to be the only way forward to reduce overall greenhouse gas emissions.

Inquiry: Global efforts to reduce climate change

Carry out some research into the aims of one of the following global agreements or international organizations and present your findings back to your class. In each case consider the guiding questions included.

1. The Kyoto Protocol
   - **a)** Which countries have not signed up, and why?
   - **b)** What is the clean development mechanism (CDM)?
   - **c)** What are certified emission reductions (CERs), and how are they used?
   - **d)** Has any progress been made in reducing greenhouse gas emissions to date?
2. Asia-Pacific Partnership on Clean Development and Climate (APP)
   - **a)** Which countries have signed this agreement?
   - **b)** What are its aims?
   - **c)** How will its objectives reduce climate change?
3. The Inter-Governmental Panel on Climate Change (IPCC)
   - **a)** How is this body organized, and who funds it?
   - **b)** What are its aims?
   - **c)** What has it achieved so far?
4. The United Nations Framework Convention on Climate (UNFCC)
   - **a)** What is the role of this body within the United Nations?
   - **b)** What has it achieved?
   - **c)** How does it support efforts to reduce global warming?

The following websites should be useful, but try to use several sources in your research.

- The United Nations: www.un.org/
- UNFCC: unfccc.int/2860.php
- IPCC: www.ipcc.ch/
- APP: www.asiapacificpartnership.org/
- Kyoto Protocol: unfccc.int/kyoto_protocol/items/2830.php (accessed July 2009)

End of chapter summary

Chapter 20 has two main themes: global warming and the enhanced greenhouse effect. The list below summarises the knowledge and skills that you should be able to undertake after having studied this chapter. Further research into more detailed explanations using other resources and/or further practice at solving problems in all these topics is recommended – particularly the items in bold.

Global warming

- Describe some possible models of global warming, describe the evidence that links global warming to increased levels of greenhouse gases and outline some of the mechanisms that may increase the rate of global warming.
- Define *coefficient of volume expansion*.
- State that one possible effect of the enhanced greenhouse effect is a rise in mean sea level and outline possible reasons for this.

Enhanced greenhouse effect

- Identify climate change as an outcome of the enhanced greenhouse effect, **solve problems related to the enhanced greenhouse effect**.
- Discuss international efforts to reduce the enhanced greenhouse effect.

Chapter 20 questions

1 This question is about the evidence that links increased levels of greenhouse gases to global warming.

a) The Vostok and similar ices cores provide evidence of the climate in the past. Two useful measurements have been atmospheric carbon dioxide levels and temperature variations. With reference to how the carbon dioxide is trapped, briefly outline how historical carbon dioxide levels can be determined. [3]

b) Describe any relationship found between carbon dioxide levels and temperature variation. [2]

c) The records indicate several previous ice ages and inter ice age warm periods. We are currently in a warmer period. Explain why many climate scientists think that this inter ice age warm period is anthropogenic, and not a natural occurrence as has happened in the past. [2]

d) Some scientists think that global warming is caused by natural phenomena. State two such natural phenomena, and explain how they may cause global warming. [4]

(Total 11 marks)

2 This question is about the effects of climate change, and about international efforts to reduce global warming.

a) One prediction of global warming is that sea levels will rise. State three different causes of sea level rise, and indicate which one is predicted to cause the biggest effect. [4]

b) Two important international efforts to reduce global warming are the Kyoto Protocol and the Asia-Pacific Partnership on Clean Development and Climate. Compare and contrast the approaches of these two international efforts in the attempt to reduce global warming. [6]

(Total 10 marks)

References

Barnola, J.-M., Raynaud, D., Lorius, C. and Barkov, N.I. 2003. Historical CO_2 record from the Vostok ice core. In *Trends: A Compendium of Data on Global Change*. Oak Ridge, TN, USA. Carbon Dioxide Information Analysis Center, Oak Ridge National Laboratory, US Department of Energy.

IPCC. 2007. *Fourth Assessment Report,* February. Geneva. Intergovernmental Panel on Climate Change.

Lugina, K.M., Groisman, P.Ya., Vinnikov, K.Ya., Koknaeva, V.V. and Speranskaya, N.A. 2006. In *Trends Online: A Compendium of Data on Global Change*. Oak Ridge, TN, USA. Carbon Dioxide Information Analysis Center, Oak Ridge National Laboratory, US Department of Energy.

Marland, G., Boden, T.A. and Andres, R.J. 2006. Global, regional, and national CO_2 emissions. In *Trends: A Compendium of Data on Global Change*. Oak Ridge, TN, USA. Carbon Dioxide Information Analysis Center, Oak Ridge National Laboratory, US Department of Energy.

Petit, J.R., Raynaud, D. Lorius, C., Jouzel, J., Delaygue, G., Barkov, N.I. and Kotlyakov, V.M. 2000. Historical isotopic temperature record from the Vostok ice core. *Trends: A Compendium of Data on Global Change*. Oak Ridge, TN, USA. Carbon Dioxide Information Analysis Center, Oak Ridge National Laboratory, US Department of Energy.

21 Technologies to reduce climate change

Improved efficiencies in the use of fossil fuels can have a significant effect in reducing greenhouse gas emissions.

We can improve energy efficiency in two ways:

- by reducing the amount of energy that we need to support our needs through technological improvements, such as generating more electricity per tonne of coal burnt, or achieving more kilometres of distance travelled by a car per litre of fuel so that less primary energy is required, and
- by changing our behaviour so that we use and waste less energy.

In addition, we need to reduce our dependence on fossil fuels. If any savings are achieved then these can be further invested in research into new technologies to generate further savings.

Energy-use projections (see Figure 3 on page 283 and Figure 7 on page 287) would, however, suggest we are still likely to be heavily dependent on fossil fuels for some time to come. All improvements in efficiency will go some way to reduce the effects on climate change, but other techniques are also needed to reduce carbon dioxide emissions. One promising technique has been to capture and store carbon dioxide at the point of production and prevent its release into the environment.

Efficiency is the key

The idea behind trying to improve the efficiencies of thermal power stations and vehicles is to get more useful energy out for less energy put in. Not only will our fossil fuel resources last longer, but less carbon dioxide will be produced in the combustion process to generate each MJ of energy.

Improvements in end-use efficiencies are likely to have an even bigger impact on overall fuel reduction. For example, energy-efficient buildings could reduce energy needs and hence contribute significantly to the reduction of carbon dioxide emissions.

It can be shown (see pages 157–8 in Chapter 11) that the maximum obtainable efficiency η of a generator is given by

$$\eta = 1 - \frac{T_2}{T_1}$$

where T_1 is the absolute temperature of combustion and T_2 is the ambient or air temperature (measured in kelvin). This maximum

1 Steam is heated, under pressure, to 540 °C in a thermal power station.

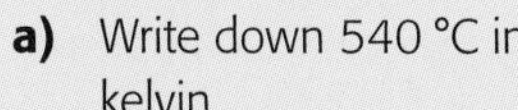

a) Write down 540 °C in kelvin.

b) Calculate the maximum possible efficiency when the ambient temperature is 30 °C.

One way of increasing the efficiency of a heat engine is to increase the furnace temperature, T_1, or reduce the ambient temperature, T_2. As the air temperature cannot be easily controlled, a sensible solution would be to design generators that operate at different temperatures. (This also works for vehicle engines.)

2 **a)** Try the calculation in question **1 b)** again, with an increased value of T_1 and a lower value of T_2.

b) Summarize the relationship between efficiency and operating temperature.

value of efficiency is reduced significantly by further energy losses in overcoming resistive forces, heat losses, and noise.

Gas-fuelled generators

The use of gas turbines to generate electricity is becoming increasingly popular. They can operate at higher temperatures than coal-fired and oil-fired power stations, and so potentially can be more efficient. Also, the thermal energy produced in the combustion of an equivalent mass of gas is higher than that for coal and oil. Carbon dioxide emissions produced from the different fossil fuels are compared in Table 1 using the unit, the **toe** – tonne of oil equivalent. The toe is the amount of fuel needed to produce the same energy as 1 tonne of crude oil. As well as being more efficient, the combustion of 1 toe of gas produces less carbon dioxide than 1 toe of oil or coal.

Table 1 Comparing production of carbon dioxide from fossil fuels

Fuel	CO_2 produced when burning 1 toe of fuel/t
coal	3.8
fuel oil	2.9
gas	2.1

Explain why the increased use of natural gas for electrical energy generation and transport has the potential to reduce global warming.

The technology involved in the design of gas turbines is more complex than for conventional oil and coal generators, mainly because of the higher temperatures and pressures involved. This has involved designing turbine blades of low-density, high-strength alloys into very precise aerofoil shapes. The gas is drawn in through one set of turbines (fan blades), which rapidly increase the gas pressure (Figure 1).

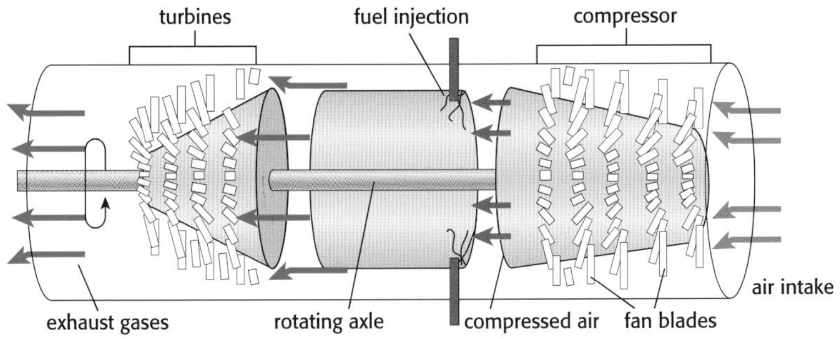

Figure 1 A gas turbine

The gas is then ignited in the combustion chamber and expands rapidly, turning the second set of turbine blades on its way to the exhaust. These blades turn the generator.

Another advantage of gas as a fuel is its relative abundance. New gas fields have recently been discovered, and the largest in the world – just off the coast of Northern Qatar in the Arabian Gulf – only started commercial production in the early years of this century. By 2011 Qatar could very likely supply one-third of the world's liquefied natural gas (LNG) needs (http://www.state.gov/r/pa/ei/bgn/5437.htm accessed July 2009).

With gas now recognized as a cleaner alternative to coal and oil, its use is projected to increase over the coming years, and it is expected to generate an increasing proportion of our electrical energy in the future.

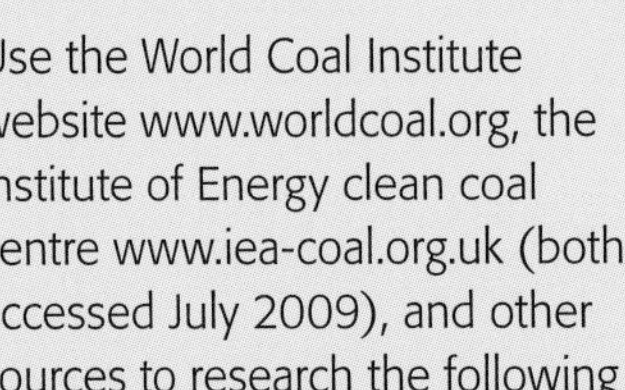

Inquiry: Making coal more efficient

Use the World Coal Institute website www.worldcoal.org, the Institute of Energy clean coal centre www.iea-coal.org.uk (both accessed July 2009), and other sources to research the following questions:

1. Coal is pulverized and ground into a powder before being burnt. Explain why this increases the combustion efficiency.
2. Early coal-fired power stations had efficiencies of only 25%, but it seems likely that this can be increased to greater than 45% in the future. Outline some of the techniques used to achieve these greater efficiencies.

Physics issues: improving our energy use

The following section discusses three examples of different ways in which energy use can be improved. This is achieved either by improving the efficiency of the process involved, or by improving the environmental impact of a process.

Energy-efficient vehicles

Vehicles, run on fossil fuels, are another major contributor to the enhanced greenhouse effect through the emission of carbon dioxide. Motor manufacturers are designing cars and trucks that are increasingly efficient, based on improvements in aerodynamic design. Similarly, combustion at higher temperatures can allow for increased efficiencies. However, conventional petrol (gasoline) or diesel engines still operate at very low efficiencies.

The best petrol engines rarely transfer more than 13% of the energy input to the wheels; 87% of the energy produced from burning the fuel is dissipated as thermal energy or noise in the engine and other moving parts, as well as additional losses due to running air conditioners, etc. Of the remaining 13%, over half is used to heat up the tyres, leaving only 6% available to produce useful kinetic energy, which is lost as thermal energy in the brakes every time the car stops. The vast majority of this remaining 6% is the kinetic energy of the car itself, and only a very small proportion, less than 1%, is the kinetic energy of the driver – the original intention of the car in the first place.

Outline, with reference to Newton's second law, how the development of lightweight alloys can be used to reduce the energy consumption of a car.

New and innovative ways are being devised to overcome these losses, including:

- the use of lightweight alloys and advanced polymer composite materials
- the use of electric vehicles
- the use of hybrid vehicles (see below), particularly the plug-in hybrid electric vehicle (PHEV).

Purely electric vehicles do not provide many energy savings. The electric motor is more efficient than a thermal engine, but is not able to transfer as much power to the wheels. Huge battery packs are needed to provide the electrical energy required, significantly increasing the mass of the vehicle. The speeds and maximum range available are currently much less than those obtainable from a conventional petrol-powered car. Electric vehicles also require regular charging, and the energy for this, of course, comes mainly from thermal power stations. The data in Figure 2 show that total carbon dioxide emissions over the lifetime of a purely electric vehicle can be only slightly less than those produced directly by petrol-powered cars, if the power station is coal burning.

Although carbon dioxide emissions in the manufacture of the different types of vehicle are fairly similar, the production of electric vehicles does involve more greenhouse gas production, because the scale of production is smaller than for petrol-powered cars. The use of hybrid vehicles leads to the emission of far less carbon dioxide than gasoline (petrol) vehicles during their lifetime, but the emissions from electric vehicles depend entirely on the source of electricity. If coal is used to generate the electrical energy then carbon dioxide emissions will be almost as high as using petrol directly. The use of liquefied natural gas (LNG) to generate the electricity leads to significantly lower emissions, and hydroelectric power generation leads to extremely low emissions.

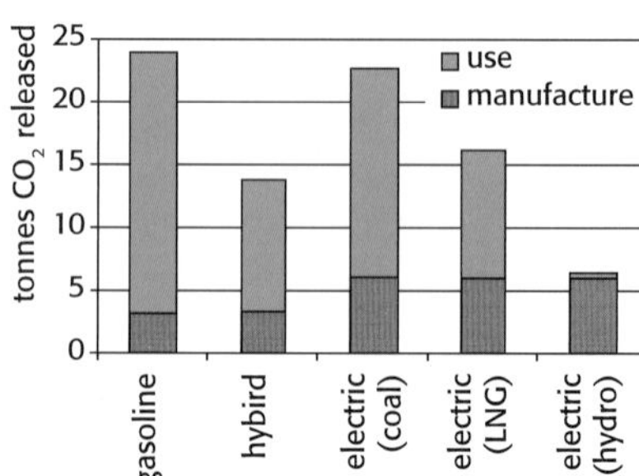

Figure 2 Total carbon dioxide emissions over the lifetimes of gasoline, hybrid, and electric cars. The electric car is shown three times, depending on the method of generating electricity: coal, liquefied natural gas, or hydroelectric.

Hybrid vehicles aim to utilize the best of both worlds. These cars use electric motors, with a petrol engine as back-up to provide additional power when necessary, such as for rapid acceleration, going uphill, or when the batteries have run down. Sophisticated computerized systems have been devised that can switch from the electric motor to the petrol engine and back as required to achieve the best possible efficiency.

The next possible step is the development of **plug-in hybrid electric vehicles** or PHEVs (Figure 3). Manufacturers claim that these may well operate in cities with fuel consumptions over 100 miles per gallon, about 32 km L^{-1}. This compares with about 21 km L^{-1} for the most efficient hybrid vehicle on the market in 2006 and 10 km L^{-1} for the average petrol-powered car. They can be recharged by plugging into the ac mains.

Figure 3 This plug-in Toyota Prius routinely gets more than 130 mpg. The black circle on the left of the back bumper is a plug socket, allowing an easy connection to the ac mains electrical supply.

1. Explain why the maximum efficiency obtainable from electric motors will be higher than that for a comparable petrol (gasoline) engine.
2. List what factors should be the same to ensure a fair comparison of the two motors in the previous question.
3. State some situations when a hybrid engine will need to switch from the electric motor to the petrol (gasoline) motor.

Hydrogen also has potential as a clean fuel; its combustion produces water as the only waste product. The German car manufacturer BMW has a prototype hydrogen-fuelled car available on the market, although only the very wealthy will be currently able to afford this technology. There are some interesting limitations: with a range of only 200 km you would have to plan your journeys very carefully. An additional slight inconvenience is that at the time of writing there are only five hydrogen filling stations in the world. One more concern is that there is a risk of explosion if the car is parked in a garage!

Figure 4 The Elcho combined heat and power plant in Poland, which is capable of supplying 226 MW of electricity and up to 500 MW of district heat

1. State the waste product of burning hydrogen.
2. Briefly outline some safety concerns when using hydrogen as a fuel.

Combined heat and power (CHP)

One major area of focus in the attempt to reduce our energy use has been to try and find a use for the waste heat from electrical energy generation. One technique is to use this thermal energy to heat buildings and homes in colder climates such as Europe or North America (Figure 4), or for the desalination of sea water in countries with limited fresh water supplies, such as the Arabian Gulf and North East Africa (Figure 5). Such strategies increase the overall efficiency of the system up to between 70% and 90%. In cooler countries, cars can also use the waste heat from the engine to heat the vehicle interior.

Figure 5 The Ras Abu Fontas combined power and desalination plant, near Doha, Qatar. The whole plant generates over 1600 MW of power using several gas turbines and produces about 100 ML per day of fresh water from sea water using the waste heat from the gas turbines.

Carbon capture and storage

Improved efficiencies and new technologies will have an impact on our fuel demand, and will certainly help to reduce carbon dioxide emissions. But the planet will continue to produce vast quantities of greenhouse gases, which will continue to build up in our atmosphere. Another solution, which is in the prototype stage in many countries, is to capture the carbon dioxide and bury it, safely underground, in rock formations such as coal seams, or in saline aquifers, where it cannot escape. This technique is known as carbon capture and storage or carbon sequestration (Figure 6).

Outline how the increased use of CHP electrical generation can reduce global warming.

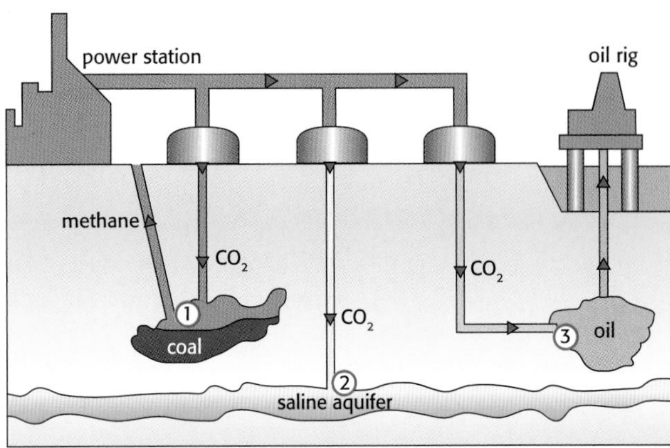

Figure 6 Carbon capture and storage

This is a particularly promising technique, because the necessary technology is already available, and it has the potential to reduce carbon dioxide emissions into the atmosphere significantly in a relatively short time. One added advantage is that the carbon dioxide can be pumped down into depleted oil reservoirs to help recover further oil.

One obstacle to the full-scale development of carbon capture and storage is that it is very difficult and the costs are prohibitive to "retro-fit" existing power plants with the technologies required to store carbon dioxide underground effectively. Only newly designed plants will be able to do this, and the design of a new power plant takes time.

The next few years

Critical decisions will be made in the next few years that may have a permanent impact on how our climate emerges from this current crisis. The adoption of these new emission-reducing technologies comes at a massive financial cost, but this is a very small cost compared with the potential impact of climate change on global society. A UK government report produced in late 2006 by Sir Nicholas Stern has suggested that an immediate increase in spending by governments of 1% annually to reduce emissions of carbon dioxide and other greenhouse gases could offset the potential costs of climate change, which were estimated to be as high as 20% of the world's total spending.

End of chapter summary

Chapter 21 has one main theme: possible solutions to climate change. Further research into more detailed explanations using other resources and/or further practice at solving problems in this topic is recommended.

Chapter 21 questions

This question is based on the Ras Abu Fontas combined power and desalination plant near Doha, Qatar.

Data for sea water

density = 1.03×10^3 kg m^{-3}

boiling point = 101 °C

specific heat capacity = 3990 J kg^{-1} K^{-1}

1 litre = 10^3 cm^3

The whole plant generates over 1600 MW of power using several gas turbines.

a) Outline some advantages of using gas rather than oil or coal in an effort to reduce global warming. [4]

As well as generating electricity, the Ras Abu Fontas plant produces fresh water from sea water using the waste heat from the gas turbines. The water is for the nearby city of Doha.

b) Show that the amount of energy needed to raise the temperature of 1 litre of sea water from 25 °C to the boiling point is about 310 kJ. Use the data provided. [3]

c) Show that the power needed to desalinate 100 ML day^{-1} of sea water is greater than 360 MW. [3]

To reduce the energy required, the pressure above the boiling sea water is reduced in a vacuum vessel so that the boiling point is lowered to 40 °C.

d) Estimate the power needed to desalinate 100 ML day^{-1} at this lower boiling point. [2]

22 Renewable energy

Renewable sources of energy (see page 282) provide us with one of the greatest causes for optimism in the whole energy debate. With appropriate design, natural energy sources can be harnessed to produce a sustainable supply of electrical energy without making any significant demands on the Earth's dwindling resources. In many cases the energy source is effectively free, and, as an added bonus, some systems ensure that no pollutants or greenhouse gases are pumped back into the environment. It seems almost too good to be true: all we need do is to build enough renewable generating capacity to provide for our energy needs and we have solved both the current "fuel crisis" and the climate change crisis.

The combustion of biofuels and waste material to generate electricity is slightly different from the other renewable sources, because both release significant amounts of carbon dioxide into the atmosphere, and waste burning releases other pollutants as well. Neither source makes significant demands on the Earth's resources, but they are not necessarily environmentally friendly, and their use does not contribute to a reduction in greenhouse gas emissions.

Biofuels are produced from plants, which absorb carbon dioxide from the atmosphere through photosynthesis. They may then either be burnt directly (wood, for example) or converted into an energy-rich fuel such as ethanol, for example by fermentation, which is then burnt. In both cases the carbon dioxide absorbed in photosynthesis is released back into the atmosphere, so there is no net effect on atmospheric carbon dioxide concentrations.

Of the other renewable sources, hydroelectricity is the most used, but wind power is perhaps the most promising source of clean energy for the future. In 2003, 1% of all electricity globally was generated by wind turbines, and this is likely to increase dramatically. In some countries significant amounts of electricity are already generated in this way, and as technology improves its use is likely to expand further.

Figure 1 A biofuel production plant

It is only comparatively recently that the focus in energy research has switched to renewables. The initial investment required to set up wind farms, dams, and tidal schemes can be prohibitive, and the efficiency of energy conversion into electrical energy is low. This means that large investments in capacity have produced only small returns in terms of electrical energy produced compared with nuclear power, for example. The location of renewable sources means that the energy often has to be generated in remote places, leading to additional costs in order to transport it to where it is most needed: the cities. These remote locations are often areas of outstanding natural beauty, and any human impact affects the environment with visual or noise pollution.

Hydroelectricity

Hydroelectricity is generated when water stored behind a dam is allowed to fall and drive a turbine. It may also be generated by pumped storage or from tidal energy in suitable river estuaries.

The Three Gorges Dam

One of the most ambitious hydroelectric schemes in history has been the construction of the Three Gorges Dam on the Yangtze River near Yichang in Hubei Province, Central China (Figure 2). The Yangtze River is the longest in Asia and the third longest in the world. It flows over 5000 km, from the edge of the Tibetan plateau to the East China Sea near the rapidly expanding city of Shanghai. The river itself has been a lifeline in central China for centuries, providing much-needed irrigation and a convenient means of transport for trade and travel in a region of mountainous and difficult terrain.

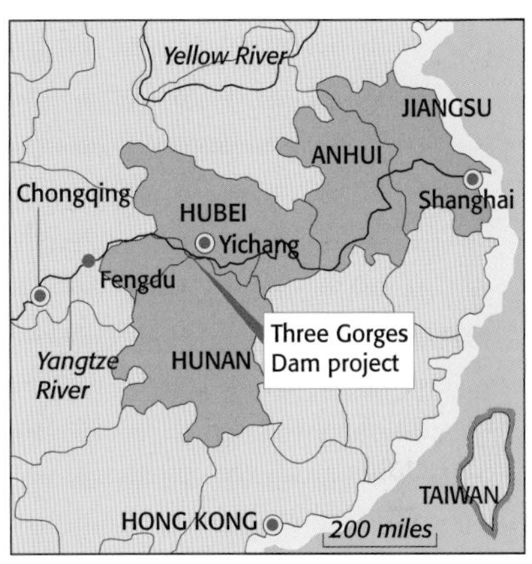

Figure 2 Location of the Three Gorges Dam

Use the data in Table 1 to answer the following questions.

Table 1 Data for the Three Gorges Dam

Economic	
Type	Concrete gravity dam
Construction dates	1993–2008 (original target date was 2009)
Official cost	US$22.5 billion (180 billion yuan at 2006 exchange rate)
Physical	
Height of dam	181 m
Length of dam	2.309 km
Power generation	
Left (north) bank	14 × 700 MW turbine generators (operating from 2003)
Right (south) bank	12 × 700 MW turbine generators (operating by 2008)
Underground	6 × 700 MW turbine generators (operating by 2011)
Reservoir data	
Length	660 km
Width	1.1 km (mean)
Area	1084 km^2
Land submerged	632 km^2
Water volume	39 billion m^3
Original river level	65 m above sea level
Final reservoir level	175 m above sea level

1 **a)** Estimate the mean drop of water in the reservoir, when it has risen to full height, as it falls through the turbines.

b) State what assumptions you have made.

Hint: Fill in all the relevant heights above sea level on a sketch diagram like that shown in Figure 3.

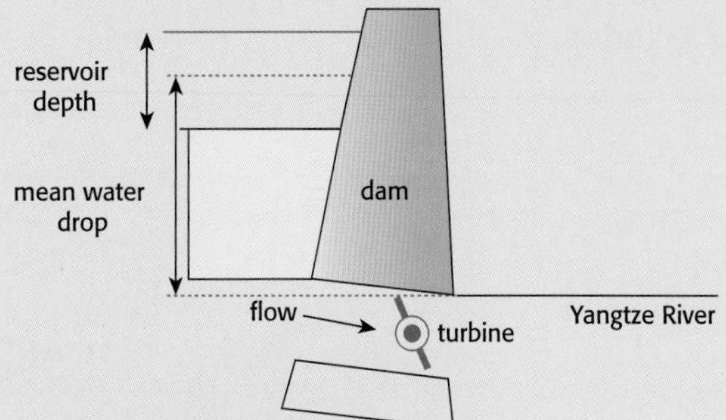

Figure 3 Sketch of dam cross-section

2 Show that the maximum power generated is approximately 22 GW.

3 A turbine converts kinetic energy in the water into electricity, which is distributed through China's electricity grid. If the turbines operate at 90% efficiency, calculate the kinetic energy per second of water flow required to generate power at 22 400 MW.

(If we ignore any resistive energy loses – a big assumption – it is reasonable to assume that all the potential energy stored in the water behind the dam is converted into kinetic energy as the water flows through the turbines.)

4 Use the data to estimate the speed of the water as it flows through the turbines.

5 Estimate the total flow rate of the water through the turbines.

The dam will form part of the largest hydroelectric scheme in the world. It is designed to generate at least 22 GW at peak capacity in 2011, which is equivalent to approximately 22 conventional coal-fired power stations. It was originally intended to generate up to 10% of China's electricity, but because of China's rapid economic growth this figure has been revised down to 3%. This could still have a huge effect in reducing China's total carbon dioxide emissions, but it would be a mistake to assume that such hydroelectric schemes have zero impact on global warming. Reservoirs and similar bodies of water release small amounts of methane – another greenhouse gas – into the atmosphere produced in the biological decomposition of organic material.

Data is a powerful tool of propaganda. It should always be viewed critically and never taken at face value. Information that is left out is just as important as the information that is included. The Chinese government would be keen to show how it is considering the energy needs of its fast-developing country with careful consideration of the environment and the need to reduce carbon dioxide emissions, and to control the regular flooding of the Yangtze. Environmental groups may be concerned about the effect on local people (it is estimated that more than 1.4 million will be displaced by the project), the changes in lifestyles caused by the loss of so much fertile farm land, the effects on the river downstream owing to increased siltation (deposition of sand and small rock particles) behind the dam caused by reduced flow, and the impact on flora and fauna in the Three Gorges area (which is designated by UNESCO as an area of outstanding natural value).

(a)

(b)

Figure 4 (a) The Three Gorges Dam (b) Landscape in the surrounding area

Tidal energy

Tidal energy schemes use a dam across a suitable river estuary. Sluice gates in the dam are opened as the tide rises and sea water fills the estuary. At high tide the gates are shut and the water is trapped at this high level behind the dam. At low tide, there is a significant height difference between the water level trapped behind the dam and the sea water outside. The trapped water is now allowed to run out through water turbines, generating electrical energy. Although the tide rises and falls only twice a day, a significant amount of power may be generated in this way owing to the huge volume of water involved. The environmental impact of such a scheme needs to be carefully considered. Currently the only tidal plant in Europe is on the Rance estuary in northern France. A useful website to start research on this is http://home.clara.net/darvill/altenerg/tidal.htm (accessed July 2009).

Pumped storage

Pumped storage adds no overall electrical energy into the grid. In fact it takes energy. Electrical energy is very difficult to store, so pumped storage provides a useful way of producing electricity when it is needed. At peak times when electrical demand is high a water turbine may be switched into the national grid. Such water turbines may reach peak generating capacity in under a minute, which ensures that the electrical supply can adjust to rapidly fluctuating demand. At such peak times many electrical companies charge consumers more per unit of electricity consumed than at off-peak times, when demand is lower. During off-peak times the action of the water turbine is reversed, and excess capacity is used to pump water back up to the upper body of water ready to be used again at peak times. The Dinorwig power station in Wales has a good website for research: http://www.fhc.co.uk/dinorwig.htm (accessed July 2009).

Figure 5 Taum Sauk pumped storage plant

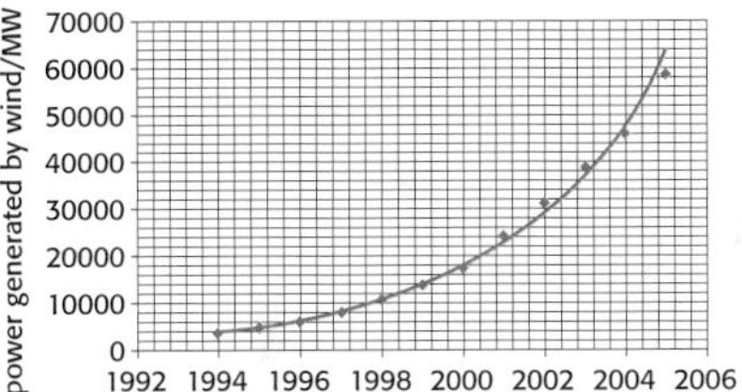

Figure 6 Recent growth in wind power. Source: Global Wind Energy Council (http://www.gwec.net accessed July 2009)

Wind power

Wind power has the potential to generate a significant proportion of the world's electrical energy needs. Recent improvements in technology and materials used have already led to the development of larger and cheaper wind turbines. The graph in Figure 6 indicates the growth of wind power in recent years.

Exponential increase

If the graph in Figure 6 can be described by an exponential function, then it should show a constant doubling time.

Time to grow from:

10 000 to 20 000 MW	~2.6 years
20 000 to 40000 MW	~2.6 years
30 000 to 60 000 MW	~2.5 years

It seems reasonable to assume that the growth in wind power generation is approximately exponential, with a mean doubling time of about 2.6 years.

We can use this data to project the wind power generated in 2020.

Power generated by the wind in 2004 = 48 000 MW

Number of doubling times from 2004 to 2020 = 16/2.6 = 6.2

1. Total world power demand is projected to be 5.4 TW in 2020. Discuss whether wind power can achieve the target of 34% of global electrical energy production, as suggested by the Global Wind Energy Council.
2. Use the data in Figure 6 to plot a graph of ln(power generated by wind) against time in years.
3. Does the graph plotted in question 2 suggest an exponential relationship?

$\therefore$ projected power in 2020 $\approx 48\,000 \times 2^{6.2}$

$= 3.5\ 3\ 10^6$ MW

$= 3.5$ TW

Wind capacity

Wind turbines of different sizes can generate a range of powers, from 100 kW (enough for a village) up to 2 MW (adequate for a small town). The turbine blades may vary in size, from 30 m up to 80 m in length, and are designed to rotate at anything from 10–30 rev/min. Modern turbines can generate electricity at wind speeds as low as 4 m s^{-1}, reaching maximum capacity at about 15 m s^{-1}. Hurricane or typhoon winds can damage wind turbines, so a braking system is incorporated into the design so that they stop turning at speeds in excess of 30 m s^{-1}. With careful planning, wind power can be fed into any national grid system, backed up by conventional thermally generated power when the wind is low. Over a whole country it is often true that the wind is always blowing *somewhere*, so that wind power can be fed into the electricity grid fairly continuously.

Figure 7 Wind turbines

1. Show that the maximum power P_0 that may be extracted from the wind per unit area is given by the formula $P_0 = \frac{1}{2}\rho v^3$, where v is the wind speed and ρ is the air density, assuming that the air stops as it hits the blades.
   (Hint: Start by calculating the kinetic energy of air per second passing through the turbine blades.)
2. If the turbine blade rotates once every 3.0 seconds, compare the blade speed at the tip, 20 m from the axis, with the speed at the other end, only 0.50 m away from the same axis.
   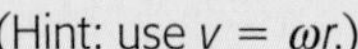
   (Hint: use $v = \omega r$.)
3. Show that the maximum power per m^2 in a wind of speed 7.0 m s^{-1} is 210 W m^{-2}. Assume that the density of air = 1.225 kg m^{-3}
4. Calculate the power per m^2 for an 8.0 m s^{-1} wind, and comment on why a small increase in wind speed causes a significantly larger increase in power.

Inquiry: Sources of renewable energy

Hydroelectricity and wind power are arguably the most important sources of renewable energy: hydroelectricity because, of all the renewable sources of energy, it currently generates the most electricity, and wind power because it is likely to be the leading renewable form of generation in the future.

Other promising renewable ways of generating electrical energy include:

- **Geothermal energy.** This uses thermal energy from "hot rocks" under the ground to produce steam, which can then drive turbines to generate electricity.
- **Solar energy.** Photovoltaic cells (otherwise known as solar cells) can convert solar energy directly into electrical energy.

  Unfortunately a typical photovoltaic cell produces a very small voltage and is not able to provide much current and hence overall power. Thus they are used to run electric devices that do not require a great deal of energy. An alternative approach to capturing energy radiated from the Sun is an active solar heater (otherwise known as a solar panel). Typically these devices are used to directly warm water and thus save on the use of electrical energy.

  Both devices rely on capturing as much energy as possible from the incident radiation from the Sun. The solar power incident per unit area of the Earth's surface varies as a result of many factors including the latitude of the location, the season of the year, the time of the day and prevailing weather. Thus they are not suitable for all locations.
- **Wave power.** An oscillating water column (OWC) ocean-wave energy converter changes the rising and falling motion of ocean waves into rotational motion, which is used to turn turbines. The total energy available within ocean waves is huge, but so is the cost of creating such schemes far out to sea.

Carry out some research to make sure you fully understand the basic principles of using each source, and their advantages and disadvantages.

To estimate the amount of energy available in wave power we can model water waves by considering a square wave of amplitude A, speed v, and wavelength λ as shown below:

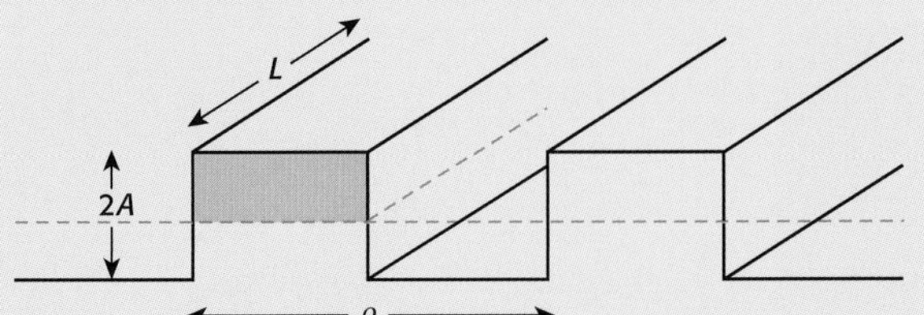

Absorbing all this wave's energy would result in a flat body of water. This would be equivalent to moving the shaded top part of the wave down into the trough. The loss of gravitational PE as a result would be $(A^2\lambda L \rho g)/2$

By considering the number of waves passing a point in unit time, this model suggests that the maximum power available per unit length is $\frac{1}{2}A^2\rho g v$

End of chapter summary

Chapter 22 has four main themes: solar power, hydroelectric power, wind power and wave power. The list below summarises the knowledge and skills that you should be able to undertake after having studied this chapter. Further research into more detailed explanations using other resources and/or further practice at solving problems in all these topics is recommended – particularly the items in bold.

Hydroelectric power

- Distinguish between different hydroelectric schemes.
- Describe the main energy transformations that take place in hydroelectric schemes.
- **Solve problems involving hydroelectric schemes.**

Wind power

- Outline the basic features of a wind generator.
- Determine the power that may be delivered by a wind generator, assuming that the wind kinetic energy is completely converted into mechanical kinetic energy, and explain why this is impossible.
- **Solve problems involving wind power.**

Solar power

- Distinguish between a photovoltaic cell and a solar heating panel.
- **Outline reasons for seasonal and regional variations in the solar power incident per unit area of the Earth's surface.**
- **Solve problems involving specific applications of photovoltaic cells and solar heating panels.**

Wave power

- Describe the principle of operation of an oscillating water column (OWC) ocean-wave energy converter.
- Determine the power per unit length of a wavefront, assuming a rectangular profile for the wave and solve problems involving wave power.

Chapter 22 questions

1 This question is about wind turbines.

a) State **two** factors that affect the maximum theoretical power output of a wind turbine. [2]

A wind farm is to be built to supply electrical energy to a small town. The following data is available.

Energy consumption for the town for 1 year = 5.0×10^7 kWh

Length of turbine blade = 20.0 m

Average wind speed = 8.0 m s^{-1}

Density of air = 1.1 kg m^{-3}

1 year = 3.2×10^7 s

b) Deduce from this data that approximately 16 wind turbines are required. [5]

c) State **three** reasons why in fact more than 16 turbines will be needed. [3]

2 A nuclear power station has an output power of 200 MW. It is proposed to replace the power station with a series of wind turbines, each with an energy output of 750 kW. State and discuss **one** advantage and **one** disadvantage that such a change might bring if this proposal was to take place. [4]

23 Nuclear power

As an alternative to fossil fuel generation, nuclear power has had a controversial press. It has become a major source of energy in many countries across the world: the UK, USA, and France, for example, could not meet their current energy supply targets without it. The use of nuclear power to generate electrical energy is fairly widespread (Figure 1).

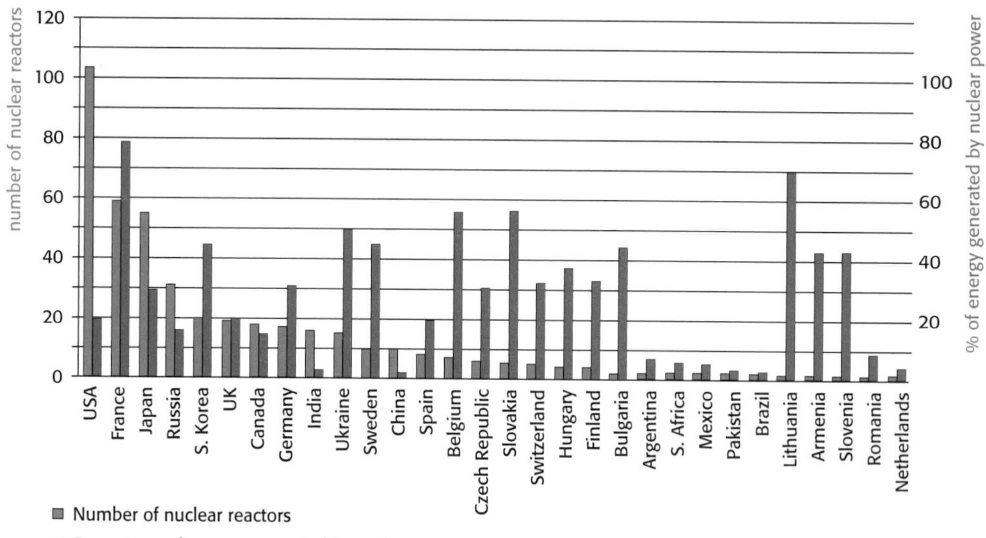

Figure 1 Use of nuclear power to generate electrical energy. Data from International Atomic Energy Authority, January 2007.

The thinking on the use of nuclear energy to generate electricity has changed since the 1950s, when it first became commercially available. The timeline in Figure 2 illustrates some trends in the development of nuclear power.

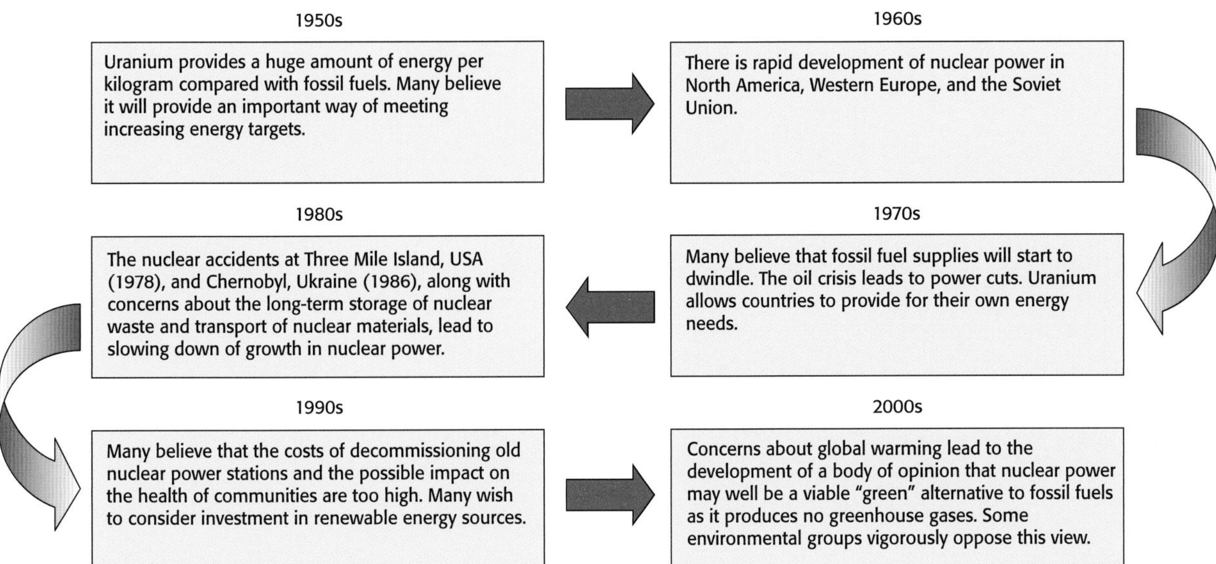

Figure 2 Timeline: the development of nuclear power

As an IB learner you will need to weigh up the evidence and form your own opinion on whether nuclear power is a safe, reliable, and efficient technology, and whether we should consider increasing investment in it. Alternatively, nuclear power may have such a negative impact on the environment in other ways that it could be wiser to invest more heavily in renewable technologies, such as wind, tidal, and wave power, to achieve our energy needs with minimal environmental impact.

The questions (right) aim to encourage you to think critically about nuclear power generation worldwide and to draw conclusions about the different circumstances and needs of individual countries.

Of the 442 operational nuclear reactors worldwide (2006), more than half of them are found in only four countries: the USA, France, Japan, and Russia. The first three of these have a very high demand for electricity, and rely heavily on nuclear power to achieve the required level of supply. France achieves a very large proportion of its electricity generating capacity through nuclear means. Table 1 enables you to compare the average rate of energy generation in these countries. The table also includes China and Lithuania for contrast.

1 Compare the number of nuclear reactors in OECD and non-OECD countries and outline your findings.

2 Outline some possible reasons that might explain why there are a relatively high number of reactors in Eastern Europe, previously the Communist bloc.

3 **a)** State which three countries have the largest number of nuclear reactors.

b) Suggest reasons to explain this.

4 Outline what other data would be helpful in order to get a full picture of a country's nuclear power capacity.

5 Describe what the data tells us about nuclear power in the developing world.

Table 1 Comparison between six countries' nuclear power reliance

	Average total electricity power production* / MW	Number of nuclear reactors	Percentage of total electrical power produced from nuclear energy	Average electricity power production by nuclear power / MW
USA	460 889	103	19.3	
France	62 694	59	78.5	
Japan	109 250	55	29.3	
Russia	99 292	31	15.8	
China	282 500	10	2.0	
Lithuania	1689	1	69.6	

*Estimates.

Source: Data from www.iaea.org, October 2006.

1 Calculate the average rate of electrical power production from nuclear energy for each country in Table 1 and so complete the table.

2 Outline how the following factors may affect the growth of nuclear power in these countries.

a) Rising fossil fuel prices.

b) Protests and demonstrations by environmental groups at the source of uranium.

c) Environmental protests in the end-user country.

d) A small nuclear accident or leak of radioactive materials into the environment.

e) A transport strike in the generating country.

Inside the reactor

The enrichment process

Most naturally occurring uranium ores contain a mixture of two isotopes: uranium-235 and uranium-238. Most of the ore (99.3%) is uranium-238; the uranium-235 required for fission makes up only 0.7%. Modern reactors require enriched uranium, containing up to 3.5% of the isotope uranium-235.

In the **enrichment** of uranium the gas uranium fluoride is fed into a **centrifuge**. This contains a series of evacuated cylinders, each with a rotor between 1 and 2 m long with a diameter of 15–20 cm (Figure 3). The mixture of isotopes is spun round at rates of up to 70 000 rev/min. The heavier uranium-238 requires a larger **centripetal** force to stay in a circular path and thus moves to the outside of the centrifuge. The uranium-235 requires a smaller force and collects together, closer to the centre (Figure 3). The uranium-238 can then be removed, as **depleted uranium**, and the remaining mixture is then centrifuged several more times, in a cascade of centrifuges, before the concentration of uranium-235 has increased to the required 3.5%. The whole process is complicated, and involves some very technical and challenging engineering, partly because of the tremendous forces involved.

Moderation and control

In order for a nuclear reactor to continue to release thermal energy at a steady rate, the reaction must be **self-sustaining**. For every one neutron that successfully collides with a uranium-235 atom, another one has to be produced that will definitely induce another fission. If too many neutrons are available for further fissions the rate of production of thermal energy in the reactor may increase in such a way that the reaction becomes uncontrollable, leading to an explosion. This uncontrolled fission, or **thermal runaway**, is the mechanism by which a nuclear bomb works. A great many precautions are used to ensure that this does not happen in a nuclear reactor.

1 Estimate the mass of one molecule of uranium hexafluoride, UF_6, in kg.

(Relative atomic mass U = 238, F = 19.0, 1 amu = 1.66×10^{-27} kg)

2 Assume the starting conditions in a centrifuge are as follows:

Initial orbital radius r = 1.5 m

Rate of rotation = 70 000 rev/min.

Calculate:

a) the number of revolutions per second.

b) the period of rotation.

c) the angular velocity.

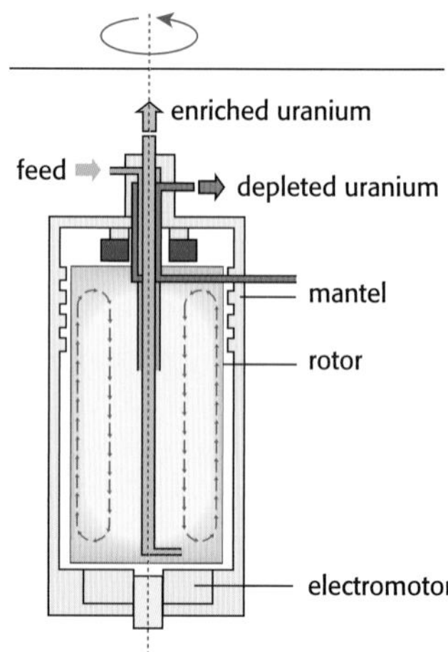

Figure 3 Structure of a uranium centrifuge. The rotor spins around a vertical axis. The uranium-235 collects towards the axis and the uranium-238 towards the outer wall of the rotor.

The nuclear energy spectrum in Figure 4 illustrates that most neutrons produced in a reactor are **fast neutrons** in the range 1–2 MeV. These have too much energy to cause further uranium-235 fission, and are far more likely to be **scattered** by uranium-238. **Thermal neutrons** are more likely to cause further fission. These neutrons have energies of about 0.02 eV. **Moderators** are used to "slow" down these fast neutrons to thermal energies to increase the probability of further fission. **Control rods** are used to remove any excess neutrons to ensure the reaction continues safely.

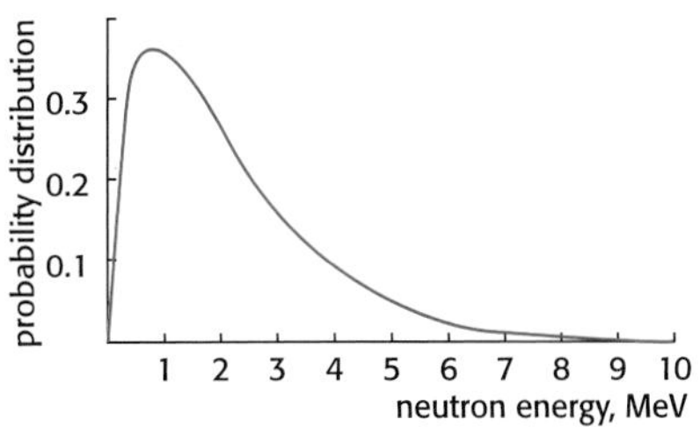

Figure 4 Neutron energy spectrum from a fission reactor core

Scattering cross-sections and probability

Typical nuclear fission reactions include:

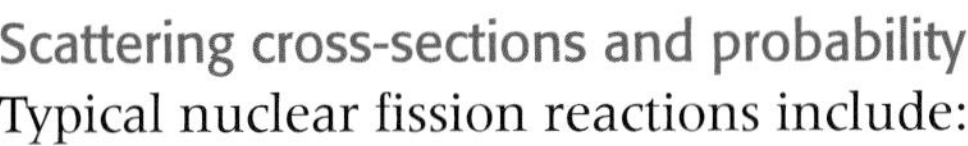

$$^{1}_{0}n + {}^{235}_{92}U \rightarrow {}^{141}_{56}Ba + {}^{92}_{36}Kr + 3\,{}^{1}_{0}n$$

$$^{1}_{0}n + {}^{235}_{92}U \rightarrow {}^{138}_{55}Cs + {}^{96}_{37}Rb + 2\,{}^{1}_{0}n$$

The fission equations show that for every inducing neutron that has a successful collision with a nucleus of uranium-235, two or three neutrons are produced in the resulting fission. In fact, the mean overall number of neutrons resulting from each fission is about 2.5.

These neutrons may:

- leave the fuel rod through the external surface
- collide inelastically with a uranium-238 nucleus and be absorbed
- collide inelastically with a uranium-235 nucleus, inducing fission
- collide elastically with either a nuclide or the moderator and control rod molecules, transferring energy only.

The laws of probability are used to determine the chance of each outcome occurring. These calculations enable appropriate conditions to be set up within the reactor to ensure that criticality – a self-sustaining fission reaction – is maintained. As we are discussing many billions of events, the laws of probability work very well.

The concept of **scattering cross-sectional** area is used. A large scattering cross-sectional area implies a large probability of a particular event occurring (Figure 5).

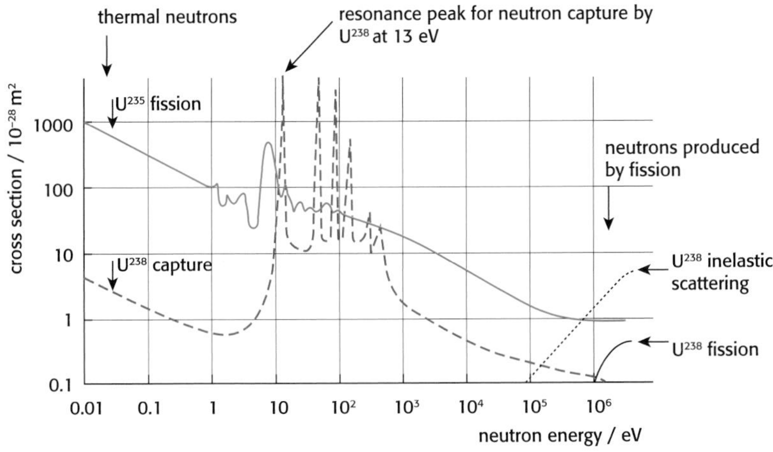

Figure 5 Scattering cross-sections for uranium nuclides

The curves are rather complicated, but show that

- the scattering cross-section and hence the probability of a fission for uranium-235 increases as the neutron energy is reduced
- the scattering cross-section for uranium-238 neutron capture is particularly large in the 10–100 eV energy range. The sharp peaks in this energy range are known as **resonances** (see page 95).

The uranium-235 and uranium-238 nuclides present a similar geometric cross-sectional area to a neutron, but their scattering cross-sections depend on the neutron energy.

1 Consider a neutron moving along a tube of area A towards a nuclide of area a, as shown in Figure 6a. Write down the probability of the neutron colliding with the nuclide. (You can assume the neutron is a point object.)

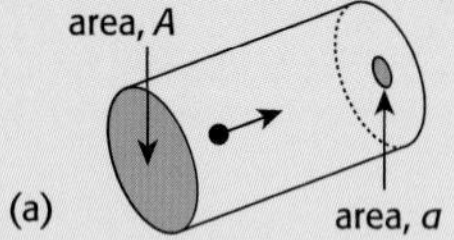

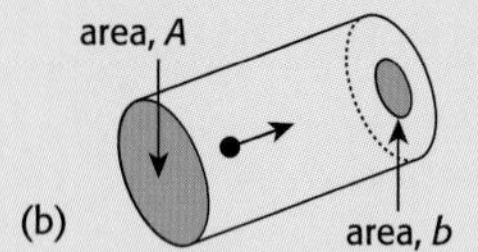

Figure 6 Calculating probabilities of nuclear collisions

2 Now assume that the tube contains a different nuclide of area b (Figure 6b). Write down the probability of a collision with the second nuclide.

3 **a)** Write down the ratio of your answer to question 1 to your answer to question 2.
b) Explain what this answer means in terms of probabilities.

4 Assume the neutron is a thermal neutron of energy 0.01 eV. Using the graph in Figure 5, determine the ratio of the probability of uranium-235 fission to uranium-238 capture.

5 Repeat question 4 for a faster neutron of energy 13eV.

6 Write a short paragraph to explain to a younger student how cross-sectional areas can be imagined to change in particle experiments.

Thinking about science: *Uranium enrichment*

Nuclear power features frequently in the news. At the time of writing, both Iran and North Korea have featured regularly in the international news because of their nuclear programmes, but for different reasons. Iran has started the enrichment of uranium-235, and North Korea exploded a small test nuclear device in October 2006. It is interesting that so much attention should be focused on these two countries when many countries in the world have far more developed and advanced nuclear programmes. In addition, the USA, France, and Russia possess nuclear weapons, yet we rarely hear as much discussion or concern in the international press about these countries as we do about North Korea or Iran.

As well as relying on the use of radio, television, and newspapers to distribute their message, modern news agencies also use the internet: websites, email, and podcasting. Many of these press agencies attempt to give a balanced view, and have no apparent political alignment. Some press agencies do, however, align themselves with political philosophies, parties, or organizations. In some countries newspapers have restrictions placed on what they can report. Other news organizations are part of profitable business concerns and can be obliged to represent the interests of their owners.

Often the press tends to present a viewpoint or opinion in order to make the news item more interesting or "newsworthy". Emotive or subjective language may be used to support an opinion, rather than relying on objective facts and evidence. Readers must remember to separate fact from opinion, and draw their own conclusions based on objective observation, just as should take place in an experiment.

As an IB learner you are encouraged to be thinkers, principled, and open-minded (see page 3). You should **think critically** about what you are reading, listening to, or watching. Ask yourself questions such as:

- What is the evidence to justify this opinion?
- What sources were used to produce the article?
- How does the type of language used distort or influence my viewpoint? (You will almost certainly have discussed this in your group 1 subject lessons.)
- What is the truth, and how do I know this?

In order to make a judgment, you may need to apply **principles** based upon your own moral and

ethical standpoint in order to formulate your own personal opinions. By keeping **open-minded** you may be able to separate fact from misinformation, and even be able to use a blatantly distorted article to develop a more accurate picture of the truth because you are aware of the biases behind any article that you read.

Many news agencies report events from a prejudicial viewpoint. This may be the result of illogical argument or irrationality. It may be deliberate or unintentional. Prejudice consists of judging an issue or holding an opinion on it without rationally examining the evidence for it. Prejudice may well be based on preconceived false, or inaccurate, stereotypes.

Apply your critical thinking skills to the following situations. Try to give a balanced answer, considering all possible viewpoints. These questions are probably better to use in a class debate rather than as written answers. Discuss:

a) whether developing countries have a moral responsibility to develop nuclear power.

b) whether any country has the right to develop nuclear weapons for self-defence.

c) whether it is better to isolate a country thought to be developing nuclear technology or to engage it with technological and logistical support.

Inquiry: Types of nuclear reactor

Each reactor is unique, in that it is designed and engineered to fit a range of particular criteria. However, there are a few designs or templates that have popular appeal for reasons of cost, safety, and efficiency.

The most common type of reactor is the **thermal reactor**, so called because it is powered by the fission of uranium-235 with thermal neutrons. Although there are important differences between each type of thermal reactor, there are some common features. A block diagram for a nuclear power station has much in common with a fossil fuel power station (see page 286) – the key difference is that the fossil fuel furnace is replaced by the nuclear reactor (Figure 7).

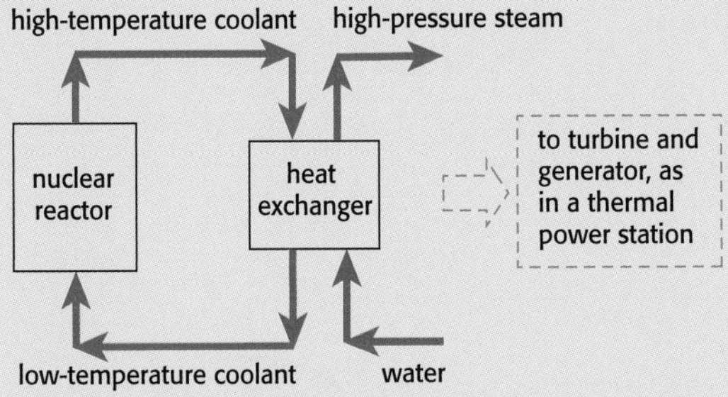

Figure 7 Block diagram for a nuclear reactor generating electricity. The source of thermal energy is illustrated.

The main types of thermal reactor in commercial use today are the:

- pressurized water reactor, PWR
- boiling water reactor, BWR
- pressurized heavy water reactor, PHWR
- gas-cooled reactor, GCR
- light water cooled, graphite moderated reactor, LWGM (also called RMBK in Russia)

There are important differences in design that effect the efficiency, the cost, and even the safety and risk involved. Some of the common features are described below.

- **Fuel rods.** All thermal reactors utilize the fission of uranium-235 to release thermal energy. The proportion of uranium-235 in the rods varies from 0.7% to 3.5% depending on the reactor.
- **Moderator.** Most neutrons released in fission are fast neutrons, so a moderator is used to reduce their energy down to thermal levels to ensure that the fission is self-sustaining. The moderators used vary: for example, solid graphite in the LWGM and steam in the BWR.
- **Coolant.** This is used to transfer the thermal energy released from the reactor core to raise steam. This steam then drives the turbines. The BWR is unique in that the coolant, steam, also drives the turbines directly. Both the BWR and PWR use water, in different states, as the coolant and moderator.
- **Control rods.** These are inserted among the fuel rod assembly and are used to control the reactor temperature, thus preventing it from overheating. If **criticality** is exceeded, neutron-absorbing rods made of cadmium or boron steel are lowered into the reactor core and any excess thermal neutrons are effectively absorbed.
- **Containment building.** The reactor core is usually surrounded by a steel vessel and a concrete structure, several metres thick, to absorb any excess radiation and help contain any potentially catastrophic event, such as an explosion, should all other safeguards fail.

The other main type of commercial reactor is the **fast breeder reactor**, FBR, which is quite different from thermal reactors.

To develop your **personal skills**, work in teams to research and present your findings on one of the four inquiries suggested below. It would be a useful exercise to work with students whom you do not normally partner so that team work can be practised ready for the **Group 4 project**, when these skills will be formally assessed. Strong team members are able to:

- offer ideas and advice to the group
- listen to the views of others
- agree and persevere on a specific task with a clear goal in mind
- work ethically and consider any environmental impacts, e.g. excess use of paper and the printer.

Some useful websites for the research include:

www.iaea.org
www.world-nuclear.org
aua.org.au (all accessed July 2009)

1 Compare and contrast the PWR and the BWR.
You should consider the following:
    - the proportion of uranium-235 in the fuel rods
    - the role of the coolant
    - the type of coolant and moderator used
    - operating efficiencies
    - safety features and possible risks
    - average power output of each reactor.

2 Find out about some of the advantages and disadvantages of the PHWR compared to the PWR and consider why many different countries have selected this design.
In your research you could consider
    - the average power output
    - the efficiency of heavy water as a moderator compared with normal water
    - an advantage of using heavy water in terms of safety
    - why this reactor has been so successfully exported worldwide.

3 Find out about the FBR and how it works. In your research you should answer the following questions.
    - Is a moderator needed? Explain your answer.
    - Describe how this reactor is able to "breed" further plutonium-239 fuel.
    - Why are liquid metals suitable for use as coolants? Why would these coolants be inappropriate for a thermal reactor such as a PWR?
    - At the current rate of use, uranium may well run out in the next 60 years. How will FBRs affect this estimate?
    - Outline any ethical concerns involved when countries develop FBRs.

4 Find out what you can about the LWGM reactors used in Eastern Europe, and compare then with the PWR. The Chernobyl accident happened in a LWGM reactor. Try and find out what safety improvements have been put in place since this disaster took place in 1986.
You should consider how:
    - the LWGM is cooled and moderated
    - the accident happened
    - the design of subsequent generations of LWGM has improved safety.

Physics issues: Is nuclear power safe?

Nuclear accidents have occurred in the recent past, and may occur again. Because of the historical links between nuclear research and nuclear weapons, many people are concerned about the use of nuclear power for electricity generation because of the possible misuse of nuclear waste from civilian reactors to produce weapons-grade plutonium.

These are all serious concerns, but we can question whether the apparent fear of nuclear power is rational, and investigate whether there is any scientific evidence to justify such fears.

Figure 8 Radiation hazard symbol

Risk

On page 105 you saw that risk may be assessed using the equation

$$\text{risk} = \text{probability} \times \text{consequence}$$

To fully assess the risks involved in using nuclear power we need to consider all the possible hazards involved, statistically analyse the probability of an event occurring, and assess the consequences of such an accident. The effect of radiation on biological material is studied in topic 16 (see pages 253–6).

Some possible risks of and consequences from a nuclear reactor are listed in Table 2. These risks are fairly meaningless without knowing the probability of the risk occurring. The extremes are easier to understand. Risk 1 is certainly possible, but has little consequence, whereas risk 5 is extremely improbable, but the consequences are very serious.

Commercial nuclear reactors have a very good safety record in most countries; more deaths can be attributed to accidents involved in mining for coal for use in conventional power stations, than to nuclear power. Long-term coal mining often leads to pneumoconiosis (a lung disease that may cause death), but there is limited statistical evidence among workers in the nuclear industry to suggest they are at a higher risk of developing cancer. Inevitably, any nuclear accident – however small and inconsequential – tends to make the news, and press headlines tend to remain in our memory. For accidents to be prevented, they need to be reported. How well known is it in North America and Europe, for example, that every year several hundred coal miners die in tragic and preventable circumstances in small and often illegally operated coalmines in rural China?

Table 2 Some risks of radiation

Risk	Consequence
1 Low-grade radiation leak over a short time	Very little
2 Low-grade radiation leak over a long time	Possible ill-health in the long term, including cancers
3 High-grade radiation leak over a short time; exposure to a small area of the body	Redness of the skin, localized "burns"
4 High-grade radiation leak over a short time; exposure to a large area of the body	Radiation sickness and death
5 Supercritical reactor, leading to thermal meltdown and explosion	Possible radiation sickness and death near to the reactor. Possible ill-health in the long term, including cancer for others

Nuclear hazards

There are many possible hazards associated with the use of uranium in the nuclear power industry, starting with its mining and ending with the disposal of waste. The level of actual risk is extremely low, however, as long as rigorous safety procedures are enforced. However, it is entirely possible that such safety standards are more rigorous in some countries than in others.

The stages in this uranium cycle are:

1 uranium mining
2 enrichment
3 fuel rod production
4 use in the reactor
5 waste disposal.

Use the World Nuclear Association website (www.world-nuclear.org accessed July 2009) and other sources to:

- research what happens in each stage of the uranium cycle
- list the possible hazards at each stage
- outline safety procedures to minimize risk at each stage.

Nuclear waste

One of the greatest concerns is management of the waste left over from the fission process. Although much of the waste, notably the fission products, has a relatively short half-life, some material will be significantly hazardous for many generations to come, and so it is the responsibility of the nuclear industry (and the governments who run it) to ensure that a reliable method is used for the safe disposal of waste.

The major constituents of nuclear waste are as follows:

Fission fragments are generally beta decay emitters, with half-lives up to 10 years.

Transuranic elements are elements beyond uranium in the Periodic Table. They include neptunium, plutonium, and americium, and are formed as shown in the decay chains below:

$${}^{238}_{92}\mathrm{U} + {}^{1}_{0}\mathrm{n} \rightarrow {}^{239}_{92}\mathrm{U} \quad \text{(neutron capture)}$$

$${}^{239}_{92}\mathrm{U} \rightarrow {}^{239}_{93}\mathrm{Np} + {}^{0}_{-1}\mathrm{e} + \bar{\nu}_{\mathrm{e}}$$

$${}^{239}_{93}\mathrm{Np} \rightarrow {}^{239}_{94}\mathrm{Pu} + {}^{0}_{-1}\mathrm{e} + \bar{\nu}_{\mathrm{e}}$$

Further neutron capture can produce plutonium-**241**:

$${}^{239}_{94}\mathrm{Pu} + {}^{1}_{0}\mathrm{n} \rightarrow {}^{240}_{94}\mathrm{Pu}$$

$${}^{240}_{94}\mathrm{Pu} + {}^{1}_{0}\mathrm{n} \rightarrow {}^{241}_{94}\mathrm{Pu}$$

This is unstable, and decays into americium-**241**:

$${}^{241}_{94}\mathrm{Pu} \rightarrow {}^{241}_{95}\mathrm{Am} + {}^{0}_{-1}\mathrm{e} + \bar{\nu}_{\mathrm{e}}$$

These elements decay with half-lives up to 1000 years or more.

Between them, the transuranic elements and fission products account for up to 6% of the thermal energy produced in the reactor.

Activation products are radioisotopes produced by the absorption of neutrons by any other material, including the steel containment vessel, coolant, and moderator. These may include tritium, a form of hydrogen: hydrogen-3.

Once the spent fuel rods are removed from the reactor they will continue to emit thermal energy, because of radioactive decay, at a rate of about 10 kW per tonne, but this falls to about 1 kW per tonne after 10 years. During this time the waste is stored under water in large cooling ponds.

Solidification processes have been developed in France, the UK, the USA, and Germany over the past 35 years. Liquid high-level wastes are evaporated, mixed with glass-forming materials, melted, and poured into robust stainless steel canisters, which are then sealed by welding.

The graph in Figure 9 illustrates the decay of the various components of nuclear waste. The graph has a log scale on both axes, and covers very large ranges of activity and time. This may make the graph appear difficult to interpret.

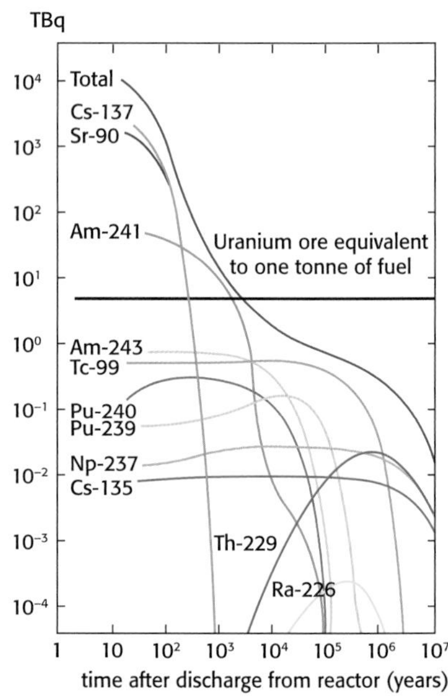

Figure 9 Activity of high-level waste from 1 tonne of spent fuel

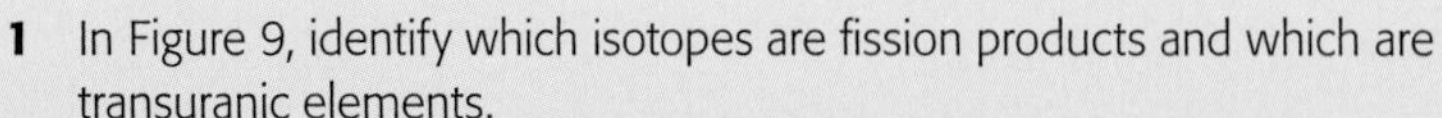

1 In Figure 9, identify which isotopes are fission products and which are transuranic elements.

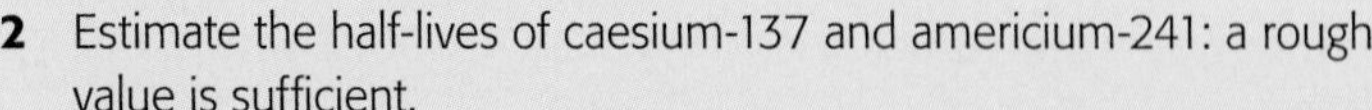

2 Estimate the half-lives of caesium-137 and americium-241: a rough value is sufficient.

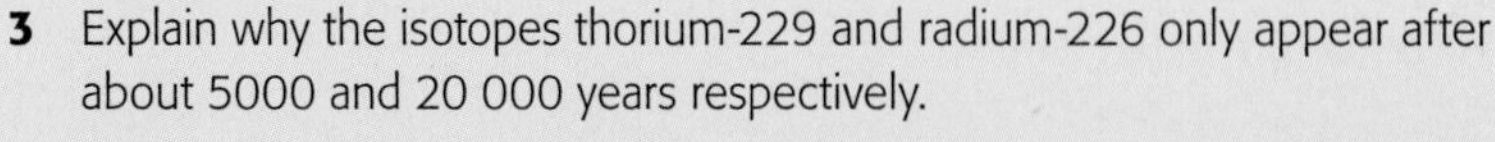

3 Explain why the isotopes thorium-229 and radium-226 only appear after about 5000 and 20 000 years respectively.

(Hint: Consider the possible decay chains in the waste fuel, starting with the alpha decay of uranium-238.)

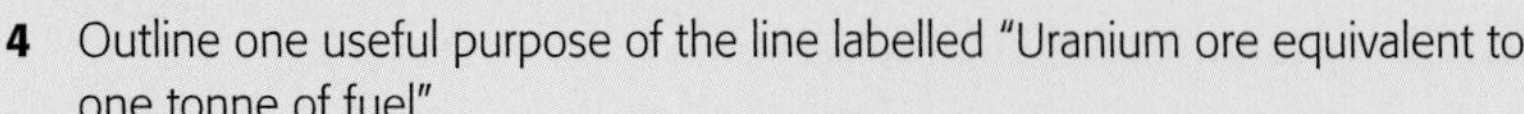

4 Outline one useful purpose of the line labelled "Uranium ore equivalent to one tonne of fuel".

5 Does the waste fuel have an overall half-life? If so, is it possible to estimate its value?

Reactor accidents

In common with other sectors of the electricity generation industry, there have been many small-scale accidents at nuclear power stations. There have been very few that have had any serious effect on either the workers or nearby residents of the power station, although long-term environmental damage is much more difficult to gauge. The two most serious accidents were in 1979 at Three Mile Island, in Pennsylvania off the East Coast of the USA, and in 1986 at Chernobyl, in the Ukraine.

Three Mile Island was a PWR, now the most popular reactor worldwide, and Chernobyl was a LWGM, still widely used in Russia. Both had potentially disastrous consequences but only in Chernobyl was there a serious radioactive leak into the environment. In this case the leak was so serious that the effects were measured as far away as Finland to the North and Wales to the west. These tragic events led to major increases in safety precautions and any subsequent accidents in the last twenty years or so have been minor. Compared to other electricity generating sectors, nuclear power has a very safe record.

Development of nuclear weapons

Modern nuclear weapons use the uncontrolled fission of uranium-235 or plutonium-239 to generate tremendous amounts of energy. They require almost pure pieces of the fissile isotope, which involves a highly sophisticated and technical engineering capability to produce. They work on a very simple principle. The fissile material is converted into a critical mass by use of a chemical explosive. One method is to force two smaller pieces of uranium together, and another is to compress a piece of plutonium to increase its density (Figure 10). In both cases, the critical mass is exceeded and an uncontrolled fission chain reaction follows.

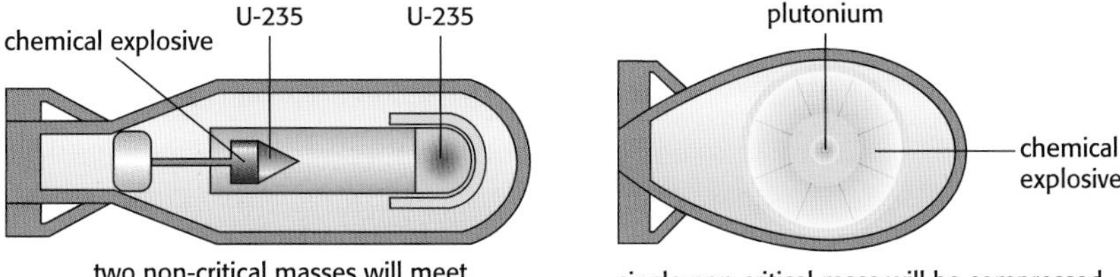

Figure 10 Two methods of achieving critical mass in a weapon

Pure uranium-235 does not occur in nature, and plutonium is not found naturally. Uranium-235 concentrations for conventional reactors are increased by centrifuging (see page 322), and this process may be continued to generate the higher concentrations needed for a weapon. Plutonium-239 is produced in the nuclear waste of most conventional reactors, and may be chemically separated out and purified. Fast breeder reactors use plutonium, but also "breed" further plutonium.

Nuclear reactors are therefore necessary in order to manufacture weapons. One of the risks of nuclear power is the possibility that weapons-grade fissile material may be produced as a by-product. Many governments have openly used a civilian nuclear programme to generate material for nuclear weapons; others have done it

?

This question is about one of the challenges for nuclear engineers caused by the low number of thermal neutrons produced in fission. Without any further intervention, the fast neutrons will be captured by uranium-238 and the reaction will come quickly to a halt.

Moderators are used to slow down the fast neutrons to thermal energies in the process of moderation. The fast neutrons undergo a series of elastic collisions with the atoms or molecules making up the moderator. The formula below (which can be derived from conservation of momentum considerations) can be used to calculate the transfer of kinetic energy from the neutron of mass m and energy E_1 to the moderating particle of mass M and kinetic energy E_2 using:

$$\frac{E_2}{E_1} = \frac{4mM}{(M + m)^2}$$

In order to reduce the energy of fast neutrons down to thermal levels as quickly as possible a moderator is used whose constituent particles have a mass, M, as close as possible to the neutron mass, m.

For the questions below use the following relative masses:

neutron = 1, water molecule = 18, deuterium oxide molecule = 20

a) Show that for
  i) water moderation E_2/E_1 = 0.20, and
  ii) deuterium oxide moderation E_2/E_1 = 0.18. [4]

b) Calculate the proportion of its original kinetic energy retained by the neutron in each case. [3]

c) Estimate the number of collisions a neutron of energy 1 MeV has to undergo in a water moderator to reach an energy of 0.02 eV. [2]

covertly. However, it is clear that nuclear reactors may be used for the totally ethical generation of electrical power without necessarily leading to the production of these weapons of mass destruction.

It is unlikely, but certainly a possibility, that terrorist groups could get hold of sufficiently pure material to make an effective fission weapon. Another fear is the use of so-called "dirty" bombs. These contain conventional, chemical explosives surrounded with highly radioactive waste. The explosive is used to disperse the waste over a large area, thus leading to large-scale radiation exposure and greatly increasing the number of fatalities.

Security at nuclear facilities is extremely high, at all stages of the nuclear cycle, in an attempt to reduce the risk of such awful weapons ever being produced.

The above section may make for depressing reading but there is much room for optimism. Despite continued fears, the nuclear generating industry has a very good overall safety record, and produces up to 15% of the world's much-needed electricity without producing greenhouse gases or other pollution from its generators. This is potentially very good news indeed.

The bigger picture: The nuclear debate

This final section in the chapter encourages you to weigh up the evidence for yourselves and decide whether the continued use of nuclear power is desirable and necessary or undesirable and avoidable. Much has been made of the potential for nuclear power to generate increased amounts of electricity without further adding to greenhouse gas emissions. However, there are strong arguments to increase the use of renewable energy sources, particularly wind power, in order to meet our electrical needs without compromising the climate.

In the nuclear debate you have to consider whether governments should:

- increase nuclear power generating capacity
- increase the use of renewable energy sources to generate electricity
- increase the use of both sources.

Which is going to be most beneficial option in the long term with the least costs? In order to get the debate started, two articles are presented below.

The first article comes from Greenpeace, a global environmental organization that campaigns against nuclear power, and argues as follows:

Greenpeace has always fought – and will continue to fight – vigorously against nuclear power because it is an unacceptable risk to the environment and to humanity.

The only solution is to halt the expansion of all nuclear power, and for the shutdown of existing plants.

We need an energy system that can fight climate change, based on renewable energy *and energy efficiency. Nuclear power already delivers less energy globally than renewable energy, and the share will continue to decrease in the coming years.*

Despite what the nuclear industry tells us, building enough nuclear power stations to make a meaningful reduction in greenhouse gas emissions would cost trillions of dollars, create tens of thousands of tons of lethal high-level radioactive waste, contribute to further proliferation of nuclear weapons materials, and result in a Chernobyl-scale accident once every decade. Perhaps most significantly, it will squander the resources necessary to implement meaningful climate change solutions.

Things are moving slowly in the right direction. In November 2000 the world recognized nuclear power as a dirty, dangerous and unnecessary technology by refusing to give it greenhouse gas credits during the UN Climate Change talks in The Hague. Nuclear power was dealt a further blow when a UN Sustainable Development Conference refused to label nuclear a sustainable technology in April 2001.

The risks from nuclear energy are real, inherent and long-lasting.

Safety: *No reactor in the world is inherently safe. All operational reactors have inherent safety flaws, which cannot be eliminated by safety upgrading. Highly radioactive spent fuel requires constant cooling. If this fails, it could lead to a catastrophic release of radioactivity. They are also highly vulnerable to deliberate acts of sabotage, including terrorist attack.*

Waste: *From the moment uranium is mined, nuclear waste on a massive scale is produced. There is no secure, risk-free way to store nuclear waste. No country in the world has a solution for high-level waste that stays radioactive for hundreds of thousands of years. The least damaging option at this current time is for waste to be stored above ground, in dry storage at the site of origin, but this option also presents major challenges and threats.*

Weapons proliferation: *The possession of nuclear weapons by the US, Russia, France, the UK and China has encouraged the further proliferation of nuclear technology and materials. Every state that has a nuclear power capability has the means to obtain nuclear material usable in a nuclear weapon. Basically this means that the 44 nuclear power states could become 44 nuclear weapons states. Many nations that have active commercial nuclear power programs, began their research with two objectives – electricity generation and the option to develop nuclear weapons. Also, nuclear programs based on reprocessing plutonium from spent fuel have dramatically increased the risk of proliferation, as the creation of more plutonium means more nuclear waste, which in turn means more materials available for the creation of dirty bombs.*

Source: www.greenpeace.org/international/campaigns/nuclear (accessed July 2009)

James Lovelock is a respected scientist and author, creator of the Gaia theory (see the biology course companion in this series of books), and a leading member of Environmentalists for Nuclear Energy. In an article published in the British newspaper *The Independent* in May 2004, entitled "Nuclear power is the only green solution", he makes the following points.

What makes global warming so serious and so urgent is that the great Earth system is trapped in a vicious circle of positive feedback. Extra heat from any source, whether from greenhouse gases, the disappearance of Arctic ice or the Amazon forest, is amplified, and its effects are more than additive. It is almost as if we had lit a fire to keep warm, and failed to notice, as we piled on fuel, that the fire was out of control and the furniture had ignited. When that happens, little time is left to put out the fire before it consumes the house. Global warming, like a fire, is accelerating and almost no time is left to act.

So what should we do? We can just continue to enjoy a warmer 21st century while it lasts, and make cosmetic attempts, such as the Kyoto Treaty, to hide the political embarrassment of global warming, and this is what I fear will happen in much of the world. When, in the 18th century, only one billion people lived on Earth, their impact was small enough for it not to matter what energy source they used.

But with six billion, and growing, few options remain; we cannot continue drawing energy from fossil fuels, and there is no chance that the renewables, wind, tide and water power can provide enough energy and in time. If we had 50 years or more we might make these our main sources. But we do not have 50 years; the Earth is already so disabled by the insidious poison of greenhouse gases that even if we stop all fossil fuel burning immediately, the consequences of what we have already done will last for 1000 years. Every year that we continue burning carbon makes it worse for our descendants and for civilization.

Opposition to nuclear energy is based on irrational fear fed by Hollywood-style fiction, the Green lobbies and the media. These fears are unjustified, and nuclear energy from its start in 1952 has proved to be the safest of all energy sources. We must stop fretting over the minute statistical risks of cancer from chemicals or radiation. Nearly one third of us will die of cancer anyway, mainly because we breathe air laden with that all pervasive carcinogen, oxygen. If we fail to concentrate our minds on the real danger, which is global warming, we may die even sooner, as did more than 20 000 unfortunates from overheating in Europe last summer.

Even if the greens are right about its dangers, and they are not, its worldwide use as our main source of energy would pose an insignificant threat compared with the dangers of intolerable and lethal heat waves and sea levels rising to drown every coastal city of the world. We have no time to experiment with visionary energy sources; civilization is in imminent danger and has to use nuclear – the one safe, available, energy source – now or suffer the pain soon to be inflicted by our outraged planet."

Source: James Lovelock, Nuclear Power is the only green solution, The Independent, May 24, 2006

Carefully consider the above two articles and write down the arguments for and against the use of nuclear power. Analyse the effect of the style of language used. Is the language objective or is it emotional? Do the articles just state facts or are they effective propaganda?

You should also consider the views of:

- the workers at nuclear power stations, and the trade unionists who represent them
- the local communities living near nuclear reactors and waste disposal units.

If possible, organize a formal debate in your teaching group or senior school on the nuclear issue.

End of chapter summary

Chapter 23 has one main theme: nuclear power (fission and fusion). The list below summarises the knowledge and skills that you should be able to undertake after having studied this chapter. Further research into more detailed explanations using other resources and/or further practice at solving problems in all these topics is recommended – particularly the items in bold.

- Describe how neutrons produced in a fission reaction may be used to initiate further fission reactions (chain reaction) and distinguish between controlled nuclear fission (power production) and uncontrolled nuclear fission (nuclear weapons).
- Describe the main energy transformations that take place in a nuclear power station and what is meant by *fuel enrichment*.
- Discuss the role of the moderator, the control rods and the heat exchanger in a thermal fission reactor.
- Describe how neutron capture by a nucleus of uranium-238 (^{238}U) results in the production of a nucleus of plutonium-239 (^{239}Pu) and describe the importance of ^{239}Pu as a nuclear fuel.
- Discuss safety issues and risks associated with the production of nuclear power.
- Outline the problems associated with producing nuclear power using nuclear fusion **and solve general problems on the production of nuclear power**.

Chapter 23 questions

1 This question is about nuclear power.

a) A fission reaction taking place in the core of a nuclear power reactor is

$$^{1}_{0}n + ^{235}_{92}U \rightarrow ^{144}_{56}Ba + ^{89}_{36}Kr + 3^{1}_{0}n$$

i) State **one** form in which energy is released in this reaction. [1]

ii) Explain why, for fission reactions to be maintained, the mass of the uranium fuel must be above a certain minimum amount. [2]

iii) The neutrons produced in the fission reaction are fast moving. In order for a neutron to fission U-235 the neutron must be slow moving. Name the part of the nuclear reactor in which neutrons are slowed down. [1]

iv) In a particular reactor approximately 8.0×10^{19} fissions per second take place. Deduce the mass of U-235 that undergoes fission per year. [3]

b) The thermal power from the reactor is 2400 MW, and this is used to drive (operate) a heat engine. The mechanical power output of the heat engine is used to drive a generator. The generator is 75% efficient and produces 600 MW of electrical power. This is represented by the energy flow diagram below.

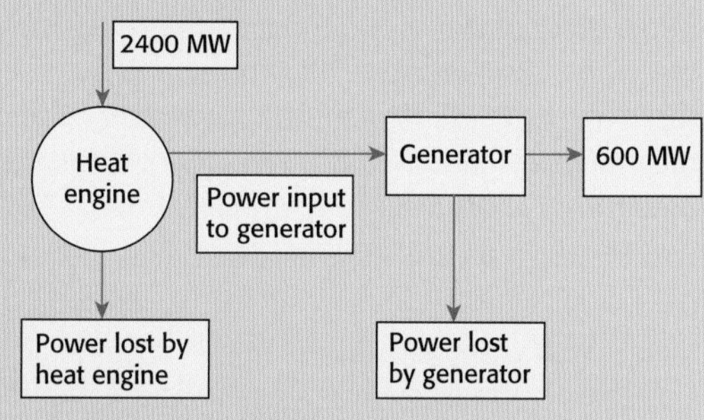

i) Calculate the power input to the generator. [1]

ii) Calculate the power lost from the generator. [1]

iii) Calculate the power lost by the heat engine. [1]

iv) State the name of the law of physics that prohibits **all** of the 2400 MW of input thermal power from being converted into mechanical power. [1]

v) Deduce that that the efficiency of the heat engine is 33%. [1]

IB Physics SL paper 3, May 2005 TZ1

24 Digital technology

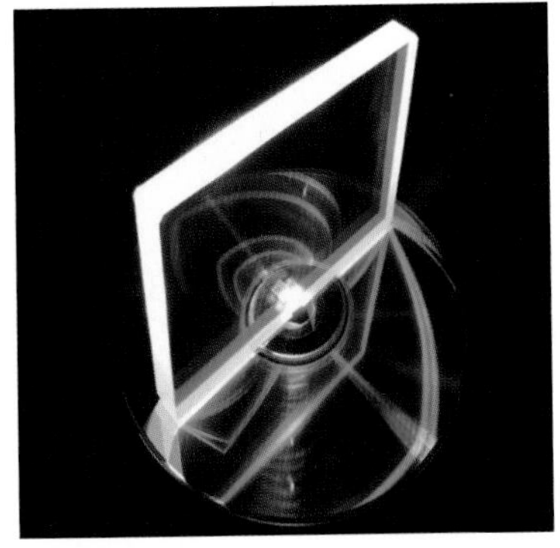

Digital technology affects the way we work, learn, and use our leisure time. The internet has proved to be an extremely useful tool for gaining knowledge, learning, and communicating. Its growth will continue, along with the development of new uses. However, although many of us take access to this technology for granted, there are many parts of the world where computers, mobile phones, and MP3 players are hardly known, and internet access is still something for the future.

This chapter will consider the use of compact discs (CDs) and digital versatile discs (DVDs) to store data, and of charged-coupled devices (CCDs) to create digital images.

Make a list of electronic devices you have used in the last week, and sort the list into things used for work/learning, those used for entertainment, and those for any other uses.

Thinking about science: The internet

The internet has increased access to data, and hence to knowledge, in many parts of the world. The amount of knowledge available is difficult to quantify but it is estimated that in 2002 the internet contained more than 530 000 terabytes (10^{12} bytes) of data – though much of this is not accessible or useful – and that between 2007 and 2010 the amount of digital data available is forecast to increase by 57% a year.

Knowledge is good

Education relies on the acquisition of knowledge and skills, and then on the application of these to help understand a range of different situations, contexts, or problems. Most, if not all, of this knowledge is now available from databases stored on servers all round the world, and is accessible from the internet. This vast quantity of knowledge can cause difficulties. Not only does the sheer volume lead to inefficiencies, because time is wasted searching for what you need, but much information is replicated, and some is wrong. Accurate sources must be distinguished from misinformation and propaganda. It is important to be discerning when you are acquiring the knowledge you need to complete a task.

Ethics of the internet

There is no one code of ethics governing the internet. Essentially, anything goes. Websites that are considered immoral or unacceptable by some groups of people are seen as acceptable by others. For example, paedophiles, extremist political parties, criminals, and terrorists can post information on the internet just as easily as transnational corporations and governments. Is it possible to set out a moral argument that correctly asserts that particular websites are immoral and others are not? Probably not.

We each have our own moral code, based on our cultural upbringing and conscience; in other words, we all have our own set of standards. Public opinion also changes over time.

The internet exemplifies the concept of freedom of speech, a principle closely upheld by many democratic organizations, including the United Nations. If this principle is acceptable on moral grounds, then it can be used to justify one's right to pervert, lie, and offend, which leads to an ethical dilemma, particularly in countries that have libel laws protecting the individual

from false accusations. Many would argue that a private opinion voiced between adults is very different from a public statement posted on the internet.

Material that can offend needs careful consideration. Many would argue, particularly educators, that a basic aim should be to ensure that children and young learners can use the internet safely as a valuable educational resource. Parents and teachers trust that students are safe from the potential corrosive social effects and harm that many websites can cause. This can easily be achieved by the use of appropriate filtering software to remove any harmful or offensive material, which can therefore provide a safe refuge for learning.

Legal difficulties also arise when considering how to prosecute inappropriate web postings. Which country's laws apply? The country where the offensive article was posted, the country of the server where the website is stored, or the country where the material is accessed? Solutions have been achieved, however. International cooperation has been able to virtually outlaw the posting of child pornography for the use of paedophiles throughout the world. Uploading or downloading such material can lead to prosecution and lengthy prison terms by anyone involved.

The internet in society

The internet grew from developments at CERN, the European Centre for Nuclear Research based in Switzerland. CERN is an example of international cooperation between governments when no one country could afford the vast expenses of such research. A prototype of the internet was developed to share data quickly with similar projects across the world. Nowadays email, chat sites, and internet communications have greatly improved and reduced the cost of local and international conversations.

As well as bringing people together, the technology also has the potential to divide societies. Within the richer countries of the world, although most people have access, significant proportions of the population do not. See www.internetworldstats.com/stats.htm (accessed July 2009).

For example, nearly 75% of the population of the USA – the most internet-connected country in the world – were estimated to have internet access in July 2009. But this means that 25%, or over 80 million people, did not have access. The internet can divide societies by creating an underclass of people without such access to information. At the time of writing, Sierra Leone in West Africa, for example, has the lowest internet connectivity, with only 0.3% of its population being able to access the internet. A lack of internet access has the potential to further impoverish societies.

Analogue and digital signals

An analogue signal varies continuously between maximum and minimum values. A microphone, for example, produces an **analogue** signal that represents the sounds it detects. The **continuous** electrical signal generated contains all values of voltage between the maximum and minimum allowed values. The graph in Figure 1 represents an analogue signal that varies continuously between +6 and −6 V.

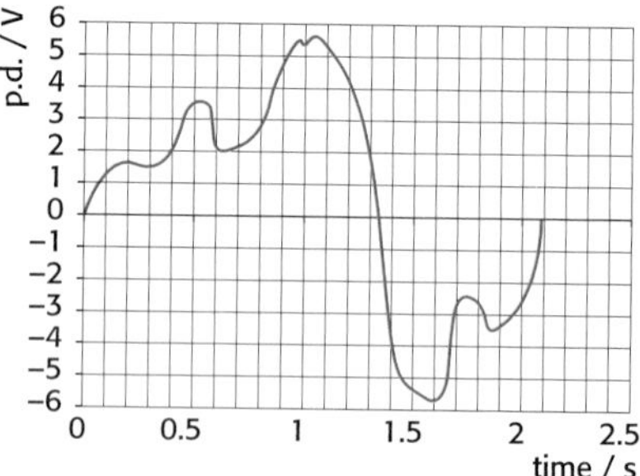

Figure 1 An analogue signal

A **digital** signal, by contrast, contains **discrete** (or quantized) values only. The signal in Figure 2 is a digital signal which has two values only; +6 or 0 V, representing a binary 1 or 0 (see page 345). Each 1 or 0 lasts for the same time: 0.2 seconds in this case.

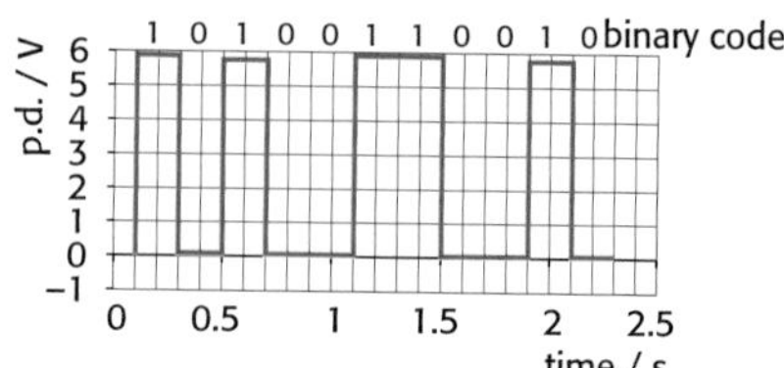

Figure 2 A digital signal

The photographs in Figure 3 show two ammeters reading the same value of current: Figure 3a is an analogue ammeter and Figure 3b is a digital ammeter. The analogue meter displays all values of current from the minimum to the maximum value allowable on the scale. Intermediate values can be achieved by interpolating (reading between) the gradations of the scale. If the position of the needle is in between two marks on the scale you can judge whether the current is closer to the lower or the higher value. In the picture below the needle is clearly between 0.20 A and 0.21 A and closer to 0.20 A. It is reasonable to estimate that the position of the needle in this case is about one quarter of the way between the two values:

that is, 0.203 A ±0.003. The digital meter is accurate to ±0.01 A. The value displayed is somewhere between 0.20 A and 0.21 A. However, there is no way of knowing exactly what the reading is. In other words, the digital reading is quantized in steps of 0.01 A. The analogue meter can therefore be used to give a more precise reading than the digital meter.

Figure 3 (a) An analogue ammeter, (b) a digital ammeter

Binary code

Computers use a binary code at their simplest, most basic level of operation.

The most common number system in general use is a base 10 system. It uses 10 digits represented by the characters, 0, 1, 2, 3, 4, 5, 6, 7, 8, 9. Numbers are written in columns, representing powers of 10:

$\mathbf{10^3}$ $\mathbf{10^2}$ $\mathbf{10^1}$ $\mathbf{10^0}$

So, for example, the number three thousand, seven hundred and forty-one is written 3741 and represents $(3 \times \mathbf{10^3}) + (7 \times \mathbf{10^2}) + (4 \times \mathbf{10^1}) + (1 \times \mathbf{10^0})$.

Numbers can be written to any base number (this is covered in the HL mathematics course), but base 10 emerged as the most useful system – presumably because humans have 10 fingers, and this was a convenient way to count.

Binary code is written in base 2, and so involves just two digits: 1 and 0. This is the simplest number system. Table 1 shows the first 16 numbers in base 10 and the corresponding binary numbers.

Table 1 Conversion table for binary numbers

Base 10	2^3	2^2	2^1	2^0	Binary
0	0	0	0	0	0
1	0	0	0	1	1
2	0	0	1	0	10
3	0	0	1	1	11
4	0	1	0	0	100
5	0	1	0	1	101
6	0	1	1	0	110
7	0	1	1	1	111
8	1	0	0	0	1000
9	1	0	0	1	1001
10	1	0	1	0	1010
11	1	0	1	1	1011
12	1	1	0	0	1100
13	1	1	0	1	1101
14	1	1	1	0	1110
15	1	1	1	1	1111

Outline the differences between an analogue and a digital signal.

1 Write the following base 10 numbers in base 2:
16, 17, 45, 63, 100, 127.

2 Write the following binary numbers in base 10:
11110, 101010, 111000, 110111, 1010101.

One **binary digit**, a 1 or a 0, occupies one **bit** of memory in a computer or any data storage device. The right-hand digit is called the **least significant bit**; when counting sequentially it changes most frequently, and represents the smallest value. The left-hand digit is called the **most significant bit**; this changes least often when counting, and represents the largest value. A binary signal containing a series of several 1s and 0s is called a **string**. An 8-bit number occupies 1 **byte** of memory.

The **hexadecimal** system, which is used in computing, uses a base of 16, with the digits 0–9 and the letters A–F. It can also be used to represent a byte. For example, the decimal numeral 93, whose binary representation is 01011101, is 5D in hexadecimal (5 = 0101, D = 1101). Each hexadecimal character represents a 4-bit string. The two strings are written down side by side to give an 8-bit string – a byte. Table 2 shows the conversion between decimal, binary and hexadecimal numbers.

8-bit numbers can be used to represent the basic word processing characters. The Notepad word processor in Windows uses an 8-bit string to represent each character. To check the size of a byte open Windows Notepad (or an equivalent program): in Windows XP the file path is *Start/All programs/Accessories*. Type the phrase "physics course companion". Save the file as *companion.txt* and close the file. The size of the file can be determined by right clicking on the file name in Windows Explorer and then clicking *Properties* (or by just holding the cursor over the file icon); it should show 24 bytes. The phrase uses 24 characters: 22 letters and two spaces, which are also considered characters. Windows Notepad stores text in its simplest form, so memory is used only to store the characters; there is no formatting (such as **bold** or *italic*) and hence no additional memory use, as is required from more sophisticated programs such as Microsoft Word.

The relationship between each character and its binary number is written in ASCII code, which may be researched further on www.asciitable.com (accessed July 2009).

Values stored on a CD are represented by a 2-byte number.

1 How many different numbers can be represented by
   a) 1 byte
   b) 2 bytes?
2 Computer memory is measured in kilobytes (kB), megabytes (MB), gigabytes (GB), terabytes (TB), and so on. Carry out some research to find out how many bytes there are in each of these values.
3 Suggest why the phrase "physics course companion" typed in Microsoft Word needs over 23 kB (kilobytes) of memory.

Write the following hexadecimal numbers **a)** in binary code and **b)** as decimal numbers:

11, FF, A5, 5A, 97, 79, CD, DC.

Table 2 Conversion table for hexadecimal numbers

Decimal	Binary	Hexadecimal
0	0000	0
1	0001	1
2	0010	2
3	0011	3
4	0100	4
5	0101	5
6	0110	6
7	0111	7
8	1000	8
9	1001	9
10	1010	A
11	1011	B
12	1100	C
13	1101	D
14	1110	E
15	1111	F

The bigger picture: Analogue-to-digital conversion

The process of converting an analogue signal to a digital signal involves some complex electronic components; this is covered in more detail in the communications option (Chapter 28). The basic steps in digitization involve:

1 sampling the analogue signal at fixed time intervals
2 assigning a quantum level to the value of the sample

3 converting the quantum level into binary code

The key components are illustrated in Figure 4.

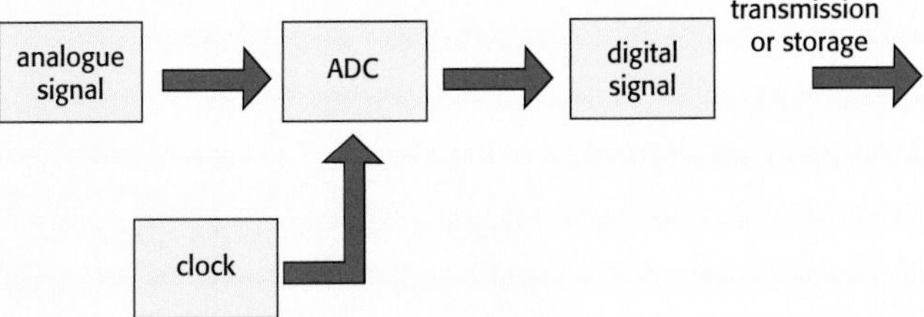

Figure 4 Analogue-to-digital conversion

The analogue signal is fed into the ADC (analogue-to-digital converter). This is a chip that samples the signal at a predetermined rate, assigns a quantum level to each sample, and converts these values into binary code. The clock produces a regular pulse to ensures that the 1s and 0s are produced at regular intervals. The digital signal produced is then either transmitted (e.g. a digital TV signal) or stored (e.g. on a CD).

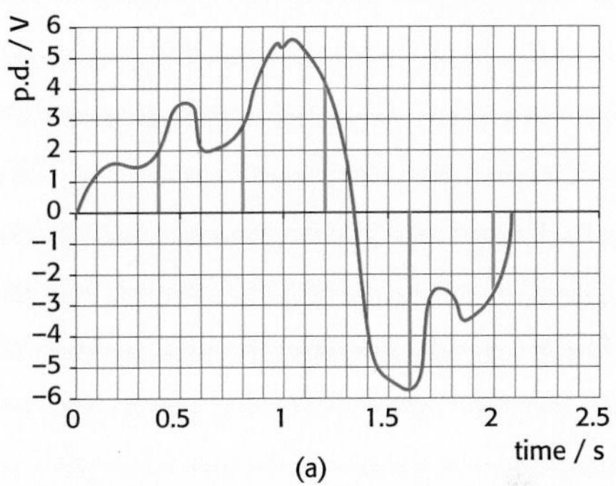

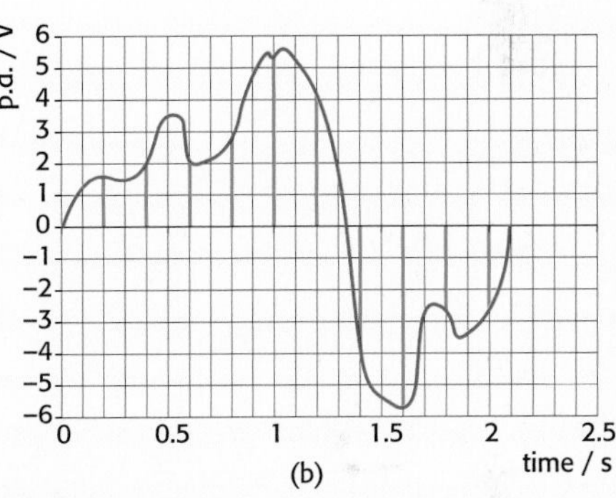

Figure 5 (a) Sampling every 0.4 s; (b) sampling every 0.2 s

Sampling

The analogue signal is read or sampled at specific time intervals, as illustrated in Figure 5.

The signal illustrated in Figure 5a is only sampled every 0.4 s. This means that significant amounts of information about the original signal will be lost in the conversion. To reduce this loss of information the number of samples each second – the **sampling frequency** – can be increased, as illustrated in Figure 5b. Sampling every 0.2 s allows more information to be retained, but increases the amount of data that has to be transmitted.

To ensure that enough data is transmitted, so that the original signal can be reproduced accurately, the signal must be sampled at a minimum rate of twice the maximum signal frequency. Humans are able to hear sounds in the frequency range of about 10 Hz to 20 kHz. The sine wave illustrated in Figure 6 has a frequency of 20 kHz. In general, the signal is sampled at least every half cycle for it to be fully reproduced after transmission.

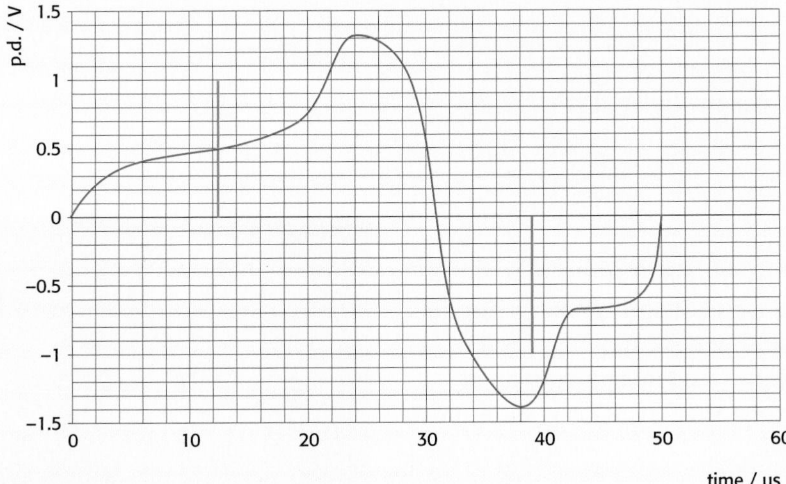

Figure 6 A 20 kHz sine wave sampled every half cycle

1 Record the sampled values and corresponding times in a table for:

a) the data in Figure 5a;

b) the data in Figure 5b.

2 Plot the data obtained in question 1 in two graphs, one for each table. Join up the points appropriately.

3 **a)** Outline how you decided to join up the points in the graphs in question 2.

b) Discuss which graph represents the original signal more accurately.

Quantum levels and coding

The voltage values of the analogue signal samples are assigned quantum levels, as illustrated in Figure 7. These quantum levels are simply discrete numbers, and have nothing to do with the quantum levels assigned to electrons or nucleons in nuc lear physics.

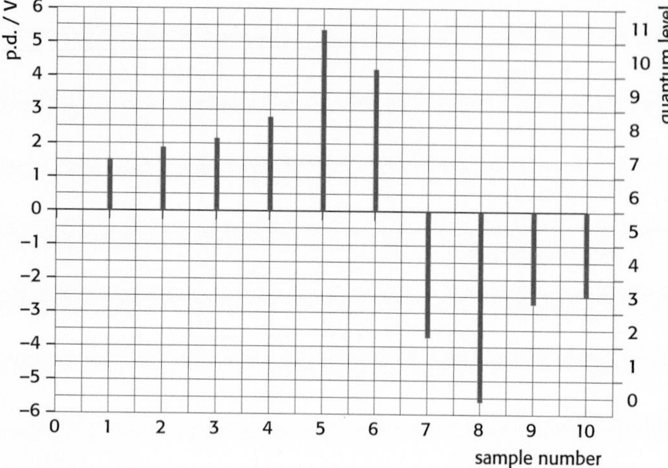

Figure 7 Quantum levels are assigned to samples from an analogue signal. The table illustrates the quantum level of each sample and the corresponding 4-bit binary code.

Sample no	Quantum level	Binary code
1	7	0111
2	7	0111
3	8	1000
4	8	1000
5	11	1011
6	10	1010
7	2	0010
8	0	0000
9	3	0011
10	3	0011

The sampling rate used to store music on CDs is 44 100 samples per second, written as 44.1 kHz. Explain why this rate is suitable.

The signal illustrated in Figure 5b has been sampled every 0.2 s; each sample is shown with a thick line. The quantum levels of each sample are recorded in the table. Each quantum level is converted into the corresponding binary number. In the example above there are 12 quantum levels, labelled 0–11. These can be represented with a 4-bit binary code. The binary code is then transmitted as a series of 1s and 0s. To ensure that each sample is assigned an accurate quantum level, there need to be as many such levels available as possible. In the example illustrated in Figure 7, samples 1 and 2 are both given a quantum level of 7, yet their voltage values are clearly different. Level 7 represents all voltages from 1 V to 2 V. A 4-bit code does not allow the digital signal to accurately represent the original signal.

In order to re-create the original signal, the digital signal (taken from the storage medium that is being used) is fed into a DAC (digital-to-analogue converter). The binary signals are converted back into voltages. The DAC "estimates" the values of the signal in between the digitized values and reproduces an analogue signal.

1 a) State how many quantum levels can be achieved with a 4-bit binary code.

b) Explain why this is not suitable for storing music on CDs.

2 Samples are stored as 16-bit (2-byte) binary numbers on CDs. Calculate how many quantum levels this represents.

Advantages of digital signals

Although analogue signals can accurately represent the source of the signal, they usually pick up noise in the transmission process. **Noise** is a form of interference that distorts the signal. It may be added to the signal during transmission, or when the signal is being played back. Depending on your own level of tolerance, noise can spoil the playback of a piece of music, for example.

The use of the term "noise" refers to modifications of the actual signal, not just to background noise, although this can be equally frustrating. Vinyl records or LPs (LP stands for long player) were a

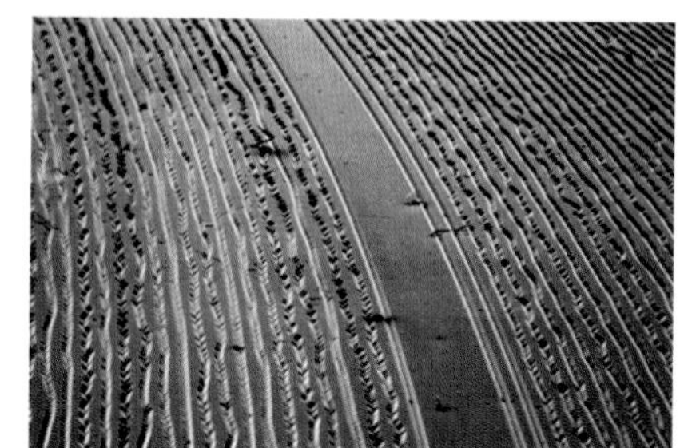

Source: www.audiography.com.au

Figure 8 Magnified view of grooves in an LP record

popular format for marketing music prior to the introduction of compact discs or CDs in the early 1980s. The musical signal in an LP is stored in a groove, physically cut into the record by a sharp stylus (Figure 8). The shape of the groove is an analogue representation of the musical signal. The LP is played back by placing another stylus into the groove, and then rotating the record. The stylus vibrates in the groove and produces an electrical signal, which is played back through an amplifier and speakers so that it can be heard. Every time the LP is played, the stylus slightly damages and changes the shape of the groove. In addition, dust and other small particles settle in the groove. The stylus vibrates in this distorted and dust-filled groove, and so noise is added to the signal.

Use the internet and other sources to find out how the following store data and, in each case, whether it is in an analogue or digital format.

a) cassette tape
b) floppy disk
c) hard disk drive
d) video (VHS) tape.

Digital signals, made up of binary strings of ON (1) and OFF (0) values may also pick up noise in transmission, but they are still recognizable as a series of 1s and 0s. Figure 2 on page 344 illustrates a binary signal where 1 = +6 V and 0 = 0 V. Music is stored on a compact disc in binary code. A digital receiver is programmed with a set cut-off voltage, often about 2 V. Any signal higher or equal to 2 V is identified as a 1 but anything below is read as a 0. A certain amount of distortion can therefore be tolerated by the system.

The choice, then, in deciding how best to store and transmit data is between an analogue signal, where the representation of the original variations is stored in a way that directly matches the original signal, and the use of an agreed code to translate between the original variations and a set of digital numbers. These digital numbers can be transmitted with effectively no corruption, but they then need to be translated back into the original analogue signal.

Compact discs

CDs have become one of the most popular formats for storing digital sound recordings. Large computer files can also be saved on CDs using CD writers (CD-R). Currently CDs can store about 0.75 GB (gigabytes) of data, and so are very convenient for transferring and storing resources.

Structure

CDs are just under 12 cm in diameter. The data is stored in a track of microscopic "pits" and "bumps", which are moulded into a thin layer of transparent plastic. These indentations are then covered with a layer of reflective aluminium (Figure 9).

The pits and bumps are arranged one after the other in a spiral track, starting from the centre of the disc, as illustrated in Figure 9b.

(a)

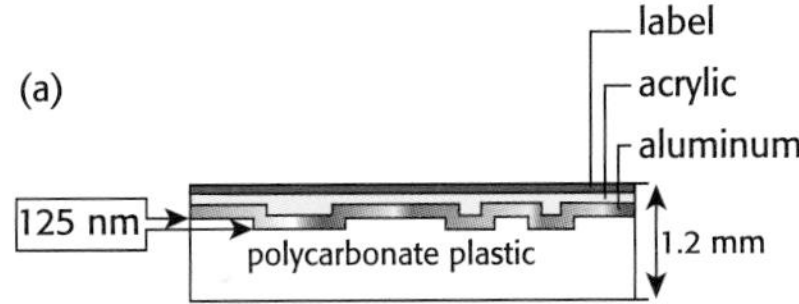

(b)

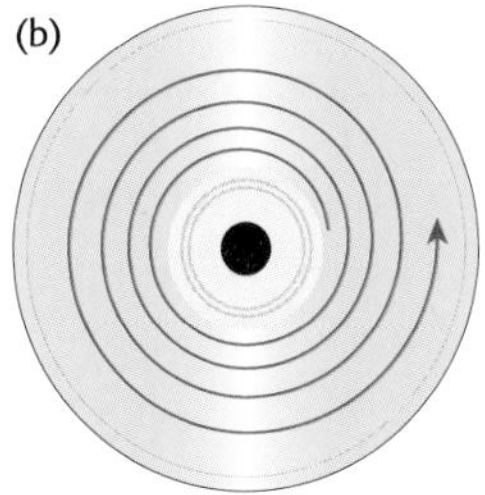

Source: howstuffworks.com.

Figure 9 (a) Cross-section through a CD; (b) the spiral.

(a)

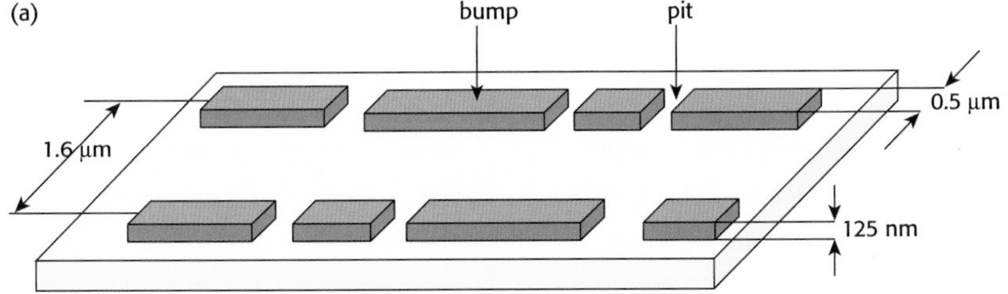

(b)

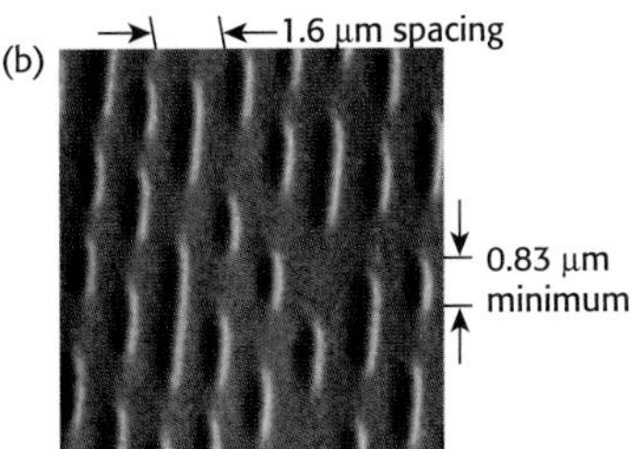

Source: Computer Desktop Encyclopedia

Figure 10 (a) Sketch showing the arrangement of "pits" and "bumps" in a track on a CD (not to scale). (b) Highly magnified plan view of a CD.

The diagram in Figure 10a illustrates the dimensions and arrangement of some "pits" and "bumps" in two consecutive arcs of the spiral.

The disk rotates at about 500 revolutions per second, and a focused laser beam "reads" the data by reflecting off the undulating track. The whole mechanism is very precise, and the laser is focused down to a spot of diameter 1.7 μm. The laser starts reading the data from the centre and moves outward along a radial line as the disk rotates rapidly next to it. One of the big advantages of a CD over an LP is that there is no mechanical contact between the laser and the disc, so it can be played and replayed without damage or the introduction of noise.

To replay the music perfectly from a CD, the laser must read the data at a constant rate. In order to do this the rotation rate of the disc must gradually change. Explain why this change is necessary, and outline how the speed changes while a CD is being played.

Reading the data

The laser beam is focused on the CD surface, and the reflected light is detected by a photodiode. The incident laser beam in reality is incident almost normal to the disc surface, but is shown in Figure 11 at an acute angle so that the incident and reflected beams are clearly visible. If the beam is entirely incident either on a bump or a pit then all parts of the reflected beam are in phase with each other. Constructive interference takes place, and a strong signal is detected by the photodiode.

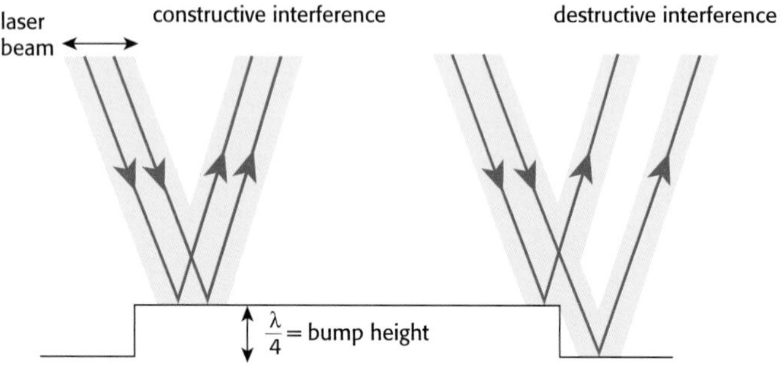

Figure 11 Reflection of the laser beam from the bump surface, and from the bump/pit transition

If part of the beam is incident on a pit and part on a bump – that is, at the transition between the pit and the bump – then there is a path difference between the two parts of the reflected beam. If the pit height and laser beam wavelength, λ, are selected appropriately, this can lead to destructive interference at the photodiode. The signal is reduced in intensity. (See pages 112–13 for more details of interference.) As the beam reflects off the spiral track, the destructive and constructive interference conditions mean that a digitally varying signal is detected at the photodiode. This signal, along with an accurate timing pulse, is converted into a series of 1s and 0s. Finally a digital-to-analogue converter (DAC) converts this digital signal back into an analogue signal. The width of the pits and bumps are varied to reproduce exactly the right binary code for conversion back into a recognizable analogue signal.

a) The "bumps" for each digital bit on the surface of a CD are laid down along a spiral track around the disc. Dimensions are as follows:

Length of bump along track = 0.97 μm
Width of track = 2.1 μm

Show, using an estimate for the area for storage, that the disc can store about 4 gigabits (4×10^9 bits) of digital information. [2]

b) Sound is stored in digital form by sampling the analogue waveform 44 000 times each second, and encoding each sample into a 16-bit number. Explain why 44 000 samples each second is an appropriate sampling rate. [2]

c) Normally two stereo channels are recorded. Use this fact and the other information above to estimate the playing time available on a 4 gigabit music CD. [3]

Adapted from Edexcel Salters Physics, A2, paper 4, June 2001

Inquiry: DVDs and future technologies

Information technologies change rapidly as new techniques are discovered to store data on ever smaller devices and transfer data at ever increasing rates. At the time of writing (April 2007) DVDs are a very popular way of storing very large amounts of data – more than seven times the data capacity of a CD. This means that a DVD can hold the

equivalent of about seven music albums or at least one full length movie. DVD writers (DVD-R) have been available for computers for a few years, and have now replaced the standard video (VHS) recorders used to record TV programmes and manage home movies.

In a small group, carry out some research to find out about DVDs and how they work. In particular, include:

- details of the structure and scale
- a comparison with a CD and explanation of why DVDs can store more data
- a brief outline of how a rewritable DVD-R works

Your findings should be presented in a format agreed with your teacher in advance (e.g. poster/article for school magazine/web page/podcast/essay/presentation). You should ensure that all ideas obtained in your research are properly cited.

Present your research to your class in a suitable way.

You may also want to include some details on the next possible technological breakthrough, which may change the way we store and access data.

1 Explain why the bump height is equal to ¼ of the incident beam's wavelength.

2 Calculate the wavelength of the laser beam for a bump height of 125 nm.

3 It has been suggested that a CD could be designed to store more data if a shorter-wavelength laser was used. Outline how the bump height would change.

4 A stereo music CD can store up to 74 minutes' worth of music. The sampling rate is 44 100 Hz, and each sample is stored as a 16-bit code. Show that a CD can store just under 800 MB of data. (Note: a stereo signal has two samples stored at the same time, one for each speaker.)

Data capture and the CCD

CCD (charge-coupled device) and CMOS (complementary metal oxide semiconductor) sensors have revolutionized the way images can be stored and captured. They are used in digital cameras, camcorders, fax machines, scanners, and astronomical telescopes. Both sensors operate on the same basic principle: light falling on the sensor results in an accumulation of electric charge. The value and position of this charge is then converted into a digital signal. CCDs produce better-quality images with less noise but are more expensive than CMOS sensors. Improvements in design, however, mean that at the time of writing (2007) it is not clear which type of sensor is going to be preferred for everyday applications. This section will focus on CCDs.

CCD basics

A CCD is composed of thin layers of light-sensitive semiconducting material arranged in small independent elements called **pixels** (**pic**ture **el**ements). A typical CCD is 2 cm × 2 cm in dimension and contains an array of 1000 × 1000 pixels (Figure 12).

The semiconducting material in each pixel is sandwiched between two thin electrodes. In a camera, light from the image is focused onto the CCD. Each pixel in a CCD behaves like a **solar** (photovoltaic) **cell** and converts light energy into electrical energy The incident photons free electrons from the semiconductor through the **photoelectric effect** (see pages 262–3 for more details). The amount of charge that collects in each pixel is proportional to the number of photons or light intensity incident upon it. More charge collects in a pixel where the image is brighter than in a region where it is dim.

In this way the image is built up on the CCD: pixels with varying amounts of charge correspond to the different light intensities in parts of the image. The potential difference (p.d.) across each pixel increases with its charge. The value of the p.d and the position of the pixel are converted into a digital signal in binary code. The signal can be stored and then used later to reconstruct the image.

(a)

Source: NASA

(b)

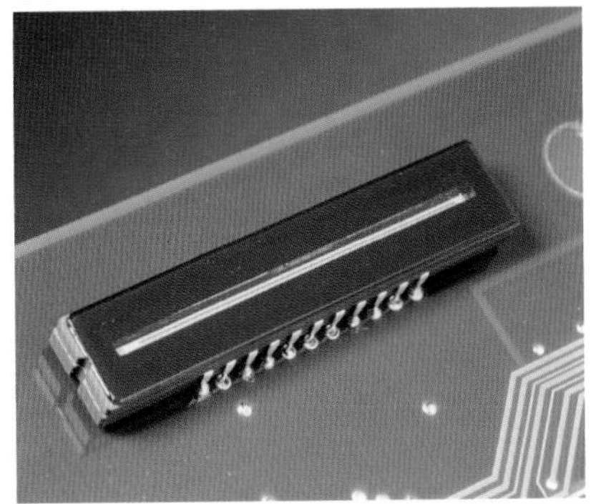

Source: http://www.betterlight.com/how_they_work.html

Figure 12 (a) A typical CCD, as used in a digital camera. (b) A linear CCD, as used in a scanner or fax machine

Capacitance

Each pixel in a CCD stores electric charge in a device called a **capacitor**. In its simplest form a capacitor consists of two parallel, conducting plates separated by a small insulating gap filled with a material called a **dielectric** (Figure 13). Real capacitors are generally quite small, about 1 cm or less in length, and generally have the plates and dielectric wrapped round in a cylindrical shape. Some typical capacitors are shown in Figure 14.

Capacitors may be charged either by contact with a charged object or by connecting them to a dc power supply. A charge, $-q$, builds up on one plate, inducing an equal but opposite charge of $+q$ on the other. If, for example, a capacitor is charged so that a charge of +250 nC collects on one plate and −250 nC on the other, it is conventional to say that the charge q on the capacitor is 250 nC.

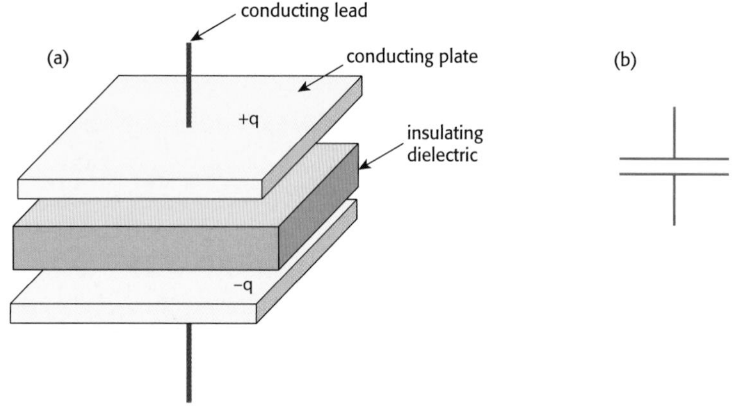

Figure 13 (a) Schematic illustration of a capacitor; (b) capacitor circuit symbol

The circuit illustrated in Figure 15 can be used to charge a capacitor. Current flows for a short time until the potential difference across the capacitor is equal to the emf of the cell.

Experiments show that the amount of charge stored on a capacitor, q, is proportional to the charging potential difference, V (Figure 16).

$$q \propto V$$

The constant of proportionality is defined as the **capacitance** of the capacitor.

$$q = CV$$

$$\Rightarrow C = \frac{q}{V}$$

where
C is the capacitance in F
q is the charge stored in the capacitor in C
V is the p.d. across the capacitor in V

The unit of capacitance is the **farad**, F, where $1\ F = 1\ C\ V^{-1}$. In reality the farad is an extremely large unit, so most commonly used capacitors have a capacitance measured in picofarads (pF), nanofarads (nF), or microfarads (μF).

1 State what is meant by a semiconducting material.
2 A 2 cm × 2 cm CCD contains 1000 × 1000 pixels.
   a) State the number of pixels in this CCD.
   b) Calculate the length of one side of a pixel.
3 Briefly outline the photoelectric effect (revision exercise).
4 a) Compare the amount of charge that collects in a pixel that is brightly illuminated with one that is not illuminated.
   b) Describe how an image is represented on a CCD.
5 Explain why there is a difference in shape between a CCD used in a camera and one used in a scanner.

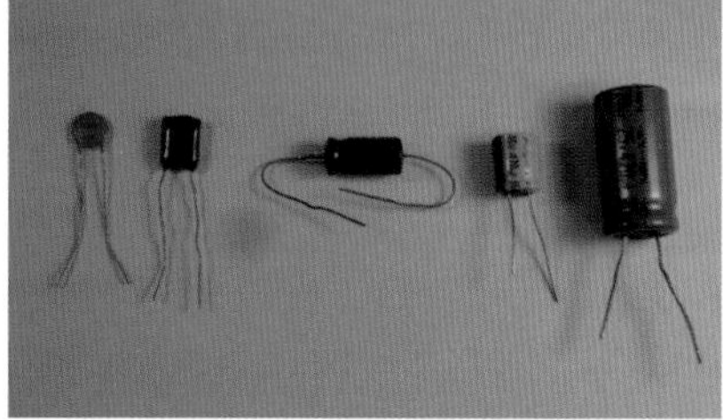

Figure 14 Miscellaneous capacitors

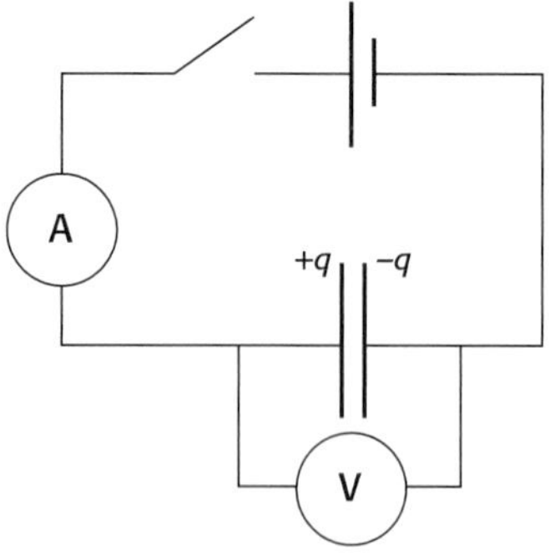

Figure 15 A circuit to charge a capacitor

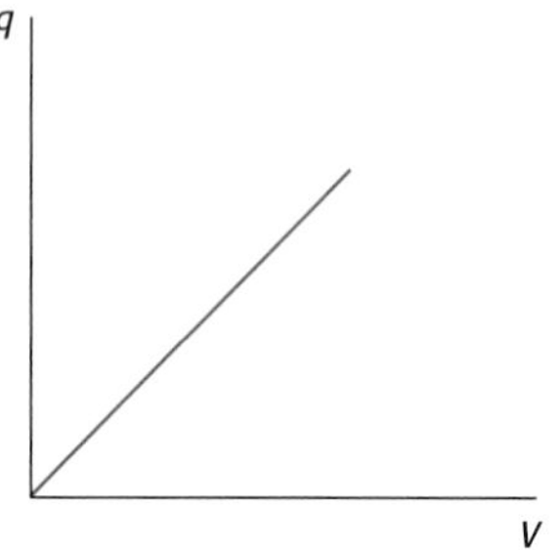

Figure 16 Sketch graph showing the variation of charge with p.d. across a capacitor

1 Use a spreadsheet or otherwise to produce a graph of the data in Table 3, which shows the variation with voltage of the charge stored in a capacitor.

  a) Add appropriate error bars to the graph.

  b) Determine the capacitance of the capacitor.

  c) Suggest an appropriate uncertainty for your answer in **b)** using maximum and minimum gradients or otherwise.

Table 3

P.D. / mV ± 1 mV	200	190	180	170	160	150	140	130	120	110	100	90	80	70	60	50	40	30	20	10	0
Charge / nC ± 5 nC	200	190	180	170	160	149	140	129	120	110	100	90	80	70	60	50	40	30	20	10	0

2 A capacitor is charged with a 6 V dc supply. Calculate the capacitance of the capacitor when the maximum charge on the plates is

  a) 30 nC **b)** 300 μC.

3 A 20 μF capacitor is connected to a battery of emf 8 V. Calculate the charges on the plates.

4 Calculate the potential difference needed to charge the plates of a 10 μF capacitor to ±2.5 mC.

5 A constant 20 μA charging current is used to fully charge a capacitor in 20 s. During this time the potential difference across the capacitor rises from 0.0 V to 5.0 V. Calculate the capacitance of the capacitor.

6 If the circuit illustrated in Figure 15 is used to charge a capacitor, then the current falls from a maximum value to zero fairly quickly.

  a) Outline in terms of electron flow how current can flow for this short time even though the capacitor is effectively an insulator.

  b) Explain why the current falls from a maximum value to zero.

Quantum efficiency

The quantum efficiency of a CCD is defined as the ratio of the number of photoelectrons emitted in a pixel to the number of incident photons. That is:

$$QE = \frac{N_e}{N_p}$$

where

QE is the quantum efficiency

N_e is the number of photoelectrons emitted in a pixel

N_p is the number of photons incident on the pixel

In an ideal situation QE = 1: that is, each incident photon liberates an electron from the semiconductor. However, this is rarely obtainable.

In effect, the quantum efficiency is a measure of the CCD's sensitivity and can be related to the number of incident photons that the chip is actually able to detect. The quantum efficiency varies with the wavelength of the incident light, and is generally higher for back-lit CCDs than those that are illuminated from the front (see Figure 17).

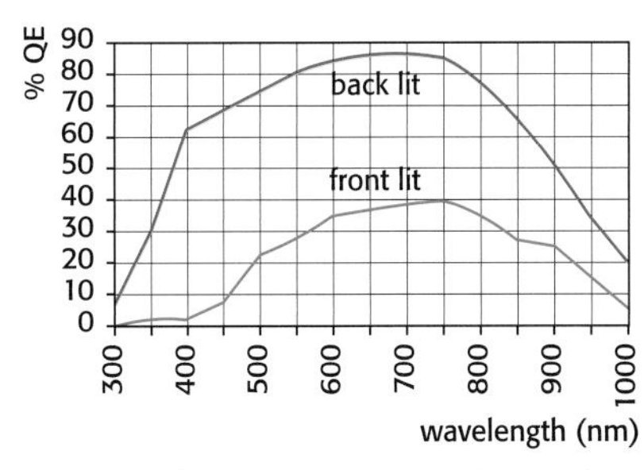

Source: http://www.ccd.com

Figure 17 Graphs illustrating the quantum efficiency of front-lit and back-lit CCDs over the range 300–1000 nm

Back-lit CCDs have relatively high quantum efficiencies, being greater than 70% for wavelengths of 450–800 nm, which corresponds to most of the visible waveband. Photographic film, by contrast, has a sensitivity of only about 10% – that is, only about 1 in 10 of the incident photons are recorded.

Resolution

The resolution of a CCD is a measure of the smallest detail that the sensor is able to detect. The higher the resolution, the finer and more detailed the image becomes. The resolution of a CCD depends on the number of pixels. The more pixels there are in a given area, the greater the resolution. A typical chip with a 1024 × 1024 pixel array is said to have a size of 1 megapixel, although the actual number of pixels is slightly more than 1 million. At the time of writing, top-end professional cameras use CCDs with more than 10 megapixels.

Two adjacent points on an object are said to be just resolved by the CCD if the images of these points form on two pixels that are separated by at least one pixel in between. The Hubble Space Telescope was launched in April 1990 into orbit around the Earth. Being located above the atmosphere it is able to record much clearer, almost undistorted, images of distant astronomical objects, compared with telescopes on Earth. The image in Figure 18 is a globular cluster, NGC 2808, located in our Milky Way galaxy. This particular cluster contains more than a million stars.

The two adjacent stars, labelled A and B in the figure, are resolved. This means that they can be seen clearly as two separate objects. In order for this to happen the light from each star must fall on separate pixels on the CCD. The stars in region C are unresolved. They cannot be seen as separate objects, because light from several stars falls on one pixel.

In order to determine whether two separate objects in an image can be resolved or not, the magnification of the lens system that focuses light on the CCD also needs to be considered.

Magnification

When light from an object is focused onto a CCD, the image size is adjusted to fit appropriately onto the chip. Images of people, for example, are smaller than the people themselves, whereas pictures of, for example, animal cells are generally increased in size to allow detail to be studied. The **linear magnification** M of a lensing system is defined as

$$M = \frac{\text{image length}}{\text{object length}}$$

Digitizing the image

In order to store the image in digital format, at least two sets of data are required for each pixel:

- the value of the voltage
- the position of the pixel.

This question is about a CCD.

a) A digital camera is used to take a photograph of a freshwater crab. The CCD in the camera has 5.8×10^6 square pixels. Each pixel has an area of 5.8×10^{-10} m^2. A claw of the crab has an area of 1.5 cm^2. The image of the claw formed on the CCD is 0.10 cm^2. Two spots on the claw are separated by 0.80 mm. Deduce that the images of the two spots will be resolved. [4]

b) Light is incident on the image collection area for a time of 50 ms. The number of photons incident on one pixel is 4.5×10^4. Each pixel has a quantum efficiency of 85% and a capacitance 20 pF.

- **i)** State what is meant by *quantum efficiency.* [1]
- **ii)** Estimate the change in potential difference across each pixel. [4]

c) Outline how the variation in potential difference across individual pixels enables a black-and-white image to be produced by a digital camera. [2]

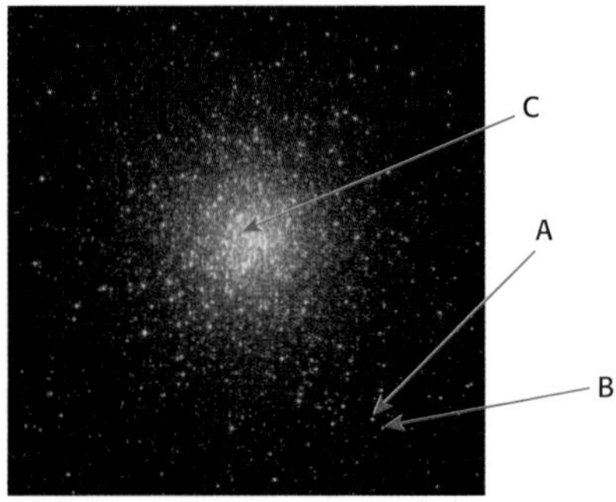

Source NASA, ESA. A. Sarajedini (University of Florida) and G. Piotto (University of Padua [Padova])

Figure 18 Globular cluster NGC 2808 taken by the Hubble Space Telescope in May 2005

The voltage across each pixel is proportional to the intensity of the light falling on that pixel: the brighter the light, the higher the voltage. A high value represents a "white" part of the image, whereas a low value represents "dark grey" or "black". When the data for all pixels is combined, the image can be reconstructed.

A complex electronic process is used to record the value of voltage across each pixel and the pixel position. CCDs are designed so that these electronic components are mainly arranged around the edge of the chip. Some components do obscure the front of the CCD slightly, however – this is why backlit CCDs have a higher quantum efficiency. A sensor is not required for each pixel, so very little of the incident light is blocked out. CMOS chips, by contrast, detect the voltage directly at each pixel, which means that there are a lot more wires in front of the sensor. These block out some of the incident light, hence reducing the sensitivity of the chip.

A digital photograph is taken of a person whose eyes are 0.060 m apart. The distance between the eyes of the image incident on the CCD is 7.1×10^{-4} m. The CCD has an area of 4×10^{-4} m^2 and contains 5.0×10^6 pixels.

a) Calculate the magnification of the camera.

b) Calculate the length of one side of a pixel.

c) By comparing the distance between two adjacent pixels and the distance between the eyes in the image on the CCD, determine whether the eyes are resolved.

Colour and bit depth

The CCD as described above produces grey scale images. Most chips use an 8-bit code, which means that there are 2^8 or 256 shades of grey available. A special filter, containing the primary colours of blue, red, and green, can be placed over the CCD so that the intensity of each specific colour can be detected. A special kind of filter called a **Bayer array filter** ensures that, in a group of four adjacent pixels, two detect green light, one red and the other blue (Figure 19); our eyes are more sensitive to green light, so this array simulates human vision reasonably accurately.

Figure 19 A Bayer array filter used with four adjacent pixels

The pds developed over each such group of pixels are combined to create a true colour image at each pixel.

If an 8-bit code is used to encode the intensity for each primary colour, this gives 256 different values. When all three primary colours are combined at each pixel, this gives $2^{(8 \times 3)}$ or 16 777 216 different colours. This is known as **true colour imaging**. This number of colours may seem a bit excessive, considering that the human eye can only distinguish about 10 million colours. The number of bits used to store the digital signal is called the **bit depth**.

Carry out some research and make a list of some of the advantages and disadvantages of using CCDs to produce images compared with film.

End of chapter summary

Chapter 24 has two main themes: Analogue and digital signals, data capture and digital imaging using charge-coupled devices (CCDs). SL candidates studying Option C (digital technology) need to study both topics and some additional material from Chapter 28. The list below summarises the knowledge and skills that you should be able to undertake after having studied this chapter. Further research into more detailed explanations using other resources and/or further practice at solving problems in all these topics is recommended – particularly the items in bold.

Analogue and digital signals

- Solve problems involving the conversion between binary numbers and decimal numbers **and describe different means of storage of information in both analogue and digital forms**.
- Explain how interference of light is used to recover information stored on a CD and calculate an appropriate depth for a pit from the wavelength of the laser light.
- **Solve problems on CDs and DVDs related to data storage capacity.**
- **Discuss the advantage of the storage of information in digital rather than analogue form and the implications for society of ever increasing capability of data storage.**

Data capture and digital imaging using charge-coupled devices (CCDs)

- Define *capacitance*.
- Outline how the image on a CCD is digitized by describing the structure of a CCD and explaining how incident light causes charge to build up within a pixel.
- State that two points on an object may be just resolved on a CCD if the images of the points are at least two pixels apart.
- Define *quantum efficiency* of a pixel and *magnification* and discuss their effects (along with resolution) on the quality of the processed image.
- Describe a range of practical uses of a CCD, and list some advantages compared with the use of film.
- Outline how the image stored in a CCD is retrieved **and solve problems involving the use of CCDs**.

Chapter 24 questions

1 It has been said that, using high-speed digital signal transmission. the complete texts of all of Shakespeare's plays can be transmitted in about *one quarter of a second.*

A typical "Complete Plays of Shakespeare" in book form has about *1100 pages*. Make an order-of-magnitude estimate of the bit rate of the signal that would be needed. State any quantities that you are estimating, and show clearly your reasoning. [5]

Adapted from Edexcel Salters Physics, A2, paper 4, June 2000

2 a) Give two reasons why compact discs (CDs) are now preferred to LPs by most people who buy recorded music. [2]

b) Figure 23 shows a CD with a scale alongside. Make suitable measurements, and then calculate the area of the disc on which digital information can be stored. [3]

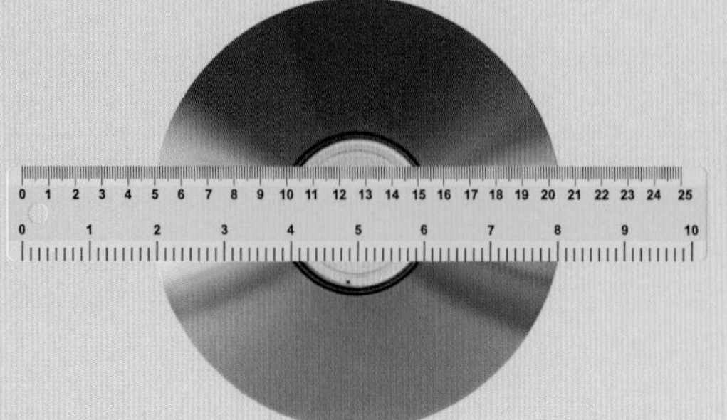

c) The "bumps" for each digital bit on the surface of a CD are laid down along a spiral track around the disc. Dimensions are as follows:

Length of bump along track = 0.97 μm

Width of track = 2.1 μm

Show, using your answer for the area for storage, that the disc can store about 4 gigabits (4×10^9 bits) of digital information. [2]

d) Sound is stored in digital form by sampling the analogue waveform 44 000 times each second, and encoding each sample into a 16-bit number. Explain why 44 000 samples each second is an appropriate sampling rate. [2]

e) Normally two stereo channels are recorded. Use this fact and the other information above to estimate the playing time available on a 4 gigabit music CD. [3]

Adapted from Edexcel Salters Physics, A2, paper 4, June 2001

3 The voltage signal below is produced by a sound synthesizer.

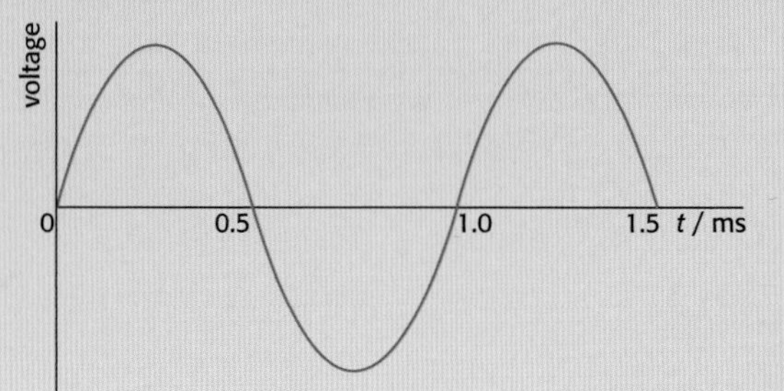

a) Calculate the frequency of the signal illustrated above. [2]

b) The signal is being sampled eight times in one cycle. Calculate the sampling rate in hertz [1]

c) What is the maximum frequency of signal that your calculated value of sampling rate could be effective in encoding? [1]

d) This signal is to be encoded using four bits per value. Each bit takes 10^{-8} s to transmit. How many different signals sampled in the same way could be transmitted along a single fibre? [3]

e) State one advantage and one disadvantage of using more than four bits to encode each value. [2]

Adapted from Edexcel Salters Physics, A2, paper 4, June 2002

4 This question is about a CCD.

a) A digital camera is used to take a photograph of a freshwater crab. The CCD in the camera has 8.2×10^6 square pixels. Each pixel has an area of 3.2×10^{-10} m^2. A claw of the crab has an area of 1.5 cm^2. The image of the claw formed on the CCD is 0.10 cm^2. Two spots on the claw are separated by 0.50 mm. Deduce that that the images of the two spots will be resolved. [4]

b) Light is incident on the image collection area for a time of 50 ms. The number of photons incident on one pixel is 4.5×10^4. Each pixel has a quantum efficiency of 85% and a capacitance 20 pF.

i) State what is meant by *quantum efficiency*. [1]

ii) Estimate the change in potential difference across each pixel. [4]

c) Outline how the variation in potential difference across individual pixels enables a black-and-white image to be produced by a digital camera. [2]

25 The extended essay

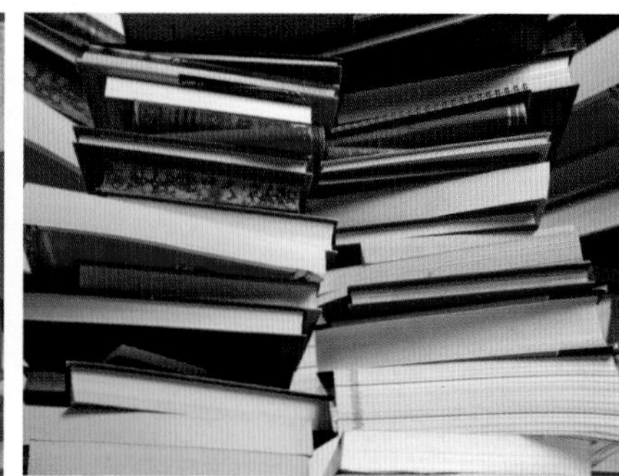

One of the unique aspects of the IB Diploma Programme is the **extended essay**. Every student has to submit this formal piece of independent research or investigation, which is the result of approximately 40 hours of work. The essay (which should contain no more than 4000 words) is externally assessed, and, in combination with the theory of knowledge grades, can contribute up to three extra bonus points to the total IB Diploma score.

The essay needs to be associated with one specific Diploma Programme subject, and an appropriate teacher has to be available to act as a supervisor through the process in order to provide advice and guidance. Although many of the principles of this chapter apply to all extended essay subjects, it has been written with a physics extended essay in mind. Students familiar with the IB Middle Years Programme (MYP) will find that the extended essay is a natural progression from the MYP personal project.

A physics extended essay is a piece of **new research** with a clear emphasis on physics, and will be judged only on its physics content. New, in this context, means new to you, the student. Practical data needs to be linked to a physical model or hypothesis for any given situation. This also means that great care needs to be taken with an interdisciplinary area such as materials science, and ideally this should be avoided. Whatever topic is chosen, clearly any experimental work must be safe, and this may rule out some possible areas of focus.

Types of extended essay

There are several different approaches that can be used when writing a physics extended essay. The most successful essays are usually experimental in nature: the author designs and carries out an experiment and analyses the data appropriately.

In this situation, it is important to keep the initial experiments as straightforward and uncomplicated as possible. An essay that relies on the complicated construction of apparatus before any readings are available has the potential to waste a great deal of time. The effort expended in building the apparatus does not gain any credit in the assessment of the essay.

Other non-experiment approaches can still score highly but care must be taken to ensure that the essay is still a valid piece of *new research*, and that the analysis and the conclusions of the essay are the author's and nobody else's. For example, it would be inappropriate to attempt to use the essay to summarize current research on a particular area of physics. This does not constitute new research, and such an essay, if submitted, would score poorly. An essay on "black holes", for example, could be nothing more than a literature review based on the writing of other authors, unless you are able to collect and analyse your own data.

If secondary sources were used as the source of data, then an important aspect of the essay would be to assess the reliability of the sources and to challenge assumptions and statements with alternative interpretations. As a general rule, an essay that is based solely on material found in textbooks or on the internet is unlikely to provide sufficient scope for critical analysis, and so is unlikely to score highly.

The key to a good extended essay is a well-focused, well-defined, and realistic research question that allows for an in-depth treatment within the limitations of your school science laboratories. It is recommended that the situation being studied can be understood without having to study new areas of physics beyond those already required by the IB Diploma Programme, though this does not mean that new areas cannot be considered. The idea for the essay should be yours, not your supervisor's. If you are unable to come up with a suitable research question in physics then it would be sensible to choose another subject for your essay.

Formal presentation of the extended essay

The vast majority of essays are word-processed. All extended essays must have the following:

- **Title page**. This provides a clear indication of the focus of the essay. It should be precise, and the title of the best essay is often written as a research question. The title "Force/extension characteristics", for example, is rather dull and extremely wide ranging. A much more focused title could be: "Do conditioners really soften your hair? A physical analysis of human hair involving force/extension graphs".
- **Abstract**. This is a separate page that consists of a couple of paragraphs not exceeding 300 words in total. It presents an overview or summary of the extended essay. It needs to include the precise research question, the process that was undertaken, and the conclusion of the essay. It is one of the last things to be written.
- **Contents**. All pages in the essay should be numbered, and this allows the different sections to be identified with subheadings that are easily located.
- **Introduction**. This is a summary section that places the topic of the essay into context, and links the research question to existing physics ideas. The research aim or question should be clearly and succinctly stated here, either as a direct question or as a statement or proposition for discussion. There is no need to "pad out" the

essay with irrelevant information. Some poor essays involve the student writing all they can find out about their topic without any specific focus.

- **Body (development/methods/results).** This is the core of the essay. The knowledge and understanding demonstrated should build on physics that you have already studied. All graphs, diagrams, and summary tables of data should be incorporated within the text and must be directly relevant to the essay. They should be numbered so that specific references can be made to them in the body of the essay. All personal views and conclusions from the data need to be justified with supporting arguments. To improve the flow of the essay, it is sometimes convenient to put all your raw data into a suitably referenced appendix and include just a summary of the key data in the main body.
- Your essay needs to include an **evaluation** of the processes and/or experimental work that has been undertaken. It does not need to be a separate section, but it is an integral part of the essay.
- **Conclusion**. This summarizes the ideas that have been discussed in the body of the text. It should be based directly on your data, and could involve a quantitative statement such as a mathematical relationship between the variables under investigation. This is not the place to introduce new ideas.
- **References and bibliography.** All sources of quotations, ideas, and points of view used in the essay need to be cited properly at the appropriate point in the text, and accurately identified in a list of references in the bibliography. This is an alphabetical list of sources that were used in the production of the text. General sources of information that are not directly cited in the body of the essay should be cited in the introduction or a separate acknowledgment.
- **Appendices.** These contain additional information and data.

Citation

A very important aspect of the extended essay is the correct citation of all sources of information used. It does not matter what particular layout is used for this academic referencing, but if someone else's opinions or material is included in the essay without a proper citation then this would be **plagiarism** – an attempt to pass off other people's ideas as your own. Plagiarism is as serious as being caught cheating in an exam.

There are many possible documentation styles for being consistent with citations, and often your extended essay supervisor or coordinator will recommend a style for your school. The aim is to provide the reader with sufficient information to be able to locate the same source of information that the writer of the essay used. For a book the minimum information required is the author, the title, the publisher, and the date and place of publication. Web page citations need the home page title, the precise URL, and the date that it was accessed. The following website gives a useful overview of some of the different types of acceptable referencing system.

http://writing.wisc.edu/Handbook/Documentation.html (accessed July 2009).

General knowledge and/or concepts such as Newton's law of gravitation or precise definitions do not require any specific reference.

The concluding interview **(viva voce)**

At the very end of the process, the supervisor gives the author a short interview, which is an opportunity to reflect on the whole process and on what has been learned. The supervisor writes a report, which is submitted with the essay to the external marker.

The length of the extended essay

The upper limit is 4000 words for all extended essays. This upper limit includes the introduction, the body, the conclusion, and any quotations, but does not include:

- the abstract
- acknowledgements
- the contents page
- maps, charts, diagrams, annotated illustrations, and tables
- equations, formulas, and calculations
- citations/references (whether parenthetical or numbered)
- footnotes or endnotes
- the bibliography
- appendices.

Essays containing more than 4000 words are subject to penalties, and examiners are not required to read material in excess of the word limit. It is, however, very difficult to be precise in physics, as most essays include tables. The following conversions apply if including computer code or if writing in Japanese or Chinese.

Japanese: 1 word = approximately 2 Japanese characters
Chinese: 1 word = approximately 1.2 Chinese characters
Computer code: Each line of code of a program fragment included in the body of the essay should count as two words towards the word limit.

Timescale

Although the task can initially seem daunting, the process becomes manageable if broken down into smaller individual tasks. For this approach to be successful it is important to set yourself deadlines for when these tasks are going to be completed. Different schools have different approaches.

Some schools allow students a great deal of flexibility, whereas others choose to control the whole process in great detail. Whatever process you are working under, it is extremely important to stick to deadlines. The list below identifies 15 important milestones. You need to know (or decide) when these tasks are going to be completed.

1. Choice of subject area and identification of supervisor
2. Choice of research question and general approach to be used
3. Preliminary experiments, pilot tests, and/or literature research
4. Main experiment and/or data collection
5. Analysis of data
6. Discussion of provisional findings with supervisor
7. First draft of the main body of the essay

8 First draft of abstract, references, bibliography, acknowledgments, etc.
9 Submission of complete first draft
10 Receipt of feedback from supervisor
11 Revision work on essay
12 Submission of second draft
13 Proofreading and final corrections
14 Submission of final draft
15 Concluding interview (viva voce)

Although it is impossible to plan for the unexpected, the timeline that you create should be flexible enough to be able to cope with delays and/or unforeseen problems.

Throughout the whole process it is very important to keep your supervisor up to date with your progress, and to discuss any difficulties, or any changes of direction you are contemplating.

Assessment

The extended essay is assessed on the 11 different aspects listed below.

A **Research question** (2 marks). The question or proposition for discussion needs to be clearly stated and centred on actual physics rather than, for example, the impact of physics on society.

B **Introduction** (2 marks). This should be related to the relevant physics rather than the experimenter. Avoid the temptation to add lengthy material that just restates the context in excessive detail.

C **Investigation** (4 marks). The plan needs to include all relevant theory, and demonstrate an understanding of the uncertainties and/or limitations associated with any chosen approach.

D **Knowledge and understanding of the topic studied** (4 marks). The starting point for the essay must be at the level specified by the IB Diploma physics syllabus. The best essays take this knowledge and understanding and then apply it to a new situation. Traditional ideas are interpreted in a new and novel context.

E **Reasoned argument** (4 marks). The key to gaining these marks is for all opinions or points of view to be backed up with reasoning. For example, you might have to demonstrate why a graph can be interpreted in a particular way. The argument needs to have a logical structure.

F **Application of analytical and evaluative skills appropriate to the subject** (4 marks) When data is analysed, you need to demonstrate that you understand the analysis that is being presented. All too often technology is used to present a great deal of information that looks impressive but has no real substance. Uncertainties and the use of significant digits are important both for raw data and for information resulting from any form of data manipulation—graphing, for example.

G **Use of language appropriate to the subject** (4 marks). The key concern here is precision. Scientific language, including mathematical descriptions such as "proportional", needs to be used correctly and appropriately. Take care that standard symbols (t for time or v for velocity) are not used on several occasions to represent different quantities (the time of flight of projectile and the time to accelerate to maximum velocity).

H **Conclusion** (2 marks). The conclusion must be consistent with the data presented. Too many essays end badly, with the writer presenting what they think should be the case rather than what their research has discovered. The uncertainty range is an essential tool when deciding possible interpretations for data.

I **Formal presentation** (4 marks). Your essay needs to be as well presented as a scientific research paper: this means that references, citations and a bibliography are essential.

J **Abstract** (2 marks). The abstract is marked on how well it summarizes the research undertaken and the essay, rather than the quality of the research itself.

K **Holistic judgment** (4 marks). This final criterion relates to such things are intellectual initiative, insight and depth of understanding, and the originality and creativity shown.

Possible topics to consider

The IB Extended Essay guide presents some examples of research question essays that represent the range of approaches possible.

- Is it possible to determine the presence of a black hole at the centre of the Milky Way? (Data-based approach.)
- Do wine bottles of different shapes behave as Helmholtz resonators? (Experimental approach.)
- What will be the angular deflection of starlight by the Sun if Newton's universal law of gravitation is applied? (Theoretical approach.)
- Is the efficiency of electromagnetic damping of a moving glider a function of the initial kinetic energy of the glider? (Experimental approach.)

Alternatively, an essay may focus on a statement or a hypothesis:

- The objective is to establish theoretically the proportionality existing between the terminal velocity of a cylindrical magnet falling down a metallic pipe and the resistivity of the metal of the pipe as well as the pipe's wall thickness. (Experimental approach.)
- Water waves are observed in a long and narrow trough, and their speeds are measured. It is assumed that, for shallow water, the speed of the wave will be proportional to the square root of the depth of the water and independent of the wavelength.
- The objective is to establish the relationship between power and temperature for an incandescent lamp.
- A retractable ballpoint will be used to test the law of energy conservation.
- The objective is to establish an acoustic model of the concert flute.

End of chapter summary

Chapter 25 has one main theme: the extended essay. The list below summarises the knowledge and skills that you should be able to undertake after having studied this chapter.

- Understand the different types of possible extended essay and the importance of formal presentation.
- Understand the timescale and processes involved in writing a physics extended essay.

26 Exam techniques

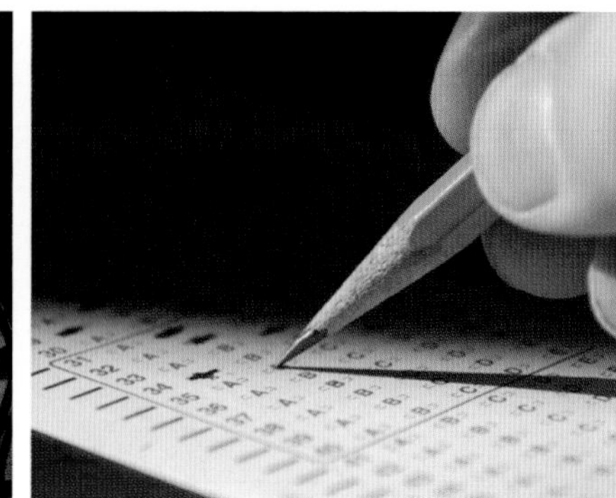

This chapter is devoted to improving your examination techniques. The two most important pieces of advice are:

- Don't panic.
- Read the question carefully.

Examination technique can be improved with practice, and the best students consolidate their knowledge and understanding at the end of a topic by attempting past examination questions. Initially these can be done without any time constraints, but as the real examinations come closer it is important to practice working within the same time as you will have in the real examination.

In every examination you have access to a clean copy of the physics Data Booklet. Apart from some basic pieces of information, such as the SI multipliers and the values of some fundamental constants, this document lists a large number of equations. It is important that you become familiar with these equations as you study the individual topics. Many schools give their students their own copy of the Data Booklet at the beginning of the course and allow them to add annotations and notes.

These notes cannot be brought into the final examinations, but they do allow students to ensure that they properly understand every formula that is contained in the booklet.

Always remember that the examiner can only give you credit if you are able to demonstrate knowledge and understanding. This means your answers must be understandable and legible. It is also important to show all the steps in your working. A wrong answer without any working must be worth zero marks. A wrong answer that allows the examiner to identify, for example, a simple mathematical slip during a calculation would lose only one mark from the total. Remember that, unlike your teacher, the examiner does not know you. In order to show excellence you must take care to craft your answers to the highest quality and not take short cuts or use abbreviations as you might do in internal tests.

Possible points to keep in mind include:

- If a question asks to you to explain how an observation is consistent with a named piece of theory, an important part of the answer would include an outline of the named theory.
- Sometimes there are several ways in which to approach a physics problem or explanation. If a question specifies the approach to take, it is very important to do just that. For example, if the question asks you to explain the expansion of a gas on heating *with reference to the kinetic theory of gases,* is very important to explain what this theory means and use it to answer the question.

Some calculations involve several steps, and these are often presented as different parts of one question: (a), (b) etc. In this case the answer in a later part of the question depends on an earlier answer. The principle of error carried forward **(ECF)** applies: if your working demonstrates that you have got the wrong a nswer to a later section because you have used an earlier wrong answer in the right way, you would be awarded the full marks for that section even if your answer was wrong.

Once you have arrived at an answer to a mathematical question, it is important to think about two additional things. First, the answer will almost certainly have units associated with it. If you forget to include these units, then your answer will not be complete and you will lose marks. Second, the number of significant digits that you quote in your answer should match the number of significant digits given in the original data.

For example, a calculator gives the result of dividing 8.3 by 1.7 as 4.882352941. The average speed of an object covering a distance of 8.3 m in a time of 1.7 s, however, is not 4.882352941 but should be correctly stated as 4.9 m s^{-1}. The units of the answer are derived from the units of the original data. The data was given to two significant digits, so the answer should be quoted to the same level of accuracy.

Command terms

The IB syllabuses for all group 4 subjects make use of the same limited number of **command terms** that define the level of detail required. The same command terms are used in the examination papers, so it is important to understand their precise meaning. The tables below lists the command terms classified by **objective level**.

The objective level is best imagined as a measure of the level of difficulty or detail of understanding required. There are three different objective levels: objective 1 is easiest and objective 3 the hardest. Every syllabus statement is categorized by objective level.

Objective 1

The syllabus statements and the questions that use these command terms require you to be able to remember information only, and so they represent the easiest tasks. You need to be able to demonstrate an understanding of facts, concepts, scientific methods and techniques, scientific terminology, and methods of presenting scientific information.

Table 1 Objective 1: command terms

Define	Give the precise meaning of a word, phrase, or physical quantity.
Draw	Represent by means of pencil lines.
Label	Add labels to a diagram.
List	Give a sequence of names or other brief answers with no explanation.
Measure	Find a value for a quantity.
State	Give a specific name, value, or other brief answer without explanation or calculation.

Objective 2

The syllabus statements and the questions that use these command terms require you to be able to apply the information that you have learned in a given situation. They represent tasks that are probably reasonably straightforward but that involve some problem-solving ability. You need to be able to demonstrate an ability to apply and use scientific facts and concepts, scientific methods and techniques, scientific terminology to communicate effectively, and appropriate methods to present scientific information.

Table 2 Objective 2: command terms

Annotate	Add brief notes to a diagram or graph.
Apply	Use an idea, equation, principle, theory, or law in a new situation.
Calculate	Find a numerical answer showing the relevant stages in the working (unless instructed not to do so).
Describe	Give a detailed account.
Distinguish	Give the differences between two or more different items.
Estimate	Find an approximate value for an unknown quantity.
Identify	Find an answer from a given number of possibilities.
Outline	Give a brief account or summary.

Objective 3

The syllabus statements and the questions that use these command terms require you to be able to apply the information that you have learned to a more complicated situation. They represent tasks and problems that involve several stages or steps before arriving at a possible solution. You need to be able to demonstrate an ability to construct, analyse, and evaluate hypotheses, research questions, predictions, scientific methods and techniques, and scientific explanations.

Table 3 Objective 3: command terms

Analyse	Interpret data to reach conclusions.
Comment	Give a judgment based on a given statement or result of a calculation.
Compare	Give an account of similarities and differences between two (or more) items, referring to both (all) of them throughout.
Construct	Represent or develop in graphical form.
Deduce	Reach a conclusion from the information given.
Derive	Manipulate a mathematical relationship(s) to give a new equation or relationship.
Design	Produce a plan, object, simulation, or model.
Determine	Find the only possible answer.
Discuss	Give an account including, where possible, a range of arguments for and against the relative importance of various factors, comparisons, or alternative hypotheses.
Evaluate	Assess the implications and limitations.
Explain	Give a detailed account of causes, reasons, or mechanisms.
Predict	Give an expected result.
Show	Give the steps in a calculation or derivation.
Sketch	Represent by means of a graph showing a line and labelled but unscaled axes but with important features (for example, intercept) clearly indicated.
Solve	Obtain an answer using algebraic and/or numerical methods.
Suggest	Propose a hypothesis or other possible answer.

Structure of exams

The three papers in the physics exam are designed to test different topics, but also to test different skills and levels of understanding. A clean copy of the *Physics Data Booklet* is required for each paper at both SL and HL.

Paper 1

This is the multiple-choice paper and addresses objectives 1 and 2 (see above). Questions can be on any topic in the common core (topics 1 to 8 for standard level and 1 to 14 for higher level students). Time is very short: SL candidates have to answer 30 questions in 45 minutes and HL candidates have 40 questions in 60 minutes. This corresponds to 90 seconds per question for reading, solving, and entering your answer on the sheet provided.

You have access to the Data Booklet but are not allowed to use a calculator. This means that the examiners cannot ask questions that involve complicated calculations. Any sum must be able to be done quickly in one's head.

Paper 2

This is the longest paper and addresses all three objectives, and a calculator is required. Typically this paper is timetabled immediately after paper 1, so you need to be prepared for a long examination session. The paper requires answers that are written on the question paper. You can always ask for additional paper if you run out of space. As for paper 1, questions can be on any topic in the common core. The paper is divided into two sections.

In section A there is a data-based question that requires students to analyse a given set of data. The remainder of section A is made up of short-answer questions. This section is all compulsory. The total number of marks is 25 for SL candidates and 45 for HL candidates. A good rule of thumb is that there is about a minute available per mark awarded.

In section B students at SL are required to answer one question (worth 25 marks) from a choice of three, and students at HL are required to answer two questions (each also worth 25 marks) from a choice of four. These extended-response questions may involve writing a number of paragraphs, solving a substantial problem, or carrying out a substantial piece of analysis or evaluation.

At the beginning of the examination you will have 5 minutes' reading time. During this time it is important to read through the questions in section B and think about which question(s) you will attempt. One possible approach is to attempt the questions in the order in which you are most confident, attempting your "best" question first and your "worst" question last. With this approach you will tend to use time more effectively.

Paper 3

Paper 3 tests knowledge of the options and addresses objectives 1, 2, and 3; again, a calculator is required. You will also get 5 minutes' reading time. If you have managed to study more than the minimum number of options, which is not recommended, then you can use this time to decide which options you wish to attempt. If you have only studied the minimum number of options, as is generally the case, then you can use this time to start thinking about how you are going to tackle the questions.

It is very important that you do not attempt an option question in a topic that you have not studied. It often happens that the first part of an option question is fairly straightforward and can be attempted successfully by most students, whatever options they have studied. However, you will soon find that you will be unable to proceed further. Every year, it is obvious when students have attempted a question in an option they have not studied by the poor quality of their answers.

Students at SL are required to answer several short-answer questions in each of the two options studied. Each option is worth 20 marks. Students at HL are required to answer several short-answer questions and an extended-response question in each of the two options studied. The questions in each HL option are worth 30 marks.

SL assessment specifications

Component	Overall weighting %)	Approximate weighting of objectives		Duration (hours)	Format and syllabus coverage	Marks awarded
		1 + 2	3			
Paper 1	20	20		¾	30 multiple-choice questions on the core syllabus	30
Paper 2	32	16	16	1¼	Section A: one data-based question and several short-answer questions on the core (all compulsory) Section B: one extended-response question on the core (from a choice of three)	25 + 25 = 50
Paper 3	24	12	12	1	Several short-answer questions in each of the two options studied (all compulsory)	20 + 20 = 40

HL assessment specifications

Component	Overall weighting (%)	Approximate weighting of objectives		Duration (hours)	Format and syllabus coverage	Marks awarded
		1 + 2	3			
Paper 1	20	20		1	40 multiple-choice questions (±15 common to SL plus about five more on the core and about 20 more on the AHL)	40
Paper 2	36	18	18	2½	Section A: one data-based question and several short-answer questions on the core and the AHL (all compulsory) Section B: two extended-response questions on the core and AHL (from a choice of four)	45 + 25 + 25 = 95
Paper 3	20	10	10	1¼	Several short-answer questions and one extended-response question in each of the two options studied (all compulsory)	30 + 30 = 60

Number of marks available

Each mark allocated to any given question corresponds to a step in a calculation or a point that is required in your answer. For example, an experimental description that is worth 4 points must be looking for four different pieces of information, and a calculation worth 3 marks will involve at least three identifiable steps before getting to the correct answer.

Many students take the number of lines available on the answer paper to be an indication of the amount of information required. For this reason the aim is to provide sufficient rather than excess space. If you need to write more information to complete your answer, make sure you ask for some continuation paper, which should be attached to the exam script. However, it is often the case that an answer that occupies a lot more space than is provided indicates that the student has not understood what is being tested.

Use of technical language

It is very important to use technical language in a precise way. For example, *speed* and *velocity* have the same meaning in everyday language, but they have different scientific definitions (speed is a scalar whereas velocity is a vector). It is correct to say that an object with constant velocity must have a zero resultant force acting on it, but it is not correct to say that an object with constant speed must have a zero resultant force. Confusions are also common with mass and weight.

Mathematical terms are often used inappropriately. See pages 19–20 for the difference between a proportional relationship and a linear relationship between two variables.

Describing graphs accurately can involve some complicated language. For example, consider an experiment in which an object is allowed to roll down a slope from rest. Measurements are taken of the velocity of the object at different times from the instant of release. The graph in Figure 1 is obtained.

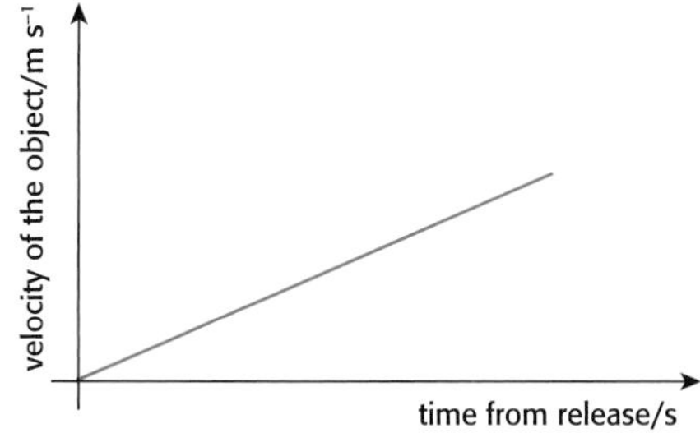

Figure 1 Is this relationship proportional or linear?

A precise way of describing this graph would be "the variation with time from release of the velocity of the object". The form that can be used to describe the graph is "the variation *with* (the variable on the x-axis) *of* (the variable on the y-axis).

Use of drawings

There are many situations where a good, labelled diagram or sketch graph can be used demonstrate, quickly and precisely, knowledge and understanding. A poor diagram, however, can be confusing. Remember the following points:

- Precise labels are essential.
- Use a sharp pencil for drawing.
- Graph lines should be either a smooth curve or a straight line drawn with a ruler.

- Think about the need for precision in your drawings. For example, unless labelled otherwise, the relative lengths of the arrows in a free body diagram will be taken to represent the relative magnitudes of the forces involved.

In conclusion

If you are approaching your final examinations, you will have completed nearly two years of challenging study. So long as you have followed an extensive and thorough revision plan over several months there is no need for you to worry about the final exams. Consistent hard work throughout the course and a final dedicated revision programme are certain to get you successfully through the final examination.

Good luck with the exams and in your future studies!

End of chapter summary

Chapter 26 has one main theme: how to improve your examination technique. The list below summarises the knowledge and skills that you should be able to undertake after having studied this chapter.

- Know the importance of command terms in the IB syllabus and exams.
- Understand the structure of the different written external assessments and appreciate the importance of answering the questions in a precise way with the correct use of technical language.

27 Option E: Astrophysics

Have you ever been lucky enough to see stars on a clear night far away from the artificial lights of a city? Many of the pinpoints of light that you see whenever you look at the night sky may in fact be huge collections of stars, called galaxies. In fact, the observable universe is made up of more than 10^{20} stars at distances of up to 10^{27} m away from the Earth, making its truly huge dimensions hard to comprehend.

The Astrophysics option aims to give you a glimpse into the processes that have shaped the development of our universe into the one we now observe. This understanding can not only be used to predict the future development of the universe, but also often results in a sense of awe and wonder at our own place in space.

This chapter aims to provide the framework for studying the astrophysics option rather than all of the detailed information required. Many of the questions in the text require some independent research and there are many excellent sources of further information available on the internet or in textbooks. Of particular note are the pages associated with space agencies that involve a great deal of international collaboration, for example NASA (the National Aeronautics and Space Administration, www.nasa.gov) and ESA (European Space Agency, http://sci.esa.int). The latter site has published teacher support material closely linked with this IB Diploma Programme Phyiscs option on their teacher pages, for example http://sci.esa.int/science-e/www/object/index.cfm?fobjectid=35713

Other possibilities include RKA (the Russian Federal Space Agency, www.federalspace.ru) or CNSA (China National Space Administration, www.cnsa.gov.cn). All websites accessed July 2009.

Figure 1 (Top) Photographs taken of stars with a long exposure time show their motion. (Middle) The Orion constellation. (Bottom) The Southern Cross constellation.

Introduction to the universe

An observer at a given location on Earth sees a fixed and essentially unchanging pattern of stars. This pattern rotates from east to west through nearly 180° during the night. Although the pattern of stars remains essentially the same, slightly different sections of the pattern are visible at different times during a year. Different observers located around the world will see different sections of the whole pattern of stars and observers in the northern hemisphere see different stars from those in the southern hemisphere.

The fixed patterns of stars has been categorized into 88 **constellations.** The stars and galaxies in a constellation are all located in the same general direction from Earth but are not necessarily close to one another. Historical observers named the constellations after mythical figures and creatures which the shapes seemed to represent.

The Moon and all the planets follow the same rotation as the stars during any given night but over time they move position with respect to the "fixed" background pattern of the stars, a change that is only noticeable over several nights.

It is not surprising that ancient civilizations explained these observations by imagining the stars and planets to be fixed on transparent spheres that rotated around the Earth.

The same observations are better explained by the more modern view of the universe that says that the stars are fixed and the Earth rotates on its own axis (time period 24 hours) and revolves around the Sun (time period 1 year).

1. Describe, using your own words and diagrams, the apparent motion of the stars over a period of a night and over a period of a year.
2. Explain the apparent motion of the stars over a period of a night and over a period of a year in terms of the rotation and revolution of the Earth.

The distribution of stars

Stars are not uniformly distributed throughout space but are gathered together by their mutual gravitational attraction. Small groups of stars that gravitationally interact with one another are called **stellar clusters**. Very large numbers of stars bound together by gravity make up a **galaxy.**

To measure the large distances between stars, astronomers use a unit called the **light-year** (ly), which is the distance travelled by a beam of light in one year. An alternative unit is the **parsec** (pc) which is defined as 3.1×10^{16} m. The nearest star to the Earth (Proxima Centauri) is approximately 4.27 ly away.

1. Show that 1 light-year is approximately equal to 9.5×10^{15} m. How many light-years are there in a parsec? How far away is Proxima Centauri in parsecs?
2. Compare, in terms of order of magnitude, the relative distances between stars within a galaxy and between galaxies.

Our own galaxy contains approximately 2×10^{11} stars and its diameter is approximately 10^5 light-years in diameter. Our Sun takes approximately 2×10^8 years to complete one orbit around the centre of the galaxy.

Research the structure of our Sun with specific reference to the following regions: core, radiative zone, convective zone, photosphere, chromosphere, and corona.

Galaxies also occur in **clusters**. There are approximately 20 other galaxies within 2.5×10^8 light-years of us including the Andromeda galaxy. Galaxy clusters also group together in **superclusters**. Despite the apparent abundance of stars, most of space is completely empty (a vacuum).

The solar system

Our Sun is a typical star. The only reason it appears to be bigger and brighter than all the other stars is that we are much closer to it than to the other stars.

1. Use Table 1 to compare the planets in terms of size and distance from the Sun.
2. Research the difference between an asteroid and a comet.
3. Research the nature and position of the asteroid belt.

Our **solar system** consists of eight planets (Mercury, Venus, Earth, Mars, Jupiter, Saturn, Uranus, and Neptune) in elliptical orbits around the Sun. The gravitational attraction between the Sun and

the planet provides the centripetal force necessary for the planet to stay in orbit. In addition to the planets, smaller masses called **planetoids** (e.g. Pluto), **asteroids**, and **comets** orbit the Sun. Many planets (including Earth) are also orbited by moons. The table below shows the approximate distances, comparative radii and masses of the planets.

The average distance from the Earth to the Sun is called the **astronomical unit (AU).** This unit is useful for talking about distances on the interplanetary scale.

Table 1

	Distance from Sun / 10^6 km	Radius / radius of Earth	Mass / mass of Earth
Mercury	57.9	0.38	0.06
Venus	108.2	0.95	0.86
Earth	149.4	1.00	1.00
Mars	227.9	0.53	0.15
Jupiter	778.3	11.19	1323
Saturn	1427.0	9.41	752
Uranus	2869.6	3.98	64
Neptune	4496.6	3.81	54

Stellar radiation and stellar types

Stars are large nuclear fusion reactors, which convert light nuclei (hydrogen, helium, etc.) into heavier nuclei (carbon, iron, etc.). The process releases a great deal of energy as thermal energy and electromagnetic radiation. See pages 379–80 for more details of the process.

Most of the atoms in a stable star are in the fourth state of matter – plasma. Many stars radiate energy whilst remaining stable for billions of years and from this we can infer that the processes going on in the star must be balanced. Firstly, the star must be in **thermal equilibrium**. As the star's temperature remains constant, the rate at which energy is being radiated must equal the rate at which nuclear energy is being converted into thermal energy.

Secondly, the star must be in **hydrostatic equilibrium** because it does not expand or contract. This is contrary to our expectations that a hot object in a vacuum would simply expand to fill the space. The only force keeping the constituent mass of the star together is gravity so at any given point in the star, the inward gravitational forces must balance the outward force. This is sometimes expressed in terms of pressure. We say that there must be equilibrium between the outward **radiation pressure** and the inward **gravitational pressure**.

Given the huge distances involved, most, if not all, of the information that we can gain about any star comes from an analysis of the electromagnetic radiation that we receive from it. The total power

(a)

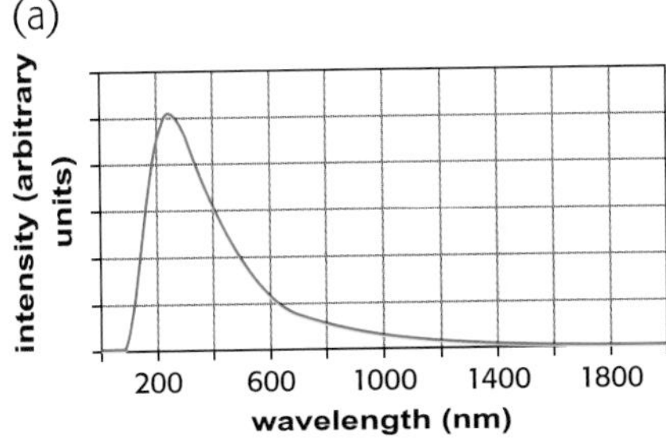

(b)

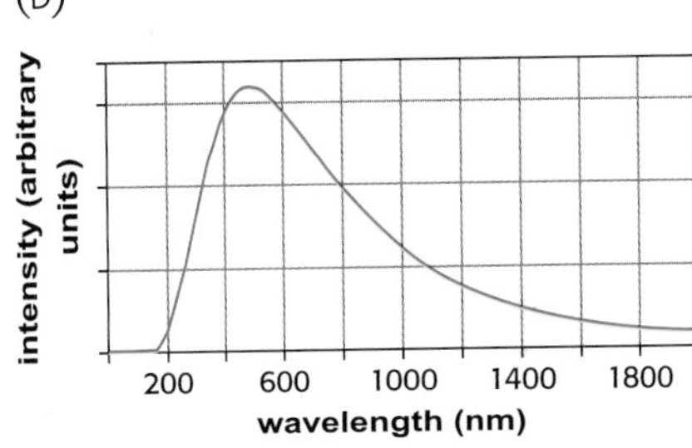

(c)

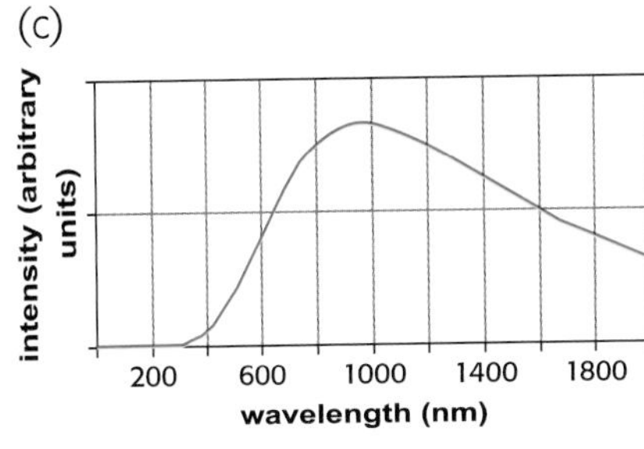

Figure 2 The black body radiation curves for stars of surface temperature (a) 12 000 K, (b) 6000 K and (c) 3000 K. Note that the intensity scale is different for each graph.

radiated by a star is called its **luminosity** L. The relative brightness of a star depends on the power received from it over a given area on the surface of Earth. This will depend on the luminosity and on how far away the star is. The **brightness** b (sometimes referred to as **apparent brightness**) of a star is defined as the power received per unit area:

$$b = \frac{L}{4\pi r^2}$$

where
L is the luminosity of the star, measured in W
b is the brightness of the star, measured in W m^{-2}
r is the distance between the star and the observer, measured in m

This relationship can be used to calculate the stellar distance once a star's luminosity and brightness have been determined. Comparisons of brightness of different stars are also often expressed in terms of a different measurement – the **magnitude**. See pages 376–7 for more details.

The radiation emitted from stars corresponds closely to the theoretical black body model for radiation introduced on page 263. This allows us to utilize two further relationships to analyse radiation from stars.

The German physicist Wilhelm Wein (1864–1928) identified a relationship that allows us to calculate the surface temperature of stars from their colour. **Wien's law** links the absolute temperature of a black body and the wavelength at which the radiated energy reaches its maximum intensity.

$$\lambda_{max} = \frac{2.90 \times 10^{-3}}{T}$$

where
T is the absolute surface temperature of the star, measured in K
λ_{max} is the peak wavelength (where radiated energy reaches a maximum intensity), measured in m

The **Stefan–Boltzmann law** can be used to calculate a star's surface area (and hence size) if we know its luminosity and its surface temperature. For any black body radiator, the total power radiated is calculated as follows:

$$L = \sigma A T^4$$

where
L is the luminosity of the black body (i.e. the star), in W
σ is the Stefan-Boltzmann constant, 5.67×10^{-8} W m^{-2} K^{-4}
A is the surface area of the black body (i.e. the star), in m^2
T is the absolute temperature of the back body (i.e. the surface temperature of the star), in K

The surface area of a star is linked to its radius r:

$$A = 4\pi r^2$$

It should be noted that luminosity can only be calculated for a star from a measurement of its brightness providing we know the

?

1 Use these three graphs to calculate the peak wavelength for a star with a surface temperature of **a)** 12 000 K, **b)** 6000 K, **c)** 3000 K.

2 Use your answers to explain why relatively cool stars are orange/red in colour, medium temperature stars are yellow/white, and hot stars are bluish white.

3 Explain why green stars do not exist.

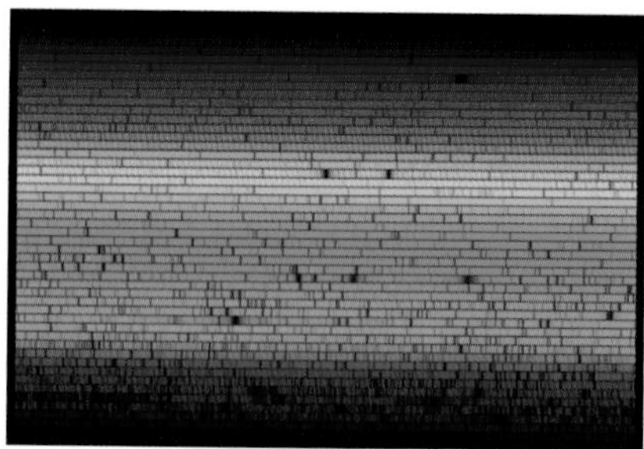

Figure 3 Visible light absorption spectrum from a star.

?

1 Calculate the luminosity of the Sun. Its radius is 6.96×10^8 m and the peak wavelength of its emitted radiation is 500 nm.

2 Use your answer to question 1 and the mass of the Sun (1.99×10^{30} kg) to calculate the power emitted per unit mass. Compare your answer with 1 W kg^{-1}, the average power emitted by a human being. What can be deduced from this information?

3 Explain whether starlight is an example of an absorption or an emission spectra. How are these spectra used to provide evidence for the chemical composition of stars?

4 Explain whether starlight allows us to deduce the elements in the interior of the star, in its outer layers, or both.

distance between the star and the observer. See pages 376–7 for further details of how stellar distances as determined.

Finally, it is possible to deduce chemical and physical data from stellar radiation by analysing the wavelengths present in the light for evidence of atomic spectra. Further detail can be deduced from the relative intensities of each of the spectral wavelengths but details of this analysis are beyond the scope of this Course Companion.

Types of star

The analysis of spectral lines allow us to categorize the different types of star that exist into different spectral classes. Each star can be allocated into one of seven categories. Each category is designated by a letter which, in descending order of surface temperature, are: O, B, A, F, G, K, M. Each type is then further subdivided into ten groups designated by a number from 9 to 0. The Sun is in spectral class G2.

Table 2

Class	Surface Temperature / K	Colour
O	28000 – 50000	blue
B	9900 – 28000	blue
A	7400 – 9900	blue-white
F	6000 – 7400	white
G	4900 – 6000	yellow
K	3500 – 4900	orange
M	2000 – 3500	red

The majority of stars are, in fact, **binary stars**. This means that the points of light visible on Earth are not single stars but pairs of stars orbiting their common centre of mass. Some binary stars can be resolved either by eye or by telescope and are called **visual binaries** but the existence of the majority of binary stars has been deduced from the analysis of their stellar radiation. These systems are called **spectroscopic binaries**.

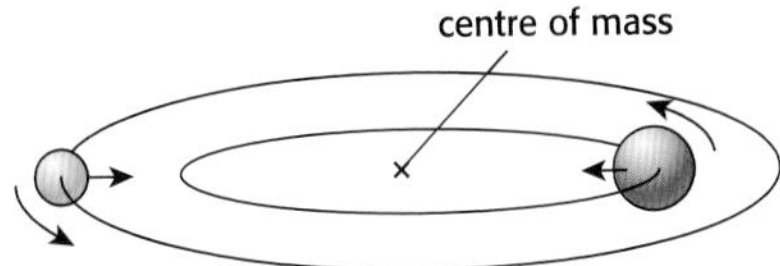

Figure 4 Binary stars – two stars in orbit around their common centre of mass

The existence of spectroscopic binaries is deduced from the periodic Doppler shift that can be seen in the spectra of the two stars that make up the binary system. If both stars are identical, over a period of time each spectral line can be seen to "split" into two separate wavelengths as a result of their orbital motion. During the "split" one of the stars in the binary pair has a component of its velocity moving towards the observer and thus has a Doppler blueshift whilst the other has a component moving away from the observer and has a Doppler redshift.

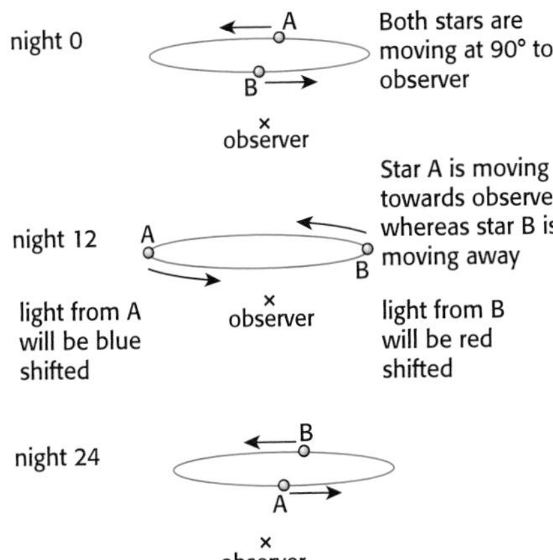

Figure 5 A spectroscopic binary system

Eclipsing binary stars are deduced from a periodic variation in the received brightness. In this situation, the orbits must be edge-on when viewed from Earth so that when one star passes in front of the other, the overall intensity received at Earth is reduced.

Sometimes individual stars are observed to have a regular variation in brightness. One class of this type of star are the **Cepheid variables**. They are named after the star delta Cephei in the Cepheus constellation. Their brightness oscillations are believed to result from an oscillation in the star's physical size. As the size changes so does its surface temperature and its luminosity. See pages 378–9 for information about the use of this type of star in calculating stellar distances.

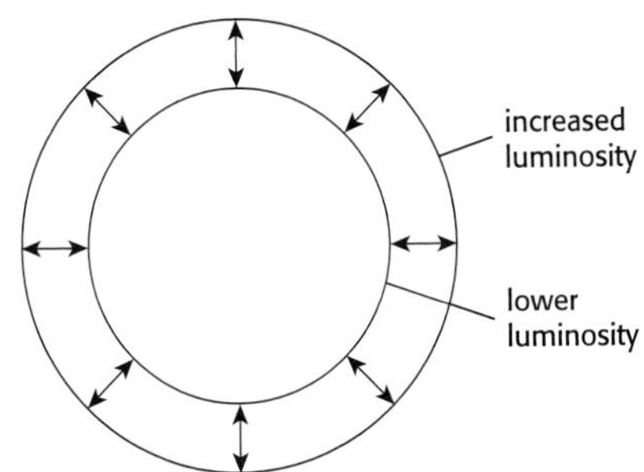

Figure 6 Periodic expansion and contraction of a Cepheid variable

Some stars are categorized as **giants** or even **supergiants**, depending on their size and mass. The **red giant** stars are in stellar class K and M and have low surface temperatures (in the range 3000 to 4000 K). At the other end of the size variation are the **dwarf** stars. **White dwarf** stars are known to have a high surface temperature (typically in stellar class B or A but sometimes at lower temperatures), their mass is of the same order of magnitude as our Sun but their volume is similar to that of the Earth. See page pages 379–80 for a description of how they can be formed.

Different stars are observed to have very different luminosities and surface temperatures. The range of surface temperature is shown by the stellar categories. Typically, star luminosities can range from ones that are a million times more luminous than our sun to ones that are ten thousand times less luminous. A very useful visual categorization of the properties of different stars is the **Hertzsprung–Russell diagram.**

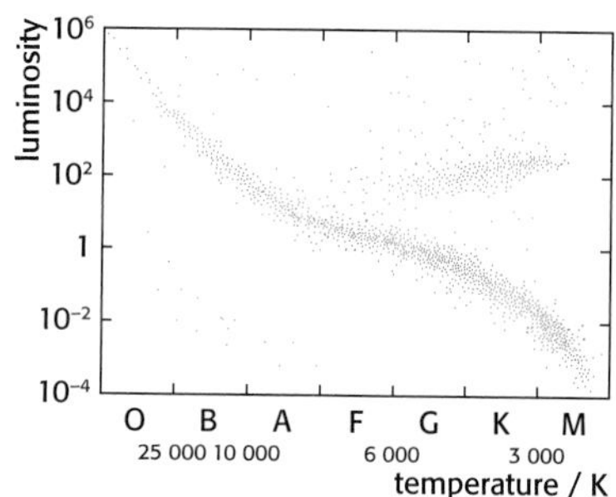

Figure 7 Hertzsprung – Russell diagram

On this diagram we plot increasing luminosity on the y-axis and decreasing temperature on the x-axis. Both scales are logarithmic. Most of the stars that we see in the night sky (approximately 90%) are located in the band that goes from the top left-hand corner of this diagram to the bottom right-hand end. This band is known as the **main sequence** and any star whose luminosity and surface temperature place it on this band is called a **main sequence star**. Stars that are not on the main sequence are more rare and most of these are either red giants or white dwarfs.

Note that different versions of this diagram exist with different (but related) variables on each axis. An alternative variable for the y-axis is called the **absolute magnitude** of the star. This quantity is defined below and ranges from +15 (which represents a very low luminosity) to −10 (which represents a very high luminosity). An alternative variable for the x-axis is the spectral class of the star going from O to M.

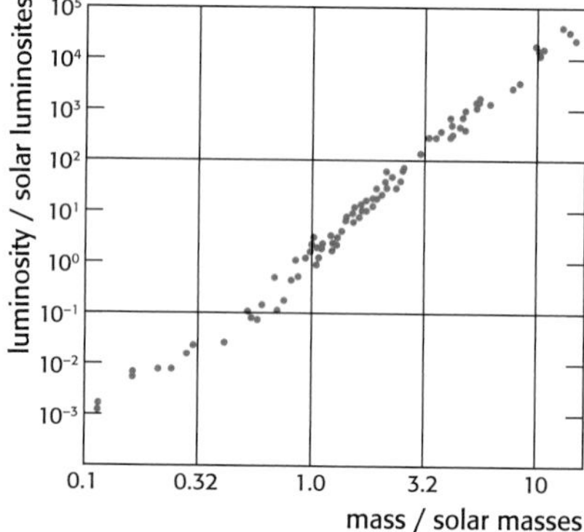

Figure 8 Variation with mass of the luminosity of main sequence stars

The largest mass stars are towards the top of the diagram and this decreases down the main sequence from top left to bottom right. The variation with mass, m of the luminosity, L of main sequence stars is shown in Figure 8. The mathematical relationship between m and L is $L \propto m^n$

Where $3 < n < 4$.

Stellar distances

A key piece of information necessary for the analysis of the light from a star is the distance from observer to star. Unfortunately, this is not necessarily an easy measurement to achieve. Consider the difference between a distant star with high luminosity and a closer star with a lower luminosity. They will both appear to have the same brightness.

Astronomers use three different methods to find the distance to stars, each with an associated degree of uncertainty. They are: **stellar parallax, spectroscopic parallax**, and calculations involving **Cepheid variables**. Once the distance to a star has been found, measurements of brightness allow us to calculate its luminosity. The brightness of stars are usually compared using the magnitude scale.

Magnitude scale

For historical reasons, the brightness of stars is compared on the **magnitude scale**. The brightness of a star as seen by an observer on Earth is recorded as a magnitude, m. On this scale a bright star is magnitude +1 and a dim star (just visible with the naked eye) is magnitude +6. The bright star ($m = +1$) is roughly 100 times brighter than the dim star, corresponding to a difference of 5 magnitudes

Table 3 gives luminosity and surface temperature data for some stars. For each star, locate its position on the HR diagram and use this to categorize the stellar type.

Table 3

Star	Surface temperature / K	Luminosity / luminosity of the Sun
Aldebaran	4700	200
Betelgeuse	3000	20000
Regulus	14000	200
Rigel	13000	20000
Sirius B	20000	0.002

(i.e. $5 = 6 - 1$). This means that each step of the magnitude scale is equivalent to a change of brightness by a factor of $100^{0.2} = 2.512$.

If the distance to stars is known, then their luminosities can be compared directly using the concept of **absolute magnitude**, *M*. This is defined as the magnitude that the star would have if it were placed at a distance of 10 pc from Earth. When stars' absolute magnitudes are compared then we are directly comparing their luminosities. Each step on the absolute magnitude scale is equivalent to change of luminosity by a factor of 2.512.

The closest star to the Earth (apart from the Sun), Proxima Centauri, is 1.3 pc away and most stars are much further away. This means the majority of stars have a smaller numerical value for absolute magnitude than for apparent magnitude. Some stars would be so bright if placed 10 pc away that they have negative absolute magnitudes. A star of absolute magnitude −4.0 is five magnitudes brighter than a star with absolute magnitude +1.0 and thus its luminosity is 100 times greater.

The following formulae can be used to convert between absolute magnitudes and apparent magnitudes.

$$m - M = 5\log\left(\frac{d}{10}\right) \qquad (1)$$

If we rearrange this we get:

$$d = 10^{\frac{(m-M+5)}{5}} \text{ pc} \qquad (2)$$

where
m is the apparent magnitude of the star
M is the absolute magnitude of the star
d is the distance between the Earth and the star in pc

The first equation is included in the IB data booklet. It is also possible to use a difference in apparent magnitudes to compare stars' brightnesses (or absolute magnitudes to compare stars' luminosities). The conversions are summarized here:

$$m_2 - m_1 = -2.5\log\left(\frac{b_2}{b_1}\right) \qquad (3)$$

$$M_2 - M_1 = -2.5\log\left(\frac{L_2}{L_1}\right) \qquad (4)$$

where
m_1 and m_2 are the apparent magnitudes of star 1 and star 2 (respectively)
M_1 and M_2 are the absolute magnitudes of star 1 and star 2
b_1 and b_2 are the brightnesses of star 1 and star 2
L_1 and L_2 are the luminosities of star 1 and star 2

Stellar parallax

The distance to nearby stars can be calculated using **stellar parallax**. As the Earth moves in its orbit around the Sun, a close star will appear to change in position when compared with the fixed background of distant stars.

?

Star A and star B have the same brightness but star A is known to have 100 times the luminosity of star B. What is the distance of star A, d_A, in terms of the distance of star B, d_B?

?

1. The star Pollux in the Gemini constellation has an apparent magnitude of +1.2 and lies at a distance of 12 pc from the Earth. Calculate its absolute magnitude.
2. The star Sirius in the Canis Major constellation has an apparent magnitude of −1.46 and lies at a distance of 2.64 pc from the Earth. Calculate its luminosity as compared with the Sun. The absolute magnitude of the Sun is +4.77
3. (AHL only). Using the inverse square relationship and the definitions of magnitude and brightness, derive each of the four formulae in the section entitled "Magnitude scale."

The observed parallax angle is half the size of maximum angular deviation over a period of 6 months, as shown in Figure 9.

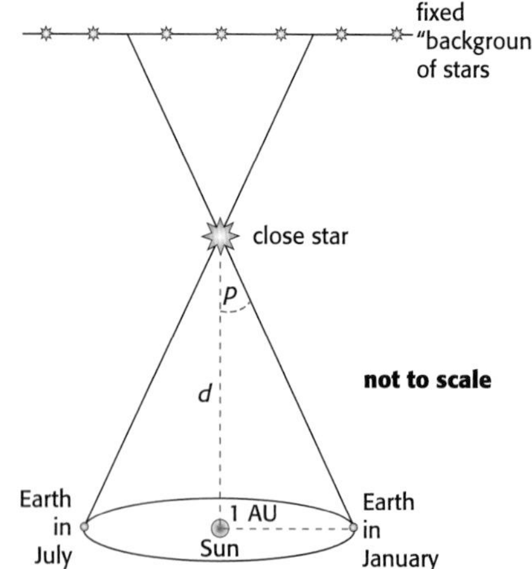

Figure 9 Stellar parallax

In reality, parallax angles are tiny and can be measured in fractions of one second of arc (arcsec). One arcsec is $\frac{1}{3600}$ of one degree. The parallax angle p is inversely proportional to the distance to the star d.

$$d = \frac{1}{p}$$

where
d is the distance from the sun to the star, measured in pc
p is the parallax angle, measured in arcsec

A star that has a parallax angle of 1 second is defined to be at a distance of 1 parsec (pc).

Parallax angles of stars measured from Earth are limited by atmospheric refraction to ~0.01 arc seconds. The Hipparcos Satellite, launched by the ESA, can measure parallax angles as small as 0.001 arc seconds.

SI prefixes are often used in conjunction with the parsec and kpc (= 1000 pc) or Mpc (= 10^6 pc) are common.

1. Show that
   1 pc = 2.06×10^5 AU
2. The star Arcturus in the Bootes constellation has a parallax angle of 0.09 arcsec. Determine its distance from the Earth.
3. Explain why the method of stellar parallax is limited to measuring stellar distances less than a few hundred parsecs.

Spectroscopic parallax

For distances greater than approximately 1000 pc, using stellar parallax is no longer feasible. The method used is called **spectroscopic parallax** but it has nothing to do with the parallax angle. In this method, the relative intensities and profile of different lines in a star's emission spectrum are used to estimate its temperature and spectral type. Using the HR diagram we can estimate its luminosity. Measurements of the star's brightness mean that we can find its apparent magnitude and then equation (2) on page 377 can be used to calculate the distance to the star. This method is only reliable for distances less than about 10 Mpc.

Cepheid variables

Beyond about 10 Mpc, the measurement of stellar distances discussed so far become prone to large uncertainties. In order to calculate the distance to a star, it is necessary to have a source of known luminosity nearby. Such a source is described as a **standard candle** by astronomers. One such standard candle is a Cepheid variable star.

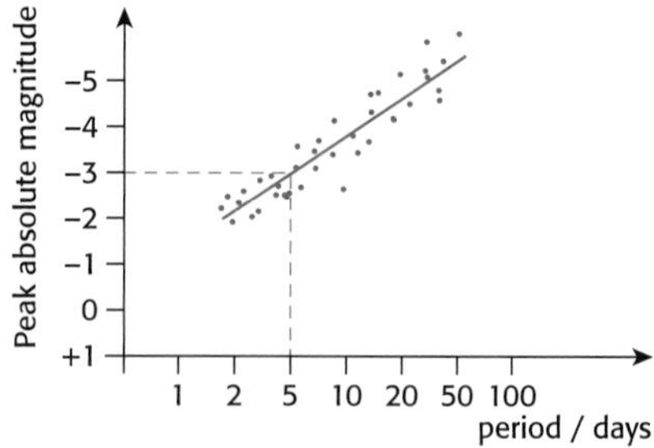

Figure 11 Magnitude–period relationship for Cepheid variable stars

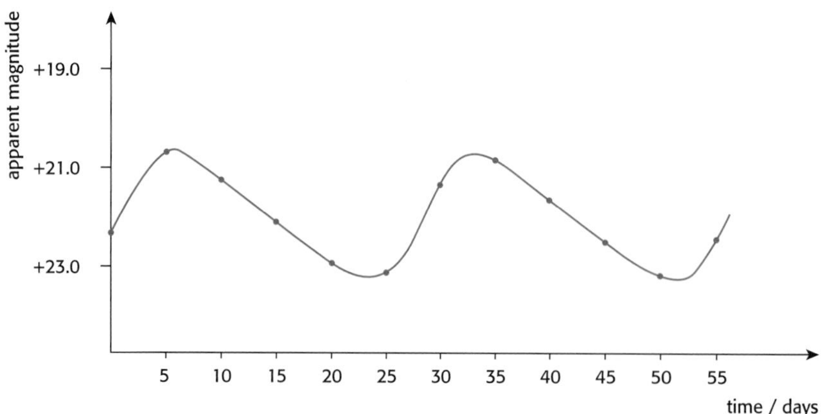

Figure 10 The apparent magnitude of Cepheid variable stars oscillates over a period of time

Figure 10 shows the variation with time of the apparent magnitude of a Cepheid variable star. Estimate using Figure 10:

a) its absolute magnitude and

b) its distance away.

Measurements on Cepheids whose distances are already known have shown that the longer the period of this oscillation in brightness, the greater the absolute magnitude (and thus average luminosity) of the star.

If a Cepheid is identified in a distant galaxy, measurements of its period of brightness oscillation and the general luminosity-period relationship for Cepheids can be used to estimate the luminosity of the star. Measurements of its brightness then allow the distance to the Cepheid (and hence the galaxy) to be calculated.

Stellar processes and stellar evolution

The nuclear fusion taking place in stars involves interactions between atomic nuclei. This can only happen at very high temperatures, so that the strong nuclear force (active at distances of 10^{-15} m or less) can overcome the Coulomb repulsion. The greater the density, the more likely a reaction is to take place. How did the stars become hot enough and dense enough for nuclear fusion reactions to start?

A star begins its life as a large cloud of gas. The mutual gravitational attraction causes the cloud to begin to collapse. The gravitational potential energy is converted into kinetic energy as the material accelerates, and the temperature of the gas cloud increases. If the initial mass of gas is sufficiently large it is possible for the gas cloud to get hot enough for nuclear reactions to start. Once they are underway, an equilibrium is created, with the energy released in the nuclear reactions balancing the energy radiated away from the star.

A main sequence star fuses hydrogen into helium. This is sometimes referred to as **hydrogen burning**, even though burning is a chemical reaction, rather than a nuclear reaction. The overall process can be viewed as the conversion of four protons into a helium nucleus, two positrons, and two neutrinos, as represented by the following equation:

$$4{}^1_1\text{H} \rightarrow {}^4_2\text{He} + 2{}^0_1\text{e}^+ + 2{}^0_0\upsilon$$

When there is not enough hydrogen left in a star's core to maintain the hydrogen burning required to support the star's equilibrium, the core collapses under gravity. This gravitational collapse causes the temperature of the core to rise still further and for most stars, the core is eventually hot enough for helium fusion reactions to take place.

When this process begins, typically helium nuclei are fusing in the core of the star to form beryllium, carbon, and oxygen in a process called the **triple alpha process**. There is still a thin shell of hydrogen fusion reactions around the helium burning core. The energy released by the collapse of the core causes a large expansion of the outer layer of the star, causing its structure to change. The radius greatly increases and most of the star is relatively low in temperature and density. It has become a **red giant** star.

The final fate of the star depends on its mass. Low mass stars, e.g. type G (up to about four times the mass of the Sun), are not massive enough to initiate any further fusion processes after helium burning. In the last stages of the process helium and hydrogen fusion takes place in shells

1 Two possible mechanisms for the fusion of hydrogen into helium are the proton-proton chain (the p-p chain) and (for large mass stars) the carbon-nitrogen-oxygen (CNO) cycle. Research each of these series of reactions.

2 Use the masses provided in Table 4 to calculate the energy released in the production of one helium nucleus.

Table 4

Particle	Mass / kg
proton	1.673×10^{-27}
helium nucleus	6.6465×10^{-27}
positron	9.11×10^{-31}
neutrino	negligible

3 The fusion of four hydrogens into a helium nucleus releases about 3.9×10^{-12} J per helium nucleus. The luminosity of the Sun is 3.8×10^{26} W. Calculate:

a) The number of reactions that are taking place each second,

b) the mass of hydrogen converted into helium each second, and

c) the mass converted into energy by the Sun each second.

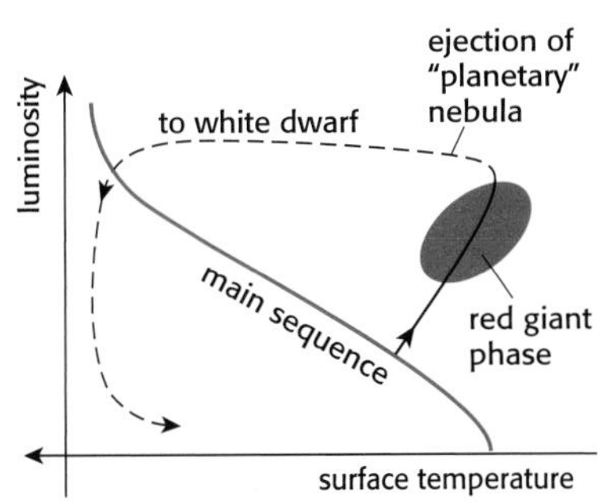

Figure 12 Evolution of a low-mass star

around the contracting core. This causes the outer layers to be ejected as a spherical shell of cooling matter called a **planetary nebula**.

The remnant that is collapsing inward will keep on getting smaller until the matter is so densely packed together that **electron degeneracy pressure** prevents any further increase in density. In this condition it is known as a **white dwarf** star. The matter is extremely hot, extremely dense, and continues to radiate energy as it cools down but no further nuclear reactions are taking place so after billions of years it will cool and stop emitting energy.

The maximum possible mass of a white dwarf star is known as the **Chandrasekhar limit** and is calculated to be 1.4 solar masses. The original star that created this remnant can be up to about 4 solar masses.

Larger mass stars, e.g. type A, form **red supergiants** and are capable of sustaining further nuclear fusions. Carbon fusions can take place to produce neon, magnesium, and oxygen. Other reactions can take place in very large stars and even silicon can be involved in fusion processes and produce iron. Iron has the highest binding energy per nucleon (see page 248) so no more energy production stages are possible after this (Figure 13).

In the final stages of a red supergiant, the core of the star is composed of different layers each fusing different nuclei of increasing mass. Further fusion reactions would take in energy rather than release it and so do not happen inside the star. In the absence of any nuclear reactions occurring to sustain the equilibrium in the star, it collapses inwards.

If the mass of the core is greater than the Chandrasekhar limit, electron degeneracy pressure is insufficient to halt the collapse. Electrons and protons are crushed together to form neutrons and the density continues to rise. Eventually a process called **neutron degeneracy pressure** prevents any further collapse and the core becomes a **neutron star**. Any material falling inwards bounces off the inner core, creating a shockwave that causes the outer layers of the star to be expelled in a huge release of energy. This explosion is called a **supernova**.

Spinning neutron stars have been identified as the cause of **pulsars**. Pulsars are objects that are seen to emit pulses of radio waves with a regular time period (of about one second or less). The model astronomers have for the neutron star hypothesizes the creation of a very strong magnetic field. Protons and electrons on the surface of the neutron star are accelerated towards the magnetic poles and this creates a narrow beam of EM radiation. As the neutron star rotates, this beam of radiation sweeps around with the star. When the beam is pointing towards the Earth we receive the radiation. The period of the pulses then suggests that neutron stars have an extremely high rotational speed.

Even neutron degeneracy pressure is not sufficient to halt the compression of the core due to gravity for very high mass stars. In this case the core is above the Oppenheimer–Volkoff limit for a neutron star of about 2.5 or 3.0 stellar masses. The core of such a star collapses and a black hole is formed. The density of a black hole is so high that even light is unable to escape.

Use the binding energy per nucleon graph (page 248) to explain why red supergiants cannot continue to fuse elements heavier than iron in their cores.

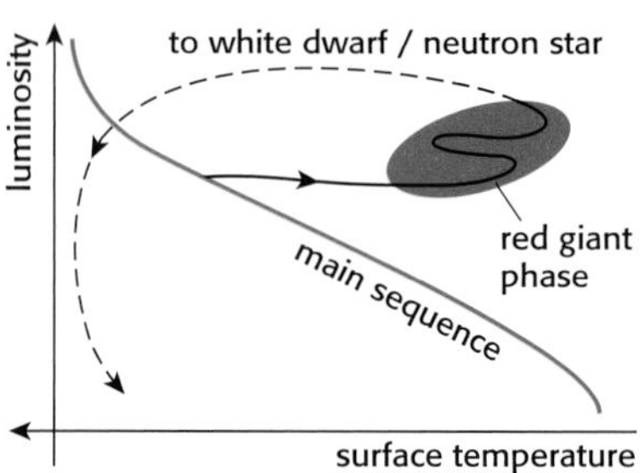

Figure 13 Evolution of a high-mass star

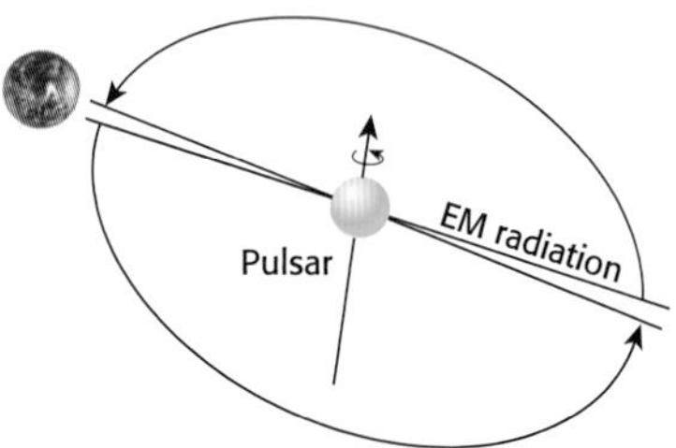

Figure 14 A spinning neutron star is a pulsar

1. Research the properties of a neutron star and pulsars.
2. Research how a physics student, Jocelyn Bell, first discovered pulsars.
3. Research what is meant by the following terms as applied to a black hole:
   - singularity
   - Schwarzchild radius
   - event horizon.

Cosmology

Given that the stars and galaxies that we observe at night do not appear to be moving it seems reasonable to assume, as Newton did, that the universe is both infinite and static.

However, the German astronomer Heinrick Olbers (1758–1840) identified a paradox that forced physicists to reconsider this model of the universe. The paradox states that if Newton's model of the universe were true then the night sky would be bright in all directions. This is not the case – so a new model was needed.

There are three important statements that can be made about the observable universe. Firstly, the universe seems to be made up of the same kind of matter wherever we look; so we say that the universe is **homogeneous**. Secondly, it seems to be more or less the same in whatever direction we consider; the universe is **isotropic**. Finally, the physical laws that we observe to apply in our region of the universe seem to be the same throughout the universe; physical laws are **universal**.

Research how the brightness of stars is expected to vary with distance, and how the number of stars is expected to vary with distance. Use your answers to show that the night sky might be expected to appear uniformly bright.

To resolve Olbers' paradox physicists have hypothesized that the universe has not been around forever. The idea that the universe has a finite age can be linked to the expansion of the universe.

If a star is moving away from us, its light will be red shifted. If the star is moving towards us, it will be blue shifted (see pages 127–8). When light is analysed from different galaxies, the vast majority of galaxies are red shifted and so we conclude that they must be moving away from us. In addition, the speed of recession is generally greater for the more distant galaxies. See page 383 for more details.

It is not correct to imagine the expansion of the universe as matter moving outwards from a point, filling up empty space. A better way is to imagine that space itself is expanding taking the matter with it. A good analogy is to imagine space as a sheet of rubber being pulled in all directions.

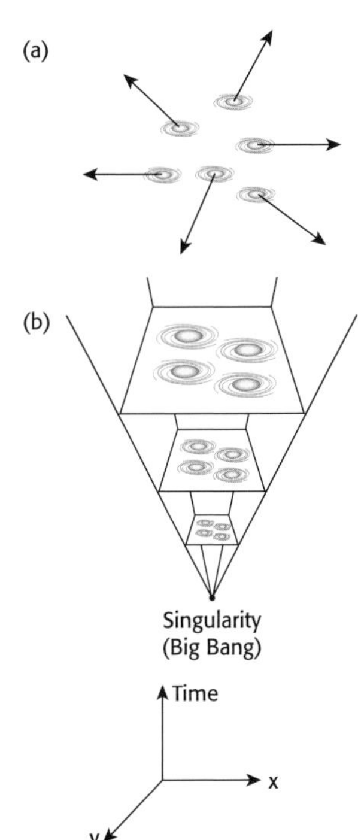

Figure 15 The expansion of the universe. **a)** Physicists believe that the model of matter expanding in to empty space is incorrect. **b)** Space itself is expanding from a singularity in this model, which is believed to be correct.

If the galaxies are all moving apart from one another (receding), they must have been closer together at some point in the past. The observations are consistent with a **Big Bang**. This theory says that all the observable matter in the universe was concentrated into one single point in space approximately 15 000 million years ago. In the Big Bang theory, space and time were created simultaneously at a single point, a singularity, of infinitely high temperature and density. After this creation, the matter expanded outwards and the matter has cooled down and eventually formed the universe we observe today.

Olbers' analysis assumes that no matter how distant a star is from Earth, its light will contribute to the brightness of the night sky but if the universe is only 15 000 million years old then we will only see light from stars less than 15 000 million light years away. This is the solution to Olbers' paradox.

Einstein was able to show that space and time should not be considered as separate dimensions but rather considered together as an entity called the **space-time continuum**. The creation of space at the Big Bang also means the creation of time. From this viewpoint it does not make sense to ask what existed before the Big Bang – there is no such thing as "before".

The universe is expanding, but we do not know if this expansion will continue on forever. The rate of expansion will slow down due to gravitational attraction amongst the matter in the observable universe, but we do not know for certain if there is enough matter to eventually halt this expansion.

If the average density of the universe is sufficiently high, then the force of gravity will slow down the expansion to zero. The universe must then start to contract and all the matter will come back together (sometimes referred to as the **big crunch**). This is known as a **closed** universe.

If the average density of the universe is not sufficiently high enough for a closed universe, then, although the force of gravity will continue to slow down the expansion, it will never stop. The universe will continue to expand forever and this model is known as an **open** universe.

If the average density of the universe is at a certain value, **the critical density**, then the force of gravity will continue to slow down the rate of expansion but will take an infinite length of time to get to zero. This is known as a **flat** universe. The critical density works out to be approximately 1.9×10^{-26} kg m^{-3} or a few hydrogen atoms in every m^3.

Although current scientific evidence suggests that the universe is open, an accurate determination of the average density is extremely difficult. Observations of stars and galaxies can only identify matter that gives off light but the vast majority of the universe is already known to be **dark matter**, that is, matter which cannot be observed directly.

An estimation can be made of all matter present in a galaxy by estimating the number of stars it contains. This calculation is always several orders of magnitude smaller than the amount of mass calculated from its speed of rotation. Estimates that include all the known sources of mass (planets, asteroids, and so on) still fall short of explaining the matter that must be present in a galaxy. At the time of writing most of the mass of the universe is currently unexplained.

Another piece of evidence supporting the Big Bang model of creation is the discovery of cosmic microwave background (CMB) radiation by Penzias and Wilson. Research their discovery and explain how it is consistent with the Big Bang model. Your answer should compare the spectrum of microwave radiation with the expected characteristics of black body radiation and explain the significance of the CMB temperature of 3 K.

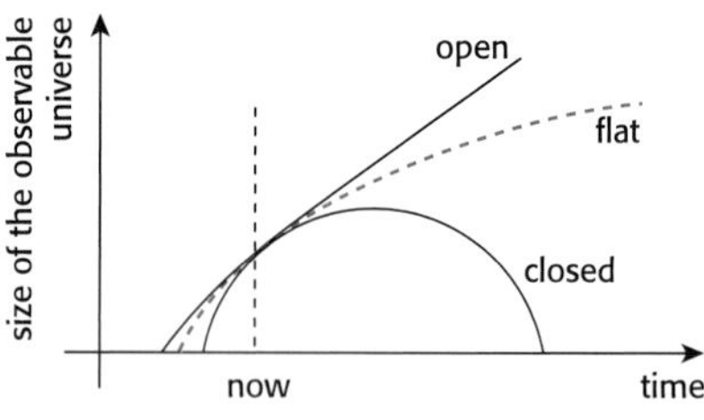

Figure 16 Different models for the development of the universe

1. Research current possible theories that attempt to explain why so much dark matter exists. Possible explanations include MACHOs and WIMPs; find out the meaning of these terms.
2. Much of the current fundamental research that is being undertaken in astrophysics involves close international collaboration and the sharing of resources. Research one current example and present a summary of your findings. If possible you should try to include all of the following information:
   - the title of the project
   - an outline of the experimental procedures being followed
   - the countries that are involved

- the timescale of the project
- the expected outcomes/benefits of the project.

3 All countries have a limited budget available for the scientific research that that country undertakes. Some would argue that research into the nature of the universe is one of the most fundamental, interesting, and important areas for mankind as a whole and it therefore deserves to be properly resourced. Others would argue that it is much more worthwhile to invest limited resources into medical research. This offers the immediate possibility of saving lives and improving the quality of life for some sufferers. List as many specific arguments as you can for and against investing significant resources into researching the nature of the universe. What is your view?

Galaxies and the expanding universe

The observed redshift for different galaxies depends on their distance from Earth. In general, the further away a galaxy is, the greater its redshift and hence the greater its recessional velocity.

Hubble's law states that the recession velocity v for a galaxy is proportional to its distance away r.

$$v \propto r \qquad \text{or} \qquad v = H_0 r \qquad (5)$$

where
v is the recessional velocity of the galaxy, measured in km s^{-1}
r is the distance to the galaxy, measured in Mpc
H_0 is the Hubble constant, measured in km s^{-1} Mpc^{-1}

The value of the Hubble constant has a great deal of uncertainty associated with it, as the measurements we use in order to calculate it (distance and recessional velocity) suffer from a great deal of uncertainty.

A galaxy at a distance x then will have a recessional velocity of $H_0 x$. If it has been travelling at a constant speed since the beginning of the universe (which is clearly not true), then the time taken will be:

$$\text{time} = \frac{\text{distance}}{\text{speed}} = \frac{x}{H_0 x} = \frac{1}{H_0}$$

The Big Bang model for the development of the universe imagines space expanding from a very hot, dense beginning to the present day. At the very high temperatures of the early universe (say up to 10^{-5} seconds after the Big Bang – the quark-lepton era), the universe was so hot that matter could not exist in the form that we see it today. Only elementary (fundamental) particles could exist. As the universe expanded it cooled, which allowed the formation of protons and neutrons to take place (from 10^{-5} to 10^{-2} seconds after the Big Bang – the hadron era).

The universe would have cooled sufficiently for helium nuclei to be formed from 10^{-2} to 10^{2} s after the Big Bang but for a significant time after this (10^{2} s to 300 000 years) the temperature of the universe was still so high that individual atoms could not exist. For these 300 000 years, all matter was in the form of a plasma with electrons, protons, neutrons, helium nuclei, and photons all interacting

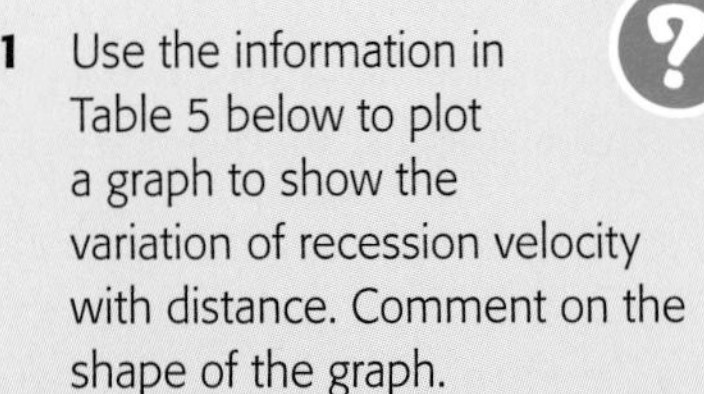

1 Use the information in Table 5 below to plot a graph to show the variation of recession velocity with distance. Comment on the shape of the graph.

Table 5

Name	Distance/ Mpc	Recessional velocity/ km s-1
NGC 55	1.99	129
NGC 157	21.5	1652
NGC 185	0.61	−202
SMC	0.06	158
Sculptor	0.08	243
M74	7.36	657
NGC 720	27.6	1745
NGC 891	9.81	528
NGC 908	18.4	1509
NGC 936	23	1430
LMC	0.05	278
UGC 4305	3.68	142
M82	3.68	203
LEO I	0.25	285
whirlpool	7.1	463
M 107	12.6	1048
NGC 1433	12.3	1075
NGC 2841	12.3	638
NGC 3198	14.4	663
M95	10.1	778
M96	10.1	897

2 Use the graph that you drew for question 1 to calculate a value for the Hubble constant.

3 Estimate the age of the universe from a value for the Hubble constant of 75 km s^{-1} Mpc^{-1}.

4 Explain why this must be an upper limit for the age of the universe.

5 Identify where the age of the universe as calculated from appears on Figure 16 on page 382, which shows the possible future development of the universe.

(the plasma/radiation era). Finally at about 300 000 years after the Big Bang, the universe was cool enough for the nuclei to capture electrons and for atoms to form.

At this point the universe was essentially 75% hydrogen and 25% helium. The elements that are more massive than helium were created when the first stars started to shine approximately 1000 million years after the Big Bang. A human being contains carbon, oxygen, and many other elements that are more massive than hydrogen and helium. Most of the atoms in your body must have been created as a result of fusion that took place in a star that shone many millions of years ago.

End of chapter summary

Chapter 27 has six main themes, two of which are only relevant to those studying higher level physics. The four themes common to all levels are: an introduction to the universe, stellar radiation and stellar types, stellar distances and cosmology. The two HL themes are: stellar processes and evolution, and galaxies and the expanding universe. The list below summarises the knowledge and skills that you should be able to undertake after having studied this chapter. Further research into more detailed explanations using other resources and/or further practice at solving problems in all these topics is recommended – particularly the items in bold.

Introduction to the universe (solar system and beyond)

- Outline the general structure of the solar system distinguishing between a stellar cluster and a constellation.
- Define the *light year* **and use it to compare the relative distances between stars within a galaxy and between galaxies, in terms of order of magnitude**.
- Describe the apparent motion of the stars/constellations over a period of a night and over a period of a year, and explain these observations in terms of the rotation and revolution of the Earth.

Stellar radiation and stellar types (energy source, luminosity, Wien's law and the Stefan-Boltzmann law, stellar spectra, types of star, the Hertzsprung-Russell diagram)

- State that fusion is the main energy source of stars and explain that, in a stable star, there is an equilibrium between radiation pressure and gravitational pressure.
- Define the *luminosity* and *apparent brightness* and state how they are measured.
- State and apply the Stefan – Boltzmann law and Wien's (displacement) law to different stars.
- Explain how atomic spectra may be used to deduce chemical and physical data for stars and describe the overall classification system of spectral classes for different types of star identifying the general regions of star types on a Hertzsprung – Russell (HR) diagram.
- Discuss the characteristics of spectroscopic and eclipsing binary stars.

Stellar distances (parallax method, absolute and apparent magnitudes, spectroscopic parallax, Cepheid variables)

- Define the parsec and describe the stellar parallax method of determining the distance to a star and use it to solve problems, explaining why this method is limited to measuring stellar distances less than several hundred parsecs.
- Define the apparent magnitude scale and absolute magnitude **and solve problems involving distance and apparent brightness**.
- State that the luminosity of a star may be estimated from its spectrum and explain how stellar distance may be determined using apparent brightness and luminosity explaining why the method of spectroscopic parallax is limited to measuring stellar distances less than about 10 Mpc.
- **Solve problems involving stellar distances, apparent brightness and luminosity.**

- Outline the nature of a Cepheid variable and state the relationship between period and absolute magnitude that it has and how this relationship allow them to be used as *standard candles.*
- **Determine the distance to a Cepheid variable using the luminosity – period relationship.**

Cosmology (Olbers' paradox, the Big Bang model, the development of the universe)

- Describe Newton's model of the universe and explain how this model leads to Olbers' paradox.
- Suggest that the red-shift of light from galaxies indicates that the universe is expanding and how both space and time originated with the Big Bang.
- Describe the discovery of cosmic microwave background radiation by Penzias and Wilson and explain this discovery is consistent with the Big Bang model (which provides a resolution to Olbers' paradox).
- Define the terms *open, flat* and *closed* when used to describe the development of the universe and discuss how the density of the universe (and the *critical density*) determines its development.
- **Discuss problems associated with determining the density of the universe** but state that current scientific evidence suggests that the universe is open.
- **Discuss an example of the international nature of recent astrophysics research and evaluate arguments related to investing significant resources into researching the nature of the universe.**

In addition HL candidates should be able to:

Stellar processes and stellar evolution (nucleosynthesis, evolutionary paths of stars and stellar processes)

- Describe the conditions that initiate fusion in a star and state the effect of a star's mass on the end product of nuclear fusion.
- Outline the changes that take place in nucleosynthesis when a star leaves the main sequence and becomes a red giant.
- Apply the mass – luminosity relation for main sequence stars and explain how the Chandrasekhar and Oppenheimer – Volkoff limits are used to predict the fate of stars of different masses in particular comparing the fate of a red giant and a red supergiant by drawing evolutionary paths of stars on an HR diagram.
- Outline the characteristics of pulsars.

Galaxies and the expanding universe (Galactic motion, Hubble's law).

- Describe the distribution of galaxies in the universe, explain the redshift of light from distant galaxies **and solve problems involving red-shift and the recession speed of galaxies**.
- State Hubble's law, discuss its limitations **and explain how the Hubble constant may be determined**.
- **Solve problems involving Hubble's law** including an explanation of how the Hubble constant may be used to estimate the age of the universe.
- Explain how the expansion of the universe made possible the formation of light nuclei and atoms.

Chapter 27 questions

1 This question is about the nature of certain stars on the Hertzsprung-Russell diagram and determining stellar distance.

The grid below is a Hertzsprung-Russell (H-R) diagram on which the positions of the Sun and four other stars A, B, C and D are shown.

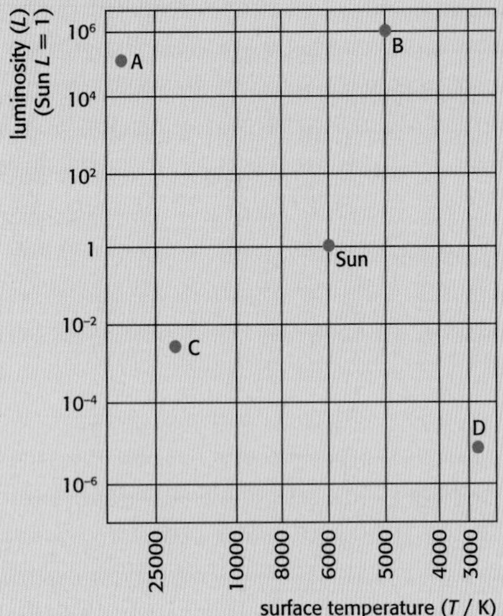

a) State an alternative unit for axes,

i) *x*-axis [1]

ii) *y*-axis [1]

b) Complete Table 6.

Table 6

Star	Type of star
A	
B	
C	
D	

[4]

c) Explain, using information from the H-R diagram, and without making any calculations, how astronomers can deduce that star **B** is larger than star **A**. [3]

d) Using the following data and information from the H-R diagram, show that star **B** is at a distance of about 700 pc from Earth.

Apparent visual brightness of the Sun $= 1.4 \times 10^3$ Wm^{-2}

Apparent visual brightness of star **B** $= 7.0 \times 10^{-8}$ Wm^{-2}

Mean distance of the Sun from Earth $= 1.0$ AU

1 parsec $= 2.1 \times 10^5$ AU [4]

e) Explain why the distance of star **B** from Earth cannot be determined by the method of stellar parallax. [1]

(Total 14 marks)

2 This question is about the possible evolution of the universe.

The sketch graph below shows three possible ways in which the size of the universe might change with time.

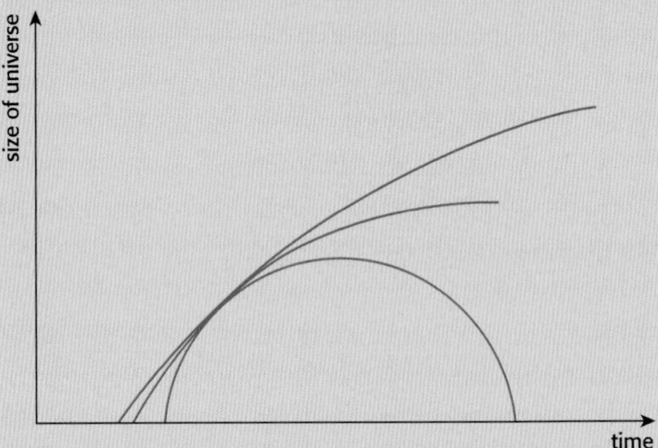

Depending on which way the size of the universe changes with time, the universe is referred to either being open or *flat* or *closed*.

a) Identify each type of universe on the graph above.

b) Complete the table below to show how the mean density ρ of each type of universe is related to the critical density ρ_0.

Table 7

Type of universe	Relation between ρ and ρ_0
Open	
Flat	
Closed	

[3]

(Total 6 marks)

3 This question is about some of the properties of Barnard's star.

Barnard's star, in the constellation Ophiuchus, has a *parallax angle* of 0.549 arc-second as measured from Earth.

a) With the aid of a suitable diagram, explain what is meant by *parallax angle* and outline how it is measured. [6]

b) Deduce that the distance of Barnard's star from the Sun is 5.94 ly. [2]

c) The ratio $\frac{\text{apparent brightness of Barnard's star}}{\text{apparent brightness of the sun}}$ **is** 2.6×10^{-14}.

i) Define the term *apparent brightness*. [2]

ii) Determine the value of the ratio

$$\frac{\text{luminosity of Barnard's star}}{\text{luminosity of the sun}}$$

($1 \text{ ly} = 6.3 \times 10^{4}$ AU). [4]

d) The surface temperature of Barnard's star is about 3 500 K. Using this information and information about its luminosity, explain why Barnard's star cannot be

i) a white dwarf [1]

ii) a red giant [1]

(Total 16 marks)

4 This question is about Cepheid variables.

a) Define

i) *luminosity.*

ii) *apparent brightness.* [1]

b) State the mechanism for the variation in the luminosity of the Cepheid variable. [1]

The variation with time t, of the apparent brightness b, of a Cepheid variable is shown here.

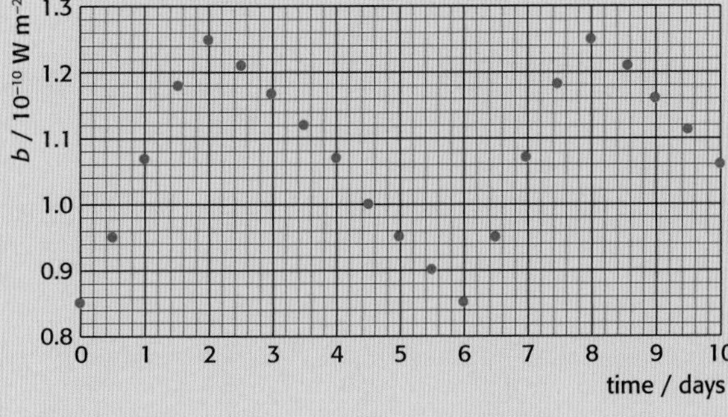

c) i) Assuming that the surface temperature of the star stays constant, deduce whether the star has a larger radius after two days or after six days. [2]

ii) Explain the importance of Cepheid variables for estimating distances to galaxies. [3]

d) i) The maximum luminosity of this Cepheid variable is 7.2×10^{29} W. Use data from the graph to determine the distance of the Cepheid variable. [3]

ii) Cepheids are sometimes referred to as "standard candles". Explain what is meant by this. [3]

(Total 13 marks)

5 This question is about the Big Bang model.

a) Describe what is meant by *cosmic background radiation.* [2]

b) Explain how cosmic background radiation is evidence in support of the Big Bang model of the universe. [1]

c) State **one** other piece of evidence in support of the Big Bang model. [1]

d) A student makes the statement that *"as a result of the Big Bang, the universe is expanding into a vacuum"*. Discuss whether the student's statement is correct. [2]

(Total 7 marks)

6 This question is about various bodies in the universe.

a) Briefly describe the nature of a star. [2]

b) Distinguish between a constellation and a galaxy. [4]

(Total 6 marks)

7 This question is about white dwarfs and neutron stars.

a) State the property that determines whether a star ends its life as a white dwarf **or** as a neutron star. [1]

b) Define the *Chandrasekhar limit* and use this concept to explain the difference between a white dwarf and neutron star. [3]

c) State the name given to a rotating neutron star. [1]

(Total 5 marks)

8 This question is about galactic redshift, the Hubble constant and the age of the universe.

a) State how the observed redshift of light from many distant galaxies is explained. [1]

b) Sketch a graph to show how the recessional speed v between galaxies varies with the distance d between them. (*Please note this is a sketch graph; you do not need to add any numerical values.*)

c) State how the Hubble constant is determined from such a graph. [1]

d) A possible value for the Hubble constant is 100 $\text{kms}^{-1}\ \text{Mpc}^{-1}$. Use this value to estimate the age of the universe in years. (1 Mpc $\approx 3 \times 10^{19}$ km, 1 year $\approx 3 \times 10^{7}$ s) [2]

(Total 5 marks)

9 This question is about the evolution of stars.

a) Outline the process that provides the source of energy for stars while on the main sequence. [2]

b) State the conditions required for the above process to take place. [1]

c) State the reason why stars leave the main sequence. [1]

d) Main sequence stars eventually evolve to form red giants. With reference to the Chandrasekhar limit, describe and distinguish between the ***subsequent*** evolutionary paths of ***red giant*** stars that have evolved from main sequence stars of mass

i) about two times that of the Sun [3]

ii) about ten times that of the Sun [3]

(Total 10 marks)

28 Option F: Communications

We are all part of the digital revolution – a push for ever more efficient means of transmitting information. The 20th century saw improvements in the use of analogue signals, and in our ability to send these over large distances using cables and radio waves. With the advent of digitization we are now looking to develop better ways of getting such messages from one part of the world to another, whether this is over networks of cables and optical fibres, being carried by electromagnetic waves within our atmosphere, or by "bouncing" these signals off satellites in orbit around our planet.

Modern-day communications technology is a result of advances necessitated by the growth of civilization. Such developments have variously been driven by a historical desire to cooperate with, trade with, fight, and entertain each other better. This chapter aims to introduce you to some of the physics involved in communication.

Many things need to be considered in the building of an effective communications system. These include, but are not limited, to:

- **Speed**: How fast can you get your information from A to B?
- **Distance**: How far apart can A and B be?
- **Quality**: How similar is the received message to the original information that was sent?
- **Security**: In a world where personal privacy has become so important, how easy is it for others to access your information?
- **Mobility**: How important it is to be able to move during communications?
- **Cost**: Whether we are talking about hardware, about software, or in human terms, there are many economic and commercial demands that must be met.

It is impossible to optimize all these factors. We often have to balance demands in one area against needs in another. Before starting this chapter it would be a good idea to review the difference between analogue and digital signals (see pages 344–5).

?

Construct a timeline that summarizes some of the important steps in the history of communications. We suggest that you begin around 1800 (after electricity came under people's control), though going back further may allow you to appreciate the global nature of communications development. Along with the achievements of such well-known people as Edison, Marconi, and Bell, you should aim to include other milestones, including the invention of the first cathode ray tube, the first transistor, and the first magnetic recordings, as well as identifying when the first television signal was transmitted, what ARPAnet was, and when exactly the World Wide Web was legally born. Identify where a significant improvement was made to any of the factors listed above that contribute to a good communications system.

Channels of communication

A channel of communication is essentially any way of physically getting your message, coded or not, from one place to another. There have been some novel channels of communication developed over the ages.

Modern-day communications systems use one of three basic channels in order to transport data:

- electromagnetic waves including radio waves
- electrically conducting wires
- optical fibre.

Radio communication

One of the key trends in communication is mobility. Of course, the idea of being able to connect without the need for wires and cables is not new. The need to send messages to places where the use of a permanent channel is impossible has been with us for many centuries, made necessary by humankind's relentless exploration of the planet we inhabit. Military demands have often pushed communication technology, and the advent of naval warfare and – more recently – the use of airplanes demonstrated the importance of electromagnetic waves as a means for carrying information. Visual signals – flags, lamps and so on – have been superseded by the application of radio technology.

Modulation

The block diagram in Figure 1 shows the basic sequence of events in an amplitude-modulated radio transmission.

The key to all this is the concept of **modulation**, which is a way of combining two electrical waves. One wave is the information that you wish to send – the signal; the other is the **carrier** wave. The key feature of the carrier is its frequency, which essentially does not change. This means that it is very easy to detect the carrier and "tune" in to it to facilitate retrieval of the information. The combination process takes place in a device called a **modulator**. At the other end after receipt of the transmission the information must be stripped from the carrier using a **demodulator**.

There are two forms of modulation. Both involve combining a signal at relatively low frequencies – what are termed **audio frequencies**, a.f. (human speech patterns vary between about 20 Hz and 20 000Hz) – with a significantly higher-frequency carrier oscillating at what is termed a **radio frequency** or r.f.

The frequencies used for carrier purposes are determined by the practicality of transmission and reception aerial size. Table 1 indicates how radio waves are classified.

Table 1 Frequencies and wavelengths for the classification of radio waves

Classification	Frequency range	Wavelength range
VLF (very low)	3–30 kHz	100–10 km
LF (low)	30–300 kHz	10–1 km
MF (medium)	300–3000 kHz	1000–100 m
HF (high)	3–30 MHz	100–10 m
VHF (very high)	30–300 MHz	10–1 m
UHF (ultrahigh)	300–3000 MHz	100–10 cm
SHF (superhigh)	3–30 GHz	10–1 cm
EHF (extremely high)	30–300 GHz	10–1 mm

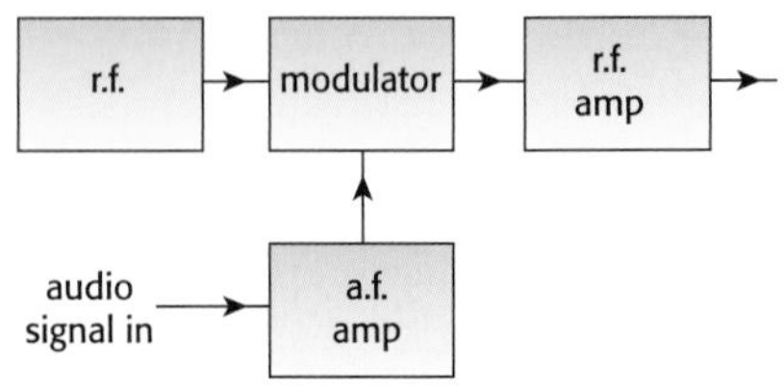

Figure 1 Sequence of events in amplitude-modulated radio transmission

Amplitude modulation (AM) is the more simple process. The amplitude of the carrier is altered depending upon the amplitude of the signal at different instances in time. If the signal amplitude is positive, then the amplitude of the carrier is increased; if the signal amplitude is negative then the carrier amplitude is decreased (Figure 2).

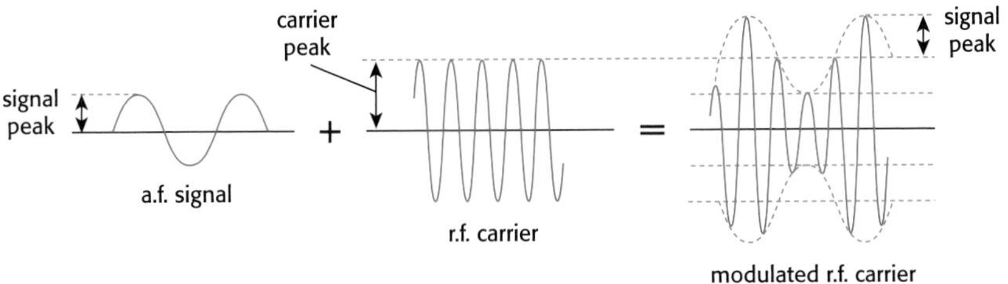

Figure 2 Amplitude modulation

The process is perfectly acceptable but does have a lot of problems in terms of quality being reduced due to the susceptibility of the transmission to noise. Also due to the need to increase **bandwidth** (covered later), the full range of audio frequencies is often not transmitted, so this limits the accuracy of the received information.

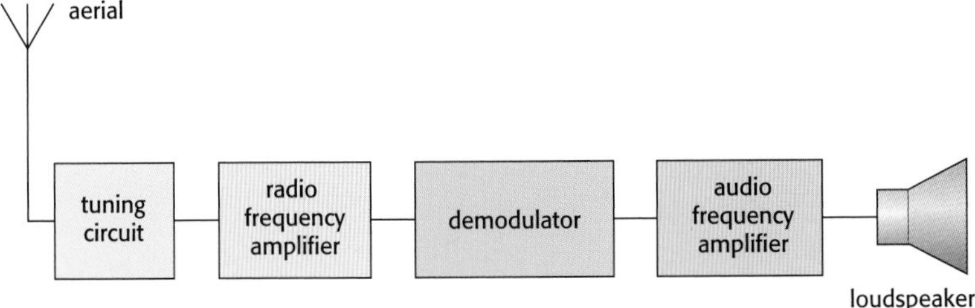

Figure 3 Block diagram of radio receiver

Frequency modulation (FM) involves altering the frequency of the carrier. When the a.f. signal amplitude is positive then the carrier frequency experiences a small proportional increase. When the a.f. amplitude goes negative, then the r.f. frequency is decreased (Figure 4).

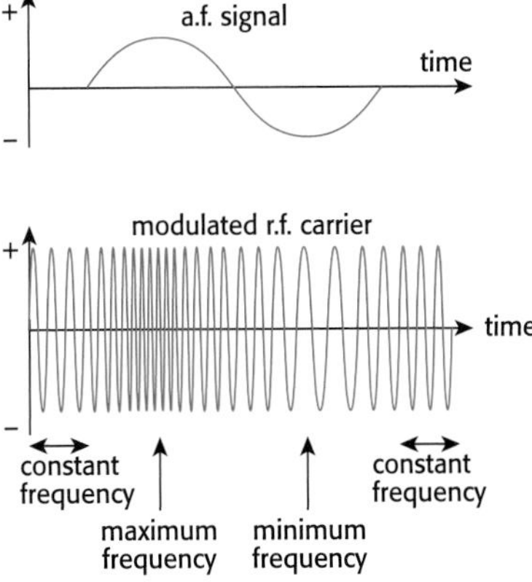

Figure 4 Frequency modulation

Tuning in

How does a receiver actually pick up the carrier wave? A full understanding of this requires some knowledge of how aerials work but mainly an appreciation of what are known as **tuned circuits**. The basic principle is that an **RC circuit** (R = resistance, C = capacitance) can cause electric charge to oscillate. Anything that is able to oscillate will, of course, have a natural or **resonant frequency** (see page 95).

By changing the values of either the resistance or the capacitance (usually the latter) in an RC circuit it is possible to select a particular resonant frequency. All oscillations of every frequency that an aerial is affected by will move the charges in the RC circuit, but only a very narrow range of frequencies will be truly effective in doing this. In Figure 5 anything below f_1 or above f_2 is not going to have much driving effect on the circuit.

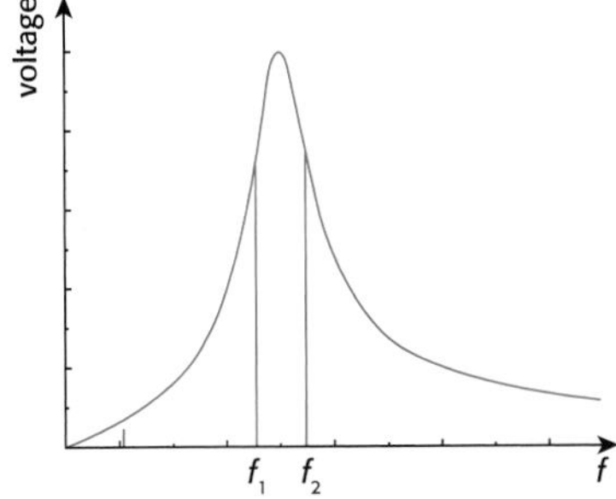

Figure 5 Resonance response of an RC circuit

Bandwidth

Certain queries should arise from all of this. For example, with frequency modulation, how much can the carrier frequency be allowed to change before we lose information because of poor tuning? Does the amount of data being transferred affect the modulation or tuning in any way?

Analogue signals, and the way the term "bandwidth" applies to them, may come across as being slightly older technology, but they are vital now as we move towards greater mobility with our communications.

AM bandwidth works like this. Because the r.f. carrier has been modulated, it is no longer sinusoidal in nature. From work covered in the waves units (see *The bigger picture: Fourier analysis* on pages 91–2) you will know that a wave such as that shown in Figure 6 can be made by superimposing several waves of different frequencies. The range of frequencies is called the **bandwidth**.

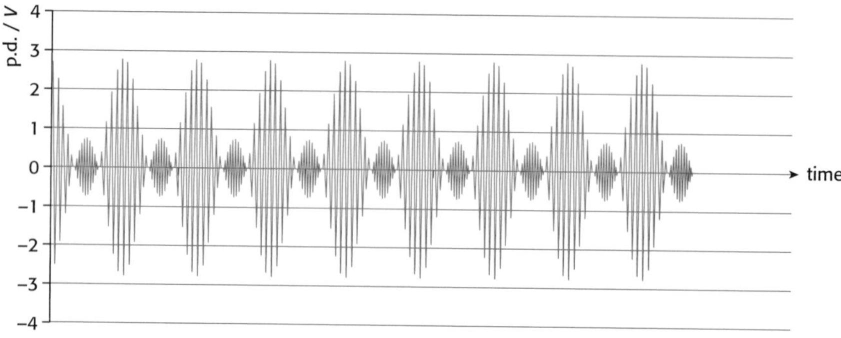

Figure 6 Complex a. t. signal involving a range of frequencies

It is quite easy to ascertain what is often called **digital bandwidth** but may more accurately be referred to as digital capacity or **bit rate**. This is just a direct measure of how many bits per second can be transferred, and is probably best understood in relation to internet connection "speeds" – a broadband connection typically offers nothing less than 100 megabits per second, for example (see page 395).

Power spectrums

It can be shown that, for an r.f. carrier of frequency f_0 being modulated by a single-frequency signal of frequency f', the overall signal can be split into three sinusoidal waveforms of the same amplitude with frequencies f_0, $f_0 + f'$ (upper side) and $f_0 - f'$ (lower side) (see Figure 7a). The difference between the upper side and lower side frequencies is the **bandwidth**.

For a normal signal that varies in frequency, as long as you know the range of a.f. frequencies, you can predict what are termed **sidebands**. These are shown in Figure 7b, which is a **power spectrum** diagram.

If we wish to transmit digital information by radio, then we need as high a signal frequency as possible. This means that a large r.f. bandwidth also becomes big. CD-quality digitally transmitted sound requires a sampling (and therefore signal) frequency of 44.1 kHz, and therefore a bandwidth of around 88 kHz. Radio stations broadcasting on AM wavelengths are limited to a bandwidth of around 9 kHz by agreement so sound quality is limited.

FM bandwidth is different. In this case the r.f. carrier has its frequency altered in direct proportion to the **frequency** of the signal. A normal a.f. signal will produce many frequencies, which suggests that the bandwidth for a normal FM transmission would need to be extremely wide. However, if we calculate all the component frequencies we see

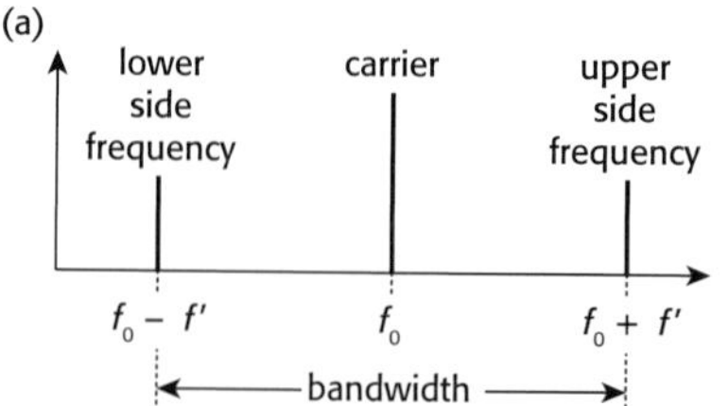

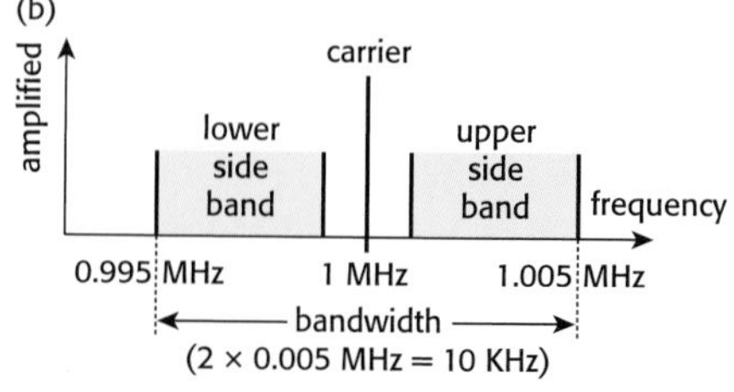

Figure 7 Power spectrum for AM **a)** single frequency and **b)** carrier wave of 1 MHz with audio frequencies (up to 5kHg)

a trend to lower amplitudes the greater the amount by which the component signal frequency differs from that of the carrier.

In practice, FM bandwidth is larger than needed for AM. By international agreement FM deviation must be no more than ±75 kHz, which provides plenty of bandwidth to deal with digital audio signals.

Bigger picture: Demodulator

At this level communications gets treated with a "black box" approach. A detailed understanding of the exact electronics or circuitry involved in doing a particular job is not normally considered necessary. This short section describes the basics of how a demodulator works, to show how some of the physics described in earlier sections "comes to life".

Consider the circuit in Figure 8(a).

As we mentioned on page 390, a demodulator is a circuit that is used to recover information from the carrier wave of a signal.

The diode rectifies the modulated carrier signal (essentially cutting out anything with negative potential). If the voltage through the diode is higher than the voltage stored on the capacitor, then the capacitor will charge up. At the same time current will flow through the resistor. When the diode is blocking reverse current in the cycle the capacitor discharges through the resistor. The resulting waveform is the original signal. The second capacitor simply prevents any dc passing on from this stage. This entire process is demodulation.

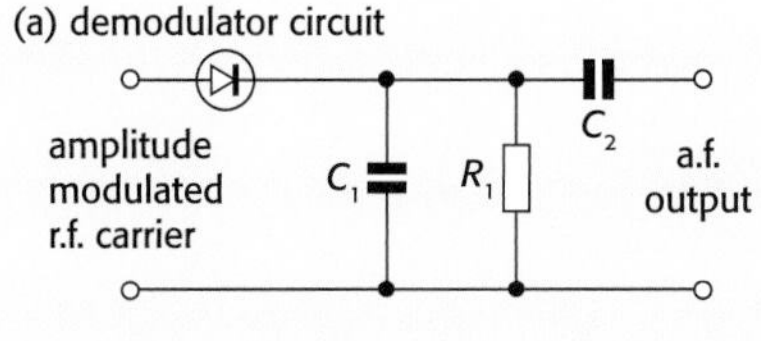

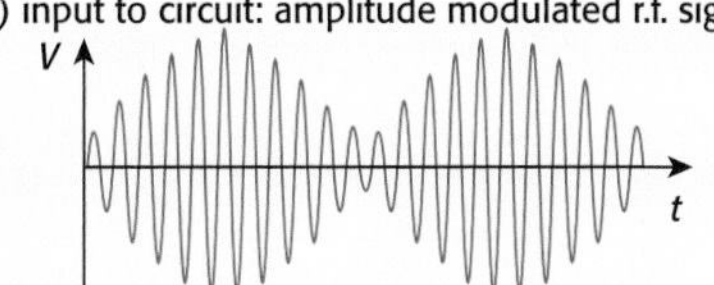

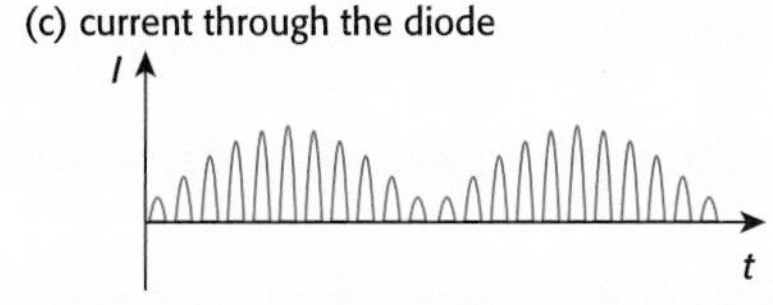

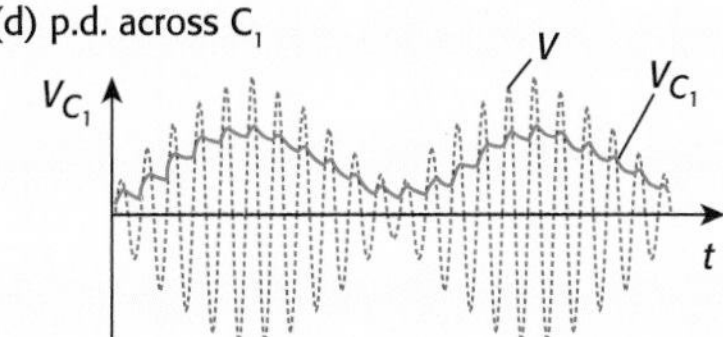

Figure 8 Demodulation

Digital signals

In most communications systems a message has to be **encoded**. A code is a way of transferring a message in a format that enables successful receipt at the other end. Some codes are used for preservation of privacy, but this is only one aspect of them. Encoding a signal often makes it easier to send it from one place to another, or assists in making communication more effective in another way.

As discussed on pages 344–5 digital signals involve just two values – 0 (OFF) and 1 (ON). Digital signals can be more accurately conveyed in terms of avoiding **noise**. However, we have to accept a certain loss of information; what we receive will never be an exact replica of the original. Additional advantages of digital over analogue include the ease with which such information can be stored, and the ease with which signals can be transmitted and received. One of the biggest issues concerning modern communication is how we balance the quality of a transmitted digital signal against other factors such as speed and security.

Block diagrams

Define or explain the following terms: digital, analogue, binary, bit, and byte.

When discussing how any complicated system works, it is often easiest to use a block diagram approach and break the whole system down into its various stages. To do this with communications systems we need to refer to certain common devices. Examples of these include:

- **encoder**: converts the information you wish to send into a useful form
- **decoder**: extracts the information that has been sent
- **transmitter**: sends the communication signal
- **receiver**: receives the communication signal
- **input transducer**: converts the information to be sent into a form that can be processed (usually electrical)
- **output transducer**: converts the received signal back into a usable form
- **amplifiers**: designed to boost the strength of either signal or carrier
- **storage devices**: store information for use at a later time, e.g. disk drives, magnetic tape, CD, DVD.

Figure 9 shows a typical radio communications block diagram:

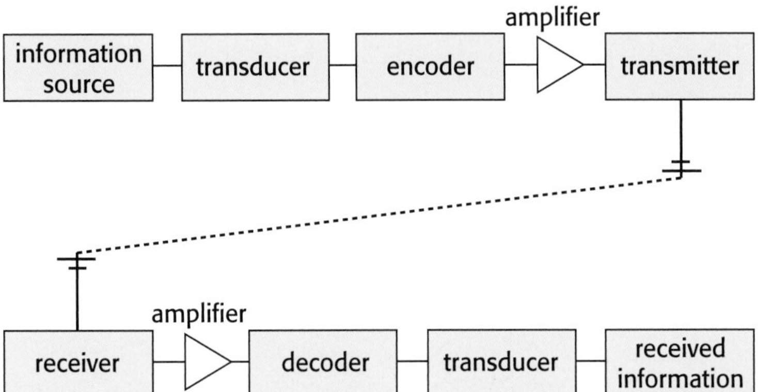

Figure 9 Block diagram of a typical communications system. This example uses radio waves, but the link could be by wire in a telephone system, for example.

A digital communications system includes some devices additional to those already mentioned. They are listed below.

- **Sample-and-hold**: some way of measuring an input analogue voltage level and storing this temporarily
- **Clock**: controls the speed at which the system can process information
- **Analogue-to-digital converter (ADC)**: changes an analogue voltage to a digital code
- **Shift register (parallel to serial converter)**: takes many digital bits of information that are produced at once and puts them in a logical order ready for transmission
- **Shift register (serial to parallel converter)**: takes a series of data bits that come in one after another and then sends them all out at once down separate wires
- **Digital-to-analogue converter**: changes a digital signal into an analogue voltage.

a) Identify the system blocks used in the following circuit which takes an analogue input signal and converts it into a serial digital output signal.

b) Construct a block diagram for the second half of a digital communications system based on the circuit above.

c) Construct a block diagram for a digital communications system using these listed devices:

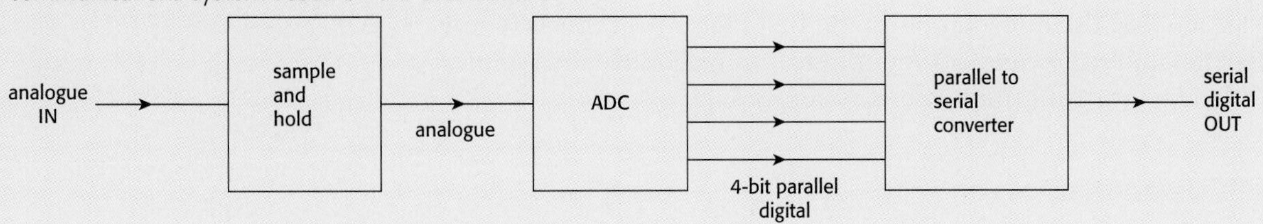

Bits and bit rate

"640k ought to be enough for anybody." Bill Gates, 1981

Generally speaking, the more detail you need to send in a digital signal, the more bits you need to represent each part of the signal; but the more bits you have, the longer it will take to transmit the information. Another important factor is the channel of communication that you use to get your message across. These issues all involve the concept of bandwidth. In the digital world bandwidth means how many bits of information you can transmit every second. This is also called the bit rate.

- **Written text**: 8-bit binary is more than enough to cover the necessary range of letters, numbers, and symbols commonly found on a standard keyboard. This allows us 2^8 or 256 different values. Each symbol can be provided with a unique code in order to identify it. If all the text on this page were sent without any formatting in an email, then there would be roughly 3000 characters $\times$ 8 bits per character—in other words, 24 000 bits of information (or 3 kB).
- **Pictures**: Even with *compression* technology (the ability to save digital information at a lower memory than initially required – this is principally done using mathematical techniques) a good print-quality digital photograph is unlikely to be less than 2 MB or $2 \times 10^6 \times 8 = 1.6 \times 10^7$ bits of digital information.
- **Audio**: A standard 60-minute CD typically contains around 700 MB of digital information. This is considered to be high quality. Many audio formats, such as MP3, will be less detailed, and therefore require fewer bits.
- **Video**: A standard European TV screen refreshes 25 times per second, with a screen resolution in the region of 720 $\times$ 486 pixels, each of which requires information on colour as well as brightness. Add to this the desire for a cinema-quality surround sound audio track, and it starts to become clear why, at the time of writing, this level of communication is still being experimented with in terms of mass media availability.

There are, of course, many other things to consider. There is an obvious trade-off in terms of quality. An MP3 recording cannot compete with an original CD in terms of accuracy to the original sound, but users are often happy with this, because it means they can carry more music around with them. You do not need to send *all* the information in a video stream 25 times every second, as much of the detail in any one frame is often similar to that in the previous one.

There are many tricks that can be employed to keep the amount of data down. However, that does not stop us from searching for faster ways of sending the information in the first place. And this is where we start to appreciate some of the physical as well as the economic limitations to our communications systems.

Research the following terms or phrases and explain what they mean in terms of digital communication. You may be asked to cover one or more of these, and you could share your research with the class in some way:

a) sampling, quantization, pulse code modulation

b) internet telephony, radio and streaming audio

c) shift registers

d) companding

e) aliasing

f) Nyquist theorem.

Time division multiplexing

Whatever the information, a digital signal is only ever a very long series of 1s and 0s that need to be sent from one place to another. The simplest method would be to have a long piece of cable. Unfortunately, for global communications there will often be thousands of individual signals, which need to share one **channel of communication**. The physical limitations on how many signals we can send down one wire will be dealt with later. However, the process by which it is done is called **multiplexing**. With digital signals we shall deal with what is called **time division multiplexing**.

Suppose we have three signals from different sources (A, B, and C); each signal is 20 bits long, and they all have to travel along the same cable. You could send all of signal A, then all of signal B, and finally all of signal C. The people interested in signal A would receive the information quickly, but those looking for signal C would have to wait. The way round this is to allow each signal a share of the total time available to that channel.

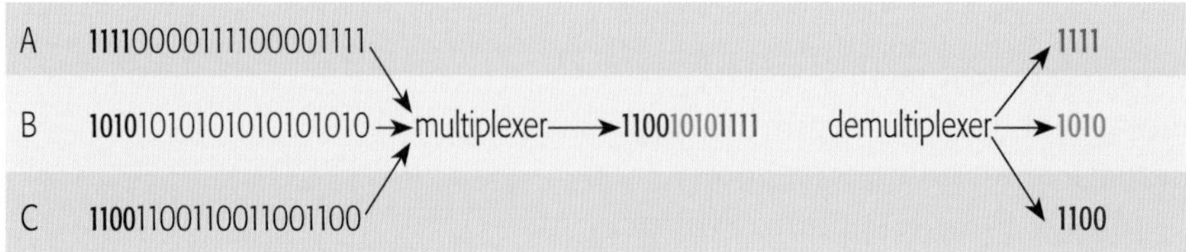

Figure 10 Principle of time division multiplexing

This is done by passing all three signals into a **multiplexer**, which cuts them up into equal-length slices of information (see Figure 10). For example, we might take the first 4 bits of signal A and then follow this with the first 4 bits of signal B and the first 4 bits of signal C after that. If we repeat this pattern, then everybody receives their signal at a constant rate, but at a lower bit rate than if one signal had the cable all to itself.

Prepare some arguments for a debate entitled "The development of better communications systems is increasing the economic divide between rich western nations and poorer developing countries." You should research arguments for and against this statement and try to draw some conclusions.

At the end of the communication channel there is a **demultiplexer**, which cuts the data stream into, in this example, 4-bit slices and then combines all the relevant signals back together again.

Any channel of communication can only cope with a limited number of messages at any one time, and the more signals the system needs to cope with, the slower the data transmission will be. The greater the capacity that a channel has, the more data it can accommodate – and the more expensive it will be! internet traffic is making ever-increasing use of data in the form of images, audio and video, so there is a need for ever more efficient, high-volume channels of communication. This will continue to push communications engineering to the limit.

There are many consequences of this demand for better internet systems. These can be simple annoyances such as roads being dug up so that new cables can be laid but can also involve some contentious decisions. There are moral, ethical, economic and environmental issues arising from access to the internet.

Wires

Traditional wiring is increasingly being replaced for long-distance communications; it now tends to be used only at a very local level. Coaxial cable is superior to simple pairs of wires, but is still limited in how much information it can carry. The physics of how signals are transferred along cables is similar to that of the other two channels of communication covered in this section. Traditional wiring is increasingly being replaced by optical fibre.

Optical fibre

There are several types of optical fibre. They vary widely in terms of their cost and characteristics, but they all operate on the same basic principles of light transmission and **total internal reflection** (see page 108).

The basic construction of any optical channel is the same. A central core of extremely pure glass is surrounded by a **cladding** material of slightly lower refractive index. The whole structure is then protected, usually by a plastic outer layer. At one end of the fibre there will be a light emitting diode or laser that can be switched on and off. Digital signals are transmitted as pulses of light that indicate a binary level 1 when the light is on. When the light is not on a binary level 0 is assumed to exist. In analogue signals the intensity of the light varies.

Multimode fibre

Multimode fibres allow rays of light to follow many different paths. In this section we will look at **step indexed** fibres. These have a relatively large core of one material surrounded by a cladding material with a different refractive index. When a ray of light hits the boundary between core and cladding there is a distinct "step" or change in the refractive index. The rays that make up a beam of light will typically follow several different paths, as shown in Figure 11.

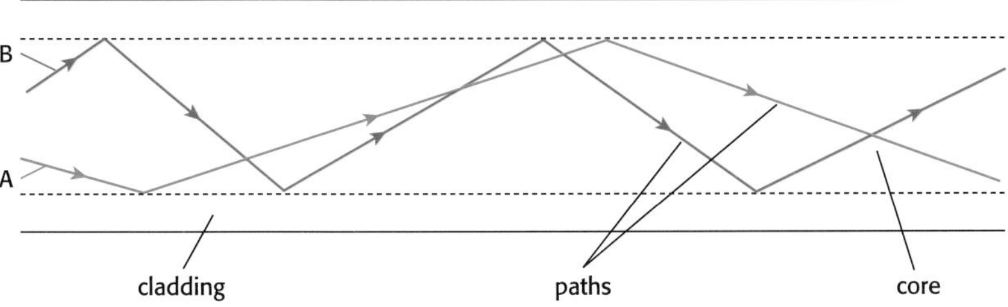

Figure 11 Two different ray paths along a multimode optical fibre

The ray labelled A will arrive at the end in a shorter space of time than B, because it has to travel a shorter overall distance. While the light travels more slowly in the glass than in a vacuum it still travels very quickly ($\sim 2 \times 10^8\ \mathrm{m\,s^{-1}}$). The high frequency switching of the transmission light source leads to a problem called **dispersion**, which is illustrated in Figure 12. An ON–OFF–ON sequence (binary 1–0–1) may start at one end of a cable, looking like Figure 12a.

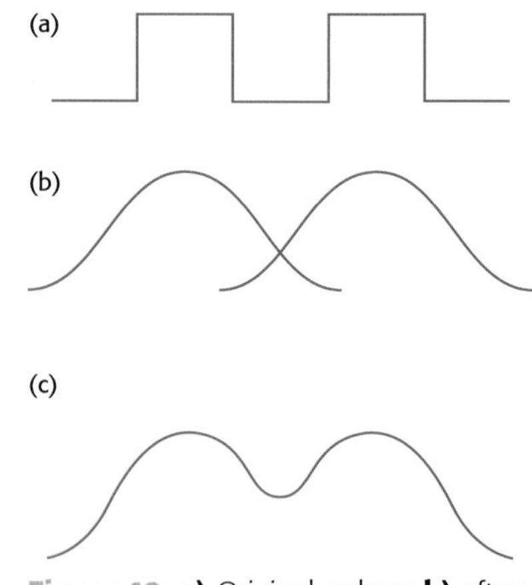

Figure 12 **a)** Original pulses; **b)** after dispersion; **c)** combined effect

But by the time it reaches the other end the information will have been spread out, as in Figure 12b, losing the middle OFF in the process. Dispersion effects can therefore limit the bit rate, especially over large distances.

Monomode fibres

These have a much smaller diameter core. This means that there are significantly fewer possible paths for rays to follow, so dispersion more or less ceases to be a problem, and therefore higher bit rates can be attained (Figure 13).

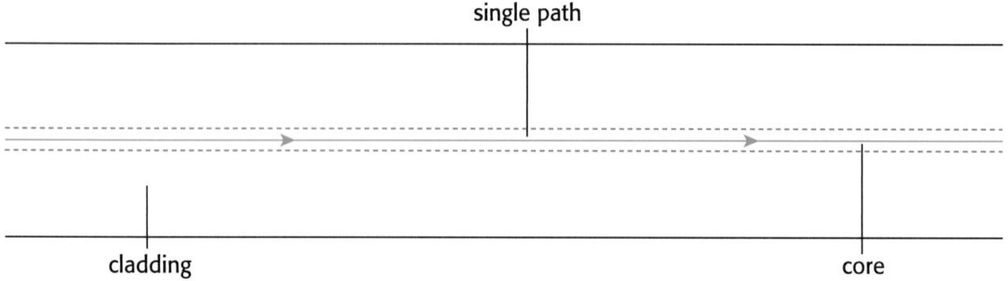

Figure 13 Monomode fibre

Attenuation

One problem for all channels of communication is the degradation of **signal strength**. Energy is being transferred from one place to another in the form of the communications signal, but some of that energy will be "lost" along the way. This process is called **attenuation**. Glass fibres are no exception. In a fibre optic channel the power of the signal decreases exponentially according to the equation

$$P = P_0 e^{-\alpha x}$$

where P is the instantaneous signal power, and P_0 is the original signal power; α is called the attenuation coefficient of the fibre – this is a measure of how good the fibre is at transmitting light; and x is the distance along the fibre (usually measured in km).

The amount of attenuation is measured on a logarithmic scale with units of decibels (dB)

$$\text{attenuation} = 10 \log \left(\frac{\text{output Power (W)}}{\text{input power (W)}} \right)$$

It is common to measure the attenuation per unit length. A typical optical fibre may have an attenuation of $-4\ \mathrm{dB\,km^{-1}}$.

However well an optical fibre is manufactured there will always be some level of impurity, which causes scattering and absorption of light travelling along the cable. There are additional problems with aging of materials, because of impurities leaching into the glass over time, as well as with how individual lengths of fibre are joined together.

> The input signal power to a 5 km length of optical fibre of attenuation $-4\ \mathrm{dB\,km^{-1}}$ is 200m W. Calculate the output power.

One of the big advantages of digital transmission is that, even though the signal attenuates and disperses over long distances it will still maintain the basic ON–OFF structure as originally sent. At some point the signal will degrade so much it will become impossible to decipher and so the signal must be periodically tidied up. This is done using an amplification device called a **regenerator** (which helps overcome attenuation problems) combined with something called a **reshaper** (which helps overcome dispersion issues) so that as the signal travels along the cable it will be kept as close to the original transmission as possible (Figure 14).

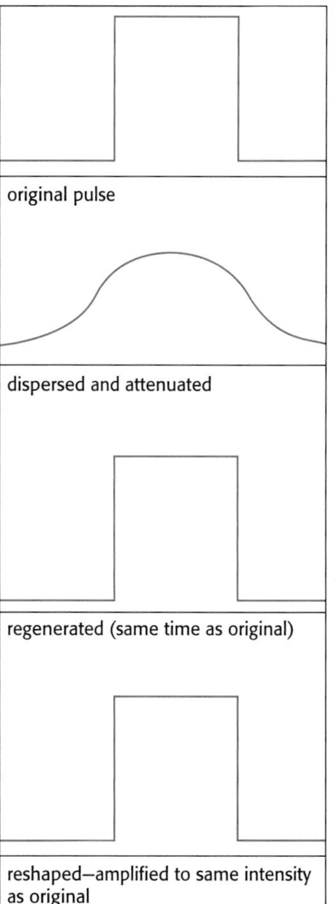

Figure 14 Regeneration and reshaping of an optical pulse

The various wavelengths of light are affected in different ways within a glass medium. Ideally, the signal will be carried by long wavelengths, because short wavelengths (higher-frequency light) will experience greater dispersion. Typically, red light or infrared is used. LEDs or tiny lasers transmit the pulses of light. Ideally the transmission will be from a monochromatic source of radiation in order to avoid **spectral spreading**, another dispersion effect caused by the fact that different wavelengths will travel at different speeds along the same channel (Figure 15).

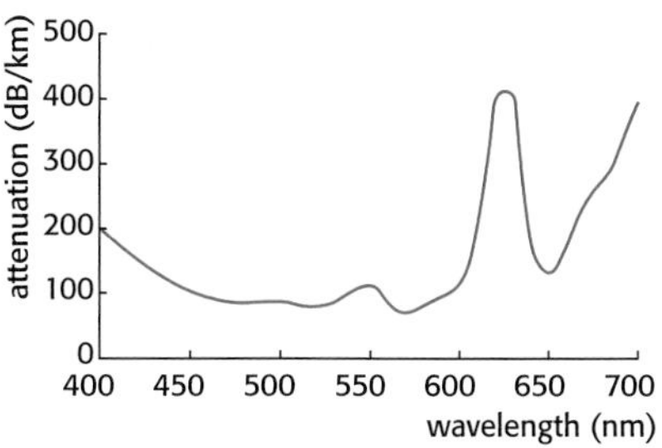

Figure 15 Typical spectral attenuation in a glass fibre

Getting the message across – Channels of Communication

This last section of the core material focuses on the practicalities of physically moving information from one place to another. The different options for data transfer are called **communication channels**. Each channel has advantages and disadvantages associated with them. These are summarized in the table below:

Table 2

Channel	Advantages	Disadvantages
Wire pair	Low cost, easy installation	Break easily, susceptible to interference, low security, low bandwidth, large distances require regular amplification.
Coaxial cable	Compared to wire pair they have higher bandwidth and are less prone to interference	Greater bulk, expensive, amplification still needed over distance

(Table continued)

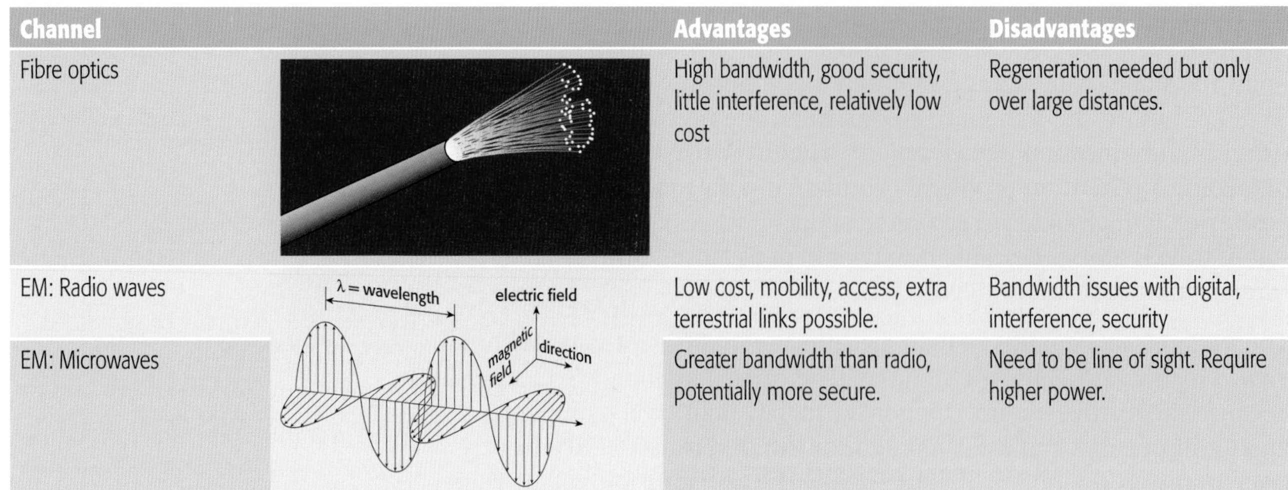

Channel		Advantages	Disadvantages
Fibre optics		High bandwidth, good security, little interference, relatively low cost	Regeneration needed but only over large distances.
EM: Radio waves	λ = wavelength; electric field; magnetic field; direction	Low cost, mobility, access, extra terrestrial links possible.	Bandwidth issues with digital, interference, security
EM: Microwaves		Greater bandwidth than radio, potentially more secure.	Need to be line of sight. Require higher power.

Electromagnetic waves can be split into three groups depending on how they travel from transmitter to receiver.

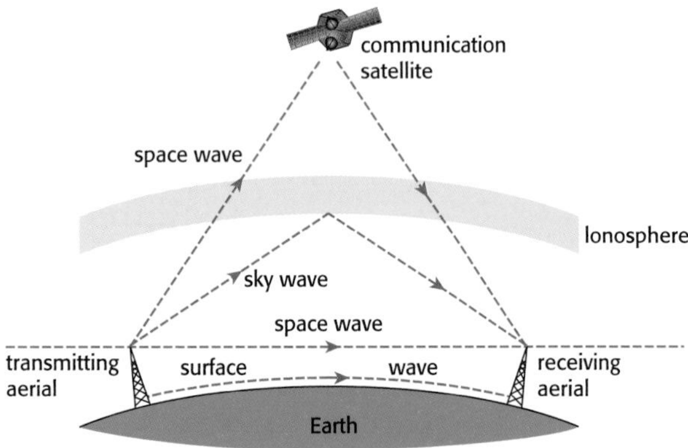

Figure 16 transmission of radio waves

Ground Waves – As the name suggests these travel in close proximity to the surface of the Earth. Energy carried by the wave is relatively quickly dissipated with different terrain affecting different levels of absorption. Such waves can, for example, travel 1500 km over water if of low frequency. In general Long Wave has a higher range than the much higher frequency VHF. Additionally the longer wavelengths are more easily diffracted around buildings and raised terrain.

Sky Waves – If below about 30 MHz radio waves can be 'bounced' off the ionosphere and back to the Earth's surface. This process can be repeated several times thus avoiding problems of line of sight due to curvature of the earth. Ranges are typically thousands of kilometers. Problems include dual reception of surface and sky waves causing interference. Also the reflection effect of the ionosphere can vary depending on the time of day and solar activity.

Space Waves – VHF and above radio transmissions are not affected by the ionosphere and so have limitations similar to those of microwaves. They can, however, pass through the entire atmosphere if powerful enough not to be completely attenuated. Once in 'outer space' there is nothing to significantly reduce the signal strength other than distance and the waves can be relatively easily received by a communications satellite, amplified and then beamed back to Earth again.

Table 3

Frequency	Waveband	Surface	Sky	Space	Use
30 – 300 kHz	LF – low frequency	Medium range; long enough to diffract around hills and buildings	Long range possible but ground wave often has higher amplitude		Communications
300 kHz – 3 MHz	MF – Medium frequency	Local sound broadcasts; long enough to diffract around hills and buildings. Range up to 1000 km possible.	Distant sound broadcast		Sound broadcasts
3 – 30 MHz	HF – High frequency (SW - short wave)	Range up to 100 km. Some signal loss due to interference with reflected space wave of same frequency	Distant sound broadcast; communication; Line of sight transmission possible	Refection off ionosphere increases possible range.	Sound broadcasts
30 – 300 MHz	VHF – Very High frequency	Range approximately 100 km;	Line of sight;	Pass through ionosphere so must be reflected off satellites	FM sound broadcasts; TV; mobile systems
300 MHz – 3 GHz	UHF – Ultra High frequency	Range less than 100 km;	Line of sight;	Pass through ionosphere so must be reflected off satellites	TV
Above 3 GHz	SHR – Super High Frequency	Range less than 100 km;		Radar; communication via satellites; microwave links for telephone etc.	Satellite Radar; telecommunications

Satellites

There are many types of satellite orbit and each has advantages and disadvantages. Different possible orbits can be specified in terms of the angle that the satellite's orbit makes with Earth's equatorial plane. This is known as the satellite's **inclination** (see Figure 17).

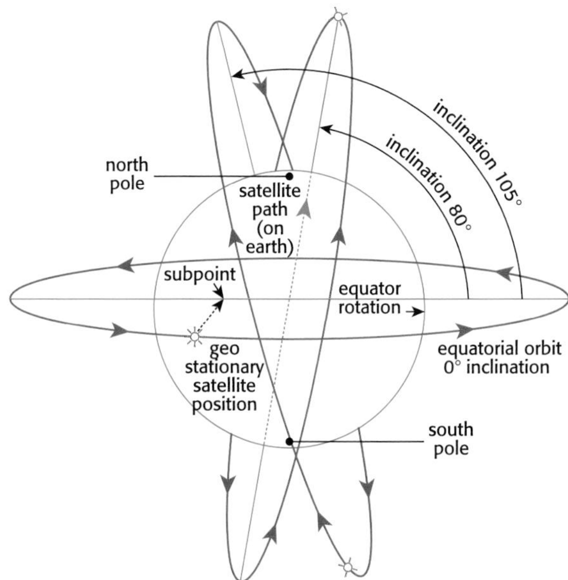

Figure 17

Geostationary satellites – All geostationary satellites have an inclination of 0 degrees and occupy a very specific **geosynchronous** orbit. The radius of this orbit is such that their orbital period is exactly

24 hours. This means that the satellite always appears to be stationary above one point (the subpoint shown in Figure 17) on the Earth's equator. They are extremely useful for communications because transmitting and receiving equipment does not need to track the satellite. This greatly reduces ground based costs and any need for complex equipment.

Unfortunately, geostationary satellites have certain limitations. At high latitudes on the Earth's surface a particular satellite could be below the horizon which makes communication with it impossible. If it is close to the horizon difficulties can still arise due to multi-path interference, where waves reflected from the ground superpose with those directly from the sky.

It is also necessary for the uplink and downlink frequencies to be different. This is due to that fact that any incoming signal will be of extremely low power. If the transmitter is close to the receiver (which will always be the case on the satellite but not necessarily so on Earth) then the transmitting frequency could swamp the incoming frequency. In addition, operating at the same frequency could cause significant interference due to internal resonance. The satellite would be transmitting to itself and feedback would occur. While not difficult to overcome this does significantly affect bandwidth as two sets of frequencies are required for each satellite communication channel.

It is an expensive proposition to put a satellite in geostationary orbit. The large distance between Earth and the geostationary orbit also demands a higher power signal. Such equipment also has a limited lifetime as the orbit will be degraded over time due to orbital perturbations due mainly to solar radiation pressure. While changes in the orbit can be corrected using thrusters, the propellant used in such devices will eventually run out. It is not economically feasible to correct such problems by visiting the satellite so at the end of its life the scrap metal must be moved into a 'junk orbit'.

Polar Satellites – these satellites are rarely used for communications (as their position in the sky will continually vary) but are very powerful tools for mapping and reconnaissance. A Polar satellite has an inclination of 90 degrees. Typically they can be put into low orbits and so have quite short orbital periods. They will travel through many different longitudes during one orbit (see Figure 18).

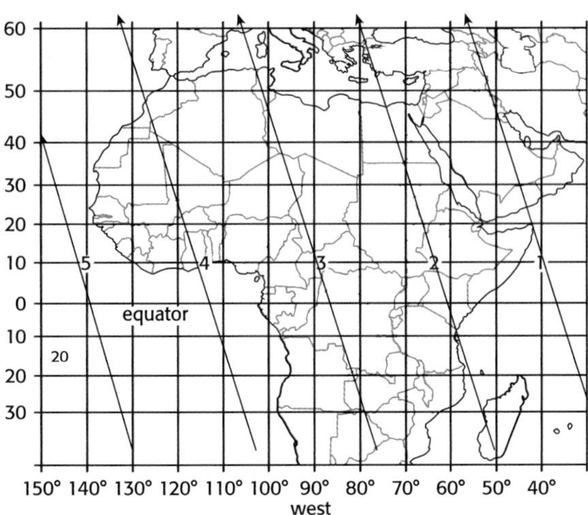

Figure 18

Mathematical Physics: The Clarke Belt

Science fiction writer Arthur C. Clarke is famously cited as the 'inventor' of the geostationary orbit. The mathematics draws from work on circular motion and gravitation.

M = mass of Earth, m = mass of satellite. For the satellite, the centripetal force is provided by the gravitational attraction between the Earth and the satellite:

Gravitational Force = Centripetal Force

$$\frac{GMm}{r^2} = \frac{mv^2}{r}$$

Rearrange and $\times \frac{1}{r}$ $\quad \frac{v^2}{r^2} = \frac{GM}{r^3}$

But angular velocity ω is defined as: $\omega = \frac{v}{r}$

So: $\omega^2 = \frac{v^2}{r^2}$

i.e. $\omega^2 = \frac{GM}{r^3}$

Thus $r = \sqrt[3]{\frac{GM}{\omega^2}}$

For the satellite to remain over exactly the same spot this requires an angular velocity, ω, equal to that of the Earth's surface:

1 Research values for G and M and then calculate, using the Earth's period of rotation as 24 hours, the radius r for all geostationary satellites.

As well as obvious costs there are many other issues worth thinking about.

- The ability to put satellites in orbit has, until relatively recently, been limited to a small number of rich nations who effectively 'control' space.
- Since humankind began throwing things into space in 1949 a lot of debris has built up which could be potentially hazardous in the future. Who, if anybody, should be held accountable?
- Could our need for satellite communications ever be compromised by an act of terror and what can nations do to safeguard against this?

You may be able to come up with some other ideas worth further research and present your ideas to the class.

Advanced Topic: Electronics

In order to study electronics, you need to be familiar with the properties of capacitors, resistors and diodes but these components alone are not sufficient. Central to many modern electronic systems is a component known as the transistor. Miniaturisation has allowed huge numbers of transistors to be combined with other components into a single **integrated circuit** or **(micro)chip**.

These discreet system blocks, can be mass produced cheaply and combined together in new and novel electronic solutions. Electronics is the study of how these systems can be usefully combined. One extremely usefully system block is the **operational amplifier** or **op-amp**.

The Operational Amplifier

A full understanding of how the complex combination of transistors and other devices work within a chip is not required; You just need to understand its overall effect. The circuit symbol for an op-amp is shown below.

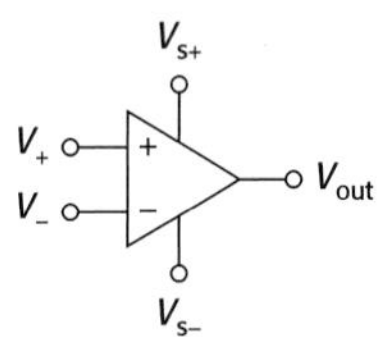

Figure 19 the full electrical symbol for an op-amp (the power supply connections are often not included)

The various labels refer to the following:

V_{s+} is the positive supply voltage

V_{s-} is the negative supply voltage

V_+ is called the non-inverting input

V_- is called the inverting input

V_{out} is the output voltage

The function of an op-amp is to give an output that depends on the combination of the two input voltages. Specifically (up to a limit called **saturation**), the output is a fixed multiple of the difference between the inputs. The op-amp behaves as a **differential voltage amplifier**.

The terms "inverting input" and "non-inverting input", refer to the polarity of the resulting output voltage:

If $V_- > V+$ (meaning V_- is more positive) then V_{out} will be positive.

If $V_+ > V_-$ then V_{out} will be negative.

A general expression for the output voltage is:-

$V_{out} = A_o (V_+ - V_-)$

A_o is called the **open-loop gain**. The term **gain** describes the amplifying or multiplying effect. **Open-loop** gain is defined as amplification between the input and output without any **feedback** (when some portion of the output is connected back into the input). A typical open-loop gain for a real op-amps is in the region of $A_o = 10^5$. Thus when the output of an op-amp not saturated, the difference between the two inputs is very tiny. An idealised op-amp is considered to have **infinite open-loop gain**.

Another property of an ideal operational amplifier is to have **infinite input impedance**. The term impedance may be new to you. Impedance is the a.c. equivalent of resistance and also measured in ohms. Infinite input impedance just means that the op-amp does not allow any electric current to flow into it (on either input).

As far as the output is concerned, there would be no limit to the current that an ideal op-amp could supply so it is said to have **zero output impedance**. A typical op-amp output impedance is approximately 100 Ω.

You may be wondering how the op-amp manages to do all this as the huge increase in voltage and current seems to be breaking the law of conservation of energy! Remember that, although the power supply connections will often not appear in circuit diagrams, external power being supplied to any op amp. As a result, output voltages are always limited to be smaller in magnitude that the power supply voltages. When the output of an op-amp reaches these limits it is said to be **saturated**. For example, an op-amp operating with supply voltages of +15V and − 15V would have its output restricted to, say, a value between −13V and +13V.

Four important circuits involving op-amps are the non-inverting amplifier, the inverting amplifier, the comparator and the Schmitt trigger.

Non-Inverting Amplifier

This circuit introduces the idea of feedback. The non-inverting input has a varying external signal, V_{in}, fed into it. In this case, part of the output signal is returned in to the inverting input. This is an example of **negative feedback**.

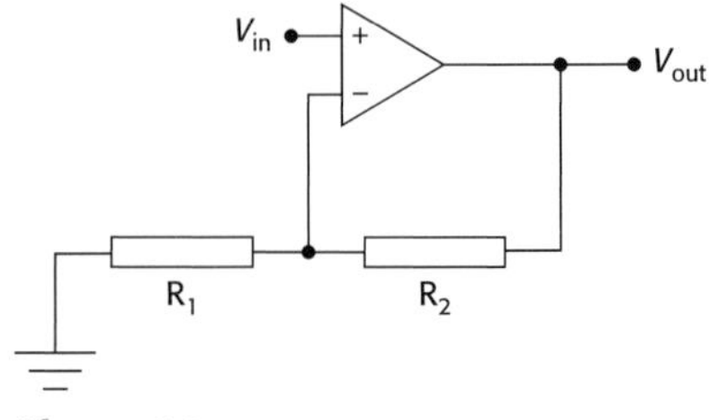

Figure 20

The huge open-loop gain of the op-amp means that, providing it is not saturated, the potential difference between the inverting input and the non-inverting input will be very small.

Effectively $V_- = V_{in}$

Thus

the (potential difference across R_1) $= V_{\text{in}}$

the (potential difference across R_2) $= (V_{\text{out}} - V_{\text{in}})$

Since the op-amp has zero input impedance, no current flows into the non-inverting input and the current through R_1 = current through $R_2 = \frac{V_{in}}{R_1}$

$$V_{\text{out}} = \text{pd across } R_1 + \text{pd across } R_2 = \frac{V_{in}}{R_1} + \left(\frac{V_{in}}{R_1} \times R_2\right) = V_{in} \times \left(1 + \frac{R_2}{R_1}\right)$$

$$\text{Closed loop gain} = \frac{V_{out}}{V_{in}} = \left(1 + \frac{R_2}{R_1}\right)$$

This circuit amplifies the input signal and the output will be **in phase** with the input. The output is a fixed multiple of the input and the size of this multiple is depended only on the values of R_1 and R_2.

Inverting Amplifier

In this case the output is still fed back to the inverting input (so this is still negative feedback) but note on the diagram below how the V_+ and V_- inputs have been switched around.

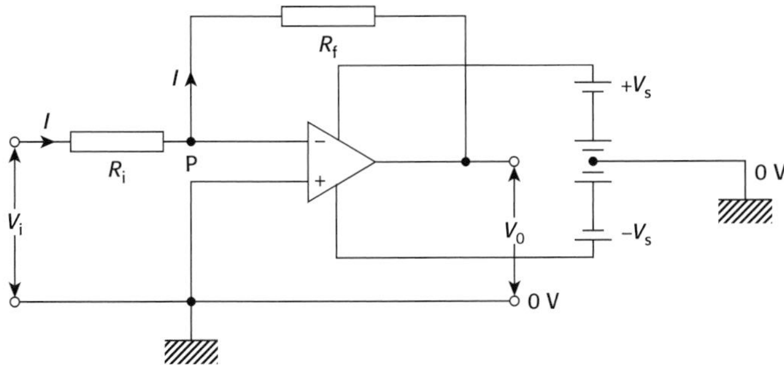

Figure 21 The inverting amplifier circuit

As the non-inverting input is connected to zero (or Earth), then the high gain means that the non-inverting input must also be at the same potential. , Point P can be considered to be a **virtual earth** or at a potential of 0V. The fact that an op-amp does not draw any current allows us to simplify the above circuit to this below:

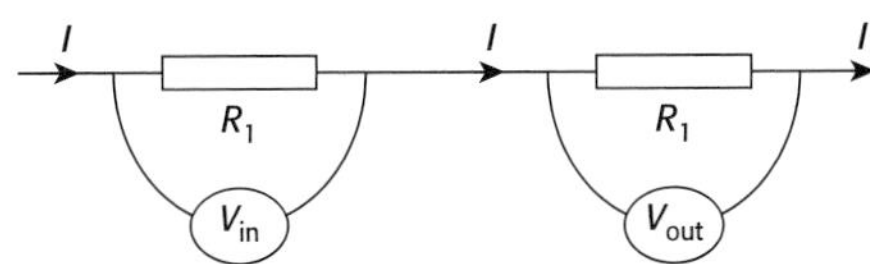

Figure 22 The virtual Earth equivalent circuit for an inverting amplifier

$I = V_{in}/R_1 = -V_{out}/R_2$

$$\text{Closed loop again} = \frac{V_{out}}{V_{in}} = -\left(\frac{R_2}{R_1}\right)$$

In other words, once again we have output that is a fixed multiple of the input and the size of this multiple is depended only on the values of R_1 and R_2 but this time the output is in **antiphase** with the input.

Comparator

When used as a comparator both the inverting and non-inverting inputs are connected to input signals. With no feedback loops the circuit simply looks like this:

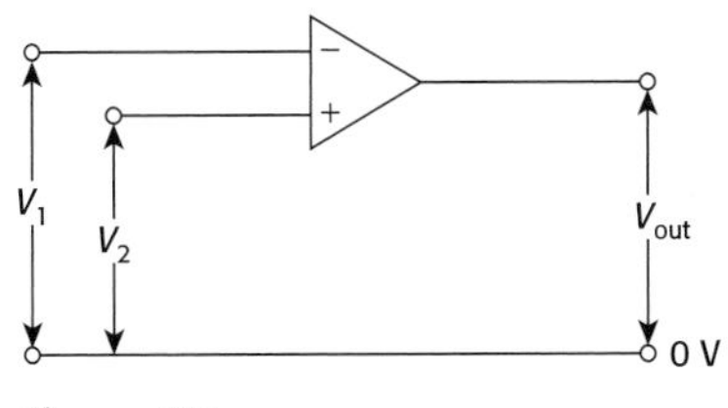

Figure 23

The open-loop gain A_0 is very large so when we apply the equation $V_{out} = A_0(V_+ - V_-)$, the output becomes **saturated** for even a small difference between the two input voltages. When the output saturates it is either at a value close to that of the positive supply voltage if $V_+ > V_-$ or to the negative supply voltage if $V_- > V_+$. The op-amp effectively transforms analogue signals with many possible values into a signal that only has two possible outputs.

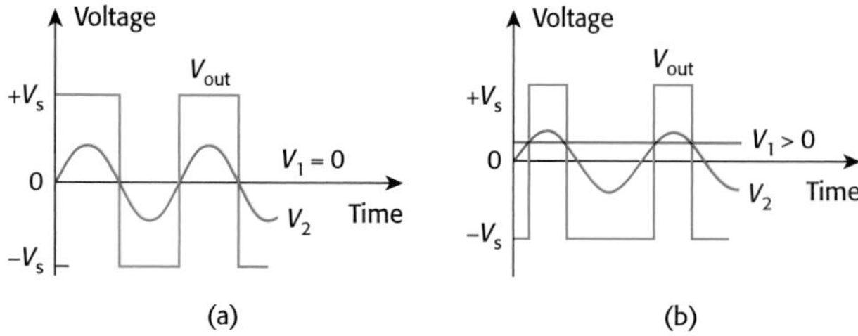

Figure 24 (a) A sinusoidal input to a Comparator with its non-inverting input set at 0V (b) A sinusoidal input to a Comparator with its non-inverting input set at a value greater than 0V

Schmitt trigger

An op-amp Schmitt trigger circuit is shown below. In this case, part of the output signal is returned in to the non-inverting input. This is an example of **positive feedback** and thus the output will either be positive or negative saturation.

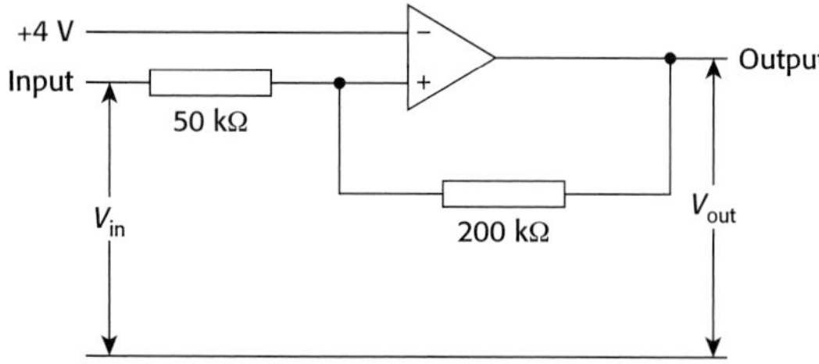

Figure 25

We begin the analysis of this circuit with an input value $V_{in} = -10$ V and the output on negative saturation (−13V). With these values, the two resistors fix the non-inverting at a value between −10V and −13V which maintains the output on negative saturation as its value is below +4V.

As the input rises, it will eventually reach a value that causes the output to switch to positive saturation. This happens when the non-inverting input gets to +4V. At this instant, the potential difference across the 200 kΩ resistor is from +4V down to −13 V i.e. 17V.

Current through 200 kΩ resistor $= \frac{17}{200} = 0.085mA$

Since the input impedance of the op-amp is very high, the same current must flow through the 50 kΩ resistor so

Potential difference across 50 kΩ resistor = 0.085 $mA \times 50$ kΩ = 4.25 V

The value of V_{in} that causes the non-inverting input to get to 4 V is thus:

$V_{in} = 4 + 4.25 = +8.25$ V.

This is the upper switch over point and when it is reached, the output switches to positive saturation (+`13V). This value now holds the non-inverting at a value between +8.25V and +13V which maintains the output on positive saturation as its value is above +4V. If the input falls a little below 8.25 V the output will remain at positive saturation. In order to identify the lower switch over point the following calculation is necessary.

The lower switch over happens when the non-inverting input returns to +4V. At this instant, the potential difference across the 200 kΩ resistor is from +13V down to +4 V i.e. 9V.

Current through 200 kΩ resistor $= \dfrac{9}{200} = 0.045mA$

The same current must flow through the 50 kΩ resistor so

Potential difference across 50 kΩ resistor = 0.045 mA × 50 kΩ = 2.25 V

The value of V_{in} that causes the non-inverting input to get to 4 V is thus:

$V_{in} = 4 - 2.25 = 1.75$ V.

This is the lower switch over point and when it is reached, the output switches back to negative saturation (−`13V).

Overall this Schmitt trigger circuit has two different switchover points which are significantly different.

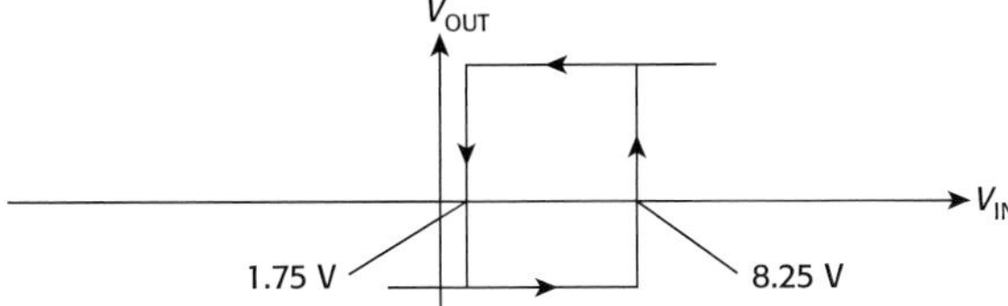

Figure 26 Schmitt trigger has two different switch-over trip points for switching between output states

In Figure 27 note how V_{out} switches to a high value when V_{in} reaches a certain point (the Upper Trip Point) but stays at this high value until V_{in} falls below a certain value (the Lower Trip Point). This should be compared to a comparator circuit (Figure 27). Comparators have a single switch-over point that is defined by the voltage on the non-inverting input. Schmitt trigger circuits perform a similar function but once the input voltage has reached a high enough value to move the output from negative saturation to positive saturation, it needs to fall to a significantly lower value before switching back. This can be used to shape digital pulses that have been subjected to noise or dispersion.

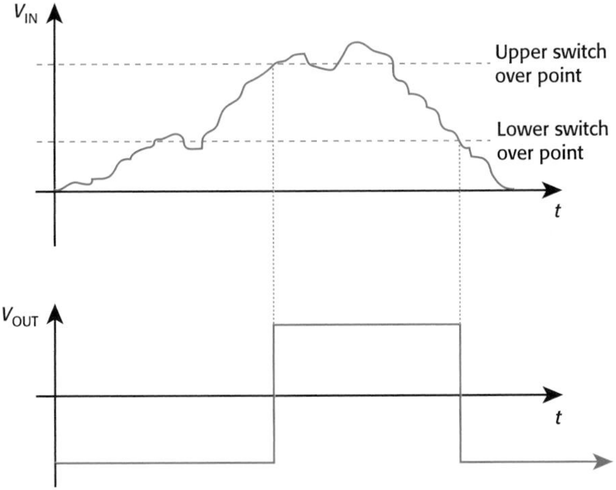

Figure 27

Advanced Topic: The Mobile Phone System

Mobile phone technology is synonymous with what modern communication is all about. Along with the internet it has had a great impact on the lives of millions of people in recent years and yet few people understand even the basics of how our voice, text or multimedia communications actually get from one phone to another

Cells

A region where mobile telephone calls can be made is divided up into a number of different **cells**. At the centre of each cell is a **base station** with a transmitting/receiving unit to which is allocated a range of frequencies for the communication channels it can use. The signals from these base stations will overlap but the overall effect is to produce a pattern of cells that is effectively hexagonal.

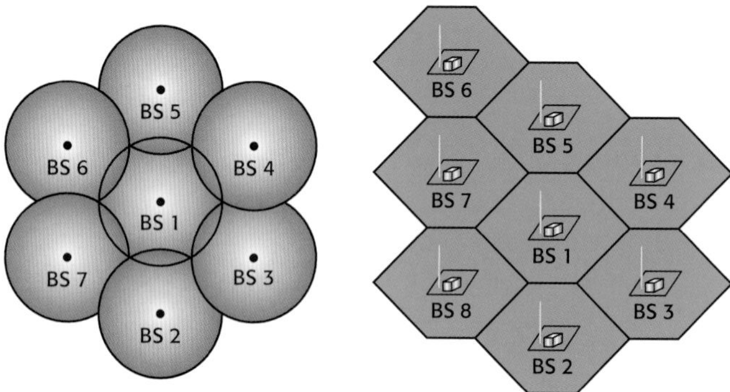

Figure 28 The overlapping signals from different base stations (BS1, BS2 etc). The effective pattern produced by overlapping signals can be considered as hexagonal.

Which region of the electromagnetic spectrum do mobile phones operate in?

The size of individual cells depends on the context. In a densely populated city, a cell can be as small as several hundred metres wide. In more rural areas, however, cells can be up to 30 km or so wide. Each cell will up and down link information to and from phones within its area of influence by using a certain range of frequencies. In order for several cells to function without interference between the base stations, no two adjacent cells use the same range of frequencies. This also allows the same frequency carrier to be used by many different users in different locations thus enabling an increased volume of traffic. All frequencies used are around 900 MHz though this is slightly different in North America where such frequencies were already allocated for other uses prior to the advent of cellular telephony.

At all times when a mobile phone is switched on it will be in transmitting and receiving information with the nearest base station in order to identify itself in case an incoming call arrives. When an outgoing call is made a connection is created with the closest receiver on its particular carrier frequency. The base station then allocates a particular channel for the mobile phone to use. Stations within a network will normally be linked together either by cabling or radio links. They are also connected to the public switched telephone network (PSTN) allowing calls to and from landlines to be made as well as connections between users on different networks.

When a mobile user moves from the influence of one cell into another the phone will automatically establish a different connection with the new base station and the call can continue without interruption. This is called 'handover'. During handover, even though the base station and associated frequency may change, the part of the connection which is via the PSTN will, of course, remain unchanged.

Standards

All engineers always work to agreed standards. This doesn't mean they want their work to be of a certain quality but rather that they agree on certain ways of doing things. With so many solutions possible to any one particular problem it is advantageous to all concerned that agreed international standards be adhered to when designing products like mobile phones.

There are over a dozen accepted mobile phone standards because in the early days of any new technology many companies will try and establish their own solution to the problem. If you can get your answer accepted as the best way to do things it can mean a lot in business terms. This is why there are so many arguments amongst companies and even at government level when trying to establish a standard.

GSM and GPRS are two of the most well known and used cell phone standards. Any phone manufactured that claims to be GPRS will be able to do a certain number of things and will be able to communicate over certain networks. GPRS is part of what has been called the 3G or 3rd generation in mobile phone technology. Unsurprisingly the next generation of phones is already waiting to be sold to you the consumer and there is currently a lot of talk as to how 4G will operate.

Inquiry: Health Implications of Mobile Phones

One of the recurring issues regarding mobile phone technology is the concern over health impacts surrounding the carrier frequencies used, especially amongst young people. There has been extensive research carried out to try and establish links between emissions from headsets and all manner of illnesses including brain tumours. Such issues are often emotive and frequently sensationalized by the media. With massive profits to protect, mobile phone companies naturally want to convince the public that there is no threat to its health. Current thinking is inconclusive at best but the arguments that abound do show how important accurate scientific communication can be. See pages 228–30 for discussion of a related safety issue.

Use the internet to identify a piece of research that has been undertaken relating to the health impacts related to the use of mobile phones. Try to access the raw data associated with the study and summarise its findings for others in your class.

References

Advanced Physics (OUP)

Duncan, T (xxxx) *Success in Electronics* (2nd edn). John Murray.

Salters A2 book

End of chapter summary

Chapter 28 has six main themes, two of which are only relevant to those studying higher level physics. The four themes common to all levels are: radio communication, digital signals, optic fibre transmission and channels of communication. The two HL themes are: electronics and the mobile phone system. SL candidates studying Option C (digital technology) also need to study these two HL themes. The list below summarises the knowledge and skills that you should be able to undertake after having studied this chapter. Further research into more detailed explanations using other resources and/or further practice at solving problems in all these topics is recommended – particularly the items in bold.

Radio communication

- Describe what is meant by the modulation of a wave, distinguish between a carrier wave and a signal wave and describe the nature of amplitude modulation (AM) and frequency modulation (FM).
- **Solve problems based on the modulation of the carrier wave in order to determine the frequency and amplitude of the information signal.**
- Sketch and analyse graphs of the power spectrum of a carrier wave that is amplitude-modulated by a single – frequency signal, define what is meant by *sideband frequencies* and bandwidth **and solve problems involving them**.
- **Describe the relative advantages and disadvantages of AM and FM for radio transmission and reception.**
- Describe, by means of a block diagram, an AM radio receiver.

Digital signals

- Distinguish between analogue and digital signals and solve problems involving the conversion between binary numbers and decimal numbers.
- Describe, using block diagrams, the principles of the transmission and reception of digital signals and state the advantages of this method, as compared to an analogue method, for the transmission of information.
- **Explain the significance of the number of bits and the bit-rate on the reproduction of a transmitted signal.**
- Describe what is meant by time – division multiplexing and solve problems involving analogue-to-digital conversion.
- **Describe the consequences of digital communication and multiplexing on worldwide communications and discuss the moral, ethical, economic and environmental issues arising from access to the Internet.**

Optic fibre transmission

- Explain what is meant by critical angle and total internal reflection and solve problems involving refractive index.
- Apply the concept of total internal reflection to the transmission of light along an optic fibre and describe the effects of material dispersion and modal dispersion.
- Explain what is meant by attenuation **and solve problems involving attenuation measured in decibels (dB)**.
- Describe the variation with wavelength of the attenuation of radiation in the core of a monomode fibre.
- State what is meant by noise in an optic fibre and describe the role of amplifiers and reshapers in optic fibre transmission.
- **Solve problems involving optic fibres.**

Channels of communication

- Outline different channels of communication, including wire pairs, coaxial cables, optic fibres, radio waves and satellite communication and discuss their uses and the relative advantages and disadvantages.
- State what is meant by a geostationary satellite and the order of magnitude of the frequencies used for communication with them, explaining why the up-link frequency and the down-link frequency are different.
- Discuss the relative advantages and disadvantages of the use of geostationary and of polar-orbiting satellites for communication and the moral, ethical, economic and environmental issues arising from satellite communication.

In addition HL candidates should be able to:

Electronics

- State the properties of an ideal operational amplifier (op-amp) and draw circuit diagrams for both inverting and non-inverting amplifiers (with a single input) incorporating op-amps.

- Derive an expression for the gain of an inverting amplifier and for a non-inverting amplifier.
- Describe the use of an operational amplifier circuit as a comparator and of a Schmitt trigger for the reshaping of digital pulses.
- **Solve problems involving circuits incorporating operational amplifiers.**

The mobile phone system

- State that any area is divided into a number of cells (each with its own base station) to which is allocated a range of frequencies and describe the role of the cellular exchange and the public switched telephone network (PSTN) in communications using mobile phones.
- **Discuss the use of mobile phones in multimedia communication and the moral, ethical, economic, environmental and international issues arising from the use of mobile phones.**

Chapter 28 questions

1 This question is about refractive index and critical angle.

The diagram below shows the boundary between glass and air.

glass | air

a) Draw a ray of light to illustrate what is meant by critical angle. Mark the critical angle with the letter "c". [3]

A straight optic fibre has length 1.2 km and diameter 1.0 mm. Light is reflected along the fibre as shown below.

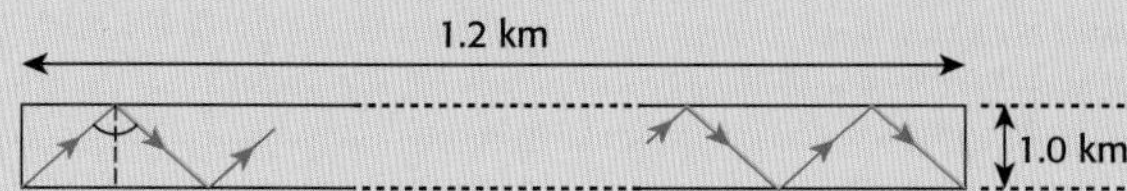

At each reflection, the angle of incidence is equal in value to the critical angle. The refractive index of the glass of the fibre is 1.5.

b) Deduce that the length of the light path along the optic fibre is about 1.8 km. [4]

The speed of light in the fibre is 2.0×10^8 m s^{-1}.

c) Calculate the time for a pulse of light to travel the length of the fibre when its path is

i) along the axis of the fibre. [1]

ii) as calculated in (b). [1]

(Total 9 marks)

2 This question is about amplitude-modulated radio waves.

The diagram below shows a sketch graph of signal voltage against time for an amplitude-modulated radio wave.

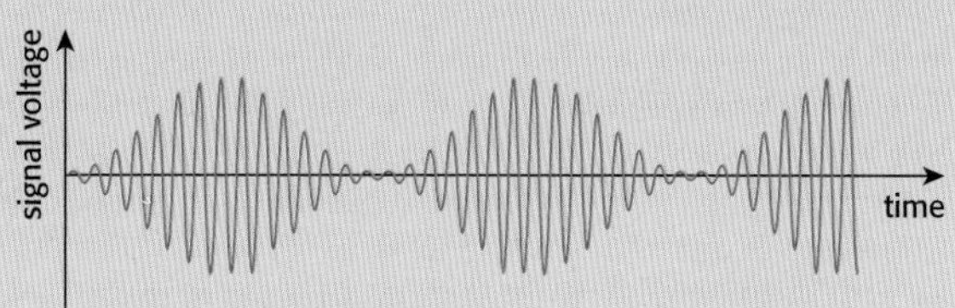

a) The information signal consists of a continuous single frequency sine wave. The frequency of the carrier wave is 18 kHz.

i) Determine the frequency of the information signal. [3]

ii) Draw the power spectrum for the amplitude-modulated wave. (*Numerical values are not required on the power axis.*) [3]

b) The block diagram below shows the principal systems in a radio that receives an amplitude-modulated signal. The unlabelled boxes represent amplifiers.

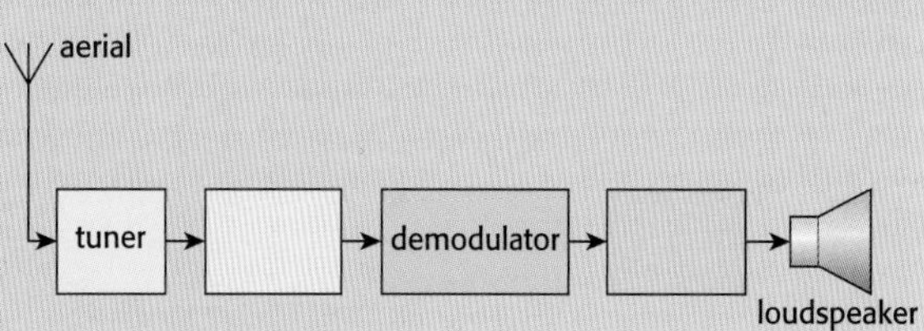

i) Label the blank boxes with the type of amplifier used. (1)

ii) State the function of the demodulator. [1]

(Total 8 marks)

3 The graph below shows the variation with time t of the voltage V of an analogue signal.

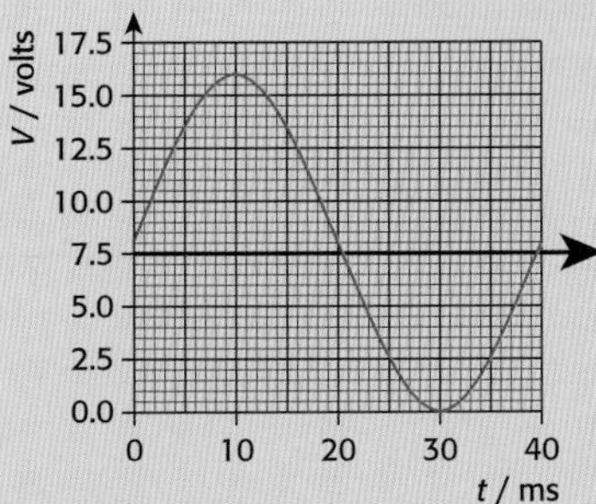

The signal is sampled at a frequency of 200 Hz and digitized using a three-bit analogue to digital converter (ADC). The first sample is taken at $t = 0$.

The possible outputs of the ADC are given below.

Analogue signal/volts	ADC binary output
	111
14	
	110
12	
	101
10	
	100
8	
	011
6	
	010
4	
	001
2	
	000
0	

a) Calculate the time at which the fourth sample is taken. [2]

b) Determine the binary output of the fourth sample. [2]

c) The ADC output is fed into a three-bit digital to analogue converter (DAC). State, and explain, whether the output of the DAC will be a faithful reproduction of the original analogue signal. [2]

(Total 6 marks)

4 This question is about communication channels.

a) State the order of magnitude of the frequencies used for communication with geostationary satellites. [1]

b) A voice communication channel is to be established between a scientific base in the northern hemisphere and its headquarters in the southern hemisphere.

For this communication channel, state and explain **one** advantage of using

i) a geostationary satellite. [2]

ii) a polar orbiting satellite. [2]

c) State **one** reason why the up-link frequency and the down-link frequency for communication satellites are different. [1]

(Total 6 marks)

5 The diagram below shows an inverting amplifier circuit.

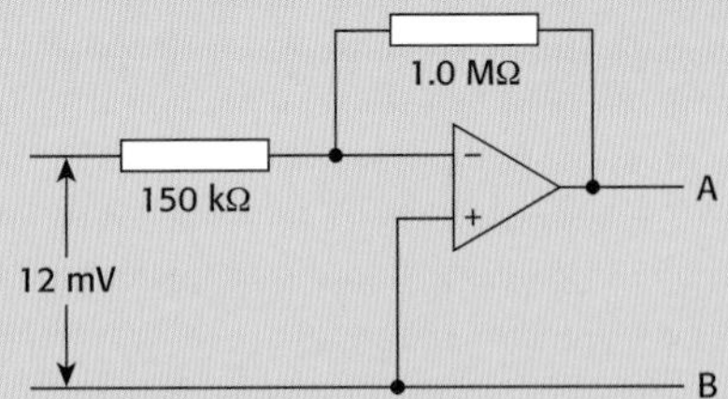

a) State what is meant by an inverting amplifier circuit. [1]

b) The input voltage is 12 mV. Calculate

i) the current in the 150 kΩ resistor. [2]

ii) the potential difference between A and B. [1]

c) Outline any assumptions that you have made in (b). [2]

(Total 6 marks)

6 In a mobile phone system an area is divided into a large number of cells.

a) State what is meant by a cell. [2]

b) Suggest **two** advantages of organizing the mobile phone system in this way. [2]

(Total 4 marks)

29 Option G: Electromagnetic waves

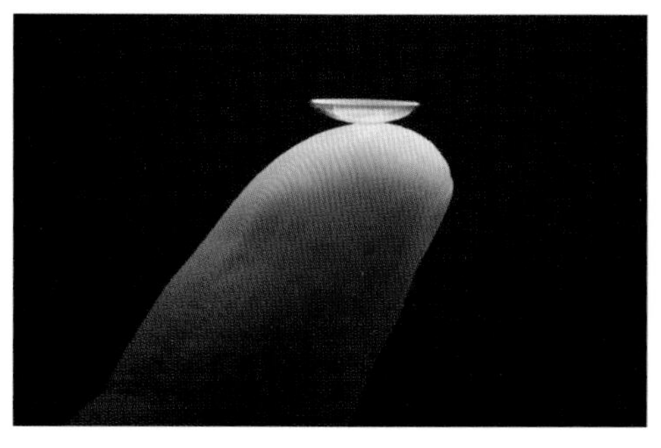

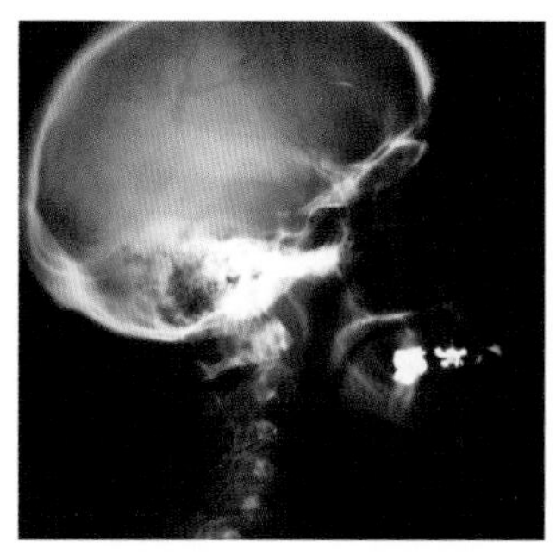

Much of this chapter focuses on applications and ideas involving the hugely important region of the electromagnetic spectrum of waves known as visible light. The later sections extend the ideas of diffraction and interference introduced in Chapter 8, to consider applications within the visible region with X-rays. Before starting work on this chapter, it would be worth familiarising yourself with the general principles introduced in Chapters 8 and 9.

The nature of EM waves

Electromagnetic (EM) waves involve oscillations of electric and magnetic fields. These fields are always at right-angles to one another and also to the direction of energy transfer (the direction of **propagation** of the wave). Thus all EM waves are transverse waves. The diagram below represents a "snap-shot" of the fields at one instant in time. At some later time, the pattern will have moved on along the direction of propagation.

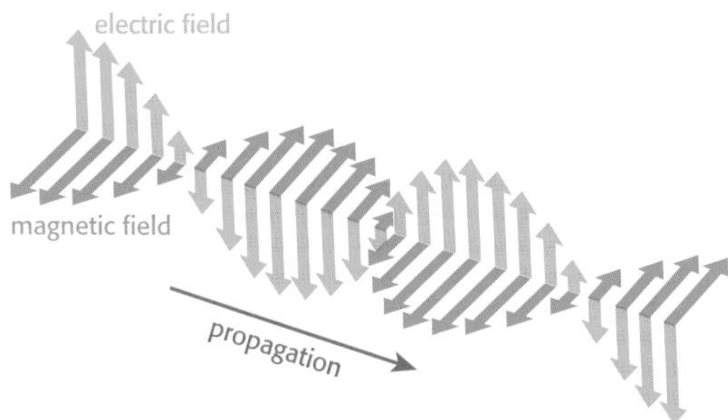

Figure 1

The varying electric and magnetic fields are caused by oscillating electric charges and the resultant EM waves all propagate at the same speed in a vacuum (3×10^8 m s^{-1}). The very different properties of the different regions of the EM spectrum depend on the particular wavelength and frequency of the EM waves under consideration.

As the diagram below shows, our eyes are only sensitive to a very small section of the EM spectrum that we call visible light. The different wavelengths are perceived by us as different colours. When all these wavelengths arrive at the same time, this is perceived as white light. Any process by which white light is split

> With colleagues, produce your own version of the EM spectrum chart to be displayed in the physics lab or along a corridor in your school/college. Try to choose examples that will be understood by younger pupils and/or non scientists.

into its component colours is called **dispersion**. The most familiar example of this is a prism. Each colour is refracted by a different amount and as a result the refractive index is dependent on wavelength.

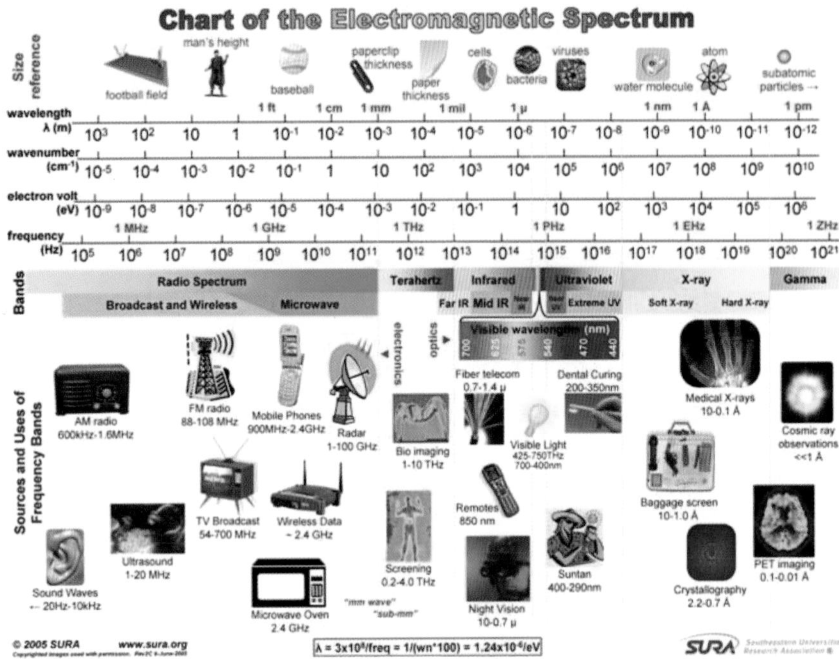

Figure 2

When light travels through a medium, energy is transferred from source to observer. Most of the time, the energy received after transmission is not the same as the energy emitted. Two processes account for this: **absorption** and **scattering**.

Absorption is the process by which the wave energy is received by the medium through which the light is travelling. Typically, particular frequencies are absorbed as a result of resonance (see page 95), but it can also be a result of breaking bonds. For example, the ozone layer in the upper atmosphere (approximately 35 km from the surface of the Earth) preferentially absorbs ultraviolet energy emitted from the Sun. A range of wavelengths, known as UV-C (from 280 to 100 nm), which would be very harmful to humans, are absorbed. Atmospheric absorption plays a significant role in the greenhouse effect (see page 302).

Scattering is the process by which energy is sent in all directions, reducing the amount of energy that is transmitted. For example, when visible light travels through the atmosphere, blue light tends to be preferentially scattered by the molecules (by absorption and re-radiation). As a result, the sky appears blue during the day. This process also explains why sunsets and sunrises appear red – the light from the Sun is travelling through the maximum length of atmosphere before reaching the observer, so the preferential scattering of the blue end of the spectrum results in the light that arrives at the observer being red.

Figure 3

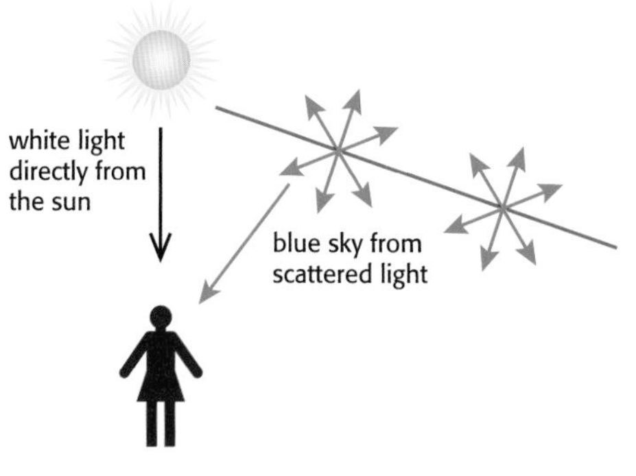

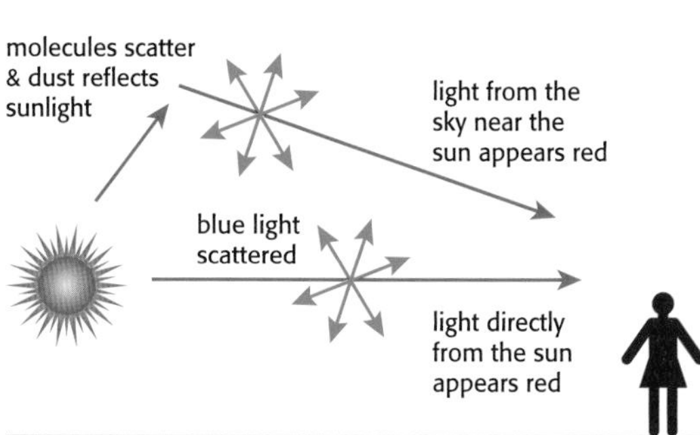

Figure 4

Lasers

The light from a laser is often emitted in a very narrow beam that hardly spreads out at all, even after having travelled a significant distance. The property arises from its method of production and is extremely useful in many applications, as it allows the transmitted energy to be focused into a very small area.

The method of production of laser light is described in its name – **L**ight **A**mplification by the **S**timulated **E**mission of **R**adiation. Light is emitted in "packets" or photons. Photons of light are produced when atomic electrons fall from a higher energy level to a lower energy level (see page 269). This process will happen naturally with any atom that is given energy – the increase in energy promotes the electrons up their energy levels and their subsequent falls result in photons being emitted. The production of laser light involves a process that promotes a large number of electrons into a higher energy level than would normally be expected. This is known as a **population inversion** and the details of how this is achieved are complex. With a population inversion established, electrons can be stimulated to fall from the higher energy level to the lower energy level with a photon of exactly the same energy as the energy "gap". The net result is that an incident photon of the correct frequency causes many more photons to be produced. This amplification of light is repeated as a result of many reflections between carefully aligned mirrors.

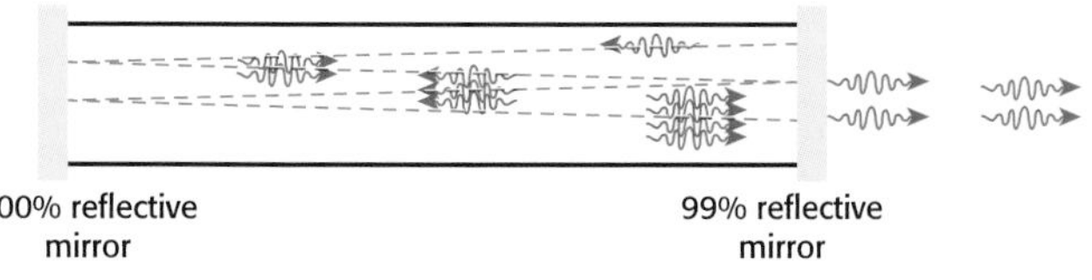

Figure 5

Laser light has many different applications, including:

- Medical application
- Communications
- Technology (bar-code scanners, laser disks)
- Industry (surveying, welding and machining metals, drilling tiny holes in metals)
- Production of CDs
- Reading and writing of CDs, DVDs, etc.

Undertake your own research into one application. Your findings should be presented in a format agreed with your teacher in advance (e.g. poster/article for school magazine/web page/podcast/essay/presentation). Remember to be aware of your target audience (what technical knowledge do they have?) and to ensure that any information that you present is properly cited.

Two other notable properties of laser light, which result from this method of production, are that it is both monochromatic and coherent.

Monochromatic literally means "one colour" and laser light contains only a very narrow band of frequencies. By contrast, a normal light bulb will emit many different frequencies, which combine together to make the light seem white.

Two sources of light are said to be **coherent**, if there is a fixed phase relationship between the oscillations involved. Normally the photons emitted from a source will be independent from one another and the different photons that a source emits will have a random phase difference between them. However, the photons of laser light are all of the same phase.

Optical instruments

This section considers three important optical instruments involving convex lenses. A simple magnifying glass is used to understand image creation and magnification and then combinations of lenses are used to create a microscope and telescope. The non-perfect behaviour of lenses, or their aberrations, brings the section to a close.

Simple magnifying glass

The terminology associated with a convex lens is outlined below:

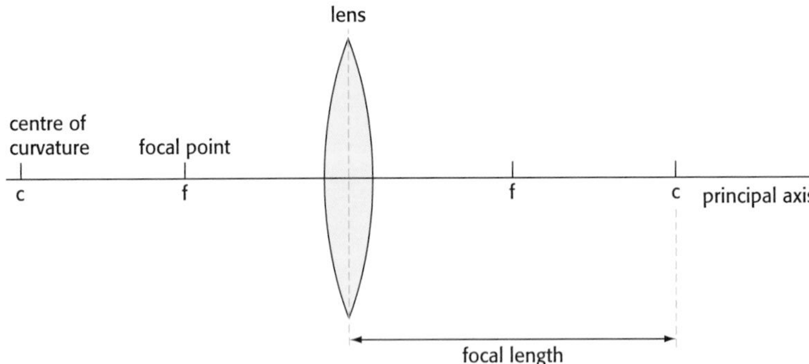

Figure 6

principle axis	The principle axis is the line going directly through the middle of the lens. Technically, it joins the centres of curvatures of the two surfaces.
focal point	The focal point of a lens is the point on the principal axis to which rays that were parallel to the principal axis are brought into focus after passing through the lens.
focal length	The focal length is the distance between the centre of the lens and the focal point.
centre of curvature	Each surface of a spherical lens is a small part of a larger sphere. The centre of curvature of a lens's surface is the point that is at the centre of this sphere.
lens	The power of a lens is the reciprocal of the focal length: $P = \frac{1}{f}$ P is the power of the lens in *dpt* f is the focal length of the lens in m. The dioptre (dpt) is the unit for the power of a lens. 1 dpt = 1 m^{-1}

Light from an object passes through the lens and forms an image. The type, magnification, orientation and position of this image depend on the object's location. In order to locate the image correctly we need to construct a scale diagram using particular rays whose paths we can predict. The paths of all the other rays can be inferred from where these rays meet (or from where they appear to diverge). The special rays are:

- Any ray that was travelling parallel to the principle axis will be refracted towards the focal point on the other side of the lens.
- Any ray that travelled through the focal point (or comes from that direction) will be refracted parallel to the principal axis.
- Any ray that goes through the centre of the lens will be undeviated.

Figure 7 shows the ray diagram for an object placed some distance away from a lens, with these special rays highlighted. This type of image is known as a **real image**, because rays of light actually pass through the point identified. If we placed a photographic film at this point, we could capture an image.

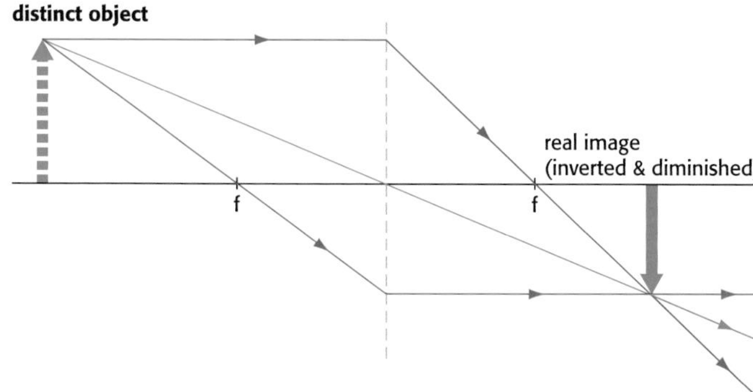

Figure 7

The other type of possible image is a **virtual image**. An example of the formation of a virtual image is a convex lens used as a magnifying glass. In this situation, the object is closer to the lens than the focal length, as shown in figure 8. In a virtual image, the rays of light do not actually pass through the point identified, but just appear to come from that place. If we placed a photographic film at this point, we would not be able to capture an image.

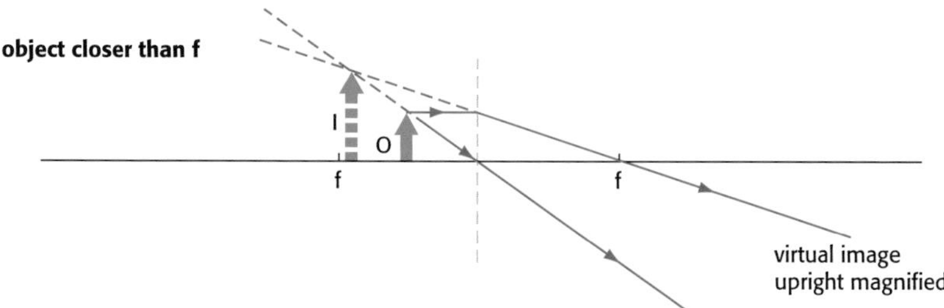

Figure 8

A full image description not only includes its type (real or virtual) location (exactly where it is) and orientation (upright or inverted), but will also calculate the magnification. Linear magnification is the ratio between the size of the image and the size of the object. It has no units.

A convex lens has a focal length of 20 cm. Use four different ray diagrams to locate and fully describe the images formed when an object is place at the following distances from the lens:

a) 50 cm

b) 30 cm

c) 10 cm

d) 20 cm

A magnification greater than 1 means that the image is bigger than the object. Less than 1 means that the image is smaller than the object.

$$m = \frac{h_i}{h_o} = -\frac{v}{u}$$

m is the linear magnification (no units)
h_i is the height of the image in m
h_o is the height of the object in m
v is the image distance in m
u is the object distance in m

There are many applications available on the internet to help understand image formation, for example, www.physics.uoguelph.ca/applets/Intro_physics/kisalev/java/clens/index.html or www.walter-fendt.de/ph14e/imageconvlens.htm (accessed Aug 2009). Researching others would be useful.

A mathematical alternative method of locating the image formed by a convex lens exists – the thin lens formula:

$$\frac{1}{f} = \frac{1}{v} + \frac{1}{u}$$

f is the focal length of the lens in m
v is the image distance in m
u is the object distance in m

The "real is positive" convention for this thin lens formula is that the distances are always measured out from the lens, so positive values of u and v correspond to object and image being on different sides of the lens, i.e. a real image. A virtual image is represented by a negative value of v, in other words it will be on the same side of the lens as the object.

A convex lens has a focal length of 20 cm. Use the thin lens formula to locate and fully describe the images formed when an object is placed at the following distances from the lens:

a) 50 cm
b) 30 cm
c) 10 cm
d) 20 cm

Compare your answers with those resulting from ray diagrams.

As we get closer to an object, it occupies more of our field of vision and we are able to see it in more detail. However, there is a limit to how close we can get and still focus on the object. The **near point** is the closest that an object that can be brought into focus by the unaided eye. The distance from the eye to the near point is called the **least distance of distinct vision**. For normal vision, the near point is taken to be 25 cm away from the observer. The **far point** is the furthest that an object that can be brought into focus by the unaided eye. For normal vision, the far point is taken to be an infinite distance away from the observer.

An important point to note is that when relaxed, our eyes are able to focus on objects that are far away (i.e. "at infinity"). The rays of light that come from any one point on the object must be slightly diverging from one another but are effectively parallel. A ray diagram that produces parallel rays from a point object can be interpreted as a virtual image formed at infinity.

Magnifying glasses allow us to see objects in more detail by allowing an image to be formed that can be focused upon whilst placing the object nearer to our eye than the near point. In this situation an important quantity is the **angular magnification**, M.

The angular magnification is the ratio of the angle subtended by the object for the unaided eye to the angle subtended by the object with the use of lenses. The angle subtended by the object for the unaided eye is often assumed to be its maximum possible value, i.e. when the object is placed at the near point.

$$M = \frac{\theta_i}{\theta_o}$$

M is the angular magnification (no units)
θ_i is the angle subtended by the image in °
θ_o is the angle subtended by the object in °

Figures 9a and 9b show the angular magnification of a simple magnifying glass for an image formed at the near point and at infinity.

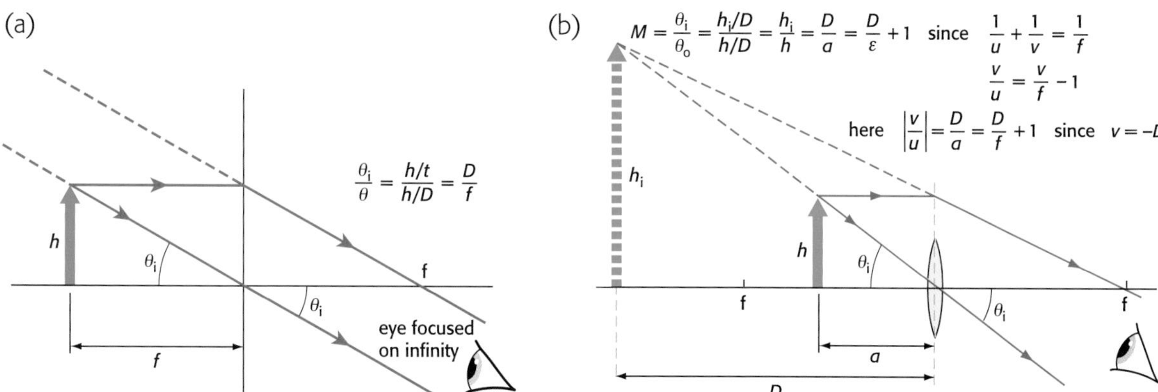

Figure 9

Compound microscope

A compound microscope uses two lenses (the objective lens and the eyepiece lens) to form a final virtual but magnified image of the object being viewed. In normal arrangement the angular magnification for this image is arranged to be as large as possible, by locating the image at the near point of the eye. The arrangement works by placing the object between one and two focal lengths away from the first lens (the objective). This arrangement will form a real but magnified intermediate image. The eyepiece lens is then used as a standard magnifying lens by arranging for the intermediate image formed by the objective lens to be less than the focal length away from the eyepiece lens. The result is a final virtual but magnified image.

The following set of instructions is designed to construct a ray diagram for the arrangement of lenses in a compound microscope:

1 Start with a sheet of graph paper (at least A4 in size) in landscape format. The arrows and lines representing the object and objective lens should be positioned far to the left of sheet for the diagram to fit into the sheet.
2 Choose a suitable scale that can represent at least 28 cm width along the page.
3 The objective lens has focal length 2.8 cm and the object (of height h) is placed 2.4 cm away. Draw two suitable rays (A & B) to locate the image (of height h_1). The image will be formed at the focal length.
4 The eyepiece lens has a focal length of 5.6 cm and is placed a further 4.0 cm to the right of the intermediate image.

5 Two construction lines can now be drawn from the top of the intermediate image to the eyepiece lens (as if they were rays), in order to locate the final image. One line is to the centre of the eyepiece lens – this goes undeviated through the eyepiece lens. The second line is parallel to the principle axis and this goes through the focal point on the other side of the eyepiece lens.
6 The two construction lines can be extrapolated back to locate the virtual image (of height h_2).
7 The rays A & B (no. 3 above) can now be continued from the intermediate image to the eyepiece lens. Their paths after refraction will be coming from the top of the virtual image (no. 6 above).
8 The position of the eye can be added to the diagram.

The angular magnification of a compound microscope is the ratio of the angle subtended by the virtual image at the near point to the angle that would be subtended by the object if placed at the near point. Thus:

$$M = \frac{\theta_i}{\theta_o} = \frac{\frac{h_2}{D}}{\frac{h}{D}} = \frac{h_2}{h}$$

M is the angular magnification of the microscope (no units)
h_2 is the height of the final image in m
h is the height of object in m

Astronomical telescope

An astronomical telescope also uses two lenses, the objective lens and the eyepiece lens to form a final virtual image of the object being viewed. In a normal arrangement, the location of this final image is arranged to be at infinity so that the observer's eye is relaxed. The arrangement works by using the object lens to produce a real but diminished intermediate image. The eyepiece lens is then used as a standard magnifying lens by arranging for the intermediate image formed by the objective lens to be formed at the focal length away from the eyepiece lens. The result is a final virtual image at infinity.

The following set of instructions is designed to construct a ray diagram for the arrangement of lenses in an astronomical telescope:

1 Start with a sheet of graph paper (at least A4 in size) in landscape format. The arrows and lines representing the objective lens should be positioned towards the left-hand side of sheet for the diagram to fit into the sheet.
2 Choose a suitable scale that can represent 40 cm width along the page.
3 The objective lens has focal length $f_o = 16$ cm and parallel rays (at an angle θ_o to the principal axis) arrive from the distant object at the lens. The image will be formed at the focal length draw suitable rays (A & B) to locate the image (of height h_1).
4 The eyepiece lens has a focal length of $f_e = 7$ cm and is placed 7 cm to the right of the intermediate image.
5 A construction line can now be drawn from the top of the intermediate image to the centre eyepiece lens (as if it was a ray), making an angle θ_i to the principal axis final image – this goes undeviated through the eyepiece lens.

6 The construction line can be extrapolated towards infinity to locate the virtual image.
7 The rays A & B (no. 3 above) can now be continued from the eyepiece lens. They will be refracted so as to be parallel to the construction line (no. 6 above).
8 The position of the eye can be added to the diagram.

The angular magnification of the telescope is the ratio of the angle subtended by the image to the angle that subtended by the object. Thus:

$$M = \frac{\theta_i}{\theta_o} = \frac{\frac{h_1}{f_e}}{\frac{h_1}{f_o}} = \frac{f_o}{f_e}$$

M is the angular magnification for the telescope (no units)
f_o is the focal length of the objective lens in m
f_e is the focal length of the eyepiece lens in m

Since the focal points of the objective lens and the eyepiece lens have been arranged to be in the same place, the total length between the two lenses is just $f_o + f_e$.

Aberrations

Simple lenses suffer from two basic problems, or aberrations, which mean that the images produced will not be perfect.

Spherical aberration is the term used to describe the fact that rays striking the outer regions of a spherical lens will be brought to a slightly different focus point from those striking the inner regions of the same lens. In general, a point object will focus into a small circle of light rather than a point image. This problem can be reduced either by using lenses that are not spherical (meaning that a particular shaped lens would only work for objects that were a particular distance away) or by decreasing the aperture in front of the lens and not using the outer sections. The disadvantage to this latter solution is that less total light is allowed into the lens and the effect of diffraction will be worsened.

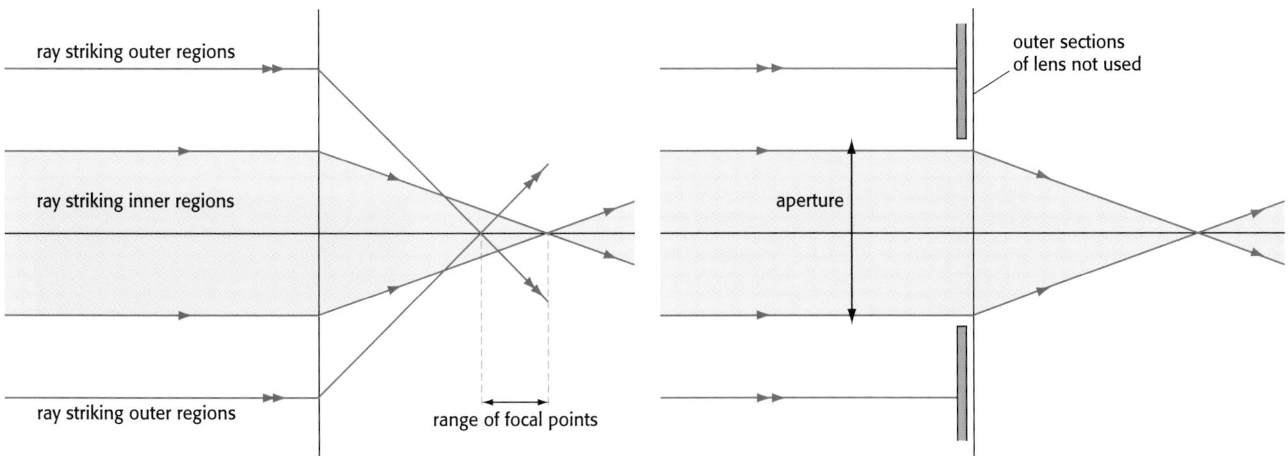

Figure 10 Spherical aberration

Chromatic aberration is the term used to describe the fact that rays of different colours will be brought to a slightly different focus point by the same lens. In general, a white point object will focus into a blurred image of different colours. This problem can be reduced by using two different materials to make up a compound lens or **achromatic doublet**.

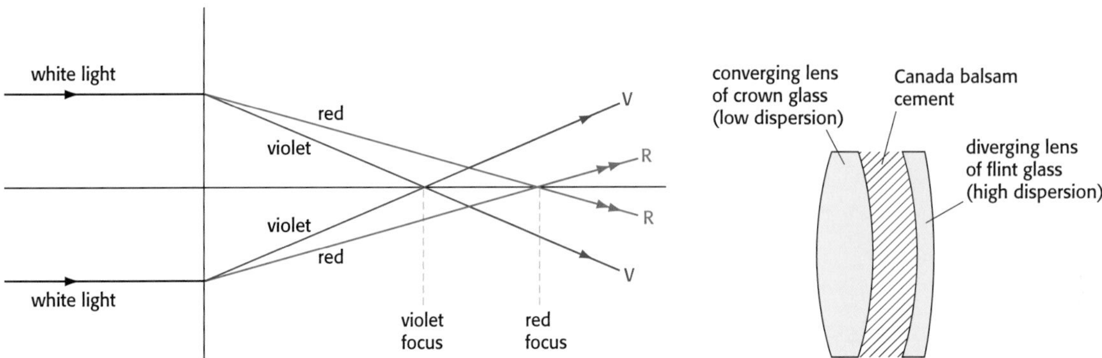

Figure 11

Two-source interference of waves

Rediscovering physics: Diffraction and interference together

Young's double slit experiment provides an example of a situation involving both the diffraction and interference of light.

Diffraction takes place at the slits, making each one a separate source of light. Light waves from these two sources interfere. The light and dark regions that are observed on a screen placed a short distance from the slits correspond to different interference conditions that occur at different points in space. At any given point, the overall result will depend on the relative phase of the two waves being added, and this in turn depends on several variables, including the distance from each source to the point being considered.

Consider two sources of waves S_1 and S_2 thwat are always in phase with each other. Figure 13 shows the individual wave patterns that are formed and the path differences between some points of interest.

The solid black circles represent the points where the waves from both sources are at their maximum displacements (e.g. two crests for transverse waves), the open white circles represent the points where both waves are at their maximum displacements in the opposite direction (e.g. two troughs) and the half black/half white circles represent the points where one wave is at a maximum in one direction and the other wave is at a maximum in the opposite direction (e.g a crest and a trough).

All the points on the line marked A_0 are equidistant from each source. In other words, the **path difference** between the two waves is zero. This means that the waves from S_1 and the waves from S_2 must always arrive in phase. This is known as **constructive interference** and anywhere on the line A_0 will oscillate with double the amplitude of either of the sources.

Constructive interference also takes place at any point where the path *difference* is equal to a whole number of wavelengths. In these situations, the waves also arrive in phase. All the points on the line A_1 are exactly one whole wavelength closer to one of the sources than the other, thus they still arrive in phase. All the points on the line A_2 have a path difference of two whole wavelengths, and so on.

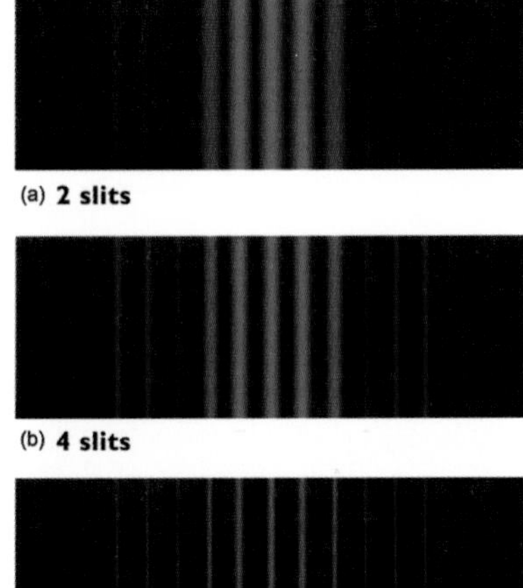

Figure 12 Patterns produced by laser light scattering from diffraction gratings

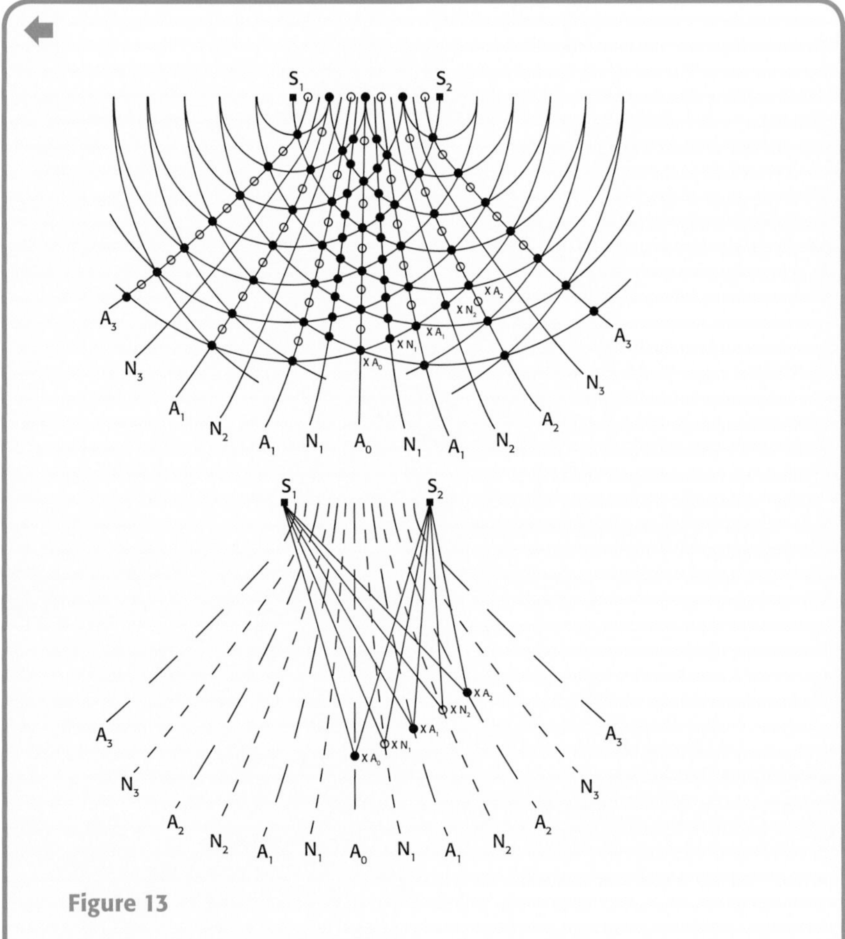

Figure 13

Destructive interference takes place when the waves arrive exactly half a wavelength out of phase. The oscillations from one source exactly cancel the oscillations from the other source and the overall result is no oscillation at all. This takes place on all the lines marked N e.g. N_1, N_2 etc.

There are several equivalent equations for the analysis of the double-slit fringe pattern:

$$s = \frac{\lambda D}{d}$$

$$\sin\theta = \frac{n\lambda}{d}$$

For a bright fringe:

$$\frac{x}{D} = \frac{n\lambda}{d}$$

For a dark fringe:

$$\frac{x}{D} = \left(n + \frac{1}{2}\right)\frac{\lambda}{d}$$

s is the fringe separation in m
λ is the wavelength of light in m
D is the distance from slits to screen in m
d is the separation between the slits in m
θ is the angle between the straight through direction and the bright fringe under consideration in °

n is an integer (1, 2, 3, etc.)
x is the distance on the screen from the central maximum to the fringe under consideration in m.

Diffraction grating (multiple-slit diffraction)

A diffraction grating is a series of parallel slits that have a regular separation. Increasing the total number of slits does not affect the separation of the bright fringes but it does make them sharper and brighter.

The diffraction grating formula is:

$$d \sin\theta = n\lambda$$

d is the separation between the slits in m
λ is the wavelength of light in m
θ is the angle between the straight through direction and the bright fringe under consideration in °
n is an integer (1, 2, 3, etc.)

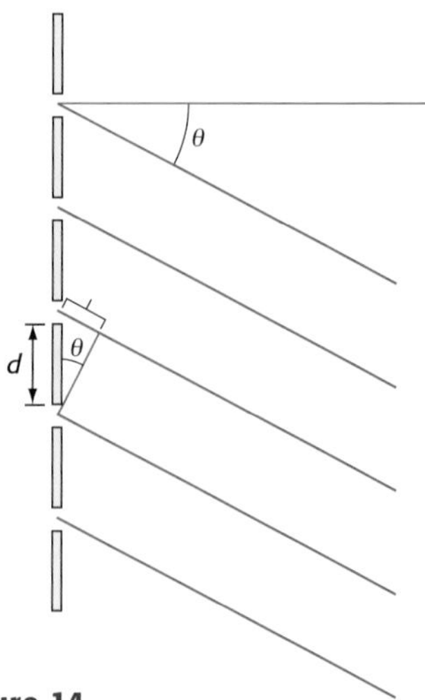

Figure 14

This question aims to derive the diffraction grating formula applied for normal incidence.

a) On the diagram, identify the path difference between two adjacent slits.

b) State the value of the path difference identified in (a), in terms of d and θ.

c) For constructive interference to take place, state the value of the path difference in terms of the wavelength λ and an integer n (= 1, 2, 3, etc.).

d) By equating your answers to (b) and (c), show that the diffraction grating formula applies for constructive interference.

X-rays and their diffraction

The previous sections of this chapter have concentrated on the small region of the EM spectrum known as visible light. All the other regions will, of course, show the same wave properties of diffraction and interference, providing the gratings and obstacles are of an appropriate size relative to the wavelength under consideration. X-rays are high energy EM waves with wavelengths that are of the same order of magnitude as the distance between atoms. This means that X-ray diffraction can be used to determine the internal structure of matter.

A Coolidge X-ray tube produces X-rays that result from the collisions between fast moving electrons and a metal target. The electrons are emitted from a heated cathode (thermionic emission) and are accelerated towards the target by a p.d. The target is cooled. The intensity of the X-rays is increased by increasing the heated current to the cathode. The hardness of an X-ray beam measures its penetration power and higher frequencies are harder.

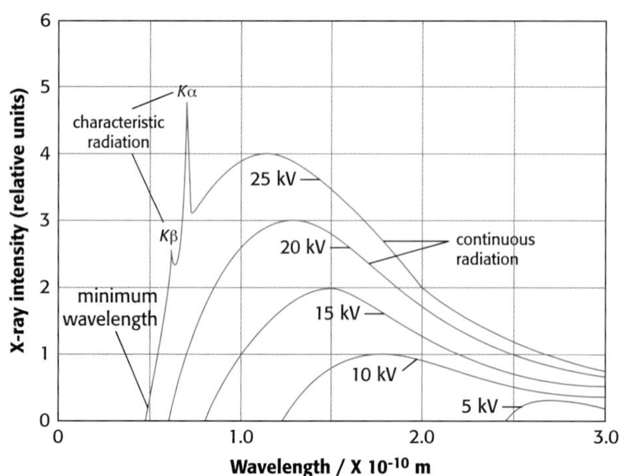

Figure 15 X-ray spectrum

A typical X-ray spectrum contains a range of wavelengths above a minimum wavelength. These X-rays are produced as the incoming electrons collide with the target atoms and are decelerated. The energy of the X-ray photon depends on the energy lost by the electron in the collision. It often also contains high intensities of some wavelengths that are specific to the target's element. These X-rays are produced as the incoming electrons collide with the target atoms and cause inner orbital electrons of the target atom to be promoted up to higher energy levels. As these promoted electrons fall back down, they emit X-rays of a particular wavelength.

The minimum wavelength limit of an X-ray spectrum corresponds to a collision where all of the incoming electron's energy has been converted into an X-ray photon.

$$\lambda_{min} = \frac{hc}{eV}$$

λ_{min} is the minimum wavelength of emitted X-rays in m
h is Plank's constant (6.63×10^{-34} J s)
c is the speed of light (3×10^{8} m s^{-1})
e is the charge on an electron (1.6×10^{-19} C)
V is the accelerating potential difference in V

X-ray diffraction is the term used to describe the scattering of X-rays by a crystal. For a given wavelength of X-rays, the majority of X-rays will pass straight through, but there will also be particular directions in which a high intensity of X-rays is recorded. These directions correspond to directions where constructive interference of the X-rays scattered from different planes of atoms in the crystal takes place. Typically a crystal structure contains many different constructive possibilities and the resulting X-ray pattern can be analysed in order to deduce the atomic arrangements in the crystal. The structure of DNA was discovered by means of X-ray diffraction.

The Bragg scattering equation is

$$2d \sin\theta = n\lambda$$

d is the distance between the atomic planes in m
θ is the half the angle of deviation by the beam in °
λ is the wavelength of light in m
n is an integer (1, 2, 3, etc.)

This question aims to derive the Bragg scattering equation.

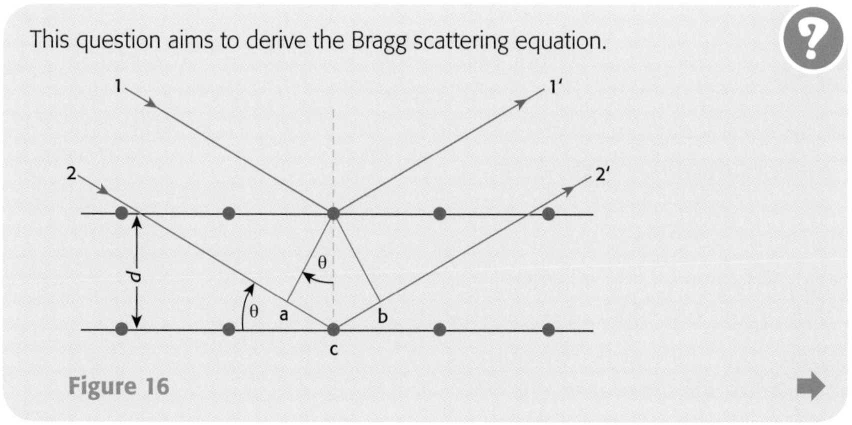

Figure 16

a) Show that the beam (paths 1 & 2) has been deviated by 2θ.

b) State the value of the length AC in terms of d and θ.

c) On the diagram, identify the path difference between two possible paths (1 & 2).

d) State the value of the path difference identified in (c) in terms of d and θ.

e) For constructive interference to take place, state the value of the path difference in terms of the wavelength λ and an integer n (= 1, 2, 3, etc.).

f) By equating your answers to (d) and (e), show that the Bragg scattering equation applies for constructive interference.

Thin-film interference (Wedge films and Parallel films)

Interference patterns occur as a result of path differences that are introduced by thin sections of media (or "films"). The pattern that results depends on whether the film thickness changes (i.e. a wedge film) or whether the film has a constant thickness (a parallel film).

An air wedge is the gap between the two surfaces of two glass plates that are at a small angle to one another (e.g. two microscope slides separated on one edge by a thick piece of paper). Light shone at right-angles to the bottom plate will be reflected back and a path difference is introduced, depending on whether the light is reflected from the bottom of the top surface or from the top of the bottom surface.

Calculating the conditions for constructive and destructive interference are complicated by the fact that when light is reflected back from an optically denser medium, there is a phase change of π, whereas when light is reflected back from an optically less dense medium, there is no phase change. An optical flat is a region that is designed to be completely smooth and thin-film interference can be used to test whether or not this is the case.

The conditions for interference in wedge films are:

Constructive interference:

$$2nt = m\lambda$$

Destructive interference:

$$2nt = \left(m + \frac{1}{2}\right)\lambda$$

n is the refractive index of the medium in the wedge (no units)
t is the thickness of the wedge at the point being considered in m
λ is the wavelength of light in m
m is an integer (1, 2, 3, etc.)

Interference takes place in parallel sided films as a result of a path difference being introduced between reflections that take place at the top surface of the film and reflections that take place at the bottom surface of the film. Once again the mathematics is quite involved, because of the phase differences that are introduced at some surfaces. The conditions for interference in parallel films are:

Constructive interference:

$$2nt \cos \phi = m\lambda$$

Destructive interference:

$$2nt \cos \phi = \left(m + \frac{1}{2}\right)\lambda$$

n is the refractive index of the medium in the film (no units)
t is the thickness of the film in m
ϕ is angle of refraction (the angle between the refracted ray and the normal) in the film in °.
λ is the wavelength of light in m
m is an integer (1, 2, 3, etc.)

Examples of a parallel film that can cause interference include oil on the surface of water and soap films (e.g. in a bubble). When white light is reflected from the film in both of these situations, coloured fringes can be seen as a result of the constructive and destructive interference that are taking place for the different wavelengths of light that are included in white light.

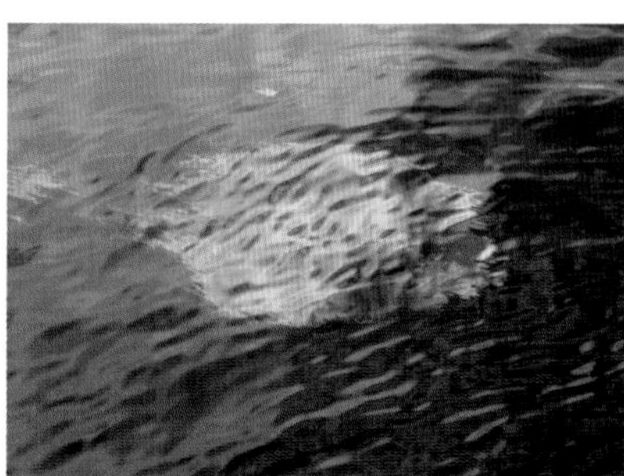

Figure 17

Applications of parallel thin films include:

1 The design of non-reflecting radar coatings for military aircraft. By choosing the appropriate thickness of the outside coating on an aircraft, it is more likely for radar signals to destructively interfere when they reflect from both surfaces, thus preventing reflection and allowing the aircraft to go undetected.
2 Measurements of the thickness of oil slicks caused by spillage can be made from the amount of interference taking place when EM waves are reflected from both surfaces of the oil.
3 Blooming is a thin coating that is added to lenses, solar panels or solar cells to increase the amount of light entering the device. The width of the coating is such that, for typical wavelengths, destructive interference takes place between rays that are reflected from the top and bottom surfaces of the blooming. This maximises the amount of light energy that is transmitted.

Figure 18

End of chapter summary

Chapter 29 has six main themes, two of which are only relevant to those studying higher level physics. The four themes common to all levels are: the nature of EM waves, optical instruments, two-source interference of waves and diffraction gratings. The two HL themes are: X-rays and thin-film interference. The list below summarises the knowledge and skills that you should be able to undertake after having studied this chapter. Further research into more detailed explanations using other resources and/or further practice at solving problems in all these topics is recommended – particularly the items in bold.

The nature of EM waves and light sources
(including lasers)

- Outline the nature of electromagnetic (EM) waves and describe the different regions of the electromagnetic spectrum.
- Describe what is meant by the dispersion of EM waves and in terms of the dependence of refractive index on wavelength.
- Distinguish between and discuss examples of the transmission, absorption and scattering of EM radiation.
- Explain the terms *monochromatic* and *coherent* and identify laser light as a source of coherent light.

- Outline the mechanism for the production of laser light and an application of its use.

Optical instruments (simple magnifying glass, compound microscope, astronomical telescope and aberrations)

- Define the terms *principal axis, focal point, focal length, linear magnification* and *power* of a *convex lens* and the *dioptre*.
- Construct ray diagrams to locate the image formed by a convex lens and distinguish between a real image and a virtual image.
- Apply the convention "real is positive, virtual is negative" for a single convex lens using the thin lens formula **and so solve problems**.
- Define the terms *angular magnification* and *far point* and *near point* for the unaided eye.
- **Derive an expression for the angular magnification of a simple magnifying glass for an image formed at the near point and at infinity.**
- Construct a ray diagram for a compound microscope and for an astronomical telescope in normal adjustment.
- State the equation relating angular magnification to the focal lengths of the lenses in an astronomical telescope in normal adjustment **and solve problems involving the compound microscope and the astronomical telescope**.
- Explain the meaning of *spherical aberration* and of *chromatic aberration* as produced by a single lens and describe how they may be reduced.

Two-source interference of waves

- State the conditions necessary to observe interference between two sources and explain, by means of the principle of superposition, the interference pattern produced by waves from two coherent point sources.
- Outline a double-slit experiment for light and draw the intensity distribution of the observed fringe pattern **and solve related problems**.

Diffraction grating (multiple-slit diffraction)

- Describe the effect on the double-slit intensity distribution of increasing the number of slits.
- **Derive the diffraction grating formula for normal incidence and outline how a diffraction grating is used to measure wavelengths and solve problems involving a diffraction grating.**

In addition HL candidates should be able to:

X-rays and their diffraction

- Outline the experimental arrangement for the production of X-rays, draw and annotate a typical X-ray spectrum and explain the origins of its features.
- **Solve problems involving accelerating potential difference and minimum wavelength.**
- Explain how X-ray diffraction arises from the scattering of X-rays in a crystal, **derive the Bragg scattering equation and solve problems using it**.
- **Outline how cubic crystals may be used to measure the wavelength of X-rays and how X-rays may be used to determine the structure of crystals.**

Thin-film interference (wedge films and parallel films)

- Explain the production of interference fringes by a thin air wedge and how wedge fringes can be used to measure very small separations.
- Describe how thin-film interference is used to test optical flats **and solve problems involving wedge films**.
- State the condition for light to undergo either a phase change of π, or no phase change, on reflection from an interface and so describe how a source of light gives rise to an interference pattern when the light is reflected at both surfaces of a parallel film and state the conditions for constructive and destructive interference.
- Explain the formation of coloured fringes when white light is reflected from thin films, such as oil and soap films.
- Describe the difference between fringes formed by a parallel film and a wedge film.
- Describe applications of parallel thin films **and solve problems involving them**.

Chapter 29 questions

1 This question is about light and the electromagnetic spectrum.

a) Outline the electromagnetic nature of light. [2]

b) The diagram below is a representation of the electromagnetic spectrum.

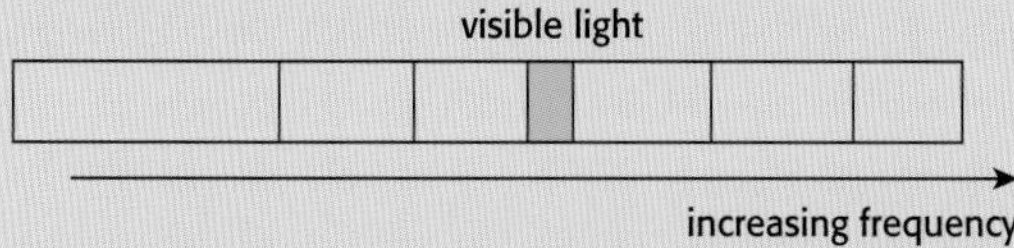

In the diagram the region of visible light has been indicated. Indicate on the diagram above the approximate position occupied by

i) infrared waves (label this I); (1)

ii) microwaves (label this M); (1)

iii) gamma rays (label this G). (1)

(Total 5 marks)

2 This question is about optical dispersion.

The graph below shows the variation with wavelength λ of the speed v of light in one type of glass.

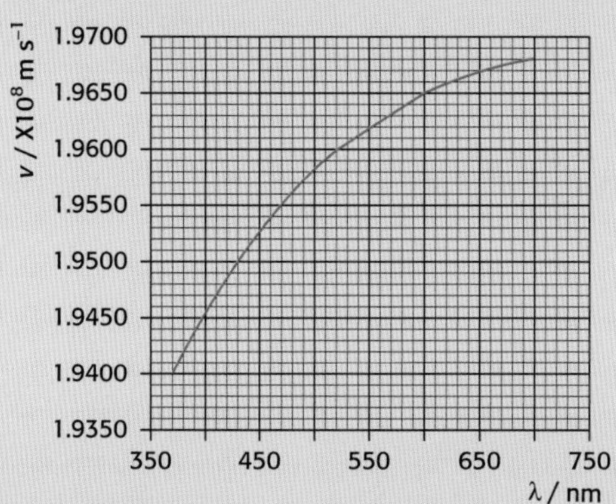

a) Use data from the graph to determine, to the correct number of significant digits, the refractive index for blue light of wavelength 400 nm in this type of glass (free space speed of light $c = 2.9979 \times 10^8$ m s^{-1}) [2]

b) The refractive index of red light of wavelength 650 nm in this type of glass is about 1.52. Use this fact and your answer in (a) to explain optical dispersion. [2]

(Total 4 marks)

3 This question is about laser light.

a) State **two** differences between the light emitted by a laser and that emitted by a filament lamp.

1

2 [2]

b) The production of laser light relies on population inversion. Outline the meaning of the term population inversion. [2]

(Total 4 marks)

4 This question is about converging lenses.

a) The diagram shows a small object O represented by an arrow placed in front of a *converging* lens L. The focal points of the lens are labelled F.

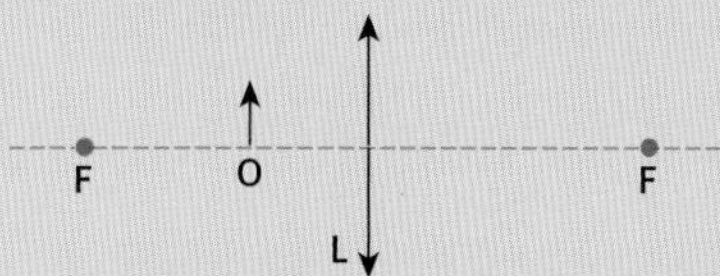

i) Define the *focal point* of a converging lens. [2]

ii) On the diagram above, draw rays to locate the position of the image of the object formed by the lens. [3]

iii) Explain whether the image is real or virtual. [1]

b) A convex lens of focal length 6.25 cm is used to view an ant of length 0.80 cm that is crawling on a table. The lens is held 5.0 cm above the table.

i) Calculate the distance of the image from the lens. [2]

ii) Calculate the length of the image of the ant. [2]

(Total 10 marks)

5 This question is about an astronomical telescope.

(a) Define the focal point of a convex (converging) lens. [2]

The diagram below shows two rays of light from a distant star incident on the objective lens of an astronomical telescope. The paths of the rays are also shown after they pass through the objective lens and are incident on the eyepiece lens of the telescope.

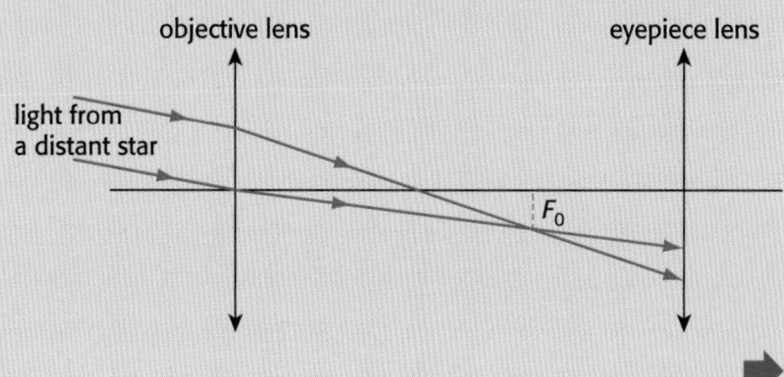

The principal focus of the objective lens is F_O

b) On the diagram above, mark

i) the position of principal focus of the eyepiece lens (label this F_E); [1]

ii) the position of the image of the star formed by the objective lens (label this I). [1]

c) State where the final image is formed when the telescope is in normal adjustment. [1]

d) Complete the diagram above to show the direction in which the final image of the star is formed for the telescope in normal adjustment. [1]

The eye ring of an astronomical telescope is a device that is placed outside the eyepiece lens of the telescope at the position where the image of the objective lens is formed by the eyepiece lens. The diameter of the eye ring is the same as the diameter of the image of the objective lens. This ensures that all the light passing through the telescope passes through the eye ring.

e) A particular astronomical telescope has an objective lens of focal length 98.0 cm and an eyepiece lens of focal length 2.00 cm (ie f_o = 98.0 cm, f_e = 20.0 cm). Determine the position of the eye ring. [4]

(Total 11 marks)

6 This question is about image formation by lenses

The diagram below shows the positions of two convex lenses L_1 and L_2 used in an optical instrument. F_1 and F_2 are the principal foci of L_1 and L_2 respectively. The object O is viewed through the two lenses.

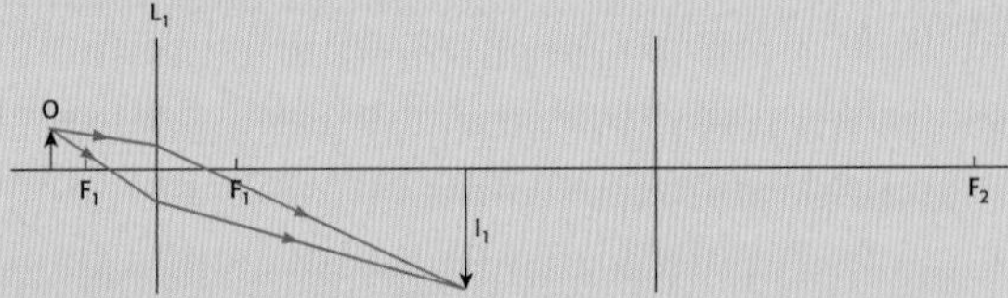

The diagram also shows two rays from the object O to the position of the image I_1 produced in the lens L_1.

a) i) Mark the position of the other principal focus of lens L_2. Label this position F_2. [1]

ii) The image I_1 acts as an object for the lens L_2. Draw **two** construction rays to locate the position of the image I_2 formed by lens L_2. Label this image I_2. [3]

b) State and explain whether the image I_2 is real or virtual. [1]

c) State the name of this optical instrument. [1]

d) State

i) the change, if any, in the positions of the lenses so that the final image in (a) (ii) is formed at infinity; [2]

ii) why the image, formed at infinity, is magnified. [1]

(Total 9 marks)

7 This question is about image formation.

a) A converging lens L has principal foci at F. An object O is placed in front of the lens as shown below.

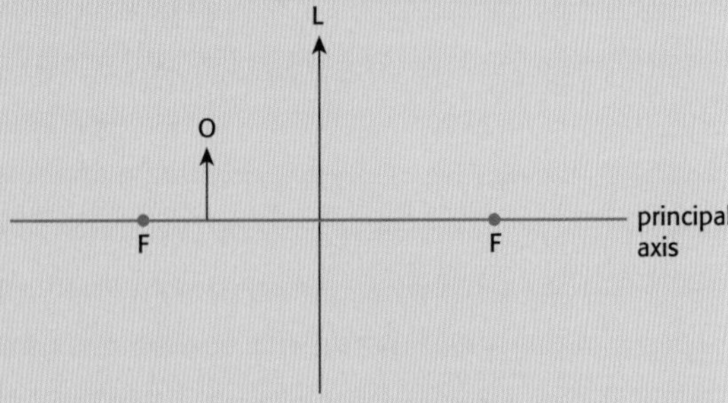

i) Define *principal axis* and *principal foci*. [2]

ii) Construct rays to locate the position of the image formed by the lens. [2]

iii) Explain whether the image is real or virtual. [2]

b) The image is formed at a distance of 25 cm from the lens. The angular magnification produced is 6.0.

i) Determine the distance of object O from the lens. [3]

ii) State the advantage of using the lens with the image formed at the near point of the eye. [1]

(Total 10 marks)

8 This question is about image formation by a converging lens.

a) Define the *principal focus* (*focal point*) of a converging lens. [2]

b) An object is placed 30 cm in front of a converging lens of focal length 15 cm. The object is moved 5.0 cm closer to the lens. Determine the displacement of the image. [4]

(Total 6 marks)

9 This question is about two-source interference.

A double slit is illuminated normally with coherent light. The interference pattern is observed on a screen. The apparatus is shown below.

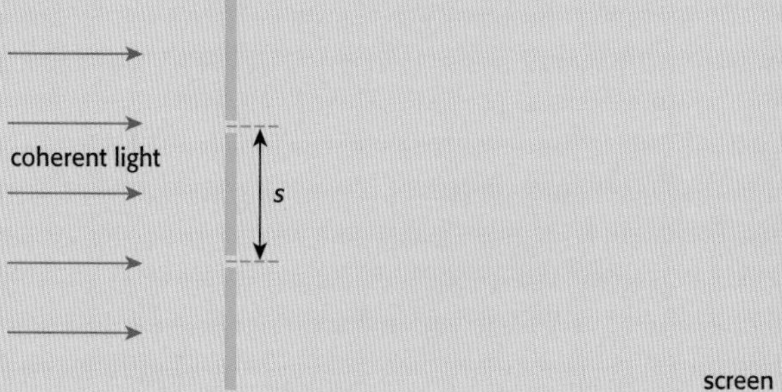

The width of both slits in the double slit arrangement is increased without altering the separation s.

Describe and explain the effect, if any, of this change on

a) the number of fringes observed; [2]

b) the intensity of the fringes. [3]

(Total 5 marks)

10 This question is about a diffraction grating.

Light of wavelength 590 nm is incident normally on a diffraction grating, as shown below.

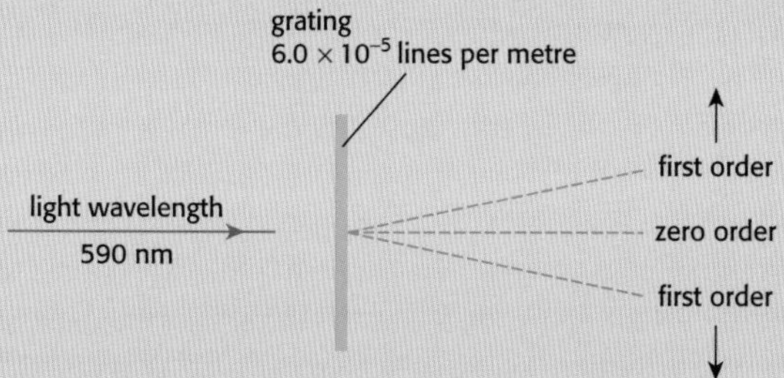

The grating has 6.0×10^5 lines per metre.

a) Determine the **total** number of orders of diffracted light, including the zero order, that can be observed. [4]

b) The incident light is replaced by a beam of light consisting of two wavelengths, 590 nm and 589 nm.

State **two** observable differences between a first order spectrum and a second order spectrum of the diffracted light. [2]

(Total 6 marks)

11 This question is about X-ray spectra.

The diagram shows the X-ray spectrum produced when electrons are accelerated from rest through a potential difference of 25 kV and are then incident on a metal target.

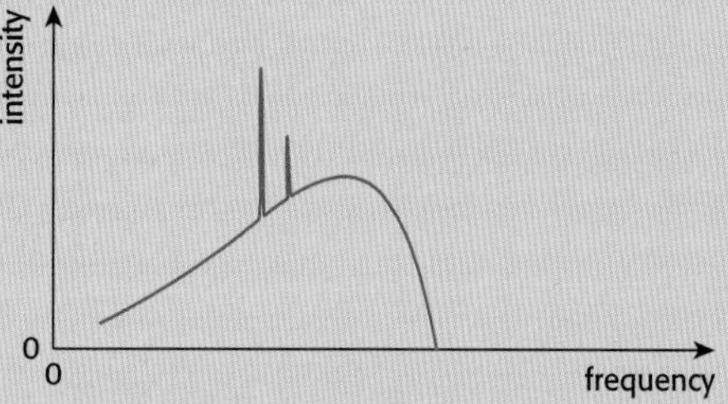

a) Calculate the minimum X-ray wavelength. [3]

b) The electrons are now accelerated through a potential difference of 50 kV. On the diagram above draw the new X-ray spectrum. [2]

(Total 5 marks)

12 This question is about X-rays.

a) Draw a labelled diagram of an experimental arrangement for the production of X-rays. [4]

b) Suggest how each of the following may be controlled.

i) Intensity of the X-ray beam [2]

ii) Minimum wavelength of X-ray photons [2]

c) State, and explain, the origin of the continuous part of an X-ray spectrum. [2]

(Total 10 marks)

13 This question is about the formation of coloured fringes when white light is reflected from thin films.

a) Name the wave phenomenon that is responsible for the formation of regions of different colour when white light is reflected from a thin film of oil floating on water. [1]

b) A film of oil of refractive index 1.45 floats on a layer of water of refractive index 1.33 and is illuminated by white light at normal incidence.

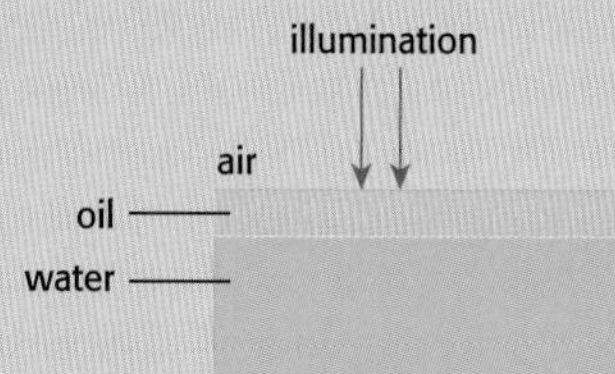

When viewed at near normal incidence a particular region of the film looks red, with an average wavelength of about 650 nm. An equation relating this dominant average wavelength λ, to the minimum film thickness of the region t, is $\lambda = 4nt$.

i) State what property n measures and explain why it enters into the equation. [2]

ii) Calculate the minimum film thickness. [1]

iii) Describe the change to the conditions for reflection that would result if the oil film was spread over a flat sheet of glass of refractive index 1.76, rather than floating on water. [2]

(Total 6 marks)

14 This question is about thin film interference.

Two flat glass plates are in contact along one edge and are separated by a piece of thin metal foil placed parallel to the edge, as shown below.

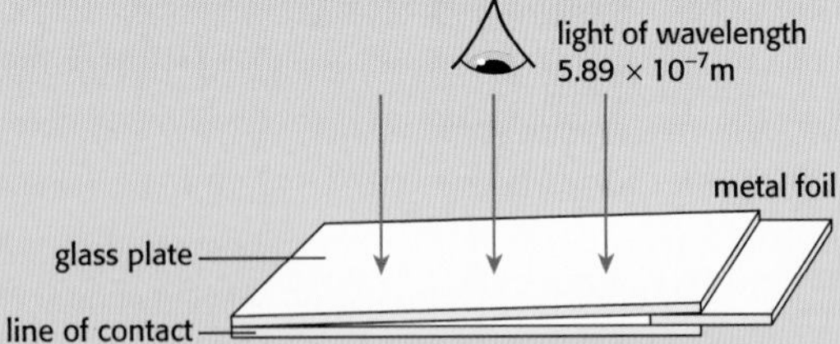

Air is trapped between the two plates. The gap between the two plates is viewed normally using reflected light of wavelength 5.89×10^{-7} m.

A series of straight fringes, parallel to the line of contact of the plates is seen.

a) State what can be deduced from the fact that the fringes are straight and parallel. [1]

b) Explain why a dark fringe is observed along the line of contact of the glass plates. [3]

c) The distance between the line of contact of the plates and the edge of the metal foil is 9.0 cm. The dark fringes are each separated by a distance of 1.4 mm. Calculate the thickness of the metal foil. [3]

The lenses used in astronomical telescopes are frequently "bloomed". This means that a thin film is deposited on the lens in order to reduce the intensity of unwanted light reflected by the lens. Destructive interference occurs between the light reflected from the upper and the lower surfaces of the film. The reflections at both surfaces for one incident ray are shown in the diagram.

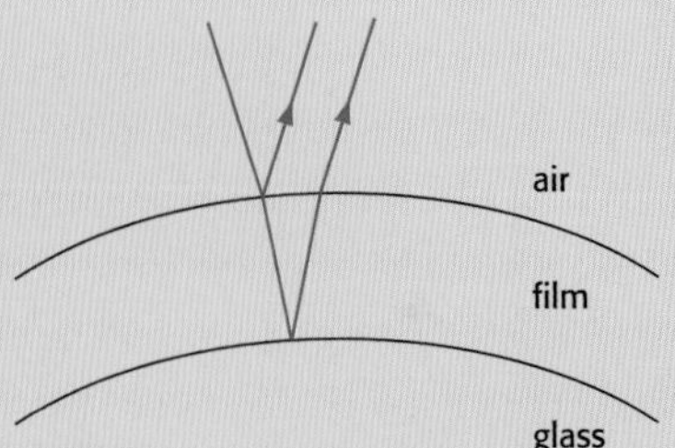

d) i) State why complete destructive interference of all the reflected light does not occur. [1]

ii) With reference to your answer in (i), suggest why the film appears to be coloured. [2]

(Total 10 marks)

15 This question is about a wedge film.

A lensmaker will sometimes check the shape of a lens surface by resting the lens on a flat sheet of glass and then viewing the arrangement in monochromatic light at near-normal incidence, as shown in the diagram below.

● light source

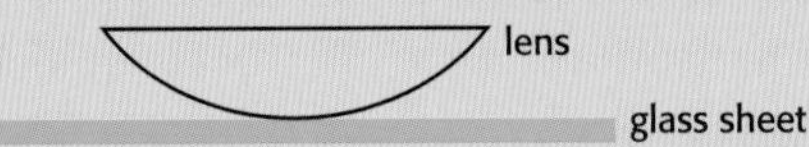

This produces a circular interference pattern with a dark centre, as shown below.

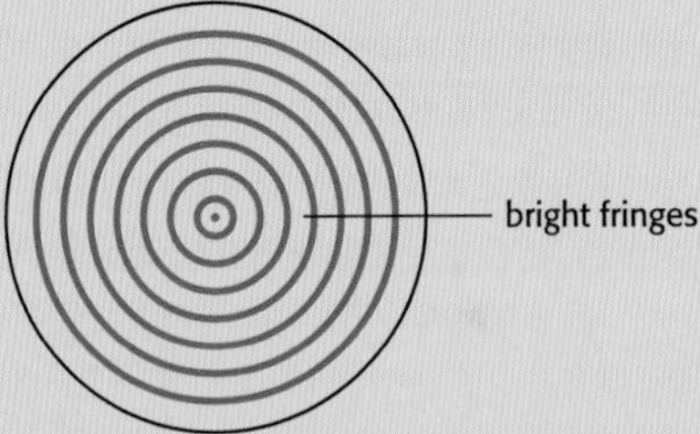

Explain how **one** of the bright fringes arises. You may draw on the diagram if you wish. [4]

(Total 4 marks)

30 Option H: Relativity

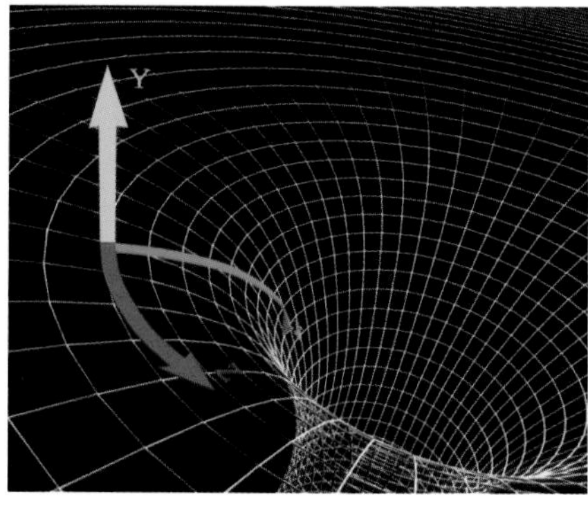

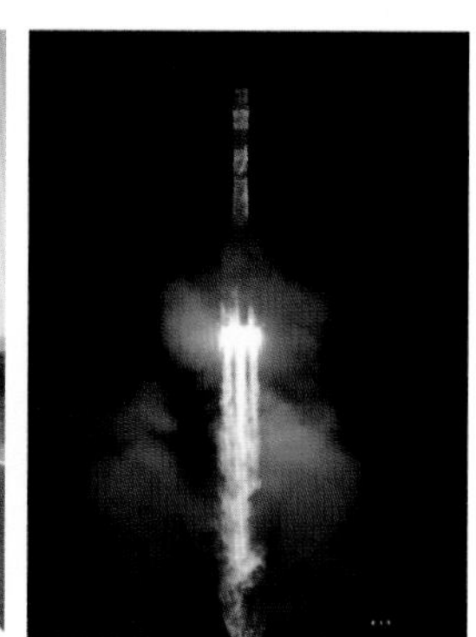

This chapter introduces Einstein's two theories of relativity: the special theory and the general theory. The special theory of relativity forces us to re-examine how we view the world around us and the very nature of space and time. The general theory of relativity extends these ideas to change how we view the force of gravity.

The principles of relativity

It is said that when trains were introduced people feared that "the human body was not made to withstand such extreme speeds". Nowadays, airplanes in smooth level flight move at about 900 km h^{-1} relative to the ground, but as you may have experienced, this does not cause any inconvenience when pouring a drink or walking down the aisle. In fact, without extra external information (e.g. looking out of the window), a passenger in the airplane cannot distinguish if they are moving uniformly or not moving at all. Even Newton's laws of motion cannot distinguish whether an object is moving uniformly or not at all, as both states involve zero acceleration.

Uniform motion can only be described relative to something else. We call this "something else" the **frame of reference**: the airplane is moving at 900 km h^{-1} in the frame of reference of a person on the ground, while a passenger is at rest in their own frame of reference. Formally, a frame of reference is the co-ordinate system in which we measure the position of objects.

Unlike uniform motion, accelerated motion *can* be felt. When the airplane is taking off we feel the force between the seat and our backs. Pouring a drink becomes tricky when the airplane is accelerating randomly due to air turbulence! Newton's laws do not work in an accelerated frame of reference, so we define an **inertial frame of reference** as one in which Newton's Laws of Motion work. An inertial frame of reference is not accelerating and is therefore either stationary or moving at a constant velocity in a straight line.

A basic principle of relativity is that if you are inside an inertial reference frame (with no external information), you cannot distinguish it from being inside any other inertial reference frame.

?

According to an observer on the side of a long straight road, a car that is moving at a constant velocity of 25 m s^{-1} is passed by another car moving at a constant velocity of 30 m s^{-1}. What is the relative velocity of the faster car as measured in the slower car's frame of reference?

We say that uniform motion is relative. Galilean relativity is essentially using relative velocity (see page 33) to find velocities of bodies in different reference frames.

The problem of light and the Michelson-Morley experiment

The problem of light is that the Galilean equations for relative velocity do not seem to apply for a moving observer – the speed of light is independent of the speed of the observer.

In the 19th century, the wave theory of light was predominant, and it seemed that, like other waves, light needed a medium to displace for light waves to exist. This suggested that the speed of light would depend on the motion of the observer through this medium, the "aether", (sometimes spelt "ether"), just as the speed of sound changes with the motion of the observer through the air. This suggests that it would be possible to measure the speed of the Earth through space by measuring the speed of the "aether wind" due to the Earth's motion around the Sun.

In its orbit around the Sun, the Earth travels at 30 km s^{-1}. Michelson and Morley's classic experiment, performed in 1887, was designed to detect the small difference in the speed of light travelling with and across the aether wind that this speed would produce.

Use your calculator to confirm this. The mean radius of the orbit is 1.5×10^{11} m and the time to complete one orbit is 1 year.

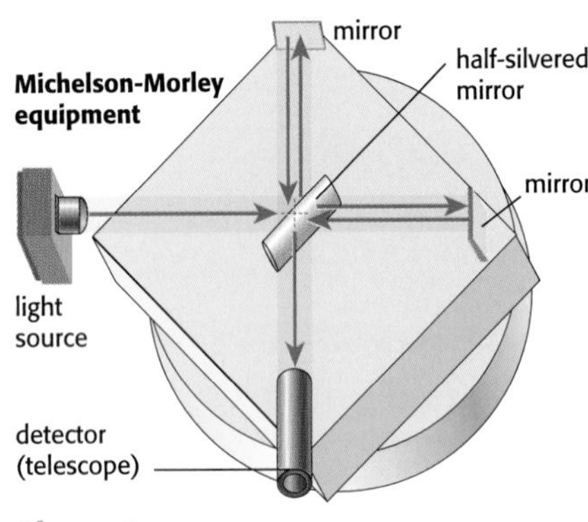

Figure 1

The length of the arm in the 1887 experiment was 11 m. If the aether wind is in the direction of the inline mirror to the telescope, the time for the upwind-downwind return trip is longer than for the crosswind return trip to the perpendicular mirror. As the apparatus is rotated (it was floated on mercury to allow this), the travel time for the light along the two arms would change continuously, resulting in a change in the position of the *interference pattern* observed in the telescope. The modern version, using a laser light source, could use a screen rather than a telescope to observe the interference.

The experiment is famous for its null result. No aether wind speed could be detected, even though the precision of the measurements easily allowed for this, given the Earth's orbital speed. Repeating the experiment at different times of the year (when the direction of the Earth's velocity would be different) had no effect. There is no aether wind. We measure the same speed of light regardless of our velocity relative to the sun.

Mathematical physics: *Calculating the time difference in the perpendicular arms of the Michelson-Morley Experiment*

An analogy to the motion of light down the two arms of the Michelson-Morley experiment is a favourite problem in physics textbooks (see the chapter on velocity vectors). It concerns a boat crossing a river perpendicular to the current. In order to cross a river perpendicularly to the shore, the boat needs to head upstream into the current. The boat's resultant velocity relative to the shore is the addition of its velocity relative to the water and the velocity of the water relative to the shore.

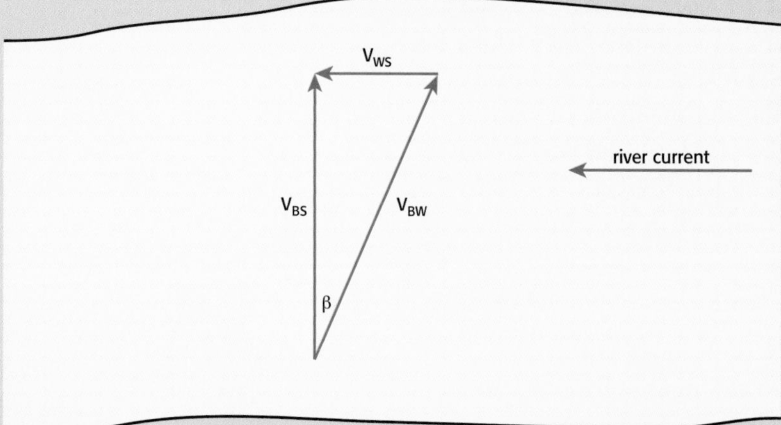

Figure 2

$\mathbf{V}_{BW}$ is the velocity of the **B**oat relative to the **W**ater
$\mathbf{V}_{BS}$ is the **B**oat velocity relative to the **S**hore
$\mathbf{V}_{WS}$ is the **W**ater velocity relative to the **S**hore.

If the boat in Figure 2 can achieve 5 m s^{-1} relative to the water, the current is 3 m s^{-1}, and the crossing is made perpendicular to the current:

a) What is the speed of the boat relative to the shore?

b) How long will the boat take to cross the river, if the river is 100 m wide?

c) Show that the time for a return trip, going 100 m upstream and back again downstream, is longer than the time for the return trip across the river and back again.

There are many good pages on the internet that analyse in detail how the Michelson-Morley experiment was expected to work, for example, http://galileoandeinstein.physics.virginia.edu/more_stuff/flashlets/mmexpt6.htm (accessed July 2009) has a simulation of the experiment allowing the "ether wind speed" to be changed, showing how it affects the travel times for the light in the two perpendicular directions

The problem of EM waves in general: fast moving pions

We expect that the speed of a wave is defined relative to its medium, so that if the observer moves relative to the medium, a change in wave speed is observed. The null result of the Michelson-Morley experiment shows that this is not the case for light. But in quantum physics we treat light as a stream of photons, particles of light. Particles in classical physics emitted by a moving source have a speed that is the sum of the velocity of the source and the emission velocity with respect to the source: if a gun moving at v_g shoots a bullet with muzzle velocity v_m, then the bullet's velocity is $v_g + v_m$. Does the speed of light particles depend on the speed of the source? In general, the problem of EM waves is that not only do the Galilean equations not apply for a moving observer, but they also do not work for a moving source!

A neutral pion (π^0) is one of a family of particles produced when a high energy proton collides with other matter. It decays into two gamma photons, which are like light but with much higher energy. In an experiment performed at the CERN in 1964, Torsten Alväger measured the speed of the gamma rays emitted by π^0 particles travelling at more than 99% of the speed of light in the laboratory frame. He showed with high precision that in the laboratory frame the gamma rays move at the same speed, c, that we always find for light. The speed of light does not depend on the speed of the source.

The postulates of relativity

Einstein starts with two basic ideas or **postulates**, and develops their consequences:

Postulate 1: The laws of physics are the same in all inertial reference frames.
Postulate 2: The speed of light in a vacuum is the same in all inertial reference frames.

Postulate 1 seems a completely reasonable assumption and we have seen that Postulate 2 is experimentally verified. These two ideas are combined to show how our everyday ideas of space and time need to be adapted if these postulates are both valid.

Time dilation: the light clock

In Newtonian physics, time "flows equably" for everyone. This is an instinctive assumption that we make without thinking about it, so it comes as a surprise that the Postulates of Relativity force us to change it. We consider what happens to a moving clock. To help the thought experiment, the clock is very simple and just involves a beam of light bouncing between two mirrors:

One tick t is the time between the two events: a light pulse bounces off the lower mirror, and a light pulse bounces off the lower mirror again. In the clock frame, the two events happen at the same place. In this case, we call the time interval t_0, the **proper time**. Clearly, $t_0 = \frac{2d}{c}$ or $2d = ct_0$.

Next consider the time for one tick as measured from a frame in which the clock is moving with a constant velocity v:

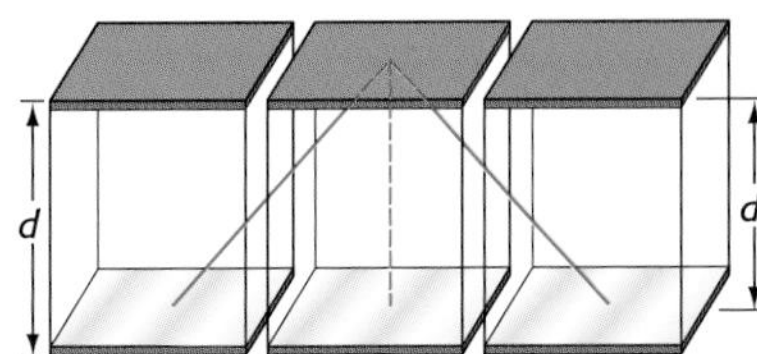

Figure 3 Light path for moving clock

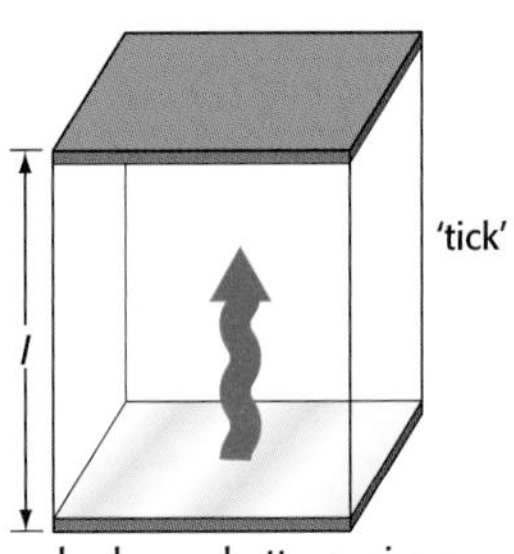

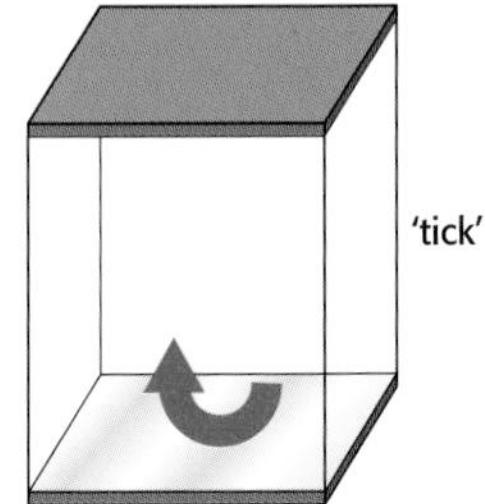

Figure 4 Light path for stationary clock

Using Pythagoras' Theorem, we find the distance travelled by the light pulse during one tick to be:

$$\left(2d^2 + v^2\left(\frac{t}{2}\right)^2\right)^{\frac{1}{2}}$$

Crucially, the speed of light is the same in both inertial frames of reference, so

$$t = \frac{distance}{speed} = \frac{2[\sqrt{(d]^2 + v^2\left(\frac{t}{2}\right)^2})}{c}$$

or $$t^2 = \frac{4d^2}{c^2} + 4v^2\frac{\left(\frac{t}{2}\right)^2}{c^2} = t_0^2 + \frac{v^2t^2}{c^2},$$

giving $$t = \frac{t_0}{\left(1 - \frac{v^2}{c^2}\right)^{\frac{1}{2}}}$$

We see that the time for one tick of the clock t, measured in the frame in which the clock is moving, is longer than the proper time t_0,

since the factor $\frac{1}{\left(1 - \frac{v^2}{c^2}\right)^{\frac{1}{2}}}$ is greater than 1. This "**Lorentz factor**"

$\frac{1}{\left(1 - \frac{v^2}{c^2}\right)^{\frac{1}{2}}}$ is often written as γ, so $t = \gamma t_0$. We call this effect **time dilation**. A moving clock runs slower than one that is at rest. Note also that the interval t is not a *proper* time interval according to the definition above, as the events that start and end the interval happen at different places when the clock is moving.

Use your graphic display calculator to plot the graph of γ against the relative velocity v. What happens as v approaches c?

There are many good animations available on the internet to help visualise what is going on, for example, http://www.phys.unsw.edu.au/einsteinlight/jw/module4_time_dilation.htm (accessed July 2009).

Time dilation is real: the problem of decaying muons

Muons are fundamental particles belonging to the lepton family (see Chapter 32). They are unstable and decay with a half life of 2.2 μs (see pages 276–8 for the mathematics of decay curves and half life). Muons were first discovered among the products of high energy particle collisions resulting from cosmic rays (typically high energy protons coming from space) interacting with the upper atmosphere. These muons travel towards the surface of the Earth at speeds close to the speed of light. Measurements done in the 1940s and subsequently repeated many times, involve counting the rate of arrival of muons of a particular speed at a high altitude, for example, at the top of a mountain, and at sea level. Suppose that the height difference is 4000 metres. Muons travelling at 0.99 c take 13.5 μs to travel this distance, which is more than 6 half lives as measured in the Earth reference frame, so only $1/2^6$ or less than 2% of muons observed at 4000 m would be observed at sea level by this reasoning.

However, the proper time for the travel must be measured in the muon reference frame, and will be equal to $13.5/\gamma\,\mu s$, where $\gamma = (1 - 0.99^2)^{-1/2} = 7.1$, i.e. the muons take 1.9 μs of proper time. Using the exponential decay equation, we calculate that the proportion of muons observed at sea level compared to muons observed at 4000 m would be about 55%, which is confirmed by many measurements. The "muon clock" runs slow as seen from the Earth reference frame.

Use the data in the text to verify that we would expect about 55% of the muons to arrive at the surface of the Earth.

Time dilation seems impossible: the twin paradox

The **twin paradox** is an attempt to show that time dilation is logically inconsistent. Suppose that one of a pair of twins sets off on an interstellar journey, travelling close to light speed to a distant star, and then travels back to Earth. The earthbound twin measures, say, 20 years for the return trip but calculates that because time passes more slowly for the travelling twin, only 1 year of spaceship time has passed, so when they are reunited the travelling twin will be younger. This in itself seems strange, but it is not the paradox.

If all inertial reference frames are equivalent, then while travelling at constant velocity the spaceship frame can be taken as our reference, so that the Earth is doing the travelling while the spaceship is at rest. Thus it is the earthbound twin that ages more slowly and will be younger when they are reunited. This is the paradox: they cannot *both* be younger.

There are several ways to resolve the paradox but the basic error is to say that there are two frames of reference involved when in fact there are three: the Earth, the spaceship travelling away from the Earth, and the spaceship returning to the Earth. The spaceship changes reference frame at the turnaround point, so the twins are not in symmetrical situations and thus are not equivalent.

Figure 5 Since the Earth is always moving in the rocket's frame of reference, time will run slow for the twin on the Earth and the twin on the Earth will be younger when reunited

In order to return to Earth, the twin that returns must have accelerated. It turns out that the twin in the rocket really will age less than the twin left on Earth. This prediction has been experimentally verified using sensitive atomic clocks (see Hafele-Keating experiment on pages 446–7).

Length contraction

If two observers disagree on their measurements of an interval of time, will they agree on their measurements of length? Going back to the muon decay experiment (see page 437), we think about what we would observe from the muon reference frame: in this frame, the Earth would now be rushing towards the muon at 0.99c, taking 1.9 μs to cover the distance from the mountain top to sea level. Clearly the distance cannot be 4000 m as measured in the Earth reference frame, but instead is $0.99c \times 1.9\,\mu s$ or about 560 m. As seen in the muon reference frame, the Earth is contracted – squashed – in the direction of relative motion.

Figure 6 Since the Earth is always moving in the rocket's frame of reference, time will run fast for the twin in the rocket and the twin on the Earth will be younger when reunited

The **proper length** is measured between two objects in the reference frame in which they are at rest. In this case the mountain top and sea level are at rest in the Earth frame, so the proper length L_0 is 4000 m. The contracted length $L = L_0/\gamma$.

Length contraction seems impossible: the barn and pole paradox

This problem attempts to show that length contraction cannot be real because it leads to a paradox. A barn has a door at each end, and a pole is made so that it is slightly too long to fit inside the barn. However, if the pole moves fast enough, it will contract and it will be possible for a brief moment to close both doors simultaneously, with the pole inside the barn. But looking at the situation from the pole frame of reference, the barn is moving and contracting, so the pole cannot possibly fit inside. Explain!

The explanation involves the realisation that events that are simultaneous in one reference frame are not necessarily simultaneous in another. This idea is explored in the next section.

Simultaneity: events that happen at the same time

In this section we will consider in detail how the Second Postulate of Relativity, that the speed of light in free space is the same in any inertial reference frame, affects our ideas about a basic aspect of time: what it means to say is that two events happen at the same time.

Our thought experiment will be about exploding stars. Very massive stars, when they run out of fuel at the end of their lives, explode as supernovae, extremely powerful explosions that can be seen across the universe. By observing distant galaxies with powerful telescopes, we can see these explosions going on all the time. *Nearby* supernovae are seen on average every few centuries - the second most recent was observed in 1604, when a star some 20 000 light years from the Earth was seen to explode. Note that while the explosion was seen in 1604, the light from the explosion had been travelling towards the Earth for 20 000 years.

Let's suppose that by an amazing coincidence light from *two* exploding stars reaches the Earth on the same day, coming from opposite directions. That is, light from exploding star A and light from exploding star B reached the Earth simultaneously. The

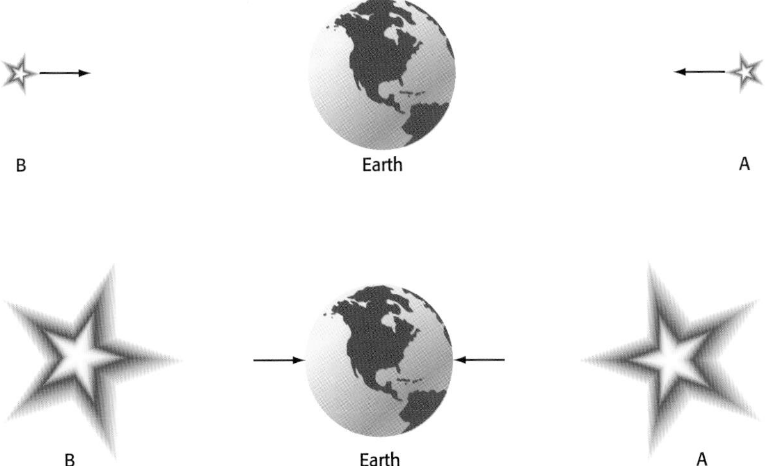

Figure 7 Light from simultaneous events can arrive at the observer at the same time – diagrams are as perceived in the Earth's frame of reference

simultaneity of the arrival times of the light is an observed fact, beyond dispute. What about the actual explosions? Would they also have been simultaneous? If the two explosions occurred at equal distances from the Earth, then the explosions must have been simultaneous as well. Light, from the explosions, travels towards the Earth from opposite sides. Given these conditions we would calculate when the explosions occurred and conclude that they happened at time, D/c, before they were seen from the Earth, where c is the speed of light and D the distance. If D was 10 light years, then the explosions both occurred 10 years before the day that their light reached the Earth, and the explosions were simultaneous.

How would this situation be viewed by another inertial frame that is moving in the Earth's frame of reference? We imagine a starship

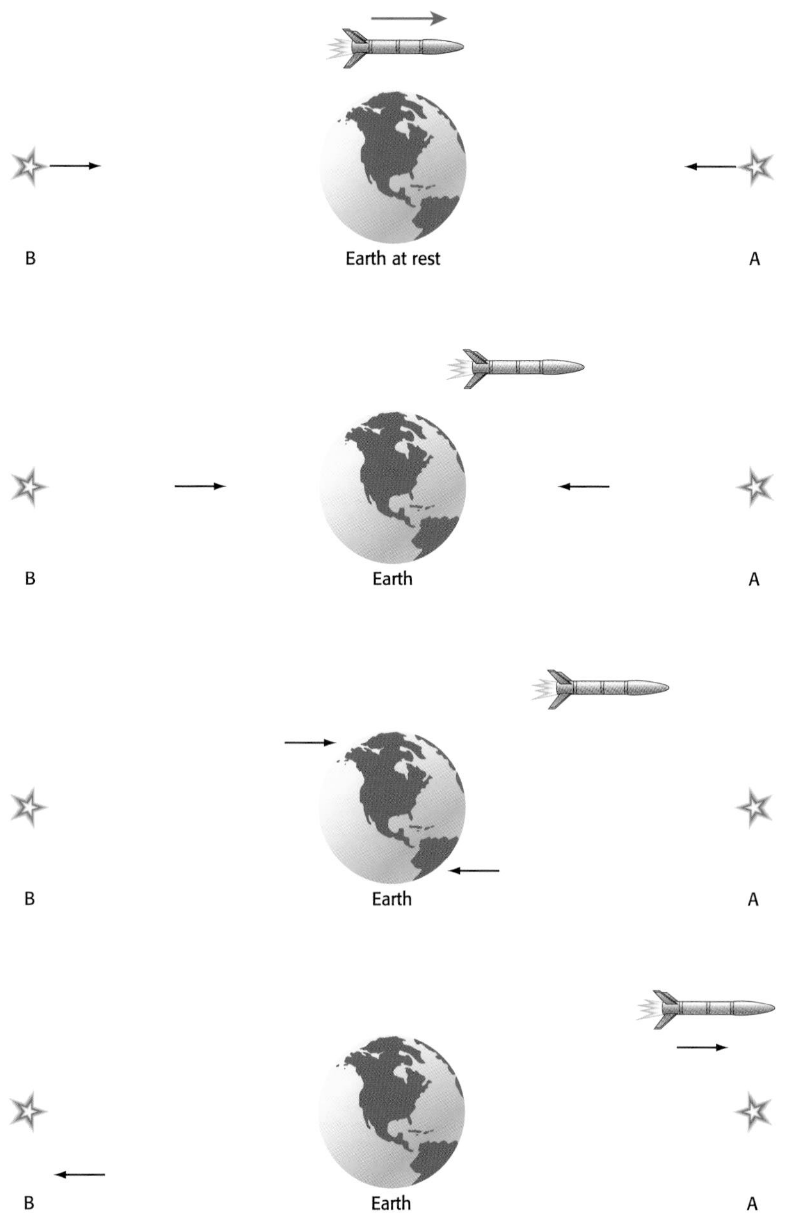

Figure 8 Light from simultaneous events arrive at the observer at different times – diagrams are as perceived in the Earth's frame of reference

travelling from star B to star A, coasting along at 50% of the speed of light. The simplest case to think about is the one where the starship coasts past the Earth on the day (10 years before in our example above) when Earth astronomers calculate that the stars exploded. The starship crew will not see the light from the two explosions arrive at the same time: by the time the light from B reaches the starship, the crew will already have met the light coming from A as it's travelling away from B and towards A. From the Earth's frame of reference this is what we expect: we know that the explosions were simultaneous, but the distance the light travelled from A to the starship was a lot less than the distance travelled from B.

Now we look at the view from the starship frame of reference. The starship crew would measure the two explosion events as having happened at equal distances from their starship,

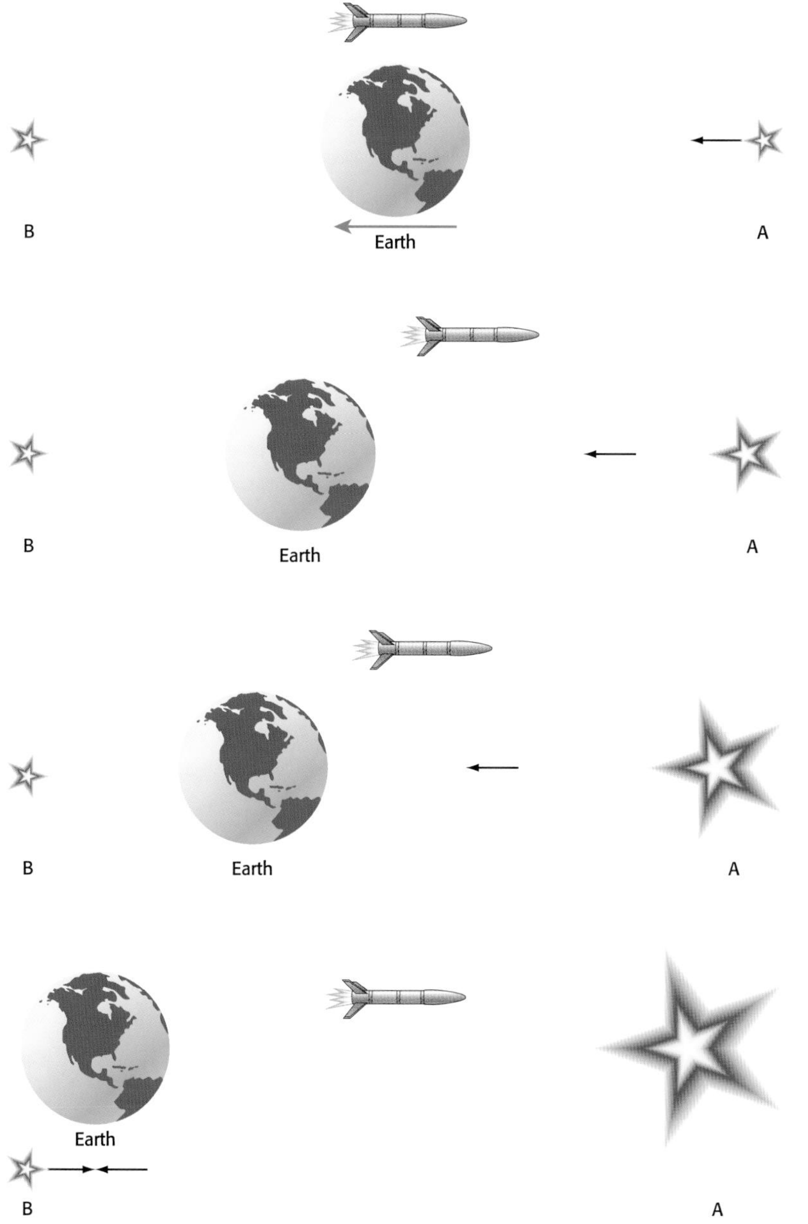

Figure 9 Light from non-simultaneous events can arrive at a moving observer at the same time – diagrams are as perceived in the starship's frame of reference

just when the *Earth* was coasting by at 50% of the speed of light, heading towards the location of B's explosion. Sometime later the Earth astronomers report the simultaneous arrival of the light from the A and B explosions. Nobody can argue with that: it's what they saw. However, what does it mean about the A and B explosion events? The Earth is rushing from A towards B, and it was in the middle as it passed the starship, so the crew conclude that when the light from the two explosions finally reaches the Earth, the light from A has travelled quite a lot further than the light from B, so there is no way that the two explosions could have been simultaneous.

The crew does not see the light reach them at the same time from the explosions, and they are in the middle, at equal distances from the events. This is expected because in their frame of reference the stars *did not explode simultaneously*. The direct consequence of the Principle of Relativity is that our idea that events that are simultaneous must be so, regardless of where we measure them from, must be modified.

Two little warnings: the simultaneity argument usually has two light sources, whether lamps, lightning flashes (as in Einstein's original thought experiment) or exploding stars. Don't confuse the object (lamp, star…), which may be at rest in one of the frames, with the events, which are like points in space and time, and should not be thought of as belonging to one reference frame or the other. In our thought experiment, the observers must measure where the explosions occurred, and these points are fixed in their reference frame, regardless of how the stars that exploded may have been moving. The other problem that can confuse is to think that the argument applies only if the observers are at the midpoint. In fact the Earth and the starship could be anywhere relative to the explosion events and we could show that if the explosions are simultaneous in one of the frames of reference they cannot be simultaneous in the other. It's just that we would have more calculations to do as there would be other distance differences to take account of.

Remind yourself of the barn and pole paradox (page 439). Use the concept of simultaneity to explain why the barn appears to be able to contain the moving pole.

Thinking about science: *Relativity vs. Relativism*

"Everything is relative". How often have you heard that statement? Sometimes it is used to end a discussion where there is no agreement, sometimes to justify a different view. A *truth relativist* claims that something can be "true for you" but not for someone else. An *ethical relativist* holds that right and wrong are meaningful only in relation to one's culture.

Relativity Theory is not "relativist". Indeed the two postulates that we started with are statements about what is true for all inertial observers, and while we have seen that some quantities become frame dependent, the theory is also seeking to identify the *invariant* quantities that are not. These are regarded as most important.

Whether science can be relativist is a different and controversial question. On the one hand, we would say that the results of an experiment are *facts*, which are *objectively true*. On the other hand, it can be argued that the idea of *paradigms* as the basis of science means that true and false are judged within a limited context. For example, something that is true in Newtonian physics may not be true in Relativistic physics. An important idea here is that Relativity does not abolish Newtonian physics, but rather sets the limits in which it can be applied. Relativity formulas must turn into Newtonian formulas when the relative velocity is much less than the speed of light.

Mass, momentum and energy in special relativity

In this section we explore the consequences of the mass-energy relationship, $E = mc^2$. This famous equation states that mass and energy are equivalent. The derivation of the relationship is left for a later course.

The formula for the Lorentz factor suggests that there should be something odd about a situation where $v > c$, since γ then involves the square root of a negative number. In fact, it is well established experimentally in particle accelerators that accelerating a particle with a constant force results in a decreasing acceleration as the speed of light is approached:

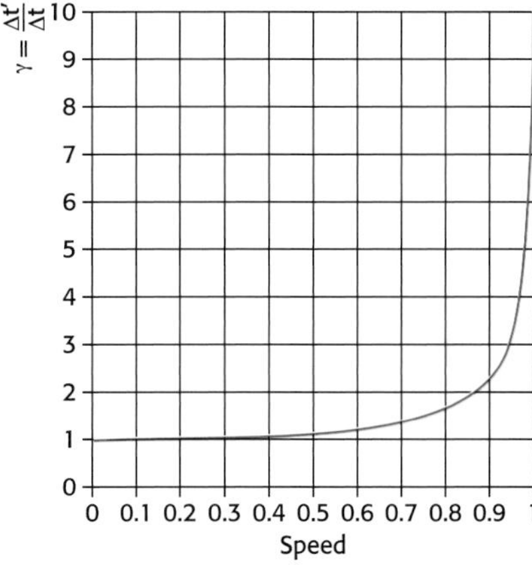

Figure 10

The work done by the accelerating force continues to increase the kinetic energy of the particle, but this takes the form of an increase in mass: mass and energy are equivalent. The overall result is that nothing can go faster that the speed of light.

The **rest energy** of a particle $E_0 = m_0c^2$, where m_0 is the mass measured in the rest frame of the particle

The **total energy** of the particle $E = mc^2$, where $m = \gamma m_0$

The **kinetic energy** of the particle is the difference between total and rest energies

$$KE = E - E_0 = (\gamma - 1)m_0c^2$$

It can be shown that when $v << c$, this becomes approximately the Newtonian kinetic energy ½ mv^2, but in problems with relativistic speeds, the Newtonian formula cannot be used.

Relativistic momentum is defined as $p = \gamma m_0 v$

The momentum and total energy of a particle are related by

$$E^2 = p^2c^2 + m_0^{\,2}c^4$$

The velocity addition formula

The previous section identified the speed of light as the ultimate speed, but consider the situation when starship A leaves the Earth at 0.9c and starship B leaves the Earth travelling in the opposite direction at 0.5c. Isn't the velocity of A in B's frame of reference 1.4c greater than the speed of light?

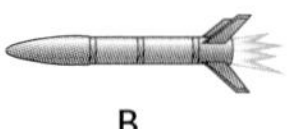

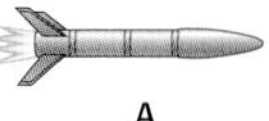

Figure 11

This is of course exactly what Galilean calculations say, but in relativity we have a different way of adding velocities, which allows us to move between frames of reference:

$$u'_x = \frac{u_x - v}{1 - \frac{u_x v}{c^2}}$$

In our example, u'_x is the velocity of A in B's frame of reference, u_x is the velocity of A in the Earth's frame of reference (−0.9c), and v is the velocity of B's frame of reference relative to the Earth (+0.5c). This gives

$$\frac{-0.9c - 0.5c}{1 - (-0.9 \times 0.5)} = -0.966c$$

as the velocity of A in B's frame. Note the importance of keeping track correctly of the plus and minus signs, i.e. the directions of the velocities. The net result is that the speed of light is not exceeded in any frame of reference.

Spacetime as an alternative approach

In 1908, Hermann Minkowski proposed the concept of **spacetime**, a four-dimensional co-ordinate system, with proper time added to the usual three-space dimensions. Just as points are located in three-dimensional space, so events are located in four-dimensional spacetime.

The distance from the origin, d, to a point x, y, z in space is given by $d^2 = x^2 + y^2 + z^2$, according to Pythagoras' Theorem. The spacetime interval between an event at the origin and an event at x,y,z,t, is defined as $I^2 = x^2 + y^2 + z^2 - c^2t^2$.

This interval is invariant: it does not change when we change reference frame. Time dilation and length contraction turn out to be like the apparent change in the length of a rod in normal space, when the rod is viewed from two different frames of reference.

Spacetime turns out to be a very useful idea, which Einstein used in his theory of gravity.

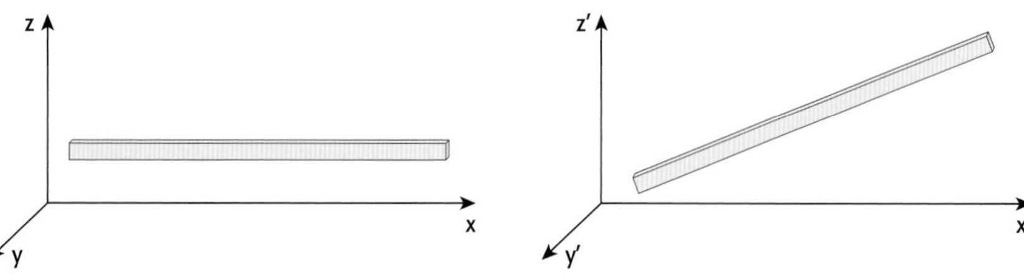

Figure 12

The Equivalence Principle

So far we have been dealing with "Special Relativity", so called because it deals with the special case of inertial reference frames, as contained in Einstein's 1905 paper. "General Relativity" was the content of Einstein's 1916 paper and is his theory of gravity. An important step on the way to this theory is the Equivalence Principle, which confronts a mystery of Newtonian Physics that we are so used to that we don't even notice it.

At the beginning of this chapter, it was stated that we can tell when we are in an accelerated (non-inertial) reference frame because we feel a force. This is so in every case but one: gravitational acceleration.

In free fall, we say that we are accelerating at g, but inside a freely falling box we have no sensation of force. In fact, the free-fall reference frame perfectly satisfies the requirements for an inertial frame of reference. However, when our box is parked on the surface of the planet, we do feel a force pushing us up, just as if we were accelerating upwards at acceleration g. We say this because in order to stay at rest in the box, the force of gravity must be countered. From what we feel inside the box, there is no difference between the accelerating "upwards", far from any planet, and sitting on the surface of the planet.

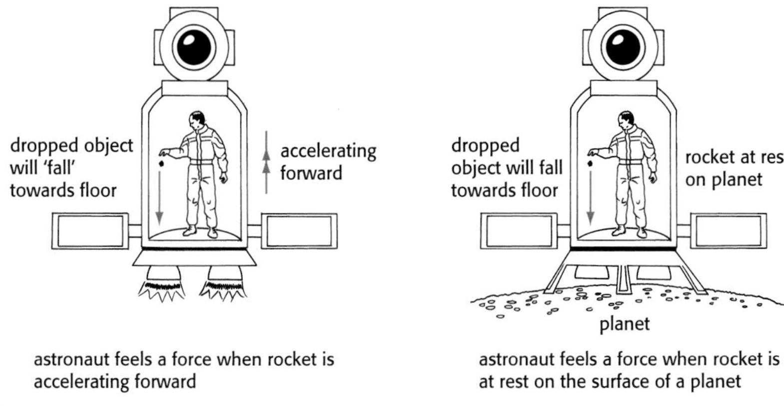

Figure 13

We cannot feel any difference between the two because they are the same: (uniform) gravity and acceleration are equivalent.

Gravitational redshift and time dilation

The Equivalence Principle leads directly to the concept of **gravitational redshift**: the frequency of light changes when it travels up or down in a gravitational field, implying that there is a slowing of time lower down. To see this we consider what happens to the light wave as viewed from an accelerating reference frame, and then by the equivalence principle, conclude that the same must be true in a gravity field.

Because the rocket is accelerating, the receiver of the light pulse is moving upwards faster than the emitter was moving, so there will be a Doppler shift towards longer wavelengths: the light is redshifted. Similarly, light emitted from the top to a receiver at the bottom of the rocket is blueshifted. By the Equivalence Principle, anything caused by acceleration is caused also by gravity, so if A and B are at the top and bottom of a tower resting on the surface of the planet in a gravity field, the same redshift would be observed.

If the source A is a clock that ticks by emitting light flashes, then an observer B at the top will see the clock ticking slower by the same argument. There is a gravitational time dilation. If we switch the positions of A and B, then the higher clock is seen to run faster: the effect is not symmetrical like the time dilation of Special Relativity. Time is slower lower down in a gravity field.

The formula for the gravitational redshift is

$$\frac{\Delta f}{f} = \frac{g\Delta h}{c^2}$$

where Δh is the change in height in constant gravitational field of strength g, and Δf is the frequency shift produced.

Show that the formula for gravitational redshift fits the idea that the photon energyw changes by $h\Delta f$ as a result of a change in gravitational potential energy $mg\Delta h$, where the original photon energy $hf = mc^2$. There is an approximation involved here: under what conditions can it be used?

Experimental evidence for general relativity

1 *Pound-Rebka experiment: a test of the gravitational redshift*
In 1959, Pound and Rebka devised an experiment to test the gravitational redshift over just 22.5 m, the height of the "tower" at the Jefferson Laboratory at Harvard University, by detecting the change in the frequency of gamma rays emitted by ^{57}Fe nuclei. The frequency shift prevents the gamma rays from being reabsorbed by a receiving sample of ^{57}Fe: a precise frequency match is needed for absorption to occur. The source was oscillated up and down at variable speeds, to impose an additional Doppler shift on the gamma rays, opposite to the expected gravitational shift, and the speed when the two shifts cancelled was noted by observing when the receiving sample absorbed the gamma rays. The percentage change in the frequency over such a small change in height is extremely small, and the experiment was only possible because of the Mossbauer Effect, which limits the recoil of the emitting and absorbing nuclei, so that emission and absorption frequencies are sufficiently precise.

2 *The Shapiro time delay*
When a radar signal passes close to the Sun, it is slowed in accordance with the gravitational time dilation, as it will pass through a region of lower gravitational potential near the Sun. This test was performed in 1964 by the American astronomer Irwin Ira Shapiro, by timing a radar signal bounced off the planets Mercury and Venus as they passed on the far side of the Sun.

3 *The Hafele-Keating experiment*
In 1971, a test of the predictions of the time dilations predicted by the kinematic time dilation of Special Relativity and the gravitational time dilation predicted by the Equivalence Principle was performed by flying caesium beam atomic clocks around the world on commercial airliners. Such clocks have nanosecond precision, which is sufficient to detect the predicted time differences when the travelling clocks are compared with laboratory clocks at the end of the journeys.
The calculations must take into account the rotation of the Earth: if we take as our reference the centre of the Earth, then the lab clock is travelling round in an eastbound direction as the Earth rotates. A clock travelling east is going round faster, while a clock travelling west is going round in an easterly direction too, but slower. Then there is the gravitational effect: a clock on the

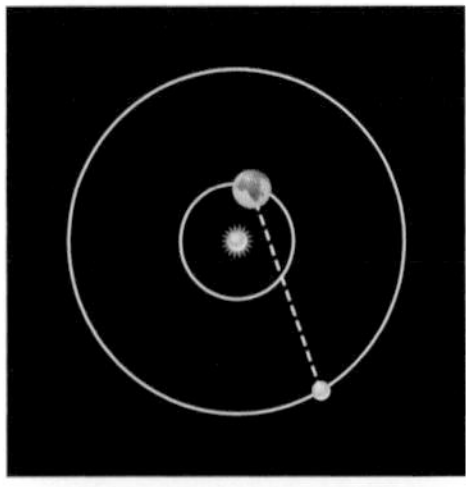

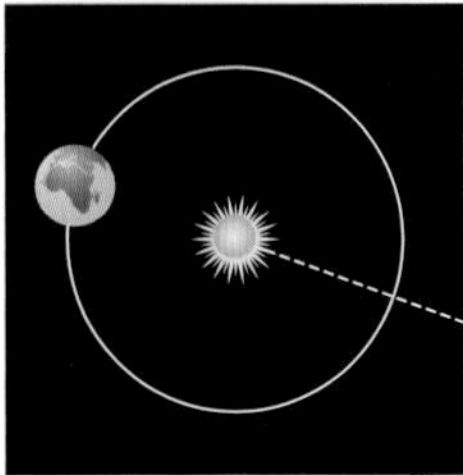

Figure 14 The Shapiro time delay

surface of the Earth runs slower than a clock 10 km above the surface (the cruising altitude for airliners), according to the gravitational time dilation. The westbound clock spends more time in flight, so it will be affected more by this effect.

The published results were as follows:

Table 1

	Nanoseconds gained			
	Predicted			Measured
	Gravitational	Kinematic	Total	Total
Eastbound clock	144 ± 14	−184 ± 18	−40 ± 23	−59 ± 10
Westbound clock	179 ± 18	96 ± 10	275 ± 21	273 ± 7

J.C. Hafele and R.E. Keating, Science 177, 166 (1972) as quoted in http://hyperphysics.phy-astr.gsu.edu/HBASE/Relativ/airtim.html accessed July 2009

GPS measurements work by comparing the times for signals from different satellites orbiting the Earth at a height of 20 000 km to reach the receiver, so high precision time measurement is necessary, comparing orbiting and earthbound clocks. The effects tested in the Hafele-Keating experiment are routinely incorporated in GPS measurements, so as to get correct results.

Einsteinian gravity: warped spacetime and the bending of light

In Einsteinian gravity we use the concept of curved spacetime: mass warps the spacetime around it. A body in free fall simply follows the shortest path in spacetime, which turns out to be curved because of the distortion. Because we cannot visualise four dimensions, we often picture a flat sheet warped by a mass pulling it down, with a small light ball rolling round in an orbit, but be aware of the limitations of this picture.

Mass tells spacetime how to curve; curved spacetime "tells" mass how to move. This results in messy calculations, and before computers were widely available to do numerical analysis, the applications of Einstein's gravity were few. When the curvature is slight, Newton's Law of Gravity is close enough, but it was known already for some decades that calculations for the orbit of Mercury, the planet closest to the Sun and therefore passing at closest approach (**perihelion**) through more extreme spacetime curvature, has features not explained precisely by Newton's Laws (exactly how the axis of the ellipse of the orbit swings around the Sun). An important success of the new theory was to provide a precise calculation for the orbit of Mercury.

The idea that light particles would be affected by gravity dates back to the 18th century, but it moved to the background when the wave model of light became widespread. The Equivalence Principle by itself brings back the idea that Newtonian gravity would deflect light, since a light beam curves when viewed from an accelerated frame, and therefore also in a uniform gravity field. Einstein originally predicted that starlight passing close to the edge of the Sun would be deflected by 0.875 arc seconds, which, while it could not be observed under normal conditions (the Sun is too bright for stars to be visible!), could be noticed as a shift in the apparent position of stars near the Sun during a total eclipse, when the moon blocks out the sunlight. Such an eclipse was to occur in 1919, but in the intervening years the full theory of gravity was developed, in which spacetime curvature causes additional deflection, to give 1.75 arc seconds in total, so that the observation of the effect becomes a test to distinguish between Newtonian and Einsteinian theories of gravity. (For an animation of these effects, see: www.astro.ucla.edu/~wright/deflection-delay.html - consulted August 2009.)

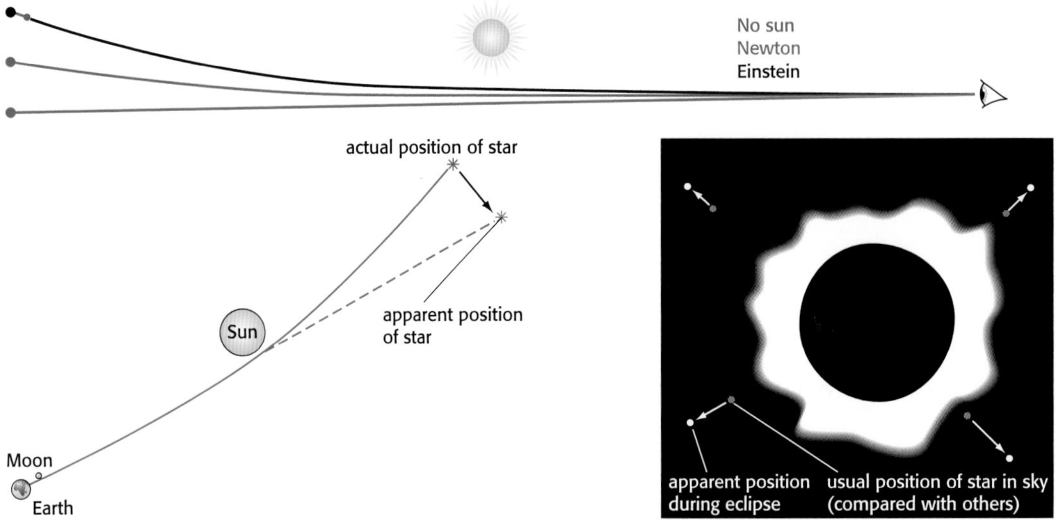

Figure 15 Differences between the predictions of Newtonian and Einsteinian theories of gravity

Famously, the astronomer and physicist Sir Arthur Eddington led an expedition to the Island of Principe, which lay in the path of totality of the eclipse, to photograph the stars near the Sun. Although the results have been disputed, it is now generally accepted that Eddington's measurements confirmed the correctness of the predictions of Einsteinian gravity.

Thinking about science: *Scientific collaboration*

The Eddington-Einstein collaboration was notable for having continued through the bitterly fought First World War, when the two physicists were on opposite sides. However, physics is not always detached from politics. Researching how one affects the other can provide good case studies for ToK Presentations.

Eddington's results were disputed by another team observing the eclipse in Brazil, who reported results more closely agreeing with the Newtonian prediction. The dispute over who was right depended on technical details about how a telescope overheated by the Sun would behave, but has been taken as an example of how theory, observation and ethics interact in complicated ways.

Black holes

If you have studied Chapter 27 on Astrophysics, you will be familiar with black holes as the final state of collapse of very massive stars, when the remnant is so dense that gravity prevents the escape of light and everything else. In terms of Einsteinian gravity, a black hole is the result of such extreme spacetime curvature that even the path of light becomes a closed orbit.

The **Schwarzschild radius** R_s is the distance from the centre of the black hole to the point at which the escape velocity is equal to the speed of light. Effectively it is the "edge" of the black hole, also called the **event horizon**. Although Newtonian gravity does not work close to a black hole, we do get the correct expression for the Schwarzschild radius by setting escape velocity equal to c in the Newtonian expression:

$$R_s = \frac{2GM}{c^2}$$

This is true, provided that the black hole is perfectly spherical.

Calculate the Schwarzschild radius for a 10 kg black hole. Could such an object exist? Do an internet search for "micro black holes".

The gravitational time dilation near a black hole is much more extreme than previously discussed. If Δt_0 is the time for one tick of a clock situated at distance r from the centre of a black hole, then an observer situated far from the black hole (where the spacetime curvature is negligible) observes the time as

$$\Delta t = \frac{\Delta t_0}{\sqrt{1 - \frac{R_s}{r}}}$$

Note that at the event horizon, time stops altogether, Δt becomes infinite.

Gravitational lensing

Quasars are objects that caused much controversy in cosmological circles. They are star-like objects, but their light shows such extreme redshift that if this is due to Hubble's Law (see the Astrophysics Option), then they must be placed right at the edge of the observable universe. If true, then their luminosity is similar to that of an entire galaxy, and yet they appear point-like, like stars. An interesting feature of quasars is that they sometimes appear in pairs, or multiples, all showing the same spectral features and fluctuations in brightness, usually delayed with respect to each other, as if we were seeing double or multiple images of the same object.

Gravitational lensing provides an explanation: a massive object such as a galaxy between the Earth and the quasar bends the light from the quasar. A spherical object exactly in line would cause a ring of light to be seen, but the more complicated real situation produces double or multiple images more often. The importance of gravitational lensing, the debate between rival cosmological theories, is that it confirms that the quasars must indeed lie at great distances from the Earth.

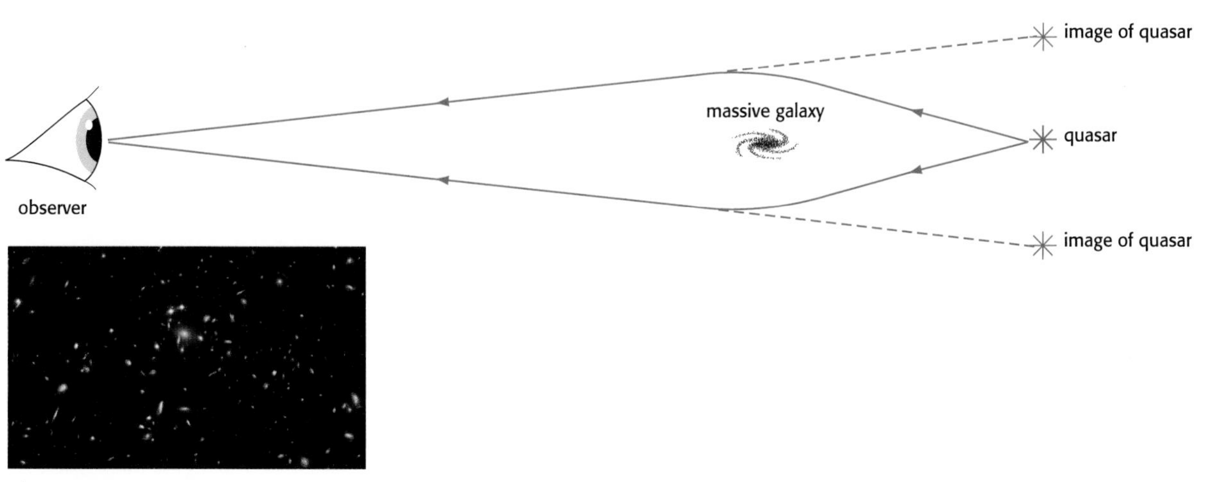

Figure 16

End of chapter summary

Chapter 30 introduces Einstein's two theories of relativity – special relativity and general relativity. The section on special relativity has six main themes: frames of reference, concepts and postulates, relativistic kinematics, consequences, evidence and relativistic momentum and energy. The section on general relativity has two main themes: concepts and evidence. SL candidates studying Option D (relativity and particle physics) need to study the three themes marked with an asterisk* in addition to some material from Chapter 32. The list below summarises the knowledge and skills that you should be able to undertake after having studied this chapter. Further research into more detailed explanations using other resources and/or further practice at solving problems in all these topics is recommended – particularly the items in bold.

Frames of reference*

- Describe what is meant by a *frame of reference* **and a *Galilean transformation***.
- **Solve problems involving relative velocities using the Galilean transformation equations.**

Concepts and postulates of special relativity*

- Describe what is meant by an inertial frame of reference, state the two postulates of the special theory of relativity and discuss the concept of simultaneity.

Relativistic kinematics (time dilation and length contraction)*

- Describe the concept of a light clock.
- Define *proper time interval* and *proper length*.
- Describe the phenomena of *time dilation* and *length contraction*.
- Derive the time dilation formula **and use it to sketch and annotate a graph showing the variation with relative velocity of the Lorentz factor**.
- **Solve problems involving time dilation and length contraction.**

Some consequences of special relativity (the twin paradox, velocity addition, mass and energy)

- Describe how the concept of time dilation leads to the "twin paradox" and discuss the Hafele – Keating experiment.
- **Solve one-dimensional problems involving the relativistic addition of velocities.**
- State the formula representing the equivalence of mass and energy and define *rest mass*.
- Distinguish between the energy of a body at rest and its total energy when moving, determine the total energy of an accelerated particle and explain why no object can ever attain the speed of light in a vacuum.

Evidence to support special relativity

- Discuss **and solve problems** involving the muon decay experiment.
- Outline the Michelson – Morley experiment and discuss its result and implication.
- Outline an experiment that indicates that the speed of light in vacuum is independent of its source.

Relativistic momentum and energy

- Apply the relations for the relativistic momentum and for the kinetic energy of particles **and use them to solve problems involving relativistic momentum and energy**.

General relativity (the equivalence principle, space-time, black holes and gravitational red-shift)

- Explain the difference between the terms gravitational mass and inertial mass.
- Describe and discuss Einstein's principle of equivalence and use it to deduce the predictions of the bending of light rays in a gravitational field and that time slows down near a massive body.
- Describe the concept of space-time and state that moving objects follow the shortest path between two points in space-time.
- Explain gravitational attraction in terms of the warping of space-time by matter and use this idea to describe black holes.
- Define the term *Schwarzschild radius*, calculate its value **and solve problems involving time dilation close to a black hole**.
- Describe the concept of gravitational redshift **and solve problems involving either frequency shifts between different points in a uniform gravitational field or the gravitational time dilation formula**.

Evidence to support general relativity

- Outline experiments that provide evidence for the bending of electromagnetic waves by a massive object and for gravitational redshift.
- Describe *gravitational lensing*.

Chapter 30 questions

1 This question is based upon a thought experiment first proposed by Einstein.

a) Define the terms *proper time* and *proper length*. [2]

In the diagram below Miguel is in a railway carriage that is travelling in a straight line with uniform speed relative to Carmen who is standing on the platform.

Miguel is midway between two people sitting at opposite ends A and B of the carriage.

According to Miguel, at the moment that Miguel and Carmen are directly opposite each other, the person at end A of the carriage strikes a match as does the person at end B of the carriage. These two events take place simultaneously.

b) i) Discuss whether the two events will appear to be simultaneous to Carmen. [4]

ii) Miguel measures the distance between A and B to be 20.0 m. However, Carmen measures this distance to be 10.0 m. Determine the speed of the carriage relative to Carmen. [2]

iii) Explain which of the **two** observers, if either, measures the correct distance between A and B? [2]

(Total 10 marks)

2 This question is about electrons travelling at relativistic speeds.

A beam of electrons is accelerated in a vacuum through a potential difference *V*.

The sketch-graph below shows how the speed *v* of the electrons, as determined by non-relativistic mechanics, varies with the potential *V*, (relative to the laboratory). The speed of light *c* is shown for reference.

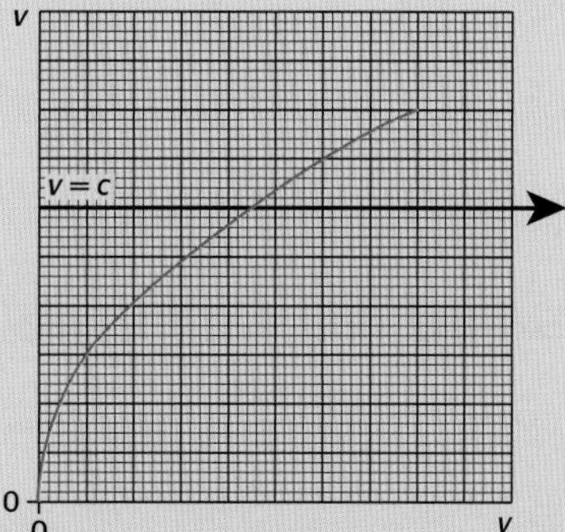

a) Draw a graph to show how the speed of the electrons varies over the same range of *V* as determined by relativistic mechanics.
(This only needs to be a sketch graph; you do not need to include any numerical values.)

b) Explain briefly, the general shape of the graph that you have drawn. [3]

c) When electrons are accelerated through a potential difference of 1.50×10^6 *V*, they attain a speed of 0.97c relative to the laboratory.

Determine, for an accelerated electron,

i) its mass; [3]

ii) its total energy. [2]

(Total 10 marks)

3 The total energy of a particle is always given by $E = mc^2$. Calculate the speed at which a particle is travelling if its total energy is equal to three times its rest mass energy.

(Total 3 marks)

4 This question is about the postulates of special relativity.

a) State the **two** postulates of the special theory of relativity. [2]

b) Two identical spacecraft are moving in opposite directions each with a speed of 0.80 *c* as measured by an observer at rest relative to the ground. The observer on the ground measures the **separation** of the spacecraft as increasing at a rate of 1.60 *c*.

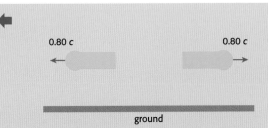

i) Explain how this observation is consistent with the theory of special relativity. [1]

ii) Calculate the speed of one spacecraft relative to an observer in the other. [3]

(Total 6 marks)

5 This question is about muon decay.

Muons, created in the upper atmosphere, travel towards the Earth's surface at a speed of 0.994 c relative to an observer at rest on the Earth's surface.

A muon detector at a height above the Earth's surface of 4150 m, as measured by the observer, detects 2.80×10^4 muons per hour. A similar detector on the Earth's surface detects 1.40×10^4 muons per hour, as illustrated below.

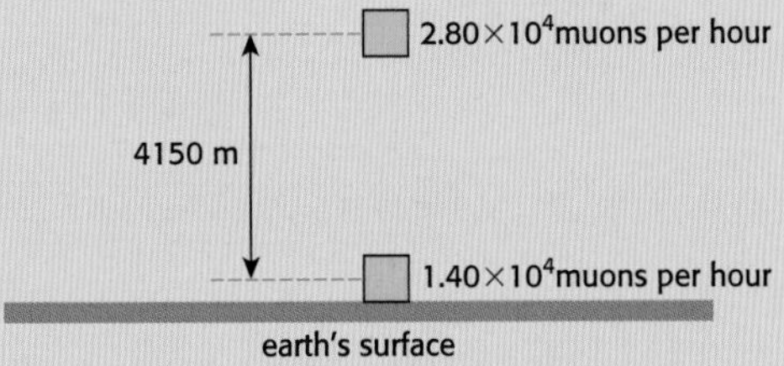

The half-life of muons as measured in a reference frame in which the muons are at rest is 1.52 μs.

a) Calculate the half-life of the muons, as observed by the observer on the Earth's surface. [2]

b) Calculate, as measured in the reference frame in which the muons are at rest,

i) the distance between the detectors; [1]

ii) the time it takes for the detectors to pass an undecayed muon. [1]

c) Use your answers to (a) and (b) to explain the concepts of

i) time dilation; [2]

ii) length contraction. [2]

(Total 8 marks)

6 This question is about mass-energy.

a) Distinguish between the rest mass-energy of a particle and its total energy. [2]

b) The rest mass of a proton is 938 MeV c^{-2}. State the value of its rest mass-energy. [1]

c) A proton is accelerated from rest through a potential difference V until it reaches a speed of 0.980 c. Determine the potential difference V as measured by an observer at rest in the laboratory frame of reference. [4]

(Total 7 marks)

7 This question is about relativistic momentum and energy.

A proton is accelerated from rest through a potential difference of 2.0×10^9 V. Calculate the final momentum of the proton in units of MeV c^{-1}.

(Total 7 marks)

8 This question is about a relativistic decay.

Particle X has rest mass 3520 MeV c–2. It decays at rest in a laboratory, into two identical particles Y. The two particles move in opposite directions with momentum 1490 MeV c–1 with respect to the laborary as shown in the diagram.

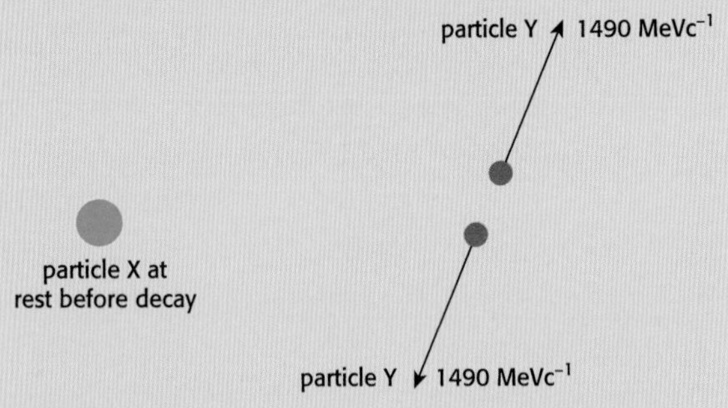

Determine the rest energy, in units of MeV, of particle Y.

(Total 4 marks)

31 Option I: Medical physics

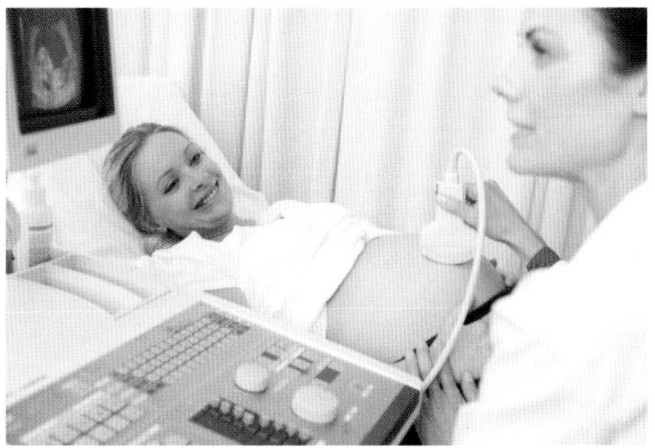

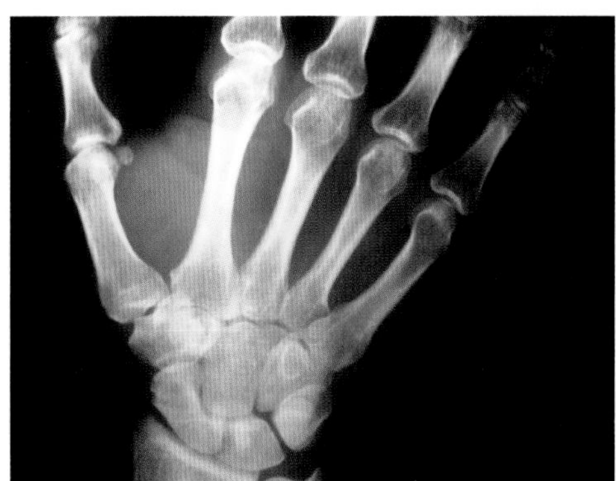

This chapter deals with a selection of medical applications that rely on physics. In general these applications can be categorised as either **diagnostic** or **therapeutic**. Diagnostic applications are those that are used to determine information about a patient, whereas therapeutic applications are those that are used to treat medical conditions. The first section concentrates on how we hear sounds. This is followed by a section that considers a range of imaging techniques. Finally the role of radioactive sources is considered from both the diagnostic and therapeutic point of view.

There are many good pages on the internet that can be used to support your research. One starting point would to search for information on each application mentioned in the text. Alternatively many medical schools or physics institutions have good pages that support self-learning, for example, http://www.teachingmedicalphysics.org.uk/ (accessed September 2009).

The ear and hearing

Sound is a longitudinal wave and a normal human ear can respond to frequencies in the range from 20 Hz up to 20 kHz, but is most sensitive to sounds of about 2 kHz. In order for a person to hear a sound, the oscillations of the air molecules need to be converted into the electrical impulses that are analysed in our brains. This conversion is complex and involves several stages. Firstly, the oscillations in the air in the **outer ear** are converted into mechanical oscillations in the **middle ear** (in three small bones). These oscillations are then converted into oscillations in the fluid trapped inside the **inner ear**, where the oscillations are converted into electrical signals that are sent along an **auditory nerve** to the brain. The reasons for this seemingly over-complex process are explained below:

The shape of the ear arranges for sound waves to be transmitted down the ear canal onto the **tympanic membrane** (the ear drum) and the variation in air pressure movement of the air causes the ear drum to oscillate. This oscillation is transmitted along three

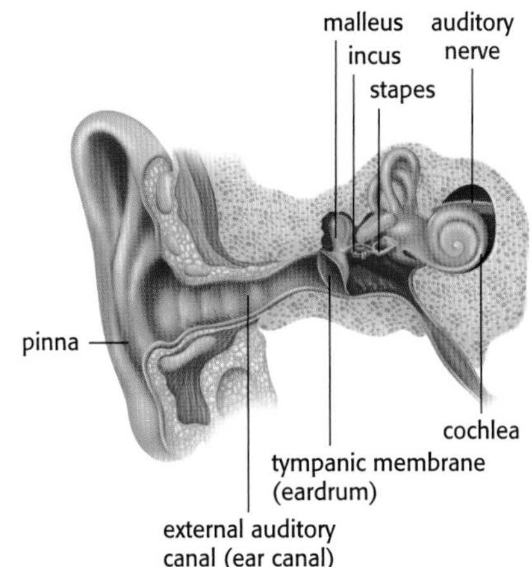

Figure 1 The ear

very small bones, the **malleus** (shaped like a hammer), the **incus** (shaped like an anvil) and the **stapes** (shaped like a stirrup). These bones are collectively known as the **ossicles**. The oscillations arrive at the **oval window**, which is the bridge between the middle ear and the inner ear and which is in contact with the fluid in the inner ear.

This structure successfully transmits sound energy into oscillations in the fluid in the inner ear. Without the ossicles, little sound energy would make it into the inner ear. In particular, the arrangements of the bones act as a set of levers and increase the force transmitted to the oval window. In addition, the area of the stapes and the oval window is much smaller than the area of the eardrum, resulting in an increase in pressure.

The fluid in the inner ear is therefore set in oscillation by the movement of the oval window. This fluid is trapped inside a spiral shape known as the **cochlea**.

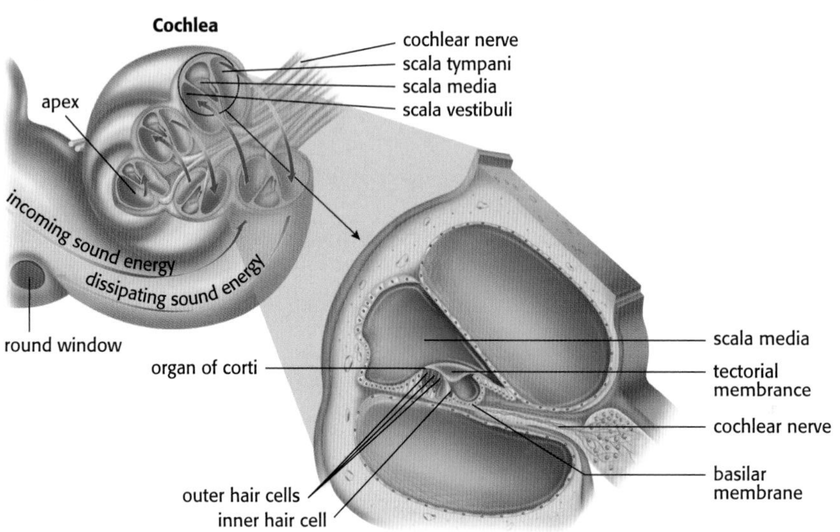

Figure 3 The inner ear

The movement means that a longitudinal wave (a pressure wave) starts from the oval window and travels in the fluid up and down the spiral. As the wave travels, small hair-like structures within a structure called the organ of Corti convert vibrations in the fluid into electrical signals in the auditory nerve. The operation of these structures in not fully understood but regions along the cochlea are involved in detecting different frequencies of sound.

Sound intensity, intensity level and loudness

The intensity of a sound is defined as the power per unit area that is received by the observer. This quantity is different to the loudness of the sound, which is a measure of how loud the sound appears to the listener:

$$I = \frac{p}{A}$$

I is the intensity in W m^{-2}
p is the power received in W
A is the area at right-angles (normal) to the wave that receives the p.

(a)

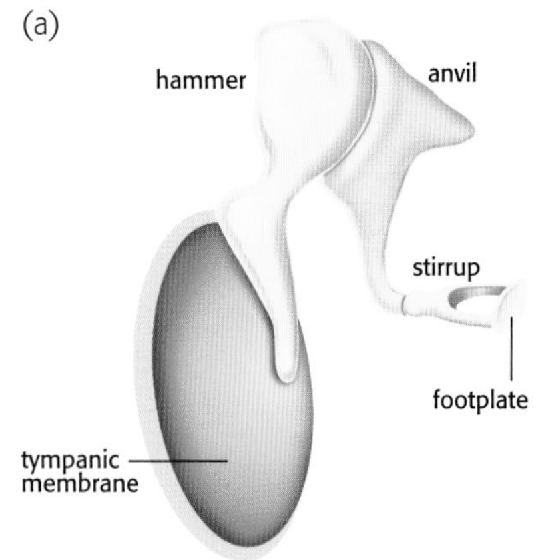

(b)

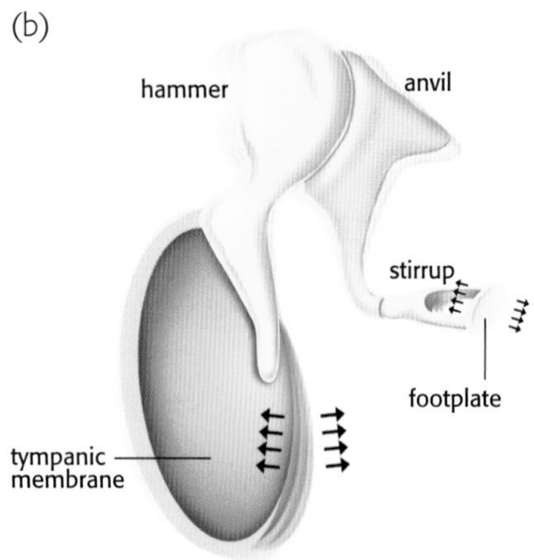

Figure 2 Ossicles

Textbooks will also refer to two other parts of the ear's structure that not involved in the sensing of sound but do provide other useful functions. Research the structure and function of (a) the semicircular canals and (b) the Eustachian tube.

A healthy human can just perceive a sound intensity of 1.0×10^{-12} W m^{-2} (at 1 kHz), whereas the onset of pain takes place when intensity is around 1 W m^{-2}. Between these two extremes, the loudness of a sound is just a subjective quantity that depends on the observer's hearing. Experiments suggest that normally there is a logarithmic response of the ear to intensity of sound.

As a result, sounds are usefully compared on a scale of sound intensity level (*IL*), which uses the decibel scale (a decibel is one tenth of a bel). Intensity level is defined using the following equation:

$$IL = 10 \lg \frac{I}{I_0}$$

IL is the intensity level measured in decibels
lg is the logarithm to the base 10
I is the intensity of the received sound in W m^{-2}.
I_0 is the minimum audible intensity of sound (1.0×10^{-12} W m^{-2}).

Use the information given to show that the range of normal hearing in terms of intensity level is from 0 dB to 120 dB.

Do some research to estimate typical sound intensity levels for a range of sounds, from very loud to very quiet. Use your results to draw up a decibel scale of noise levels.

Sound perception is a complicated situation as a normal ear responds differently at different frequencies, as shown in the diagram below. Each line represents the intensity of sounds that are perceived as being of equal loudness. The top line is the loudest and is on the threshold of pain. The bottom line represents the lower limit of audibility. Note that the frequency scale is logarithmic.

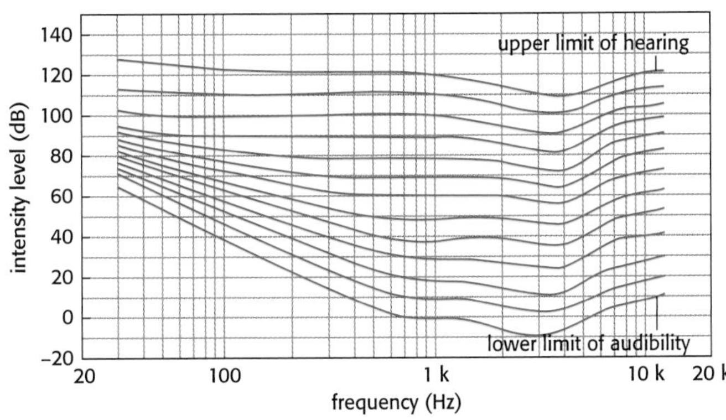

Figure 4 Equal loudness curves (based on Fletcher and Munson)

It should also be noted that the above graphs represent how a normal ear responds. Hearing can be affected for many different reasons, including:

- Blockages in the outer ear
- Damage to the outer or middle ear due to accidents
- Disease (where the ossicles can be prevented from moving)
- Age (the ossicles tend to become solidified)
- Short- and long-term exposure to loud noise.

A plot comparing normal and defective hearing is known as an **audiogram**. The curves shown in Figure 5 represent the average hearing loss as a function of age. Hearing aids are available that preferentially amplify the higher frequencies.

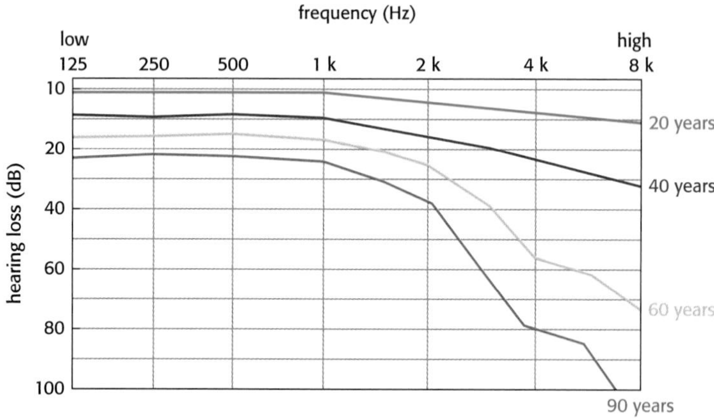

Figure 5 Standard hearing loss with age audiogram

Medical imaging

This section considers four different diagnostic imaging techniques: X-rays, Ultrasound, Nuclear Magnetic Resonance (NMR) and lasers. Some of these techniques (X-rays, Ultrasound and lasers) can also be adapted for therapeutic use.

X-rays

X-Rays (also known as Röntgen Rays, after the discoverer) are electromagnetic waves with shorter wavelengths than ultraviolet radiation, having wavelengths from about 10^{-9} to 10^{-11} m. These high energy photons are capable of penetrating and ionising matter. The production and characteristics of X-rays are discussed on pages 424–5.

The use of X-rays entails an inherent health risk, because they are a form of ionising radiation, and the dose of radiation received must be monitored. The **attenuation** of X-rays (the percentage that is absorbed) as they pass through matter depends on the density of the material and how well it absorbs the X-ray energy. The first X-ray images showing the bones inside a living person were produced in the 1890s and caused great excitement. Such pictures are the shadows of the denser bones, produced by shining the X-rays from a point-like source through the body, and recording the shadow on photographic film.

Little has changed in the principles involved in obtaining this kind of shadow image, except in the technology available for recording the image. In modern X-ray film, the photographic film is sandwiched between intensifying screens of fluorescent material that produce light from X-rays, thus amplifying the image, but photographic film has now been replaced by CCD (see Chapter 24) arrays to produce direct digital images, with the important additional advantage that these require a much lower intensity of X-rays to get a usable image, so that the dose of radiation is lower, and real-time images can be viewed.

The **contrast** of the image will be poor if the organ to be imaged absorbs X-rays in the same way as the surrounding tissue. Thus, X-rays are good for imaging bones, which attenuate X-rays much more than surrounding tissue. In some situations a contrast medium can be used to get a better image: to get X-ray images of the bowel, the patient is fed a "barium meal" so that the digestive tract is filled with a relatively opaque material.

?

Explain why somebody suffering from age-related hearing loss would not benefit from a hearing-aid that amplifies all frequencies by the same amount.

Draw an audiogram to represent the hearing loss that might be experienced by a short-term exposure to loud sounds (e.g. music from a loud concert) where the whole frequency range is affected.

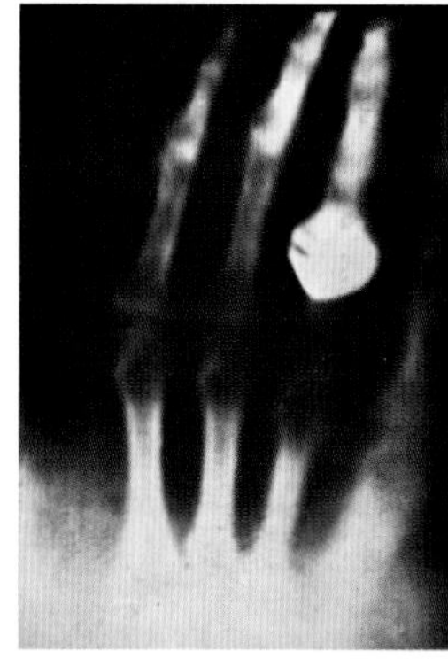

Figure 6 The first x-ray image

Most clinical X-rays are used for imaging, but at higher doses X-rays can also be used in a therapeutic way to target cancers.

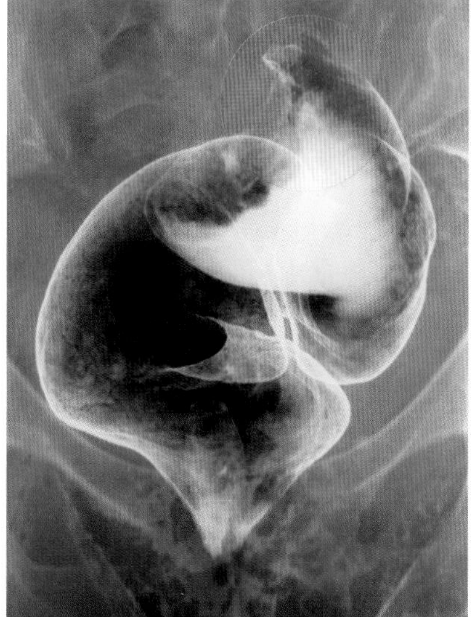

Figure 7

Mathematical physics: *Attenuation of X-rays by matter*

The intensity of the X-ray beam decreases exponentially with the distance travelled through the material, so the same mathematics of exponential decay that is found, for example, in decay of radioactive materials applies, with a distance variable (instead of a time variable) along the *x*-axis. The fractional change in the intensity is proportional to the thickness of the material, *x*:

$$\frac{\Delta I}{I} \propto -x \text{ or } \frac{\Delta I}{I} = -\mu x$$

where μ is the **attenuation coefficient** of the material. The negative sign indicates that the intensity is reducing with increasing *x*.

This gives intensity:

$$I = I_0 e^{-\mu x}$$

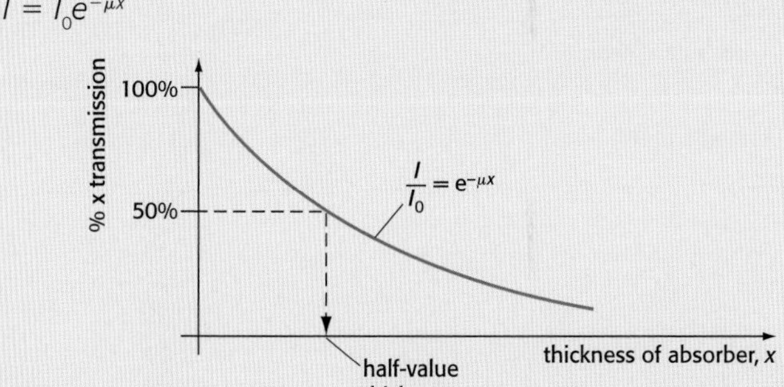

Figure 8

The half value thickness of the material is the thickness *x* that will reduce the intensity to half the value on entry:

$$\frac{I}{I_0} = \frac{1}{2} = e^{-\mu x}$$

so

$$ln2 = \mu x_{\frac{1}{2}}$$

or half value thickness

$$x_{\frac{1}{2}} = \frac{ln2}{\mu}$$

CT Tomography

Tomography means making an image of a slice. If the X-ray source and film are moved in a co-ordinated way, it is possible to sharpen the image at a particular "depth" as all other regions are blurred out of focus. Computed tomography is the use of computation to produce images of selected slices through the patient from the attenuation of a narrow beam of X-rays passing through in many different directions. CT X-ray scanning can be used to create three-dimensional images.

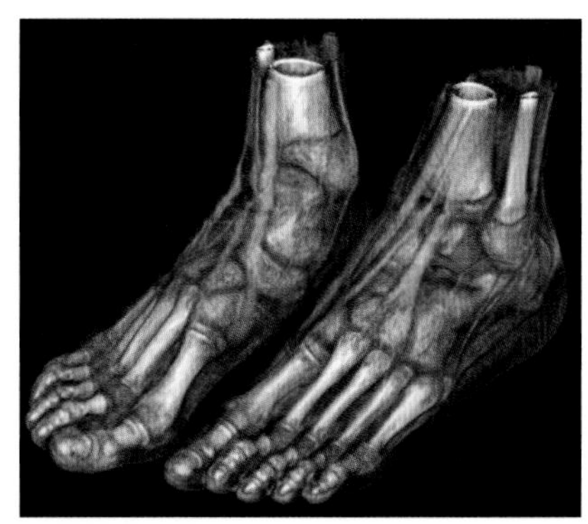

Figure 9 CT image of feet

Ultrasound

Whereas X-rays make shadow pictures, ultrasound uses echo-location, relying on the reflections that take place at the boundaries between different kinds of tissue. The imaging device sends a pulse of high frequency (in the MHz region) sound into the body and measures the time taken for reflections to return from different boundaries. A direct simple distance measurement can be calculated or a more complex scanning method can be used to build up an image.

At the high frequencies used for ultrasound, a piezoelectric device must be used to turn electrical signals into mechanical vibrations, and back again. This exploits the property of some materials, quartz crystals being a common example, which vibrate when an alternating potential difference is applied across them, and conversely, vibrations of the crystal produce an alternating voltage. The same piezoelectric crystal is used as both emitter and detector of the ultrasound pulses and reflections.

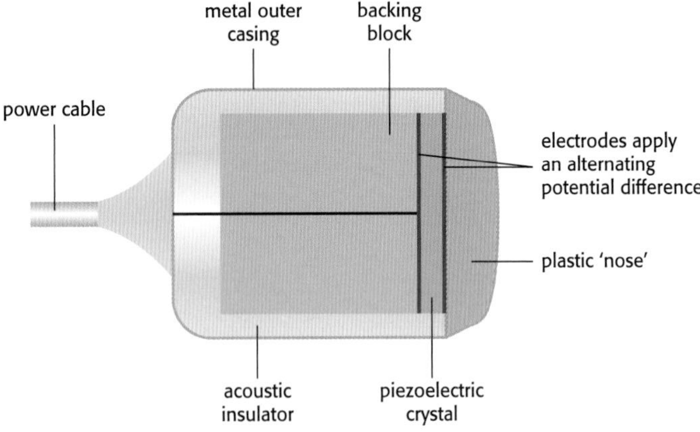

Figure 10 Ultrasound imaging device

An important quantity that can be used to determine the strength of the reflection that comes back from a boundary is the acoustic impedance Z of any material. This is defined as the density ρ multiplied by the speed of sound c in the material:

$$Z = \rho c$$

The strength of the reflection that comes back from a boundary between difference types of tissue depends on the change in the acoustic impedance. One consequence of this is that the air gap between the probe and the skin must be eliminated, as there is a big change in Z going from air to skin and most of the energy reflects off the surface so very little enters the body. This is avoided by using a gel of similar Z to the body tissue between the probe and the skin, therefore getting most of the energy of the vibration in through the skin.

Table 1

material	density ρ (Kg m^{-3})	speed c (m/s)	impedance Z (kg m^{-2} s^{-1})
air (20 °C)	1.20	343	412
water (37 °C)	1.10^3	1525	1.52×10^6
brain	1.02×10^3	1530	1.56×10^6
Muscle	1.04×10^3	1580	1.64×10^6
fat	0.92×10^3	1450	1.33×10^6
trabecular bone	0.9×10^3	1540	1.39×10^6
cortical bone	1.9×10^3	4040	7.68×10^6

For most ultrasound purposes, frequencies of between 1 MHz and 10 MHz are used. The choice of frequency is a compromise between **resolution** and **attenuation**:

- A smaller wavelength allows the ultrasound to resolve smaller details, so the higher the frequency the more detailed the information available and the higher the resolution.
- Shorter wavelength ultrasound is also more scattered by molecules and its energy dissipated by heating the tissue. The result is that the higher the frequency the less the distance that the ultrasound penetrates into the tissue – it is more attenuated.

Different types of ultrasound scans: A- and B-scans

The simplest kind of display from the probe would be a graph showing the intensity of the echo plotted against time. This would quickly show the diagnostician the distance from the probe to the various reflecting boundaries. This A-scan (amplitude scan) is useful when the anatomy of the zone being examined is clear, but a distance measurement is needed. For example, in the case of a suspected brain injury or tumour, the position of the boundary between the two brain hemispheres can be checked to see if it has been pushed off centre or the dimensions of an eyeball can be quickly measured for a lens implant.

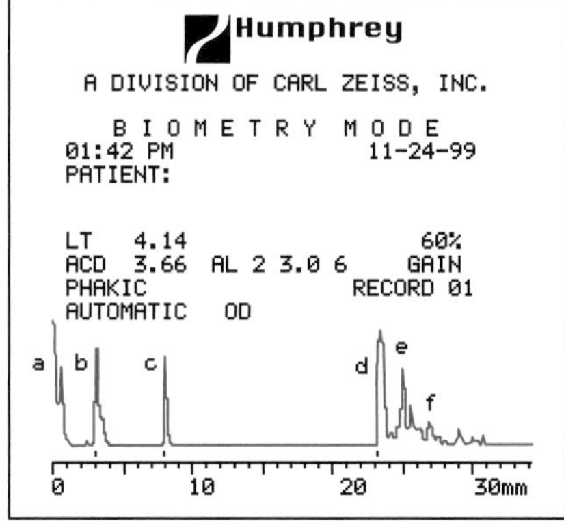

Figure 11 A-scan of eye

In a B-scan (brightness scan), instead of peaks showing the strength of the reflections, a dot is plotted, with the strength of reflection shown in the brightness of the dot on the screen. This allows the operator to move the probe either linearly across the body or by changing orientation, to build up an image. To build up the image, the computer must combine the information from the probe with information about the position and orientation of the probe. In this way the two-dimensional image of a section of the body can be obtained.

A further refinement of the B-scan is to use an array of hundreds of transducers in the same probe. Wavefronts at different angles can be generated by firing the transducers in different phases, so the probe can send and receive ultrasound in different directions without the need for the probe to be moved. Thus it can collect information to build up a two-dimensional image without relying on the operator to move a single transducer. A well-known use of B-scans is to image the foetus in the womb, now a standard procedure.

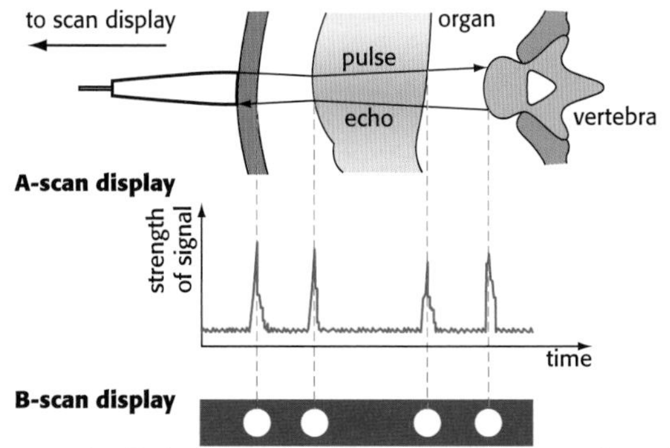

Figure 12 B-scan procedure

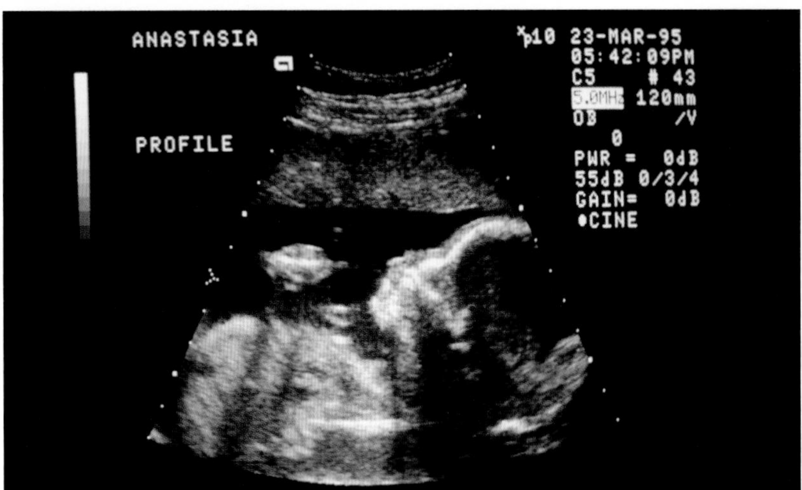

Figure 13 Ultrasound image of a fetus

Ultrasound has several advantages when compared with other imaging methods. Most importantly, no ionising radiation is used, so the procedures are safer than X-rays, and no discomfort is caused to the patient. Equipment can be made portable. Good contrast is obtained with soft tissue as well as bone.

Disadvantages include the fact that lungs cannot be imaged because the very strong reflection from gas wipes out other information. For the same reason gas bubbles in the digestive tract are problematic when imaging organs in the abdomen. Because of attenuation by bone it is difficult to image deep inside the body or inside the skull. Finally, successful ultrasound imaging can depend a lot on the skill of the operator. One "trick" often adopted is for the operator to get the pregnant woman to drink a lot of water before having a scan – this helps reduce reflections from the bladder.

Nuclear magnetic resonance (NMR) imaging

The phenomenon called nuclear magnetic resonance was used for the first time to form an image of the organs of a person in 1977. Only in the past 20 years or so has the method become widely used as the equipment needed remains very expensive. Nucleons are magnetic dipoles – protons and neutrons can be thought of as magnets with north and south poles. When placed in a uniform magnetic field they line up in the direction of the field in such a way that their spin axis precesses around the field direction, like a spinning top with its axis off the vertical.

Spinning proton precessing

The frequency at which the precession occurs is known as the Larmor frequency, and is in the radio part of the electromagnetic spectrum. When nucleons in a strong magnetic field are subjected to radio waves at the Larmor frequency, a **resonance** occurs and they absorb energy, jumping into a higher energy precession state. After the pulse of radio waves ends, the nucleons return to the lower energy state, re-emitting the energy at radio frequency. The time that they take to return to the initial state, called the **relaxation time**, depends on the environment of the nucleon, as well as the applied magnetic field.

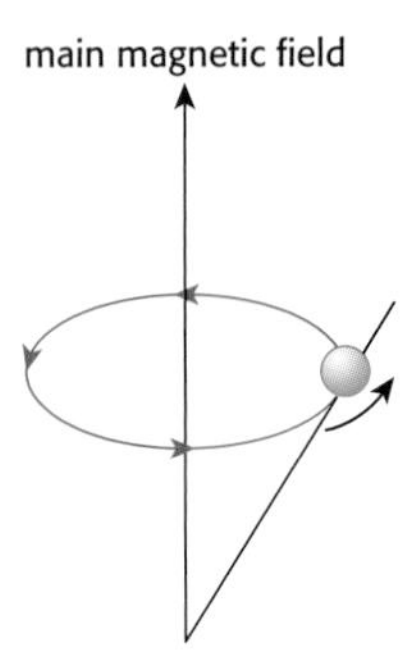

Figure 14 Spinning proton precessing

In practice, magnetic resonance imaging (MRI) is done using the protons that form the nuclei of the hydrogen atoms in water molecules. These are very abundant in body tissue and have a strong magnetic moment. The relaxation time varies between different kinds of tissue, providing the **contrast** needed to form the image.

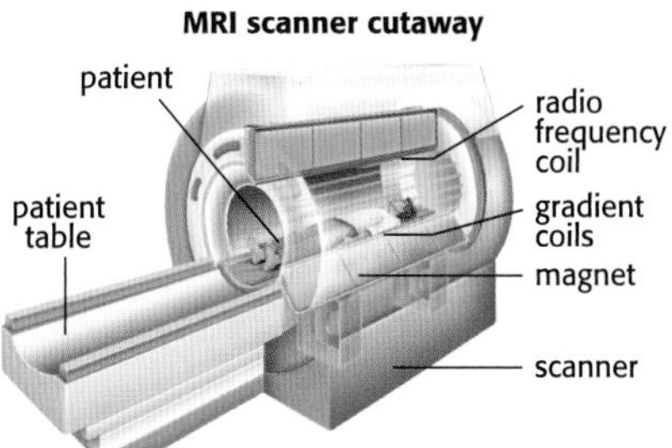

Figure 15 MRI scanner

The patient is placed in a very strong magnetic field. Fields of up to 3T are used – compare this with the strongest magnet that can typically be found in an IB school lab, which has a field of about 200 mT between the poles. A large superconducting coil is used, with liquid helium as the coolant to maintain the low temperatures needed. For good image definition the field must be uniform to a high degree. In addition to the coil for producing the main field, there is a coil for transmitting the radio frequency pulse, and for receiving the protons re-emitting the energy. Finally there are the gradient coils, which are used to make small changes in the main field, at well defined locations. This is important because small changes in the magnetic field result in different frequencies emitted by the protons, so the system can identify their position by emitted frequency.

In effect, the MRI system divides the body into small cells and determines what kind of tissue surrounds the protons in each cell. The response comes in the form of the frequencies emitted (combined with a knowledge of the gradient fields) and the relaxation time of the signal. There are different relaxation times along different directions, and they behave differently in different tissue. Contrast between different kinds of tissue depends on the magnetic field strength and the combination of relaxation times used, so it is possible to programme for better contrast according to the kind of tissue being imaged, many different contrast settings being available. Once the tissue data has been stored in the computer memory, images corresponding to whatever section (slice) of the body is needed can be displayed.

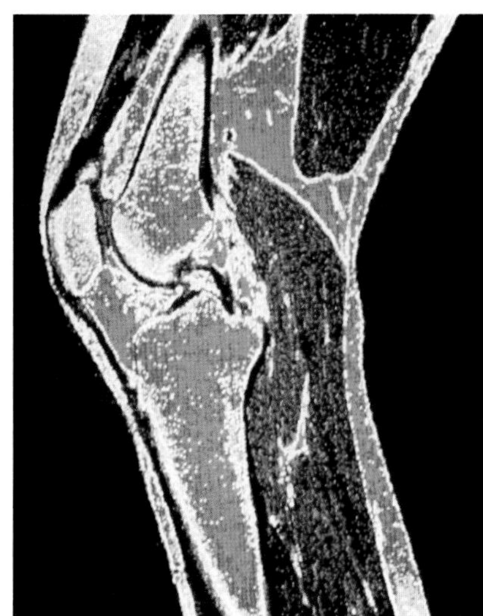

Figure 16 MRI scan of a knee

Compared to X-rays, there are two important advantages. Firstly, as no ionising radiation is used, and there is no known biological risk associated with exposure to strong magnetic fields, the process is safer. However, the radio pulse can cause heating of the body, and the intensity used must be limited.

Secondly, the contrast available for imaging different kinds of tissue makes MRI much more versatile than either X-ray tomography or ultrasound. However, the equipment needed is more expensive than an ultrasound system.

The main danger in the MRI room is from ferromagnetic objects accidentally brought within range of the magnet. These experience huge forces and become deadly projectiles! Patients have to be checked for ferromagnetic implants, and the radio pulse can interfere with pacemakers.

In recent years, MRI has opened up the possibility of monitoring brain activity in real time, while the subject performs different tasks. The active areas of the brain can be identified because of difference in blood flow and metabolism, providing an important new tool for the

Do some web research on this application of MRI and consider the topic for a ToK presentation on how two very different areas of knowledge may try to answer the same question.

neuroscientist. Questions that were the exclusive domain of areas like philosophy, as for example how we reach ethical decisions, can be further enlightened by systematically investigating which parts of the brain are active when we answer different types of question, and mapping the function of different areas of the brain.

Light and lasers

From the point of view of medical applications, the two most important properties of laser light are its ability to be concentrated into a fine beam and the fact that it is of a single frequency.
The production of laser light, the details of which do not need to be known for this option, is described in more detail on pages 415–6. Four procedures are introduced in the section below, but many others exist.

Pulse oximetry is a diagnostic process that measures the proportion of haemoglobin (the chemical in the blood that transports oxygen) that is saturated with oxygen. It is a **non-invasive** technique, which means that the measurement is achieved without the need for any injections or cuts. The principle involved is that blood in which the haemoglobin is carrying oxygen (a form called oxyhaemoglobin) is a different colour (a brighter red) when compared with blood that is not carrying oxygen. Light of two different frequencies is shone through a part of the patient's body that allows some light to pass (e.g. a finger or an ear lobe). One frequency will be red light and the other is typically in the Infrared part of the EM spectrum. The difference in the amount of absorption of these two frequencies can be used to calculate the relative percentage of the two different forms of haemoglobin in the blood.

Endoscopes are devices which use fibre optics to image internal organs for diagnostic purposes (see Chapter 8). In addition to simple imaging, additional medical instruments can be attached to the endoscope to allow for sampling or therapeutic procedures. In essence, an endoscope is a tube that is inserted into a patient.
The tube contains a bundle of fibre optics and light travels down some of the fibres and illuminates the object being observed. Reflected light is transmitted back to an imaging system, which allows medical staff to observe structures inside the body. Additional equipment can be attached to the endoscope to allow samples to be taken for further analysis (biopsies to check for cancer), blockages to be removed or even for laser surgery to take place (see below).

The insertion of a tube probably seems a highly unpleasant idea but the procedure is what is known as **minimally invasive**. Local anaesthetics can help and the procedure is certainly preferable to full surgery. Endoscopy can provide information on the gastrointestinal track (via mouth or rectum), the respiratory track (via the nose), the urinary track (male and female), the ear and the female reproductive system – including during pregnancy. Regions of the body that are not accessible can also be imaged by inserting an endoscope through a very small incision.

Laser beams can concentrate a large amount of energy into a small region, allowing tissue to be cut in the same way that a sharp knife (**scalpel**) is used during surgery. Providing the intensity of the laser beam remains constant, the depth of cut (incision) from a laser scalpel will not vary as it moves and tissue nearby will be unaffected. Another advantage is that concentrated heat generated by the laser that achieves the cut will also seal off any small blood vessels and thus prevent bleeding. This process is called **cauterizing**.

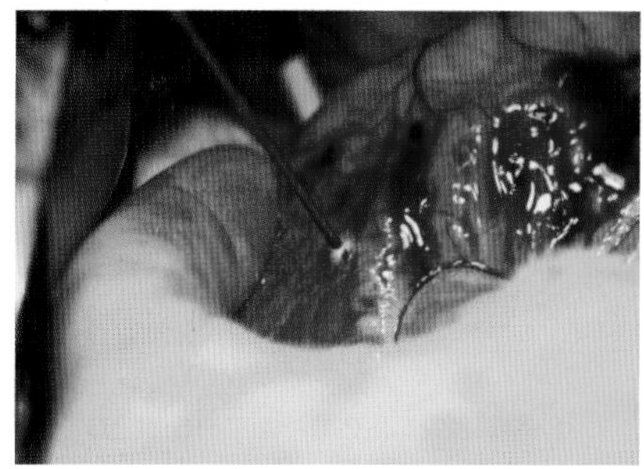

Figure 17 Cauterizing blood vessels during surgery

A laser can be also used as a **coagulator**. Coagulation is the name given to the formation of a blood clot. In this technique, a defocused laser produces heat and this is absorbed by haemoglobin in the blood and causes the blood to clot. This results in blood vessels being sealed off.

Radiation in medicine

The medical use of radioisotopes exposes a patient to ionising radiation and thus must involve an extra risk. Some of the biological effects of ionising radiation have already been introduced on pages 456–7. Given these dangers, it seems strange that radiations and radioactive sources are widely used for both diagnostic and therapeutic procedures. The important point to keep in mind is that refraining from using radiations and radioisotopes also exposes the patient to risk as the condition might either be undiagnosed or untreated. **Balanced risk** is the concept that both the use and the non-use of radiations should be considered when deciding upon what procedures to use.

> Do some research to investigate other medical uses for lasers. When presenting your findings, try to summarise the procedure as succinctly as possible and make it clear if the application is diagnostic or therapeutic. Possible uses to investigate include dentistry, cleaning arteries, cosmetic treatments, laser therapy and eye surgery.

In any situation, the risk depends on the exposure or dose received. Any radiation that causes ionisation could be dangerous, so different quantities need to be considered when assessing different situations. The most important quantities are the exposure, X; the absorbed dose, D; and dose equivalent, H. Of these three quantities, it is the dose equivalent that gives the best measure of the biological damage that may take place as a result of an exposure to radiation, but in all cases a higher number indicates a greater risk.

These are defined as follows:

1 The exposure is the total amount of ionisation produced as a result of ionising radiation:

$$X = \frac{Q}{m}$$

X is the exposure in C kg^{-1}
Q is the total charge produced as the radiation passes through air in C
m is the mass of air in kg.

2 Absorbed dose is the energy absorbed per unit mass of tissue:

$$D = \frac{E}{m}$$

D is the absorbed does in Gy
E is the total energy absorbed by the tissue in J
m is the mass of tissue in kg.

3 Dose equivalent is a measure of the radiation damage that occurs in tissues. It recognises the fact that even if two sources of radiation (e.g. one alpha and one gamma) provide a body with the same absorbed dose, then the alpha source will cause more damage. It is defined as:

$$H = QD$$

H is the dose equivalent in Sv
Q is the quality factor or relative biological effectiveness for different radiations (see table below) – no unit
D is the absorbed dose in Gy.

Table 2

Radiation	Q
X-rays	1
γ-rays	1
β-particles	1
α-particles	20

Radiation precautions

The three basic methods of protecting oneself from radiation exposure are shielding (putting something in between yourself and the source), distance (moving away from the source) and time of exposure (reducing the time next to the source). The effectiveness of these three techniques depends on the nature of the radiation involved and in particular its penetration power.

A film badge is used to monitor the exposure received by those who work with radiation. It is a piece of photographic film held within a plastic container and not exposed to light. Different filters are placed either side of the film so as to be able to determine the nature of the ionizing radiation that has affected the film.

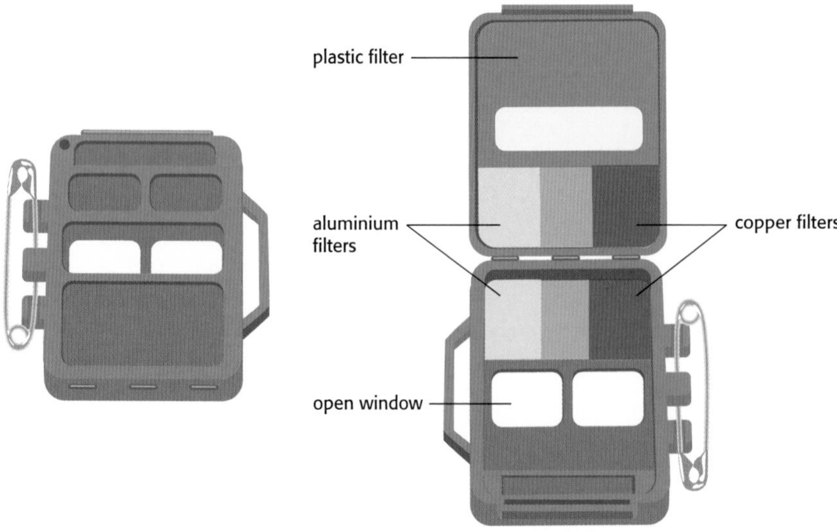

Figure 18 Radiation film badge closed (left) and open (right)

1 A radioactive source needs to be injected into a patient in a hospital. Outline some of the precautions that need to be taken by the hospital to ensure that medical staff are not put at risk. Outline any differences in procedure that should be considered, depending on whether the source concerned emits alpha, beta or gamma radiation.

2 It is normal practice of the operator of an X-ray machine to stand behind a protective screen during the short time that the patient is being exposed to the X-rays. How would you explain this procedure to a patient who asks why the X-rays are okay for her but not for the operator?

3 Research one make of film badge and explain how the different absorbers help identify the radiation in different situations.

Radiation calculations

If a patient is exposed to radiation for a long time, then this will increase the risk. One very important factor to be considered when contemplating the introduction of a radioactive chemical into the body is its half-life. The process of radioactive decay means that the sample's activity will decay exponentially (see pages 276–8).

At the same time, it is likely that normal biological processes will also tend to remove the chemical from the body. These biological processes result in a separate exponential decrease in the amount of chemical in the body. It is the combination of these two effects that regulate the effective half-life of the radioactive isotope in the body.

The physical half-life of an isotope is the time taken for the number of nuclei concerned to halve as a result of radioactive decay, whereas the biological half-life of a chemical is the time taken for the number of molecules of a chemical in the body to halve as a result of biological processes. The overall effective half-life of a radioactive isotope in the body is the time taken for the number of atoms of the radioactive isotope in the body to halve as a result of both biological processes and radioactive decay. This can be calculated using the following relationship:

$$\frac{1}{T_E} = \frac{1}{T_P} + \frac{1}{T_B}$$

T_E is effective half-life of a radioactive isotope in the body measured in s
T_P is physics half-life of a radioactive isotope measured in s
T_B is biological half-life of a radioactive isotope in the body measured in s.

Applications of radiation

Diagnostic applications of radioisotopes often target particular organs in the body. A tracer element is introduced into the body and the uptake of the particular radioactive chemical by a given organ can be traced by sensing (from outside the body) where the radiation has reached. For the radiation to be measurable outside the body, the radioisotope would need to be a gamma-emitter. A gamma camera detects gamma rays coming from the body. By appropriately shielding the many sensors inside the gamma camera, it is possible to detect where gamma rays are coming from and so build up a picture of the sources of radioactivity within the body. For example, a bone scan can be made using the isotope

A radioisotope has a physical half-life of 5 days and a biological half-life of 10 days. Estimate the percentage amount of radioisotope that remains after a month.

technetium-99m. The radioisotope preferentially identifies regions where bone growth it taking place.

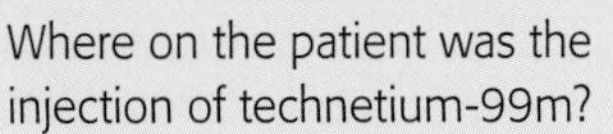

Where on the patient was the injection of technetium-99m?

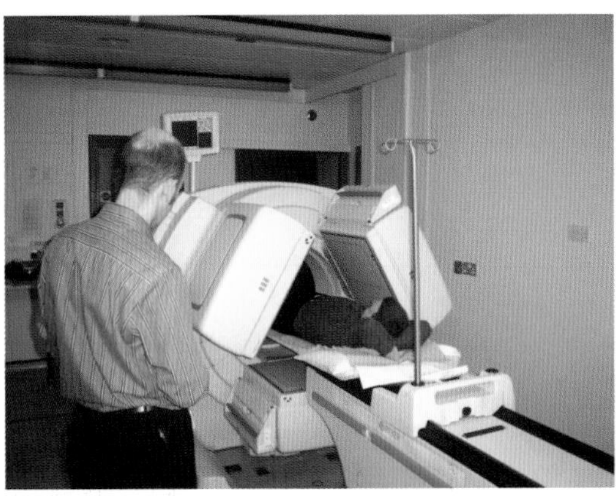
Figure 19

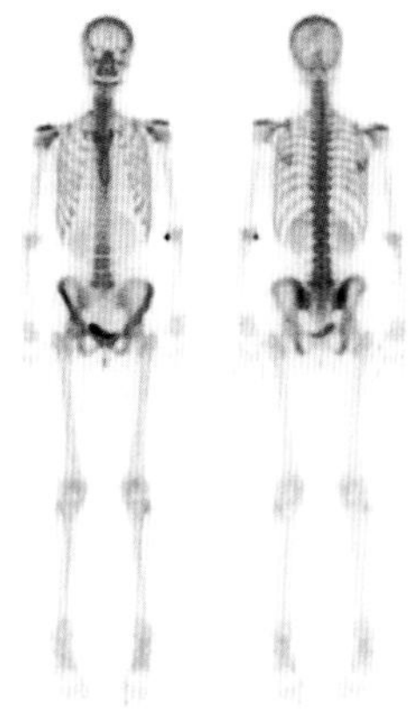
Figure 20

Another example of radioactive tracing is a ventilation scan, where the patient breathes radioactive xenon or krypton gas. The resulting image can be analysed to see if there is a physical blockage preventing oxygen getting into one or both of the lungs.

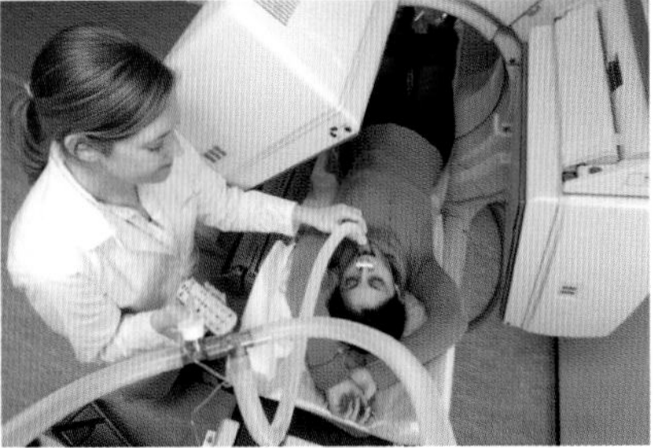
Figure 21

Radiation can also be used in a therapeutic way. Large doses of radiation will damage cells and this can be used to target cancerous cells within the body. Radiotherapy aims to preferentially target malignant cancerous cells in preference to normal healthy cells. The skill of the medical physics team is used to minimise the dose received by healthy cells and maximise the dose received by the cancerous cells. This can be extremely challenging, as the two types of cells will be next to one another.

Luckily, malignant cells are slightly more susceptible to damage from radiation compared with normal healthy cells and once damaged, the rate of repair of cancerous cells is slower than healthy cells.

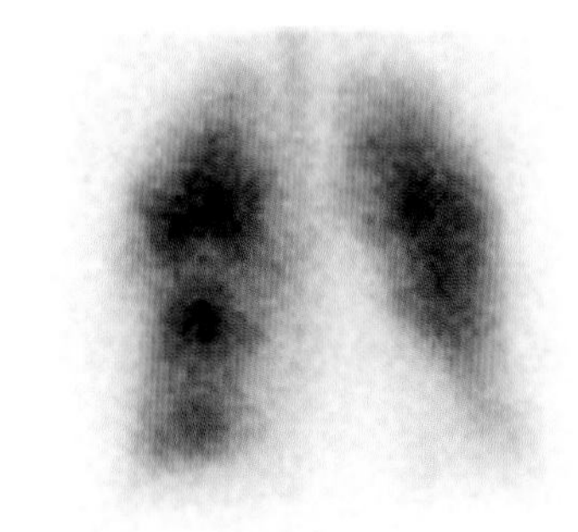
Figure 22

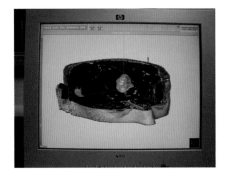

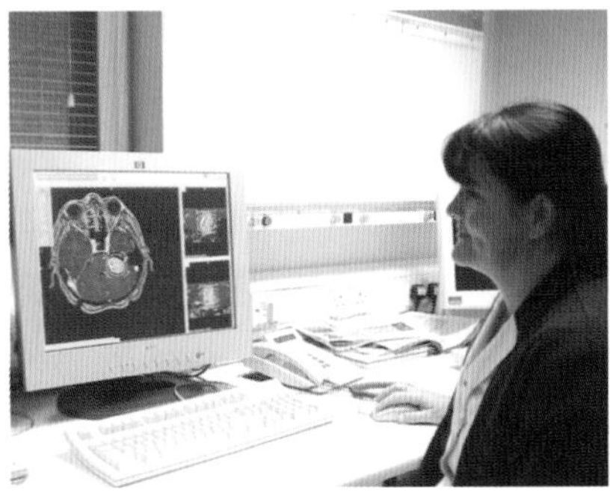
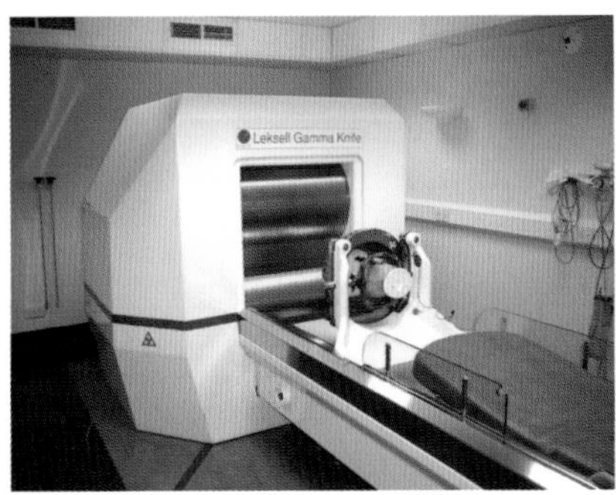
Figure 23 Computer imaging

Research the use of some commonly used procedures using radioisotopes. Your summary should include:

a) whether the procedure is diagnostic or therapeutic;

b) the radioisotope used and its properties (half-life, radiation emitted, etc.);

c) the organ being studied;

d) any appropriate precautions. Possible sources include cobalt-60, technetius-99m, Iodine-123, iodine-131 and xenon-133.

End of chapter summary

Chapter 31 has three main themes: the ear and hearing, medical imaging and radiation in medicine. The list below summarises the knowledge and skills that you should be able to undertake after having studied this chapter. Further research into more detailed explanations using other resources and/or further practice at solving problems in all these topics is recommended – particularly the items in bold.

The ear and hearing

- Describe the basic structure of the human ear and state the range of audible frequencies experienced by a person with normal hearing.
- State and explain how sound pressure variations in air are changed into larger pressure variations in the cochlear fluid and that a change in observed loudness is the response of the ear to a change in intensity and the ear has a logarithmic response.
- Define *intensity* and *intensity level* (*IL*) and state the approximate magnitude of the IL at which discomfort is experienced by a person with normal hearing **and solve IL problems**.
- Describe the effects on hearing of short-term and long-term exposure to noise, analyse **and give a simple interpretation of graphs where IL is plotted against the logarithm of frequency for normal and for defective hearing**.

Medical imaging (X-rays, ultrasound, NMR and lasers)

- Define the terms *attenuation coefficient* and *half-value thickness*, derive the relation between these two **quantities and solve problems using the X-ray attenuation equation**.
- Describe X-ray detection, recording and display techniques and explain standard imaging techniques used in medicine including the principles of computed tomography (CT).
- Describe the principles of the generation and the detection of ultrasound using piezoelectric crystals, define *acoustic impedance* **and use it to solve ultrasound problems**.
- Outline the differences between A-scans and B-scans and identify factors that affect the choice of diagnostic frequency.
- Outline the basic principles of nuclear magnetic resonance (NMR) imaging and describe examples of the use of lasers in clinical diagnosis and therapy.

Radiation in medicine

- State the meanings of the terms *exposure*, *absorbed dose*, *quality factor* (relative biological effectiveness) and *dose equivalent* as used in radiation dosimetry and discuss the precautions taken in situations involving different types of radiation.
- Discuss the concepts of balanced risk, physical half-life, biological half-life and effective half-life and **solve problems involving radiation dosimetry**.
- Outline the basis of radiation therapy for cancer.
- **Solve problems, including the suitable choice of radio-isotope, for a particular diagnostic or therapeutic application.**

Chapter 31 questions

1 This question is about sound intensity.

a) Define *sound intensity level.* [2]

b) The earphone of a personal radio produces 2.8×10^{-7} W of sound power. This power may be assumed to be incident uniformly on the eardrum of area 1.9×10^{-5} m^2. Calculate the sound intensity level at the eardrum. [3]

c) Comment on your answer to (b). [1]

(Total 6 marks)

2 This question is about sound intensity levels.

a) Distinguish between *sound intensity* and *loudness*. [2]

b) An engine generates 2.4 W of sound power that is emitted uniformly in all directions. For health reasons, the intensity level at the ear must not exceed 82 dB. Calculate the minimum distance that any person must be from the engine unless wearing ear protection. (The surface area of a sphere of radius r is $4\pi r^2$) [5]

(Total 7 marks)

3 This question is about sound and hearing.

The sound intensity level is defined by the equation

$$\text{intensity level (dB)} = 10\lg\left(\frac{I}{1.0 \times 10^{-1}}\right)$$

where I is the intensity of the sound.

a) State what the number 1.0×10^{-12} represents. [1]

b) A person is listening to a sound that has an intensity of $1.0 \times 10^{-6}\,\text{Wm}^{-2}$ at the ear. The intensity of the sound at the ear is then increased by a factor of 3. Determine the change in intensity level at the person's ear. [2]

c) The person then detects a change in loudness that corresponds to a 20 dB change in intensity level at the ear. Determine the factor by which the intensity at the ear has increased. [2]

d) A young person with normal hearing has a hearing test. The results of the test are shown below.

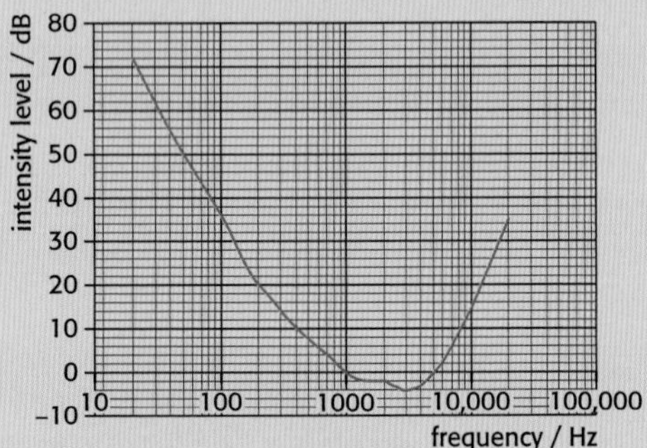

Using the same axes, draw a sketch graph to show the results of a hearing test for an elderly person. [3]

(Total 8 marks)

4 This question is about medical imaging.

a) State and explain which imaging technique is normally used

i) to detect a broken bone. [2]

ii) to examine the growth of a fetus. [2]

The graph below shows the variation of the intensity I of a parallel beam of X-rays after it has been transmitted through a thickness x of lead.

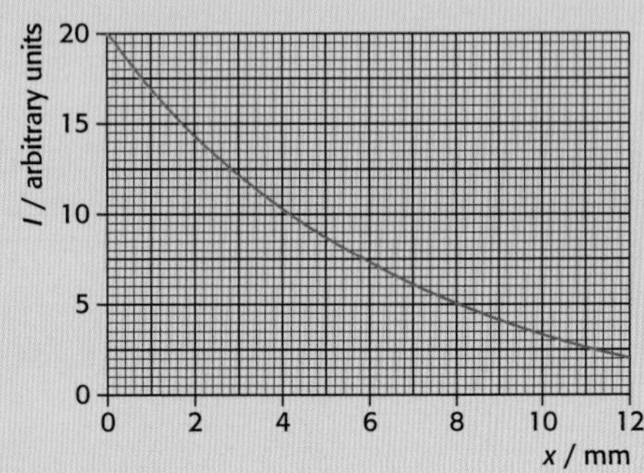

b) i) Define *half-value thickness*, $x_{\frac{1}{2}}$. [2]

ii) Use the graph to estimate $x_{\frac{1}{2}}$ for this beam in lead. [2]

iii) Determine the thickness of lead required to reduce the intensity transmitted to 20% of its initial value. [2]

iv) A second metal has a half-value thickness $x_{\frac{1}{2}}$ for this radiation of 8 mm. Calculate what thickness of this metal is required to reduce the intensity of the transmitted beam by 80%. [3]

(Total 13 marks)

5 This question is about medical imaging.

a) State the approximate range of ultrasound frequencies used in medical imaging. [1]

b) Distinguish between an A-scan and a B-scan.

A-scan:

B-scan: [1]

c) State **one** advantage and **one** disadvantage of using ultrasound at a frequency in the upper part of the range stated in (a). [2]

d) A parallel beam of X-rays of a particular energy is used to examine a bone. At this energy, the half-value thickness of bone is 0.012 m and of muscle is 0.040 m. The beam passes through bone of thickness 0.060 m and through muscle of thickness 0.080 m. Determine the ratio

$$\frac{\text{decrease in intensity of beam produced by bone}}{\text{decrease in intensity of beam produced by muscle}}$$

[3]

e) Suggest, using your answer to (d), why this beam is suitable for identifying a bone fracture. [1]

(Total 8 marks)

6 A patient of mass 60 kg receives a dose equivalent of 3 30 μSv during a chest X-ray. The quality factor (relative biological effectiveness) of X-rays is 1.

a) i) Calculate the absorbed dose received by the patient. [1]

ii) Estimate the total energy received by the patient. [2]

b) Outline **two** precautions that an X-ray machine operator should take to minimize his/her exposure to X-rays. [2]

c) State **two** possible biological effects for an X-ray machine operator of not taking suitable precautions. [2]

X-rays can also be used in radiation therapy. The X-rays used are of a much higher intensity than those used to take chest X-rays.

d) i) Suggest a situation in which radiation therapy might be used. [1]

ii) Outline the basis of X-radiation therapy. [2]

(Total 10 marks)

7 This question is about radiation used in medicine.

a) Define the terms *exposure* and *absorbed dose*.

b) Explain, with reference to α and γ radiation, the distinction between absorbed dose and dose equivalent. [3]

c) Explain why, when using radioactive tracer elements in the treatment of cancer, it is better to use radioactive isotopes that have a long physical half-life and a short biological half-life. [2]

(Total 7 marks)

8 This question is about the biological effectiveness of radiation.

a) Explain the term *quality factor* (*relative biological effectiveness*). [1]

b) A beam of protons is directed at a tumour of mass 0.015 kg. In order to kill the tumour, a dose equivalent of 240 J kg^{-1} is required.

Using the data below, determine the exposure time, assuming all the incident protons are absorbed within the tumour.

Energy of each proton = 4.0 MeV.

Number of protons incident on the tumour per second = 1.8×10^{10}.

Quality factor for protons of this energy = 14.

[4]

(Total 5 marks)

9 When X-rays are used for diagnostic purposes, beam energies of about 30 keV are used. This results in good contrast on the radiogram because the most important attenuation mechanism is not simple scattering.

a) Outline the most important attenuation mechanism that is taking place at this energy. [2]

b) Explain the following terms.

i) *Attenuation coefficient* [2]

ii) *Half-value thickness* [2]

c) The attenuation coefficient at 30 keV varies with the atomic number Z as shown below.

Attenuation coefficient $\propto Z^3$

The data given below list average values of the atomic number Z for different biological materials.

biological material	atomic number Z
fat	5.9
muscle	7.4
bone	13.9

i) Calculate the ratio

$$\frac{\text{attenuation coefficient for bone}}{\text{attenuation coefficient for muscle}}$$ [2]

ii) Suggest why X-rays of 30 keV energy are useful for diagnosing a broken bone but a different technique must be used for examining a fat-muscle boundary. [4]

(Total 12 marks)

10 This question is about dosimetry.

a) Define *absorbed dose*. [1]

b) Outline why, when assessing the biological effects of radiation, dose equivalent is used, rather than absorbed dose. [1]

c) A patient undergoes computerized tomography (CT scanning) and receives a dose equivalent of 2.2 mSv. The increased health risk as a result of exposure to this radiation is estimated to be 5% per Sv.

i) Calculate the increased health risk as a result of this CT scan. [1]

ii) Suggest why patients are prepared to accept this increased health risk. [1]

(Total 4 marks)

32 Option J: Particle physics

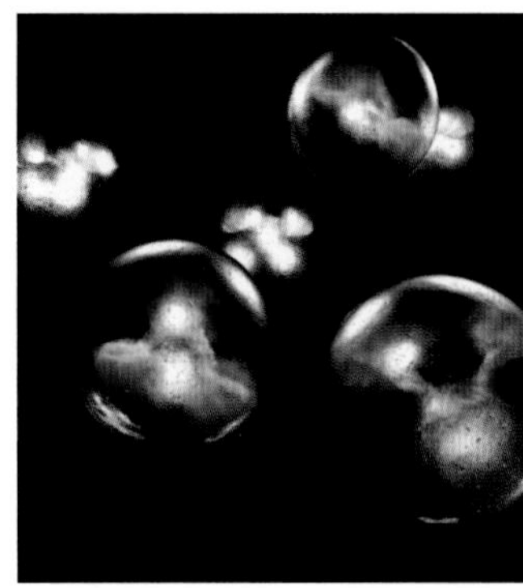

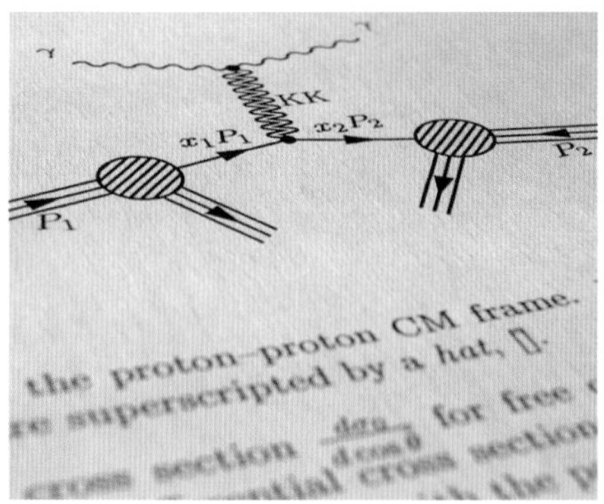

Particles and interactions

An atom is the smallest piece of an element that can exist and still be recognized as that element. Nearly all of these elements, such as the oxygen you breathe and the carbon in your skin, were made in stars at least 5 billion years ago (around the time that the Earth was forming). Hydrogen and helium are even older, most hydrogen having been made soon after the Big Bang.

A hundred years ago atoms were thought to be small impenetrable objects, like miniature versions of billiard balls. Today we know that each atom has a rich labyrinth of inner structure. At its centre is a dense, compact nucleus, which accounts for almost all of the atom's mass and carries positive electrical charge. The nucleus consists of **protons** and **neutrons**. In the outer regions of the atom there are tiny lightweight particles known as **electrons**.

However, protons and neutrons are themselves **composite particles**. Magnify one a thousand times and you will find that they too have a rich internal structure. They are clusters of smaller particles called **quarks**.

There is nothing smaller than these quarks, and so they are called **elementary particles**. There is another class of elementary particles, which includes electrons, which are called the **leptons**.

Table 1 The six quarks arranged in three families

	u	c	t
charge/e mass/m_p	+2/3 1/3	+2/3 1.7	+2/3 186
	d	**s**	**b**
charge/e mass/m_p	−1/3 1/3	−1/3 0.5	−1/3 4.9

m_p is the mass of a proton. $m_p = 0.938$ GeV/$c^2 = 1.67 \times 10^{-27}$ kg.

Note that it is conventional in particle physics to fix the unit of charge as +1 so that the actual charge on an up quark is $2/3 \times 1.6 \times 10^{-19} = 1.07 \times 10^{-19}$C. This convention is followed from now on.

Table 2 The six leptons arranged in three families

	e	μ	τ
charge/e mass/m_p	−1 0.0005	−1 0.1	−1 1.9
	ν_e	ν_μ	ν_τ
charge/e mass/m_p	0 *negligible*	0 *negligible*	0 *negligible*

Elementary particles

Quarks and leptons are thought not to have any internal structure and hence are thought to be elementary particles. There are six known types (**flavours**) of each (see Tables 1 and 2). Exchange particles (see page 472) are also elementary, as is a hypothetical particle known as the Higgs boson (see page 495).

Quantum numbers

All particles have intrinsic physical properties, for example, their rest mass (mass when they are stationary or moving at non-relativistic speeds). There is another set of intrinsic properties that all particles possess, called their quantum numbers. They are not necessarily directly related to physical properties but describe quantities that are conserved in interactions. The quantum numbers used to identify particles include: electric charge/spin, strangeness, charm, color, lepton number and bayon number.

Antimatter

In addition, every particle has an antimatter partner with the same mass, but its other properties are opposite (electric charge, for example).

In 1932 Carl Anderson discovered the first evidence for antimatter. He observed a track on a cloud chamber photograph that had been left by a cosmic ray particle with the same mass as an electron, but a positive electric charge. He had discovered the **positron** – the antimatter partner of the electron.

In 1995 CERN announced that a small experiment using LEAR (Low-Energy Antiproton Ring) had managed to create atoms of anti-hydrogen (a negatively charged anti-proton with a positively charged positron in orbit).

Understanding antimatter is incredibly important in cosmology, because of the so-called asymmetry between matter and antimatter in the universe. See page 498 for more details.

Figure 1 The first cloud chamber photograph of a positron. The line across the centre is a 6 mm lead sheet. The chamber was in a magnetic field which caused the track to be curved. The positron was moving upwards on the photograph.

Spin and the Pauli exclusion principle

A very important quantum number for any particle (either elementary or composite) is its **spin**. As the name suggests it has some things in common with the classical concept of bodies spinning on their own axis and their associated **angular momentum**. In classical physics, an isolated object's angular momentum is measured in units of JS, is always conserved and can have any value depending on the object's shape, mass, size and speed of rotation.

The spin quantum number used in particle physics is, however, strictly associated with the rules of quantum mechanics and can only have well defined values. It also has units of JS but a particle's spin is expressed in units of $\frac{h}{2\pi}$ where h is Planck's constant (6.63×10^{-34} Js). This is often written as $\hbar$ (pronounced "h – bar"). The smallest quantum of spin is $\frac{\hbar}{2}$.

All elementary particles either have half-integral spin values $\left(\pm\frac{\hbar}{2}\right)$ or they have integral multiples of spin ($0, \hbar, 2\hbar$). The half-integral spin possibilities are known as **spin up** $\left(+\frac{\hbar}{2}\right)$ or **spin down** $\left(-\frac{\hbar}{2}\right)$. When elementary particles combine to produce composite particles, the spin of the composite particle (in its lowest energy state) is just the addition of the spins of the constituent elementary particles but needs to take into account the fact that particles will tend to align so as to form pairs of spin up and spin down particles.

Particles that have half-integral spin are called **fermions** where as particles that have integral spin are called **bosons**. The basic constituents of an atom, protons, neutrons and electrons are all fermions.

Thinking about science: Conservation laws

Conservation laws place limits on the events that can take place in nature. The two most well known conservation laws, those of energy and momentum, state that whenever objects collide, explode, stick together, or repel each other, the total energy and momentum in the universe must remain the same.

The essence of such a conservation law is to compare the total amount of some quantity, such as energy, before an event has taken place with the total amount afterwards. If the total has not changed then the quantity is conserved. Conservation of energy and momentum are important in particle physics (always remembering to use the relativistic equations), but there are many other rules that introduce some new physics. By studying the quantities that are conserved in reactions and decays, particle physicists can learn more about the fundamental forces driving these processes. Examples of additional quantities conserved in all particle interactions include charge, baryon number and family lepton number (see pages 493–4 for more details).

The fundamental forces

A fundamental force is one that cannot be explained by the action of another force. For example, friction is not a fundamental force as it is actually due to the electromagnetic forces between the atoms in one object and those in another.

There are four known fundamental forces that are believed to govern all interactions in the universe. They are the electromagnetic, strong nuclear, weak nuclear, and gravitational forces. On the scale of the elementary particles the gravitational force is generally considered to be so weak its effects are negligible. The electromagnetic and the weak interactions are treated separately but are, in fact, two aspects of a single interaction called the electroweak force.

Table 3 The four fundamental forces

Force	Felt by	Range	Relative strength*
Gravity	Any particle with mass	infinite	10^{-38}
Weak nuclear	Any particle	10^{-18} m	10^{-5}
Electromagnetic	Any charged particle	infinite	1
Strong nuclear	Only quarks†	10^{-15} m	100

* The strength of forces relative to the electromagnetic force has been estimated for two protons that are just touching.

† Technically, the strong force between nucleons is the remnant of the color force between quarks.

Exchange particles

The four fundamental interactions that are possible between the particles that make up our universe are described in terms of the exchange of another group of particles – the **exchange particles**. A simple model of the role of exchange particles involves a brother and sister both on roller skates. If they throw identical objects between them then the emission the object followed by receiving another one back does not result in an overall change for either of them. As a result, however, they are now both moving – an interaction has taken place. The exchange object has been the "carrier" of the force.

Table 4 Exchange particles in the standard model

Force	Exchange particle	Charge/e	Mass	Spine/$\frac{h}{2}\pi$
Electromagnetic	photon (γ)	0	0	1
Strong nuclear	gluon (g)	0	0	1
Weak nuclear	W^+	+1	$89m_p$	1
	W^-	−1	$89m_p$	1
	Z^0	0	$99m_p$	1
Gravitational	gravitation	0	0	2

The graviton is the theoretical carrier of the gravitational force, but their existence has not been experimentally verified.

"Anything that is not expressly forbidden by Nature is compulsory."

Richard Feynman

The Heisenberg uncertainty principle

A potential problem exists when imagining interactions in terms of the exchange of particles of given quanta of energy – the law of conservation of energy. In classical physics, a particle can only emit a certain packet of energy if it already possesses enough total energy. If you were in a car weighing 1 tonne and suddenly 89 tonnes were ejected, you would complain that something was wrong! If this principle applied to particle physics, then many observed interactions would be disallowed.

In quantum physics, however, it is possible for the conservation of energy to be broken for a very short amount of time provided that the energy imbalance is corrected within a strict time limit expressed by a principle called the **Heisenberg uncertainty principle**.

Werner Heisenberg (1901–1976) realised that all physics measurements are subject to some degree of uncertainly. These uncertainties do not just result from a lack of precision in experimental techniques, but result from fundamental principles involved whenever two related measurements are attempted. Two such related quantities are energy and time and it turns out that the shorter the time interval (Δt) considered, the greater must be the uncertainty in the measured energy (ΔE). His uncertainty principle allows the value of these uncertainties to be quantified.

$$\Delta E \Delta t \geq \frac{\hbar}{2}$$

The fact that the energy for these exchange particles has been "borrowed" and then "paid back" within a very short period of time means that associated particle must have been emitted and absorbed within the time proscribed by Heisenberg's uncertainty principle. This means that the exchange particle can never be detected. Thus it is known as a **virtual particle**.

What is the maximum energy that can be "borrowed" for one second?

Virtual particles

Table 4 lists the exchange particles for three of the four fundamental forces.

The particles that mediate the weak interaction, the W^+, W^-, and Z^0 were predicted in the late 1960s by Sheldon Glashow, Steven Weinberg and Abdus Salaam in their theory of the electroweak force (unified theory of the electromagnetic and weak nuclear forces at high energies, see pages 494–5).

Heisenberg's uncertainty principle can be used to estimate the range R of a given interaction if the mass m of the relevant virtual particle is known (or vice versa). The energy that will be exchanged is $\Delta E = mc^2$. This energy is available for a time that is limited by the uncertainty principle:

$$\Delta E \Delta t \approx \frac{h}{4}\pi$$

$$\Delta t \approx \frac{h}{4}\pi \Delta E \approx \frac{h}{4}\pi mc^2$$

The fastest the virtual particle can move is the speed of light c. If moving at this speed, then the range R will be related to the time taken to travel Δt by:

$$\Delta t = \frac{R}{c}$$

An exchange particle for the weak interaction, the W boson, has a mass of about 80 GeV c^{-2}. Estimate the range of the weak interaction.

Combining these equations gives:

$$R \approx \frac{h}{4}\pi mc$$

The discovery of a massive W boson in CERN in 1983 is considered to be a major achievement of the Standard Model of particle physics (see page 496).

Feynman diagrams

In the 1950s the physicist Richard Feynman developed a method for dealing with the electromagnetic force between charged particles that has since been adapted for use with the other fundamental forces. The approach is very mathematical and requires an understanding of advanced quantum theory. However, the outline of his technique can be seen in his **Feynman diagrams**.

Consider a simple situation, such as two electrons scattering off each other. The event is completely described by stating the energy and momentum of each electron before and after. These are the only measurements that can be made without affecting the reaction.

In Figure 2 this is symbolized by the shaded "bubble of ignorance" in the middle. All we know for certain is what went in (*before*) and what came out (*after*).

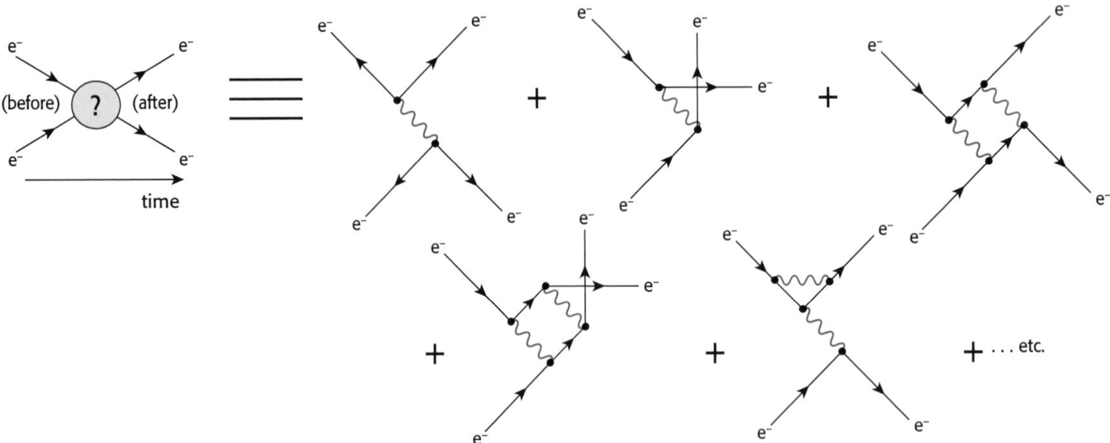

Figure 2 Feynman diagrams "fill in" the region between the start and end of an interaction. Each diagram represents a term in an infinite mathematical series; only the total of the series accurately represents the interaction.

Feynman worked out how to calculate the likelihood of a given initial situation leading to a given result. He did this by imagining all the possible things that might have happened inside the "bubble" to get from the initial to the final state. He represented each of the possible events as a "doodle", or Feynman diagram. Figure 2 shows the first five simplest Feynman diagrams that can be drawn for electron-electron scattering.

In a Feynman diagram the straight lines represent the electrons and the wavy line is the energy and momentum being passed from one to the other in the reaction. The reaction is happening because of the electromagnetic force between the charged electrons so the energy and momentum is transferred by a virtual photon. Each Feynman diagram represents a precise mathematical formula.

In fact, for any given interaction there is an infinite sum of possible Feynman diagrams. So why are they useful? It turns out that for each exchange particle involved, the probability of the reaction taking that particular pathway decreases by a known factor. For the electromagnetic force, each photon exchanged reduces the probability by a factor of $(137)^2$. The contribution to the sum of diagrams with increasing exchange particles decreases rapidly, and in this limit it is possible to use mathematical techniques to calculate the sum.

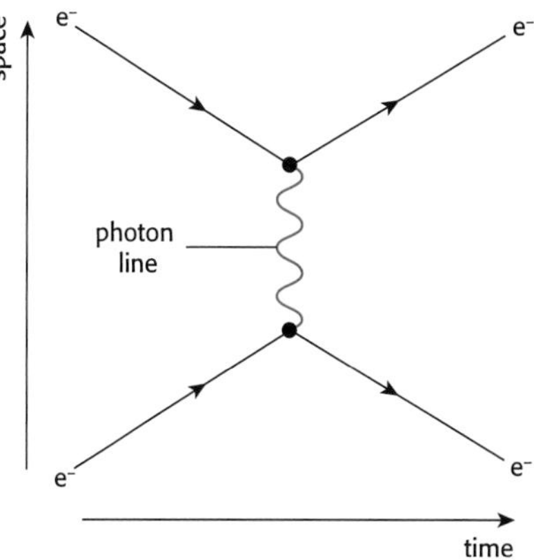

Figure 3 A Feynman diagram. In this diagram the photon appears to violate Einstein's theory of special relativity and travel at an infinite speed. However, remember that the diagram is just a representation of a mathematical expression and as such is not really physically meaningful.

Annihilation and pair production

Feynman diagrams allow the connections between seemingly unconnected interactions to become clear – they turn out to be the same pattern of lines that have just been rotated or stretched. This allows more complicated interactions to be constructed very easily so long as we stick to the same basic rules for each change or "vertex" on the diagram.

An electromagnetic interaction is shown in Figure 4. Remember that time is represented from left to right on the diagram and space is represented from bottom to top, so this picture shows an electron emitting a photon and changing direction as a result.

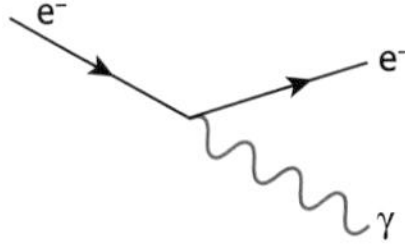

Figure 4 An electron emits a photon and changes direction

An equivalent picture (Figure 5) can be created by reflection. In this diagram, the arrows moving from right to left. This seems to represent an electron moving backwards in time but actually shows a positron moving forward in time.

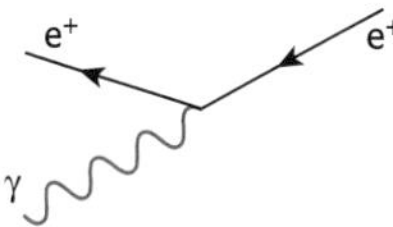

Figure 5 A positron absorbs a photon and changes direction

Another equivalent picture (Figure 6) can result from a rotation of the same diagram so that the photon is goes from left to right. In this case, the diagram represents an photon changing into an electron and positron.

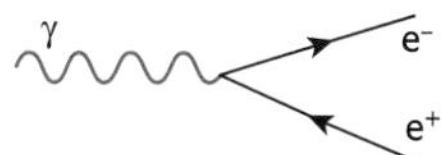

Figure 6 A photon produces an electron–positron pair

Similar interaction vertexes can be drawn for all interactions and providing that the junctions are all linked by appropriate exchange particles then we have a Feynman diagram that represents a possible interactions. Further examples are shown in Figures 7 and 8.

In summary, any interaction vertex will always show:

- a solid line representing the leptons or the quarks involved and
- a curly (gluons) or wavy line (photons, W^+, W^- or Z) representing the exchange particle and
- one arrow heading into the vertex and one heading out.

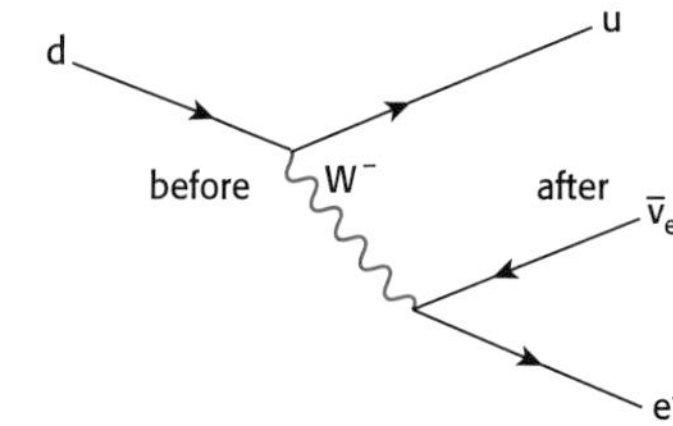

Figure 7 Beta decay – a down quark changes into an up quark with the emissions of a W^- particle that decays into an electron and an antineutrino

Lines going from left to right represent particles, lines from right to left represent antiparticles. The vertices that are involved in a Feynman diagram, the more unlikely it is to occur.

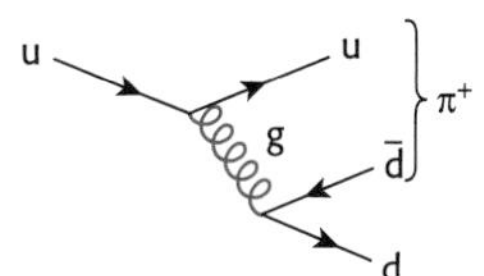

Figure 8 The up quark in a proton emits a gluon which decays into a down quark and an antidown quark

Particle accelerators and detectors

Energy requirement

A γ photon of sufficient energy can produce a particle – antiparticle pair, for example an electron and positron.

$$\gamma\ (\text{photon}) \ / \ e^+ + e^-$$

The rest mass of an electron or positron is 0.51 Mev c^{-2} so the minimum photon energy required (from the equation $E = hf$, see page 262) for such a reaction is 2×0.51 MeV = 1.02 MeV. If the electron and positron are to have energy to move then more energy is required:

$$E = mc^2 + E_K$$

where
E is the total energy of a particle
m is the rest mass of a particle
c is the speed of light
E_K is the kinetic energy of the particle.

In order to create an antiproton (p^-) the pair production p^+/p^- would require a minimum photon energy of 2×938 MeV c^{-2} = 1900 MeV c^{-2}, almost 2000 times the energy required for the creation of an e^-/e^+ pair.

Can you explain why the law of conservation of momentum means more energy is required?

Actually, in order to create the γ photon, a beam of particles is often used to strike a stationary (target) particle. For conservation of momentum, the required initial kinetic energy of the incident particle must be about 5600 MeV (5.6 GeV).

Another reason to aim for high-energy particles is the requirement dictated by the need for high resolution.

Resolution of small targets

How close can two point sources of visible light be to each other and still be resolved with the human eye? Light diffraction is a limiting physical parameter (see pages 128–31). Rayleigh suggested the following criterion for circular apertures of diameter b:

$$\sin\theta = 1.22\frac{\lambda}{b} \quad (1)$$

where
θ is the minimum angular separation of the two sources
λ is the wavelength of the visible light measured by the observer.

The shorter the wavelength, the smaller the minimum angle of resolution and so the greater the resolving power. The same principle applies to a telescope looking at stars or to a microscope looking at bacteria.

By decreasing the wavelength of the light observed, the resolution power increases proportionally. The smallest object that can be resolved in a visible light microscope is close to 1.2×10^{-7} m. In an ultraviolet microscope, targets of 5×10^{-10} m can be distinguished (the size of an atom!) since ultraviolet light has a wavelength of 10^{-7} m – 10^{-9} m. As a rule of thumb, the size of a resolvable object is in the order of the wavelength of the light used.

Elementary particles are much smaller than an atom. For example, the cross-section of the proton is about 10^{-15} m. Gamma rays have a wavelength of the order 10^{-11} m, however, much shorter wavelengths are required to resolve individual protons. In order to study the structure of particles, the use of extremely high momentum/short wavelength particles is necessary. De Broglie's hypothesis (page 264) assigns wavelength to all particles that have momentum.

(a)

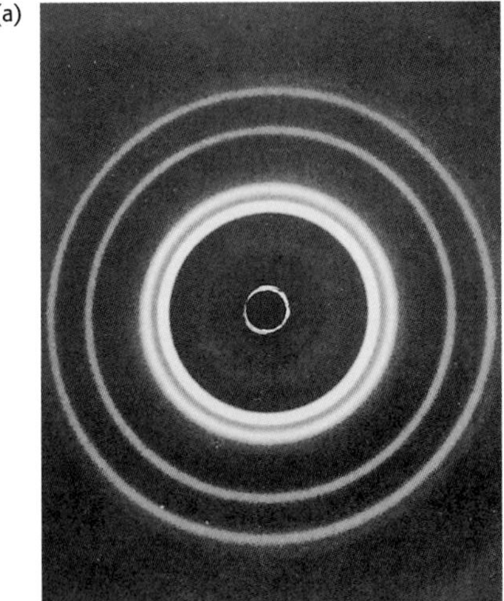

(b)

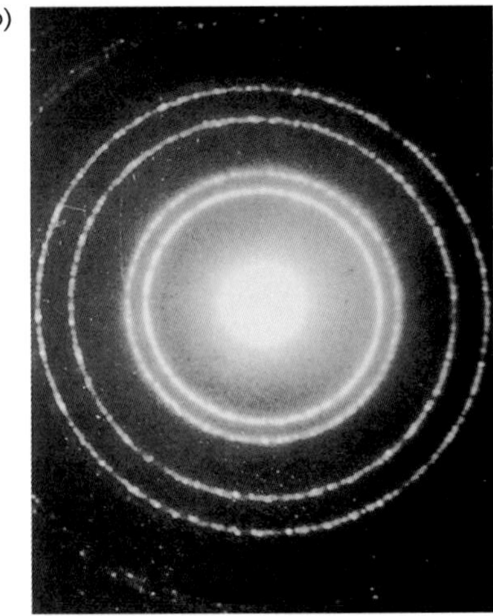

Figure 9 (a) diffraction pattern made by a beam of X-rays of wavelength 0.71×10^{-10} m from an aluminium foil target. (b) diffraction pattern made by 600 eV electrons from an aluminium foil target.

A 110 eV electron will have an associated wavelength of 1.2×10^{-10} m in the order of the wavelength of X-rays. This is also the order of magnitude of an atom. Electron diffraction patterns have been produced as predicted and they look like X-ray diffraction patterns.

To probe structures smaller than the size of an atom, electrons are accelerated so they have very large values of velocity and momentum.

1 Estimate the wavelength associated with 30 keV electrons.
2 In a given accelerator, protons of 10 GeV are produced. Using the equation for the momentum of a particle $p = E/c$, estimate the wavelength associated with these protons.

The linear accelerator (linac)

A charged particle could acquire 20 MeV of kinetic energy by moving through a single potential difference of 20×10^6 V.

However, such high voltage cannot be used effectively in the laboratory. A linear accelerator can produce high-energy charged particles by accelerating them through successive application of low electric potential differences.

The operation of a linear accelerator

Figure 10 is a sketch of a typical linear accelerator. The drift tubes are connected to an alternating voltage. This analysis starts by considering a proton that is leaving tube 1 and heading towards tube 2.

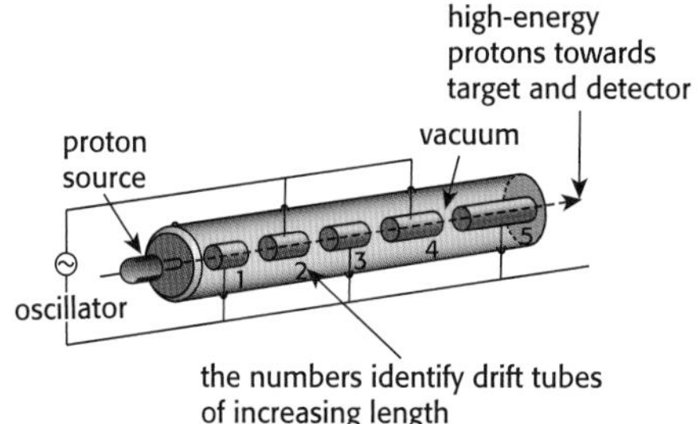

Figure 10 Linear acceleration

Stage 1 Tubes 1, 3, and 5 are positively polarized by the oscillator, Tubes 2 and 4 are negatively polarized. The protons are accelerated in the gap between tube 1 and 2, then drift inside tube 2 at a constant speed during half-period.

Stage 2 When the protons reach the end of the (now positive) tube 2, they are accelerated again towards tube 3 (now negative).

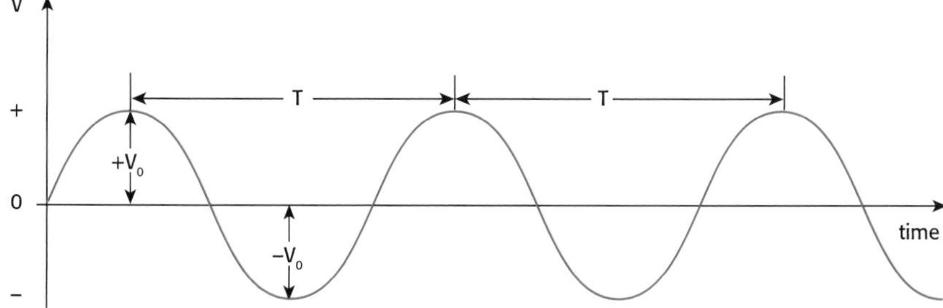

Figure 11 The voltage alternates so that the proton is always accelerated between drift tubes

Stage 3 The protons drift inside tube 3. The length of tube 3 is greater than the length of tube 2.

This process continues until the kinetic energy of the protons reaches the expected maximum value. The high-energy protons are then directed towards the target inside a detector.

Figure 12 SLAC (the Stanford Linear Accelerator Laboratory) in California

1 State the reason why the main chamber is a vacuum.
2 Explain whether or not the time of flight of the protons inside a tube change from one tube to another.
3 Derive an expression giving the final kinetic energy of the proton at the end of the accelerator as a function of the total number of "gaps" n and of the *effective* voltage between successive tubes V_{eff}.
4 Determine whether such an accelerator can be used with electrons. What about heavy ions? If so, will all these charged particles have the same final kinetic energy? Will they reach the same final speed (assuming non-relativistic effect)?
5 **a)** Discuss whether the oscillator should operate at low or high frequency to minimize the length of the tubes.
b) A typical oscillator in a linac uses a radio frequency (RF) of about 2×10^8 Hz. What is the corresponding period of the signal?
6 Estimate the speed of the proton at the end of the first potential jump if the effective accelerating voltage is 25 keV. Assume its initial speed is effectively zero.

The special theory of relativity indicates that the speed of a particle can get very close to the speed of light c but will never reach or exceed it. So what would happen if energy were still given to the proton when the speed of the particle is very close to the speed of light?

The particle's speed will not change much (see Table 5), but its mass will increase by an amount Δm proportional to the given energy increment ΔE:

$$\Delta m = \frac{\Delta E}{c^2} \qquad (2)$$

There is equivalence between mass and energy. When a particle becomes relativistic[1] in a linear accelerator, its speed remains constant. Consequently, there is no need to change the length of the drift tubes so the design of a linear accelerator is much simpler for relativistic particles.

The constant length of these equal drift tubes is equal to $c/2f$, where f is the frequency of the oscillator.

Table 5 Kinetic energy of different particles and their speed

Kinetic energy	Electron speed*	Proton speed*
0.1 MeV	0.553	0.0146
1 MeV	0.941	0.0462
10 MeV	0.9988	0.145
100 MeV	0.999987	0.429
1 GeV	0.99999987	0.876
10 GeV	0.9999999987	0.9963
100 GeV	0.999999999987	0.999957

* All speeds are expressed as fractions of the speed of light.

1 Using Table 5, plot a graph of the ratio v/c against E_k.
2 Explain why the accelerated electron reaches relativistic speeds at lesser kinetic energies than the accelerated proton.
3 Calculate the speed of an electron with a kinetic energy of 1.0 MeV, assuming Newtonian mechanics. Compare your results to the corresponding value in Table 5.

The cyclotron

Dependence of speed of particles on their kinetic energy

In the cyclotron, a relatively low potential is used to accelerate a beam of particles in circular trajectories inside a homogeneous magnetic field.

Trajectories of charged particles inside a homogeneous magnetic field

The behaviour of charged particles inside a homogeneous field can be deduced from knowledge of circular motion. Figure 13 illustrates the motion of a positive particle moving under a constant magnetic force $\boldsymbol{F}_m$ acting as the centripetal force. The field $\boldsymbol{B}$ is perpendicular and directed into the page. The radius of curvature is r and the velocity is $\boldsymbol{v}$.

[1] Generally, a particle moving at a speed v = 0.10 c or more is considered relativistic.

The vectors $\boldsymbol{B}$, $\boldsymbol{F}_m$, and $\boldsymbol{v}$ are always perpendicular to each other. As a consequence, the direction of the velocity of the particle changes constantly but not its magnitude. No work is done by the magnetic field.

The magnetic force $\boldsymbol{F}_m$, acting on the charge q of mass m, is the centripetal force:

$$F_m = \frac{mv^2}{r} = Bqv \quad (3)$$

$$r = \frac{mv}{qB} \quad (4)$$

The product mv is the momentum p, so for constant q and B the radius of curvature r is a direct measure of the momentum p of the particle. The momentum is related to the kinetic energy E_k if the particle is non-relativistic:

$$E_k = \frac{p^2}{2m} \quad (5)$$

Since p = rqB by equation 4:

$$E_k = \left[\frac{q^2B^2}{2m}\right]r^2 \quad (6)$$

The kinetic energy of the particle is proportional to the square of the radius of curvature. If the radius of the particle's path is increased by a factor of three, the energy of the particle increases by a factor of nine.

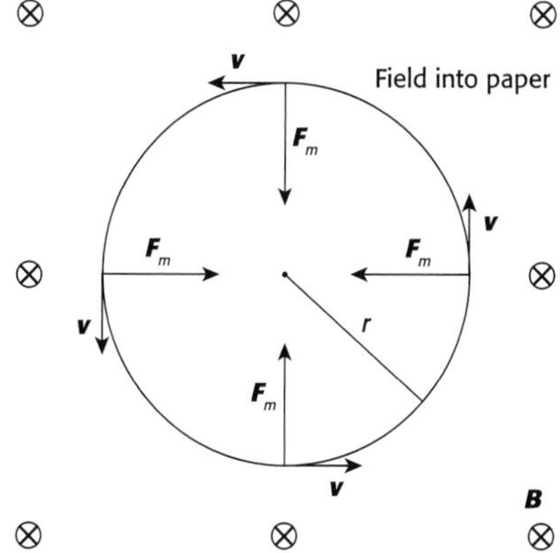

Figure 13 The field, velocity and force vectors are all perpendicular. The particle is positive.

Deduce that the period T of one orbit is given by

$$T = \frac{2\pi m}{qB} \quad (7)$$

Interestingly, the period given by this equation is **neither** a function of the radius r **nor** of the speed v. How can this be? What happens to r when v increases?

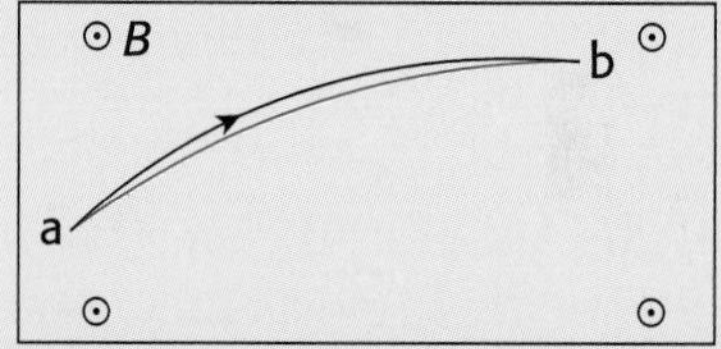

Figure 14 Analysis of a simple particle track

Figure 14 illustrates the trajectory of a proton moving from a to b in the plane of the page. The field (1.4 T) is perpendicular to the page and directed out of it.

1 State the sign of the charge.
2 Explain how you would go about measuring the radius of curvature of the trajectory.
3 The radius of curvature of the proton is 10 cm. Determine:
   a) the momentum
   b) the speed
   c) the kinetic energy of this proton.

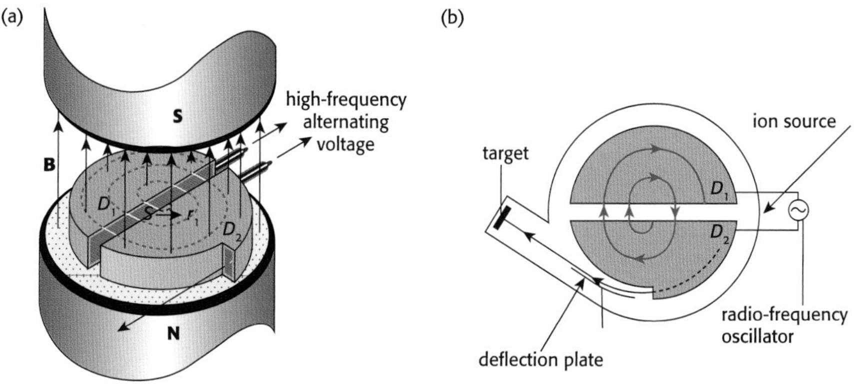

Figure 15 The design of a cyclotron

A homogeneous magnetic field B is created between the poles of an electromagnet. Two hollow semicircular electrodes are inside a vacuum chamber (because of their shapes, they are referred to as Dees). An ion source at the centre injects the particle between the two electrodes.

If D_2 is positive and D_1 negative, the particle will be accelerated towards D_1. Inside the Dee, the particle will not feel the electric field due to the charges on the surface of the electrode (see page 197), but will curve under the influence of the magnetic field. The time required to come back to the gap between the two Dees is not a function of the radius of the curvature nor of the speed of the ion (see equation 7). During that time, the polarity of the two Dees reverses just in time for the particle to be accelerated again during its second passage through the gap.

After this second acceleration, the radius of curvature inside D_2 will be greater. If the particle remains non-relativistic, the time of travel inside the second Dee will be the same as it was inside the first. Under this condition, the frequency of the oscillator is constant and the accelerated particle remains in phase with the oscillator (accelerating field). The particle will follow a circular path with an increasing radius (spiral), until a deflecting plate sends it towards the target.

The frequency f of the oscillator is given by

$$f = \frac{1}{T} = \frac{qB}{2\pi m} \tag{8}$$

The energy gained at each gap ΔE_k is

$$\Delta E_k = V_{eff}\, q \tag{9}$$

where V_{eff} is the effective voltage of the oscillator.

The maximum value of the radius of the curvature is close to the actual radius of the Dee chambers.

Values for a typical cyclotron

Particle: proton
Radius: 0.40 m
Magnetic field: 1.5 T
Oscillator frequency: 22 MHz
Maximum kinetic energy: 17 MeV
Alternating voltage: V_{eff} = 200 kV

1 The proton moves at 19% of the speed of light. Assuming the proton is non-relativistic, determine how many times will it pass between the Dees before reaching its maximum energy. Assume that the initial kinetic energy of the injected proton is negligible.

2 If a deuteron ion is used, instead of a proton, state the new frequency of the oscillator.

The synchrotron

Theory

Bremsstrahlung (breaking radiation)

Unfortunately the above analysis does not take into consideration the fact that when any charged particle accelerates, it radiates electromagnetic energy. The general term for the radiation emitted by accelerating charges is **bremsstrahlung** radiation where the German word means "slowing down" or "braking".

This term was first used to describe the continuous X-ray spectrum produced when electron are decelerated by hitting a stationary target. Figure 13 illustrates and electron moving the Coulomb field of the nucleus of an atom. The electric force must change the path of the electron and thus the electron must be accelerating. As is consequence it must emit a photon of electromagnetic radiation. This loss of energy means it must slow down.

The particles in a synchrotron are also being accelerated. Of course, this happens when they move across the accelerating gaps but it also happens when they are being bent by a magnetic field into following a circular path (circular motion involves constant centripetal acceleration). The lost energy during circular motion needs to be continually replaced and many synchrotrons have very large diameters to reduce the energy lost.

In the synchrotron, the radius of the orbit followed by the accelerated particles remains constant. Figure 16 shows a proton circulating in an evacuated tube in a magnetic field.

A series of accelerating stations (only one is shown in Figure 16) situated around the circumference of the synchrotron accelerate the particle. At each gap, the increase in kinetic energy of the particle is given by $V_{eff}\, q$.

As the speed increases, the magnetic field must be proportionally adjusted.

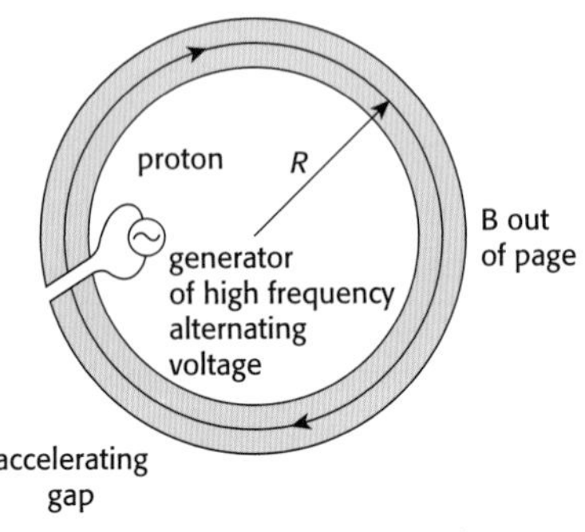

Figure 16 Principle of a Synchrotron

The equations $f = \dfrac{qB}{2\pi m_{rel}}$ and $f = \dfrac{v}{2\pi r}$

must both be satisfied during the acceleration of the particle around the ring. This is done by adjusting the magnetic field B and the resonance frequency f. As v tends to be the speed of light c, the frequency remains constant and B must be adjusted for increasing relativistic mass, m_{rel}.

The particle will stay in orbit of radius R if the magnetic field strength has the value given by

$$B = \frac{m_{rel} v}{qR} \qquad (10)$$

where the values of v and m_{rel} increase with time.

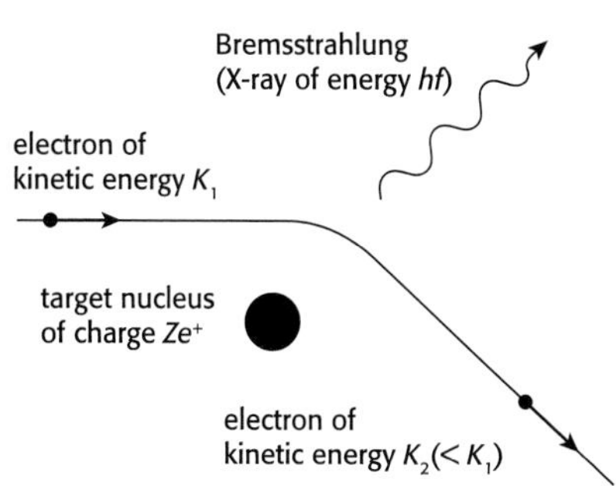

Figure 17

Mathematical Physics: *The energy required to create a set of particles at rest from the collision of an incoming particle with a stationary target*

A relativistic incoming proton collides with a target proton. This scattering event produces a neutral pion π^0

$$p + p \rightarrow p + p + \pi^0$$

The rest mass of the pion is 135 MeV c^{-2}. More than 135 MeV is in fact required since the incoming proton has momentum. Some of the particles must carry momentum after the collision in order to conserve momentum, and thus kinetic energy, away.

To simplify the mathematics we swap our viewpoint, or "frame of reference" to look at the event in the frame of reference of the centre of mass (CM) where, before the collision, the two protons move towards each other with opposite and equal velocities and the momentum of the system is zero. In the CM frame, the minimum energy required to create the neutral pion with the two protons, all at rest, equals the sum of the rest mass energy of these three particles. This sum must be equal to the total relativistic energy in the CM frame:

$$2(mc^2) = 2\,(m_p c^2) + m_\pi c^2$$

where

m is the relativistic mass of each proton in the CM frame

m_p is the rest mass of a proton

m_π is the rest mass of a pion.

The speed of the proton v in the CM frame can be determined by:

$$m = \frac{m_p}{\sqrt{(1 - v^2/c^2)}}$$

m_p: 938 MeV c^{-2}
m_π: 135 MeV c^{-2}
v: speed of each proton in the CM frame

Calculations yield $v = 0.36$ c.

However, in order to obtain a meaningful result we must look at the event from the laboratory frame of reference (where the target proton is at rest).

+0.36 c —— −0.36 c

• → × ← •

Centre of mass frame

• ⟶ •

+0.64 c

Lab frame

In the lab FR, V can be calculated by using the relativistic addition of velocities equation:

$$V_{tot} = \frac{V_1 + V_2}{1 + \dfrac{V_1 V_2}{c^2}}$$

The result is: $V = 0.64$ c (not 0.72 c!)

This is the velocity of the incoming proton. Its relativistic mass ($m = \gamma m_p$) is 1.3 m_p, and its kinetic energy ($E_k = \Delta m\, c^2$) is near to 280 MeV.

$$\text{The ratio } \frac{\text{energy of proton}}{\text{rest mass of pion}} = \frac{280}{135} \approx 2$$

The proton must have an energy of more than twice the rest mass of the pion to produce it. Collisions with a stationary target are not efficient, since a significant amount of kinetic energy is "used up" to conserve momentum.

If we generalize this approach we obtain

$$E_a^2 = 2Mc^2E + (Mc^2)^2 + (mc^2)^2$$

where E_a is the total rest mass energy of the product particles. This is the minimum energy that must be available for the event.

E: total relativistic energy (threshold energy) of the incoming particle (lab FR)

M: rest mass of the stationary target

m: rest mass of incoming particle

Experimentally the value of E_a is first determined, then its value is introduced into the equation to find the value of *E*.

If for a different event, the total rest mass energy of the particles after collision is larger than the above, what effect does this have on the efficiency of this collision (with a stationary target) increase?

For example, consider the event where an antiproton $\bar{p}$ and a proton p are produced in the collision of an incoming proton with a stationary proton:

$$p + p \rightarrow p + p + \bar{p} + p$$

Similar calculations show the incoming proton must have a kinetic energy of 6 $m_p c^2$ (total energy of 7 $m_p c^2$). Three times more energy than the rest mass energy is required. We can see that when higher energies are required, the efficiency drops.

The colliding-beam accelerator

The high-energy particles emerging from a synchrotron collide with a target to create new particles. Unfortunately, the total energy of the incident particle (kinetic energy + rest mass energy) is not entirely available when the target is fixed.

When colliding particles are moving towards each other at relativistic speeds much more energy is available for the creation of new particles.

Considering the amount of extra energy available for the creation of new particles it is clear that if both the "beam" and "target" protons are moving we are at a considerable advantage. However, aligning two protons so that they collide head-on (the radius of a proton is about 10^{-15} m.) is very challenging.

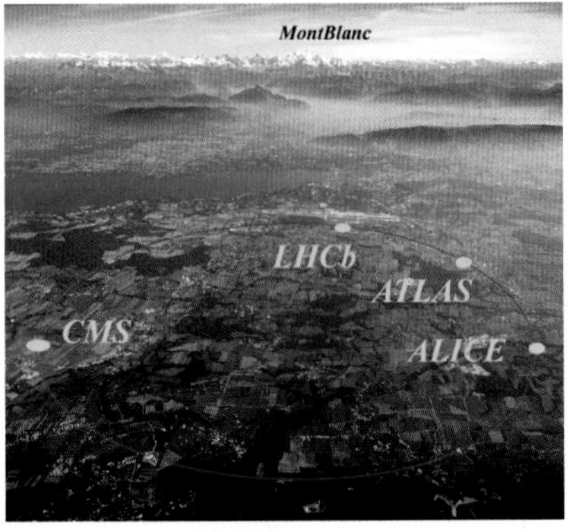

Figure 18 The Large Hadron Collider (LHC) at CERN in Geneva

Comparative look at a relativistic collision between two moving protons

Two accelerators each produce a beam of protons (of rest mass m_p), moving at a speed v = 0.98 c relative to the laboratory frame of reference to which the accelerator tube is attached. The protons are aimed at each other.

In the lab frame, the gain in relative speed of approach is not great (0.0198 c). However, in the same frame, the relativistic energy is about 8 times greater than its Newtonian value. The energy gained in using colliding beams is significant. The principle of operation of the **colliding-beam** accelerator takes advantage of this fact. The probability of having two protons colliding head-on is not very high but the energy gain is important.

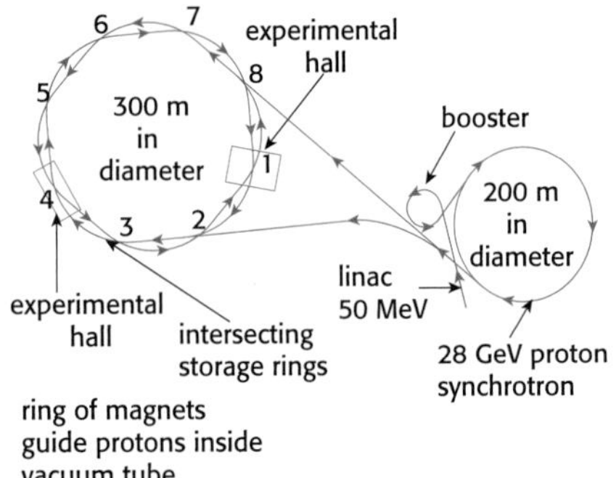

Figure 19 Proton synchrotron at CERN

Figure 19 shows the proton synchrotron at CERN, which uses storage rings.

The proton beam from the synchrotron is divided in two separate beams moving in opposite directions, thus creating two beams intersecting at different points where the proton beams collide (almost head-on) with an energy equivalent to that produced by a single 1700 GeV accelerator.

Investigation

The Large Hadron Collider (LHC)

Using the page found here: http://public.web.cern.ch/Public/en/LHC/LHC/-en.html prepare a poster about the LHC accelerator. Include information about the type of accelerator, the detectors located around it, and some of its goals.

At the time of writing, the Large Hadron Collider was planned to be operational in CERN by autumn 2009. In this accelerator phyicists will be able to recreate the state of matter and energy a fraction of a second after the Big Bang.

Comparison of different particle accelerators

Accelerator type	Advantages	Disadvantages
Linear	Low energy losses as Bremsstrahlung radiation is kept to a minimum.	A fixed target means that proportion of energy available for particle production is low. Low relative final energies as a result of long lengths that would be required. Time of collision cannot be altered after acceleration has started.
Cyclotrons	Compact size and relatively low cost make small scale research and applications possible.	A fixed target means that proportion of energy available for particle production is low. Compact size places limit on final energies. Time of collision cannot be altered after acceleration has started.
Synchrotrons	Able to accelerate particles to extremely high energies. Colliding beams means that proportion of energy available for particle production is high. Use of storage rings allows for the control of the timings of collisions.	High energy losses as a result of Bremsstrahlung radiation. Technical difficulty of arranging collisions to take place. Very high cost means often means that facilities rely on international collaboration.

Thinking about science: ***Can we justify the expense of particle physics research centres?***

Consider the following points of view:

- In 1905, Einstein published three papers on light, special relativity, and the sizes and movements of molecules. These theories help our understanding of matter and the universe but, also, lay down the foundations for the technology and economy of today. The television, GPS, the mobile phone, and the electron microscope are examples of such applications.
- In the 18th century, Faraday, a brilliant scientist, was investigating magnetic fields and their effect on coils. Was his work then productive for his society? One day, Prime Minister Gladstone asked him what was the use of electricity "Some day you will tax it," answered Faraday. Today, electricity is an intrinsic part of our life thanks to his work and that of others.
- Colliding particles in accelerators produce new particles that are used for biological research and therapy. Synchrotron radiation has a large range of applications including biology, medicine, and technology.
- The sharing of large research costs by developing countries is a wise approach that permits reasonable scientific endeavour. The sharing of the design of accelerators and detectors as well as the sharing of research and data with developing countries encourage cooperation between people of different cultures and beliefs. It also feeds the progress of these nations.
- Without experimental physics, theoretical physics stagnates in pure speculation.
- Without fundamental research some new technology will not materialize.

Above all, curiosity is an essential part of the human mind.

Particle detectors

When the high-energy particles from accelerators strike a target, rapid reactions occur. These particles are invisible to the human eye so physicists design instruments that make it possible to "see" the path of charged particles in a medium. Three of these devices are described here: the bubble chamber, the spark chamber, and the photomultiplier. In each case, the ionizing power of charged particles moving in the medium is the dominating effect.

The bubble chamber

If pressure is exerted on a liquid, it can be superheated. By reducing the pressure suddenly, the liquid vaporizes. At that instant, incoming charged particles can ionize the liquid along its track, which creates gas bubbles around the ions. These bubbles reflect light and can be photographed by two cameras.

Figure 20 The bubble chamber

Liquid hydrogen is generally used in bubble chambers because it provides a dense target of protons. The analysis of events is also facilitated by the fact that the proton is the simplest nucleus available. However, the temperature at which the hydrogen is superheated is $-246\,°C$, so a cooling system is required.

A magnetic field of about 1 T curves the tracks of charged particles.[2] Events unfold very rapidly so that the picture can reveal a series of cascading tracks across the chamber. Uncharged particles can be "seen" indirectly through their interaction with protons, which, after impact, will leave an ionic trail of their own.

[2] As a comparison, a typical value of the field at the Earth's surface is 5.5×10^{-5} T.

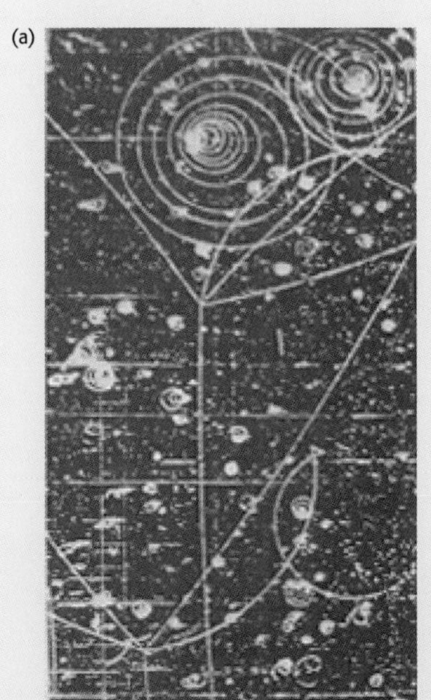

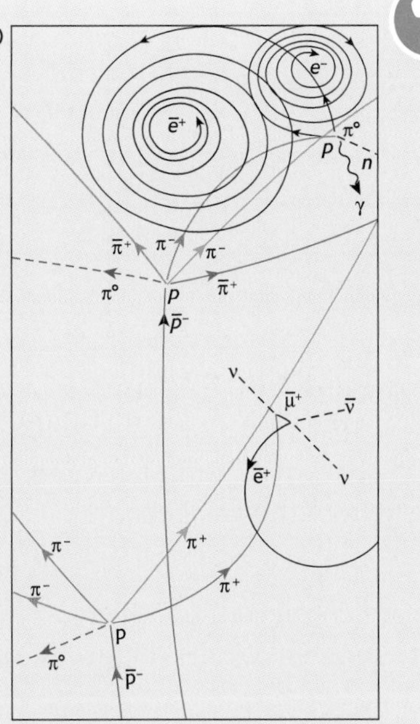

Figure 22 (a) Photograph of antiproton reactions in a hydrogen bubble chamber showing meson pairs in the kind of reaction that first led to the identification of the ω meson. (b) Legend diagram showing the unchanged trackless particles as dotted lines.

1 At the top of Figure 22a, the spiralling trajectories of an electron and a positron are shown. Which particle carries less energy? What are your assumptions?

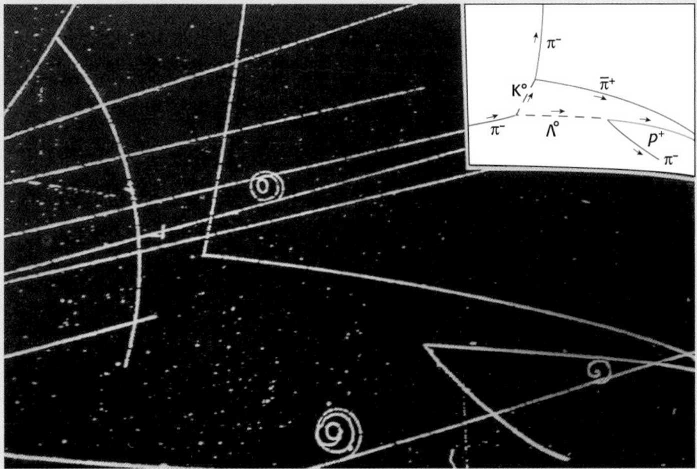

Figure 23 Bubble-chamber photograph with π^-, p, K°, Λ°, $\bar{\pi}^+$, and π^- tracks, all in one related event. Legend diagram at upper right.

2 Neutral particles do not leave any tracks in the chamber. Explain why this is the case.

Figure 21 A 4 MeV electron slows down in a liquid hydrogen bubble chamber immersed in a homogeneous magnetic field. The beaded trajectory is characteristic of electron tracks.

The spark chamber

Figure 24 shows two thin metallic plates immersed in a noble gas (neon or argon). A high voltage is applied to the plates. If the path of a charged particle runs between the plates, a trail of ion pairs will be produced in the space between the plates (Figure 24a).

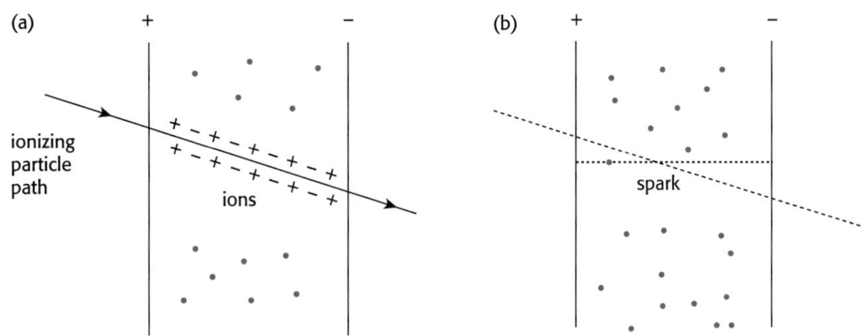

Figure 24

Because of the high potential difference, a spark will jump between the plates at the location of the ion pairs, where the gas becomes temporarily conductive (Figure 24b).

In a spark chamber, a series of electrodes are connected to a high-voltage source. A few millimetres separate successive foils (Figure 25).

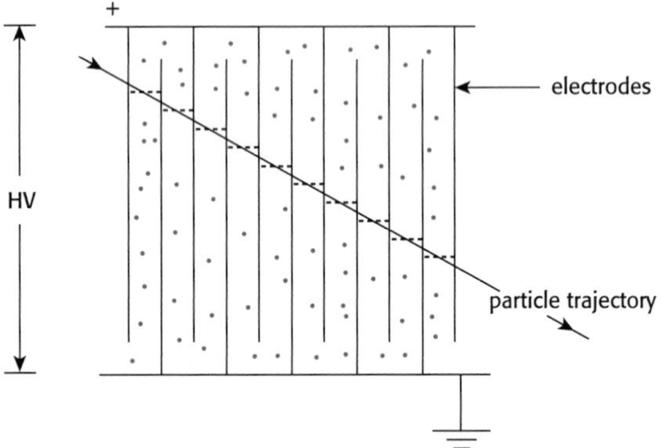

Figure 25

When the potential difference is applied, a succession of sparks will reveal the trajectory of the particle through the chamber, since the charged particle will leave a trail of ions in the space between the electrodes. The series of sparks can be photographed or electronically counted (Figure 27).

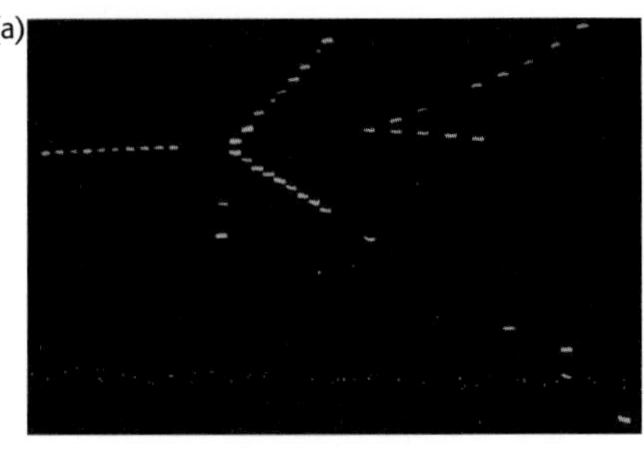

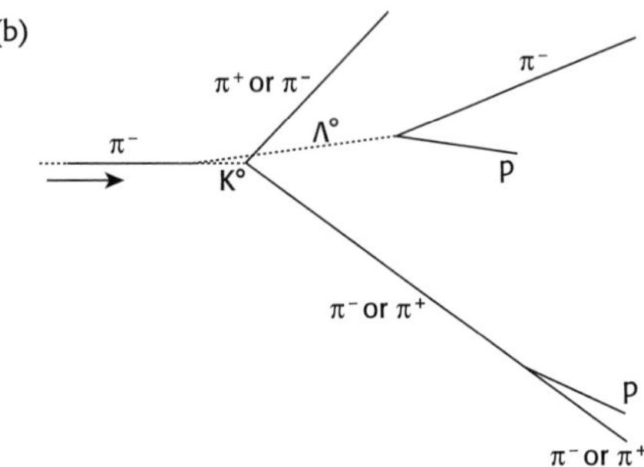

Figure 26 (a) shows appearance of particle tracks. No magnetic field is applied. (b) identifies the charged particles and their tracks as well as the implied tracks of the neutral particles.

The photomultiplier

Gamma rays produce very little ionization in a gas or liquid. They do not leave trails in a bubble chamber or in a spark chamber but a photomultiplier tube can be used to detect gamma rays and to measure their energy.

A single gamma ray photon collides with an atom inside a transparent crystalline solid called a scintillator, producing light in the visible region. A reflector sends this light towards a photosensitive layer. The action of the light on this layer triggers the emission of a single photoelectron, i.e. an electron that has been liberated from its atom by a photon, which is accelerated towards an anode (+). Upon collision, a series of secondary electrons are liberated and accelerated towards a second anode, which is at a higher potential than the first one. At this anode, even more secondary electrons are ejected. This avalanche effect continues through a series of anodes called dynodes. The final signal is amplified

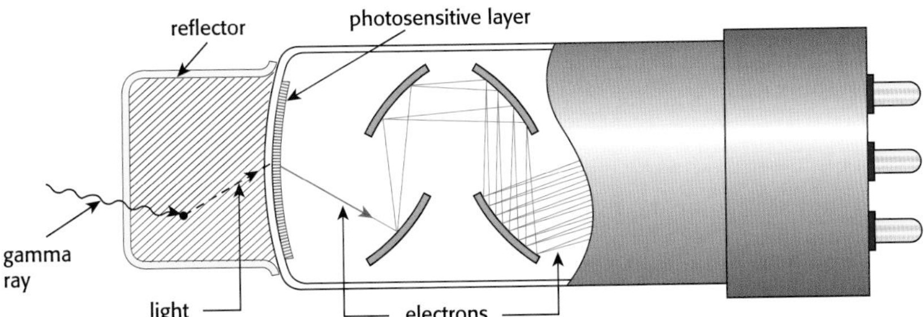

Figure 28 Scintillation detector coupled to a photomultiplier

electronically so the pulses can be counted. These pulses are also a measure of the energy of the incoming gamma ray. Ultimately, an energy spectrum of the photons entering the scintillation detector can be produced. The advantage of the photomultiplier is that individual photons of relatively low energy can be counted.

The particle physics international community

Particle research laboratories

In 1949, at the European Cultural Conference, in Lausanne, Switzerland, the French scientist Louis de Broglie proposed the idea for a European research laboratory where "greater resources than those available to the national laboratories and could then embark upon tasks whose magnitude and nature preclude them from being done by the latter on their own."[3] A series of meetings under the auspices of UNESCO brought about the foundation of CERN. Today, its membership comprises 20 European countries and more than half the world's particle physicists collaborate on the work done at CERN.

Scientists do fundamental research in the name of scientific endeavour, independently of their nationalities, politics, culture, and religion. This cultural melting pot is at the heart of the activities of these facilities. "… based in their own countries, members of collaborations not only provide most of the ambitious experimental apparatus, but they also contribute to a novel, global, powerful information and communication infrastructure using their own countries' industries and talents in a fair and constructive partnership. And the motivation for all this: cutting-edge physics."[4]

The Large Electron Positron collider (LEP) which ran from 1989–2000 contributed to the current Standard Model of particles and their interactions.

The new LHC project comprising the accelerator and four major detectors involves the efforts of more than 400 research laboratories from dozens of countries around the world. It is due to switch on soon and physicists hope to detect evidence for the Standard Model.

The International Linear Collider (ILC) is currently under development. It goes beyond the regional concept to a global one. The idea is to form a partnership between American, Asian, and European countries, which will share the design, construction, and

Examples of some other big national laboratories:

- USA: Fermilab, SLAC
- Germany: DESY
- Japan: KEK
- Canada: TRIUMF

[3] *CERN Courier*, November 2004
[4] *CERN Courier*, March 2005

management of the facility. Emphasis will be on the research and development of detectors. The goals of the ILC are:

- to create a research centre where strong international cooperation flourishes
- to ensure access to the facilities
- to include the broad community in the decision-making process: "access to materials, land use, ecology, and economic impacts due to the resources that are required."[5]
- to make the data available to the broader community
- to consider new forms of detector collaboration.

Scientists work together as colleagues and friends in a way that does not reflect the external tensions and hostilities existing between some countries and cultures. Even in times of conflicts, scientific collaboration continues to play a role in penetrating political barriers, as it was between Western and Soviet societies during the Cold War.

Research one of the following further examples of international collaboration in the particle physics community:

- JINR
- SESAME
- The Berlin Declaration
- The Pugwash conference
- ICSC
- The Erice statement
- ICTP

Quarks

Quarks are elementary particles. The six different types of quark are shown in Table 1 (page 470). Any particle that is made up of quarks is called a **hadron**.

According to the best theories at the moment, it is impossible for quarks to exist on their own; they must always be bound into particles by the color force that exists between them. As it happens there are only three stable combinations that the color force will allow.

Table 6 Allowed combinations of quarks and antiquarks

Combination	Symbol	Name
three quarks	qqq	baryon
three antiquarks	$\bar{q}\bar{q}\bar{q}$	antibaryon
a quark and an antiquark	$q\bar{q}$	meson

Distinguish the category of hadrons into which the following particles can be placed.

1. uud (a proton)
2. udd (a neutron)
3. $u\bar{d}$ (a π^+ particle)
4. $\bar{u}d$ (a π^- particle)
5. $u\bar{u}$ (a π^0 particle)

The three π particles have very similar masses, and belong to a family known as the pions

Baryons, antibaryons, and mesons were discovered long before quarks and their properties were well known. The success of the quark idea was to draw together and explain their properties in terms of the properties of the quarks inside them.

Baryon number

Baryon number is a property of hadrons that was discovered by studying various reactions taking place in particle accelerators. It is possible to give every hadron a value of baryon number (B) based on a simple scheme:

Any baryon $B = 1$
Any antibaryon $B = -1$
Any meson $B = 0$

e.g. protons and neutrons have $B = 1$, an antiproton has $B = -1$, all pions have $B = 0$. In all reactions baryon number is found to be a conserved quantity. The baryon number of the various hadrons can be explained if the quarks are given values of baryon number shown in Table 7.

Table 7 Quarks and antiquarks

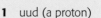

	Baryon number	Charge	Type of quark
Matter	1/3	+2/3	u c t
		−1/3	d s b
Antimatter	−1/3	−2/3	$\bar{u}$ $\bar{c}$ $\bar{t}$
		+1/3	$\bar{d}$ $\bar{s}$ $\bar{b}$

[5] *CERN Courier*, October 2005

Antiquarks

Quarks all have antiquark partners. They have the same mass but opposite charge and **baryon numbers**.

Consider the baryon numbers of the quarks and antiquarks. You should be able to convince yourself that the sum of the baryon numbers of the quarks will produce the required baryon numbers of the hadrons.

Spin

Quarks all have an intrinsic angular momentum of $\hbar/2$, or in the usual shorthand, "spin ½". Spin can be directed "up" (spin +½) or "down" (spin −½).

Quantum numbers are, in general, additive. This means that when elementary particles combine to form a new system, the quantum numbers of the particle that is formed are just the sum of the numbers of the constituent parts. Spin systems will tend to align so as to form pairs of spin "up" and spin "down" particles. When a proton is formed from three quarks, there will always be one spin "up" quark and one spin "down" quark. The third quark can be aligned in either of the two spin states so that the proton will have an overall spin of ±½ in its lowest energy state.

Determine the possible values of spin of

a) baryons

b) antibaryons

c) mesons

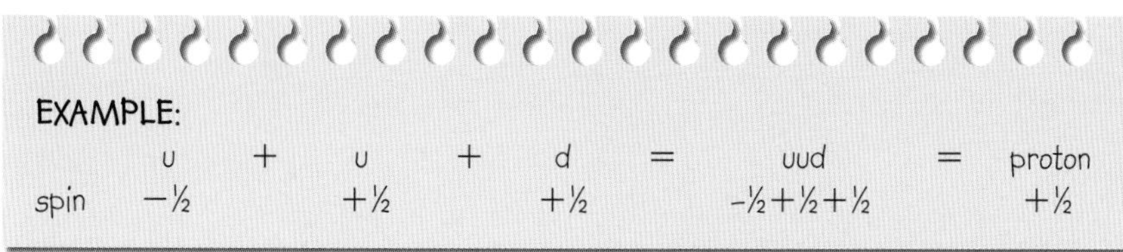

EXAMPLE:
u + u + d = uud = proton
spin −½ +½ +½ -½+½+½ +½

Conservation of baryon number

Baryon number was discovered when physicists noticed that some reactions that conserved electrical charge were not taking place:

$$p + p \not\to p + \pi^+$$
$$p + n \not\to p + \pi^0$$
$$p + p \not\to \pi^+ + \pi^+$$

It seems that the total number of baryons must remain the same. Pions, which can appear in any number given enough energy, are mesons whilst protons and neutrons are baryons. This suggested a simple scheme in which the proton and neutron have $B = 1$ and the pions have $B = 0$. It is then easy to show that the reactions do not happen because they do not conserve baryon number:

$$p + p \not\to p + \pi^+$$
$$B \quad 1 + 1 \neq 1 + 0$$
$$p + n \not\to p + \pi^0$$
$$B \quad 1 + 1 \neq 1 + 0$$
$$p + p \not\to \pi^+ + \pi^+$$
$$B \quad 1 + 1 \neq 0 + 0$$

Baryon number conservation indicates that the fundamental forces must all keep the net number of quarks (i.e. the number of quarks minus the number of antiquarks) constant.

Strangeness

In the early 1950s, particles with large masses and long lifetimes started to be discovered in cosmic ray experiments. In 1954 these particles were categorized and given a new property – **strangeness**. The odd lifetimes of the strange particles (heavy particles generally decay more rapidly than light ones) could then be explained by assuming that strangeness was conserved in strong force interactions but not conserved by the weak force.

Faced with the need for an arbitrary choice, physicists chose the K^+ meson to have strangeness $S = +1$, and categorized other particles relative to this. It is now understood that the strangeness of a baryon indicates the presence of **strange quarks** within it. An unfortunate consequence of the choice of the K^+ meson to have strangeness $S = +1$ is that the strange quark itself has $S = -1$, but the choice is arbitrary so it is not important. The antistrange quark has $S = +1$.

The other heavy quarks (charm, top, and bottom) are all defined to have their own associated quantum number in a similar fashion (see Table 8).

Table 8 Quark properties

Quark	charge/e	baryon number	strangeness	charm	bottom	top
u	−1/3	1/3	0	0	0	0
d	+2/3	1/3	0	0	0	0
s	−1/3	1/3	−1	0	0	0
c	+2/3	1/3	0	+1	0	0
b	−1/3	1/3	0	0	−1	0
t	+2/3	1/3	0	0	0	+1

Table of quark content of some hadrons (baryons and mesons)

	Name	Symbol	Charge / e	Baryon no.	Strangeness	Charm	Quarks
Baryons	proton	p	+1	+1	0	0	uud
	neutron	n	0	+1	0	0	udd
	lambda	Λ	0	+1	−1	0	uds
	sigma plus	Σ^+	+1	+1	−1	0	uus
	sigma minus	Σ^-	−1	+1	−1	0	dds
	sigma zero	Σ^0	0	+1	−1	0	uds
	xi minus	Ξ^-	−1	+1	−2	0	dss
	xi zero	Ξ^0	0	+1	−2	0	uss
	omega minus	Ω^-	−1	+1	−3	0	sss
	charmed lambda	Λ_c	1	+1	0	1	udc
Anti Baryons	anti proton	$\bar{p}$	−1	−1	0	0	$\bar{u}\bar{u}\bar{d}$
	anti neutron	$\bar{n}$	0	−1	0	0	$\bar{u}\bar{d}\bar{d}$
	anti lambda	$\bar{\Lambda}$	0	−1	+1	0	$\bar{u}\bar{d}\bar{s}$

	Name	Symbol	Charge / e	Baryon no.	Strangeness	Charm	Quarks
Mesons	pi zero	π^0	0	0	0	0	$u\bar{u}$ / $d\bar{d}$
	pi plus	π^+	+1	0	0	0	$u\bar{d}$
	pi minus	π^-	−1	0	0	0	$\bar{u}d$
	K zero	K^0	0	0	+1	0	$d\bar{s}$
	K plus	K^+	+1	0	+1	0	$u\bar{s}$
	K minus	K^-	−1	0	−1	0	$\bar{u}s$
	J / Psi	J/Ψ	0	0	0	0	$c\bar{c}$
	D zero	D^0	0	0	0	1	$c\bar{u}$
	D plus	D^{+1}	+1	0	0	1	$c\bar{d}$
	upsilon	Y	0	0	0	0	$b\bar{b}$

- A quark's spin is $\pm\frac{1}{2}$ making it fermion that obeys the Pauli exclusion principle.
- The three quarks in a baryon means that all baryons will have a spin of $\frac{1}{2}$ (two aligned and one opposite) or $\frac{3}{2}$ (all aligned) and must also be fermions. Baryons also obey the Pauli exclusion principle.
- The quark/anti-quark pair in a meson means that all mesons have a spin of 0 (opposite spins) or 1 (aligned spins) and are thus bosons. Mesons do not obey the Pauli exclusion principle.

Using conservation laws

Consider the following reaction in which X is a newly discovered particle:

$$K- + p \rightarrow K^0 + K+ + X$$

K particles are mesons. What can be deduced about the properties of X by applying the various conservation laws that we know?

1 Conservation of electric charge shows that X must carry a negative charge:

$$\begin{array}{lccccccccc} & K^- & + & p & \rightarrow & K^0 & + & K^+ & + & X \\ Q & (-1) & + & (+1) & \rightarrow & (0) & + & (+1) & + & (-1) \end{array}$$

2 Conservation of baryon number shows that X must be a baryon:

$$\begin{array}{lccccccccc} & K^- & + & p & \rightarrow & K^0 & + & K^+ & + & X \\ B & (0) & + & (+1) & \rightarrow & (0) & + & (0) & + & (+1) \end{array}$$

3 The next thing to decide is whether strangeness is conserved in this reaction. The reaction comes about due to the interaction of a K meson with a proton so it is highly likely that the strong force is responsible. Strangeness must then be conserved:

$$\begin{array}{lccccccccc} & K^- & + & p & \rightarrow & K^0 & + & K^+ & + & X \\ S & (-1) & + & (0) & \rightarrow & (+1) & + & (+1) & + & (-3) \end{array}$$

This means that X is a very strange baryon indeed. In fact, X is the Ω^- particle (three strange quarks) whose existence was predicted by Murray Gell-Mann.

Color

The Ω^- is three apparently identical fermions bound in the same system, which suggests that the Pauli exclusion principle (see page 471) is being violated, so we must propose a new quantum number for quarks.

It turns out that there are three distinct varieties, or **flavors** of strong charge and to distinguish among them we call them red (R), green (G), and blue (B). As such they are known as color charges, though this is nothing to do with colour in its familiar sense – it is just a naming convention.

Quarks cannot exist in isolation but must be bound in colorless combinations. This can only be achieved in one of two ways:

1 **Baryons**: 3 quarks each of which has a different color charge from the other two. Antibaryons are essentially the same but each antiquark possesses a different anticolor from the other two.
e.g. a proton u u d
(red) (green) (blue) = colorless

2 **Mesons**: a quark of one color and an antiquark of the corresponding anticolor.
e.g. a π^+ u $\bar{d}$
(green) (antigreen) = colorless

Gluons

The quarks inside a given hadron are exchanging gluons all this time. As a result individual quarks are changing their colour but overall the colourless nature of the hadron is maintained. If one of the quarks in a given hadron is pulled away from its neighbours, the colour field "stretches" between that quark and its neighbours. The force between quarks increases with their separation so more and more energy needs to be added as the quarks are pulled apart.

At some point, it is energetically favourable for a new quark-antiquark pair to be produced rather than a further increase in the energy of the colour field. The energy of the colour field is converted into the mass of the new quarks, and the colour field can "relax" back to an unstretched state. The overall effect is that it is never possible to observe an isolated quark (or gluon). This is known as **quark confinement**.

The strong interaction between nucleons is the residual colour interaction between the quarks in the nucleons. This residual interaction is short-ranged.

Leptons and the standard model

In 1911–12 Victor Hess (1936 Nobel Prize winner) made a series of balloon flights to measure radiation in the atmosphere. He found that the intensity of radiation increased to five times the amount at sea level by about 5000 m altitude; hence the radiation must be coming from space. Robert Millikan coined the name **cosmic rays** for this penetrating and ionizing radiation, and believed it was electromagnetic in nature. However, but studying the intensity of radiation at different latitudes, Compton showed that the radiation was being deflected by the Earth's magnetic field and so must be charged.

Most of the cosmic rays striking the Earth are protons. Some are the nuclei of heavy elements (uranium nuclei have been detected) and there are also some very high-energy electrons. Typically they have energies between 1 and 10 GeV, but occasionally extraordinarily high-energy particles (up to 2×10^7 GeV) are detected, smashing into protons and producing hundreds of particles. The nature and origin of these cosmic rays is still a mystery.

The muon

Since its discovery the muon has been extensively studied. It is known to be unstable, and decays with a lifetime of 2.19703×10^{-6} seconds. This sort of time-scale is characteristic of a weak-force decay. Muons appear to be unstable "heavy electrons". It is possible to create atoms in which a muon replaces an electron in orbit around the nucleus. The biggest mystery that remains concerning the muon is why nature needs such a particle to exist.

The tau

In 1974 another "heavy electron", the **tau** (τ) particle, was discovered. This discovery came from the SPEAR collider (Stanford Positron-Electron Asymmetric Ring) built on a car park at SLAC, the Stanford Linear Accelerator. The particle was totally unexpected but it did not take long to confirm that it was another unstable lepton. However, the tau's enormous mass (3.5 million times that of the electron) and very short lifetime (3.3×10^{-13} seconds) means that it is very unlikely to be seen in cosmic rays.

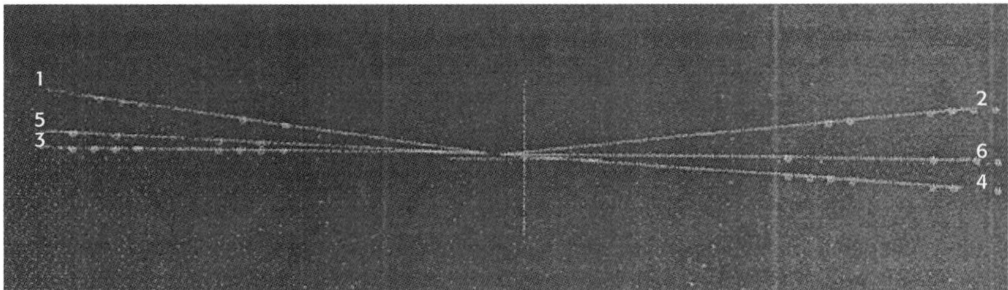

Figure 29 A typical tau decay. In this electronic track reconstruction a tau and antitau have been produced, travelled a short distance, and each decayed into three pions (numbered 1–6) and a neutrino. The small gap at the centre shows the distance travelled by the taus.

Lepton number and generations

There are three particles; the electron, the muon, and the tau, that appear to be versions of the same particle, but with increasing mass. They all carry the same electric charge and they all react via the weak and electromagnetic forces, but not the strong force. In each case there appears to be a partner: a neutral, almost massless object called a neutrino.

In order to categorize these particles physicists invented a labelling system called the **lepton number**, L, analogous to the baryon number of quarks.

Lepton number turns out to be more than simply a convenient way of tabulating the particles – it is also a conserved quantity. In any reaction involving leptons the total number in each generation must always remain the same.

State and explain which of the following reactions conserves lepton number.

a) $\nu_e + \mu^- \rightarrow e^- + \nu_\mu$

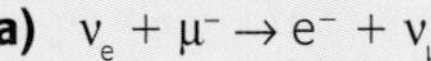

b) $\nu_e + n \rightarrow p + \mu^-$

Table 9 Lepton generations

First generation, $L_e = 1$	Second generation, $L_\mu = 1$	Third generation, $L_\tau = 1$
electron • found in atoms • involved in electrical and thermal conduction • produced in β^- radioactive decay	**muon** • produced in large numbers when cosmic rays collide with nuclei in the upper atmosphere—most cosmic rays that reach the surface are muons	**tau** • so far only observed in laboratories
electron neutrino • produced in β^- radioactive decay • produced in large numbers by nuclear reactors • produced in huge numbers by nuclear reactions in the sun	**muon neutrino** • produced by atomic reactors • produced in the upper atmosphere by cosmic rays • produced in the sun by nuclear reactions	**tau neutrino** • so far only observed in laboratories

Conservation laws – a summary

In all particle reactions, the following quantities are always conserved:

- Electric charge
- Total energy
- Total momentum (linear and angular)
- Baryon number
- Family lepton number
- Colour

Since a strong (colour) interaction involves gluons which only change the color of component quarks (Red, Green, Blue) as opposed to the individual quark type (u, d, s, c, t, b), the total number of any given quark type must be conserved. This means that the following quantities are always conserved in strong (and electromagnetic) interactions:

- Charm, Strangeness, Top & bottom

Since a weak interaction involves W^+, W^- or Z (which can change the quark type) these quantities will not necessarily be conserved in weak interactions.

Unification

Modern-day particle physicists are searching for beauty in the idea that the four fundamental forces can be united into one theory that shows them to be different aspects of one force (just in the same way that electricity and magnetism can be brought together in electromagnetism). In 1979 Glashow, Salam, and Weinberg received the Nobel Prize for bringing this goal a step nearer with their combined theory of the weak and electromagnetic forces.

The electroweak force

In 1961 Sheldon Glashow showed how the weak force could be explained by Feynman diagrams and linked with the electromagnetic force. For 30 years physicists had known about the weak **charged current** reactions such as:

$\nu_e + n \rightarrow p + e^-$ W^+ exchange   (Figure 30)
$\bar{\nu}_e + p \rightarrow n + e^+$ W^- exchange   (Figure 31)

Glashow predicted the existence of another weak interaction (the **neutral current**) that involved the Z^0 exchange particle:

$\bar{\nu}_\mu + e^- \rightarrow \bar{\nu}_\mu + e^-$ Z^0 exchange   (Figure 32)

Direct evidence for all three particles came from a giant CERN bubble chamber in 1983. Protons and antiprotons were collided. The W^+, W^- or Z^0 particles that were created immediately decayed:

$Z^0 \rightarrow e^+ + e^-$ $Z^0 \rightarrow \mu^+ + \mu^-$
$W^+ \rightarrow e^+ + \nu_e$ $W^+ \rightarrow \mu^+ + \nu_\mu$
$W^- \rightarrow e^- + \bar{\nu}_e$ $W^- \rightarrow \mu^- + \bar{\nu}_\mu$

Once the neutral current had been established the link with the electromagnetic force was more promising – in many ways the Z^0 and the photon are similar objects.

The final problem was that the W and Z exchange particles are very massive and the photon is massless. Any theory seemed to require all the exchange particles to be massless.

The next step was taken in 1967 when Steven Weinberg and Abdus Salam applied an idea, first suggested by Peter Higgs, to the electroweak theory. Higgs's idea was to introduce another type of force field (now called the Higgs field) into the theory. By carefully selecting the right sort of field to add, Weinberg and Salam made the Higgs field interact with the W and Z in such a way as to give them mass.

The Higgs field is very important – without it you cannot explain why the W and Z exchange particles have mass. If the Higgs field exists then **Higgs bosons** should exist as well.

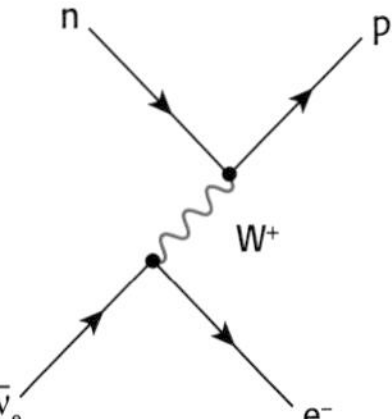

Figure 30 W^+ interaction

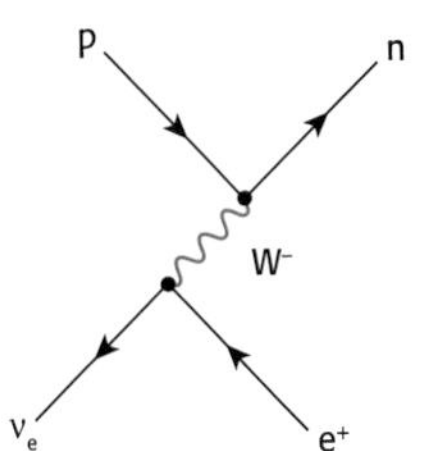

Figure 31 W^- interaction

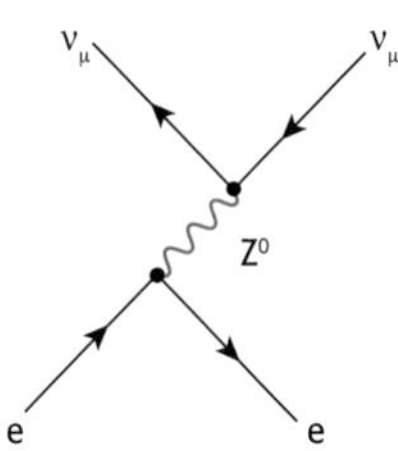

Figure 32 Z^0 in interaction. An electron and a neutrino scatter off each other.

Higgs bosons

The Higgs field is currently the best explanation for the origin of the mass of not only the W and Z but all massive particles. So far there is no experimental evidence for Higgs particles (Higgs bosons). The next generation of particle accelerators (e.g. the Large Hadron Collider at CERN) have been designed specifically to find them. If they are not found, theoretical particle physicists will have a new challenge – to find a new explanation for the origin of mass.

Experimental evidence for the quark and standard models

Deep inelastic scattering

In the late 1960s, 20 GeV electrons were collided with a proton target. As a result of these collisions the electrons scattered off at some angle to the target, and the detectors used were able to measure the angle.

What the results showed was that at low energies the electrons collided "softly" with the protons and scattered away at quite small angles; the protons simply recoiled from the collision. However, once the energy of the electrons passed a certain minimum value they started to see much "harder" collisions – the electrons scattered off at greater angles and the protons shattered into a stream of hadrons rather than simply recoiling. This effect became known as **deep inelastic scattering**.

A careful study of the results, notably by Bjorken and Feynman in 1968, suggested that the protons contained three very small charged objects. At high energies they were able to "resolve" the very small objects inside the protons – in effect the electrons were starting to diffract off the quarks inside the protons. As a result, one of the quarks was deflected away from the other two and shattered the proton into a stream of hadrons.

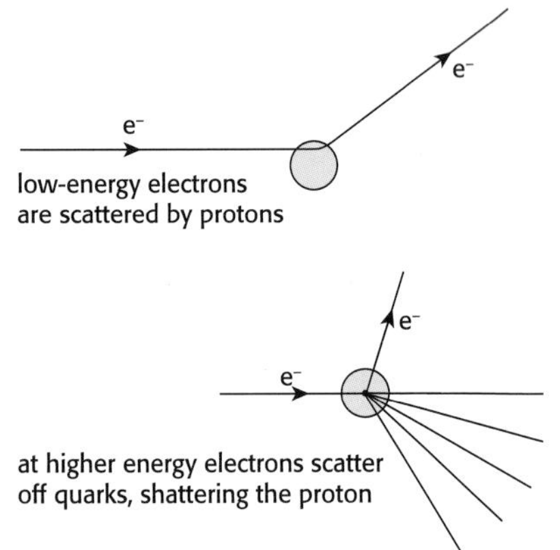

Figure 33 Deep inelastic scattering one particle to probe the structure of another

The idea is very similar to that of the Rutherford scattering experiment with α particles that revealed the presence of the nucleus inside the atom (Figure 34).

Detailed analysis of the experimental data was also able to determine the electrical charge of objects within the proton. They were shown to be $\pm 1/3^e$ or $\pm 2/3^e$. Evidence also existed for other neutral particles inside the proton (the gluons) and for the different types of objects within the proton (different colour quarks.)

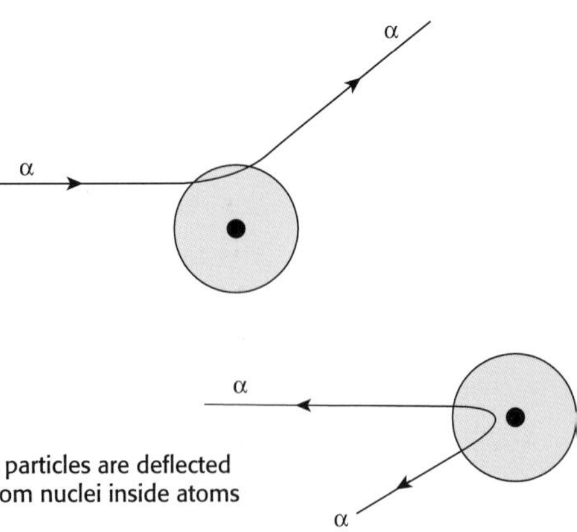

Figure 34 Rutherford scattering

The SLAC results have since been refined using neutron targets and also neutrino beams. Physicists have been able to show that the three objects have the charges expected of the u and d quarks, and in the correct ratios.

Asymptotic freedom

One final observation from the deep inelastic scattering experiments was that as the electron beam that was probing the proton was increased in energy, the more the electrons were being scattered as if they were encountering particles that were moving freely within the proton. The quarks that were being indentified appeared to be behaving more like free particles than ones that were bound inside the proton. This seemed to contradict the notion of quark containment and the phrase **asymptotic freedom** is used to describe the way quarks seemed to interact more feebly at higher energies.

This phenomenon was surprising because the highest-energy collisions probe the strong force at the shortest distances, where researchers expected the interaction to be at its strongest. This view of how the colour interaction strength decreases with the energy of the virtual gluon is now a central part of the standard model.

The Standard Model of particle physics

It seems that mass is the essential feature distinguishing the corresponding members of the three generations of basic particles (both quarks and leptons). Each generation contains two quarks and two leptons.

This "full set" of fundamental particles and the forces between them make up a complete description of nature, called the Standard Model of particle physics. There are many theories that propose further modifications but so far physicists have found no experimental evidence for any of them.

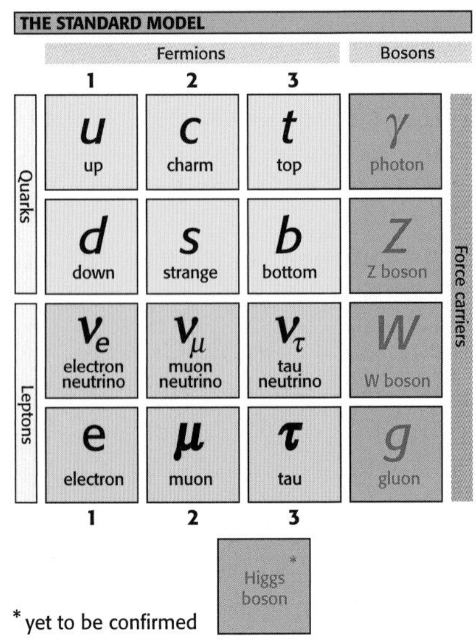

Figure 35 The Standard Model

Cosmology and strings

The Big Bang

The universe as we know it began in a hot Big Bang approximately 15 billion years ago. The temperature of the universe immediately following the Big Bang is thought to have vastly exceeded 10^{30} K.

At this temperature particles had huge energies and so exotic particles would have existed that are no longer common today. As the universe cooled, these exotic particles would not have survived but would have annihilated or decayed to produce radiation and common fundamental particles (quarks and leptons). Within a millionth of a second, quarks had grouped to produce hadrons such as neutrons and protons and after about 3 minutes Helium and some light nuclei (Beryllium, Boron) had formed as the temperature was low enough for protons and neutrons to remain bound in these nuclei. It was, however, still too hot for electrons to be trapped by these nuclei to form neutral atoms.

Some 300,000 years later, the ambient temperature had fallen below 10^4 degrees, that is, similar or cooler than the outer regions of our sun today. At these energies the negatively charged electrons were at last able to be held fast by the electrical attraction to the positively charged atomic nuclei whereby they combined to form neutral atoms. Electromagnetic radiation was set free and the universe became transparent as light could roam unhindered across space.

In today's universe, the once hot electromagnetic radiation now forms a black body spectrum with an effective temperature of 2.7 K (2.7 degrees above absolute zero). The discovery of this **cosmic microwave background** by Penzias and Wilson half a century ago is one of the great pieces of support for the Big Bang theory. Today precision measurements of the spectrum by instruments in satellites reveal small fluctuations in this radiation. These give hints of proto galaxies forming in the early universe.

It is possible to estimate the temperature of the early universe from the energies involved with different particle interactions that were taking place at different times. An important equation to use is the Boltzmann equation which links the average kinetic energy of particles in a medium to its absolute temperature:

$$E_K = \frac{3}{2}kT$$

Where E_K is the average kinetic energy of particles in the medium (in J)

T is the medium's absolute temperature (in K)

k is the Boltzmann's constant 1.38×10^{-23} J K^{-1}

As the universe expanded, it cooled down. By applying the Boltzmann equation to the cooling universe after the Big Bang, we can use particle physics to identify some important temperatures and times in the development of the early universe. The creation of a matter-antimatter pair of particles from a photon depends on the energy of the photons that are available which in turn depends on the temperature of the universe

Early in the universe's history, the temperature was so hot that helium nuclei could not form because the kinetic energy of the particles was so high that any helium nucleus that formed would be immediately broken apart. Calculate the temperature of the Universe when the average kinetic energy of the particles had just fallen below the energy needed to break apart a nucleus of Helium (28 MeV).

From 10^{-43} s after the Big Bang to 10^{-4} s (temperatures from 10^{32} K $\rightarrow 10^{12}$ K), the particle energies are so high that pair production is taking place all the time. The early universe is filled with different types of massive elementary particles and contained almost equal numbers of particles and antiparticles. Towards the end of this period (10^{-6} s after the Big Bang) protons and neutrons are beginning to be formed. Most of the matter and antimatter created during this time is annihilated leaving just a small remnant of matter that makes up the current observable universe (see below).

From 10^{-4} s after the Big Bang to 10 s (temperatures from 10^{12} K $\rightarrow 10^{10}$ K), the particle energies are not high enough for massive particles to be formed but lighter particles (e.g. electrons and positrons) can still be formed. Neutrons can be created from protons and electrons.

From 10 s after the Big Bang to 10^{13} s [10^{6} years] (temperatures from 10^{10} K $\rightarrow$3000 K), light nuclei (e.g. helium, beryllium and lithium) can be created. In fact this process can only happen for a very short time (1000 s) at the beginning of this period. These nuclei are not able to capture electrons to form atoms until the temperature has fallen low enough which does not take place until the end of the period.

From 10^{13} s [10^{6} years] after the Big Bang to now (temperatures from 3000 K $\rightarrow$2.7 K), the universe is transparent to radiation which means that the photons that exist stop having to interact with matter. The matter can come together under gravity form star and all the other elements in the universe can be created.

Matter/antimatter asymmetry

The early universe contained equal quantities of matter and antimatter. If they had remained equally balanced we would expect that the particles and antiparticles would completely annihilate each other, leaving no matter. This is not the case, and we observe vastly more matter than antimatter in the universe. Physicists have observed a subtle difference in the behaviour of matter and antimatter that might explain this imbalance.

In the 1960s a type of weak decay was observed that did not conserve a quantity called "CP", which was previously thought to always be conserved in weak interactions. Since the Standard Model does not predict this **CP violation**, it is now thought that there might be gaps in the Standard Model, paving the way for new theories, known as **physics beyond the Standard Model**. Supersymmetry and string theory are two well-known examples of these types of theories.

String theory perceives the fundamental particles as strings of energy, rather than points, using more than the four dimensions of space and time that are used in the Standard Model.

The extra dimensions (typically an extra 6) are not noticeable to us but fundamental particles are no longer to be perceived as points but extended objects (strings) in these extra dimensions. The fundamental particles that are known to us have different energies. These are seen as just different modes of vibration of these strings. The different modes of vibration are similar to the harmonics of an ordinary vibrating string where different modes have different energies. String theories have been developed as a result of the failure to reconcile gravitation with quantum theory.

End of chapter summary

Chapter 32 has six main themes, two of which are only relevant to those studying higher level physics. The two themes common to all levels are: particles and interactions, and quarks. The four HL themes are: accelerators and detectors, leptons and the standard model, experimental evidence and cosmology and strings. SL candidates studying Option D (relativity and particle physics) need to study the two themes marked with an asterisk* in addition to some material from Chapter 30. The list below summarises the knowledge and skills that you should be able to undertake after having studied this chapter. Further research into more detailed explanations using other resources and/or further practice at solving problems in all these topics is recommended – particularly the items in bold.

Particles and interactions (description and classification of particles, fundamental interactions and Feynman diagrams)*

- State what is meant by an elementary particle and identify them.
- Describe particles in terms of mass and various quantum numbers including spin.
- State what is meant by an antiparticle and the Pauli exclusion principle.
- List the fundamental interactions and describe them in terms of exchange particles.
- Discuss the uncertainty principle for time and energy in the context of particle creation.
- Describe what is meant by a Feynman diagram and discuss how one may be used to calculate probabilities for fundamental processes.
- Describe what is meant by virtual particles and apply the formula for the range R for interactions involving the exchange of a particle.
- **Use Feynman diagrams to describe pair annihilation (and pair production) and predict particle processes.**

Quarks*

- List the six types of quark.
- State the content, in terms of quarks and antiquarks, of hadrons (and deduce their spin structure). In particular the quark content of the proton and the neutron needs to be remembered.
- Define *baryon number* **and apply the law of conservation of baryon number**.
- Explain the need for colour in forming bound states of quark and state the colour of quarks and gluons.
- Outline the concept of strangeness **and discuss quark confinement**.
- Discuss the interaction that binds nucleons in terms of the colour force between quarks.

Particle accelerators and detectors

- Explain the need for high energies in order to produce particles of large mass and to resolve particles of small size.
- Outline the structure and explain the operation of a linear accelerator, a cyclotron and a synchrotron and compare their relative advantages and disadvantages.
- State what is meant by bremsstrahlung (braking) radiation **and solve problems related to the production of particles in accelerators**.
- Outline the structure and operation of the bubble chamber, the photomultiplier and the wire chamber.
- Outline international aspects of research into high-energy particle physics **and discuss the economic and ethical implications of high-energy particle physics research**.

Leptons and the standard model

- State the three-family structure of quarks and leptons in the standard model and identify the lepton number of the leptons in each family.
- Solve problems involving conservation laws in particle reactions and evaluate the significance of the Higgs particle (boson).

Experimental evidence for the quark and standard models

- State what is meant by deep inelastic scattering and analyse the results of deep inelastic scattering experiments.
- Describe what is meant by *asymptotic freedom* and *neutral current*.
- Describe how the existence of a neutral current is evidence for the standard model.

Cosmology and strings

- State the order of magnitude of the temperature change of the universe since the Big Bang and solve problems involving particle interactions in the early universe.

- State that the early universe contained almost equal numbers of particles and antiparticles suggest a mechanism by which the predominance of matter over antimatter has occurred.
- Describe qualitatively the theory of strings.

Chapter 32 questions

1 a) Possible particle reactions are given below. They **cannot** take place because they violate one or more conservation laws. For each reaction identify **one** conservation law that is violated.

i) $\mu^- \rightarrow e^- + \gamma$ [1]

ii) $p + n \rightarrow p + \pi^0$ [1]

iii) $p \rightarrow \pi^+ + \pi^-$ [1]

b) State the name of the exchange particle(s) involved in the strong interaction. [1]

(Total 4 marks)

2 This question is about deducing the quark structure of a nuclear particle.

When a K^- meson collides with a proton, the following reaction can take place.

$K^- + p \rightarrow K^0 + K^+ + X$

X is a particle whose quark structure is to be determined.

The quark structure of mesons is given below.

particle	quark structure
K^-	$s\bar{u}$
K^+	$u\bar{s}$
K^0	$d\bar{s}$

a) State and explain whether the original K^- particle is a hadron, a lepton **or** an exchange particle. [2]

b) State the quark structure of the proton. [2]

c) The quark structure of particle X is sss. Show that the reaction is consistent with the theory that hadrons are composed of quarks. [2]

(Total 6 marks)

3 This question is about fundamental particles and conservation laws.

Nucleons are considered to be made of quarks.

a) State the name of

i) the force (interaction) between quarks; [1]

ii) the particle that gives rise to the force between quarks. [1]

b) Outline in terms of conservations laws, why the interaction $\bar{v} + p = n + e^+$ is observed but the interaction $v + p = n + e^+$ has never been observed. (*You may assume that mass-energy and momentum are conserved in both interactions.*) [3]

(Total 5 marks)

4 This question is about the decay of a neutron.

The diagram below illustrates a neutron decaying into a proton by emitting a β^--particle.

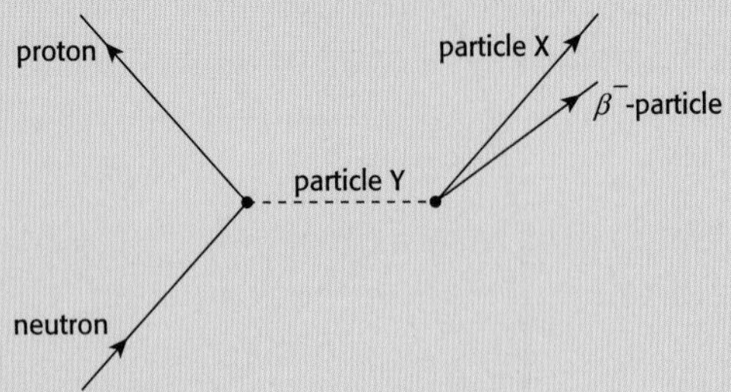

State the name of

a) the force involved in this decay; [1]

b) the particle X; [1]

c) the exchange particle Y involved in the decay. [1]

(Total 3 marks)

5 This question is about radioactive decay.

The decay process of a neutron is given by the following equation.

$$n \rightarrow p + \bar{e} + \bar{v}_e$$

a) Complete the table below.

particle	n	p	e^-	$\bar{v}_e$
baryon number				
lepton number				

[2]

b) Baryon number and lepton number are both conserved in this decay process. State **one** other property that is conserved. [1]

(Total 3 marks)

6 This question is about fundamental particles.

Particle production and annihilation are subject to conservation laws. Two of these laws are conservation of mass-energy and conservation of momentum.

a) State the names of **three** other conservation laws. [3]

b) Free neutrons are unstable. A neutron may decay to become a proton with the emission of an electron. A student represents the decay by the following equation.

$^1_0n \rightarrow {}^1_1p + {}^0_{-1}e$

i) State, by reference to conservation laws, why the student's equation is not correct. [1]

ii) Write down the correct decay equation. [1]

(Total 5 marks)

7 This question is about fundamental interactions.

a) The Feynman diagram below represents a β^- decay via the weak interaction process.

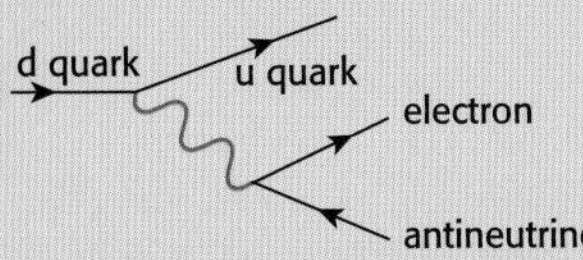

The exchange particle in this weak interaction is a virtual particle.

i) State what is meant by a virtual particle. [1]

ii) Detemine whether the virtual particle in the process represented by the Feynman diagram is a W^+, a W^- or a Z^0 boson. [2]

b) The order of magnitude of the mass of the $W^{\pm}$ and Z^0 bosons is 100 GeV c^{-2}. Estimate the range of the weak interaction. [3]

(Total 6 marks)

8 This question is about a proton.

The proton particle is made out of three quarks.

a) Explain why the three quarks in the proton do not violate the Pauli exclusion principle. [2]

b) Quarks have spin $\frac{1}{2}$. Explain how it is possible for the proton to also have spin $\frac{1}{2}$. [2]

(Total 4 marks)

9 This question is about the synchrotron and particle production.

a) In a synchrotron ring, a beam of protons and another beam of antiprotons move in opposite directions through regions of electric and magnetic fields as they circle the ring.

Describe the purpose of the

i) electric fields. [1]

ii) magnetic fields. [1]

b) Explain why the magnitudes of the magnetic fields in a synchrotron must be increased as the energy of the accelerated particles increases. [3]

c) The neutral lambda baryon Λ and its antiparticle may be produced in proton-antiproton collisions according to the following reaction.

$$p + \overline{p} \rightarrow \Lambda + \overline{\Lambda}$$

The minimum energy required to produce the Λ and the $\overline{\Lambda}$ is 2240 MeV. The rest mass of the proton is 938 MeV c^{-2}.

Calculate the **minimum** kinetic energy, E_K, of the antiproton, in order to produce the $\overline{\Lambda}$ and Λ particles when

i) the proton and the antiproton are each accelerated to a kinetic energy E_K; [1]

ii) the antiproton is accelerated to a kinetic energy E_K and collides with a stationary proton. [3)

d) By reference to your answers to (c), state an advantage of collisions between protons and antiprotons in a synchrotron compared with collisions between stationary protons and moving antiprotons. [1]

(Total 10 marks)

10 a) Outline

i) what is meant by a deep inelastic scattering experiment. [2]

ii) how deep inelastic scattering experiments give evidence in support of the existence of quarks and gluons. [4]

b) Deep inelastic scattering experiments indicate that the quarks inside hadrons behave as free particles. Suggest a reason for this. [2]

c) State **two** fundamental differences between the standard model for quarks and leptons and the theory of strings. [2]

(Total 10 marks)

33 Option overview

All IB Diploma Programme candidates need to study at least two optional topics from the list of available subjects. This material is tested in the paper 3 of the external examinations and counts for either 20% (HL) or 24% (SL) of your final performance.

Higher level candidates need to choose **two** optional topics from the following list:

Option E: Astrophysics (covered in chapter 27)

Option F: Communications (covered in chapter 28)

Option G: Electromagnetic waves (covered in chapter 29)

Option H: Relativity (covered in chapter 30)

Option I: Medical physics (covered in chapter 31)

Option J: Particle physics (covered in chapter 32)

Standard level candidates also need to choose **two** optional topics. Their list has some overlaps with the HL list but it is different:

Option A: Sight and wave phenomena (covered in chapter 9 and sections of this chapter)

Option B: Quantum physics and nuclear physics (covered in chapter 17)

Option C: Digital technology (covered in chapter 24)

Option D: Relativity and particle physics (covered in sections of chapters 30 and 32)

Option E: Astrophysics (covered in sections of chapter 27)

Option F: Communications (covered in sections of chapter 28)

Option G: Electromagnetic waves (covered in sections of chapter 29)

Apart from option A (Sight and wave phenomena), all the material needed to study the SL options are already contained in other chapters. The extra material needed for option A is included here and then the rest of this chapter just identifies which sections of this book are relevant to particular SL options.

Chapter 33: Sight and wave phenomena

Most of this SL option concerns itself with wave phenomena. This topic area needs to be studied by all HL students as a part of the additional higher level material. The exception is a small topic concerning the human eye.

The eye and sight

Looking into somebody's eyes, the coloured part, the **iris**, has a dark central gap, the **pupil**, that changes in size to control the amount of light that enters the eye. As shown in Figure 1, light entering the eye passes through the **cornea** (where it is refracted) and then the **aqueous humour** before passing through the pupil and going through the **lens**. Further refraction takes place at the lens whose shape is controlled by the **ciliary muscles** and the **suspensory ligaments**. The light then passes through the **vitreous humour** before striking the **retina** where light sensitive cells (**rods** and **cones**) respond to the light by sending electrical signals via the **optic nerve** to the brain. A part of the retina is occupied by the optic nerve resulting in a **blind spot** that is not sensitive to light.

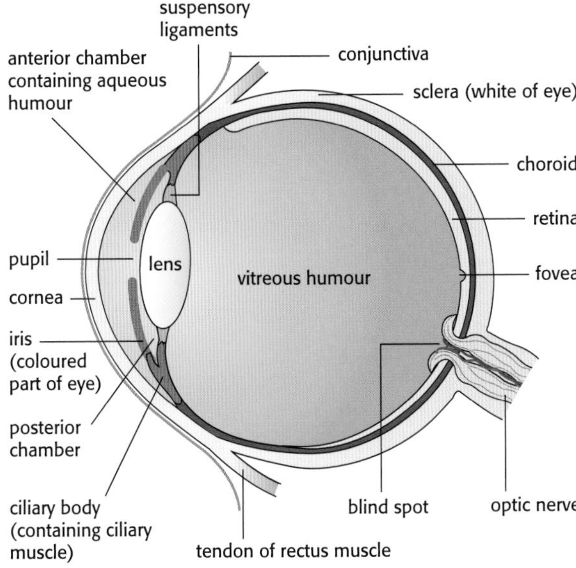

Figure 1 The eye

The depth of vision is the range of distances that are in focus at any given instant. Normal vision has the eye focused on infinity when the eye is relaxed and this results in the greatest depth of vision. **Accommodation** is the name given to the process by which the eye can focus on different objects. The lens changes shape as a result of changes in tension in the suspensory ligaments and the ciliary muscle.

The near point is the closest point that can be focused upon without straining or optical aids. In normal vision, the distance to the near point is approximately 25 cm. The far point is the furthest point that can be focused upon without straining or optical aids. In normal vision, the distance to the near point is taken as infinite.

Rods are light sensitive cells responsible for scotopic vision (the black and white vision that takes place in dim light). Rods are mainly located away from the centre of the retina. The density of the rods on the retina peak approximately 20° away from the centre.

Cones are light sensitive cells responsible for photopic vision (the colour vision that takes place at normal light levels). There are three different types of cone which have a different peak wavelength sensitivity. Colour blindness is often caused by the failure of one or more types of cone to respond.

Cones are mainly located in the centre of the retina.

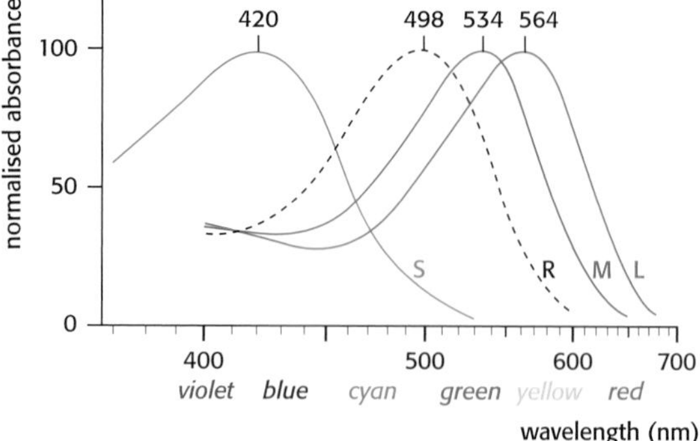

Figure 2 Spectral response graphs of cones in the eye

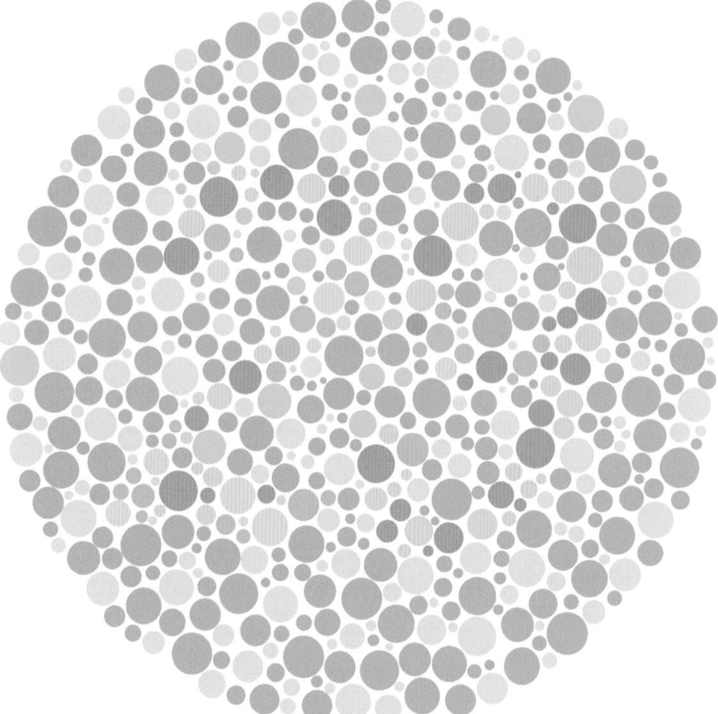

Figure 3 Normal vision should read the number 29. Red-green deficiencies should read the number 70. Total colour blindness should not read any numeral.

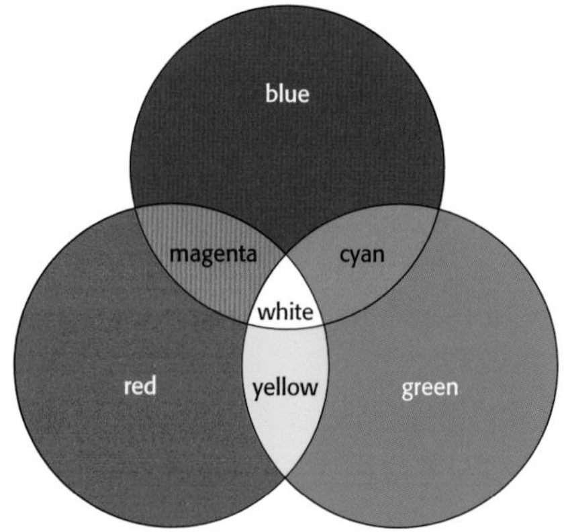

Figure 4 Combining two primary colours of light gives one of the secondary colours. Combining all three primary colours of light gives white light.

The primary colours are red, green and blue. By adding combinations of these three wavelengths at different intensities, the other colours of the rainbow can be perceived. The secondary colours are yellow, magenta and cyan. They result from mixing together two primary colours

Tricking the eye is all too easy to achieve either as an optical illusion or simply using the effect of light, dark and colour to alter how we perceive different objects. Examples of this include:

- Architectural effects can be created by the use of light and shadow – deep shadow gives the impression of massiveness.
- Glow can be used to give the impression of "warmth" (e.g. blue tints are cold) or to change the perceived size of a room (e.g. light-coloured ceilings appear to heighten the room)

Figure 5 An optical illusion. How many legs does the elephant have?

End of chapter summary

Chapter 33 has two main themes: the structure of the different SL and HL options and the eye and sight (for Option A). The list below summarises the knowledge and skills that you should be able to undertake after having studied this chapter. Further research into more detailed explanations using other resources and/or further practice at solving problems in all these topics is recommended – particularly the items in bold.

Structure of different SL and HL options

- Understand the structure of optional part of the physics diploma programme and how this material is tested in the different written external assessments.

The eye and sight

- Describe the basic structure of the human eye and state (and explain) the process of depth of vision and accommodation.
- State that the retina contains rods and cones, describe the variation in density across the surface of the retina and describe their function in photopic and scotopic vision.
- Describe colour mixing of light by addition and subtraction **and discuss the effect of light and dark, and colour, on the perception of objects**.

Index

Index entries are arranged in letter-by-letter sequence. Page numbers in bold refers to tables; page numbers in italics refers to figures.